Monografie Matematyczne

Instytut Matematyczny Polskiej Akademii Nauk (IMPAN)

Volume 70
(New Series)

Founded in 1932 by
S. Banach, B. Knaster, K. Kuratowski,
S. Mazurkiewicz, W. Sierpinski, H. Steinhaus

Volumes 31–62 of the series
Monografie Matematyczne were published by
PWN – Polish Scientific Publishers, Warsaw

Leonid Positselski

Homological Algebra of Semimodules and Semicontramodules

Semi-infinite Homological Algebra of Associative Algebraic Structures

Appendix C in collaboration with Dmitriy Rumynin
Appendix D in collaboration with Sergey Arkhipov

 Birkhäuser

Leonid Positselski
Sector of Algebra and Number Theory
Institute for Information Transmission Problems
Bolshoy Karetny per. 19 str. 1
Moscow 127994
Russia
e-mail: posic@mccme.ru

Sergey Arkhipov
Department of Mathematics
University of Toronto
40 St. George Street
Toronto, Ontario
Canada M5S 2E4
e-mail: hippie@math.toronto.edu

Dmitriy Rumynin
Mathematics Department
University of Warwick
Coventry, CV4 7AL
UK
e-mail: D.Rumynin@maths.warwick.ac.uk

2000 Mathematics Subject Classification: 16E05, 16E30, 16E35, 16E40, 17B56, 17B65, 18E25, 18E30, 18G10, 18G15, 18C20, 20G15, 22D12, 22E50

ISBN 978-3-0346-0435-2   e-ISBN 978-3-0346-0436-9
DOI 10.1007/978-3-0346-0436-9

Library of Congress Control Number: 2010929656

Cover design: deblik, Berlin

Printed on acid-free paper

Springer Basel AG is part of Springer Science+Business Media

www.birkhauser-science.com

To the memory of my father

Contents

Appendices

Preface*

The subject of this book is Semi-Infinite Algebra, or more specifically, Semi-Infinite Homological Algebra. The term "semi-infinite" is loosely associated with objects that can be viewed as extending in both a "positive" and a "negative" direction, with some natural position in between, perhaps defined up to a "finite" movement. Geometrically, this would mean an infinite-dimensional variety with a natural class of "semi-infinite" cycles or subvarieties, having always a finite codimension in each other, but infinite dimension and codimension in the whole variety [37]. (For further instances of semi-infinite mathematics see, e.g., [38] and [57], and references below.)

Examples of algebraic objects of the semi-infinite type range from certain infinite-dimensional Lie algebras to locally compact totally disconnected topological groups to ind-schemes of ind-infinite type to discrete valuation fields. From an abstract point of view, these are ind-pro-objects in various categories, often endowed with additional structures. One contribution we make in this monograph is the demonstration of another class of algebraic objects that should be thought of as "semi-infinite", even though they do not at first glance look quite similar to the ones in the above list. These are *semialgebras* over coalgebras, or more generally over corings – *the* associative algebraic structures of semi-infinite nature.

The subject lies on the border of Homological Algebra with Representation Theory, and the introduction of semialgebras into it provides an additional link with the theory of corings [23], as the semialgebras are the natural objects dual to corings. The author's main interests belong to Homological Algebra, and so the main body of the monograph consists of the formal development of the homological theory of corings and semialgebras, while the representation-theoretic (and other) examples and applications are relegated to appendices.

One such application worth mentioning here is related to the duality between complexes of representations of an infinite-dimensional Lie algebra with the complementary central charges, e.g., c and $26 - c$ for the Virasoro algebra [39, 77]. We interpret it as a particular case of a very general homological phenomenon related to coalgebras, which we call the *comodule-contramodule correspondence*. The latter is a coalgebra version of the Serre–Grothendieck duality – covariant, noncommutative, and not depending on any finiteness assumptions (the coalgebra $\mathcal{C}$ itself plays the role of the dualizing complex; cf. [65, 71]). This allows us to formulate the duality for infinite-dimensional Lie algebra representations as a (covariant) equivalence of triangulated categories.

On a less ambitious level, with the formal neighborhood of a closed subgroup in an algebraic group one can associate a semialgebra of (roughly speaking) dis-

---

*What follows is very speculative and should be taken with a grain of salt.

tributions on it, and the category of Harish-Chandra modules over an algebraic Harish-Chandra pair is the category of semimodules over this semialgebra. For further applications to Representation Theory, see [16], [17], and [45].

Another important area that Semi-Infinite Algebra and Geometry are related to is Mathematical Physics. The author of this monograph stands at the receiving end of a long chain of interpretative work through which the ideas originating in the interaction of Mathematics with Quantum Field Theory or String Theory are transferred to the heart of Algebra. We are not in a position to comment here on the possibilities of applications of the content of this book to Mathematical Physics, so we will restrict ourselves to a couple of references and some very general remarks. The semi-infinite homology of Lie algebras are closely related to what the physicists call the BRST construction [10, 11]; for a discussion of the significance of the semi-infinite homology in String Theory, see [41] and the introduction to [42].

The field of functions on the formal circle, which is the field of Laurent power series, is a very simple example of a semi-infinite algebraic object; and much more complicated algebraic or geometric objects built on the basis of the formal circle often have very visible semi-infinite structures. This includes the Virasoro and affine Kac-Moody Lie algebras, the varieties [58] and groups of formal loops, the semi-infinite flag variety [37], etc. The formal circle is obviously important for Conformal Field Theory [12], hence the significance of such objects of study as the semi-infinite homology [10, 40], the semi-infinite de Rham complex [58, 8], or the chiral differential operators [10, 59] in Mathematical Physics.

Things semi-infinite play a role in Class Field Theory [83] and the Langlands Program [11, 40] for the very same reason. A much more detailed discussion of the links between the semi-infinite cohomology and various other mathematical and physical disciplines can be found in the introduction to [84].

Another class of algebraic objects prominently featured in this monograph is that of *contramodules*. Their definition, introduced originally in the case of coalgebras or corings [33], can be extended to certain topological rings, topological Lie algebras, certain topological groups ... These are modules with infinite summation operations, but of the kind that cannot be interpreted as any sort of limit with respect to a topology. Here one finds an approach to the infinite summation entirely different from the one most common to Analysis.

Typically, for an abelian category of "discrete", "smooth", or "torsion" modules there is an accompanying abelian category of contramodules. The latter contains all kinds of objects "dual" to the objects of the former, and some other objects in addition. For example, the category of "weakly l-complete abelian groups" appearing in the continuous étale cohomology theory [54] is simply the category of contramodules over the l-adic integers. While not "semi-infinite" in themselves, contramodules always come up whenever one wishes to pass from a semi-infinite *homology* to a semi-infinite *cohomology* theory.

One area where our approach is inspired by, still essentially different from, the classical one is Relative Homological Algebra. While the classical theory [47, 33, 80, 35] emphasizes relative derived functors of nonexact functors that may be quite conventional and not necessarily "relative" in themselves, here we are mainly interested in *absolute* derived functors, but the nonexact functors that we derive and the categories where they are defined are essentially relative by their nature. We always want our derived functors to assign long exact sequences of cohomology to arbitrary short exact sequences of complexes in the arguments, not only to short exact sequences that are split over some base. Still the base (or even two bases, one over the other) are built into the definitions of the categories and functors we work with.

One thing we cannot pretend to explain, still cannot avoid mentioning here, are the *exotic derived categories*. These are variations on the theme of the unbounded derived category. Their names are the (conventional unbounded) derived, the *coderived*, the *contraderived*, and the mixed, or *semiderived* categories. Historically, these first occurred in the derived nonhomogeneous Koszul duality theory [61], but from a wider point of view, the coderived and contraderived categories appear to be intrinsic to the comodules and contramodules (while occasionally useful for modules, too). For a definitive treatment of the exotic derived categories and their role in Koszul duality, we refer the reader to the long paper [76]. As to the nonhomogeneous Koszul duality itself, it is developed and used in this book as a strong technical tool.

An object of the contraderived category can be thought of as a complex having, in addition to the conventional cohomology at finite degrees, some kind of "cohomology in the degree $+\infty$". Analogously, a complex in the coderived category can be viewed as having a "cohomology in the degree $-\infty$". This is essential, in particular, for the construction of the comodule-contramodule correspondence, as the latter can well transform irreducible modules into acyclic complexes (i.e., those with no cohomology anywhere but "at infinity") and back. For example, an acyclic, but nontrivial object in the contraderived category of contramodules can be represented by an acyclic, unbounded complex of projective contramodules, and the latter thought of as a "left projective resolution of something living in the degree $+\infty$".

We also propose a very simple, bordering on self-evident, still apparently not widely known, approach to derived functors of two arguments, which allows us to obtain double-sided derived functors for free. It wouldn't get one too far without the exotic derived categories, though. The concrete double-sided derived functors we are interested in are the SemiExt and SemiTor over semiassociative semialgebras, and the semi-infinite (co)homology of Tate Lie algebras and locally compact totally disconnected topological groups. The semimodule-semicontramodule correspondence connects these with the more conventional one-sided Ext and CtrTor.

The functors to be derived are the *semitensor product* and *semihomomor-phisms* in the semiassociative case, and the *semiinvariants* and *semicontrainvari-ants* in the Lie algebra or topological group case. These neither left, nor right exact functors are naturally associated with certain semi-infinite algebraic struc-tures, and particularly with semialgebras. Still they are nontrivial enough even for finite-dimensional Lie algebras and finite groups.

To end these preparatory notes, let us say a few words about the state of the subject after this monograph. It appears that the question of *defining* the semi-infinite homology and cohomology generally, and in the case of associative algebraic structures specifically, has been now worked out and understood to a very significant extent. Compared to this development, our knowledge of the ways of *computing* the semi-infinite cohomology is next to nonexistent outside of the classical Lie algebra case. The only example where the semi-infinite cohomology of associative algebras has been computed as of now is that of the small quantum group with its triangular decomposition [4, 17]. The methods used for this computation have so far resisted, essentially, all attempts of transfer to other situations or generalization. Computing the semi-infinite cohomology remains a challenge for future researchers to take on.

It is our special pleasure to finish these most cursory remarks with a reference to B. Feigin's paper [36] that introduced the semi-infinite homology and the very term *semi-infinite*.

Introduction

This monograph grew out of the author's attempts to understand the definitions of semi-infinite (co)homology of associative algebras that had been proposed in the literature and particularly in the works of S. Arkhipov [2, 3] (see also [17, 79]). Roughly speaking, the semi-infinite cohomology is defined for a Lie or associative algebra-like object which is split in two halves; the semi-infinite cohomology has the features of a homology theory (left derived functor) along one half of the variables and a cohomology theory (right derived functor) along the other half.

In the Lie algebra case, the splitting in two halves only has to be chosen up to a finite-dimensional space; in particular, the homology of a finite-dimensional Lie algebra only differs from its cohomology by a shift of the homological degree and a twist of the module of coefficients. So one can define the semi-infinite homology of a Tate (locally linearly compact) Lie algebra [10] (see also [5]); it depends, to be precise, on the choice of a compact open vector subspace in the Lie algebra, but when the subspace changes it undergoes only a dimensional shift and a determinantal twist. For Lie superalgebras already, there is no such shift/twist phenomenon, and the dependence on the choice of a compact open subspace or subalgebra is very substantial [10]. Let us emphasize that what is often called the "semi-infinite cohomology" of Lie algebras should be thought of as their semi-infinite *homology*, from our point of view. What we call the semi-infinite *cohomology* of Tate Lie algebras is a different and dual functor, defined in this book (see Appendix D).

In the associative case, people usually considered an algebra A with two subalgebras N and B such that $N \otimes B \simeq A$ and there is a grading on A for which N is positively graded and locally finite-dimensional, while B is nonpositively graded. To this data, under certain assumptions, one assigns another graded algebra $A^\#$ with the same subalgebras N and B such that $B \otimes N \simeq A$. An attempt to understand this construction was the very starting point of the present research. We show that both the grading and the second subalgebra B are redundant; all one needs is an associative algebra R, a subalgebra K in R, and a coalgebra $\mathcal{C}$ dual to K. Certain flatness/projectivity and "integrability" conditions have to be imposed on this data. If they are satisfied, the tensor product $\mathbf{S} = \mathcal{C} \otimes_K R$ has a *semialgebra* structure and all the machinery described below can be applied.

Furthermore, we propose the following general setting for semi-infinite (co)homology of associative algebraic structures. Let $\mathcal{C}$ be a coalgebra over a field k. Then $\mathcal{C}$-$\mathcal{C}$-bicomodules form a tensor category with respect to the operation of cotensor product over $\mathcal{C}$; the categories of left and right $\mathcal{C}$-comodules are module categories over this tensor category. Let $\mathbf{S}$ be a ring object in this tensor category; we call such an object a *semialgebra* over $\mathcal{C}$ (due to it being "an algebra in half of the variables and a coalgebra in the other half"). One can consider module objects over $\mathbf{S}$ in the module categories of left and right

$\mathcal{C}$-comodules; these are called left and right **S**-*semimodules*. The categories of left and right semimodules are abelian if **S** is an injective right and left $\mathcal{C}$-comodule, respectively; let us suppose that it is. There is a natural operation of *semitensor product* of a right semimodule and a left semimodule over **S**, denoted by

$$\lozenge_{\mathbf{S}} \colon \text{simod-}\mathbf{S} \times \mathbf{S}\text{-simod} \longrightarrow k\text{-vect};$$

it can be thought of as a mixture of the cotensor product $\square_{\mathcal{C}}$ over $\mathcal{C}$ and the tensor product in the direction of **S** relative to $\mathcal{C}$. This functor is neither left, nor right exact. Its double-sided derived functor SemiTor is suggested as the associative version of semi-infinite *homology* theory.

Before describing the functor SemiHom (whose derived functor SemiExt provides the associative version of semi-infinite *cohomology*), let us discuss a little bit of abstract nonsense. Let E be an (associative, but noncommutative) tensor category, M be a left module category over it, N be a right module category, and K be a category such that there is a pairing between the module categories M and N over E taking values in K. This means that there are multiplication functors

$$\mathsf{E} \times \mathsf{E} \longrightarrow \mathsf{E}, \quad \mathsf{E} \times \mathsf{M} \longrightarrow \mathsf{M}, \quad \mathsf{N} \times \mathsf{E} \longrightarrow \mathsf{N}, \quad \text{and} \quad \mathsf{N} \times \mathsf{M} \longrightarrow \mathsf{K}$$

and associativity constraints for ternary multiplications $\mathsf{E} \times \mathsf{E} \times \mathsf{E} \to \mathsf{E}$, $\mathsf{E} \times \mathsf{E} \times \mathsf{M} \to \mathsf{M}$, $\mathsf{N} \times \mathsf{E} \times \mathsf{E} \to \mathsf{N}$, and $\mathsf{N} \times \mathsf{E} \times \mathsf{M} \to \mathsf{K}$ satisfying the appropriate pentagonal diagram equations. Let A be a ring object in E. Then one can consider the category $_A\mathsf{E}_A$ of A-A-bimodules in E, the category $_A\mathsf{M}$ of left A-modules in M, and the category N_A of right A-modules in N. If the categories E, M, N, and K are abelian, there are functors of tensor product over A, making $_A\mathsf{E}_A$ into a tensor category, $_A\mathsf{M}$ and N_A into left and right module categories over $_A\mathsf{E}_A$, and providing a pairing

$$\mathsf{N}_A \times {}_A\mathsf{M} \longrightarrow \mathsf{K}.$$

These new tensor structures are associative whenever the original multiplication functors were right exact.

Suppose that we want to iterate this construction, considering a coring object C in $_A\mathsf{E}_A$, the categories of C-C-bicomodules in $_A\mathsf{E}_A$ and C-comodules in $_A\mathsf{M}$ and N_A, etc. Since the functors of tensor product over A are not left exact in general, the cotensor products over C will be only associative under certain (co)flatness conditions. If one makes the next step and considers a ring object S in the category of C-C-bicomodules in $_A\mathsf{E}_A$, one discovers that the functors of tensor products over S are only partially defined. Considering partially defined tensor structures, one can indeed build this tower of module-comodule categories and tensor-cotensor products in them as high as one wishes. In this book, we restrict ourselves to 3-story towers of *semialgebras* over *corings* over (ordinary) rings, mainly because we don't know how to define unbounded (co)derived categories of (co)modules for any higher levels (see below).

Now let us introduce *contramodules*. The functor $(V, W) \longmapsto \mathrm{Hom}_k(V, W)$ makes the category opposite to the category of vector spaces into a module category over the tensor category of vector spaces. A contramodule over an algebra R or a coalgebra $\mathcal{C}$ is an object of the category opposite to the category of modules or comodules in $k\text{-vect}^{\mathrm{op}}$ over the ring object R or the coring object $\mathcal{C}$ in $k\text{-vect}$. One can easily see that an R-contramodule is just an R-module, while the vector space of k-linear maps from a $\mathcal{C}$-comodule to a k-vector space provides a typical example of $\mathcal{C}$-contramodule. Setting $\mathsf{E} = \mathsf{M} = k\text{-vect}$ and $\mathsf{N} = \mathsf{K} = k\text{-vect}^{\mathrm{op}}$ in the above construction, one obtains a right module category $\mathcal{C}\text{-contra}^{\mathrm{op}}$ over the tensor category $\mathcal{C}\text{-comod-}\mathcal{C}$ together with a pairing $\mathrm{Cohom}_\mathcal{C}^{\mathrm{op}} \colon \mathcal{C}\text{-comod} \times \mathcal{C}\text{-contra}^{\mathrm{op}} \longrightarrow k\text{-vect}^{\mathrm{op}}$. Given a semialgebra $\mathbf{S}$ over $\mathcal{C}$, one can apply the construction again and obtain the category of $\mathbf{S}$-*semicontramodules* and the functor

$$\mathrm{SemiHom}_{\mathbf{S}}^{\mathrm{op}} \colon \mathbf{S}\text{-simod} \times \mathbf{S}\text{-sicntr}^{\mathrm{op}} \longrightarrow k\text{-vect}^{\mathrm{op}}$$

assigning a vector space to an $\mathbf{S}$-semimodule and an $\mathbf{S}$-semicontramodule. Though comodules and contramodules are quite different, there is a strong duality-analogy between them on the one hand, and an equivalence of their appropriately defined (exotic) unbounded derived categories on the other hand (see below).

Let us explain how we define double-sided derived functors. While the author knows of no natural way to define a derived functor of one argument that would not be either a left or a right derived functor, such a definition of derived functor *of two arguments* does exist in the balanced case. Namely, let $\Theta \colon \mathsf{H}_1 \times \mathsf{H}_2 \longrightarrow \mathsf{K}$ be a functor and $\mathsf{S}_i \subset \mathsf{H}_i$ be localizing classes of morphisms in categories H_1 and H_2. We would like to define a derived functor

$$\mathbb{D}\Theta \colon \mathsf{H}_1[\mathsf{S}_1^{-1}] \times \mathsf{H}_2[\mathsf{S}_2^{-1}] \longrightarrow \mathsf{K}.$$

Let F_1 be the full subcategory of "flat objects in H_1 relative to Θ" consisting of all objects $F \in \mathsf{H}_1$ such that the morphism $\Theta(F, s)$ is an isomorphism in K for any morphism $s \in \mathsf{S}_2$. Let F_2 be the full subcategory in H_2 defined in the analogous way. Suppose that the natural functors

$$\mathsf{F}_i[(\mathsf{S}_i \cap \mathsf{F}_i)^{-1}] \longrightarrow \mathsf{H}_i[\mathsf{S}_i^{-1}]$$

are equivalences of categories. Then the restriction of the functor Θ to the subcategory $\mathsf{F}_1 \times \mathsf{H}_2$ of the Cartesian product $\mathsf{H}_1 \times \mathsf{H}_2$ factorizes through $\mathsf{F}_1[(\mathsf{S}_1 \cap \mathsf{F}_1)^{-1}] \times \mathsf{H}_2[\mathsf{S}_2^{-1}]$ and therefore defines a functor on the category $\mathsf{H}_1[\mathsf{S}_1^{-1}] \times \mathsf{H}_2[\mathsf{S}_2^{-1}]$. The same derived functor can be obtained by restricting the functor Θ to the subcategory $\mathsf{H}_1 \times \mathsf{F}_2$ of $\mathsf{H}_1 \times \mathsf{H}_2$. This construction can be even extended to partially defined functors of two arguments Θ (see 2.7).

For this definition of the double-sided derived functor to work properly, the localizing classes in the homotopy categories have to be carefully chosen (see 0.2.3). That is why our derived functors SemiTor and SemiExt are not defined on the conventional derived categories of semimodules and semicontramodules, but on their *semiderived categories.* The semiderived category of $\mathbf{S}$-semi(contra)modules is a mixture of the usual derived category in the module direction (relative to $\mathcal{C}$) and the *co/contraderived* category in the $\mathcal{C}$-co/contramodule direction. More precisely, one defines the semiderived category of $\mathbf{S}$-semimodules as the quotient category of the homotopy category of complexes of $\mathbf{S}$-semimodules by the thick subcategory formed by those complexes that, *considered as complexes of $\mathcal{C}$-comodules, vanish in the coderived category of $\mathcal{C}$-comodules.* The coderived category of $\mathcal{C}$-comodules is equivalent to the homotopy category of complexes of injective $\mathcal{C}$-comodules, and analogously, the contraderived category of $\mathcal{C}$-contramodules is equivalent to the homotopy category of complexes of projective $\mathcal{C}$-contramodules. So the distinction between the derived and co/contra/semiderived categories is only relevant for unbounded complexes and only in the case of infinite homological dimension.

A notable attempt to develop a general theory of semi-infinite homological algebra was undertaken by A. Voronov in [84]. Let us point out the differences between our approaches. First of all, Voronov only considers the semi-infinite homology of Lie algebras, while we work with associative algebraic structures. Secondly, Voronov constructs a double-sided derived functor of a functor of one argument and the choice of a class of resolutions becomes an additional ingredient of his construction, while we define double-sided derived functors of functors of two arguments and the conditions imposed on resolutions are determined by the functors themselves. Thirdly, Voronov works with graded Lie algebras and his functor of semivariants is obtained as the image of the invariants with respect to one half of the Lie algebra in the coinvariants with respect to the other half, while we consider ungraded Tate Lie algebras with only one subalgebra chosen, and our functor of *semiinvariants* is constructed in a much more delicate way (see below). Finally, no exotic derived categories appear in [84].

Another approach to the semi-infinite cohomology (but not homology) was developed in [17]. The definition of the semi-infinite cohomology of finite-dimensional associative algebras proposed in [17] agrees with the one given in this book when the algebra has a grading satisfying the restrictive conditions under which the main argument of [17] applies, but is not readily comparable to our definition in the general case. Indeed, for finite-dimensional algebras and modules the object $\mathrm{SemiExt}_{\mathbf{S}}(\mathcal{M}, \mathfrak{P})$, being dual to the object $\mathrm{SemiTor}^{\mathbf{S}}(\mathfrak{P}^{*}, \mathcal{M})$, is a complex of profinite-dimensional (compact) vector spaces; while the semi-infinite cohomology as defined in [17] are discrete vector spaces. However, there is a natural map from the latter to the former.

The coderived category of $\mathcal{C}$-comodules and the contraderived category of $\mathcal{C}$-contramodules turn out to be naturally equivalent. This equivalence can be thought of as a covariant analogue of the contravariant functor $\mathbb{R}\operatorname{Hom}_{R}(-, R)$: $\mathsf{D}(R\text{-mod}) \longrightarrow \mathsf{D}(\text{mod-}R)$ on the derived category of modules over a ring R. Moreover, there is a natural equivalence between the semiderived categories of $\mathbf{S}$-semimodules and $\mathbf{S}$-semicontramodules. The functors $\mathbb{R}\Psi_{\mathbf{S}} \colon \mathsf{D}^{\mathsf{si}}(\mathbf{S}\text{-simod}) \longrightarrow \mathsf{D}^{\mathsf{si}}(\mathbf{S}\text{-sicntr})$ and $\mathbb{L}\Phi_{\mathbf{S}} \colon \mathsf{D}^{\mathsf{si}}(\mathbf{S}\text{-sicntr}) \longrightarrow \mathsf{D}^{\mathsf{si}}(\mathbf{S}\text{-simod})$ providing this equivalence are defined in terms of the spaces of homomorphisms in the category of $\mathbf{S}$-semi-modules and the operation of *contratensor product* of an $\mathbf{S}$-semimodule and an $\mathbf{S}$-semicontramodule. The latter is a right exact functor which resembles the functor of tensor product of modules over a ring. This equivalence of triangulated categories transforms the functor SemiExt$_{\mathbf{S}}$ into the functors Ext in either of the semiderived categories (and the functor SemiTor$^{\mathbf{S}}$ into the left derived functor CtrTor$^{\mathbf{S}}$ of the functor of contratensor product). We call this kind of equivalence of triangulated categories the *comodule-contramodule correspondence* or the *semimodule-semicontramodule correspondence*.

The duality-analogy between semimodules and semicontramodules partly breaks down when one passes from homological algebra to the structure theory. Comodules over a coalgebra over a field are simplistic creatures; contramodules are quite a bit more complicated, though still much simpler than modules over a ring, the structure theory of a coalgebra over a field being much simpler than that of an algebra or a ring. We construct some relevant counterexamples. There is an analogue of Nakayama's Lemma for contramodules, a description of contramodules over an infinite direct sum of coalgebras, etc. These results can be extended to contramodules over certain topological rings (much more general than the topological algebras dual to coalgebras). Contramodules over topological Lie algebras can also be defined; and an isomorphism of the categories of contramodules over a topological Lie algebra and its topological enveloping algebra can be proven under certain assumptions.

A *coring* $\mathcal{C}$ over a ring A is a coring object in the tensor category of bimodules over A. (In a different terminology, this is called a *coalgebroid*.) A *semialgebra* $\mathbf{S}$ over a coring $\mathcal{C}$ is a ring object in the tensor category of bicomodules over $\mathcal{C}$; for this definition to make sense, certain (co)flatness conditions have to be imposed on $\mathcal{C}$ and $\mathbf{S}$ to make the cotensor product of bicomodules well-defined and associative. Throughout this monograph (with the exception of Chapter 0 and the appendices) we work with corings $\mathcal{C}$ over noncommutative rings A and semialgebras $\mathbf{S}$ over $\mathcal{C}$. Mostly we have to assume that the ring A has a finite homological dimension – for a number of reasons, the most important one being that otherwise we don't know how to define appropriately the unbounded (co)derived category of $\mathcal{C}$-comodules. No assumptions about the homological dimension of the coring and the semialgebra are made. Besides, we mostly have to suppose that $\mathcal{C}$ is a flat left and right A-module and $\mathbf{S}$ is a coflat left and right $\mathcal{C}$-comodule, and even certain (co)projectivity conditions have to be imposed in order to work with contramodules.

All kinds of relative adjustness (flatness, projectivity, injectivity) properties are considered in this monograph, but their definitions differ from the ones typical of the classical relative homological algebra in one important respect. Specifically, we define relative adjustness in terms of complexes or exact triples *adjusted* over the base, rather than those *split* over the base. Our relative conditions tend to be weaker than the ones defined in the classical manner, more delicate, and better behaved with respect to the exact sequences that do not split over the base. When we need to consider relative adjustness properties defined in ways more resembling the classical approach, we insert the word "quite" into our terms for such properties.

Nonhomogeneous quadratic duality [73, 75] establishes a correspondence between nonhomogeneous Koszul algebras and Koszul CDG-algebras. This duality has a relative version with a base ring, assigning, e.g., the de Rham DG-algebra to the filtered algebra of differential operators (the base ring being the ring of functions, in this case). For a number of reasons, it is advisable to avoid passing to the dual vector space/module in this construction, working with CDG-coalgebras instead of CDG-algebras; in particular, this allows one to include infinitely (co)generated Koszul algebras and coalgebras [74, 76]. In the relative case, this means considering the graded coring of polyvector fields, rather than the graded algebra of differential forms, as the dual object to the differential operators. The relevant additional structure on the polyvector fields (corresponding to the de Rham differential on the differential forms) is called a *quasi-differential structure*. Another important version of relative nonhomogeneous quadratic duality uses base coalgebras in place of base rings. This situation is simpler in some respects, since one still obtains CDG-coalgebras as the dual objects. As a generalization of these two dualities, one can consider nonhomogeneous Koszul semialgebras over corings and assign Koszul quasi-differential corings over corings to them. The Poincaré–Birkhoff–Witt theorem for Koszul semialgebras claims that this correspondence is an equivalence of categories.

The relative nonhomogeneous Koszul duality theorem provides an equivalence between the semiderived category of semimodules over a nonhomogeneous Koszul semialgebra and the coderived category of quasi-differential comodules over the corresponding quasi-differential coring, and an analogous equivalence between the semiderived category of semicontramodules and the contraderived category of quasi-differential contramodules. In particular, for a smooth algebraic variety M and a vector bundle E over M with a global connection ∇, there is an equivalence between the derived category of modules over the algebra/sheaf of differential operators on M acting in the sections of E and the coderived category (and also the contraderived category, when M is affine) of CDG-modules over the CDG-algebra $\Omega(M, \mathrm{End}(E))$ of differential forms with coefficients in the vector bundle $\mathrm{End}(E)$. The differential d in $\Omega(M, \mathrm{End}(E))$ is the de Rham differential depending on ∇ and the curvature element $h \in \Omega^2(M, \mathrm{End}(E))$ is the curvature of ∇.

Natural examples of semialgebras and semimodules come from Lie theory. Namely, let $(\mathfrak{g}, H)$ be an algebraic Harish-Chandra pair, i.e., $\mathfrak{g}$ is a Lie algebra over a field k and H is a smooth affine algebraic group over k corresponding to a finite-dimensional Lie subalgebra $\mathfrak{h} \subset \mathfrak{g}$. Let $\mathcal{C}(H)$ be the coalgebra of functions on H. Then the category $\mathsf{O}(\mathfrak{g}, H)$ of Harish-Chandra modules is isomorphic to the category of left semimodules over the semialgebra $\mathsf{S}(\mathfrak{g}, H) = U(\mathfrak{g}) \otimes_{U(\mathfrak{h})} \mathcal{C}(H)$. If the group H is unimodular, the semialgebra $\mathsf{S} = \mathsf{S}(\mathfrak{g}, H)$ has an involutive anti-automorphism. In general, the opposite semialgebras $\mathbf{S}$ and $\mathbf{S}^{\mathrm{op}}$ are Morita-equivalent in some sense; more precisely, there is a canonical left $\mathbf{S} \otimes_k \mathbf{S}$-semimodule $\mathcal{E} = \mathcal{E}(\mathfrak{g}, H)$ such that the semitensor product with $\mathcal{E}$ provides an equivalence between the categories of right and left $\mathbf{S}$-semimodules. Geometrically, $\mathcal{E}(\mathfrak{g}, H)$ is the bimodule of distributions on an algebraic group G supported in its subgroup H and regular along H. So the semitensor product of $\mathbf{S}$-semimodules can be considered as a functor on the category $\mathsf{O}(\mathfrak{g}, H) \times \mathsf{O}(\mathfrak{g}, H)$. This functor factorizes through the functor of tensor product in the category $\mathsf{O}(\mathfrak{g}, H)$ and is closely related to the functor of $(\mathfrak{g}, H)$-*semiinvariants* $\mathcal{M} \longmapsto \mathcal{M}_{\mathfrak{g}, H}$ on the category of $(\mathfrak{g}, H)$-modules. The semiinvariants are a mixture of invariants over H and coinvariants along $\mathfrak{g}/\mathfrak{h}$.

More generally, let $(\mathfrak{g}, H)$ be a *Tate Harish-Chandra pair*, that is $\mathfrak{g}$ is a Tate Lie algebra and H is an affine proalgebraic group corresponding to a compact open subalgebra $\mathfrak{h} \subset \mathfrak{g}$. Let $\varkappa \colon (\mathfrak{g}', H) \longrightarrow (\mathfrak{g}, H)$ be a morphism of Tate Harish-Chandra pairs with the same proalgebraic group H such that the Lie algebra map $\mathfrak{g}' \longrightarrow \mathfrak{g}$ is a central extension whose kernel is identified with k; assume also that H acts trivially in $k \subset \mathfrak{g}'$. One example of such a central extension of Tate Harish-Chandra pairs comes from the canonical central extension $\mathfrak{g}^\sim$ of $\mathfrak{g}$; we denote the corresponding morphism by $\varkappa_0$. There is a semialgebra

$$\mathbf{S}_\varkappa(\mathfrak{g}, H) = U_\varkappa(\mathfrak{g}) \otimes_{U(\mathfrak{h})} \mathcal{C}(H)$$

over the coalgebra $\mathcal{C}(H)$ such that the category of left semimodules over $\mathbf{S}_\varkappa = \mathbf{S}_\varkappa(\mathfrak{g}, H)$ is isomorphic to the category of discrete $(\mathfrak{g}', H)$-modules where the unit central element of $\mathfrak{g}'$ acts by the identity (Harish-Chandra modules with the central charge $\varkappa$). Left semicontramodules over the opposite semialgebra $\mathbf{S}_\varkappa^{\mathrm{op}}$ can be described in terms of compatible structures of $\mathfrak{g}'$-contramodules and $\mathcal{C}(H)$-contramodules. These are called *Harish-Chandra contramodules* with the central charge $-\varkappa$; the dual vector spaces to Harish-Chandra modules with the central charge $\varkappa$ can be found among them.

The semialgebras $\mathbf{S}_\varkappa$ and $\mathbf{S}_{-\varkappa_0 - \varkappa}^{\mathrm{op}}$ are naturally isomorphic, at least, when the pairing $U(\mathfrak{h}) \otimes_k \mathcal{C}(H) \longrightarrow k$ is nondegenerate in $\mathcal{C}(H)$. In view of the semimodule-semicontramodule correspondence theorem, it follows that the semiderived categories of Harish-Chandra modules with the central charge $\varkappa$ and Harish-Chandra contramodules with the central charge $\varkappa + \varkappa_0$ over $(\mathfrak{g}, H)$ are naturally equivalent. So the well-known phenomenon of correspondence between complexes of modules with complementary central charges over certain infinite-dimensional Lie algebras can be formulated as an equivalence of triangulated categories using the notions of contramodules and semiderived categories. Besides, it follows that the category of right semimodules over $\mathbf{S}_\varkappa$ is isomorphic to the category of Harish-Chandra modules with the central charge $-\varkappa - \varkappa_0$. When the proalgebraic group H is prounipotent (and $\mathfrak{h}$ is exactly the Lie algebra of H), the object

$$\mathrm{SemiTor}^{\mathbf{S}_\varkappa}(\mathbf{N}^\bullet, \mathcal{M}^\bullet)$$

of the derived category of k-vector spaces is represented by the complex of semi-infinite forms over $\mathfrak{g}$ with coefficients in the $\mathfrak{g}^\sim$-module $\mathbf{N}^\bullet \otimes_k \mathcal{M}^\bullet$. This provides a comparison of our theory of SemiTor with the semi-infinite homology of Tate Lie algebras. The semi-infinite cohomology of Lie algebras, whose coefficients are contramodules over (the canonical central extensions of) Tate Lie algebras, is related to SemiExt in the analogous way.

To a topological group G with an open profinite subgroup H and a commutative ring k one can assign a semialgebra $\mathbf{S}_k(G, H)$ over the coring $\mathfrak{C}_k(H)$ of k-valued locally constant functions on H such that the categories of left and right semimodules over $\mathbf{S}_k(G, H)$ are isomorphic to the category of smooth G-modules over k. So the category of semimodules over $\mathbf{S}_k(G, H)$ does not depend on H, neither does the category of semicontramodules over $\mathbf{S}_k(G, H)$; all the semialgebras $\mathbf{S}_k(G, H)$ with a fixed G and varying H are naturally Morita equivalent. The semiderived categories of semimodules and semicontramodules over $\mathbf{S}_k(G, H)$ do depend on H essentially, however, as do the functors SemiTor and SemiExt over $\mathbf{S}_k(G, H)$. These double-sided derived functors may be called the *semi-infinite (co)homology of a group with an open profinite subgroup*. The semi-infinite homology of topological groups is a mixture of the discrete group homology and the profinite group cohomology.

Examples of corings $\mathfrak{C}$ over commutative rings A for which the left and the right actions of A in $\mathfrak{C}$ are different come from the algebraic groupoids theory, and examples of semialgebras over such corings come from Lie theory of groupoids. Namely, let (M, H) be a smooth affine groupoid, i.e., M and H are smooth affine algebraic varieties, there are two smooth morphisms $s_H, t_H \colon H \rightrightarrows M$ of source and target, and there are unit, multiplication, and inverse element morphisms satisfying the usual groupoid axioms. Let $A = A(M)$ be the ring of functions on M and $\mathfrak{C} = \mathfrak{C}(H)$ be the ring of functions on H. Then $\mathfrak{C}$ is a coring over A. Moreover, suppose that (M, H) is a closed subgroupoid of a groupoid (M, G). Let $\mathfrak{g}$ and $\mathfrak{h}$ be the Lie algebroids over the ring A corresponding to the groupoids (M, G) and (M, H), and let $U_A(\mathfrak{g})$ and $U_A(\mathfrak{h})$ be their enveloping algebras. Then there is a semialgebra $\mathbf{S} = \mathbf{S}_M(G, H) = U_A(\mathfrak{g}) \otimes_{U_A(\mathfrak{h})} \mathfrak{C}(H)$ over the coring $\mathfrak{C}$ and a canonical left $\mathbf{S} \otimes_k \mathbf{S}$-semimodule $\mathcal{E} = \mathcal{E}_M(G, H)$ providing an equivalence between the categories of right and left $\mathbf{S}$-semimodules. The semimodule $\mathcal{E}$ consists of all distributions on G twisted with the line bundle $s_G^*(\Omega_M^{-1}) \otimes t_G^*(\Omega_M^{-1})$, supported in H and regular along H (where Ω_M denotes the bundle of top forms on M).

Examples of corings over noncommutative rings come from Noncommutative Geometry [62, 63]. Noncommutative stacks are represented as quotients of noncommutative affine schemes corresponding to rings A by actions of corings $\mathfrak{C}$ over A. The cotensor product of $\mathfrak{C}$-comodules can be understood as the group of global sections of the tensor product of a right and a left sheaf over a noncommutative stack, while the tensor product of sheaves itself does not exist.

Notice that the roles of the ring and coring structures in our constructions are not symmetric; in particular, we have to consider conventional derived categories along the algebra variables and co/contraderived categories along the coalgebra variables. The cause of this difference is that the tensor product of modules commutes with the infinite direct sums, but not with the infinite products. This can be changed by passing to pro-objects; consequently one can still define versions of derived functors Cotor and Coext over a coring $\mathcal{C}$ without making any homological dimension assumptions at all by considering pro- and ind-modules (see Remarks 2.7 and 4.7). A problem remains to construct the comodule-contramodule correspondence without any homological dimension assumptions on the ring A. Here we only manage to weaken the finite homological dimension assumption to the Gorensteinness assumption.

Algebras/coalgebras over fields and semialgebras over coalgebras over fields are briefly discussed in Chapter 0. Semialgebras over corings and the functors of semitensor product over them are introduced in Chapter 1, and important constructions of flat comodules and coflat semimodules are presented there. The derived functor SemiTor is defined in Chapter 2. Contramodules over corings and semicontramodules over semialgebras are introduced in Chapter 3, and the derived functor SemiExt is defined in Chapter 4. Equivalence of exotic derived categories of comodules and contramodules is proven in Chapter 5; and the same for semimodules and semicontramodules is done in Chapter 6. Functors of change of ring and coring for the categories of comodules and contramodules are introduced in Chapter 7; functors of change of coring and semialgebra for the categories of semimodules and semicontramodules are constructed in Chapter 8. Closed model category structures on the categories of complexes of semimodules and semicontramodules are defined in Chapter 9. The construction of a semialgebra depending on three embedded rings and a coring dual to the middle ring is considered in Chapter 10. The Poincaré–Birkhoff–Witt theorem and the Koszul duality theorem for nonhomogeneous Koszul semialgebras are proven in Chapter 11. The basic structure theory of contramodules over a coalgebra over a field is developed in Appendix A. We compare our theory of SemiExt and SemiTor with Arkhipov's and Sevostyanov's semi-infinite Ext and Tor in Appendix B. Semialgebras corresponding to Harish-Chandra pairs and their Hopf algebra analogues are discussed in Appendix C. Tate Harish-Chandra pairs are considered in Appendix D, and the theorem of comparison with semi-infinite cohomology of Tate Lie algebras is proven there. Semialgebras corresponding to topological groups are discussed in Appendix E. Pairs of algebraic groupoids are considered in Appendix F.

Appendix C was written in collaboration with Dmitriy Rumynin. Appendix D was written in collaboration with Sergey Arkhipov.

One terminological note: we will generally use the words *the homotopy category of* (an additive category A) and *the homotopy category of complexes of* (objects from A) as synonyms. Analogously, *the homotopy category of complexes* (with a particular property) *over* A is a full subcategory of the homotopy category of A.

Acknowledgement. I am grateful to B. Feigin for posing the problem of defining the semi-infinite cohomology of associative algebras back in the first half of the 1990's. Even earlier, I learned about the problem of constructing a derived equivalence between modules with complementary central charges from M. Kapranov's handwritten notes on Koszul duality. S. Arkhipov patiently explained to me many times over the years his ideas about the semi-infinite cohomology, contributing to my efforts to understand the subject. In the Summer of 2000, this work was stimulated by discussions with S. Arkhipov and R. Bezrukavnikov, and my gratitude goes to both of them. I wish to thank J. Bernstein, B. Feigin, B. Keller, V. Lunts, and V. Vologodsky for helpful conversations, and I. Mirkovic for stimulating interest. Parts of the mathematical content of this monograph were worked out when the author was visiting Stockholm University, Weizmann Institute, Independent University of Moscow, Max-Planck-Institut in Bonn, and Warwick University; I was supported by the European Post-Doctoral Institute during a part of that time. The author was partially supported by grants from CRDF, INTAS, and P. Deligne's 2004 Balzan prize while writing the manuscript.

0 Preliminaries and Summary

This chapter contains some known results and some results deemed to be new, but no proofs. Its goal is to prepare the reader for the more technically involved constructions of the main body of the monograph (where the proofs are given). In particular, we do not have to worry about nonassociativity of the cotensor product and partial definition of the semitensor product here, distinguish between the myriad of notions of absolute/relative coflatness/coprojectivity/injectivity of comodules and analogously for contramodules, etc., because we only consider coalgebras over fields.

The material of Sections 0.1–0.3 roughly corresponds to that of Chapters 1–6.

0.1 Unbounded Tor and Ext

Let R be an algebra over a field k.

0.1.1 We would like to extend the familiar definition of the derived functor of tensor product $\mathrm{Tor}^R \colon \mathsf{D}^-(\mathsf{mod}\text{-}R) \times \mathsf{D}^-(R\text{-}\mathsf{mod}) \longrightarrow \mathsf{D}^-(k\text{-}\mathsf{vect})$ on the Cartesian product of the derived categories of right and left R-modules bounded from above, so as to obtain a functor on the Cartesian product of unbounded derived categories.

As always, the tensor product of a complex of right R-modules $N^\bullet$ and a complex of left R-modules $M^\bullet$ is defined as the total complex of the bicomplex $N^i \otimes_R M^j$, constructed by taking infinite direct sums along the diagonals. This provides a functor $\mathsf{Hot}(\mathsf{mod}\text{-}R) \times \mathsf{Hot}(R\text{-}\mathsf{mod}) \longrightarrow \mathsf{Hot}(k\text{-}\mathsf{vect})$ on the Cartesian product of unbounded homotopy categories of R-modules.

The most straightforward way to define the object $\mathrm{Tor}^R(N^\bullet, M^\bullet)$ of $\mathsf{D}(k\text{-}\mathsf{vect})$ is to represent it by the total complex of the bicomplex

$$\cdots \longrightarrow N^\bullet \otimes_k R \otimes_k R \otimes_k M^\bullet \longrightarrow N^\bullet \otimes_k R \otimes_k M^\bullet \longrightarrow N^\bullet \otimes_k M^\bullet,$$

constructed by taking infinite direct sums along the diagonals. One can check that this bar construction indeed defines a functor

$$\mathrm{Tor}^R \colon \mathsf{D}(\mathsf{mod}\text{-}R) \times \mathsf{D}(R\text{-}\mathsf{mod}) \longrightarrow \mathsf{D}(k\text{-}\mathsf{vect}).$$

The unbounded derived functor Tor^R can be also defined by restricting the functor of tensor product to appropriate subcategories of complexes adjusted to the functor of tensor product in the unbounded homotopy categories of R-modules. Namely, let us call a complex of left R-modules $M^\bullet$ *flat* if the complex of k-vector spaces $M^\bullet \otimes_R N^\bullet$ is acyclic whenever a complex of right R-modules $N^\bullet$

L. Positselski, *Homological Algebra of Semimodules and Semicontramodules*, Monografie Matematyczne 70, DOI 10.1007/978-3-0346-0436-9_1, © Springer Basel AG 2010

is acyclic. *Not every complex of flat R-modules is a flat complex of R-modules according to this definition.*

In particular, an acyclic complex of left R-modules is flat if and only if it is *pure*, i.e., it remains acyclic after taking the tensor product with any right R-module. So an acyclic complex of flat R-modules is flat if and only if all of its modules of cocycles are flat. On the other hand, any complex of flat R-modules bounded from above is flat. If the ring R has a finite weak homological dimension, then any complex of flat R-modules is flat.

For example, the acyclic complex $M^\bullet$ of free modules over the ring of dual numbers $R = k[\varepsilon]/\varepsilon^2$ whose every term is equal to R and every differential is the operator of multiplication with ε is not flat. Indeed, let $N^\bullet = (\cdots \to k[\varepsilon/\varepsilon^2] \to k \to 0 \to \cdots)$ be a free resolution of the R-module k; then the complex $N^\bullet \otimes_R M^\bullet$ is quasi-isomorphic to $k \otimes_R M^\bullet$ and has a one-dimensional cohomology space in every degree, even though the complex $N^\bullet$ is acyclic.

Any complex of R-modules is quasi-isomorphic to a flat complex, and moreover, the quotient category of the homotopy category $\mathsf{Hot}_{\mathsf{fl}}(R\text{-mod})$ of flat complexes of R-modules by the thick subcategory of acyclic flat complexes $\mathsf{Acycl}(R\text{-mod}) \cap \mathsf{Hot}_{\mathsf{fl}}(R\text{-mod})$ is equivalent to the derived category $\mathsf{D}(R\text{-mod})$. This result holds for an arbitrary ring [81], and even for an arbitrary DG-ring [60, 15]. The derived functor Tor^R can be defined by restricting the functor of tensor product over R to either of the full subcategories $\mathsf{Hot}(\text{mod-}R) \times \mathsf{Hot}_{\mathsf{fl}}(R\text{-mod})$ or $\mathsf{Hot}_{\mathsf{fl}}(\text{mod-}R) \times \mathsf{Hot}(R\text{-mod})$ of the category $\mathsf{Hot}(\text{mod-}R) \times \mathsf{Hot}(R\text{-mod})$.

0.1.2   The functor $\mathrm{Hom}_R \colon \mathsf{Hot}(R\text{-mod})^{\mathsf{op}} \times \mathsf{Hot}(R\text{-mod}) \longrightarrow \mathsf{Hot}(k\text{-vect})$ and its derived functor $\mathrm{Ext}_R \colon \mathsf{D}(R\text{-mod})^{\mathsf{op}} \times \mathsf{D}(R\text{-mod}) \longrightarrow \mathsf{D}(k\text{-vect})$ need no special definition: once the unbounded homotopy and derived categories are defined, so are the spaces of homomorphisms in them. For any (unbounded) complexes of left R-modules $L^\bullet$ and $M^\bullet$, the total complex of the cobar bicomplex

$$\mathrm{Hom}_k(L^\bullet, M^\bullet) \longrightarrow \mathrm{Hom}_k(R \otimes_k L^\bullet, M^\bullet) \longrightarrow \mathrm{Hom}_k(R \otimes_k R \otimes_k L^\bullet, M^\bullet) \longrightarrow \cdots,$$

constructed by taking infinite direct products along the diagonals, represents the object $\mathrm{Ext}_R(L^\bullet, M^\bullet)$ in $\mathsf{D}(k\text{-vect})$.

The unbounded derived functor Ext_R can be also computed by restricting the functor Hom_R to appropriate subcategories in the Cartesian product of homotopy categories of R-modules. Let us call a complex of left R-modules $L^\bullet$ *projective* if the complex $\mathrm{Hom}_R(L^\bullet, M^\bullet)$ is acyclic for any acyclic complex of left R-modules $M^\bullet$. Analogously, a complex of left R-modules $M^\bullet$ is called *injective* if the complex $\mathrm{Hom}_R(L^\bullet, M^\bullet)$ is acyclic for any acyclic complex of left R-modules $L^\bullet$.

Any projective complex of R-modules is flat. Any complex of projective R-modules bounded from above is projective, and any complex of injective R-modules bounded from below is injective. If the ring R has a finite left homological dimension, then any complex of projective left R-modules is projective and any complex of injective left R-modules is injective.

A complex of R-modules is projective if and only if it belongs to the minimal triangulated subcategory of the homotopy category of R-modules containing the complex $\cdots \to 0 \to R \to 0 \to \cdots$ and closed under infinite direct sums. Analogously, a complex of R-modules is injective if and only if up to the homotopy equivalence it can be obtained from the complex $\cdots \to 0 \to \operatorname{Hom}_k(R, k) \to 0 \to \cdots$ using the operations of shift, cone, and infinite direct product. The homotopy category $\operatorname{Hot}_{\mathsf{proj}}(R\text{-mod})$ of projective complexes of R-modules and the homotopy category $\operatorname{Hot}_{\mathsf{inj}}(R\text{-mod})$ of injective complexes of R-modules are equivalent to the unbounded derived category $\mathsf{D}(R\text{-mod})$. The results mentioned in this paragraph even hold for an arbitrary DG-ring [60, 15]. The functor Ext_R can be obtained by restricting the functor Hom_R to either of the full subcategories $\operatorname{Hot}_{\mathsf{proj}}(R\text{-mod})^{\mathsf{op}} \times \operatorname{Hot}(R\text{-mod})$ or $\operatorname{Hot}(R\text{-mod})^{\mathsf{op}} \times \operatorname{Hot}_{\mathsf{inj}}(R\text{-mod})$ of the category $\operatorname{Hot}(R\text{-mod})^{\mathsf{op}} \times \operatorname{Hot}(R\text{-mod})$.

0.1.3   The definitions of unbounded Tor and Ext in terms of (co)bar constructions were known at least since the 1960's. The notions of flat, projective, and injective (unbounded) complexes of R-modules were introduced by N. Spaltenstein [81] (who attributes the idea to J. Bernstein). Such complexes were called "K-flat", "K-projective", and "K-injective" in [81]; they are often called "H-projective" or "homotopy projective" etc. nowadays.

0.2   Coalgebras over fields; Cotor and Coext

The notion of a coalgebra over a field is obtained from that of an algebra by formal dualization. Since any coassociative coalgebra is the union of its finite-dimensional subcoalgebras, the category of coalgebras is anti-equivalent to the category of profinite-dimensional algebras. There are two ways of dualizing the notion of a module over an algebra: one can consider *comodules* and *contramodules* over a coalgebra. Comodules can be thought of as discrete modules which are unions of their finite-dimensional submodules, while contramodules are modules where certain infinite summation operations are defined. Dualizing the constructions of the tensor product of modules and the space of homomorphisms between modules, one obtains the functors of cotensor product and cohomomorphisms. Their derived functors are called Cotor and Coext.

0.2.1   A coassociative *coalgebra* with counit over a field k is a k-vector space $\mathcal{C}$ endowed with a *comultiplication* map $\mu_{\mathcal{C}} : \mathcal{C} \longrightarrow \mathcal{C} \otimes_k \mathcal{C}$ and a *counit* map $\varepsilon_{\mathcal{C}} : \mathcal{C} \longrightarrow k$ satisfying the equations dual to the associativity and unity equations on the multiplication and unit maps of an associative algebra with unit. More precisely, one should have $(\mu_{\mathcal{C}} \otimes \operatorname{id}_{\mathcal{C}}) \circ \mu_{\mathcal{C}} = (\operatorname{id}_{\mathcal{C}} \otimes \mu_{\mathcal{C}}) \circ \mu_{\mathcal{C}}$ and $(\varepsilon_{\mathcal{C}} \otimes \operatorname{id}_{\mathcal{C}}) \circ \mu_{\mathcal{C}} = \operatorname{id}_{\mathcal{C}} = (\operatorname{id}_{\mathcal{C}} \otimes \varepsilon_{\mathcal{C}}) \circ \mu_{\mathcal{C}}$.

A *left comodule* $\mathcal{M}$ over a coalgebra $\mathcal{C}$ is a k-vector space endowed with a *left coaction* map $\nu_{\mathcal{M}} : \mathcal{M} \longrightarrow \mathcal{C} \otimes_k \mathcal{M}$ satisfying the equations dual to the associativity and unity equations on the action map of a module over an associative algebra with unit. More precisely, one should have $(\mu_{\mathcal{C}} \otimes \mathrm{id}_{\mathcal{M}}) \circ \nu_{\mathcal{M}} = (\mathrm{id}_{\mathcal{C}} \otimes \nu_{\mathcal{M}}) \circ \nu_{\mathcal{M}}$ and $(\varepsilon_{\mathcal{C}} \otimes \mathrm{id}_{\mathcal{M}}) \circ \nu_{\mathcal{M}} = \mathrm{id}_{\mathcal{M}}$. A *right comodule* $\mathcal{N}$ over a coalgebra $\mathcal{C}$ is a k-vector space endowed with a *right coaction* map $\nu_{\mathcal{N}} : \mathcal{N} \longrightarrow \mathcal{N} \otimes_k \mathcal{C}$ satisfying the coassociativity and counity equations $(\nu_{\mathcal{N}} \otimes \mathrm{id}_{\mathcal{C}}) \circ \nu_{\mathcal{N}} = (\mathrm{id}_{\mathcal{N}} \otimes \mu_{\mathcal{C}}) \circ \nu_{\mathcal{N}}$ and $(\mathrm{id}_{\mathcal{N}} \otimes \varepsilon_{\mathcal{C}}) \circ \nu_{\mathcal{N}} = \mathrm{id}_{\mathcal{N}}$. For example, the coalgebra $\mathcal{C}$ has natural structures of a left and a right comodule over itself.

The categories of left and right $\mathcal{C}$-comodules are abelian. We will denote them by $\mathcal{C}$-comod and comod-$\mathcal{C}$, respectively. Both infinite direct sums and infinite products exist in the category of $\mathcal{C}$-comodules, but only infinite direct sums are preserved by the forgetful functor $\mathcal{C}$-comod $\longrightarrow$ k-vect (while the infinite products are not even exact in $\mathcal{C}$-comod). A *cofree* $\mathcal{C}$-comodule is a $\mathcal{C}$-comodule of the form $\mathcal{C} \otimes_k V$, where V is a k-vector space. The space of comodule homomorphisms into the cofree $\mathcal{C}$-comodule is described by the formula $\mathrm{Hom}_{\mathcal{C}}(\mathcal{M}, \mathcal{C} \otimes_k V) \simeq \mathrm{Hom}_k(\mathcal{M}, V)$. The category of $\mathcal{C}$-comodules has enough injectives; besides, a left $\mathcal{C}$-comodule is injective if and only if it is a direct summand of a cofree $\mathcal{C}$-comodule.

The *cotensor product* $\mathcal{N} \square_{\mathcal{C}} \mathcal{M}$ of a right $\mathcal{C}$-comodule $\mathcal{N}$ and a left $\mathcal{C}$-comodule $\mathcal{M}$ is defined as the kernel of the pair of maps

$$(\nu_{\mathcal{N}} \otimes \mathrm{id}_{\mathcal{M}}, \mathrm{id}_{\mathcal{N}} \otimes \nu_{\mathcal{M}}) \colon \mathcal{N} \otimes_k \mathcal{M} \rightrightarrows \mathcal{N} \otimes_k \mathcal{C} \otimes_k \mathcal{M}.$$

This is the dual construction to the tensor product of a right module and a left module over an associative algebra. There are natural isomorphisms $\mathcal{N} \square_{\mathcal{C}} \mathcal{C} \simeq \mathcal{N}$ and $\mathcal{C} \square_{\mathcal{C}} \mathcal{M} \simeq \mathcal{M}$. The functor of cotensor product over $\mathcal{C}$ is left exact.

0.2.2 The cotensor product $\mathcal{N}^\bullet \square_{\mathcal{C}} \mathcal{M}^\bullet$ of a complex of right $\mathcal{C}$-comodules $\mathcal{N}^\bullet$ and a complex of left $\mathcal{C}$-comodules $\mathcal{M}^\bullet$ is defined as the total complex of the bicomplex $\mathcal{N}^i \square_{\mathcal{C}} \mathcal{M}^j$, constructed by taking infinite direct sums along the diagonals.

We would like to define the derived functor $\mathrm{Cotor}^{\mathcal{C}}$ of the functor of cotensor product in such a way that it could be obtained by restricting the functor $\square_{\mathcal{C}}$ to appropriate subcategories of the Cartesian product of the homotopy categories $\mathsf{Hot}(\text{comod-}\mathcal{C})$ and $\mathsf{Hot}(\mathcal{C}\text{-comod})$. In addition, we would like the object $\mathrm{Cotor}^{\mathcal{C}}(\mathcal{N}^\bullet, \mathcal{M}^\bullet)$ of $\mathsf{D}(k\text{-vect})$ to be represented by the total complex of the cobar bicomplex

$$\mathcal{N}^\bullet \otimes_k \mathcal{M}^\bullet \longrightarrow \mathcal{N}^\bullet \otimes_k \mathcal{C} \otimes_k \mathcal{M}^\bullet \longrightarrow \mathcal{N}^\bullet \otimes_k \mathcal{C} \otimes_k \mathcal{C} \otimes_k \mathcal{M}^\bullet \longrightarrow \cdots, \qquad (1)$$

constructed by taking infinite direct sums along the diagonals. It turns out that a functor $\mathrm{Cotor}^{\mathcal{C}}$ with these properties does exist, but it *is not defined on the Cartesian product of conventional unbounded derived categories* $\mathsf{D}(\text{comod-}\mathcal{C})$ *and* $\mathsf{D}(\mathcal{C}\text{-comod})$.

For example, let $\mathcal{C}$ be the coalgebra dual to the algebra of dual numbers $\mathcal{C}^* = k[\varepsilon]/\varepsilon^2$, so that $\mathcal{C}$-comodules are just $k[\varepsilon]/\varepsilon^2$-modules. Let $\mathcal{M}^\bullet$ be the acyclic complex of cofree $\mathcal{C}$-comodules whose every term is equal to $\mathcal{C}$ and every differential is the operator of multiplication with ε, and let $\mathcal{N}^\bullet$ be the complex of $\mathcal{C}$-comodules whose only nonzero term is the $\mathcal{C}$-comodule k. Then the cobar complex that we want to compute $\mathrm{Cotor}^{\mathcal{C}}(\mathcal{N}^\bullet, \mathcal{M}^\bullet)$ is quasi-isomorphic to the complex $\mathcal{N}^\bullet \,\square_{\mathcal{C}}$ $\mathcal{M}^\bullet$ and has a one-dimensional cohomology space in every degree, even though $\mathcal{M}^\bullet$ represents a zero object in $\mathsf{D}(\mathcal{C}\text{-comod})$. Therefore, a more refined version of unbounded derived category of $\mathcal{C}$-comodules has to be considered.

A complex of left $\mathcal{C}$-comodules is called *coacyclic* if it belongs to the minimal triangulated subcategory of the homotopy category $\mathsf{Hot}(\mathcal{C}\text{-comod})$ containing the total complexes of exact triples $'\mathcal{K}^\bullet \to \mathcal{K}^\bullet \to \, ''\mathcal{K}^\bullet$ of complexes of left $\mathcal{C}$-comodules and closed under infinite direct sums. (By the total complex of an exact triple of complexes we mean the total complex of the corresponding bicomplex with three rows.) Any coacyclic complex is acyclic; any acyclic complex bounded from below is coacyclic. The complex $\mathcal{M}^\bullet$ from the above example is acyclic, but not coacyclic. (Indeed, the total complex of the cobar bicomplex (1) is acyclic whenever $\mathcal{M}^\bullet$ is coacyclic.) The *coderived category* of left $\mathcal{C}$-comodules $\mathsf{D}^{\mathrm{co}}(\mathcal{C}\text{-comod})$ is defined as the quotient category of the homotopy category $\mathsf{Hot}(\mathcal{C}\text{-comod})$ by the thick subcategory of coacyclic complexes $\mathsf{Acycl}^{\mathrm{co}}(\mathcal{C}\text{-comod})$.

In the same way one can define the coderived category of any abelian category with exact functors of infinite direct sum. Over a category of finite homological dimension, every acyclic complex belongs to the minimal triangulated subcategory of the homotopy category containing the total complexes of exact triples of complexes, even without the infinite direct sum closure.

The cotensor product $\mathcal{N}^\bullet \,\square_{\mathcal{C}} \mathcal{M}^\bullet$ of a complex of right $\mathcal{C}$-comodules $\mathcal{M}^\bullet$ and a complex of left $\mathcal{C}$-comodules $\mathcal{N}^\bullet$ is acyclic whenever one of the complexes $\mathcal{M}^\bullet$ and $\mathcal{N}^\bullet$ is coacyclic and the other one is a complex of injective $\mathcal{C}$-comodules. Besides, the coderived category $\mathsf{D}^{\mathrm{co}}(\mathcal{C}\text{-comod})$ is equivalent to the homotopy category $\mathsf{Hot}(\mathcal{C}\text{-comod}_{\mathrm{inj}})$ of injective $\mathcal{C}$-comodules. Thus one can define the unbounded derived functor

$$\mathrm{Cotor}^{\mathcal{C}} \colon \mathsf{D}^{\mathrm{co}}(\mathrm{comod}\text{-}\mathcal{C}) \times \mathsf{D}^{\mathrm{co}}(\mathcal{C}\text{-comod}) \longrightarrow \mathsf{D}(k\text{-vect})$$

by restricting the functor of cotensor product to either of the full subcategories $\mathsf{Hot}(\mathrm{comod}\text{-}\mathcal{C}) \times \mathsf{Hot}(\mathcal{C}\text{-comod}_{\mathrm{inj}})$ or $\mathsf{Hot}(\mathrm{comod}_{\mathrm{inj}}\text{-}\mathcal{C}) \times \mathsf{Hot}(\mathcal{C}\text{-comod})$ of the category $\mathsf{Hot}(\mathrm{comod}\text{-}\mathcal{C}) \times \mathsf{Hot}(\mathcal{C}\text{-comod})$.

0.2.3 If one attempts to construct a derived functor of cotensor product on the Cartesian product of conventional unbounded derived categories of comodules in a way analogous to 0.1.1, the result may not look like what one expects.

Consider the example of a finite-dimensional coalgebra $\mathcal{C}$ dual to a Frobenius algebra $\mathcal{C}^* = F$. Let us assume the convention that left $\mathcal{C}$-comodules are left F-modules and right $\mathcal{C}$-comodules are right F-modules. Then the functor $\square_{\mathcal{C}}$ is left exact and the functor $\otimes_F$ is right exact, but the difference between them is still rather small: if either a left (co)module M, or a right (co)module N is projective-injective, then there is a natural isomorphism $N \,\square_{\mathcal{C}} M \simeq (N \,\square_{\mathcal{C}} F) \otimes_F M$, and after one chooses an isomorphism between the left modules F and $\mathcal{C}$, the right modules N and $N \,\square_{\mathcal{C}} F$ will only differ by the Frobenius automorphism of the Frobenius algebra F.

So if one defines "coflat" complexes of $\mathcal{C}$-comodules as the complexes whose cotensor products with acyclic complexes are acyclic, then the quotient category of the homotopy category of "coflat" complexes by the thick subcategory of acyclic "coflat" complexes will be indeed equivalent to the derived category of comodules, and one will be able to define a "derived functor of cotensor product over $\mathcal{C}$" in this way, but the resulting derived functor will coincide, up to the Frobenius twist, with the functor Tor^F. (Indeed, any flat complex of flat modules will be "coflat".) When the complexes this derived functor is applied to are concentrated in degree 0, this functor will produce a complex situated in the negative cohomological degrees, as is characteristic of Tor^F, and not in the positive ones, as one would expect of $\mathrm{Cotor}^{\mathcal{C}}$.

Likewise, if one attempts to construct a derived functor of tensor product on the Cartesian product of coderived categories of modules in the way analogous to 0.2.2, one will find, in the Frobenius algebra case, that the tensor product of a complex of projective F-modules with a coacyclic complex is acyclic, the homotopy category of complexes of projective modules is indeed equivalent to the coderived category of F-modules, and one can define a "derived functor of tensor product over F" by restricting to this subcategory, but the resulting derived functor will coincide, up to the Frobenius twist, with the functor $\mathrm{Cotor}^{\mathcal{C}}$.

Nevertheless, it is well known how to define a derived functor of cotensor product on the conventional unbounded derived categories of comodules (see 0.2.10, cf. Remark 2.7).

0.2.4 The category $k\text{-vect}^{\mathrm{op}}$ opposite to the category of vector spaces has a natural structure of a *module category* over the tensor category $k\text{-vect}$ with the action functor $k\text{-vect} \times k\text{-vect}^{\mathrm{op}} \longrightarrow k\text{-vect}^{\mathrm{op}}$ defined by the rule $(V, W^{\mathrm{op}}) \longmapsto \mathrm{Hom}_k(V, W)^{\mathrm{op}}$. More precisely, there are two module category structures associated with this functor: the left module category with the associativity constraint $\mathrm{Hom}_k(U \otimes_k V, W) \simeq \mathrm{Hom}_k(U, \mathrm{Hom}_k(V, W))$ and the right module category with the associativity constraint $\mathrm{Hom}_k(U \otimes_k V, W) \simeq \mathrm{Hom}_k(V, \mathrm{Hom}_k(U, W))$. The category of *left contramodules* over a coalgebra $\mathcal{C}$ is the opposite category to the category of comodule objects in the *right* module category $k\text{-vect}^{\mathrm{op}}$ over the coring object $\mathcal{C}$ in the tensor category $k\text{-vect}$. Explicitly, a $\mathcal{C}$-contramodule $\mathfrak{P}$ is a k-vector space endowed with a *contraaction* map $\pi_{\mathfrak{P}} \colon \mathrm{Hom}_k(\mathcal{C}, \mathfrak{P}) \longrightarrow \mathfrak{P}$ satisfying the *contraassociativity* and *counity* equations $\pi_{\mathfrak{P}} \circ \mathrm{Hom}(\mathrm{id}_{\mathcal{C}}, \pi_{\mathfrak{P}}) = \pi_{\mathfrak{P}} \circ \mathrm{Hom}(\mu_{\mathcal{C}}, \mathrm{id}_{\mathfrak{P}})$ and $\pi_{\mathfrak{P}} \circ \mathrm{Hom}(\varepsilon_{\mathcal{C}}, \mathrm{id}_{\mathfrak{P}}) = \mathrm{id}_{\mathfrak{P}}$.

For any right $\mathcal{C}$-comodule $\mathcal{N}$ and any k-vector space V the space $\mathrm{Hom}_k(\mathcal{N}, V)$ has a natural structure of left $\mathcal{C}$-contramodule. The category of left $\mathcal{C}$-contramodules is abelian. We will denote it by $\mathcal{C}\text{-contra}$. Both infinite direct sums and infinite products exist in the category of contramodules, but only the infinite products are preserved by the forgetful functor $\mathcal{C}\text{-contra} \longrightarrow k\text{-vect}$ (while the infinite direct sums are not even exact in $\mathcal{C}\text{-contra}$). The category of contramodules has enough projectives. Besides, a $\mathcal{C}$-contramodule is projective if and only if it is a direct summand of a *free* $\mathcal{C}$-contramodule of the form $\mathrm{Hom}_k(\mathcal{C}, V)$ for some vector space V. The space of contramodule homomorphisms from the free $\mathcal{C}$-con-tramodule is described by the formula $\operatorname{Hom}^{\mathcal{C}}(\operatorname{Hom}_k(\mathcal{C}, V), \mathfrak{P}) \simeq \operatorname{Hom}_k(V, \mathfrak{P})$.

Let $\mathcal{M}$ be a left $\mathcal{C}$-comodule and $\mathfrak{P}$ be a left $\mathcal{C}$-contramodule. The *space of cohomomorphisms* $\operatorname{Cohom}_{\mathcal{C}}(\mathcal{M}, \mathfrak{P})$ is defined as the cokernel of the pair of maps

$$(\operatorname{Hom}(\nu_{\mathcal{M}}, \operatorname{id}_{\mathfrak{P}}), \operatorname{Hom}(\operatorname{id}_{\mathcal{M}}, \pi_{\mathfrak{P}})):$$
$$\operatorname{Hom}_k(\mathcal{C} \otimes_k \mathcal{M}, \mathfrak{P}) = \operatorname{Hom}_k(\mathcal{M}, \operatorname{Hom}_k(\mathcal{C}, \mathfrak{P})) \rightrightarrows \operatorname{Hom}_k(\mathcal{M}, \mathfrak{P}).$$

This is the dual construction to that of the space of homomorphisms between two modules over a ring. There are natural isomorphisms $\operatorname{Cohom}_{\mathcal{C}}(\mathcal{C}, \mathfrak{P}) \simeq \mathfrak{P}$ and $\operatorname{Cohom}_{\mathcal{C}}(\mathcal{M}, \operatorname{Hom}_k(\mathcal{N}, V)) \simeq \operatorname{Hom}_k(\mathcal{N} \square_{\mathcal{C}} \mathcal{M}, V)$. The functor of cohomomorphisms over $\mathcal{C}$ is right exact.

0.2.5 The complex of cohomomorphisms $\operatorname{Cohom}_{\mathcal{C}}(\mathcal{M}^\bullet, \mathfrak{P}^\bullet)$ from a complex of left $\mathcal{C}$-comodules $\mathcal{M}^\bullet$ to a complex of left $\mathcal{C}$-contramodules $\mathfrak{P}^\bullet$ is defined as the total complex of the bicomplex $\operatorname{Cohom}_{\mathcal{C}}(\mathcal{M}^i, \mathfrak{P}^j)$, constructed by taking infinite products along the diagonals. Let us define the derived functor $\operatorname{Coext}_{\mathcal{C}}$ of the functor of cohomomorphisms.

A complex of $\mathcal{C}$-contramodules is called *contraacyclic* if it belongs to the minimal triangulated subcategory of the homotopy category $\operatorname{Hot}(\mathcal{C}\text{-contra})$ containing the total complexes of exact triples $'\mathfrak{R}^\bullet \to \mathfrak{R}^\bullet \to ''\mathfrak{R}^\bullet$ of complexes of $\mathcal{C}$-contramod-ules and closed under infinite products. Any contraacyclic complex is acyclic; any acyclic complex bounded from above is contraacyclic. The *contraderived category* of $\mathcal{C}$-contramodules $\mathsf{D}^{\operatorname{ctr}}(\mathcal{C}\text{-contra})$ is defined as the quotient category of the homo-topy category $\operatorname{Hot}(\mathcal{C}\text{-contra})$ by the thick subcategory of contraacyclic complexes $\operatorname{Acycl}^{\operatorname{ctr}}(\mathcal{C}\text{-contra})$.

The complex of cohomomorphisms $\operatorname{Cohom}_{\mathcal{C}}(\mathcal{M}^\bullet, \mathfrak{P}^\bullet)$ is acyclic whenever ei-ther $\mathcal{M}^\bullet$ is a complex of injective $\mathcal{C}$-comodules and $\mathfrak{P}^\bullet$ is contraacyclic, or $\mathcal{M}^\bullet$ is coacyclic and $\mathfrak{P}^\bullet$ is a complex of projective $\mathcal{C}$-contramodules. Besides, the contra-derived category $\mathsf{D}^{\operatorname{ctr}}(\mathcal{C}\text{-contra})$ is equivalent to the homotopy category of projec-tive $\mathcal{C}$-contramodules $\operatorname{Hot}(\mathcal{C}\text{-contra}_{\operatorname{proj}})$. Thus one can define the derived functor

$$\operatorname{Coext}_{\mathcal{C}} \colon \mathsf{D}^{\operatorname{co}}(\mathcal{C}\text{-comod})^{\operatorname{op}} \times \mathsf{D}^{\operatorname{ctr}}(\mathcal{C}\text{-contra}) \longrightarrow \mathsf{D}(k\text{-vect})$$

by restricting the functor $\operatorname{Cohom}_{\mathcal{C}}$ to either of the subcategories

$$\operatorname{Hot}(\mathcal{C}\text{-comod}_{\operatorname{inj}})^{\operatorname{op}} \times \operatorname{Hot}(\mathcal{C}\text{-contra}) \quad \text{or} \quad \operatorname{Hot}(\mathcal{C}\text{-comod})^{\operatorname{op}} \times \operatorname{Hot}(\mathcal{C}\text{-contra}_{\operatorname{proj}})$$

of the Cartesian product $\operatorname{Hot}(\mathcal{C}\text{-comod})^{\operatorname{op}} \times \operatorname{Hot}(\mathcal{C}\text{-contra})$.

The contramodule version of bar construction provides a functorial complex computing $\operatorname{Coext}_{\mathcal{C}}$. Namely, for any complex of left $\mathcal{C}$-comodules $\mathcal{M}^\bullet$ and complex of left $\mathcal{C}$-contramodules $\mathfrak{P}^\bullet$ the total complex of the bicomplex

$$\cdots \longrightarrow \operatorname{Hom}_k(\mathcal{C} \otimes_k \mathcal{C} \otimes_k \mathcal{M}^\bullet, \mathfrak{P}^\bullet) \longrightarrow \operatorname{Hom}_k(\mathcal{C} \otimes_k \mathcal{M}^\bullet, \mathfrak{P}^\bullet) \longrightarrow \operatorname{Hom}_k(\mathcal{M}^\bullet, \mathfrak{P}^\bullet),$$

constructed by taking infinite products along the diagonals, represents the object $\operatorname{Coext}_{\mathcal{C}}(\mathcal{M}^\bullet, \mathfrak{P}^\bullet)$ in $\mathsf{D}(k\text{-vect})$.

0.2.6   The categories of left $\mathcal{C}$-comodules and left $\mathcal{C}$-contramodules are isomorphic if the coalgebra $\mathcal{C}$ is finite-dimensional, but in general they are quite different. However, the coderived category of left $\mathcal{C}$-comodules is naturally equivalent to the contraderived category of left $\mathcal{C}$-contramodules, $\mathsf{D}^{\mathrm{co}}(\mathcal{C}\text{-comod}) \simeq \mathsf{D}^{\mathrm{ctr}}(\mathcal{C}\text{-contra})$.

Indeed, the coderived category $\mathsf{D}^{\mathrm{co}}(\mathcal{C}\text{-comod})$ is equivalent to the homotopy category $\mathsf{Hot}(\mathcal{C}\text{-comod}_{\mathrm{inj}})$ and the contraderived category $\mathsf{D}^{\mathrm{ctr}}(\mathcal{C}\text{-contra})$ is equivalent to the homotopy category $\mathsf{Hot}(\mathcal{C}\text{-contra}_{\mathrm{proj}})$. Furthermore, the additive category of injective $\mathcal{C}$-comodules is the idempotent closure of the category of cofree $\mathcal{C}$-comodules and the additive category of projective $\mathcal{C}$-contramodules is the idempotent closure of the category of free $\mathcal{C}$-contramodules. One has $\mathrm{Hom}_{\mathcal{C}}(\mathcal{C}\otimes_k U,\ \mathcal{C}\otimes_k V) = \mathrm{Hom}_k(\mathcal{C}\otimes_k U,\ V) = \mathrm{Hom}_k(U,\ \mathrm{Hom}_k(\mathcal{C},V)) = \mathrm{Hom}^{\mathcal{C}}(\mathrm{Hom}_k(\mathcal{C},U),\mathrm{Hom}_k(\mathcal{C},V))$, so the categories of cofree comodules and free contramodules are equivalent.

To describe this equivalence of additive categories in a more invariant way, let us define the operation of contratensor product of a comodule and a contramodule.

Let $\mathcal{N}$ be a right $\mathcal{C}$-comodule and $\mathfrak{P}$ be a left $\mathcal{C}$-contramodule. The *contratensor product* $\mathcal{N}\odot_{\mathcal{C}}\mathfrak{P}$ is defined as the cokernel of the pair of maps

$$((\mathrm{id}_{\mathcal{N}}\otimes\mathrm{ev}_{\mathcal{C}})\circ(\nu_{\mathcal{N}}\otimes\mathrm{id}_{\mathrm{Hom}_k(\mathcal{C},\mathfrak{P})}),\ \mathrm{id}_{\mathcal{N}}\circ\pi_{\mathfrak{P}})\colon \mathcal{N}\otimes_k\mathrm{Hom}_k(\mathcal{C},\mathfrak{P}) \rightrightarrows \mathcal{N}\otimes_k\mathfrak{P},$$

where $\mathrm{ev}_{\mathcal{C}}$ denotes the evaluation map $\mathcal{C}\otimes_k\mathrm{Hom}_k(\mathcal{C},\mathfrak{P}) \to \mathfrak{P}$. The contratensor product functor is not a part of any tensor or module category structure; instead, it is dual to the functors Hom in the categories of $\mathcal{C}$-comodules and $\mathcal{C}$-contramodules. The functor of contratensor product over $\mathcal{C}$ is right exact. There are natural isomorphisms $\mathcal{N}\odot_{\mathcal{C}}\mathrm{Hom}_k(\mathcal{C},V) \simeq \mathcal{N}\otimes_k V$ and $\mathrm{Hom}_k(\mathcal{N}\odot_{\mathcal{C}}\mathfrak{P},\ V) \simeq \mathrm{Hom}^{\mathcal{C}}(\mathfrak{P},\mathrm{Hom}_k(\mathcal{N},V))$.

The desired equivalence between the additive categories of injective left $\mathcal{C}$-comodules and projective left $\mathcal{C}$-contramodules is provided by the pair of adjoint functors $\Psi_{\mathcal{C}}(\mathcal{M}) = \mathrm{Hom}_{\mathcal{C}}(\mathcal{C},\mathcal{M})$ and $\Phi_{\mathcal{C}}(\mathfrak{P}) = \mathcal{C}\odot_{\mathcal{C}}\mathfrak{P}$ between the categories of left $\mathcal{C}$-comodules and left $\mathcal{C}$-contramodules. Here the space $\mathrm{Hom}_{\mathcal{C}}(\mathcal{C},\mathcal{M})$ is endowed with a $\mathcal{C}$-contramodule structure as the kernel of a pair of contramodule morphisms $\mathrm{Hom}_k(\mathcal{C},\mathcal{M}) \rightrightarrows \mathrm{Hom}_k(\mathcal{C},\ \mathcal{C}\otimes_k\mathcal{M})$ (where the contramodule structure on $\mathrm{Hom}_k(\mathcal{C},\mathcal{M})$ and $\mathrm{Hom}_k(\mathcal{C},\ \mathcal{C}\otimes_k\mathcal{M})$ comes from the right $\mathcal{C}$-comodule structure on $\mathcal{C}$), while the space $\mathcal{C}\odot_{\mathcal{C}}\mathfrak{P}$ is endowed with a left $\mathcal{C}$-comodule structure as the cokernel of a pair of comodule morphisms $\mathcal{C}\otimes_k\mathrm{Hom}_k(\mathcal{C},\mathfrak{P}) \rightrightarrows \mathcal{C}\otimes_k\mathfrak{P}$.

0.2.7   The following examples illustrate the necessity of considering the exotic derived categories in the above construction of the derived comodule-contramodule correspondence. Let W be a vector space and $\mathcal{C}$ be the symmetric coalgebra of W. One can construct $\mathcal{C}$ as the subcoalgebra of the tensor coalgebra $\bigoplus_{n=0}^{\infty} W^{\otimes n}$ formed by the symmetric tensors. Consider the trivial left $\mathcal{C}$-contramodule k; it has a left projective $\mathcal{C}$-contramodule resolution of the form

$$\cdots \longrightarrow \mathrm{Hom}_k(\mathcal{C},(\textstyle\bigwedge_k^2 W)^*) \longrightarrow \mathrm{Hom}_k(\mathcal{C},W^*) \longrightarrow \mathrm{Hom}_k(\mathcal{C},k).$$

Applying the functor $\Phi_{\mathcal{C}}$ to the above complex of contramodules, one obtains the complex of injective left $\mathcal{C}$-comodules

$$\cdots \longrightarrow \mathcal{C} \otimes_k (\textstyle\bigwedge_k^2 W)^* \longrightarrow \mathcal{C} \otimes_k W^* \longrightarrow \mathcal{C}. \qquad (2)$$

When W is finite-dimensional, the complex (2) has its only nonvanishing cohomology in the degree $-\dim W$; this cohomology is a trivial one-dimensional $\mathcal{C}$-comodule naturally isomorphic to $\det(W)^* = (\bigwedge_k^{\dim W} W)^*$ as a vector space. When W is infinite-dimensional, the complex (2) is acyclic; one can think of it as an "injective resolution of a one-dimensional $\mathcal{C}$-comodule placed in the degree $-\infty$". So when $\dim W = \infty$ the equivalence of categories $\mathsf{D}^{\mathsf{co}}(\mathcal{C}\text{-comod}) \simeq \mathsf{D}^{\mathsf{ctr}}(\mathcal{C}\text{-contra})$ transforms the acyclic complex of $\mathcal{C}$-comodules (2) into the trivial $\mathcal{C}$-contramodule k considered as a complex concentrated in degree 0, and back.

Analogously, consider the trivial left $\mathcal{C}$-comodule k; it has a right injective $\mathcal{C}$-comodule resolution of the form

$$\mathcal{C} \longrightarrow \mathcal{C} \otimes_k W \longrightarrow \mathcal{C} \otimes_k \textstyle\bigwedge_k^2 W \longrightarrow \cdots$$

Applying the functor $\Psi_{\mathcal{C}}$ to this complex of comodules, one obtains the complex of projective left $\mathcal{C}$-contramodules

$$\operatorname{Hom}_k(\mathcal{C}, k) \longrightarrow \operatorname{Hom}_k(\mathcal{C}, W) \longrightarrow \operatorname{Hom}_k(\mathcal{C}, \textstyle\bigwedge_k^2 W) \longrightarrow \cdots$$

When W is finite-dimensional, the latter complex has its only nonvanishing cohomology in the degree $\dim W$; the cohomology is a trivial one-dimensional $\mathcal{C}$-contramodule naturally isomorphic to $\det(W)$ as a vector space. When W is infinite-dimensional, this complex is acyclic; one can think of it as a "projective resolution of a one-dimensional $\mathcal{C}$-contramodule placed in the degree $+\infty$". In this case, the equivalence of categories $\mathsf{D}^{\mathsf{co}}(\mathcal{C}\text{-comod}) \simeq \mathsf{D}^{\mathsf{ctr}}(\mathcal{C}\text{-contra})$ transforms the trivial $\mathcal{C}$-comodule k considered as a complex concentrated in degree 0 into this acyclic complex of $\mathcal{C}$-contramodules, and back.

The above-mentioned cohomology computations are very similar to computing $\operatorname{Ext}_R(k, R)$ for the algebra R of polynomials in a finite or infinite number of variables over a field k.

0.2.8 The functor $\operatorname{Ext}_{\mathcal{C}} \colon \mathsf{D}^{\mathsf{co}}(\mathcal{C}\text{-comod})^{\mathsf{op}} \times \mathsf{D}^{\mathsf{co}}(\mathcal{C}\text{-comod}) \longrightarrow \mathsf{D}(k\text{-vect})$ of homomorphisms in the coderived category $\mathsf{D}^{\mathsf{co}}(\mathcal{C}\text{-comod})$ can be computed by restricting the functor $\operatorname{Hom}_{\mathcal{C}} \colon \mathsf{Hot}(\mathcal{C}\text{-comod})^{\mathsf{op}} \times \mathsf{Hot}(\mathcal{C}\text{-comod}) \longrightarrow \mathsf{Hot}(k\text{-vect})$ of homomorphisms in the homotopy category $\mathsf{Hot}(\mathcal{C}\text{-comod})$ to the full subcategory $\mathsf{Hot}(\mathcal{C}\text{-comod})^{\mathsf{op}} \times \mathsf{Hot}(\mathcal{C}\text{-comod}_{\mathsf{inj}})$ of the category $\mathsf{Hot}(\mathcal{C}\text{-comod})^{\mathsf{op}} \times \mathsf{Hot}(\mathcal{C}\text{-comod})$. The complex $\operatorname{Hom}_{\mathcal{C}}(\mathcal{L}^\bullet, \mathcal{M}^\bullet)$ is acyclic whenever $\mathcal{L}^\bullet$ is a coacyclic complex of left $\mathcal{C}$-comodules and $\mathcal{M}^\bullet$ is a complex of injective left $\mathcal{C}$-comodules.

Analogously, the functor

$$\operatorname{Ext}^{\mathcal{C}} \colon \mathsf{D}^{\mathsf{ctr}}(\mathcal{C}\text{-contra})^{\mathsf{op}} \times \mathsf{D}^{\mathsf{ctr}}(\mathcal{C}\text{-contra}) \longrightarrow \mathsf{D}(k\text{-vect})$$

of homomorphisms in the contraderived category $\mathsf{D}^{\mathsf{ctr}}(\mathcal{C}\text{-contra})$ can be computed by restricting the functor $\mathrm{Hom}^{\mathcal{C}} \colon \mathsf{Hot}(\mathcal{C}\text{-contra})^{\mathsf{op}} \times \mathsf{Hot}(\mathcal{C}\text{-contra}) \longrightarrow \mathsf{Hot}(k\text{-vect})$ to the full subcategory

$$\mathsf{Hot}(\mathcal{C}\text{-contra}_{\mathsf{proj}})^{\mathsf{op}} \times \mathsf{Hot}(\mathcal{C}\text{-contra})$$

of the category $\mathsf{Hot}(\mathcal{C}\text{-contra})^{\mathsf{op}} \times \mathsf{Hot}(\mathcal{C}\text{-contra})$. The complex $\mathrm{Hom}^{\mathcal{C}}(\mathfrak{P}^{\bullet}, \mathfrak{Q}^{\bullet})$ is acyclic whenever $\mathfrak{P}^{\bullet}$ is a complex of projective $\mathcal{C}$-contramodules and $\mathfrak{Q}^{\bullet}$ is a contraacyclic complex of $\mathcal{C}$-contramodules.

The contratensor product $\mathcal{N}^{\bullet} \odot_{\mathcal{C}} \mathfrak{P}^{\bullet}$ of a complex of right $\mathcal{C}$-comodules $\mathcal{N}^{\bullet}$ and a complex of left $\mathcal{C}$-contramodules $\mathfrak{P}^{\bullet}$ is defined as the total complex of the bicomplex $\mathcal{N}^{i} \odot_{\mathcal{C}} \mathfrak{P}^{j}$, constructed by taking infinite direct sums along the diagonals. The complex $\mathcal{N}^{\bullet} \odot_{\mathcal{C}} \mathfrak{P}^{\bullet}$ is acyclic whenever $\mathcal{N}^{\bullet}$ is a coacyclic complex of right $\mathcal{C}$-comodules and $\mathfrak{P}^{\bullet}$ is a complex of projective left $\mathcal{C}$-contramodules. The left derived functor $\mathrm{Ctrtor}^{\mathcal{C}}$ of the functor of contratensor product,

$$\mathrm{Ctrtor}^{\mathcal{C}} \colon \mathsf{D}^{\mathsf{co}}(\mathsf{comod}\text{-}\mathcal{C}) \times \mathsf{D}^{\mathsf{ctr}}(\mathcal{C}\text{-contra}) \longrightarrow \mathsf{D}(k\text{-vect}),$$

is defined by restricting the functor of contratensor product to the full subcategory $\mathsf{Hot}(\mathsf{comod}\text{-}\mathcal{C}) \times \mathsf{Hot}(\mathcal{C}\text{-contra}_{\mathsf{proj}})$ of the category $\mathsf{Hot}(\mathsf{comod}\text{-}\mathcal{C}) \times \mathsf{Hot}(\mathcal{C}\text{-contra})$. Notice that the (abelian or homotopy) category of right $\mathcal{C}$-comodules does not contain enough objects adjusted to contratensor product.

The equivalence of triangulated categories $\mathsf{D}^{\mathsf{co}}(\mathcal{C}\text{-comod}) \simeq \mathsf{D}^{\mathsf{ctr}}(\mathcal{C}\text{-contra})$ transforms the functor $\mathrm{Coext}_{\mathcal{C}}$ into either of the functors $\mathrm{Ext}_{\mathcal{C}}$ or $\mathrm{Ext}^{\mathcal{C}}$ and the functor $\mathrm{Cotor}^{\mathcal{C}}$ into the functor $\mathrm{Ctrtor}^{\mathcal{C}}$.

0.2.9 A left $\mathcal{C}$-comodule $\mathcal{M}$ is called *coflat* if the functor $\mathcal{N} \longmapsto \mathcal{N} \,\square_{\mathcal{C}}\, \mathcal{M}$ is exact on the category of right $\mathcal{C}$-comodules. A left $\mathcal{C}$-comodule $\mathcal{M}$ is called *coprojective* if the functor $\mathfrak{P} \longmapsto \mathrm{Cohom}_{\mathcal{C}}(\mathcal{M}, \mathfrak{P})$ is exact on the category of left $\mathcal{C}$-contramodules. It is easy to see that an injective comodule is coprojective and a coprojective comodule is coflat. Using the fact that any comodule is a union of its finite-dimensional subcomodules, one can show that any coflat comodule is injective. Thus all the three conditions are equivalent.

A left $\mathcal{C}$-contramodule $\mathfrak{P}$ is called *contraflat* if the functor $\mathcal{N} \longmapsto \mathcal{N} \odot_{\mathcal{C}} \mathfrak{P}$ is exact on the category of right $\mathcal{C}$-comodules. A left $\mathcal{C}$-contramodule $\mathfrak{P}$ is called *coinjective* if the functor $\mathcal{M} \longmapsto \mathrm{Cohom}_{\mathcal{C}}(\mathcal{M}, \mathfrak{P})$ is exact on the category of left $\mathcal{C}$-comodules. It is easy to see that a projective contramodule is coinjective and a coinjective contramodule is contraflat. We will show in 5.2 that any coinjective contramodule is projective and (independently) in A.3 that any contraflat contramodule is projective. Thus all the three conditions are equivalent.

0.2.10 Our definition of the derived functor of cotensor product for unbounded complexes differs from the most traditional one, which was introduced (in the greater generality of DG-coalgebras and DG-comodules) by Eilenberg and Moore [34]. Husemoller, Moore, and Stasheff [52, I.4–5] call the functor defined by Eilenberg–Moore *the differential derived functor of cotensor product of the first kind* and denote it by $\text{Cotor}^{\mathcal{C},I}$ or simply $\text{Cotor}^{\mathcal{C}}$, while the functor $\text{Cotor}^{\mathcal{C}}$ defined in 0.2.2 is (the nondifferential particular case of) what they call *the differential derived functor of cotensor product of the second kind* and denote by $\text{Cotor}^{\mathcal{C},II}$.

The functor $\text{Cotor}^{\mathcal{C},I}$ is computed by the total complex of the cobar bicomplex (1), constructed by taking infinite *products* along the diagonals (while the tensor product complexes $\mathcal{N}^{\bullet} \otimes \mathcal{C} \otimes \cdots \otimes \mathcal{C} \otimes \mathcal{M}^{\bullet}$ constituting the cobar bicomplex are still defined as infinite direct sums). It is indeed a functor on the Cartesian product of conventional unbounded derived categories $\mathsf{D}(\text{comod-}\mathcal{C})$ and $\mathsf{D}(\mathcal{C}\text{-comod})$.

The unbounded derived functor Tor^R defined in 0.1.1 is a derived functor of the first kind in this terminology. Roughly, derived functors of the first kind correspond to the conventional derived categories D (which can be therefore called *derived categories of the first kind*), while derived functors of the second kind correspond to the coderived and contraderived categories D^{co} and D^{ctr} (which can be called *derived categories of the second kind*). The distinction, which is only relevant for unbounded complexes of modules (comodules, or contramodules), manifests itself also for quite finite-dimensional DG-modules (DG-comodules, or DG-contramodules).

The coderived categories of comodules were introduced by K. Lefèvre-Hasegawa [66, 61] in the context of Koszul duality; our definition is equivalent to his (the nondifferential case). They first appeared in the author's own research in the very same context. An elaborate discussion of the two kinds of derived categories and their roles in Koszul duality can be found in [76]; a proof of the equivalence of the two definitions is also given there. Contramodules were defined by Eilenberg and Moore in [33] and studied by Barr in [6]. A module version of the comodule-contramodule correspondence was developed in [65], where the homotopy categories of complexes of injectives and projectives are being used, following the approaches of [55, 64], in lieu of the coderived and contraderived categories.

All the most important results of this section can be extended straightforwardly to DG-coalgebras and even CDG-coalgebras (see [76] or 0.4.3 and 11.2.2 for the definition). Generally, the constructions of derived categories and functors of the first kind can be generalized to A_{∞}-algebras, while the constructions of derived categories and functors of the second kind can be naturally extended to CDG-coalgebras.

0.3 Semialgebras over coalgebras over fields

The notion of a semialgebra over a coalgebra is dual to that of a coring over a noncommutative ring. The similarity between the two theories only goes so far, however.

0.3.1 Let $\mathcal{C}$ and $\mathcal{D}$ be two coalgebras over a field k. A $\mathcal{C}$-$\mathcal{D}$-*bicomodule* $\mathcal{K}$ is k-vector space endowed with a left $\mathcal{C}$-comodule and a right $\mathcal{D}$-comodule structures such that the left $\mathcal{C}$-coaction map $\nu'_{\mathcal{K}} \colon \mathcal{K} \longrightarrow \mathcal{C} \otimes_k \mathcal{K}$ is a morphism of right $\mathcal{D}$-comodules, or, equivalently, the right $\mathcal{D}$-coaction map $\nu''_{\mathcal{K}} \colon \mathcal{K} \longrightarrow \mathcal{K} \otimes_k \mathcal{D}$ is a morphism of left $\mathcal{C}$-comodules. A bicomodule can be also defined as a vector space endowed with a *bicoaction* map $\mathcal{K} \longrightarrow \mathcal{C} \otimes_k \mathcal{K} \otimes_k \mathcal{D}$ satisfying the coassociativity and counity equations. The category of $\mathcal{C}$-$\mathcal{D}$-bicomodules is abelian. We will denote it by $\mathcal{C}$-comod-$\mathcal{D}$.

Let $\mathcal{C}$, $\mathcal{D}$, and $\mathcal{E}$ be three coalgebras, $\mathcal{N}$ be a $\mathcal{C}$-$\mathcal{D}$-bicomodule, and $\mathcal{M}$ be a $\mathcal{D}$-$\mathcal{E}$-bicomodule. Then the cotensor product $\mathcal{N} \,\square_{\mathcal{D}}\, \mathcal{M}$ is endowed with a $\mathcal{C}$-$\mathcal{E}$-bicomodule structure as the kernel of a pair of bicomodule morphisms $\mathcal{N} \otimes_k \mathcal{M} \rightrightarrows \mathcal{N} \otimes_k \mathcal{D} \otimes_k \mathcal{M}$. The cotensor product of bicomodules is associative: for any coalgebras $\mathcal{C}$ and $\mathcal{D}$, any right $\mathcal{C}$-comodule $\mathcal{N}$, left $\mathcal{D}$-comodule $\mathcal{M}$, and $\mathcal{C}$-$\mathcal{D}$-bicomodule $\mathcal{K}$ there is a natural isomorphism $\mathcal{N} \,\square_{\mathcal{C}}\, (\mathcal{K} \,\square_{\mathcal{D}}\, \mathcal{M}) \simeq (\mathcal{N} \,\square_{\mathcal{C}}\, \mathcal{K}) \,\square_{\mathcal{D}}\, \mathcal{M}$.

0.3.2 In particular, the category of $\mathcal{C}$-$\mathcal{C}$-bicomodules is an associative tensor category with the unit object $\mathcal{C}$. A *semialgebra* $\mathbf{S}$ over $\mathcal{C}$ is an associative ring object with unit in this tensor category; in other words, it is a $\mathcal{C}$-$\mathcal{C}$-bicomodule endowed with a *semimultiplication* map $\mathbf{m_S} \colon \mathbf{S} \,\square_{\mathcal{C}}\, \mathbf{S} \longrightarrow \mathbf{S}$ and a *semi-unit* map $\mathbf{e_S} \colon \mathcal{C} \longrightarrow \mathbf{S}$ which have to be $\mathcal{C}$-$\mathcal{C}$-bicomodule morphisms satisfying the associativity and unity equations $\mathbf{m_S} \circ (\mathbf{m_S} \,\square\, \mathrm{id_S}) = \mathbf{m_S} \circ (\mathrm{id_S} \,\square\, \mathbf{m_S})$ and $\mathbf{m_S} \circ (\mathbf{e_S} \,\square\, \mathrm{id_S}) = \mathrm{id_S} = \mathbf{m_S} \circ (\mathrm{id_S} \,\square\, \mathbf{e_S})$.

The category of left $\mathcal{C}$-comodules is a left module category over the tensor category $\mathcal{C}$-comod-$\mathcal{C}$, and the category of right $\mathcal{C}$-comodules is a right module category over it. A *left semimodule* $\mathcal{M}$ over $\mathbf{S}$ is a module object in this left module category over the ring object $\mathbf{S}$ in this tensor category; in other words, it is a left $\mathcal{C}$-comodule endowed with a *left semiaction* map $\mathbf{n_M} \colon \mathbf{S} \,\square_{\mathcal{C}}\, \mathcal{M} \longrightarrow \mathcal{M}$, which has to be a morphism of left $\mathcal{C}$-comodules satisfying the associativity and unity equations $\mathbf{n_M} \circ (\mathbf{m_S} \,\square\, \mathrm{id_M}) = \mathbf{n_M} \circ (\mathrm{id_S} \,\square\, \mathbf{n_M})$ and $\mathbf{n_M} \circ (\mathbf{e_S} \,\square\, \mathrm{id_M}) = \mathrm{id_M}$. A *right semimodule* $\mathcal{N}$ over $\mathbf{S}$ is a right $\mathcal{C}$-comodule endowed with a *right semiaction* morphism of right $\mathcal{C}$-comodules $\mathbf{n_N} \colon \mathcal{N} \,\square_{\mathcal{C}}\, \mathbf{S} \longrightarrow \mathcal{N}$ satisfying the equations $\mathbf{n_N} \circ (\mathbf{n_N} \,\square\, \mathrm{id_S}) = \mathbf{n_N} \circ (\mathrm{id_N} \,\square\, \mathbf{m_S})$ and $\mathbf{n_N} \circ (\mathrm{id_N} \,\square\, \mathbf{e_S}) = \mathrm{id_N}$.

For any left $\mathcal{C}$-comodule $\mathcal{L}$, the cotensor product $\mathbf{S} \,\square_{\mathcal{C}}\, \mathcal{L}$ has a natural left semimodule structure. It is called the left $\mathbf{S}$-semimodule *induced* from a left $\mathcal{C}$-comodule $\mathcal{L}$. The space of semimodule homomorphisms from the induced semimodule is described by the formula $\mathrm{Hom}_{\mathbf{S}}(\mathbf{S} \,\square_{\mathcal{C}}\, \mathcal{L}, \mathcal{M}) \simeq \mathrm{Hom}_{\mathcal{C}}(\mathcal{L}, \mathcal{M})$. We will denote the category of left $\mathbf{S}$-semimodules by $\mathbf{S}$-simod and the category of right $\mathbf{S}$-semimodules by simod-$\mathbf{S}$. The category of left $\mathbf{S}$-semimodules is abelian provided that $\mathbf{S}$ is an injective right $\mathcal{C}$-comodule. Moreover, $\mathbf{S}$ is an injective right $\mathcal{C}$-comodule if and only if the category $\mathbf{S}$-simod is abelian and the forgetful functor $\mathbf{S}$-simod $\longrightarrow$ $\mathcal{C}$-comod is exact.

The operation of cotensor product over $\mathcal{C}$ provides a pairing functor

$$\text{comod-}\mathcal{C} \times \mathcal{C}\text{-comod} \longrightarrow k\text{-vect}$$

compatible with the right module category structure on comod-$\mathcal{C}$ and the left module category structure on $\mathcal{C}$-comod over the tensor category $\mathcal{C}$-comod-$\mathcal{C}$. The *semitensor product* $\mathcal{N} \lozenge_\mathcal{S} \mathcal{M}$ of a right $\mathcal{S}$-semimodule $\mathcal{N}$ and a left $\mathcal{S}$-semimodule $\mathcal{M}$ is defined as the cokernel of the pair of maps

$$(\mathbf{n}_\mathcal{N} \square \mathrm{id}_\mathcal{M}, \ \mathrm{id}_\mathcal{N} \square \mathbf{n}_\mathcal{M}) \colon \mathcal{N} \square_\mathcal{C} \mathcal{S} \square_\mathcal{C} \mathcal{M} \rightrightarrows \mathcal{N} \square_\mathcal{C} \mathcal{M}.$$

There are natural isomorphisms $\mathcal{N} \lozenge_\mathcal{S} (\mathcal{S} \square_\mathcal{C} \mathcal{L}) \simeq \mathcal{N} \square_\mathcal{C} \mathcal{L}$ and $(\mathcal{R} \square_\mathcal{C} \mathcal{S}) \lozenge_\mathcal{S} \mathcal{M} \simeq \mathcal{R} \square_\mathcal{C} \mathcal{M}$. The functor of semitensor product is neither left, nor right exact.

0.3.3 The semitensor product $\mathcal{N}^\bullet \lozenge_\mathcal{S} \mathcal{M}^\bullet$ of a complex of right $\mathcal{S}$-semimodules $\mathcal{N}^\bullet$ and a complex of left $\mathcal{S}$-semimodules $\mathcal{M}^\bullet$ is defined as the total complex of the bicomplex $\mathcal{N}^i \lozenge_\mathcal{S} \mathcal{M}^j$, constructed by taking infinite direct sums along the diagonals. Assume that $\mathcal{S}$ is an injective left and right $\mathcal{C}$-comodule. We would like to define the double-sided derived functor $\mathrm{SemiTor}^\mathcal{S}$ of the functor of semitensor product.

The *semiderived category* of left $\mathcal{S}$-semimodules $\mathsf{D}^{\mathrm{si}}(\mathcal{S}\text{-simod})$ is defined as the quotient category of the homotopy category $\mathsf{Hot}(\mathcal{S}\text{-simod})$ by the thick subcategory $\mathsf{Acycl}^{\mathrm{co}\text{-}\mathcal{C}}(\mathcal{S}\text{-simod})$ of complexes of $\mathcal{S}$-semimodules that are *coacyclic as complexes of $\mathcal{C}$-comodules*. For example, if the coalgebra $\mathcal{C}$ coincides with the ground field k, and $\mathcal{S} = R$ is just a k-algebra, then the semiderived category $\mathsf{D}^{\mathrm{si}}(\mathcal{S}\text{-simod})$ coincides with the derived category $\mathsf{D}(R\text{-mod})$, while if the semialgebra $\mathcal{S}$ coincides with the coalgebra $\mathcal{C}$, then the semiderived category $\mathsf{D}^{\mathrm{si}}(\mathcal{S}\text{-simod})$ coincides with the coderived category $\mathsf{D}^{\mathrm{co}}(\mathcal{C}\text{-comod})$.

A complex of left $\mathcal{S}$-semimodules $\mathcal{M}^\bullet$ is called *semiflat* if the semitensor product $\mathcal{N}^\bullet \lozenge_\mathcal{S} \mathcal{M}^\bullet$ is acyclic for any $\mathcal{C}$-coacyclic complex of right $\mathcal{S}$-semimodules $\mathcal{N}^\bullet$. For example, the complex of $\mathcal{S}$-semimodules $\mathcal{S} \square_\mathcal{C} \mathcal{L}^\bullet$ induced from a complex of injective $\mathcal{C}$-comodules $\mathcal{L}^\bullet$ is semiflat.

The quotient category of the homotopy category $\mathsf{Hot}_{\mathsf{sifl}}(\mathcal{S}\text{-simod})$ of semiflat complexes of $\mathcal{S}$-semimodules by the thick subcategory of $\mathcal{C}$-coacyclic semiflat complexes $\mathsf{Acycl}^{\mathrm{co}\text{-}\mathcal{C}}(\mathcal{S}\text{-simod}) \cap \mathsf{Hot}_{\mathsf{sifl}}(\mathcal{S}\text{-simod})$ is equivalent to the semiderived category of $\mathcal{S}$-semimodules. The derived functor

$$\mathrm{SemiTor}^\mathcal{S} \colon \mathsf{D}^{\mathrm{si}}(\mathrm{simod}\text{-}\mathcal{S}) \times \mathsf{D}^{\mathrm{si}}(\mathcal{S}\text{-simod}) \longrightarrow \mathsf{D}(k\text{-vect})$$

is defined by restricting the functor of semitensor product over $\mathcal{S}$ to either of the full subcategories $\mathsf{Hot}(\mathrm{simod}\text{-}\mathcal{S}) \times \mathsf{Hot}_{\mathsf{sifl}}(\mathcal{S}\text{-simod})$ or $\mathsf{Hot}_{\mathsf{sifl}}(\mathrm{simod}\text{-}\mathcal{S}) \times \mathsf{Hot}(\mathcal{S}\text{-simod})$ of the category $\mathsf{Hot}(\mathrm{simod}\text{-}\mathcal{S}) \times \mathsf{Hot}(\mathcal{S}\text{-simod})$.

0.3.4 Let $\mathcal{C}$ and $\mathcal{D}$ be two coalgebras, $\mathcal{K}$ be a $\mathcal{C}$-$\mathcal{D}$-bicomodule, and $\mathfrak{P}$ be a left $\mathcal{C}$-contramodule. Then the space of cohomomorphisms $\mathrm{Cohom}_\mathcal{C}(\mathcal{K}, \mathfrak{P})$ is endowed with a left $\mathcal{D}$-contramodule structure as the cokernel of a pair of $\mathcal{D}$-contramodule morphisms $\mathrm{Hom}_k(\mathcal{C} \otimes_k \mathcal{K}, \mathfrak{P}) = \mathrm{Hom}_k(\mathcal{K}, \mathrm{Hom}_k(\mathcal{C}, \mathfrak{P})) \rightrightarrows \mathrm{Hom}_k(\mathcal{K}, \mathfrak{P})$. For any left $\mathcal{D}$-comodule $\mathcal{M}$, left $\mathcal{C}$-contramodule $\mathfrak{P}$, and $\mathcal{C}$-$\mathcal{D}$-bicomodule $\mathcal{K}$ there is a natural isomorphism $\mathrm{Cohom}_\mathcal{C}(\mathcal{K} \square_\mathcal{D} \mathcal{M}, \mathfrak{P}) \simeq \mathrm{Cohom}_\mathcal{D}(\mathcal{M}, \mathrm{Cohom}_\mathcal{C}(\mathcal{K}, \mathfrak{P}))$.

0.3.5   Therefore, the category opposite to the category of left $\mathcal{C}$-contramodules is a right module category over the tensor category of $\mathcal{C}$-$\mathcal{C}$-bicomodules with respect to the action functor $\mathrm{Cohom}_{\mathcal{C}}$. The category of *left $\mathbf{S}$-semicontramodules* is the opposite category to the category of module objects in the right module category $\mathcal{C}$-contra^{op} over the ring object $\mathbf{S}$ in the tensor category $\mathcal{C}$-comod-$\mathcal{C}$. In other words, a left semicontramodule $\mathfrak{P}$ over $\mathbf{S}$ is a left $\mathcal{C}$-contramodule endowed with a *left semicontraaction* map $\mathbf{p}_{\mathfrak{P}} \colon \mathfrak{P} \longrightarrow \mathrm{Cohom}_{\mathcal{C}}(\mathbf{S}, \mathfrak{P})$, which has to be a morphism of left $\mathcal{C}$-contramodules satisfying the associativity and unity equations $\mathrm{Cohom}(\mathrm{id}_{\mathbf{S}}, \mathbf{p}_{\mathfrak{P}}) \circ \mathbf{p}_{\mathfrak{P}} = \mathrm{Cohom}(\mathbf{m}_{\mathbf{S}}, \mathrm{id}_{\mathfrak{P}}) \circ \mathbf{p}_{\mathfrak{P}}$ and $\mathrm{Cohom}(\mathbf{e}_{\mathbf{S}}, \mathrm{id}_{\mathfrak{P}}) \circ \mathbf{p}_{\mathfrak{P}} = \mathrm{id}_{\mathfrak{P}}$.

For example, if the coalgebra $\mathcal{C}$ coincides with the ground field k and $\mathbf{S} = R$ is just a k-algebra, then left $\mathbf{S}$-semicontramodules are simply left R-modules.

For any right $\mathbf{S}$-semimodule $\mathcal{N}$ and any k-vector space V the space $\mathrm{Hom}_k(\mathcal{N}, V)$ has a natural structure of left $\mathbf{S}$-semicontramodule. For any left $\mathcal{C}$-contramodule $\mathfrak{Q}$, the space of cohomomorphisms $\mathrm{Cohom}_{\mathcal{C}}(\mathbf{S}, \mathfrak{Q})$ has a natural structure of left semicontramodule. It is called the left $\mathbf{S}$-semicontramodule *coinduced* from a left $\mathcal{C}$-contramodule $\mathfrak{Q}$. The space of semicontramodule homomorphisms into the coinduced semicontramodule is described by the formula $\mathrm{Hom}^{\mathbf{S}}(\mathfrak{P}, \mathrm{Cohom}_{\mathcal{C}}(\mathbf{S}, \mathfrak{Q})) \simeq \mathrm{Hom}^{\mathcal{C}}(\mathfrak{P}, \mathfrak{Q})$. We will denote the category of left $\mathbf{S}$-semicontramodules by $\mathbf{S}$-sicntr. The category of left $\mathbf{S}$-semicontramodules is abelian provided that $\mathbf{S}$ is an injective left $\mathcal{C}$-comodule. Moreover, $\mathbf{S}$ is an injective left $\mathcal{C}$-comodule if and only if the category $\mathbf{S}$-sicntr is abelian and the forgetful functor $\mathbf{S}$-sicntr $\longrightarrow \mathcal{C}$-contra is exact.

The functor $\mathrm{Cohom}_{\mathcal{C}}^{\mathrm{op}} \colon \mathcal{C}$-comod $\times\ \mathcal{C}$-contra$^{\mathrm{op}} \longrightarrow k$-vect^{op} is a pairing compatible with the left module category structure on $\mathcal{C}$-comod and the right module category structure on $\mathcal{C}$-contra^{op} over the tensor category $\mathcal{C}$-comod-$\mathcal{C}$. Thus one can define the *space of semihomomorphisms* $\mathrm{SemiHom}_{\mathbf{S}}(\mathcal{M}, \mathfrak{P})$ from a left $\mathbf{S}$-semimodule $\mathcal{M}$ to a left $\mathbf{S}$-semicontramodule $\mathfrak{P}$ as the kernel of the pair of maps

$$(\mathrm{Cohom}(\mathbf{n}_{\mathcal{M}}, \mathrm{id}_{\mathfrak{P}}), \mathrm{Cohom}(\mathrm{id}_{\mathcal{M}}, \mathbf{p}_{\mathfrak{P}})) \colon$$
$$\mathrm{Cohom}_{\mathcal{C}}(\mathcal{M}, \mathfrak{P}) \rightrightarrows \mathrm{Cohom}_{\mathcal{C}}(\mathbf{S} \,\square_{\mathcal{C}}\, \mathcal{M},\ \mathfrak{P}) = \mathrm{Cohom}_{\mathcal{C}}(\mathcal{M}, \mathrm{Cohom}_{\mathcal{C}}(\mathbf{S}, \mathfrak{P})).$$

There are natural isomorphisms $\mathrm{SemiHom}_{\mathbf{S}}(\mathbf{S} \,\square_{\mathcal{C}}\, \mathcal{L},\ \mathfrak{P}) \simeq \mathrm{Cohom}_{\mathcal{C}}(\mathcal{L}, \mathfrak{P})$ and $\mathrm{SemiHom}_{\mathbf{S}}(\mathcal{M}, \mathrm{Cohom}_{\mathcal{C}}(\mathbf{S}, \mathfrak{Q})) \simeq \mathrm{Cohom}_{\mathcal{C}}(\mathcal{M}, \mathfrak{Q})$. The functor of semihomomorphisms is neither left, nor right exact.

0.3.6   The complex of semihomomorphisms $\mathrm{SemiHom}_{\mathbf{S}}(\mathcal{M}^{\bullet}, \mathfrak{P}^{\bullet})$ from a complex of left $\mathbf{S}$-semimodules $\mathcal{M}^{\bullet}$ to a complex of left $\mathbf{S}$-semicontramodules $\mathfrak{P}^{\bullet}$ is defined as the total complex of the bicomplex $\mathrm{SemiHom}_{\mathbf{S}}(\mathcal{M}^i, \mathfrak{P}^j)$, constructed by taking infinite products along the diagonals. Assume that $\mathbf{S}$ is an injective left and right $\mathcal{C}$-comodule. Let us define the double-sided derived functor $\mathrm{SemiExt}_{\mathbf{S}}$ of the functor of semihomomorphisms.

The *semiderived category* $\mathsf{D}^{\mathrm{si}}(\mathbf{S}$-sicntr$)$ of left $\mathbf{S}$-semicontramodules is defined as the quotient category of the homotopy category $\mathsf{Hot}(\mathbf{S}$-sicntr$)$ by the thick subcategory $\mathsf{Acycl}^{\mathsf{ctr}\text{-}\mathcal{C}}(\mathcal{S}\text{-sicntr})$ of complexes of $\mathcal{S}$-semicontramodules that are *contra-acyclic as complexes of $\mathcal{C}$-contramodules*.

A complex of left $\mathcal{S}$-semimodules $\mathcal{M}^\bullet$ is called *semiprojective* if the complex $\mathrm{SemiHom}_\mathcal{S}(\mathcal{M}^\bullet,\mathfrak{P}^\bullet)$ is acyclic for any $\mathcal{C}$-contraacyclic complex of left $\mathcal{S}$-semicontramodules $\mathfrak{P}^\bullet$. A complex of left $\mathcal{S}$-semicontramodules $\mathfrak{P}^\bullet$ is called *semiinjective* if the complex $\mathrm{SemiHom}_\mathcal{S}(\mathcal{M}^\bullet,\mathfrak{P}^\bullet)$ is acyclic for any $\mathcal{C}$-coacyclic complex of left $\mathcal{S}$-semimodules $\mathcal{M}^\bullet$. For example, the complex of $\mathcal{S}$-semimodules $\mathcal{S}\,\square_\mathcal{C}\,\mathcal{L}^\bullet$ induced from a complex of injective $\mathcal{C}$-comodules $\mathcal{L}^\bullet$ is semiprojective. Any semiprojective complex of semimodules is semiflat. The complex of $\mathcal{S}$-semicontramodules $\mathrm{Cohom}_\mathcal{C}(\mathcal{S},\mathfrak{Q}^\bullet)$ coinduced from a complex of projective $\mathcal{C}$-contramodules $\mathfrak{Q}^\bullet$ is semiinjective.

The quotient category of the homotopy category $\mathsf{Hot}_{\mathsf{sipr}}(\mathcal{S}\text{-simod})$ of semi-projective complexes of $\mathcal{S}$-semimodules by the thick subcategory of $\mathcal{C}$-coacyclic semiprojective complexes $\mathsf{Acycl}^{\mathsf{co}\text{-}\mathcal{C}}(\mathcal{S}\text{-simod}) \cap \mathsf{Hot}_{\mathsf{sipr}}(\mathcal{S}\text{-simod})$ is equivalent to the semiderived category of $\mathcal{S}$-semimodules. Analogously, the quotient category of the homotopy category $\mathsf{Hot}_{\mathsf{siin}}(\mathcal{S}\text{-sicntr})$ of semiinjective complexes of $\mathcal{S}$-semicon-tramodules by the thick subcategory of $\mathcal{C}$-contraacyclic semiinjective complexes $\mathsf{Acycl}^{\mathsf{ctr}\text{-}\mathcal{C}}(\mathcal{S}\text{-sicntr}) \cap \mathsf{Hot}_{\mathsf{siin}}(\mathcal{S}\text{-sicntr})$ is equivalent to the semiderived category of $\mathcal{S}$-semicontramodules. The derived functor

$$\mathrm{SemiExt}_\mathcal{S}\colon \mathsf{D}^{\mathsf{si}}(\mathcal{S}\text{-simod})^{\mathsf{op}} \times \mathsf{D}^{\mathsf{si}}(\mathcal{S}\text{-sicntr}) \longrightarrow \mathsf{D}(k\text{-vect})$$

is defined by restricting the functor of semihomomorphisms to either of the full subcategories $\mathsf{Hot}_{\mathsf{sipr}}(\mathcal{S}\text{-simod})^{\mathsf{op}} \times \mathsf{Hot}(\mathcal{S}\text{-sicntr})$ or $\mathsf{Hot}(\mathcal{S}\text{-simod})^{\mathsf{op}} \times \mathsf{Hot}_{\mathsf{siin}}(\mathcal{S}\text{-sicntr})$ of the category $\mathsf{Hot}(\mathcal{S}\text{-simod})^{\mathsf{op}} \times \mathsf{Hot}(\mathcal{S}\text{-sicntr})$.

0.3.7 Assume that $\mathcal{S}$ is an injective left and right $\mathcal{C}$-comodule. Even when $\mathcal{C}$ and $\mathcal{S}$ are finite-dimensional k-vector spaces, the categories of left $\mathcal{S}$-semimodules and left $\mathcal{S}$-semicontramodules are quite different. Nevertheless, their semiderived categories are equivalent in the very general case.

One can check that the adjoint functors $\Psi_\mathcal{C}\colon \mathcal{C}\text{-comod} \longrightarrow \mathcal{C}\text{-contra}$ and $\Phi_\mathcal{C}\colon \mathcal{C}\text{-contra} \longrightarrow \mathcal{C}\text{-comod}$ transform left $\mathcal{C}$-comodules with an $\mathcal{S}$-semimodule structure into left $\mathcal{C}$-contramodules with an $\mathcal{S}$-semicontramodule structure and vice versa. Therefore, there is a pair of adjoint functors $\Psi_\mathcal{S}\colon \mathcal{S}\text{-simod} \longrightarrow \mathcal{S}\text{-sicntr}$ and $\Phi_\mathcal{S}\colon \mathcal{S}\text{-sicntr} \longrightarrow \mathcal{S}\text{-simod}$ agreeing with the functors $\Psi_\mathcal{C}$ and $\Phi_\mathcal{C}$ and providing an equivalence between the exact categories of $\mathcal{C}$-injective left $\mathcal{S}$-semimodules and $\mathcal{C}$-projective left $\mathcal{S}$-semicontramodules. To construct this pair of adjoint functors in a natural way, let us define the operation of contratensor product of a semimodule and a semicontramodule.

Let $\mathcal{N}$ be a right $\mathcal{S}$-semimodule and $\mathfrak{P}$ be a left $\mathcal{S}$-semicontramodule. The *contratensor product* $\mathcal{N} \odot_\mathcal{S} \mathfrak{P}$ is defined as the cokernel of the pair of maps

$$(\mathbf{n}_\mathcal{N} \odot \mathrm{id}_\mathfrak{P},\ \eta_\mathcal{S} \circ (\mathrm{id}_{\mathcal{N}\square_\mathcal{C}\mathcal{S}} \odot \mathbf{p}_\mathfrak{P}))\colon (\mathcal{N}\,\square_\mathcal{C}\,\mathcal{S}) \odot_\mathcal{C} \mathfrak{P} \rightrightarrows \mathcal{N} \odot_\mathcal{C} \mathfrak{P}$$

where the natural "evaluation" map $\eta_\mathcal{K}\colon (\mathcal{N}\square_\mathcal{C}\mathcal{K}) \odot_\mathcal{D} \mathrm{Cohom}_\mathcal{C}(\mathcal{K},\mathfrak{P}) \longrightarrow \mathcal{N}\odot_\mathcal{C}\mathfrak{P}$ exists for any right $\mathcal{C}$-comodule $\mathcal{N}$, left $\mathcal{C}$-contramodule $\mathfrak{P}$, and $\mathcal{C}$-$\mathcal{D}$-bicomodule

$\mathcal{K}$ and is dual to the map

$$\mathrm{Hom}_k(\eta_{\mathcal{K}}, V) = \mathrm{Cohom}_{\mathcal{C}}(\mathcal{K}, -):$$
$$\mathrm{Hom}^{\mathcal{C}}(\mathfrak{P}, \mathrm{Hom}_k(\mathcal{N}, V)) \longrightarrow \mathrm{Hom}^{D}(\mathrm{Cohom}_{\mathcal{C}}(\mathcal{K}, \mathfrak{P}), \mathrm{Cohom}_{\mathcal{C}}(\mathcal{K}, \mathrm{Hom}_k(\mathcal{N}, V)))$$

for any k-vector space V. There are natural isomorphisms $(\mathcal{R} \square_{\mathcal{C}} \mathcal{S}) \odot_{\mathcal{S}} \mathfrak{P} \simeq \mathcal{R} \odot_{\mathcal{C}} \mathfrak{P}$ and $\mathrm{Hom}_k(\mathcal{N} \odot_{\mathcal{S}} \mathfrak{P}, V) \simeq \mathrm{Hom}^{\mathcal{S}}(\mathfrak{P}, \mathrm{Hom}_k(\mathcal{N}, V))$. The functor of contratensor product over $\mathcal{S}$ is right exact whenever $\mathcal{S}$ is an injective left $\mathcal{C}$-comodule.

The adjoint functors $\Psi_{\mathcal{S}}$ and $\Phi_{\mathcal{S}}$ can be defined by the formulas $\Psi_{\mathcal{S}}(\mathcal{M}) = \mathrm{Hom}_{\mathcal{S}}(\mathcal{S}, \mathcal{M})$ and $\Phi_{\mathcal{S}}(\mathfrak{P}) = \mathcal{S} \odot_{\mathcal{S}} \mathfrak{P}$. Here the space $\mathrm{Hom}_{\mathcal{S}}(\mathcal{S}, \mathcal{M})$ is endowed with a left $\mathcal{S}$-semicontramodule structure as a subsemicontramodule of the semicontramodule $\mathrm{Hom}_k(\mathcal{S}, \mathcal{M})$, while the space $\mathcal{S} \odot_{\mathcal{S}} \mathfrak{P}$ is endowed with a left $\mathcal{S}$-semimodule structure as a quotient semimodule of the semimodule $\mathcal{S} \otimes_k \mathfrak{P}$.

The quotient category of the homotopy category of $\mathcal{C}$-injective $\mathcal{S}$-semimodules $\mathsf{Hot}(\mathcal{S}\text{-simod}_{\mathrm{inj}\text{-}\mathcal{C}})$ by the thick subcategory of $\mathcal{C}$-contractible complexes of $\mathcal{C}$-injective $\mathcal{S}$-semimodules is equivalent to the semiderived category of $\mathcal{S}$-semimodules. Analogously, the quotient category of the homotopy category $\mathsf{Hot}(\mathcal{S}\text{-sicntr}_{\mathrm{proj}\text{-}\mathcal{C}})$ of $\mathcal{C}$-projective $\mathcal{S}$-semicontramodules by the thick subcategory of $\mathcal{C}$-contractible complexes of $\mathcal{C}$-projective $\mathcal{S}$-semicontramodules is equivalent to the semiderived category of $\mathcal{S}$-semicontramodules. Thus the semiderived categories of left $\mathcal{S}$-semimodules and left $\mathcal{S}$-semicontramodules are equivalent, $\mathsf{D}^{\mathrm{si}}(\mathcal{S}\text{-simod}) \simeq \mathsf{D}^{\mathrm{si}}(\mathcal{S}\text{-sicntr})$.

When $\mathcal{S}$ is not an injective left or right $\mathcal{C}$-comodule, the exact categories of $\mathcal{C}$-injective $\mathcal{S}$-semimodules and $\mathcal{C}$-projective $\mathcal{S}$-semicontramodules are still equivalent, even though the functors $\Psi_{\mathcal{S}}$ and $\Phi_{\mathcal{S}}$ are not defined on the whole categories of all comodules and contramodules.

0.3.8   The functor $\mathrm{Ext}_{\mathcal{S}} : \mathsf{D}^{\mathrm{si}}(\mathcal{S}\text{-simod})^{\mathrm{op}} \times \mathsf{D}^{\mathrm{si}}(\mathcal{S}\text{-simod}) \longrightarrow \mathsf{D}(k\text{-vect})$ of homomorphisms in the semiderived category $\mathsf{D}^{\mathrm{si}}(\mathcal{S}\text{-simod})$ can be computed by restricting the functor $\mathrm{Hom}_{\mathcal{S}} : \mathsf{Hot}(\mathcal{S}\text{-simod})^{\mathrm{op}} \times \mathsf{Hot}(\mathcal{S}\text{-simod}) \longrightarrow \mathsf{Hot}(k\text{-vect})$ of homomorphisms in the homotopy category $\mathsf{Hot}(\mathcal{S}\text{-simod})$ to an appropriate subcategory of the Cartesian product $\mathsf{Hot}(\mathcal{S}\text{-simod})^{\mathrm{op}} \times \mathsf{Hot}(\mathcal{S}\text{-simod})$. Namely, a complex of left $\mathcal{S}$-semimodules $\mathcal{L}^{\bullet}$ is called *projective relative to* $\mathcal{C}$ ($\mathcal{S}/\mathcal{C}$-projective) if the complex $\mathrm{Hom}_{\mathcal{S}}(\mathcal{L}^{\bullet}, \mathcal{M}^{\bullet})$ is acyclic for any $\mathcal{C}$-contractible complex of $\mathcal{C}$-injective left $\mathcal{S}$-semimodules $\mathcal{M}^{\bullet}$. For example, the complex of $\mathcal{S}$-semimodules $\mathcal{S} \square_{\mathcal{C}} \mathcal{L}^{\bullet}$ induced from a complex of $\mathcal{C}$-comodules $\mathcal{L}^{\bullet}$ is projective relative to $\mathcal{C}$. The quotient category of the homotopy category $\mathsf{Hot}_{\mathrm{proj}\text{-}\mathcal{S}/\mathcal{C}}(\mathcal{S}\text{-simod})$ of $\mathcal{S}/\mathcal{C}$-projective complexes of $\mathcal{S}$-semimodules by the thick subcategory $\mathsf{Acycl}^{\mathrm{co}\text{-}\mathcal{C}}(\mathcal{S}\text{-simod}) \cap \mathsf{Hot}_{\mathrm{proj}\text{-}\mathcal{S}/\mathcal{C}}(\mathcal{S}\text{-simod})$ of $\mathcal{C}$-coacyclic $\mathcal{S}/\mathcal{C}$-projective complexes is equivalent to the semiderived category of $\mathcal{S}$-semimodules. The functor $\mathrm{Ext}_{\mathcal{S}}$ can be obtained by restricting the functor $\mathrm{Hom}_{\mathcal{S}}$ to the full subcategory $\mathsf{Hot}_{\mathrm{proj}\text{-}\mathcal{S}/\mathcal{C}}(\mathcal{S}\text{-simod})^{\mathrm{op}} \times \mathsf{Hot}(\mathcal{S}\text{-simod}_{\mathrm{inj}\text{-}\mathcal{C}})$ of the category $\mathsf{Hot}(\mathcal{S}\text{-simod})^{\mathrm{op}} \times \mathsf{Hot}(\mathcal{S}\text{-simod})$.

Analogously, the functor $\mathrm{Ext}^{\mathcal{S}} : \mathsf{D}^{\mathrm{si}}(\mathcal{S}\text{-sicntr})^{\mathrm{op}} \times \mathsf{D}^{\mathrm{si}}(\mathcal{S}\text{-sicntr}) \longrightarrow \mathsf{D}(k\text{-vect})$ of homomorphisms in the semiderived category $\mathsf{D}^{\mathrm{si}}(\mathcal{S}\text{-sicntr})$ can be computed by restricting the functor $\mathrm{Hom}^{\mathcal{S}} : \mathsf{Hot}(\mathcal{S}\text{-sicntr})^{\mathrm{op}} \times \mathsf{Hot}(\mathcal{S}\text{-sicntr}) \longrightarrow \mathsf{Hot}(k\text{-vect})$ to an appropriate subcategory of the Cartesian product $\mathsf{Hot}(\mathsf{S}\text{-sicntr})^{\mathsf{op}} \times \mathsf{Hot}(\mathsf{S}\text{-sicntr})$. A complex of S-semicontramodules $\mathfrak{Q}^\bullet$ is called *injective relative to* $\mathcal{C}$ ($\mathsf{S}/\mathcal{C}$-*injective*) if the complex $\mathsf{Hom}^{\mathsf{S}}(\mathfrak{P}^\bullet, \mathfrak{Q}^\bullet)$ is acyclic for any $\mathcal{C}$-contractible complex of $\mathcal{C}$-projective S-semicontramodules $\mathfrak{P}^\bullet$. For example, the complex of S-semicontramodules $\mathsf{Cohom}_{\mathcal{C}}(\mathsf{S}, \mathfrak{Q}^\bullet)$ coinduced from a complex of $\mathcal{C}$-contramodules $\mathfrak{Q}^\bullet$ is $\mathsf{S}/\mathcal{C}$-injective. The quotient category of the homotopy category $\mathsf{Hot}_{\mathsf{inj}\text{-}\mathsf{S}/\mathcal{C}}(\mathsf{S}\text{-sicntr})$ of $\mathsf{S}/\mathcal{C}$-injective complexes of S-semicontramodules by the thick subcategory

$$\mathsf{Acycl}^{\mathsf{ctr}\text{-}\mathcal{C}}(\mathsf{S}\text{-sicntr}) \cap \mathsf{Hot}_{\mathsf{inj}\text{-}\mathsf{S}/\mathcal{C}}(\mathsf{S}\text{-sicntr})$$

of $\mathcal{C}$-contraacyclic $\mathsf{S}/\mathcal{C}$-injective complexes is equivalent to the semiderived category of S-semicontramodules. The functor $\mathsf{Ext}^{\mathsf{S}}$ can be obtained by restricting the functor $\mathsf{Hom}^{\mathsf{S}}$ to the full subcategory $\mathsf{Hot}(\mathsf{S}\text{-sicntr}_{\mathsf{proj}\text{-}\mathcal{C}})^{\mathsf{op}} \times \mathsf{Hot}_{\mathsf{inj}\text{-}\mathsf{S}/\mathcal{C}}(\mathsf{S}\text{-sicntr})$ of the category $\mathsf{Hot}(\mathsf{S}\text{-sicntr})^{\mathsf{op}} \times \mathsf{Hot}(\mathsf{S}\text{-sicntr})$.

The contratensor product $\mathsf{N}^\bullet \odot_{\mathsf{S}} \mathfrak{P}^\bullet$ of a complex of right S-semimodules $\mathsf{N}^\bullet$ and a complex of left S-semicontramodules $\mathfrak{P}^\bullet$ is defined as the total complex of the bicomplex $\mathsf{N}^i \odot_{\mathsf{S}} \mathfrak{P}^j$, constructed by taking infinite direct sums along the diagonals. Let us define the left derived functor $\mathsf{CtrTor}^{\mathsf{S}}$ of the functor of contratensor product over S. A complex of right S-semimodules $\mathsf{N}^\bullet$ is called *contraflat relative to* $\mathcal{C}$ ($\mathsf{S}/\mathcal{C}$-*contraflat*) if the complex $\mathsf{N}^\bullet \odot_{\mathsf{S}} \mathfrak{P}^\bullet$ is acyclic for any $\mathcal{C}$-contractible complex of $\mathcal{C}$-projective S-semicontramodules $\mathfrak{P}^\bullet$. For example, the complex of S-semimodules $\mathcal{R}^\bullet \square_{\mathcal{C}} \mathsf{S}$ induced from a complex of right $\mathcal{C}$-comodules $\mathcal{R}^\bullet$ is contraflat relative to $\mathcal{C}$. A complex of right S-semimodules $\mathsf{N}^\bullet$ is contraflat relative to $\mathcal{C}$ if and only if the complex of left S-semicontramodules $\mathsf{Hom}_k(\mathsf{N}^\bullet, k)$ is injective relative to $\mathcal{C}$. The quotient category of the homotopy category $\mathsf{Hot}_{\mathsf{ctrfl}\text{-}\mathsf{S}/\mathcal{C}}(\mathsf{simod}\text{-}\mathcal{C})$ of $\mathsf{S}/\mathcal{C}$-contraflat complexes of right S-semimodules by the thick subcategory $\mathsf{Acycl}^{\mathsf{co}\text{-}\mathcal{C}}(\mathsf{simod}\text{-}\mathsf{S}) \cap \mathsf{Hot}_{\mathsf{ctrfl}\text{-}\mathsf{S}/\mathcal{C}}(\mathsf{simod}\text{-}\mathcal{C})$ of $\mathcal{C}$-coacyclic $\mathsf{S}/\mathcal{C}$-contraflat complexes is equivalent to the semiderived category of right S-semimodules. The left derived functor

$$\mathsf{CtrTor}^{\mathsf{S}} : \mathsf{D}^{\mathsf{si}}(\mathsf{simod}\text{-}\mathsf{S}) \times \mathsf{D}^{\mathsf{si}}(\mathsf{S}\text{-sicntr}) \longrightarrow \mathsf{D}(k\text{-vect})$$

is defined by restricting the functor of contratensor product to the full subcategory $\mathsf{Hot}_{\mathsf{ctrfl}\text{-}\mathsf{S}/\mathcal{C}}(\mathsf{simod}\text{-}\mathsf{S}) \times \mathsf{Hot}(\mathsf{S}\text{-sicntr}_{\mathsf{proj}\text{-}\mathcal{C}})$ of the category $\mathsf{Hot}(\mathsf{simod}\text{-}\mathsf{S}) \times \mathsf{Hot}(\mathsf{S}\text{-sicntr})$.

The equivalence of triangulated categories $\mathsf{D}^{\mathsf{si}}(\mathsf{S}\text{-simod}) \simeq \mathsf{D}^{\mathsf{si}}(\mathsf{S}\text{-sicntr})$ transforms the double-sided derived functor $\mathsf{SemiExt}_{\mathsf{S}}$ into the functor Ext in either of the semiderived categories and the double-sided derived functor $\mathsf{SemiTor}^{\mathsf{S}}$ into the left derived functor $\mathsf{CtrTor}^{\mathsf{S}}$.

0.3.9 Any semiprojective complex of S-semimodules is $\mathsf{S}/\mathcal{C}$-projective. An $\mathsf{S}/\mathcal{C}$-projective complex of $\mathcal{C}$-injective S-semimodules is semiprojective. The homotopy category of semiprojective complexes of $\mathcal{C}$-injective S-semimodules is equivalent to the semiderived category of S-semimodules.

Analogously, any semiinjective complex of S-semicontramodules is S/C-injective. An S/C-injective complex of C-projective S-semicontramodules is semiinjective. The homotopy category of semiinjective complexes of C-injective S-semicontramodules is equivalent to the semiderived category of S-semicontramodules.

Notice that our definitions of S/C-projective and S/C-injective complexes differ from the traditional ones; cf. B.3 and Remark 9.2.1.

0.3.10 Semialgebras and their generalizations have been studied under the name of "internal categories" in M. Aguiar's dissertation [1]; see also [27]. In [21, 25, 22], semialgebras over a coalgebra C were discussed under the name of "C-rings".

0.4 Nonhomogeneous Koszul duality over a base ring

This section is intended to supply preliminary material for Chapter 11 and Appendix D.

0.4.1 A graded ring $S = S_0 \oplus S_1 \oplus S_2 \oplus \cdots$ is called *quadratic* if it is generated by S_1 over S_0 with relations of degree 2 only. In other words, this means that if one considers the graded ring freely generated by the S_0-S_0-bimodule S_1 (the "tensor ring" of the S_0-S_0-bimodule S_1), i.e., the graded ring $\mathbb{T}_{S_0, S_1}$ with the components $S_1^{\otimes_{S_0} n} = S_1 \otimes_{S_0} S_1 \otimes_{S_0} \cdots \otimes_{S_0} S_1$, then the ring S should be isomorphic to the quotient ring of $\mathbb{T}_{S_0, S_1}$ by the ideal generated by a certain subbimodule I_S in $S_1 \otimes_{S_0} S_1$.

A quadratic ring S is called *2-left finitely projective* if both left S_0-modules S_1 and S_2 are projective and finitely generated. A quadratic ring is called *3-left finitely projective* if the same applies to S_1, S_2, and S_3. Further conditions of this kind are not very sensible to consider for general quadratic rings. Analogously one defines *2-right finitely projective* and *3-right finitely projective* quadratic rings.

There is an anti-equivalence between the category of 2-left finitely projective quadratic rings and the category of 2-right finitely projective quadratic rings, called the *quadratic duality*. The duality functors are defined by the formulas $R_0 = S_0$, $R_1 = \mathrm{Hom}_{S_0}(S_1, S_0)$, $R_2 = \mathrm{Hom}_{S_0}(I_S, S_0)$, $I_R \simeq \mathrm{Hom}_{S_0}(S_2, S_0)$, and conversely, $S_1 = \mathrm{Hom}_{R_0^{\mathrm{op}}}(R_1, R_0)$, $S_2 = \mathrm{Hom}_{R_0^{\mathrm{op}}}(I_R, R_0)$, $I_S \simeq \mathrm{Hom}_{R_0^{\mathrm{op}}}(R_2, R_0)$. Here we use the natural isomorphism

$$\mathrm{Hom}_{S_0}(N, S_0) \otimes_{S_0} \mathrm{Hom}_{S_0}(M, S_0) \simeq \mathrm{Hom}_{S_0}(M \otimes_{S_0} N, \, S_0)$$

for S_0-S_0-bimodules M and N that are projective and finitely generated left S_0-modules, and the analogous isomorphism

$$\mathrm{Hom}_{R_0^{\mathrm{op}}}(N, R_0) \otimes_{R_0} \mathrm{Hom}_{R_0^{\mathrm{op}}}(M, R_0) \simeq \mathrm{Hom}_{R_0^{\mathrm{op}}}(M \otimes_{R_0} N, \, R_0)$$

for R_0-R_0-bimodules M and N that are projective and finitely generated right R_0-modules.

The duality functor sends 3-left finitely projective quadratic rings to 3-right finitely projective quadratic rings and vice versa. Indeed, set $J_S = I_S \otimes_{S_0} S_1 \cap S_1 \otimes_{S_0} I_S \subset S_1 \otimes_{S_0} S_1 \otimes_{S_0} S_1$; then

$$0 \longrightarrow J_S \longrightarrow I_S \otimes_{S_0} S_1 \oplus S_1 \otimes_{S_0} I_S \longrightarrow S_1 \otimes_{S_0} S_1 \otimes_{S_0} S_1 \longrightarrow S_3 \longrightarrow 0$$

is an exact sequence of finitely generated projective left S_0-modules, and $R_3 \simeq \operatorname{Hom}_{S_0}(J_S, S_0)$, since the sequence remains exact after applying $\operatorname{Hom}_{S_0}(-, S_0)$.

0.4.2 A graded ring $S = S_0 \oplus S_1 \oplus S_2 \oplus \cdots$ is called *left flat Koszul* if it is flat as a left S_0-module and one has $\operatorname{Tor}^S_{i,j}(S_0, S_0) = 0$ for $i \neq j$. Here S_0 is endowed with the right and left S-module structures via the augmentation map $S \longrightarrow S_0$ and the second grading j on the Tor is induced by the grading of S. *Right flat Koszul* graded rings are defined in the analogous way. A left/right flat Koszul ring is called *left/right (finitely) projective Koszul*, if it is a projective (with finitely generated grading components) left/right S_0-module.

Notice that when S is a flat left S_0-module, the reduced relative bar construction

$$\cdots \longrightarrow S \otimes_{S_0} S/S_0 \otimes S/S_0 \longrightarrow S \otimes S/S_0 \longrightarrow S$$

is a flat resolution of the left S-module S_0, so one can use it to compute $\operatorname{Tor}^S(S_0, S_0)$. When S is a projective left S-module, the same resolution can be used to compute $\operatorname{Ext}_S(S_0, S_0)$. Assume that the grading components of S are finitely generated projective left S_0-modules; then it follows that S is left finitely projective Koszul if and only if $\operatorname{Ext}_S^{i,j}(S_0, S_0) = 0$ for $i \neq j$ and $\operatorname{Ext}_S^{i,i}(S_0, S_0)$ are projective right S_0-modules.

Assume that a graded ring S is a flat left S_0-module. Then S is left flat Koszul if and only if it is quadratic and for each degree n the lattice of subbimodules in $S_1^{\otimes_{S_0} n}$ generated by the $n-1$ subbimodules

$$S_1^{\otimes_{S_0} i-1} \otimes_{S_0} I_S \otimes_{S_0} S_1^{\otimes_{S_0} n-i-1}, \qquad i = 1, \ldots, n-1$$

is distributive. This means that for any three subbimodules X, Y, Z that can be obtained from the generating subbimodules by applying the operations of sum and intersection one should have $(X+Y) \cap Z = X \cap Z + Y \cap Z$. Furthermore, if S is a left finitely projective Koszul ring, then the ring R quadratic dual to S is right finitely projective Koszul, and vice versa; besides, in this case the graded ring $\operatorname{Ext}_S(S_0, S_0)$ is isomorphic to R^{op} and the graded ring $\operatorname{Ext}_{R^{\mathrm{op}}}(R_0, R_0)$ is isomorphic to S.

0.4.3 Let S be a 3-left finitely projective quadratic ring. Suppose that we are given a ring $S^\sim$ endowed with an increasing filtration $F_0 S^\sim \subset F_1 S^\sim \subset F_2 S^\sim \subset \cdots$ such that $S = \bigcup_n F_n S^\sim$ and the associated graded ring $\operatorname{gr}_F S^\sim$ is identified with S. Such a ring $S^\sim$ will be called a *3-left finitely projective nonhomogeneous quadratic ring*. If the graded ring S is left finitely projective Koszul, the filtered ring $S^\sim$ is called a *left finitely projective nonhomogeneous Koszul ring*.

Let R be the 3-right finitely projective quadratic ring dual to S. We would like to describe the additional structure on the ring R corresponding to the data of a filtered ring $S^\sim$ endowed with an isomorphism $\mathrm{gr}_F S^\sim \simeq S$.

A *CDG-ring* (*curved differential graded ring*) is a graded ring $R = \bigoplus_n R^n$ endowed with an odd derivation d of degree 1 and a "curvature element" $h \in R^2$ such that $d^2(x) = [h, x]$ for all $x \in R$ and $d(h) = 0$. A morphism of CDG-rings $'R \longrightarrow ''R$ is a pair (f, a) consisting of a morphisms of graded rings $f \colon 'R \longrightarrow ''R$ and a "change-of-connection element" $a \in ''R^1$ such that $f(d'(x)) = d''(f(x)) + [a, f(x)]$ (the supercommutator) for all $x \in 'R$ and $f(h') = h'' + d''(a) + a^2$. Composition of morphisms is defined by the rule $(g, b)(f, a) = (gf, \ b + g(a))$. Identity morphisms are the morphisms $(\mathrm{id}, 0)$.

So the *category of CDG-rings* is defined. Notice that the natural functor from the category of DG-rings to the category of CDG-rings is faithful, but not fully faithful. In other words, two DG-rings may be isomorphic in the category of CDG-rings without being isomorphic as DG-rings. Furthermore, two CDG-rings of the form $(R, d + [a, \cdot], \ h + d(a) + a^2)$ and (R, d, h) are always naturally isomorphic, the isomorphism being given by the pair (id, a).

There is a fully faithful contravariant functor from the category of 3-left finitely projective nonhomogeneous quadratic rings $S^\sim$ with a fixed ring $F_0 S^\sim$ to the category of CDG-rings (R, d, h) with the same component $R^0 = F_0 S^\sim$ such that the underlying graded ring R of the CDG-ring (R, d, h) corresponding to $S^\sim$ is the 3-right finitely projective quadratic ring dual to the ring $S = \mathrm{gr}_F S^\sim$ (in the grading $R_i = R^i$).

This functor is constructed as follows. For each 3-left finitely projective nonhomogeneous quadratic ring $S^\sim$ choose a complementary left $S_0 = F_0 S^\sim$-submodule V to the submodule $F_0 S^\sim$ in the left S_0-module $F_1 S^\sim$. This can be done, because the quotient module $S_1 = F_1 S^\sim / F_0 S^\sim$ is projective. Since V maps isomorphically to S_1, it is endowed with a structure of an S_0-S_0-bimodule. The embedding $V \longrightarrow F_0 S^\sim$ is only a morphism of left S_0-modules, however; the right actions of S_0 in V and $F_1 S^\sim$ are compatible modulo $F_0 S^\sim$. Put $q(v, s) = m(v, s) - vs$ for $v \in V$, $s \in S_0$, where $m(v, s)$ is the product in $S^\sim$ and vs denotes the right action of S_0 in V. This defines a map $q \colon V \otimes_{\mathbb{Z}} S_0 \longrightarrow S_0$.

Let $I^\sim$ be the full preimage of the subbimodule $I_S \subset S_1 \otimes_{S_0} S_1$ under the surjective map $S_1 \otimes_{\mathbb{Z}} S_1 \longrightarrow S_1 \otimes_{S_0} S_1$. Using the identification of V with S_1, we will consider $I^\sim$ as the full preimage of $F_1 S^\sim$ under the multiplication map $m \colon V \otimes_{\mathbb{Z}} V \longrightarrow S^\sim$. Let us split the map $m \colon I^\sim \longrightarrow F_1 S^\sim$ into two components $(g, -h)$ according to the direct sum decomposition $F_1 S^\sim \simeq V \oplus S_0$, so that $g \colon I^\sim \longrightarrow V$ and $h \colon I^\sim \longrightarrow S_0$.

The differentials $d_0 \colon R^0 \longrightarrow R^1$ and $d_1 \colon R^1 \longrightarrow R^2$ are defined in terms of the maps q and g by the formulas

$$\langle v, d_0(s) \rangle = q(v, s), \qquad \langle i, d_1(r) \rangle = \langle g(\tilde{\imath}), r \rangle - q(\tilde{\imath}_1, \langle \tilde{\imath}_2, r \rangle),$$

where $\langle\,,\,\rangle$ denotes the pairing of V with R^1 and of I_S with R^2, and $\tilde\imath$ is any preimage of $i \in I_S$ in $I^\sim$, written also as $\tilde\imath = \tilde\imath_1 \otimes \tilde\imath_2$. The map h factorizes through I_S, providing the curvature element in $R^2 = \mathrm{Hom}_{S_0}(I_S, S_0)$.

Finally, to a morphism of nonhomogeneous quadratic rings $f\colon S''^\sim \longrightarrow S'^\sim$ with chosen complementary submodules $V'' \subset F_1''S''^\sim$ and $V' \subset F_1'S'^\sim$ one assigns a morphism of dual CDG-rings $(g, a)\colon ('R, d', h') \longrightarrow (''R, d'', h'')$ defined as follows. The morphism of quadratic rings $g\colon {}'R \longrightarrow {}''R$ is the quadratic dual map to the associated graded morphism $\mathrm{gr}\, f\colon S'' \longrightarrow S'$, while the change-of-connection element $a \in {}''R^1 = \mathrm{Hom}_{S_0}(S_1'', S_0)$ is equal to minus the composition $V'' \longrightarrow F_1''S''^\sim \longrightarrow F_1'S'^\sim \longrightarrow S_0$ of the embedding $V'' \longrightarrow F_1''S''^\sim$, the map f, and the projection $F_1'S'^\sim \longrightarrow S_0$ along V'. In particular, for a given nonhomogeneous quadratic ring $S^\sim$, changing the splitting of $F_1 S^\sim$ by the rule $V'' = \{v' - a(v') \mid v' \in V'\}$ leads to a natural morphism of CDG-rings $(\mathrm{id}, a)\colon (R, d', h') \longrightarrow (R, d'', h'')$.

One has to make quite some computations in order to check that everything is well defined and compatible in this construction. In particular, the 3-left projectivity is actually used in the form of the duality between J_S (where some self-consistency equations on the defining relations of $S^\sim$ live) and R^3 (where the equations $d(e) = 0$ for $e \in I_R$, $d^2(r) = [h, r]$, and $d(h) = 0$ have to be verified).

The nonhomogeneous quadratic duality functor restricted to the categories of left finitely projective nonhomogeneous Koszul rings and right finitely projective Koszul CDG-ring becomes an equivalence of categories. In other words, any CDG-ring whose underlying graded ring is right finitely projective Koszul corresponds to a left finitely projective nonhomogeneous Koszul ring. This is the statement of the Poincaré–Birkhoff–Witt theorem for finitely projective nonhomogeneous Koszul rings.

0.4.4   A *quasi-differential ring* $R^\sim$ is a graded ring $R^\sim = \bigoplus_n R^{n\sim}$ endowed with an odd derivation ∂ of degree -1 with zero square such that the cohomology of ∂ vanishes (equivalently, the unit element of $R^\sim$ lies in the image of ∂). A *quasi-differential structure* on a graded ring R is the data of a quasi-differential ring $(R^\sim, \partial)$ together with an isomorphism of graded rings $\ker \partial \simeq R$.

The category of quasi-differential rings is equivalent to the category of CDG-rings. This equivalence assigns to a CDG-ring (R, d, h) the quasi-differential ring $R^\sim = R[\delta]$ with an added generator δ of degree 1, the relations $[\delta, x] = d(x)$ (the supercommutator) for $x \in R$ and $\delta^2 = h$, and the derivation $\partial = \partial/\partial\delta$ (the partial derivative in δ, meaning the unique odd derivation ∂ of $R^\sim$ for which $\partial(R) = 0$ and $\partial(\delta) = 1$). Conversely, to construct a CDG-ring structure on the kernel R of the derivation ∂ of a quasi-differential ring $R^\sim$, it suffices to choose an element $\delta \in R^{1\sim}$ such that $\partial(\delta) = 1$ and set $d(x) = [\delta, x]$, $h = \delta^2$. Choosing two different elements δ leads to two naturally isomorphic CDG-rings.

A *left CDG-module* M over a CDG-ring (R, d, h) is a graded left R-module endowed with a d-derivation d_M (that is a homogeneous map $M \longrightarrow M$ of degree 1 for which $d_M(rx) = d(r)x + (-1)^{|r|}r d(x)$ for $r \in R$, $x \in M$, where

$|r|$ denotes the degree of a homogeneous element r) such that $d_M^2(x) = hx$. A *quasi-differential left module* over a quasi-differential ring $R^\sim$ is just a graded left $R^\sim$-module (without any differential). The category of left CDG-modules over a CDG-ring (R, d, h) is isomorphic to the category of quasi-differential left modules over the quasi-differential ring $R^\sim$ corresponding to (R, d, h); this isomorphism of categories assigns to a graded $R^\sim$-module structure on a graded left R-module M the derivation $d_M(x) = \delta x$ on M.

Analogously, a *right CDG-module* N over (R, d, h) is a graded right R-module endowed with a d-derivation d_N (that is a homogeneous map $N \longrightarrow N$ of degree 1 for which $d_N(yr) = d_N(y)r + (-1)^{|y|}yd(r)$ for $y \in N$, $r \in R$) such that $d_N^2(y) = -yh$. A *quasi-differential right module* over a quasi-differential ring $R^\sim$ is just a graded right $R^\sim$-module. The category of right CDG-modules over (R, d, h) is isomorphic to the category of quasi-differential $R^\sim$-modules when $R^\sim$ corresponds to (R, d, h); this isomorphism of categories assigns to a graded $R^\sim$-module structure on a graded right R-module N the derivation $d_N(y) = (-1)^{|y|+1}y\delta$ on N.

0.4.5   CDG-modules over a CDG-ring form a *DG-category*, i.e., a category where for any two given objects there is a complex of morphisms between them. We will consider the cases of left and right CDG-modules separately.

Let L and M be two left CDG-modules over a CDG-ring (R, d, h). The complex $\operatorname{Hom}_R^\bullet(L, M)$ is defined as follows. The component $\operatorname{Hom}_R^n(L, M)$ consists of all homogeneous maps $L \longrightarrow M$ of degree n supercommuting with the R-module structures in L and M. This means that for $f \in \operatorname{Hom}_R^n(L, M)$ and $r \in R$, $x \in L$ one should have $f(rx) = (-1)^{n|r|}rf(x)$. The differential is defined by the formula $(df)(x) = d_M f(x) - (-1)^{|f|}fd_L(x)$. One has $d^2(f) = 0$, because $f(hx) = hf(x)$.

Let K and N be two right CDG-modules over (R, d, h). The component $\operatorname{Hom}_R^n(K, N)$ of the complex $\operatorname{Hom}_R^\bullet(K, N)$ consists of all homogeneous maps $K \longrightarrow N$ of degree n commuting with the R-module structures in L and M (without any signs). The differential is defined by the formula $(df)(y) = d_N f(y) - (-1)^{|f|}fd_K(y)$.

One can see that shifts and cones exist in the DG-categories of (left or right) CDG-modules, and moreover, a CDG-module structure can be twisted with any cochain in the complex of endomorphisms satisfying the Maurer–Cartan equation [19]. Explicitly, for any left CDG-module (M, d_M) over (R, d, h) and any $q \in \operatorname{Hom}_R^1(M, M)$ such that $d(q) + q^2 = 0$ the twisted differential $d_M(q) = d_M + q$ defines another CDG-module structure on M, and analogously, for any right CDG-module (N, d_N) over (R, d, h) and any $q \in \operatorname{Hom}_R^1(N, N)$ such that $d(q) + q^2 = 0$ the twisted differential $d_N(q) = d_N + q$ defines another CDG-module structure on N. It follows that the homotopy categories of CDG-modules, defined as the categories of zero cohomology of the DG-categories of CDG-modules, are triangulated.

Furthermore, one can speak about the total CDG-modules of complexes of CDG-modules, constructed by taking infinite direct sums or infinite products along the diagonals. In particular, there are total CDG-modules of exact triples of CDG-modules. This allows one to define the *coderived* and *contraderived categories of CDG-modules* over (R, d, h) as the quotient categories of the homotopy categories of CDG-modules by the minimal triangulated subcategories containing the total CDG-modules of exact triples of CDG-modules and closed under infinite direct sums and infinite products, respectively.

Notice that one *cannot* define the conventional derived category of CDG-modules, as CDG-modules have no cohomology groups.

0.4.6 Let $S^\sim$ be a left finitely projective nonhomogeneous Koszul ring and (R, d, h) be the dual CDG-ring. Assume that the ring S_0 has a finite right homological dimension. Then the Koszul duality theorem claims that the derived category of right $S^\sim$-modules is equivalent to the coderived category of right CDG-modules N over (R, d, h) such that every element of N is annihilated by R^n for $n \gg 0$. Assuming that S_0 has a finite left homological dimension, the derived category of left $S^\sim$-modules is also described as being equivalent to the contraderived category of left CDG-modules over (R, d, h) in which certain infinite summation operations are defined.

One can drop the homological dimension assumptions, replacing the derived categories of $S^\sim$-modules in the formulations of these results with certain semiderived categories relative to S_0 (see Theorem 11.8 and Remark 11.7.3). And the conventional derived category of right $S^\sim$-modules, without the homological dimension assumption on S_0, is equivalent to the quotient category of the coderived category of locally nilpotent (in the above sense) right CDG-modules over (R, d, h) by its minimal triangulated subcategory closed under infinite direct sums and containing all the CDG-modules N where R^n act by zero for all $n > 0$ and which are acyclic with respect to d_N (one has $d_N^2 = 0$, since $Nh = 0$). The latter result has an obvious analogue in the case of left CDG-modules with infinite summation operations.

0.4.7 The following example is thematic. Let M be a smooth affine algebraic variety and E be a vector bundle over M. Let $\mathrm{Diff}_{M,E}$ denote the ring of differential operators acting in the sections of E. The natural filtration of $\mathrm{Diff}_{M,E}$ by the order of differential operators makes it a left (and right) finitely projective nonhomogeneous Koszul ring. To construct the dual CDG-ring, choose a global connection ∇_E in E. Let $\Omega(M, \mathrm{End}(E))$ be the graded algebra of differential forms with coefficients in the vector bundle $\mathrm{End}(E)$ of endomorphisms of E, endowed with the de Rham differential d_∇ depending on the connection $\nabla_{\mathrm{End}(E)}$ corresponding to ∇_E and the element $h_\nabla \in \Omega^2(M, \mathrm{End}(E))$ equal to the curvature of ∇_E. The Koszul duality theorem provides an equivalence between the derived category of right $\mathrm{Diff}_{M,E}$-modules and the coderived category of right CDG-modules over $\Omega(M, \mathrm{End}(E))$. The proof of this result given in 11.8 generalizes easily to nonaffine varieties (the approach with quasi-differential structures allows one to get rid of the choice of a global connection).

These results are even valid in prime characteristic, describing the derived category of modules over the ring/sheaf of "crystalline" differential operators (those generated by the endomorphisms and vector fields with the commutation relations analogous to the zero characteristic case). Furthermore, it is not difficult to see that the quotient category of the homotopy category of finitely generated right CDG-modules over $\Omega(M, \mathrm{End}(E))$ by its minimal thick subcategory containing the total CDG-modules of exact triples of finitely generated CDG-modules is a full subcategory of the coderived category of CDG-modules. This full subcategory is equivalent to the bounded derived category of finitely generated (coherent) right $\mathrm{Diff}_{M,E}$-modules. All of this is applicable to any smooth varieties, not necessarily affine [76].

For a smooth affine variety M, the derived category of left $\mathrm{Diff}_{M,E}$-modules is equivalent to the contraderived category of left CDG-modules over $\Omega(M, \mathrm{End}(E))$.

0.4.8   Koszul algebras were introduced by S. Priddy; the standard contemporary sources are [13, 75]. Nonhomogeneous quadratic duality (the equivalence of categories of nonhomogeneous Koszul algebras and Koszul CDG-algebras) was developed in [73, 75]. Homogeneous Koszul duality (the equivalence of derived categories of graded modules over dual Koszul algebras) was established in [13]. Koszul duality in the context of A_∞-algebras and DG-coalgebras was worked out in [66, Chapitres 1 and 2]. All of these papers only consider duality over the ground field (or, in the case of [13], a semisimple algebra) rather than over an arbitrary ring, as above.

Notable attempts to define a version of derived category of DG-modules over the de Rham complex, so that the derived category of modules over the differential operators would be equivalent to it, were undertaken in [56] and [11, Section 7.2]. They were not entirely successful, in the present author's view, in that in [56] the analytic topology and analytic functions were used in the definition of an essentially purely algebraic category, while in [11] the right-hand side of the purported equivalence of categories is to a certain extent defined in terms of the left-hand side. The latter problem is also present in Lefèvre-Hasegawa's Koszul duality [66, 61].

1 Semialgebras and Semitensor Product

Throughout Chapters 1–11, k is a commutative ring. All our rings, bimodules, abelian groups, ... will be k-modules; all additive categories will be k-linear.

1.1 Corings and comodules

Let A be an associative k-algebra (with unit).

1.1.1 A *coring* $\mathcal{C}$ over A is a coring object in the tensor category of A-A-bimodules; in other words, it is a k-module endowed with an A-A-bimodule structure and two A-A-bimodule maps of *comultiplication* $\mathcal{C} \longrightarrow \mathcal{C} \otimes_A \mathcal{C}$ and *counit* $\mathcal{C} \longrightarrow A$ satisfying the coassociativity and counity equations: two compositions of the comultiplication map $\mathcal{C} \longrightarrow \mathcal{C} \otimes_A \mathcal{C}$ with the maps $\mathcal{C} \otimes_A \mathcal{C} \rightrightarrows \mathcal{C} \otimes_A \mathcal{C} \otimes_A \mathcal{C}$ induced by the comultiplication map should coincide with each other, and two compositions $\mathcal{C} \longrightarrow \mathcal{C} \otimes_A \mathcal{C} \rightrightarrows \mathcal{C}$ of the comultiplication map with the maps $\mathcal{C} \otimes_A \mathcal{C} \rightrightarrows \mathcal{C}$ induced by the counit map should coincide with the identity map of $\mathcal{C}$.

A *left comodule* $\mathcal{M}$ over a coring $\mathcal{C}$ is a comodule object in the left module category of left A-modules over the coring object $\mathcal{C}$ in the tensor category of A-A-bimodules; in other words, it is a left A-module endowed with a left A-module map of *left coaction* $\mathcal{M} \longrightarrow \mathcal{C} \otimes_A \mathcal{M}$ satisfying the coassociativity and counity equations: two compositions of the coaction map $\mathcal{M} \longrightarrow \mathcal{C} \otimes_A \mathcal{M}$ with the maps $\mathcal{C} \otimes_A \mathcal{M} \rightrightarrows \mathcal{C} \otimes_A \mathcal{C} \otimes_A \mathcal{M}$ induced by the comultiplication and coaction maps should coincide with each other and the composition $\mathcal{M} \longrightarrow \mathcal{C} \otimes_A \mathcal{M} \longrightarrow \mathcal{M}$ of the coaction map with the map $\mathcal{C} \otimes_A \mathcal{M} \longrightarrow \mathcal{M}$ induced by the counit map should coincide with the identity map of $\mathcal{M}$. A *right comodule* $\mathcal{N}$ over $\mathcal{C}$ is a comodule object in the right module category of right A-modules over the coring object $\mathcal{C}$ in the tensor category of A-A-bimodules; in other words, it is a right A-module endowed with a right A-module map of *right coaction* $\mathcal{N} \longrightarrow \mathcal{N} \otimes_A \mathcal{C}$ satisfying the coassociativity and counity equations for the compositions $\mathcal{N} \longrightarrow \mathcal{N} \otimes_A \mathcal{C} \rightrightarrows \mathcal{N} \otimes_A \mathcal{C} \otimes_A \mathcal{C}$ and $\mathcal{N} \longrightarrow \mathcal{N} \otimes_A \mathcal{C} \longrightarrow \mathcal{N}$.

1.1.2 If V is a left A-module, then the left $\mathcal{C}$-comodule $\mathcal{C} \otimes_A V$ is called the left $\mathcal{C}$-comodule *coinduced* from an A-module V. The k-module of comodule homomorphisms from an arbitrary $\mathcal{C}$-comodule into the coinduced $\mathcal{C}$-comodule is described by the formula

$$\mathrm{Hom}_{\mathcal{C}}(\mathcal{M}, \, \mathcal{C} \otimes_A V) \simeq \mathrm{Hom}_A(\mathcal{M}, V).$$

This is an instance of the following general fact, which we prefer to formulate in the tensor (monoidal) category language, though it can be also formulated in the monad language.

L. Positselski, *Homological Algebra of Semimodules and Semicontramodules*, Monografie Matematyczne 70, DOI 10.1007/978-3-0346-0436-9_2, © Springer Basel AG 2010

Lemma. *Let* E *be a (not necessarily additive) associative tensor category with a unit object,* M *be a left module category over it,* R *be a ring object with unit in* E, *and* $_R$M *be the category of R-module objects in* M. *Then the induction functor* M $\longrightarrow$ $_R$M *defined by the rule* $V \longmapsto R \otimes V$ *is left adjoint to the forgetful functor* $_R$M $\longrightarrow$ M.

Proof. For any object V and any R-module M in M, the map

$$\mathrm{Hom}_M(V, M) \longrightarrow \mathrm{Hom}_M(R \otimes V, M)$$

is a split equalizer (see [67, VI.6]) of the pair of maps

$$\mathrm{Hom}_M(R \otimes V, M) \rightrightarrows \mathrm{Hom}_M(R \otimes R \otimes V, M)$$

in the category of sets, with the splitting maps

$$\mathrm{Hom}_M(V, M) \longleftarrow \mathrm{Hom}_M(R \otimes V, M) \longleftarrow \mathrm{Hom}_M(R \otimes R \otimes V, M)$$

induced by the unit morphism of R (applied at the rightmost factor R). $\square$

We will denote the category of left $\mathcal{C}$-comodules by $\mathcal{C}$-comod and the category of right $\mathcal{C}$-comodules by comod-$\mathcal{C}$. The category of left $\mathcal{C}$-comodules is abelian whenever $\mathcal{C}$ is a flat right A-module. Moreover, the right A-module $\mathcal{C}$ is flat if and only if the category $\mathcal{C}$-comod is abelian and the forgetful functor $\mathcal{C}$-comod $\longrightarrow$ A-mod is exact. This is an instance of a general fact applicable to any monad over an abelian category. The "only if" assertion is straightforwardly checked, while the "if" part is deduced from the observations that the coinduction functor $V \longmapsto \mathcal{C} \otimes_A V$ is right adjoint to the forgetful functor and a right adjoint functor is left exact.

At the same time, for any coring $\mathcal{C}$ there are four natural exact categories [28] of left comodules: the exact category of A-projective $\mathcal{C}$-comodules, the exact category of A-flat $\mathcal{C}$-comodules, the exact category of arbitrary $\mathcal{C}$-comodules with A-split exact triples, and the exact category of arbitrary left $\mathcal{C}$-comodules with A-*pure* exact triples, i.e., the exact triples which as triples of left A-modules remain exact after the tensor product with any right A-module. Besides, any morphism of $\mathcal{C}$-comodules has a cokernel and the forgetful functor $\mathcal{C}$-comod $\longrightarrow$ A-mod preserves cokernels. When a morphism of $\mathcal{C}$-comodules has the property that its kernel in the category of A-modules is preserved by the functors of tensor product with $\mathcal{C}$ and $\mathcal{C} \otimes_A \mathcal{C}$ over A, this kernel has a natural $\mathcal{C}$-comodule structure, which makes it the kernel of that morphism in the category of $\mathcal{C}$-comodules.

Infinite direct sums always exist in the category of $\mathcal{C}$-comodules and the forgetful functor $\mathcal{C}$-comod $\longrightarrow$ A-mod preserves them. The coinduction functor A-mod $\longrightarrow$ $\mathcal{C}$-comod preserves both infinite direct sums and infinite products. To construct products of $\mathcal{C}$-comodules, one can present them as kernels of morphisms of coinduced comodules, so the category of $\mathcal{C}$-comodules has infinite products if it has kernels.

If $\mathcal{C}$ is a projective right A-module, or $\mathcal{C}$ is a flat right A-module and A is a left Noetherian ring, then any left $\mathcal{C}$-comodule is a union of its subcomodules that are finitely generated as A-modules [23, 18.16 and 19.12].

1.1.3   Assume that the coring $\mathcal{C}$ is a flat left and right A-module and the ring A has a finite weak homological dimension (Tor-dimension).

Lemma. *There exists a (not always additive) functor assigning to any $\mathcal{C}$-comodule a surjective map onto it from an A-flat $\mathcal{C}$-comodule.*

Proof. Let $G(M) \longrightarrow M$ be a surjective map onto an A-module M from a flat A-module $G(M)$ functorially depending on M. For example, one can take $G(M)$ to be the direct sum of copies of the A-module A over all elements of M. Let $\mathcal{M}$ be a left $\mathcal{C}$-comodule. Consider the coaction map $\mathcal{M} \longrightarrow \mathcal{C} \otimes_A \mathcal{M}$; it is an injective morphism of left $\mathcal{C}$-comodules; let $\mathcal{K}(\mathcal{M})$ denote its cokernel. Let $\mathcal{Q}(\mathcal{M})$ be the kernel of the composition

$$\mathcal{C} \otimes_A G(\mathcal{M}) \longrightarrow \mathcal{C} \otimes_A \mathcal{M} \longrightarrow \mathcal{K}(\mathcal{M}).$$

Then the composition of maps $\mathcal{Q}(\mathcal{M}) \longrightarrow \mathcal{C} \otimes_A G(\mathcal{M}) \longrightarrow \mathcal{C} \otimes_A \mathcal{M}$ factorizes through the injection $\mathcal{M} \longrightarrow \mathcal{C} \otimes_A \mathcal{M}$, so there is a natural surjective morphism of $\mathcal{C}$-comodules $\mathcal{Q}(\mathcal{M}) \longrightarrow \mathcal{M}$. Let us show that the flat dimension $\mathrm{df}_A \mathcal{Q}(\mathcal{M})$ of the A-module $\mathcal{Q}(\mathcal{M})$ is smaller than that of $\mathcal{M}$. Indeed, the A-module $\mathcal{C} \otimes_A G(\mathcal{M})$ is flat, hence

$$\mathrm{df}_A \mathcal{Q}(\mathcal{M}) = \mathrm{df}_A \mathcal{K}(\mathcal{M}) - 1 \leqslant \mathrm{df}_A(\mathcal{C} \otimes_A \mathcal{M}) - 1 \leqslant \mathrm{df}_A \mathcal{M} - 1,$$

because the A-module $\mathcal{K}(\mathcal{M})$ is a direct summand of the A-module $\mathcal{C} \otimes_A \mathcal{M}$ and a flat resolution of the A-module $\mathcal{C} \otimes_A \mathcal{M}$ can be constructed by taking the tensor product of a flat resolution of the A-module $\mathcal{M}$ with the A-A-bimodule $\mathcal{C}$. It remains to iterate the functor $\mathcal{M} \longmapsto \mathcal{Q}(M)$ sufficiently many times. Notice that the comodule $\mathcal{Q}(\mathcal{M})$ is an extension of $\mathcal{M}$ by a coinduced comodule $\mathcal{C} \otimes_A \ker(G(\mathcal{M}) \to \mathcal{M})$. $\square$

1.2   Cotensor product

1.2.1   The *cotensor product* $\mathcal{N} \square_\mathcal{C} \mathcal{M}$ of a right $\mathcal{C}$-comodule $\mathcal{N}$ and a left $\mathcal{C}$-comodule $\mathcal{M}$ is a k-module defined as the kernel of the pair of maps

$$\mathcal{N} \otimes_A \mathcal{M} \rightrightarrows \mathcal{N} \otimes_A \mathcal{C} \otimes_A \mathcal{M}$$

one of which is induced by the $\mathcal{C}$-coaction in $\mathcal{N}$ and the other by the $\mathcal{C}$-coaction in $\mathcal{M}$. The functor of cotensor product is neither left, nor right exact in general; it is left exact if the ring A is absolutely flat. For any right A-module V and any left $\mathcal{C}$-comodule $\mathcal{M}$ there is a natural isomorphism

$$(V \otimes_A \mathcal{C}) \square_\mathcal{C} \mathcal{M} \simeq V \otimes_A \mathcal{M}.$$

This is an instance of the following general fact.

Lemma. *Let* E *be a tensor category,* M *be a left module category over it,* N *be a right module category,* K *be an additive category, and* $\otimes\colon \mathsf{N}\times\mathsf{M}\longrightarrow\mathsf{K}$ *be a pairing functor compatible with the module category structures on* M *and* N. *Let* R *be a ring object with unit in* E, M *be an* R-*module object in* M, *and* V *be an object of* N. *Then the morphism* $V\otimes R\otimes M\longrightarrow V\otimes M$ *induced by the action of* R *in* M *is a cokernel of the pair of morphisms* $V\otimes R\otimes R\otimes M\rightrightarrows V\otimes R\otimes M$, *one of which is induced by the multiplication in* R *and the other by the* R-*action in* M.

Proof. The whole bar complex

$$\cdots\longrightarrow V\otimes R\otimes R\otimes M\longrightarrow V\otimes R\otimes M\longrightarrow V\otimes M\longrightarrow 0$$

is contractible with a contracting homotopy

$$\cdots\longleftarrow V\otimes R\otimes R\otimes M\longleftarrow V\otimes R\otimes M\longleftarrow V\otimes M$$

induced by the unit morphism of R (applied at the leftmost factor R). $\square$

1.2.2 Assume that $\mathcal{C}$ is a flat right A-module. A right comodule $\mathcal{N}$ over $\mathcal{C}$ is called *coflat* if the functor of cotensor product with $\mathcal{N}$ is exact on the category of left $\mathcal{C}$-comodules. It is easy to see that any coflat $\mathcal{C}$-comodule is a flat A-module. The $\mathcal{C}$-comodule coinduced from a flat A-module is coflat. A left comodule $\mathcal{M}$ over $\mathcal{C}$ is called *coflat relative to* A ($\mathcal{C}/A$-*coflat*) if its cotensor product with any exact triple of A-flat right $\mathcal{C}$-comodules is an exact triple. Any coinduced $\mathcal{C}$-comodule is $\mathcal{C}/A$-coflat.

The definition of a relatively coflat $\mathcal{C}$-comodule does not really depend on the flatness assumption on $\mathcal{C}$, but appears to be useful when this assumption holds.

Lemma. *The classes of coflat right* $\mathcal{C}$-*comodules and* $\mathcal{C}/A$-*coflat left* $\mathcal{C}$-*comodules are closed under extensions. The quotient comodule of a* $\mathcal{C}/A$-*coflat left* $\mathcal{C}$-*comodule by a* $\mathcal{C}/A$-*coflat subcomodule is* $\mathcal{C}/A$-*coflat; an* A-*flat quotient comodule of a coflat right* $\mathcal{C}$-*comodule by a coflat subcomodule is coflat. The cotensor product of an exact triple of coflat right* $\mathcal{C}$-*comodules with any left* $\mathcal{C}$-*comodule is an exact triple and the cotensor product of an* A-*flat right* $\mathcal{C}$-*comodule with an exact triple of* $\mathcal{C}/A$-*coflat left* $\mathcal{C}$-*comodules is an exact triple.*

Proof. All of these results follow from the standard properties of the right derived functor of the left exact functor of cotensor product on the Cartesian product of the exact category of A-flat right $\mathcal{C}$-comodules and the abelian category of left $\mathcal{C}$-comodules. One can simply define the k-modules $\mathrm{Cotor}_i^{\mathcal{C}}(\mathcal{N},\mathcal{M})$, $i=0,-1,\ldots$ as the homology of the cobar complex

$$\mathcal{N}\otimes_A\mathcal{M}\longrightarrow\mathcal{N}\otimes_A\mathcal{C}\otimes_A\mathcal{M}\longrightarrow\mathcal{N}\otimes_A\mathcal{C}\otimes_A\mathcal{C}\otimes_A\mathcal{M}\longrightarrow\cdots$$

for any A-flat right $\mathcal{C}$-comodule $\mathcal{N}$ and any left $\mathcal{C}$-comodule $\mathcal{M}$. Then

$$\mathrm{Cotor}_0^{\mathcal{C}}(\mathcal{N},\mathcal{M})\simeq\mathcal{N}\,\square_{\mathcal{C}}\,\mathcal{M},$$

and there are long exact sequences of $\mathrm{Cotor}_*^{\mathcal{C}}$ associated with exact triples of $\mathcal{C}$-co-modules in either argument, since in both cases the cobar complexes form an exact triple. Now an A-flat right $\mathcal{C}$-comodule $\mathcal{N}$ is coflat if and only if $\mathrm{Cotor}_i^{\mathcal{C}}(\mathcal{N}, \mathcal{M}) = 0$ for any left $\mathcal{C}$-comodule $\mathcal{M}$ and all $i < 0$. Indeed, the "if" assertion follows from the homological exact sequence, and "only if" holds since the cobar complex is the cotensor product of the comodule $\mathcal{N}$ with the cobar resolution of the comodule $\mathcal{M}$, which is exact except in degree 0. Analogously, a left $\mathcal{C}$-comodule $\mathcal{M}$ is $\mathcal{C}/A$-coflat if and only if $\mathrm{Cotor}_i^{\mathcal{C}}(\mathcal{N}, \mathcal{M}) = 0$ for any A-flat right $\mathcal{C}$-comodule $\mathcal{N}$ and all $i < 0$, since the cobar resolution of the comodule $\mathcal{N}$ is a complex of A-flat right $\mathcal{C}$-comodules, exact except in degree 0 and split over A. The rest is obvious. $\square$

Remark. A much more general construction of the double-sided derived functor $\mathrm{Cotor}_*^{\mathcal{C}}(\mathcal{N}, \mathcal{M})$ defined for arbitrary $\mathcal{C}$-comodules $\mathcal{M}$ and $\mathcal{N}$ will be given, in the stronger assumptions of 1.1.3 and 2.5, in Chapter 2. Using this construction, one can prove somewhat stronger results. In particular, $\mathrm{Cotor}_i^{\mathcal{C}}(\mathcal{M}, \mathcal{N}) = 0$ for any $\mathcal{C}/A$-coflat left $\mathcal{C}$-comodule $\mathcal{M}$, any right $\mathcal{C}$-comodule $\mathcal{N}$, and all $i < 0$, since the k-modules $\mathrm{Cotor}_i^{\mathcal{C}}(\mathcal{M}, \mathcal{N})$ can be computed using a left resolution of $\mathcal{N}$ consisting of A-flat right $\mathcal{C}$-comodules (see 2.8). Therefore, any A-flat $\mathcal{C}/A$-coflat $\mathcal{C}$-comodule is coflat. It follows that the construction of Lemma 1.1.3 assigns to any $\mathcal{C}/A$-coflat $\mathcal{C}$-comodule a surjective map onto it from a coflat $\mathcal{C}$-comodule with a $\mathcal{C}/A$-coflat kernel.

1.2.3 Now let $\mathcal{C}$ be an arbitrary coring. Let us call a left $\mathcal{C}$-comodule $\mathcal{M}$ *quasicoflat* if the functor of cotensor product with $\mathcal{M}$ is right exact on the category of right $\mathcal{C}$-comodules, i.e., this functor preserves cokernels. Any coinduced $\mathcal{C}$-comodule is quasicoflat. Any quasicoflat $\mathcal{C}$-comodule is $\mathcal{C}/A$-coflat.

Proposition. *Let $\mathcal{N}$ be a right $\mathcal{C}$-comodule, $\mathcal{K}$ be a left $\mathcal{C}$-comodule endowed with a right action of a k-algebra B by comodule endomorphisms, and M be a left B-module. Then there is a natural k-module map*

$$(\mathcal{N} \,\square_{\mathcal{C}}\, \mathcal{K}) \otimes_B M \longrightarrow \mathcal{N} \,\square_{\mathcal{C}}\, (\mathcal{K} \otimes_B M),$$

which is an isomorphism, at least, in the following cases:

(a) *M is a flat left B-module;*

(b) *$\mathcal{N}$ is a quasicoflat right $\mathcal{C}$-comodule;*

(c) *$\mathcal{C}$ is a flat right A-module, $\mathcal{N}$ is a flat right A-module, $\mathcal{K}$ is a $\mathcal{C}/A$-coflat left $\mathcal{C}$-comodule, $\mathcal{K}$ is a flat right B-module, and the ring B has a finite weak homological dimension;*

(d) *$\mathcal{K}$ as a left $\mathcal{C}$-comodule with a right B-module structure is coinduced from an A-B-bimodule.*

Besides, in the case (c) the cotensor product $\mathcal{N} \,\square_{\mathcal{C}}\, \mathcal{K}$ is a flat right B-module.

Proof. The map $(\mathcal{N} \,\square_{\mathcal{C}}\, \mathcal{K}) \otimes_B M \longrightarrow \mathcal{N} \otimes_A \mathcal{K} \otimes_B M$ obtained by taking the tensor product of the map $\mathcal{N} \,\square_{\mathcal{C}}\, \mathcal{K} \longrightarrow \mathcal{N} \otimes_A \mathcal{K}$ with the B-module M has equal compositions with two maps $\mathcal{N} \otimes_A \mathcal{K} \otimes_B M \rightrightarrows \mathcal{N} \otimes_A \mathcal{C} \otimes_A \mathcal{K} \otimes_B M$, hence there is a natural map $(\mathcal{N} \,\square_{\mathcal{C}}\, \mathcal{K}) \otimes_B M \longrightarrow \mathcal{N} \,\square_{\mathcal{C}}\, (\mathcal{K} \otimes_B M)$. The case (a) is obvious. In the case (b), it suffices to present M as the cokernel of a map of flat B-modules. To prove (c) and (d), consider the cobar complex

$$\mathcal{N} \,\square_{\mathcal{C}}\, \mathcal{K} \longrightarrow \mathcal{N} \otimes_A \mathcal{K} \longrightarrow \mathcal{N} \otimes_A \mathcal{C} \otimes_A \mathcal{K} \longrightarrow \mathcal{N} \otimes_A \mathcal{C} \otimes_A \mathcal{C} \otimes_A \mathcal{K} \longrightarrow \cdots \quad (1.1)$$

In the case (c) this complex is exact, since it is the cotensor product of a $\mathcal{C}/A$-coflat $\mathcal{C}$-comodule $\mathcal{K}$ with an A-split exact complex of A-flat $\mathcal{C}$-comodules $\mathcal{N} \longrightarrow \mathcal{N} \otimes_A \mathcal{C} \longrightarrow \mathcal{N} \otimes_A \mathcal{C} \otimes_A \mathcal{C} \longrightarrow \cdots$ Since all the terms of the complex (1.1), except possibly the leftmost one, are flat right B-modules and the weak homological dimension of the ring B is finite, the leftmost term $\mathcal{K} \,\square_{\mathcal{C}}\, M$ is also a flat B-module and the tensor product of this complex with the left B-module M is exact. In the case (d), the complex (1.1) is exact and split as a complex of right B-modules. $\square$

1.2.4 Let $\mathcal{C}$ be a coring over a k-algebra A and $\mathcal{D}$ be a coring over a k-algebra B. A $\mathcal{C}$-$\mathcal{D}$-*bicomodule* $\mathcal{K}$ is an A-B-bimodule in the category of k-modules endowed with a left $\mathcal{C}$-comodule and a right $\mathcal{D}$-comodule structures such that the right $\mathcal{D}$-coaction map $\mathcal{K} \longrightarrow \mathcal{K} \otimes_B \mathcal{D}$ is a morphism of left $\mathcal{C}$-comodules and the left $\mathcal{C}$-coaction map $\mathcal{K} \longrightarrow \mathcal{C} \otimes_A \mathcal{K}$ is a morphism of right B-modules, or equivalently, the right $\mathcal{D}$-coaction map is a morphism of left A-modules and the left $\mathcal{C}$-coaction map is a morphism of right $\mathcal{D}$-comodules. Equivalently, a $\mathcal{C}$-$\mathcal{D}$-bicomodule is a k-module endowed with an A-B-bimodule structure and an A-B-bimodule map of *bicoaction* $\mathcal{K} \longrightarrow \mathcal{C} \otimes_A \mathcal{K} \otimes_B \mathcal{D}$ satisfying the coassociativity and counity equations. We will denote the category of $\mathcal{C}$-$\mathcal{D}$-bicomodules by $\mathcal{C}$-comod-$\mathcal{D}$.

Assume that $\mathcal{C}$ is a flat right A-module and $\mathcal{D}$ is a flat left B-module. Then the category of $\mathcal{C}$-$\mathcal{D}$-bicomodules is abelian and the forgetful functor $\mathcal{C}$-comod-$\mathcal{D} \longrightarrow k$-mod is exact. Let $\mathcal{E}$ be a coring over a k-algebra F. Let $\mathcal{N}$ be a $\mathcal{C}$-$\mathcal{E}$-bicomodule and M be a $\mathcal{E}$-$\mathcal{D}$-bicomodule. Then the cotensor product $\mathcal{N} \,\square_{\mathcal{E}}\, M$ can be endowed with a $\mathcal{C}$-$\mathcal{D}$-bicomodule structure as the kernel of a pair of bimodule morphisms $\mathcal{N} \otimes_F M \rightrightarrows \mathcal{N} \otimes_F \mathcal{E} \otimes_F M$.

More generally, let $\mathcal{C}$, $\mathcal{D}$, and $\mathcal{E}$ be arbitrary corings. Assume that the functor of tensor product with $\mathcal{C}$ over A and with $\mathcal{D}$ over B preserves the kernel of the pair of maps $\mathcal{N} \otimes_F M \rightrightarrows \mathcal{N} \otimes_F \mathcal{E} \otimes_F M$, that is the natural map $\mathcal{C} \otimes_A (\mathcal{N} \,\square_{\mathcal{E}}\, M) \otimes_B \mathcal{D} \longrightarrow (\mathcal{C} \otimes_A \mathcal{N}) \,\square_{\mathcal{E}}\, (M \otimes_B \mathcal{D})$ is an isomorphism. Then one can define a bicoaction map $\mathcal{N} \,\square_{\mathcal{E}}\, M \longrightarrow \mathcal{C} \otimes_A (\mathcal{N} \,\square_{\mathcal{E}}\, M) \otimes_B \mathcal{D}$ by taking the cotensor product over $\mathcal{E}$ of the left $\mathcal{C}$-coaction map $\mathcal{N} \longrightarrow \mathcal{C} \otimes_A \mathcal{N}$ and the right $\mathcal{D}$-coaction map $M \longrightarrow M \otimes_B \mathcal{D}$. One can easily see that this bicoaction is counital and coassociative, at least, if the natural maps $\mathcal{C} \otimes_A \mathcal{C} \otimes_A (\mathcal{N} \,\square_{\mathcal{E}}\, M) \longrightarrow (\mathcal{C} \otimes_A \mathcal{C} \otimes_A \mathcal{N}) \,\square_{\mathcal{E}}\, M$ and $(\mathcal{N} \,\square_{\mathcal{E}}\, M) \otimes_B \mathcal{D} \otimes_B \mathcal{D} \longrightarrow \mathcal{N} \,\square_{\mathcal{E}}\, (M \otimes_B \mathcal{D} \otimes_B \mathcal{D})$ are also isomorphisms.

In particular, if $\mathcal{C}$ is a flat right A-module and either $\mathcal{D}$ is a flat left B-module, or $\mathcal{N}$ is a quasicoflat right $\mathcal{E}$-comodule, or $\mathcal{N}$ is a flat right F-module, $\mathcal{E}$ is a flat right F-module, $\mathcal{M}$ is an $\mathcal{E}/F$-coflat left $\mathcal{E}$-comodule, $\mathcal{M}$ is a flat right B-module, and B has a finite weak homological dimension, or $\mathcal{M}$ as a left $\mathcal{E}$-comodule with a right B-module structure is coinduced from an F-B-bimodule, then the cotensor product $\mathcal{N} \,\square_\mathcal{E}\, \mathcal{M}$ has a natural $\mathcal{C}$-$\mathcal{D}$-bicomodule structure.

1.2.5 Let $\mathcal{C}$ be a coring over a k-algebra A and $\mathcal{D}$ be a coring over a k-algebra B.

Proposition. *Let $\mathcal{N}$ be a right $\mathcal{C}$-comodule, $\mathcal{K}$ be a $\mathcal{C}$-$\mathcal{D}$-bicomodule, and $\mathcal{M}$ be a left $\mathcal{D}$-comodule. Then the iterated cotensor products*

$$(\mathcal{N} \,\square_\mathcal{C}\, \mathcal{K}) \,\square_\mathcal{D}\, \mathcal{M} \quad and \quad \mathcal{N} \,\square_\mathcal{C}\, (\mathcal{K} \,\square_\mathcal{D}\, \mathcal{M})$$

are naturally isomorphic, at least, in the following cases:

(a) *$\mathcal{C}$ is a flat right A-module, $\mathcal{N}$ is a flat right A-module, $\mathcal{D}$ is a flat left B-module, and $\mathcal{M}$ is a flat left B-module;*

(b) *$\mathcal{C}$ is a flat right A-module and $\mathcal{N}$ is a coflat right $\mathcal{C}$-comodule;*

(c) *$\mathcal{C}$ is a flat right A-module, $\mathcal{N}$ is a flat right A-module, $\mathcal{K}$ is a $\mathcal{C}/A$-coflat left $\mathcal{C}$-comodule, $\mathcal{K}$ is a flat right B-module, and the ring B has a finite weak homological dimension;*

(d) *$\mathcal{C}$ is a flat right A-module, $\mathcal{N}$ is a flat right A-module, and $\mathcal{K}$ as a left $\mathcal{C}$-comodule with a right B-module structure is coinduced from an A-B-bimodule;*

(e) *$\mathcal{M}$ is a quasicoflat left $\mathcal{C}$-comodule and $\mathcal{K}$ as a left $\mathcal{C}$-comodule with a right B-module structure is coinduced from an A-B-bimodule;*

(f) *$\mathcal{K}$ as a left $\mathcal{C}$-comodule with a right B-module structure is coinduced from an A-B-bimodule and $\mathcal{K}$ as a right $\mathcal{D}$-comodule with a left A-module structure is coinduced from an A-B-bimodule.*

More precisely, in all cases in this list the natural maps from both iterated cotensor products to the k-module $\mathcal{N} \otimes_A \mathcal{K} \otimes_B \mathcal{M}$ are injective, their images coincide and are equal to the intersection of two submodules $(\mathcal{N} \otimes_A \mathcal{K}) \,\square_\mathcal{D}\, \mathcal{M}$ and $\mathcal{N} \,\square_\mathcal{C}\, (\mathcal{K} \otimes_B \mathcal{M})$ in the k-module $\mathcal{N} \otimes_A \mathcal{K} \otimes_B \mathcal{M}$.

Proof. One can easily see that whenever both maps $(\mathcal{N} \,\square_\mathcal{C}\, \mathcal{K}) \otimes_B \mathcal{M} \longrightarrow \mathcal{N} \,\square_\mathcal{C}\,$ $(\mathcal{K} \otimes_B \mathcal{M})$ and $(\mathcal{N} \,\square_\mathcal{C}\, \mathcal{K}) \otimes_B \mathcal{D} \otimes_B \mathcal{M} \longrightarrow \mathcal{N} \,\square_\mathcal{C}\, (\mathcal{K} \otimes_B \mathcal{D} \otimes_B \mathcal{M})$ are isomorphisms, the natural map $(\mathcal{N} \,\square_\mathcal{C}\, \mathcal{K}) \,\square_\mathcal{D}\, \mathcal{M} \longrightarrow \mathcal{N} \otimes_A \mathcal{K} \otimes_B \mathcal{M}$ is injective and its image coincides with the desired intersection of two submodules in $\mathcal{N} \otimes_A \mathcal{K} \otimes_B \mathcal{M}$. Thus it remains to apply Proposition 1.2.3. $\qquad\square$

When associativity of cotensor product of four or more (bi)comodules is an issue, it becomes important to know that the pentagonal diagrams of associativity isomorphisms are commutative. Since each of the five iterated cotensor products of four factors of the form $\mathcal{N} \,\square_\mathcal{C}\, \mathcal{K} \,\square_\mathcal{E}\, \mathcal{L} \,\square_\mathcal{D}\, \mathcal{M}$ is endowed with a natural map into the tensor product $\mathcal{N} \otimes_A \mathcal{K} \otimes_F \mathcal{L} \otimes_B \mathcal{M}$, and the associativity isomorphisms are, presumably, compatible with these maps, it suffices to check that at least one of these five maps is injective in order to show that the pentagonal diagram commutes. In particular, if the above proposition provides all the five associativity isomorphisms constituting the pentagonal diagram and either $\mathcal{M}$ is a flat left B-module, or $\mathcal{N}$ is a flat right A-module, or both $\mathcal{K}$ and $\mathcal{L}$ as left (right) comodules with right (left) module structures are coinduced from bimodules, then the pentagonal diagram is commutative.

We will say that a multiple cotensor product of several bimodules $\mathcal{N} \square_{\mathcal{C}}$ $\cdots \square_{\mathcal{D}} \mathcal{M}$ is associative if for any way of putting parentheses in this product all the intermediate cotensor products can be endowed with bimodule structures via the construction of 1.2.4, all possible associativity isomorphisms between intermediate cotensor products exist in the sense of the last assertion of Proposition 1.2.5 and preserve bimodule structures, and all the pentagonal diagrams commute. This definition allows us to consider associativity of cotensor products as a *property* rather than an additional structure. In particular, associativity isomorphisms and bimodule structures on associative multiple cotensor products are preserved by the morphisms between them induced by any bimodule morphisms of the factors.

1.3   Semialgebras and semimodules

1.3.1   Assume that the coring $\mathcal{C}$ over A is a flat right A-module.

It follows from Proposition 1.2.5(b) that the category of $\mathcal{C}$-$\mathcal{C}$-bimodules which are coflat right $\mathcal{C}$-comodules is an associative tensor category with a unit object $\mathcal{C}$, the category of left $\mathcal{C}$-comodules is a left module category over it, and the category of coflat right $\mathcal{C}$-comodules is a right module category over this tensor category. Furthermore, it follows from Proposition 1.2.5(c) that whenever the ring A has a finite weak homological dimension, the $\mathcal{C}$-$\mathcal{C}$-bimodules that are flat right A-modules and $\mathcal{C}/A$-coflat left $\mathcal{C}$-comodules also form a tensor category, left $\mathcal{C}$-comodules form a left module category over it, and A-flat right $\mathcal{C}$-comodules form a right module category over this tensor category. Finally, it follows from Proposition 1.2.5(a) that whenever the ring A is absolutely flat, the categories of left and right $\mathcal{C}$-comodules are left and right module categories over the tensor category of $\mathcal{C}$-$\mathcal{C}$-bimodules. In each case, the cotensor product operation provides a pairing between these left and right module categories compatible with their module category structures and taking values in the category of k-modules.

A *semialgebra* over $\mathcal{C}$ is a ring object with unit in one of the tensor categories of $\mathcal{C}$-$\mathcal{C}$-bimodules of the kind described above. In other words, a semialgebra $\mathcal{S}$ over $\mathcal{C}$ is a $\mathcal{C}$-$\mathcal{C}$-bimodule satisfying appropriate (co)flatness conditions guaranteeing associativity of cotensor products $\mathcal{S} \square_{\mathcal{C}} \cdots \square_{\mathcal{C}} \mathcal{S}$ of any number of copies of $\mathcal{S}$ and endowed with two bimodule morphisms of *semimultiplication* $\mathcal{S} \square_{\mathcal{C}} \mathcal{S} \longrightarrow \mathcal{S}$ and *semiunit* $\mathcal{C} \longrightarrow \mathcal{S}$ satisfying the associativity and unity equations. Namely, two compositions $\mathcal{S} \square_{\mathcal{C}} \mathcal{S} \square_{\mathcal{C}} \mathcal{S} \rightrightarrows \mathcal{S} \square_{\mathcal{C}} \mathcal{S} \longrightarrow \mathcal{S}$ of the morphisms $\mathcal{S} \square_{\mathcal{C}} \mathcal{S} \square_{\mathcal{C}} \mathcal{S} \rightrightarrows \mathcal{S} \square_{\mathcal{C}} \mathcal{S}$ induced by the semimultiplication morphism with the semimultiplication morphism $\mathcal{S} \square_{\mathcal{C}} \mathcal{S} \longrightarrow \mathcal{S}$ should coincide with each other and two compositions $\mathcal{S} \rightrightarrows \mathcal{S} \square_{\mathcal{C}} \mathcal{S} \longrightarrow \mathcal{S}$ of the morphisms $\mathcal{S} \rightrightarrows \mathcal{S} \square_{\mathcal{C}} \mathcal{S}$ induced by the semiunit morphism with the semimultiplication morphism should coincide with the identity morphism of S.

A *left semimodule* over S is a module object in one of the left module categories of C-comodules of the above kind over the ring object S in the corresponding tensor category of C-C-bicomodules. In other words, a left S-semimodule $\mathcal{M}$ is a left C-comodule endowed with a left C-comodule morphism of *left semiaction* $S \square_C \mathcal{M} \longrightarrow \mathcal{M}$ satisfying the associativity and unity equations. Namely, two compositions $S \square_C S \square_C \mathcal{M} \rightrightarrows S \square_C \mathcal{M} \longrightarrow \mathcal{M}$ of the morphisms $S \square_C S \square_C \mathcal{M} \rightrightarrows S \square_C \mathcal{M}$ induced by the semimultiplication and the semiaction morphisms with the semiaction morphism $S \square_C \mathcal{M} \longrightarrow \mathcal{M}$ should coincide with each other, and the composition $\mathcal{M} \longrightarrow S \square_C \mathcal{M} \longrightarrow \mathcal{M}$ of the morphism $\mathcal{M} \longrightarrow S \square_C \mathcal{M}$ induced by the semiunit morphism with the semiaction morphism should coincide with the identity morphism of $\mathcal{M}$. For this definition to make sense, (co)flatness conditions imposed on S and/or $\mathcal{M}$ must guarantee associativity of multiple cotensor products of the form $S \square_C \cdots \square_C S \square_C \mathcal{M}$. *Right semimodules* over S are defined in an analogous way.

If $\mathcal{L}$ is a left C-comodule for which the multiple cotensor products $S \square_C \cdots \square_C S \square_C \mathcal{L}$ are associative, then there is a natural left S-semimodule structure on the cotensor product $S \square_C \mathcal{L}$. The left semimodule $S \square_C \mathcal{L}$ is called the left S-semimodule *induced* from a C-comodule $\mathcal{L}$. According to Lemma 1.1.2, the k-module of semimodule homomorphisms from the induced S-semimodule to an arbitrary S-semimodule is described by the formula

$$\operatorname{Hom}_S(S \square_C \mathcal{L}, \mathcal{M}) \simeq \operatorname{Hom}_C(\mathcal{L}, \mathcal{M}).$$

We will denote the category of left S-semimodules by S-simod and the category of right S-semimodules by simod-S. This notation presumes that one can speak of (left or right) S-semimodules with no flatness conditions imposed on them. If S is a coflat right C-comodule, the category of left semimodules over S is abelian and the forgetful functor S-simod $\longrightarrow$ C-comod is exact.

Assume that either S is a coflat right C-comodule, or S is a flat right A-module and a C/A-coflat left C-comodule and A has a finite weak homological dimension, or A is absolutely flat. Then both infinite direct sums and infinite products exist in the category of left S-semimodules, and both are preserved by the forgetful functor S-simod $\longrightarrow$ C-comod, even though only infinite direct sums are preserved by the full forgetful functor S-simod $\longrightarrow$ A-mod.

If S is a flat right A-module and a C/A-coflat left C-comodule and A has a finite weak homological dimension, then the category of A-flat right S-semimodules is exact. Of course, if S is a coflat right C-comodule, then the category of A-flat left S-semimodules is exact. In both cases there are exact categories of C-coflat right S-semimodules and C/A-coflat left S-semimodules. If A is absolutely flat, there are exact categories of C-coflat left and right S-semimodules. Infinite direct sums exist in all of these exact categories, and the forgetful functors preserve them.

1.3.2 Assume that the coring $\mathcal{C}$ is a flat left and right A-module, the semialgebra $\mathcal{S}$ is a flat left A-module and a coflat right $\mathcal{C}$-comodule, and the ring A has a finite weak homological dimension.

Lemma. *There exists a (not always additive) functor assigning to any left $\mathcal{S}$-semimodule a surjective map onto it from an A-flat left $\mathcal{S}$-semimodule.*

Proof. Let $\mathcal{P}(\mathcal{M}) \longrightarrow \mathcal{M}$ denote the functorial surjective morphism onto a $\mathcal{C}$-comodule $\mathcal{M}$ from an A-flat $\mathcal{C}$-comodule $\mathcal{P}(\mathcal{M})$ constructed in Lemma 1.1.3. Then for any left $\mathcal{S}$-semimodule $\mathcal{M}$ the composition of maps

$$\mathcal{S} \,\square_{\mathcal{C}}\, \mathcal{P}(\mathcal{M}) \longrightarrow \mathcal{S} \,\square_{\mathcal{C}}\, \mathcal{M} \longrightarrow \mathcal{M}$$

provides the desired surjective morphism of $\mathcal{S}$-semimodules. According to the last assertion of Proposition 1.2.3 (with the left and right sides switched), the A-module $\boldsymbol{\mathcal{P}}(\mathcal{M}) = \mathcal{S} \,\square_{\mathcal{C}}\, \mathcal{P}(\mathcal{M})$ is flat. $\square$

Remark. In the above assumptions, the same construction provides also a (not always additive) functor assigning to any $\mathcal{C}/A$-coflat right $\mathcal{S}$-semimodule a surjective map onto it from a semiflat right $\mathcal{S}$-semimodule (see 1.4.2) with a $\mathcal{C}/A$-coflat kernel. This follows from Lemma 1.2.2 and Remark 1.2.2, since the cotensor product with $\mathcal{S}$ over $\mathcal{C}$ preserves the kernel of the morphism $\mathcal{P}(\mathcal{N}) \longrightarrow \mathcal{N}$ and the kernel of the map $\mathcal{N} \,\square_{\mathcal{C}}\, \mathcal{S} \longrightarrow \mathcal{N}$ is isomorphic to a direct summand of $\mathcal{N} \,\square_{\mathcal{C}}\, \mathcal{S}$ as a right $\mathcal{C}$-comodule.

1.3.3 Assume that the coring $\mathcal{C}$ is a flat right A-module, the semialgebra $\mathcal{S}$ is a $\mathcal{C}/A$-coflat left $\mathcal{C}$-comodule and a coflat right $\mathcal{C}$-comodule, and the ring A has a finite weak homological dimension.

Lemma. *There exists an exact functor assigning to any A-flat right $\mathcal{S}$-semimodule an injective morphism from it into a coflat right $\mathcal{S}$-semimodule with an A-flat quotient semimodule. Besides, there exists an exact functor assigning to any left $\mathcal{S}$-semimodule an injective morphism from it into a $\mathcal{C}/A$-coflat left $\mathcal{S}$-semimodule.*

Proof. For any A-flat right $\mathcal{C}$-comodule $\mathcal{N}$, set $\mathcal{G}(\mathcal{N}) = \mathcal{N} \otimes_A \mathcal{C}$. Then the coaction map $\mathcal{N} \longrightarrow \mathcal{G}(\mathcal{N})$ is an injective morphism of $\mathcal{C}$-comodules, the comodule $\mathcal{G}(\mathcal{N})$ is coflat, and the quotient comodule $\mathcal{G}(\mathcal{N})/\mathcal{N}$ is A-flat. Now let $\mathcal{N}$ be an A-flat right $\mathcal{S}$-semimodule. The semiaction map $\mathcal{N} \,\square_{\mathcal{C}}\, \mathcal{S} \longrightarrow \mathcal{N}$ is a surjective morphism of A-flat $\mathcal{S}$-semimodules; let $\mathcal{K}(\mathcal{N})$ denote its kernel. The map $\mathcal{N}\square_{\mathcal{C}}\mathcal{S} \longrightarrow \mathcal{G}(\mathcal{N})\square_{\mathcal{C}}\mathcal{S}$ is an injective morphism of A-flat $\mathcal{S}$-semimodules with an A-flat quotient semimodule $(\mathcal{G}(\mathcal{N})/\mathcal{N}) \,\square_{\mathcal{C}}\, \mathcal{S}$. Let $\mathcal{Q}(\mathcal{N})$ be the cokernel of the composition

$$\mathcal{K}(\mathcal{N}) \longrightarrow \mathcal{N}\,\square_{\mathcal{C}}\,\mathcal{S} \longrightarrow \mathcal{G}(\mathcal{N}) \,\square_{\mathcal{C}}\, \mathcal{S}.$$

Then the composition of maps $\mathcal{N} \,\square_{\mathcal{C}}\, \mathcal{S} \longrightarrow \mathcal{G}(\mathcal{N}) \,\square_{\mathcal{C}}\, \mathcal{S} \longrightarrow \mathcal{Q}(\mathcal{N})$ factorizes through the surjection $\mathcal{N} \,\square_{\mathcal{C}}\, \mathcal{S} \longrightarrow \mathcal{N}$, so there is a natural injective morphism of $\mathcal{S}$-semimodules $\mathcal{N} \longrightarrow \mathcal{Q}(\mathcal{N})$. The quotient semimodule $\mathcal{Q}(\mathcal{N})/\mathcal{N}$ is isomorphic to $(\mathcal{G}(\mathcal{N})/\mathcal{N}) \,\square_{\mathcal{C}}\, \mathcal{S}$, hence both $\mathcal{Q}(\mathcal{N})/\mathcal{N}$ and $\mathcal{Q}(\mathcal{N})$ are flat A-modules.

Notice that the semimodule morphism $\mathcal{N} \longrightarrow \mathcal{Q}(\mathcal{N})$ can be lifted to a comodule morphism $\mathcal{N} \longrightarrow \mathcal{G}(\mathcal{N}) \, \square_{\mathcal{C}} \, \mathbf{S}$. Indeed, the map $\mathcal{N} \longrightarrow \mathcal{Q}(\mathcal{N})$ can be presented as the composition $\mathcal{N} \longrightarrow \mathcal{N} \, \square_{\mathcal{C}} \, \mathbf{S} \longrightarrow \mathcal{G}(\mathcal{N}) \, \square_{\mathcal{C}} \, \mathbf{S} \longrightarrow \mathcal{Q}(\mathcal{N})$, where the map $\mathcal{N} \longrightarrow \mathcal{N} \square_{\mathcal{C}} \mathbf{S}$ is induced by the semiunit morphism $\mathcal{C} \longrightarrow \mathbf{S}$ of the semialgebra $\mathbf{S}$.

Iterating this construction, we obtain an inductive system of $\mathcal{C}$-comodule morphisms

$$\mathcal{N} \longrightarrow \mathcal{G}(\mathcal{N}) \, \square_{\mathcal{C}} \, \mathbf{S} \longrightarrow \mathcal{Q}(\mathcal{N}) \longrightarrow \mathcal{G}(\mathcal{Q}(\mathcal{N})) \, \square_{\mathcal{C}} \, \mathbf{S} \longrightarrow \mathcal{Q}(\mathcal{Q}(\mathcal{N})) \longrightarrow \cdots,$$

where the maps

$$\mathcal{N} \longrightarrow \mathcal{Q}(\mathcal{N}) \longrightarrow \mathcal{Q}(\mathcal{Q}(\mathcal{N})) \longrightarrow \cdots$$

are injective morphisms of $\mathbf{S}$-semimodules with A-flat cokernels, while the $\mathcal{C}$-comodules $\mathcal{G}(\mathcal{N}) \, \square_{\mathcal{C}} \, \mathbf{S}$, $\mathcal{G}(\mathcal{Q}(\mathcal{N})) \, \square_{\mathcal{C}} \, \mathbf{S}$, ... are coflat. Denote by $\mathcal{J}(\mathcal{N})$ the inductive limit of this system; then $\mathcal{N} \longrightarrow \mathcal{J}(\mathcal{N})$ is an injective morphism of $\mathbf{S}$-semimodules with an A-flat cokernel and the $\mathcal{C}$-comodule $\mathcal{J}(\mathcal{N})$ is coflat (since the functor of cotensor product preserves filtered inductive limits).

A functorial injection $\mathcal{M} \longrightarrow \mathcal{J}(\mathcal{M})$ of any left $\mathbf{S}$-semimodule $\mathcal{M}$ into a $\mathcal{C}/A$-coflat left $\mathbf{S}$-semimodule $\mathcal{J}(\mathcal{M})$ is provided by the same construction (with the left and right sides switched). The only changes are that A-modules are no longer flat, for any left $\mathcal{C}$-comodule $\mathcal{M}$ the $\mathcal{C}$-comodule $\mathcal{G}(\mathcal{M}) = \mathcal{C} \otimes_A \mathcal{M}$ is $\mathcal{C}/A$-coflat, and therefore the $\mathbf{S}$-semimodule $\mathbf{S} \, \square_{\mathcal{C}} \, \mathcal{G}(\mathcal{M})$ is $\mathcal{C}/A$-coflat.

Both functors $\mathcal{J}$ are exact, since the kernels of surjective maps, the cokernels of injective maps, and the filtered inductive limits preserve exact triples. $\square$

1.4   Semitensor product

1.4.1   Assume that the coring $\mathcal{C}$ is a flat right A-module, the semialgebra $\mathbf{S}$ is a flat right A-module and a $\mathcal{C}/A$-coflat left $\mathcal{C}$-comodule, and the ring A has a finite weak homological dimension. Let $\mathcal{M}$ be a left $\mathbf{S}$-semimodule and $\mathcal{N}$ be an A-flat right $\mathbf{S}$-semimodule. The *semitensor product* $\mathcal{N} \lozenge_{\mathbf{S}} \mathcal{M}$ is a k-module defined as the cokernel of the pair of maps

$$\mathcal{N} \, \square_{\mathcal{C}} \, \mathbf{S} \, \square_{\mathcal{C}} \, \mathcal{M} \rightrightarrows \mathcal{N} \, \square_{\mathcal{C}} \, \mathcal{M}$$

one of which is induced by the $\mathbf{S}$-semiaction in $\mathcal{N}$ and another by the $\mathbf{S}$-semiaction in $\mathcal{M}$. Even under the strongest of our (co)flatness conditions on $\mathcal{C}$ and $\mathbf{S}$, the flatness of either $\mathcal{N}$ or $\mathcal{M}$ is still needed to guarantee that the triple cotensor product $\mathcal{N} \, \square_{\mathcal{C}} \, \mathbf{S} \, \square_{\mathcal{C}} \, \mathcal{M}$ is associative.

For any A-flat right $\mathbf{S}$-semimodule $\mathcal{N}$ and any left $\mathcal{C}$-comodule $\mathcal{L}$ there is a natural isomorphism

$$\mathcal{N} \lozenge_{\mathbf{S}} (\mathbf{S} \, \square_{\mathcal{C}} \, \mathcal{L}) \simeq \mathcal{N} \, \square_{\mathcal{C}} \, \mathcal{L}.$$

Analogously, for any A-flat right $\mathcal{C}$-comodule $\mathcal{R}$ and any left $\mathbf{S}$-semimodule $\mathcal{M}$ there is a natural isomorphism $(\mathcal{R} \, \square_{\mathcal{C}} \, \mathbf{S}) \lozenge_{\mathbf{S}} \mathcal{M} \simeq \mathcal{R} \, \square_{\mathcal{C}} \, \mathcal{M}$. These assertions follow from Lemma 1.2.1.

1.4.2 If the coring $\mathcal{C}$ is a flat right A-module and the semialgebra $\mathbf{S}$ is a coflat right $\mathcal{C}$-comodule, one can define the semitensor product of a $\mathcal{C}$-coflat right $\mathbf{S}$-semimodule and an arbitrary left $\mathbf{S}$-semimodule. In these assumptions, a $\mathcal{C}$-coflat right $\mathbf{S}$-semimodule $\mathbf{N}$ is called *semiflat* if the functor of semitensor product with $\mathbf{N}$ is exact on the abelian category of left $\mathbf{S}$-semimodules. The $\mathbf{S}$-semimodule induced from a coflat $\mathcal{C}$-comodule is semiflat.

If $\mathcal{C}$ is a flat right A-module, $\mathbf{S}$ is a coflat right $\mathcal{C}$-comodule and $\mathcal{C}/A$-coflat left $\mathcal{C}$-comodule, and the ring A has a finite weak homological dimension, one can define semiflat $\mathbf{S}$-semimodules as A-flat right $\mathbf{S}$-semimodules such that the functors of semitensor product with them are exact. Then one can prove that any semiflat $\mathbf{S}$-semimodule is a coflat $\mathcal{C}$-comodule.

When the ring A is absolutely flat, the semitensor product of arbitrary two $\mathbf{S}$-semimodules is defined without any conditions on the coring $\mathcal{C}$ and the semialgebra $\mathbf{S}$.

1.4.3 Let $\mathbf{S}$ be a semialgebra over a coring $\mathcal{C}$ over a k-algebra A and $\mathbf{T}$ be a semialgebra over a coring $\mathcal{D}$ over a k-algebra B. Let $\mathbf{K}$ denote a $\mathcal{C}$-$\mathcal{D}$-bicomodule. One can speak about $\mathbf{S}$-$\mathbf{T}$-*bisemimodule* structures on $\mathbf{K}$ if the (co)flatness conditions imposed on $\mathbf{S}$, $\mathbf{T}$, and $\mathbf{K}$ guarantee associativity of multiple cotensor products of the form $\mathbf{S} \,\square_{\mathcal{C}} \cdots \square_{\mathcal{C}}\, \mathbf{S} \,\square_{\mathcal{C}}\, \mathbf{K} \,\square_{\mathcal{D}}\, \mathbf{T} \,\square_{\mathcal{D}} \cdots \square_{\mathcal{D}}\, \mathbf{T}$. Assuming that this is so, $\mathbf{K}$ is called an $\mathbf{S}$-$\mathbf{T}$-bisemimodule if it is endowed with a left $\mathbf{S}$-semimodule and a right $\mathbf{T}$-semimodule structures such that the right $\mathbf{T}$-semiaction map $\mathbf{K} \,\square_{\mathcal{D}}\, \mathbf{T} \longrightarrow \mathbf{K}$ is a morphism of left $\mathbf{S}$-semimodules and the left $\mathbf{S}$-semiaction map $\mathbf{S} \,\square_{\mathcal{C}}\, \mathbf{K} \longrightarrow \mathbf{K}$ is a morphism of right $\mathcal{D}$-comodules, or equivalently, the right $\mathbf{T}$-semiaction map is a morphism of left $\mathcal{C}$-comodules and the left $\mathbf{S}$-semiaction map is a morphism of right $\mathbf{T}$-semimodules. Equivalently, the $\mathcal{C}$-$\mathcal{D}$-bicomodule $\mathbf{K}$ is called an $\mathbf{S}$-$\mathbf{T}$-bisemimodule if it is endowed with a $\mathcal{C}$-$\mathcal{D}$-bicomodule morphism of *bisemiaction* $\mathbf{S} \,\square_{\mathcal{C}}\, \mathbf{K} \,\square_{\mathcal{D}}\, \mathbf{T} \longrightarrow \mathbf{K}$ satisfying the associativity and unity equations.

In particular, one can speak about $\mathbf{S}$-$\mathbf{T}$-bisemimodules $\mathbf{K}$ without imposing any (co)flatness conditions on $\mathbf{K}$ if $\mathcal{C}$ is a flat right A-module and either $\mathbf{S}$ is a coflat right $\mathcal{C}$-comodule, or $\mathbf{S}$ is a flat right A-module and a $\mathcal{C}/A$-coflat left $\mathcal{C}$-comodule and A has a finite weak homological dimension, while $\mathcal{D}$ is a flat left B-module and either $\mathbf{T}$ is a coflat left $\mathcal{D}$-comodule, or $\mathbf{T}$ is a flat left B-module and a $\mathcal{D}/B$-coflat right $\mathcal{D}$-comodule and B has a finite weak homological dimension. We will denote the category of $\mathbf{S}$-$\mathbf{T}$-bisemimodules by $\mathbf{S}$-simod-$\mathbf{T}$. Besides, one can consider B-flat $\mathbf{S}$-$\mathbf{T}$-bisemimodules if $\mathcal{C}$ is a flat right A-module and $\mathbf{S}$ is a coflat right $\mathcal{C}$-comodule, while $\mathcal{D}$ is a flat right B-module, $\mathbf{T}$ is a flat right B-module and a $\mathcal{D}/B$-coflat right $\mathcal{D}$-comodule, and B has a finite weak homological dimension; and one can consider $\mathcal{D}$-coflat $\mathbf{S}$-$\mathbf{T}$-bisemimodules if $\mathcal{C}$ is a flat right A-module and $\mathbf{S}$ is a coflat right $\mathcal{C}$-comodule, while $\mathcal{D}$ is a flat right B-module and $\mathbf{T}$ is a coflat right $\mathcal{D}$-comodule.

1.4.4 Let $\mathbf{R}$ be a semialgebra over a coring $\mathcal{E}$ over a k-algebra F. Let $\mathbf{N}$ be an $\mathbf{S}$-$\mathbf{R}$-bisemimodule and $\mathbf{M}$ be an $\mathbf{R}$-$\mathbf{T}$-bisemimodule. We would like to define an $\mathbf{S}$-$\mathbf{T}$-bisemimodule structure on the semitensor product $\mathbf{N} \,\lozenge_{\mathbf{R}}\, \mathbf{M}$.

Assume that multiple cotensor products of the form $\mathbf{S}\,\square_{\mathbb{C}}\cdots\square_{\mathbb{C}}\,\mathbf{S}\,\square_{\mathbb{C}}\,\mathbf{N}\,\square_{\mathcal{E}}$ $\mathcal{R}\,\square_{\mathcal{E}}\,\mathbf{M}\,\square_{\mathcal{D}}\,\mathcal{T}\,\square_{\mathcal{D}}\cdots\square_{\mathcal{D}}\,\mathcal{T}$ are associative. Then, in particular, the semitensor products $(\mathbf{S}^{\square n}\,\square_{\mathbb{C}}\,\mathbf{N})\,\lozenge_{\mathcal{R}}\,(\mathbf{M}\,\square_{\mathcal{D}}\,\mathcal{T}^{\square m})$ can be defined. Assume in addition that multiple cotensor products of the form $\mathbf{S}\,\square_{\mathbb{C}}\cdots\square_{\mathbb{C}}\,\mathbf{S}\,\square_{\mathbb{C}}\,\mathbf{N}\,\square_{\mathcal{E}}\,\mathbf{M}\,\square_{\mathcal{D}}\,\mathcal{T}\,\square_{\mathcal{D}}\cdots\square_{\mathcal{D}}\,\mathcal{T}$ are associative. Then the semitensor products $(\mathbf{S}^{\square n}\,\square_{\mathbb{C}}\,\mathbf{N})\,\lozenge_{\mathcal{R}}\,(\mathbf{M}\,\square_{\mathcal{D}}\,\mathcal{T}^{\square m})$ have natural $\mathcal{C}$-$\mathcal{D}$-bicomodule structures as cokernels of $\mathcal{C}$-$\mathcal{D}$-bicomodule morphisms. Assume that multiple cotensor products of the form $\mathbf{S}\,\square_{\mathbb{C}}\cdots\square_{\mathbb{C}}\,\mathbf{S}\,\square_{\mathbb{C}}\,(\mathbf{N}\lozenge_{\mathcal{R}}\mathbf{M})\,\square_{\mathcal{D}}$ $\mathcal{T}\,\square_{\mathcal{D}}\cdots\square_{\mathcal{D}}\mathcal{T}$ are also associative. Finally, assume that the semitensor product with $\mathbf{S}^{\square n}$ over $\mathcal{C}$ and with $\mathcal{T}^{\square m}$ over $\mathcal{D}$ preserves the cokernel of the pair of morphisms $\mathbf{N}\,\square_{\mathcal{E}}\,\mathcal{R}\,\square_{\mathcal{E}}\,\mathbf{M}\,\rightrightarrows\,\mathbf{N}\,\square_{\mathcal{E}}\,\mathbf{M}$ for $n+m=2$, that is the bicomodule morphisms $(\mathbf{S}^{\square n}\,\square_{\mathbb{C}}\,\mathbf{N})\,\lozenge_{\mathcal{R}}\,(\mathbf{M}\,\square_{\mathcal{D}}\,\mathcal{T}^{\square m})\longrightarrow\mathbf{S}^{\square n}\,\square_{\mathbb{C}}\,(\mathbf{N}\lozenge_{\mathcal{R}}\mathbf{M})\,\square_{\mathcal{D}}\,\mathcal{T}^{\square m}$ are isomorphisms. Then one can define an associative and unital bisemiaction morphism $\mathbf{S}\,\square_{\mathbb{C}}\,(\mathbf{N}\,\lozenge_{\mathcal{R}}\,\mathbf{M})\,\square_{\mathcal{D}}\,\mathcal{T}\longrightarrow\mathbf{N}\,\lozenge_{\mathcal{R}}\,\mathbf{M}$ by taking the semitensor product over $\mathcal{R}$ of the morphism of $\mathbf{S}$-semiaction in $\mathbf{N}$ and the morphism of $\mathcal{T}$-semiaction in $\mathbf{M}$.

For example, if $\mathcal{C}$ is a flat right A-module, $\mathbf{S}$ is a coflat right $\mathcal{C}$-comodule, $\mathcal{D}$ is a flat left B-module, $\mathcal{T}$ is a coflat right $\mathcal{D}$-comodule, $\mathcal{E}$ is a flat right F-module, $\mathcal{R}$ is a flat right F-module and a $\mathcal{E}/F$-coflat left $\mathcal{E}$-comodule, and F has a finite weak homological dimension, then the semitensor product of any F-flat $\mathbf{S}$-$\mathcal{R}$-bisemimodule $\mathbf{N}$ and any $\mathcal{R}$-$\mathcal{T}$-bisemimodule $\mathbf{M}$ has a natural $\mathbf{S}$-$\mathcal{T}$-bisemimodule structure. Since the category of $\mathbf{S}$-$\mathcal{T}$-bisemimodules is abelian in this case, the bisemimodule $\mathbf{N}\,\lozenge_{\mathcal{R}}\,\mathbf{M}$ can be simply defined as the cokernel of the pair of bisemimodule morphisms $\mathbf{N}\,\square_{\mathcal{E}}\,\mathcal{R}\,\square_{\mathcal{E}}\,\mathbf{M}\rightrightarrows\mathbf{N}\,\square_{\mathcal{E}}\,\mathbf{M}$.

Proposition. *Let $\mathbf{N}$ be a right $\mathbf{S}$-semimodule, $\mathcal{K}$ be an $\mathbf{S}$-$\mathcal{T}$-bisemimodule, and $\mathbf{M}$ be a left $\mathcal{T}$-semimodule. Then the iterated semitensor products*

$$(\mathbf{N}\,\lozenge_{\mathbf{S}}\,\mathcal{K})\,\lozenge_{\mathcal{T}}\,\mathbf{M}\quad and\quad \mathbf{N}\,\lozenge_{\mathbf{S}}\,(\mathcal{K}\,\lozenge_{\mathcal{T}}\,\mathbf{M})$$

are well defined and naturally isomorphic, at least, in the following cases:

(a) *$\mathcal{C}$ is a flat right A-module, $\mathbf{S}$ is a coflat right $\mathcal{C}$-comodule, $\mathbf{N}$ is a coflat right $\mathcal{C}$-comodule, $\mathcal{D}$ is a flat left B-module, $\mathcal{T}$ is a coflat left $\mathcal{D}$-comodule, and $\mathbf{M}$ is a coflat left $\mathcal{D}$-comodule;*

(b) *$\mathcal{C}$ is a flat right A-module, $\mathbf{S}$ is a coflat right $\mathcal{C}$-comodule, $\mathbf{N}$ is a semiflat right $\mathbf{S}$-semimodule, and either*

   • *$\mathcal{D}$ is a flat right B-module, $\mathcal{T}$ is a coflat right $\mathcal{D}$-comodule, and $\mathcal{K}$ is a coflat right $\mathcal{D}$-comodule, or*

   • *$\mathcal{D}$ is a flat right B-module, $\mathcal{T}$ is a flat right B-module and a $\mathcal{D}/B$-coflat left $\mathcal{D}$-comodule, the ring B has a finite weak homological dimension, and $\mathcal{K}$ is a flat right B-module, or*

   • *$\mathcal{D}$ is a flat left B-module, $\mathcal{T}$ is a flat left B-module and a $\mathcal{D}/B$-coflat right $\mathcal{D}$-comodule, the ring B has a finite weak homological dimension, and $\mathbf{M}$ is a flat left B-module, or*

   • *the ring B is absolutely flat;*

(c) $\mathcal{C}$ *is a flat right A-module,* $\mathcal{S}$ *is a coflat right* $\mathcal{C}$*-comodule,* $\mathcal{N}$ *is a coflat right* $\mathcal{C}$*-comodule, and either*

- $\mathcal{D}$ *is a flat right B-module,* $\mathcal{T}$ *is a coflat right* $\mathcal{D}$*-comodule, and* $\mathcal{K}$ *as a left* $\mathcal{S}$*-semimodule with a right* $\mathcal{D}$*-comodule structure is induced from a* $\mathcal{D}$*-coflat* $\mathcal{C}$*-*$\mathcal{D}$*-bicomodule, or*

- $\mathcal{D}$ *is a flat right B-module,* $\mathcal{T}$ *is a flat right B-module and a* $\mathcal{D}/B$*-coflat left* $\mathcal{D}$*-comodule, the ring B has a finite weak homological dimension, and* $\mathcal{K}$ *as a left* $\mathcal{S}$*-semimodule with a right* $\mathcal{D}$*-comodule structure is induced from a B-flat* $\mathcal{C}$*-*$\mathcal{D}$*-bicomodule, or*

- $\mathcal{D}$ *is a flat left B-module,* $\mathcal{T}$ *is a flat left B-module and a* $\mathcal{D}/B$*-coflat right* $\mathcal{D}$*-comodule, the ring B has a finite weak homological dimension,* $\mathcal{K}$ *as a left* $\mathcal{S}$*-semimodule with a right* $\mathcal{D}$*-comodule structure is induced from a* $\mathcal{C}$*-*$\mathcal{D}$*-bicomodule, and* $\mathcal{M}$ *is a flat left B-module, or*

- *the ring B is absolutely flat and* $\mathcal{K}$ *as a left* $\mathcal{S}$*-semimodule with a right* $\mathcal{D}$*-comodule structure is induced from a B-flat* $\mathcal{C}$*-*$\mathcal{D}$*-bicomodule.*

More precisely, in all cases in this list the natural maps into both iterated semitensor products from the k-module $(\mathcal{N} \,\square_{\mathcal{C}}\, \mathcal{K}) \,\square_{\mathcal{D}}\, \mathcal{M} \simeq \mathcal{N} \,\square_{\mathcal{C}}\, (\mathcal{K} \,\square_{\mathcal{D}}\, \mathcal{M})$ *are surjective, their kernels coincide and are equal to the sum of the kernels of two maps from this module onto its quotient modules* $(\mathcal{N} \,\square_{\mathcal{C}}\, \mathcal{K}) \,\lozenge_{\mathcal{T}}\, \mathcal{M}$ *and* $\mathcal{N} \,\lozenge_{\mathcal{S}}\, (\mathcal{K} \,\square_{\mathcal{D}}\, \mathcal{M})$.

Proof. It follows from Proposition 1.2.5 that all multiple cotensor products of the form $\mathcal{N} \,\square_{\mathcal{C}}\, \mathcal{S} \,\square_{\mathcal{C}}\, \cdots \,\square_{\mathcal{C}}\, \mathcal{S} \,\square_{\mathcal{C}}\, \mathcal{K} \,\square_{\mathcal{D}}\, \mathcal{T} \,\square_{\mathcal{D}}\, \cdots \,\square_{\mathcal{D}}\, \mathcal{T} \,\square_{\mathcal{D}}\, \mathcal{M}$ are associative. Multiple cotensor products $\mathcal{N} \,\square_{\mathcal{C}}\, \mathcal{S} \,\square_{\mathcal{C}}\, \cdots \,\square_{\mathcal{C}}\, \mathcal{S} \,\square_{\mathcal{C}}\, (\mathcal{K} \,\lozenge_{\mathcal{T}}\, \mathcal{M})$ and $(\mathcal{N} \,\lozenge_{\mathcal{S}}\, \mathcal{K}) \,\square_{\mathcal{D}}\, \mathcal{T} \,\square_{\mathcal{D}}\, \cdots \,\square_{\mathcal{D}}\, \mathcal{T} \,\square_{\mathcal{D}}\, \mathcal{M}$ are also associative by the same proposition (here one has to notice that the semitensor product $\mathcal{N} \,\lozenge_{\mathcal{S}}\, \mathcal{K}$ is a coflat right $\mathcal{D}$-comodule whenever $\mathcal{K}$ is a coflat right $\mathcal{D}$-comodule and $\mathcal{N}$ is a semiflat right $\mathcal{S}$-semimodule). The map $\mathcal{N} \,\square_{\mathcal{C}}\, \mathcal{K} \,\square_{\mathcal{D}}\, \mathcal{M} \longrightarrow (\mathcal{N} \,\lozenge_{\mathcal{S}}\, \mathcal{K}) \,\square_{\mathcal{D}}\, \mathcal{M}$ factorizes through the surjection $\mathcal{N} \,\square_{\mathcal{C}}\, \mathcal{K} \,\square_{\mathcal{D}}\, \mathcal{M} \longrightarrow \mathcal{N} \,\lozenge_{\mathcal{S}}\, (\mathcal{K} \,\square_{\mathcal{D}}\, \mathcal{M})$, hence there is a natural map $\mathcal{N} \,\lozenge_{\mathcal{S}}\, (\mathcal{K} \,\square_{\mathcal{D}}\, \mathcal{M}) \longrightarrow (\mathcal{N} \,\lozenge_{\mathcal{S}}\, \mathcal{K}) \,\square_{\mathcal{D}}\, \mathcal{M}$. One can easily see that whenever this map and the analogous maps for $\mathcal{T}$, $\mathcal{T} \,\square_{\mathcal{D}}\, \mathcal{T}$, and $\mathcal{T} \,\square_{\mathcal{D}}\, \mathcal{M}$ in place of $\mathcal{M}$ are isomorphisms, the iterated semitensor product $(\mathcal{N} \,\lozenge_{\mathcal{S}}\, \mathcal{K}) \,\lozenge_{\mathcal{T}}\, \mathcal{M}$ is defined, the natural map $\mathcal{N} \,\square_{\mathcal{C}}\, \mathcal{K} \,\square_{\mathcal{D}}\, \mathcal{M} \longrightarrow (\mathcal{N} \,\lozenge_{\mathcal{S}}\, \mathcal{K}) \,\lozenge_{\mathcal{T}}\, \mathcal{M}$ is surjective, and its kernel is equal to the desired sum of two kernels of maps from $\mathcal{N} \,\square_{\mathcal{C}}\, \mathcal{K} \,\square_{\mathcal{D}}\, \mathcal{M}$ onto its quotient modules. Thus it remains to prove that the map $\mathcal{N} \,\lozenge_{\mathcal{S}}\, (\mathcal{K} \,\square_{\mathcal{D}}\, \mathcal{M}) \longrightarrow (\mathcal{N} \,\lozenge_{\mathcal{S}}\, \mathcal{K}) \,\square_{\mathcal{D}}\, \mathcal{M}$ is an isomorphism, i.e., the exact sequence of right $\mathcal{D}$-comodules $\mathcal{N} \,\square_{\mathcal{C}}\, \mathcal{S} \,\square_{\mathcal{C}}\, \mathcal{K} \longrightarrow \mathcal{N} \,\square_{\mathcal{C}}\, \mathcal{K} \longrightarrow \mathcal{N} \,\lozenge_{\mathcal{S}}\, \mathcal{K} \longrightarrow 0$ remains exact after taking the cotensor product with $\mathcal{M}$ over $\mathcal{D}$. This is obvious if $\mathcal{M}$ is a quasicoflat $\mathcal{D}$-comodule. If $\mathcal{N}$ is a semiflat $\mathcal{S}$-semimodule, it suffices to present $\mathcal{M}$ as a kernel of a morphism of (quasi)coflat $\mathcal{D}$-comodules. Finally, if $\mathcal{K}$ as a left $\mathcal{S}$-semimodule with a right $\mathcal{D}$-comodule structure is induced from a $\mathcal{C}$-$\mathcal{D}$-bicomodule, then our exact sequence of right $\mathcal{D}$-comodules splits. $\square$

2 Derived Functor SemiTor

2.1 Coderived categories

A complex $C^\bullet$ over an exact category [28] A is called exact if it is composed of exact triples $Z^i \to C^i \to Z^{i+1}$ in A. A complex over A is called acyclic if it is homotopy equivalent to an exact complex (or equivalently, if it is a direct summand of an exact complex). Acyclic complexes form a thick subcategory $\mathsf{Acycl}(\mathsf{A})$ of the homotopy category $\mathsf{Hot}(\mathsf{A})$ of complexes over A. All acyclic complexes over A are exact if and only if A contains images of idempotent endomorphisms [69]. The quotient category $\mathsf{D}(\mathsf{A}) = \mathsf{Hot}(\mathsf{A})/\mathsf{Acycl}(\mathsf{A})$ is called the derived category of A.

Let A be an exact category where all infinite direct sums exist and the functors of infinite direct sum are exact. By the total complex of an exact triple $'K^\bullet \to K^\bullet \to ''K^\bullet$ of complexes over A we mean the total complex of the corresponding bicomplex with three rows. A complex $C^\bullet$ over A is called coacyclic if it belongs to the minimal triangulated subcategory $\mathsf{Acycl}^{\mathsf{co}}(\mathsf{A})$ of the homotopy category $\mathsf{Hot}(\mathsf{A})$ containing all the total complexes of exact triples of complexes over A and closed under infinite direct sums. Any coacyclic complex is acyclic. Acyclic complexes are not always coacyclic (see 0.2.2). It follows from the next lemma that any acyclic complex bounded from below is coacyclic.

Lemma. *Let* $0 \to M^{0,\bullet} \to M^{1,\bullet} \to \cdots$ *be an exact sequence, bounded from below, of arbitrary complexes over* A. *Then the total complex* $T^\bullet$ *of the bicomplex* $M^{\bullet,\bullet}$ *constructed by taking infinite direct sums along the diagonals is coacyclic.*

Proof. An exact sequence of complexes $0 \to M^{0,\bullet} \to M^{1,\bullet} \to \cdots$ can be presented as the inductive limit of finite exact sequences of complexes $0 \to M^{0,\bullet} \to \cdots \to M^{n,\bullet} \to Z^{n+1,\bullet} \to 0$. The total complex $T_n^\bullet$ of the latter finite exact sequence is homotopy equivalent to a complex obtained from total complexes of the exact triples $Z^{n,\bullet} \to M^{n,\bullet} \to Z^{n+1,\bullet}$ using the operations of shift and cone. Hence the complexes $T_n^\bullet$ are coacyclic. The complex $T^\bullet$ is their inductive limit; moreover, the inductive system of $T_n^\bullet$ is obtained by applying the functor of total complex to a locally stabilizing inductive system of bicomplexes. Therefore, the construction of homotopy inductive limit provides an exact triple of complexes $\bigoplus_n T_n^\bullet \longrightarrow \bigoplus_n T_n^\bullet \longrightarrow T^\bullet$. Since the total complex of this exact triple is coacyclic and the direct sum of coacyclic complexes is coacyclic, the complex $T^\bullet$ is coacyclic. (In fact, this exact triple of complexes is split in every degree, so its total complex is even contractible.) $\square$

The category of coacyclic complexes $\mathsf{Acycl}^{\mathsf{co}}(\mathsf{A})$ is a thick subcategory of the homotopy category $\mathsf{Hot}(\mathsf{A})$, since it is a triangulated subcategory with infinite

L. Positselski, *Homological Algebra of Semimodules and Semicontramodules*, Monografie Matematyczne 70, DOI 10.1007/978-3-0346-0436-9_3, © Springer Basel AG 2010

direct sums [69, 70]. The *coderived category* $\mathsf{D}^{co}(\mathsf{A})$ of an exact category A is defined as the quotient category $\mathsf{Hot}(\mathsf{A})/\mathsf{Acycl}^{co}(\mathsf{A})$.

Remark. If an exact category A has a finite homological dimension, then the minimal triangulated subcategory of the homotopy category $\mathsf{Hot}(\mathsf{A})$ containing the total complexes of exact triples of complexes over A coincides with the subcategory of acyclic complexes. Indeed, let $C^\bullet$ be an exact complex over A and n be a number greater than the homological dimension of A. Let Z^i be the objects of cycles of the complex $C^\bullet$. Then for any integer j the Yoneda extension class represented by the extension $Z^{2jn} \to C^{2jn} \to \cdots \to C^{2jn+n-1} \to Z^{2jn+n}$ is trivial, and therefore, this extension can be connected with the split extension by a pair of extension morphisms $(Z^{2jn} \to C^{2jn} \to \cdots \to C^{2jn+n-1} \to Z^{2jn}) \longrightarrow (Z^{2jn} \to {'C}^{2jn} \to \cdots \to {'C}^{2jn+n-1} \to Z^{2jn+n}) \longleftarrow (Z^{2jn} \to Z^{2jn} \to 0 \to \cdots \to 0 \to Z^{2jn+n} \to Z^{2jn+n})$. Let $'C^\bullet$ be the complex obtained by replacing all the even segments $C^{2jn} \to \cdots \to C^{2jn+n-1}$ of the complex $C^\bullet$ with the segments $'C^{2jn} \to \cdots \to {'C}^{2jn+n-1}$ while leaving the odd segments $C^{2jn+n} \to \cdots \to C^{2(j+1)n-1}$ in place, and let $''C^\bullet$ be the complex obtained by replacing the same even segments of the complex $C^\bullet$ with the segments $Z^{2jn} \to 0 \to \cdots \to 0 \to Z^{2jn+n}$ while leaving the odd segments in place. Then the complex $''C^\bullet$ and the cones of both morphisms $C^\bullet \longrightarrow {'C}^\bullet$ and $''C^\bullet \longrightarrow {'C}^\bullet$ are homotopy equivalent to complexes obtained from total complexes of exact triples of complexes with zero differentials using the operation of cone repeatedly.

2.2 Coflat complexes

Let $\mathcal{C}$ be a coring over a k-algebra A. The cotensor product $\mathcal{N}^\bullet \square_{\mathcal{C}} \mathcal{M}^\bullet$ of a complex of right $\mathcal{C}$-comodules $\mathcal{N}^\bullet$ and a complex of left $\mathcal{C}$-comodules $\mathcal{M}^\bullet$ is defined as the total complex of the bicomplex $\mathcal{N}^i \square_{\mathcal{C}} \mathcal{M}^j$, constructed by taking infinite direct sums along the diagonals.

Assume that $\mathcal{C}$ is a flat right A-module. Then the category of left $\mathcal{C}$-comodules is an abelian category with exact functors of infinite direct sums, so the coderived category $\mathsf{D}^{co}(\mathcal{C}\text{-comod})$ is defined. When speaking about *coacyclic complexes* of $\mathcal{C}$-comodules, we will always mean coacyclic complexes with respect to the abelian category of $\mathcal{C}$-comodules, unless another exact category of $\mathcal{C}$-comodules is explicitly mentioned.

A complex of right $\mathcal{C}$-comodules $\mathcal{N}^\bullet$ is called *coflat* if the complex $\mathcal{N}^\bullet \square_{\mathcal{C}} \mathcal{M}^\bullet$ is acyclic whenever a complex of left $\mathcal{C}$-comodules $\mathcal{M}^\bullet$ is coacyclic.

Lemma. *Any complex of coflat $\mathcal{C}$-comodules is coflat.*

Proof. Let $\mathcal{N}^\bullet$ be a complex of coflat $\mathcal{C}$-comodules. Since the functor of cotensor product with $\mathcal{N}^\bullet$ preserves shifts, cones, and infinite direct sums, it suffices to show the complex $\mathcal{N}^\bullet \square_{\mathcal{C}} \mathcal{M}^\bullet$ is acyclic whenever $\mathcal{M}^\bullet$ is the total complex of an exact triple of complexes of left $\mathcal{C}$-comodules $'\mathcal{K}^\bullet \to \mathcal{K}^\bullet \to {''}\mathcal{K}^\bullet$. In this case, the triple of complexes $\mathcal{N}^\bullet \square_{\mathcal{C}} {'}\mathcal{K}^\bullet \longrightarrow \mathcal{N}^\bullet \square_{\mathcal{C}} \mathcal{K}^\bullet \longrightarrow \mathcal{N}^\bullet \square_{\mathcal{C}} {''}\mathcal{K}^\bullet$ is also exact, because $\mathcal{N}^\bullet$ is a complex of coflat $\mathcal{C}$-comodules, and the complex $\mathcal{N}^\bullet \,\square_\mathcal{C}\, \mathcal{M}^\bullet$ is the total complex of this exact triple. $\square$

If the ring A has a finite weak homological dimension, then any coflat complex of $\mathcal{C}$-comodules is a flat complex of A-modules in the sense of 0.1.1. (Indeed, if $V^\bullet$ is a complex of right A-modules such that the tensor product of $V^\bullet$ with any coacyclic complex of left A-modules is acyclic, then the tensor product of $V^\bullet$ with any acyclic complex $U^\bullet$ of left A-modules is also acyclic, since one can construct a morphism into $U^\bullet$ from an acyclic complex of flat A-modules with a coacyclic cone.) The complex of $\mathcal{C}$-comodules $V^\bullet \otimes_A \mathcal{C}$ coinduced from a flat complex of A-modules $V^\bullet$ is coflat.

Remark. The coderived category $\mathsf{D}^{\mathrm{co}}(\mathcal{C}\text{-comod})$ can be only thought of as the "right" version of exotic unbounded derived category of $\mathcal{C}$-comodules (e.g., for the purposes of defining the derived functors $\mathrm{Cotor}^\mathcal{C}$ and $\mathrm{Coext}_\mathcal{C}$, constructing the equivalence of derived categories of $\mathcal{C}$-comodules and $\mathcal{C}$-contramodules, etc.) when the ring A has a finite (weak or left) homological dimension. Indeed, what is needed is a definition of "relative coderived category" of $\mathcal{C}$-comodules such that for $\mathcal{C} = A$ it would coincide with the derived category of A-modules, while when $\mathcal{C}$ is a coalgebra over a field it would be the coderived category of $\mathcal{C}$-comodules defined above. (The same applies to the semiderived category $\mathsf{D}^{\mathrm{si}}(\mathsf{S}\text{-simod})$ of S-semimodules– it only appears to be the "right" definition when the ring A has a finite homological dimension.)

2.3 Semiderived categories

Let S be a semialgebra over a coring $\mathcal{C}$. Assume that $\mathcal{C}$ is a flat right A-module and S is a coflat right $\mathcal{C}$-comodule, so that the category of left S-semimodules is abelian. The *semiderived category* of left S-semimodules $\mathsf{D}^{\mathrm{si}}(\mathsf{S}\text{-simod})$ is defined as the quotient category of the homotopy category $\mathsf{Hot}(\mathsf{S}\text{-simod})$ by the thick subcategory $\mathsf{Acycl}^{\mathrm{co}\text{-}\mathcal{C}}(\mathsf{S}\text{-simod})$ of complexes of S-semimodules that are *coacyclic as complexes of $\mathcal{C}$-comodules*.

Remark. There is no claim that the semiderived category *exists* in the sense that morphisms between a given pair of objects form a set rather than a class. Rather, we think of our localizations of categories as of "very large" categories with classes of morphisms instead of sets. We will explain in 5.5 and 6.5 how to compute the modules of homomorphisms in semiderived categories in terms of resolutions; then it will follow that the semiderived category does exist, under certain assumptions.

2.4 Semiflat complexes

Let S be a semialgebra. The semitensor product $\mathcal{N}^\bullet \lozenge_\mathsf{S} \mathcal{M}^\bullet$ of a complex of right S-semimodules $\mathcal{N}^\bullet$ and a complex of left S-semimodules $\mathcal{M}^\bullet$ is defined as the total complex of the bicomplex $\mathcal{N}^i \lozenge_\mathsf{S} \mathcal{M}^j$, constructed by taking infinite direct sums along the diagonals. Of course, appropriate (co)flatness conditions must be imposed on $\mathbf{S}$, $\mathbf{N}^\bullet$, and $\mathbf{M}^\bullet$ for this definition to make sense.

Assume that the coring $\mathcal{C}$ is a flat right A-module, the semialgebra $\mathbf{S}$ is a coflat right $\mathcal{C}$-comodule and a $\mathcal{C}/A$-coflat left $\mathcal{C}$-comodule, and the ring A has a finite weak homological dimension. A complex of A-flat right $\mathbf{S}$-semimodules $\mathbf{N}^\bullet$ is called *semiflat* if the complex $\mathbf{N}^\bullet \lozenge_{\mathbf{S}} \mathbf{M}^\bullet$ is acyclic whenever a complex of left $\mathbf{S}$-semimodules $\mathbf{M}^\bullet$ is $\mathcal{C}$-coacyclic. Any semiflat complex of $\mathbf{S}$-semimodules is a coflat complex of $\mathcal{C}$-comodules. The complex of $\mathbf{S}$-semimodules $\mathcal{R}^\bullet \square_\mathcal{C} \mathbf{S}$ induced from a coflat complex of A-flat $\mathcal{C}$-comodules $\mathcal{R}^\bullet$ is semiflat.

If it is only known that $\mathcal{C}$ is a flat right A-module and $\mathbf{S}$ is a coflat right $\mathcal{C}$-comodule, one can define semiflat complexes of $\mathcal{C}$-coflat right $\mathbf{S}$-semimodules. Then the complex of $\mathbf{S}$-semimodules induced from a complex of coflat $\mathcal{C}$-comodules is semiflat; it is also a complex of semiflat semimodules.

Notice that *not every complex of semiflat semimodules is semiflat* (see 0.1.1). In particular, it follows from Theorem 2.6 and Lemma 2.7 below that (in the assumptions of 2.6) a $\mathcal{C}$-coacyclic complex of A-flat right $\mathbf{S}$-semimodules $\mathbf{N}^\bullet$ is semiflat if and only if its semitensor product with any complex of left $\mathbf{S}$-semimodules $\mathbf{M}^\bullet$ (or just with any left $\mathbf{S}$-semimodule $\mathbf{M}$) is acyclic. Thus a $\mathcal{C}$-coacyclic complex of semiflat $\mathbf{S}$-semimodules is semiflat if and only if all of its semimodules of cocycles are semiflat.

On the other hand, any complex of semiflat semimodules bounded from above is semiflat. Moreover, if $\cdots \to \mathbf{N}^{-1,\bullet} \to \mathbf{N}^{0,\bullet} \to 0$ is a complex, bounded from above, of semiflat complexes of $\mathbf{S}$-semimodules, then the total complex $\mathcal{E}^\bullet$ of the bicomplex $\mathbf{N}^{\bullet,\bullet}$ constructed by taking infinite direct sums along the diagonals is semiflat. Indeed, the category of semiflat complexes is closed under shifts, cones, and infinite direct sums, so one can apply the following lemma.

Lemma. *Let* $\cdots \to N^{-1,\bullet} \to N^{0,\bullet} \to 0$ *be a complex, bounded from above, of arbitrary complexes over an additive category* $\mathbf{A}$ *where infinite direct sums exist. Then the total complex* $E^\bullet$ *of the bicomplex* $N^{\bullet,\bullet}$ *up to the homotopy equivalence can be obtained from the complexes* $N^{-i,\bullet}$ *using the operations of shift, cone, and infinite direct sum.*

Proof. Let $E_n^\bullet$ be the total complex of the finite complex of complexes $0 \to N^{-n,\bullet} \to \cdots \to N^{0,\bullet} \to 0$. Then the complex $E^\bullet$ is the inductive limit of the complexes $E_n^\bullet$, and in addition, the embeddings of complexes $E_n^\bullet \longrightarrow E_{n+1}^\bullet$ split in every degree. Thus the triple of complexes $\bigoplus_n E_n^\bullet \longrightarrow \bigoplus_n E_n^\bullet \longrightarrow E^\bullet$ is split exact in every degree and the complex $E^\bullet$ is homotopy equivalent to the cone of the morphism $\bigoplus_n E_n^\bullet \longrightarrow \bigoplus_n E_n^\bullet$ (the homotopy inductive limit of the complexes $E_n^\bullet$). $\square$

2.5 Main theorem for comodules

Assume that the coring $\mathcal{C}$ is a flat left and right A-module and the ring A has a finite weak homological dimension.

Theorem. *The functor mapping the quotient category of the homotopy category of complexes of coflat $\mathcal{C}$-comodules (coflat complexes of $\mathcal{C}$-comodules) by its intersection with the thick subcategory of coacyclic complexes of $\mathcal{C}$-comodules into the coderived category of $\mathcal{C}$-comodules is an equivalence of triangulated categories.*

Proof. We will show that any complex of $\mathcal{C}$-comodules $\mathcal{K}^\bullet$ can be connected with a complex of coflat $\mathcal{C}$-comodules in a functorial way by a chain of two morphisms $\mathcal{K}^\bullet \longleftarrow \mathbb{R}_2(\mathcal{K}^\bullet) \longrightarrow \mathbb{R}_2\mathbb{L}_1(\mathcal{K}^\bullet)$ with coacyclic cones. Moreover, if the complex $\mathcal{K}^\bullet$ is a complex of coflat $\mathcal{C}$-comodules (coflat complex of $\mathcal{C}$-comodules), then the intermediate complex $\mathbb{R}_2(\mathcal{K}^\bullet)$ in this chain is also a complex of coflat $\mathcal{C}$-comodules (coflat complex of $\mathcal{C}$-comodules). Then we will apply the following lemma.

Lemma. *Let C be a category and F be its full subcategory. Let S be a class of morphisms in C containing the third morphism of any triple of morphisms s, t, and st when it contains two of them. Suppose that for any object X in C there is a chain of morphisms $X \leftarrow F_1(X) \rightarrow \cdots \leftarrow F_{n-1}(X) \rightarrow F_n(X)$ belonging to S and functorially depending on X such that the object $F_n(X)$ belongs to F for any $X \in \mathsf{C}$ and all the objects $F_i(X)$ belong to F for any $X \in \mathsf{F}$. Then the functor $\mathsf{F}[(\mathsf{S} \cap \mathsf{F})^{-1}] \longrightarrow \mathsf{C}[\mathsf{S}^{-1}]$ induced by the embedding $\mathsf{F} \longrightarrow \mathsf{C}$ is an equivalence of categories.*

Proof. It is obvious that the functor between the localized categories is surjective on the isomorphism classes of objects; let us show that it is bijective on morphisms. It follows from the condition on the class S that the functors F_i preserve it. Let U and V be two objects of F and $\phi \colon U \longrightarrow V$ be a morphism between them in the category $\mathsf{C}[\mathsf{S}^{-1}]$. Applying the functor $F_n \colon \mathsf{C} \longrightarrow \mathsf{F}$, we obtain a morphism $F_n(\phi) \colon F_n(U) \longrightarrow F_n(V)$ in the category $\mathsf{F}[(\mathsf{S} \cap \mathsf{F})^{-1}]$. The square diagram of morphisms in the category $\mathsf{C}[\mathsf{S}^{-1}]$ formed by the morphism ϕ, the isomorphism between U and $F_n(U)$, the morphism $F_n(\phi)$, and the isomorphism between V and $F_n(V)$ is commutative, since it is composed from commutative squares of morphisms in the category C. Since the other three morphisms in this commutative square lift to $\mathsf{F}[(\mathsf{S} \cap \mathsf{F})^{-1}]$, the morphism ϕ belongs to the image of the functor $\mathsf{F}[(\mathsf{S} \cap \mathsf{F})^{-1}] \longrightarrow \mathsf{C}[\mathsf{S}^{-1}]$. Now suppose that two morphisms ϕ and $\psi \colon U \longrightarrow V$ in the category $\mathsf{F}[(\mathsf{S} \cap \mathsf{F})^{-1}]$ map to the same morphism in $\mathsf{C}[\mathsf{S}^{-1}]$. Applying the functor F_n, we see that the morphisms $F_n(\phi)$ and $F_n(\psi)$ are equal in $\mathsf{F}[(\mathsf{S} \cap \mathsf{F})^{-1}]$. So we have two commutative squares in the category $\mathsf{F}[(\mathsf{S} \cap \mathsf{F})^{-1}]$ with the same vertices U, V, $F_n(U)$, and $F_n(V)$, the same morphism $F_n(U) \longrightarrow F_n(V)$, the same isomorphisms $U \simeq F(U)$ and $V \simeq F(V)$, and two morphisms ϕ and $\psi \colon U \longrightarrow V$. It follows that the latter two morphisms are equal. $\qquad\square$

Let $\mathcal{K}^\bullet$ be a complex of $\mathcal{C}$-comodules. Let $\mathcal{P}(\mathcal{M}) \longrightarrow \mathcal{M}$ denote the functorial surjective morphism onto an arbitrary $\mathcal{C}$-comodule $\mathcal{M}$ from an A-flat $\mathcal{C}$-comodule $\mathcal{P}(\mathcal{M})$ constructed in Lemma 1.1.3.

The functor $\mathcal{P}$ is not always additive, but as any functor from an additive category to an abelian one it is the direct sum of a constant functor $\mathcal{M} \longmapsto \mathcal{P}(0)$ and a functor $\mathcal{P}^+(\mathcal{M}) = \ker(\mathcal{P}(\mathcal{M}) \to \mathcal{P}(0)) = \operatorname{coker}(\mathcal{P}(0) \to \mathcal{P}(\mathcal{M}))$ sending zero objects to zero objects and zero morphisms to zero morphisms. For any $\mathcal{C}$-comodule $\mathcal{M}$, the comodule $\mathcal{P}^+(\mathcal{M})$ is A-flat and the morphism $\mathcal{P}^+(\mathcal{M}) \longrightarrow \mathcal{M}$ is surjective.

Set $\mathcal{P}_0(\mathcal{K}^\bullet) = \mathcal{P}^+(\mathcal{K}^\bullet)$, $\mathcal{P}_1(\mathcal{K}^\bullet) = \mathcal{P}^+(\ker(\mathcal{P}^0(\mathcal{K}^\bullet) \to \mathcal{K}^\bullet))$, etc. For d large enough, the kernel $\mathcal{Z}(\mathcal{K}^\bullet)$ of the morphism $\mathcal{P}_{d-1}(\mathcal{K}^\bullet) \longrightarrow \mathcal{P}_{d-2}(\mathcal{K}^\bullet)$ will be a complex of A-flat $\mathcal{C}$-comodules. Let $\mathbb{L}_1(\mathcal{K}^\bullet)$ be the total complex of the bicomplex

$$\mathcal{Z}(\mathcal{K}^\bullet) \longrightarrow \mathcal{P}_{d-1}(\mathcal{K}^\bullet) \longrightarrow \cdots \longrightarrow \mathcal{P}_1(\mathcal{K}^\bullet) \longrightarrow \mathcal{P}_0(\mathcal{K}^\bullet).$$

Then $\mathbb{L}_1(\mathcal{K}^\bullet)$ is a complex of A-flat $\mathcal{C}$-comodules and the cone of the morphism $\mathbb{L}_1(\mathcal{K}^\bullet) \longrightarrow \mathcal{K}^\bullet$ is the total complex of a finite exact sequence of complexes of $\mathcal{C}$-comodules, and therefore, a coacyclic complex.

Now let $\mathcal{L}^\bullet$ be a complex of A-flat left $\mathcal{C}$-comodules. Consider the cobar construction

$$\mathcal{C} \otimes_A \mathcal{L}^\bullet \longrightarrow \mathcal{C} \otimes_A \mathcal{C} \otimes_A \mathcal{L}^\bullet \longrightarrow \mathcal{C} \otimes_A \mathcal{C} \otimes_A \mathcal{C} \otimes_A \mathcal{L}^\bullet \longrightarrow \cdots$$

Let $\mathbb{R}_2(\mathcal{L}^\bullet)$ be the total complex of this bicomplex, constructed by taking infinite direct sums along the diagonals. Then $\mathbb{R}_2(\mathcal{L}^\bullet)$ is a complex of coflat $\mathcal{C}$-comodules. The functor $\mathbb{R}_2$ can be extended to arbitrary complexes of $\mathcal{C}$-comodules; for any complex $\mathcal{K}^\bullet$, the cone of the morphism $\mathcal{K}^\bullet \longrightarrow \mathbb{R}_2(\mathcal{K}^\bullet)$ is coacyclic by Lemma 2.1.

Finally, if $\mathcal{K}^\bullet$ is a coflat complex of $\mathcal{C}$-comodules, then $\mathbb{R}_2(\mathcal{K}^\bullet)$ is also a coflat complex of $\mathcal{C}$-comodules, since the cotensor product of $\mathbb{R}_2(\mathcal{K}^\bullet)$ with a complex of right $\mathcal{C}$-comodules $\mathcal{N}^\bullet$ coincides with the cotensor product of $\mathcal{K}^\bullet$ with the total cobar complex $\mathbb{R}_2(\mathcal{N}^\bullet)$, and the latter is coacyclic whenever $\mathcal{N}^\bullet$ is coacyclic.

We have constructed the chain of morphisms $\mathcal{K}^\bullet \longleftarrow \mathbb{R}_2(\mathcal{K}^\bullet) \longrightarrow \mathbb{R}_2\mathbb{L}_1(\mathcal{K}^\bullet)$ with the desired properties. The only remaining problem is that the functor $\mathbb{L}_1$ is not additive and therefore not defined on the homotopy category of complexes of $\mathcal{C}$-comodules, but only on the (abelian) category of complexes and their morphisms. So we have to apply Lemma 2.5 to the category C of complexes of $\mathcal{C}$-comodules, the full subcategory F of complexes of coflat $\mathcal{C}$-comodules (coflat complexes of $\mathcal{C}$-comodules) in it, and the class S of morphisms with coacyclic cones.

The corresponding localizations will coincide with the desired quotient categories of homotopy categories due to the following general fact [46, III.4.2–3]. For any DG-category DG where shifts and cones exist, the localization of the category of closed morphisms in DG with respect to the class of homotopy equivalences coincides with the homotopy category of DG (i.e., closed morphisms homotopic in DG become equal after inverting homotopy equivalences). In particular, this is true for any category of complexes over an additive category that is closed under shifts and cones.   □

Remark. Another proof of Theorem 2.5 (for complexes of coflat comodules or coflat complexes of A-flat comodules) can be found in 2.6. After Theorem 2.5 has been proven, it turns out that the functors $\mathbb{L}_1$ and $\mathbb{R}_2$ can be also applied in the reverse order: for any complex of $\mathcal{C}$-comodules $\mathcal{L}^\bullet$, the complex $\mathbb{R}_2(\mathcal{L}^\bullet)$ is a complex of $\mathcal{C}/A$-coflat $\mathcal{C}$-comodules, and for any complex of $\mathcal{C}/A$-coflat $\mathcal{C}$-comodules $\mathcal{K}^\bullet$, the complex $\mathbb{L}_1(\mathcal{K}^\bullet)$ is a complex of coflat $\mathcal{C}$-comodules (by Remark 1.2.2, which depends on Theorem 2.5).

2.6 Main theorem for semimodules

Assume that the coring $\mathcal{C}$ is a flat left and right A-module, the semialgebra $\mathbf{S}$ is a coflat left and right $\mathcal{C}$-comodule, and the ring A has a finite weak homological dimension.

Theorem. *The functor mapping the quotient category of the homotopy category of semiflat complexes of A-flat ($\mathcal{C}$-coflat, semiflat) $\mathbf{S}$-semimodules by its intersection with the thick subcategory of $\mathcal{C}$-coacyclic complexes of $\mathbf{S}$-semimodules into the semiderived category of $\mathbf{S}$-semimodules is an equivalence of triangulated categories.*

Proof. We will show that in the following chain of functors betweeen triangulated categories all the three functors are equivalences of categories. The quotient category of (the homotopy category of) semiflat complexes of $\mathcal{C}$-coflat (semiflat) $\mathbf{S}$-semimodules by (the thick subcategory of) $\mathcal{C}$-coacyclic semiflat complexes of $\mathcal{C}$-coflat (semiflat) $\mathbf{S}$-semimodules is mapped into the quotient category of complexes of $\mathcal{C}$-coflat $\mathbf{S}$-semimodules by $\mathcal{C}$-coacyclic complexes of $\mathcal{C}$-coflat $\mathbf{S}$-semimodules. The latter category is mapped into the quotient category of the homotopy category of complexes of A-flat $\mathbf{S}$-semimodules by $\mathcal{C}$-coacyclic complexes of A-flat $\mathbf{S}$-semimodules, and then into the semiderived category of $\mathbf{S}$-semimodules. Analogously, in the chain of functors mapping the quotient category of (the homotopy category of) semiflat complexes of A-flat $\mathbf{S}$-semimodules by (the thick subcategory of) $\mathcal{C}$-coacyclic semiflat complexes of A-flat $\mathbf{S}$-semimodules into the quotient category of $\mathcal{C}$-coflat complexes of A-flat $\mathbf{S}$-semimodules by $\mathcal{C}$-coacyclic $\mathcal{C}$-coflat complexes of A-flat $\mathbf{S}$-semimodules, into the quotient category of complexes of A-flat $\mathbf{S}$-semimodules by $\mathcal{C}$-coacyclic complexes of A-flat $\mathbf{S}$-semimodules, and into the semiderived category of $\mathbf{S}$-semimodules, all the three functors are equivalences of categories.

   In order to prove this, we will construct for any complex of $\mathbf{S}$-semimodules $\mathcal{K}^\bullet$ a morphism $\mathbb{L}_1(\mathcal{K}^\bullet) \longrightarrow \mathcal{K}^\bullet$ into $\mathcal{K}^\bullet$ from a complex of A-flat $\mathbf{S}$-semimodules $\mathbb{L}_1(\mathcal{K}^\bullet)$, for any complex of A-flat $\mathbf{S}$-semimodules $\mathcal{L}^\bullet$ a morphism $\mathcal{L}^\bullet \longrightarrow \mathbb{R}_2(\mathcal{L}^\bullet)$ from $\mathcal{L}^\bullet$ into a complex of $\mathcal{C}$-coflat $\mathbf{S}$-semimodules $\mathbb{R}_2(\mathcal{L}^\bullet)$, and for any $\mathcal{C}$-coflat complex of A-flat $\mathbf{S}$-semimodules (complex of $\mathcal{C}$-coflat $\mathbf{S}$-semimodules) $\mathcal{M}^\bullet$ a morphism $\mathbb{L}_3(\mathcal{M}^\bullet) \longrightarrow \mathcal{M}^\bullet$ into $\mathcal{M}^\bullet$ from a semiflat complex of A-flat (semiflat)

S-semimodules $\mathbb{L}_3(\mathcal{M}^\bullet)$ such that in each case the cone of this morphism will be a $\mathcal{C}$-coacyclic complex of S-semimodules. Then we will apply the following lemma.

Lemma. *Let* H *be a category and* F *be its full subcategory. Let* S *be a localizing (i.e., satisfying the Ore conditions) class of morphisms in* H. *Assume that for any object* X *of* H *there exists an object* U *of* F *together with a morphism* $U \longrightarrow X$ *belonging to* S *(or for any object* X *of* H *there exists an object* V *of* F *together with a morphism* $X \longrightarrow V$ *belonging to* S). *Then the functor* $F[(S \cap F)^{-1}] \longrightarrow H[S^{-1}]$ *induced by the embedding* $F \longrightarrow H$ *is an equivalence of categories.*

Proof. It is obvious that the functor between the localized categories is surjective on the isomorphism classes of objects; let us show that it is bijective on morphisms. Any morphism in the category $H[S^{-1}]$ between two objects U and V from F can be represented by a fraction $U \leftarrow X \rightarrow V$, where X is an object of H and the morphism $X \rightarrow U$ belongs to S. By our assumption, there is an object W from F together with a morphism $W \rightarrow X$ from S. Then the fractions $U \leftarrow X \rightarrow V$ and $U \leftarrow W \rightarrow V$ represent the same morphism in $H[S^{-1}]$, while the second fraction represents also a certain morphism in $F[(S \cap F)^{-1}]$. Furthermore, any two morphisms from an object U to an object V in the category $F[(S \cap F)^{-1}]$ can be represented by two fractions of the form $U \leftarrow U' \rightrightarrows V$, with the same morphism $U \rightarrow U'$ from $S \cap F$ and two different morphisms $U' \rightrightarrows V$ (since the class of morphisms $S \cap F$ in the category F satisfies the right Ore conditions). If the images of these morphisms in the category $H[S^{-1}]$ are equal, then there is a morphism $X \rightarrow U'$ from S with an object X from H such that two compositions $X \rightarrow U' \rightrightarrows V$ coincide. Again there is an object W from F together with a morphism $W \rightarrow X$ belonging to S. Since the two compositions $W \rightarrow U' \rightrightarrows V$ coincide in F, the morphisms represented by the two fractions $U \leftarrow U' \rightrightarrows V$ are equal in $F[(S \cap F)^{-1}]$. $\square$

Let $\mathcal{K}^\bullet$ be a complex of S-semimodules. Let $\mathcal{P}(\mathcal{M}) \longrightarrow \mathcal{M}$ denote the functorial surjective morphism onto an arbitrary S-semimodule $\mathcal{M}$ from an A-flat S-semimodule $\mathcal{P}(\mathcal{M})$ constructed in Lemma 1.3.2. As explained in the proof of Theorem 2.5, the functor $\mathcal{P}$ is the direct sum of a constant functor $\mathcal{M} \longmapsto \mathcal{P}(0)$ and a functor $\mathcal{P}^+$ sending zero morphisms to zero morphisms. For any S-semimodule $\mathcal{M}$, the semimodule $\mathcal{P}^+(\mathcal{M})$ is A-flat and the morphism $\mathcal{P}^+(\mathcal{M}) \longrightarrow \mathcal{M}$ is surjective.

Set $\mathcal{P}_0(\mathcal{K}^\bullet) = \mathcal{P}^+(\mathcal{K}^\bullet)$, $\mathcal{P}_1(\mathcal{K}^\bullet) = \mathcal{P}^+(\ker(\mathcal{P}^0(\mathcal{K}^\bullet) \rightarrow \mathcal{K}^\bullet))$, etc. For d large enough, the kernel $\mathcal{Z}(\mathcal{K}^\bullet)$ of the morphism $\mathcal{P}_{d-1}(\mathcal{K}^\bullet) \longrightarrow \mathcal{P}_{d-2}(\mathcal{K}^\bullet)$ will be a complex of A-flat S-semimodules. Let $\mathbb{L}_1(\mathcal{K}^\bullet)$ be the total complex of the bicomplex

$$\mathcal{Z}(\mathcal{K}^\bullet) \longrightarrow \mathcal{P}_{d-1}(\mathcal{K}^\bullet) \longrightarrow \cdots \longrightarrow \mathcal{P}_1(\mathcal{K}^\bullet) \longrightarrow \mathcal{P}_0(\mathcal{K}^\bullet).$$

Then $\mathbb{L}_1(\mathcal{K}^\bullet)$ is a complex of A-flat S-semimodules and the cone of the morphism $\mathbb{L}_1(\mathcal{K}^\bullet) \longrightarrow \mathcal{K}^\bullet$ is the total complex of a finite exact sequence of complexes of S-semimodules, and therefore, a $\mathcal{C}$-coacyclic complex (and even an S-coacyclic complex).

Now let $\mathcal{L}^\bullet$ be a complex of A-flat $\mathcal{S}$-semimodules. Let $\mathcal{M} \longrightarrow \mathcal{J}(\mathcal{M})$ denote the functorial injective morphism from an arbitrary A-flat $\mathcal{S}$-semimodule $\mathcal{M}$ into a $\mathcal{C}$-coflat $\mathcal{S}$-semimodule $\mathcal{J}(\mathcal{M})$ with an A-flat cokernel $\mathcal{J}(\mathcal{M})/\mathcal{M}$ constructed in Lemma 1.3.3. Set $\mathcal{J}^0(\mathcal{L}^\bullet) = \mathcal{J}(\mathcal{L}^\bullet)$, $\mathcal{J}^1(\mathcal{L}^\bullet) = \mathcal{J}(\operatorname{coker}(\mathcal{L}^\bullet \to \mathcal{J}^0(\mathcal{L}^\bullet)))$, etc. Let $\mathbb{R}_2(\mathcal{L}^\bullet)$ be the total complex of the bicomplex

$$\mathcal{J}^0(\mathcal{L}^\bullet) \longrightarrow \mathcal{J}^1(\mathcal{L}^\bullet) \longrightarrow \mathcal{J}^2(\mathcal{L}^\bullet) \longrightarrow \cdots,$$

constructed by taking infinite direct sums along the diagonals. Then $\mathbb{R}_2(\mathcal{L}^\bullet)$ is a complex of $\mathcal{C}$-coflat $\mathcal{S}$-semimodules and the cone of the morphism $\mathcal{L}^\bullet \longrightarrow \mathbb{R}_2(\mathcal{L}^\bullet)$ is a $\mathcal{C}$-coacyclic (and even $\mathcal{S}$-coacyclic) complex by Lemma 2.1.

Finally, let $\mathcal{M}^\bullet$ be a $\mathcal{C}$-coflat complex of A-flat left $\mathcal{S}$-semimodules. Then the complex $\mathcal{S} \,\square_\mathcal{C}\, \mathcal{M}^\bullet$ is a semiflat complex of A-flat $\mathcal{S}$-semimodules. Moreover, if $\mathcal{M}^\bullet$ is a complex of $\mathcal{C}$-coflat $\mathcal{S}$-semimodules, then $\mathcal{S} \,\square_\mathcal{C}\, \mathcal{M}^\bullet$ is a semiflat complex of semiflat $\mathcal{S}$-semimodules. Consider the bar construction

$$\cdots \longrightarrow \mathcal{S} \,\square_\mathcal{C}\, \mathcal{S} \,\square_\mathcal{C}\, \mathcal{S} \,\square_\mathcal{C}\, \mathcal{M}^\bullet \longrightarrow \mathcal{S} \,\square_\mathcal{C}\, \mathcal{S} \,\square_\mathcal{C}\, \mathcal{M}^\bullet \longrightarrow \mathcal{S} \,\square_\mathcal{C}\, \mathcal{M}^\bullet.$$

Let $\mathbb{L}_3(\mathcal{M}^\bullet)$ be the total complex of this bicomplex, constructed by taking infinite direct sums along the diagonals. Then the complex $\mathbb{L}_3(\mathcal{M}^\bullet)$ is semiflat by 2.4 and the cone of the morphism $\mathbb{L}_3(\mathcal{M}^\bullet) \longrightarrow \mathcal{M}^\bullet$ is not only $\mathcal{C}$-coacyclic, but even $\mathcal{C}$-contractible (the contracting homotopy being induced by the semiunit morphism $\mathcal{C} \longrightarrow \mathcal{S}$). $\hfill\square$

Remark. It is clear that the constructions of complexes $\mathbb{R}_2(\mathcal{L}^\bullet)$ and $\mathbb{L}_3(\mathcal{M}^\bullet)$ can be applied to arbitrary complexes of $\mathcal{S}$-semimodules, with no (co)flatness conditions imposed on them. For example, an alternative way of proving Theorem 2.6 is to show that the functors mapping the quotient category of semiflat complexes of $\mathcal{C}$-coflat (semiflat) $\mathcal{S}$-semimodules by $\mathcal{C}$-coacyclic semiflat complexes into the quotient category of complexes of $\mathcal{C}/A$-coflat $\mathcal{S}$-semimodules by $\mathcal{C}$-coacyclic complexes into the semiderived category of $\mathcal{S}$-semimodules are both equivalences of categories. Indeed, for any complex of $\mathcal{S}$-semimodules $\mathcal{L}^\bullet$ the complex $\mathbb{R}_2(\mathcal{L}^\bullet)$ is a complex of $\mathcal{C}/A$-coflat $\mathcal{S}$-semimodules by Lemma 1.3.3 and for any complex of $\mathcal{C}/A$-coflat $\mathcal{S}$-semimodules $\mathcal{K}^\bullet$ the complex $\mathbb{L}_1(\mathcal{K}^\bullet)$ is a complex of $\mathcal{C}$-coflat $\mathcal{S}$-semimodules by Remark 1.3.2 (hence the complex $\mathbb{L}_3\mathbb{L}_1(\mathcal{K}^\bullet)$ is a semiflat complex of semiflat $\mathcal{S}$-semimodules). Yet another useful approach to proving Theorem 2.6 was presented in 2.5: any complex of $\mathcal{S}$-semimodules $\mathcal{K}^\bullet$ can be connected with a semiflat complex of semiflat $\mathcal{S}$-semimodules in a functorial way by a chain of three morphisms $\mathcal{K}^\bullet \longleftarrow \mathbb{L}_3(\mathcal{K}^\bullet) \longrightarrow \mathbb{L}_3\mathbb{R}_2(\mathcal{K}^\bullet) \longleftarrow \mathbb{L}_3\mathbb{R}_2\mathbb{L}_1(\mathcal{K}^\bullet)$ with $\mathcal{C}$-coacyclic cones, and when $\mathcal{K}^\bullet$ is a semiflat complex of (A-flat, $\mathcal{C}$-coflat, or semiflat) $\mathcal{S}$-semimodules, all the complexes in this chain are also semiflat complexes of (A-flat, $\mathcal{C}$-coflat, or semiflat) $\mathcal{S}$-semimodules.

Question. Is the quotient category of $\mathcal{C}$-coflat complexes of $\mathcal{S}$-semimodules by the thick subcategory of $\mathcal{C}$-coacyclic $\mathcal{C}$-coflat complexes equivalent to the semiderived category of $\mathcal{S}$-semimodules?

2.7   Derived functor SemiTor

The following lemma provides a general approach to double-sided derived functors
of (partially defined) functors of two arguments.

Lemma. *Let H_1 and H_2 be two categories, H be a (not necessarily full) subcategory
in $H_1 \times H_2$, and S_1 and S_2 be localizing classes of morphisms in H_1 and H_2. Let
K be a category and $\Theta \colon H \longrightarrow K$ be a functor. Let F_1 and F_2 be subcategories
in H_1 and H_2. Assume that both functors $F_i[(S_i \cap F_i)^{-1}] \longrightarrow H_i[S_i^{-1}]$ induced by
the embeddings $F_i \longrightarrow H_i$ are equivalences of categories and the subcategory H
contains both subcategories $F_1 \times H_2$ and $H_1 \times F_2$. Furthermore, assume that the
morphisms $\Theta(U,t)$ and $\Theta(s,V)$ are isomorphisms in the category K for any objects $U \in F_1$, $V \in F_2$ and any morphisms $s \in S_1$, $t \in S_2$. Then the restrictions
of the functor Θ to the subcategories $F_1 \times H_2$ and $H_1 \times F_2$ factorize through their
localizations by their intersections with $S_1 \times S_2$, so one can define derived functors
$\mathbb{D}_1\Theta$, $\mathbb{D}_2\Theta \colon H_1[S_1^{-1}] \times H_2[S_2^{-1}] \longrightarrow K$ by restricting the functor Θ to these subcategories. Moreover, the derived functors $\mathbb{D}_1\Theta$ and $\mathbb{D}_2\Theta$ are naturally isomorphic to
each other and therefore do not depend on the choice of subcategories F_1 and F_2,
provided that both subcategories exist.*

Proof. Let us show that for any morphism $s \in S_1 \cap F_1$ and any object $X \in H_2$
the morphism $\Theta(s, X)$ is an isomorphism in K. By assumptions of the lemma, the
image of X in $H_2[S_2^{-1}]$ is isomorphic to the image of a certain object $V \in F_2$.
First suppose that there exists a fraction $X \leftarrow Y \rightarrow V$ of morphisms from S_2
connecting X and V. Then both morphisms of morphisms $\Theta(s, Y) \longrightarrow \Theta(s, X)$
and $\Theta(s, Y) \longrightarrow \Theta(s, V)$ are isomorphisms of morphisms, since the source and the
target of s belong to F_1. Now the morphism $\Theta(s, X)$ is an isomorphism, because
the morphism $\Theta(s, V)$ is an isomorphism. In the general case, there exist a fraction
$X \leftarrow Y \rightarrow V$ connecting X and V and two morphisms $Y' \rightarrow Y$ and $V \rightarrow V'$ such
that the morphism $Y \rightarrow X$ and two compositions $Y' \rightarrow Y \rightarrow V$ and $Y \rightarrow V \rightarrow V'$
belong to S_2. Then the compositions of morphisms of morphisms $\Theta(s, Y') \longrightarrow$
$\Theta(s, Y) \longrightarrow \Theta(s, V)$ and $\Theta(s, Y) \longrightarrow \Theta(s, V) \longrightarrow \Theta(s, V')$ are isomorphisms
of morphisms, so the morphism of morphisms $\Theta(s, Y) \longrightarrow \Theta(s, V)$ is both left
and right invertible, and therefore, is an isomorphism of morphisms. Since the
morphism of morphisms $\Theta(s, Y) \longrightarrow \Theta(s, X)$ is also an isomorphism of morphisms
and the morphism $\Theta(s, V)$ is an isomorphism, one can conclude that the morphism
$\Theta(s, X)$ is also an isomorphism.

   Thus the derived functor $\mathbb{D}_1\Theta$ is defined; it remains to construct an isomor-
phism between $\mathbb{D}_1\Theta$ and $\mathbb{D}_2\Theta$. But the compositions of the functors $\mathbb{D}_1\Theta$ and $\mathbb{D}_2\Theta$
with the functor $F_1[(S_1 \cap F_1)^{-1}] \times F_2[(S_2 \cap F_2)^{-1}] \longrightarrow H_1[S_1^{-1}] \times H_2[S_2^{-1}]$ coincide
by definition, and the latter functor is an equivalence of categories.   □

   Assume that the coring $\mathcal{C}$ is a flat left and right A-module, the semialgebra $\mathcal{S}$
is a coflat left and right $\mathcal{C}$-comodule, and the ring A has a finite weak homological
dimension.

The double-sided derived functor SemiTor^S on the Cartesian product of the semiderived categories of right and left S-semimodules is defined as follows. Consider the partially defined functor of semitensor product of complexes of S-semimodules

$$\lozenge_S \colon \mathsf{Hot}(\mathsf{simod}\text{-}S) \times \mathsf{Hot}(S\text{-}\mathsf{simod}) \dashrightarrow \mathsf{Hot}(k\text{-}\mathsf{mod}).$$

This functor is defined on the full subcategory of the Cartesian product of homotopy categories that consists of pairs of complexes $(\mathcal{N}^\bullet, \mathcal{M}^\bullet)$ such that either $\mathcal{N}^\bullet$ or $\mathcal{M}^\bullet$ is a complex of A-flat S-semimodules. Compose it with the functor of localization $\mathsf{Hot}(k\text{-}\mathsf{mod}) \longrightarrow \mathsf{D}(k\text{-}\mathsf{mod})$ and restrict to the Cartesian product of the homotopy category of semiflat complexes of A-flat right S-semimodules and the homotopy category of complexes of left S-semimodules.

By the definition, the functor so obtained factorizes through the semiderived category of left S-semimodules in the second argument, and it follows from Theorem 2.6 and Lemma 2.7 that it factorizes through the quotient category of the homotopy category of semiflat complexes of A-flat right S-semimodules by its intersection with the thick subcategory of $\mathcal{C}$-coacyclic complexes in the first argument.

Explicitly, let $\mathcal{N}^\bullet$ be a $\mathcal{C}$-coacyclic semiflat complex of A-flat right S-semimodules and $\mathcal{M}^\bullet$ be a complex of left S-semimodules. Using the constructions from the proof of Theorem 2.6, connect $\mathcal{M}^\bullet$ with a semiflat complex of A-flat left S-semimodules $\mathcal{L}^\bullet$ by a chain of morphisms with $\mathcal{C}$-coacyclic cones. Then the complexes $\mathcal{N}^\bullet \lozenge_S \mathcal{M}^\bullet$ and $\mathcal{N}^\bullet \lozenge_S \mathcal{L}^\bullet$ are connected by a chain of quasi-isomorphisms, and since the complex $\mathcal{N}^\bullet \lozenge_S \mathcal{L}^\bullet$ is acyclic, the complex $\mathcal{N}^\bullet \lozenge_S \mathcal{M}^\bullet$ is acyclic, too.

Thus we have constructed the double-sided derived functor

$$\mathrm{SemiTor}^S \colon \mathsf{D}^{\mathsf{si}}(\mathsf{simod}\text{-}S) \times \mathsf{D}^{\mathsf{si}}(S\text{-}\mathsf{simod}) \longrightarrow \mathsf{D}(k\text{-}\mathsf{mod}).$$

According to Lemma 2.7, the same derived functor can be obtained by restricting the functor of semitensor product to the Cartesian product of the homotopy category of complexes of left S-semimodules and the homotopy category of semiflat complexes of A-flat right S-semimodules, or indeed, to the Cartesian product of the homotopy categories of semiflat complexes of A-flat right and left S-semimodules. One can also use semiflat complexes of $\mathcal{C}$-coflat S-semimodules or semiflat complexes of semiflat S-semimodules instead of semiflat complexes of A-flat S-semimodules.

In particular, when the coring $\mathcal{C}$ is a flat left and right A-module and the ring A has a finite weak homological dimension, one defines the double-sided derived functor

$$\mathrm{Cotor}^{\mathcal{C}} \colon \mathsf{D}^{\mathsf{co}}(\mathsf{comod}\text{-}\mathcal{C}) \times \mathsf{D}^{\mathsf{co}}(\mathcal{C}\text{-}\mathsf{comod}) \longrightarrow \mathsf{D}(k\text{-}\mathsf{mod})$$

by composing the functor of cotensor product

$$\square_{\mathcal{C}} \colon \mathsf{Hot}(\mathsf{comod}\text{-}\mathcal{C}) \times \mathsf{Hot}(\mathcal{C}\text{-}\mathsf{comod}) \longrightarrow \mathsf{Hot}(k\text{-}\mathsf{mod})$$

with the functor of localization $\mathsf{Hot}(k\text{-mod}) \longrightarrow \mathsf{D}(k\text{-mod})$ and restricting it to the Cartesian product of the homotopy category of complexes of coflat right $\mathcal{C}$-comodules and the homotopy category of arbitrary complexes of left $\mathcal{C}$-comodules. The same derived functor is obtained by restricting the functor of cotensor product to the Cartesian product of the homotopy category of arbitrary complexes of right $\mathcal{C}$-comodules and the homotopy category of complexes of coflat left $\mathcal{C}$-comodules, or indeed, to the Cartesian product of the homotopy categories of coflat right $\mathcal{C}$-comodules and coflat left $\mathcal{C}$-comodules. One can also use coflat complexes of $\mathcal{C}$-comodules or coflat complexes of A-flat $\mathcal{C}$-comodules instead of complexes of coflat $\mathcal{C}$-comodules.

Remark. One can define a version of derived functor Cotor without making any homological dimension assumptions by considering pro-objects in the spirit of [44, 45]. Let $k\text{-mod}^\omega$ denote the category of pro-objects over the category $k\text{-mod}$ that can be represented by countable filtered projective systems of k-modules; this is an abelian tensor category with exact functors of countable filtered projective limits and a right exact functor of tensor product commuting with countable filtered projective limits. Let A be a ring object in $k\text{-mod}^\omega$; then one can consider right and left A-module objects and A-A-bimodule objects in $k\text{-mod}^\omega$, which we will simply call right and left A-modules and A-A-bimodules. Furthermore, let $\mathcal{C}$ be a coring object in the tensor category of A-A-bimodules; we will consider $\mathcal{C}$-comodule objects in the categories of right and left A-modules and call them right and left $\mathcal{C}$-comodules. Define the functor of cotensor product over $\mathcal{C}$ taking values in the category $k\text{-mod}^\omega$ in the usual way and extend it to the Cartesian product of the homotopy categories of complexes of right and left $\mathcal{C}$-comodules by taking infinite products along the diagonals in the bicomplex of cotensor products. The categories of right and left A-modules are abelian. Assume that $\mathcal{C}$ is a flat left and right A-module; then the categories of right and left $\mathcal{C}$-comodules are also abelian. Define the semiderived categories of right and left $\mathcal{C}$-comodules as the quotient categories of the homotopy categories by the thick subcategories of A-contraacyclic complexes (the contraacyclic complexes being defined in terms of countable products). Then one can use Lemma 2.7 to define the double-sided derived functor $\mathsf{ProCotor}^\mathcal{C}$ of cotensor product on the Cartesian product of the semiderived categories of right and left $\mathcal{C}$-comodules in terms of coflat complexes of $\mathcal{C}$-comodules.

In order to obtain for any complex of $\mathcal{C}$-comodules $\mathcal{M}^\bullet$ a coflat complex of $\mathcal{C}$-comodules connected with $\mathcal{M}^\bullet$ by a functorial chain of two morphisms with A-contraacyclic cones, one needs to construct a surjective morphism onto any $\mathcal{C}$-comodule $\mathcal{M}$ from an A-flat $\mathcal{C}$-comodule $\mathcal{F}(\mathcal{M})$. This construction is dual to that of Lemma 1.3.3 and uses the surjective map onto any A-module $\mathcal{M}$ from an A-flat A-module $\mathcal{G}(\mathcal{M}) = A \otimes_k^\omega M'$, where M' is a pro-k-module represented by a countable filtered projective system of flat k-modules mapping onto the pro-k-module $\mathcal{M}$ and $\otimes_k^\omega$ denotes the functor of tensor product in $k\text{-mod}^\omega$. The A-flat $\mathcal{C}$-comodule $\mathcal{F}(\mathcal{M})$ is obtained as the projective limit in $k\text{-mod}^\omega$ of the projective system of $\mathcal{C}$-comodules $\mathcal{M} \longleftarrow \mathfrak{Q}(\mathcal{M}) \longleftarrow \mathfrak{Q}(\mathfrak{Q}(\mathcal{M})) \longleftarrow \cdots$ Given a complex of A-flat $\mathcal{C}$-comodules $\mathcal{M}^\bullet$, a coflat complex of $\mathcal{C}$-comodules endowed with a morphism from the complex $\mathcal{M}^\bullet$ with an A-contractible cone is obtained as the total complex of the cobar complex of $\mathcal{M}^\bullet$, constructed by taking infinite products along the diagonals. One can also consider the category of arbitrary pro-k-modules in place of k-mod$^\omega$. Notice that for a conventional coalgebra $\mathcal{C}$ over a field $A = k$ and complexes of $\mathcal{C}$-comodules $\mathcal{N}^\bullet$ and $\mathcal{M}^\bullet$ in the category of k-vector spaces that are both bounded from above or from below the object of the derived category of k-vector spaces obtained by applying the derived functor of projective limit to the object ProCotor$^\mathcal{C}(\mathcal{N}^\bullet, \mathcal{M}^\bullet)$ of the derived category $\mathsf{D}(k\text{-vect}^\omega)$ coincides with Cotor$^{\mathcal{C},I}(\mathcal{N}^\bullet, \mathcal{M}^\bullet)$ (see 0.2.10).

2.8 Relatively semiflat complexes

We keep the assumptions and notation of 2.5, 2.6, and 2.7.

One can compute the derived functor Cotor$^\mathcal{C}$ using resolutions of a different kind. Namely, the cotensor product $\mathcal{N}^\bullet \square_\mathcal{C} \mathcal{M}^\bullet$ of a complex of A-flat right $\mathcal{C}$-comodules $\mathcal{N}^\bullet$ and a complex of $\mathcal{C}/A$-coflat $\mathcal{C}$-comodules $\mathcal{M}^\bullet$ represents an object naturally isomorphic to Cotor$^\mathcal{C}(\mathcal{M}^\bullet, \mathcal{N}^\bullet)$ in the derived category of k-modules. Indeed, the complex $\mathbb{R}_2(\mathcal{N}^\bullet)$ is a complex of coflat $\mathcal{C}$-comodules and the cone of the morphism $\mathcal{N}^\bullet \longrightarrow \mathbb{R}_2(\mathcal{N}^\bullet)$ is coacyclic with respect to the exact category of A-flat right $\mathcal{C}$-comodules, hence the morphism $\mathcal{N}^\bullet \square_\mathcal{C} \mathcal{M}^\bullet \longrightarrow \mathbb{R}_2(\mathcal{N}^\bullet) \square_\mathcal{C} \mathcal{M}^\bullet$ is a quasi-isomorphism. One can prove that the cotensor product of a complex coacyclic with respect to the exact category of A-flat $\mathcal{C}$-comodules and a complex of $\mathcal{C}/A$-coflat $\mathcal{C}$-comodules is acyclic in a way completely analogous to the proof of Lemma 2.2.

One can also compute the derived functor SemiTor$^\mathsf{S}$ using resolutions of different kinds. Namely, a complex of left S-semimodules is called *semiflat relative to A* if its semitensor product with any complex of A-flat right S-semimodules that as a complex of $\mathcal{C}$-comodules is coacyclic with respect to exact category of A-flat right $\mathcal{C}$-comodules is acyclic (cf. Theorem 7.2.2(a)). For example, the complex of S-semimodules induced from a complex of $\mathcal{C}/A$-coflat $\mathcal{C}$-comodules is semiflat relative to A, hence the complex $\mathbb{L}_3\mathbb{R}_2(\mathcal{K}^\bullet)$ is semiflat relative to A for any complex of left S-semimodules $\mathcal{K}^\bullet$. The semitensor product $\mathcal{N}^\bullet \lozenge_\mathsf{S} \mathcal{M}^\bullet$ of a complex of A-flat right S-semimodules $\mathcal{N}^\bullet$ and a complex of left S-semimodules $\mathcal{M}^\bullet$ semiflat relative to A represents an object naturally isomorphic to SemiTor$^\mathsf{S}(\mathcal{N}^\bullet, \mathcal{M}^\bullet)$ in the derived category of k-modules. Indeed, $\mathbb{L}_3\mathbb{R}_2(\mathcal{N}^\bullet)$ is a semiflat complex of right S-semimodules connected with $\mathcal{N}^\bullet$ by a chain of morphisms $\mathcal{N}^\bullet \longrightarrow \mathbb{R}_2(\mathcal{N}^\bullet) \longleftarrow \mathbb{L}_3\mathbb{R}_2(\mathcal{N}^\bullet)$ whose cones are coacyclic with respect to the exact category of A-flat $\mathcal{C}$-comodules and contractible over $\mathcal{C}$, respectively. Hence there is a chain of two quasi-isomorphisms connecting $\mathcal{N}^\bullet \lozenge_\mathsf{S} \mathcal{M}^\bullet$ with $\mathbb{L}_3\mathbb{R}_2(\mathcal{N}^\bullet) \lozenge_\mathsf{S} \mathcal{M}^\bullet$.

Analogously, a complex of left S-semimodules is called *semiflat relative to $\mathcal{C}$* if its semitensor product with any $\mathcal{C}$-contractible complex of $\mathcal{C}$-coflat right S-semimodules is acyclic. For example, the complex of $\mathbf{S}$-semimodules induced from any complex of $\mathcal{C}$-comodules is semiflat relative to $\mathcal{C}$, hence the complex $\mathbb{L}_3(\mathcal{K}^\bullet)$ is semiflat relative to $\mathcal{C}$ for any complex of left $\mathbf{S}$-semimodules $\mathcal{K}^\bullet$. The semitensor product $\mathcal{N}^\bullet \lozenge_\mathbf{S} \mathcal{M}^\bullet$ of a complex of $\mathcal{C}$-coflat right $\mathbf{S}$-semimodules $\mathcal{N}^\bullet$ and a complex of left $\mathbf{S}$-semimodules $\mathcal{M}^\bullet$ semiflat relative to $\mathcal{C}$ represents an object naturally isomorphic to $\mathrm{SemiTor}^\mathbf{S}(\mathcal{N}^\bullet, \mathcal{M}^\bullet)$ in the derived category of k-modules. Indeed, $\mathbb{L}_3(\mathcal{N}^\bullet)$ is a semiflat complex of right $\mathbf{S}$-semimodules and the cone of the morphism $\mathbb{L}_3(\mathcal{N}^\bullet) \longrightarrow \mathcal{N}^\bullet$ is a $\mathcal{C}$-contractible complex of $\mathcal{C}$-coflat right $\mathbf{S}$-semimodules. It follows that the semitensor product of a complex of left $\mathbf{S}$-semimodules semiflat relative to $\mathcal{C}$ with a $\mathcal{C}$-coacyclic complex of $\mathcal{C}$-coflat right $\mathbf{S}$-semimodules is acyclic.

At last, a complex of A-flat right $\mathbf{S}$-semimodules is called *semiflat relative to* $\mathcal{C}$ *relative to* A ($\mathbf{S}/\mathcal{C}/A$-semiflat) if its semitensor product with any $\mathcal{C}$-contractible complex of $\mathcal{C}/A$-coflat left $\mathbf{S}$-semimodules is acyclic. For example, the complex of $\mathbf{S}$-semimodules induced from a complex of A-flat $\mathcal{C}$-comodules is $\mathbf{S}/\mathcal{C}/A$-semiflat, hence the complex $\mathbb{L}_3\mathbb{L}_1(\mathcal{K}^\bullet)$ is $\mathbf{S}/\mathcal{C}/A$-semiflat for any complex of right $\mathbf{S}$-semi-modules $\mathcal{K}^\bullet$. The semitensor product $\mathcal{N}^\bullet \lozenge_\mathbf{S} \mathcal{M}^\bullet$ of an $\mathbf{S}/\mathcal{C}/A$-semiflat complex of A-flat right $\mathbf{S}$-semimodules $\mathcal{N}^\bullet$ and a complex of $\mathcal{C}/A$-coflat left $\mathbf{S}$-semimodules $\mathcal{M}^\bullet$ represents an object naturally isomorphic to $\mathrm{SemiTor}^\mathbf{S}(\mathcal{N}^\bullet, \mathcal{M}^\bullet)$ in the derived category of k-modules. Indeed, $\mathbb{L}_3(\mathcal{M}^\bullet)$ is a complex of left $\mathbf{S}$-semimodules semiflat relative to A and the cone of the morphism $\mathbb{L}_3(\mathcal{M}^\bullet) \longrightarrow \mathcal{M}^\bullet$ is a $\mathcal{C}$-contractible complex of $\mathcal{C}/A$-coflat right $\mathbf{S}$-semimodules. It follows that the semitensor product of an $\mathbf{S}/\mathcal{C}/A$-semiflat complex of A-flat right $\mathbf{S}$-semimodules with a $\mathcal{C}$-coacyclic complex of $\mathcal{C}/A$-coflat left $\mathbf{S}$-semimodules is acyclic.

The functors mapping the quotient categories of the homotopy categories of complexes of $\mathbf{S}$-semimodules semiflat relative to A, complexes of $\mathbf{S}$-semimod-ules semiflat relative to $\mathcal{C}$, and $\mathbf{S}/\mathcal{C}/A$-semiflat complexes of A-flat $\mathbf{S}$-semimodules by their intersections with the thick subcategory of $\mathcal{C}$-coacyclic complexes into the semiderived category of $\mathbf{S}$-semimodules are equivalences of triangulated categories. The same applies to complexes of A-flat, $\mathcal{C}$-coflat, or $\mathcal{C}/A$-coflat $\mathbf{S}$-semimodules. These results follow easily from either of Lemmas 2.5 or 2.6. So one can define the derived functor $\mathrm{SemiTor}^\mathbf{S}$ by restricting the functor of semitensor product to these categories of complexes of $\mathbf{S}$-semimodules as explained above.

Remark. Assuming that $\mathcal{C}$ is a flat right A-module, $\mathbf{S}$ is a coflat right and a $\mathcal{C}/A$-coflat left $\mathcal{C}$-comodule, and A has a finite weak homological dimension, one can define the double-sided derived functor $\mathrm{SemiTor}^\mathbf{S}$ on the Cartesian product of the semiderived category of A-flat right $\mathbf{S}$-semimodules and the semiderived category of left $\mathbf{S}$-semimodules. The former is defined as the quotient category of the homotopy category of complexes of A-flat right $\mathbf{S}$-semimodules by the thick subcategory of complexes that as complexes of $\mathcal{C}$-comodules are coacyclic with respect to the exact category of A-flat right $\mathcal{C}$-comodules. The derived functor is constructed by restricting the functor of semitensor product to the Cartesian product of the homotopy category of complexes of A-flat right $\mathbf{S}$-semimodules and the homotopy category of complexes of left $\mathbf{S}$-semimodules semiflat relative to A, or the Cartesian product of the homotopy category of semiflat complexes of A-flat right $\mathbf{S}$-semimodules and the homotopy category of complexes of left $\mathbf{S}$-semimodules. Assuming that $\mathcal{C}$ is a flat left and right A-module, $\mathbf{S}$ is a flat left A-module and a coflat right $\mathcal{C}$-comodule, and A has a finite weak homological dimension, one can define the left derived functor $\mathrm{SemiTor}^{\mathbf{S}}$ on the Cartesian product of the semiderived category of $\mathcal{C}/A$-coflat right $\mathbf{S}$-semimodules and the semiderived category of left $\mathbf{S}$-semimodules. The former is defined as the quotient category of the homotopy category of complexes of $\mathcal{C}/A$-flat right $\mathbf{S}$-semimodules by the thick subcategory of complexes that as complexes of $\mathcal{C}$-comodules are coacyclic with respect to the exact category of $\mathcal{C}/A$-coflat right $\mathcal{C}$-comodules (cf. Remark 7.2.2). The derived functor is constructed by restricting the functor of semitensor product to the Cartesian product of the homotopy category of complexes of $\mathcal{C}/A$-coflat right $\mathbf{S}$-semimodules and the homotopy category of $\mathbf{S}/\mathcal{C}/A$-semiflat complexes of A-flat left $\mathbf{S}$-semimodules, or the Cartesian product of the homotopy category of semiflat complexes of $\mathcal{C}$-coflat right $\mathbf{S}$-semimodules and the homotopy category of complexes of left $\mathbf{S}$-semimodules. Both of these definitions of derived functors are particular cases of Lemma 2.7.

2.9 Remarks on derived semitensor product of bisemimodules

We would like to define the double-sided derived functor of semitensor product of bisemimodules and in such a way that derived semitensor products of several factors would be associative. It appears that there are two approaches to this problem, even in the case of modules over rings. First suppose that we only wish to have associative derived semitensor products of three factors. Let $\mathbf{S}$ be a semialgebra over a coring $\mathcal{C}$ and $\mathcal{T}$ be a semialgebra over a coring $\mathcal{D}$, both satisfying the conditions of 2.6.

The semiderived category of $\mathbf{S}$-$\mathcal{T}$-bisemimodules $\mathsf{D}^{\mathsf{si}}(\mathbf{S}\text{-simod-}\mathcal{T})$ is defined as the quotient category of the homotopy category $\mathsf{Hot}(\mathbf{S}\text{-simod-}\mathcal{T})$ by the thick subcategory of complexes of bisemimodules that as complexes of $\mathcal{C}$-$\mathcal{D}$-bicomodules are coacyclic with respect to the abelian category of $\mathcal{C}$-$\mathcal{D}$-bicomodules. We would like to define derived functors of semitensor product

$$\lozenge_{\mathbf{S}}^{\mathcal{D}} \colon \mathsf{D}^{\mathsf{si}}(\text{simod-}\mathbf{S}) \times \mathsf{D}^{\mathsf{si}}(\mathbf{S}\text{-simod-}\mathcal{T}) \longrightarrow \mathsf{D}^{\mathsf{si}}(\text{simod-}\mathcal{T})$$

$$\lozenge_{\mathcal{T}}^{\mathcal{D}} \colon \mathsf{D}^{\mathsf{si}}(\mathbf{S}\text{-simod-}\mathcal{T}) \times \mathsf{D}^{\mathsf{si}}(\mathcal{T}\text{-simod}) \longrightarrow \mathsf{D}^{\mathsf{si}}(\mathbf{S}\text{-simod})$$

and prove the associativity isomorphism

$$\mathrm{SemiTor}^{\mathcal{T}}(\mathbf{N}^{\bullet} \lozenge_{\mathbf{S}}^{\mathcal{D}} \mathcal{K}^{\bullet}, \mathcal{M}^{\bullet}) \simeq \mathrm{SemiTor}^{\mathbf{S}}(\mathbf{N}^{\bullet}, \mathcal{K}^{\bullet} \lozenge_{\mathcal{T}}^{\mathcal{D}} \mathcal{M}^{\bullet}).$$

Let us call a complex of $\mathcal{C}$-coflat right $\mathbf{S}$-semimodules *quite semiflat* if it belongs to the minimal triangulated subcategory of the homotopy category of $\mathbf{S}$-semimodules containing the complexes induced from complexes of coflat right $\mathcal{C}$-comodules and closed under infinite direct sums. One can show (see Remark 7.2.2

and the proof of Theorem 8.2.2) that the quotient category of the category of quite semiflat complexes of $\mathcal{C}$-coflat $\mathbf{S}$-semimodules by its minimal triangulated subcategory containing the complexes of $\mathbf{S}$-semimodules induced from complexes of $\mathcal{C}$-comodules coacyclic with respect to the exact category of $\mathcal{C}$-coflat $\mathcal{C}$-comodules and closed under infinite direct sums is equivalent to the semiderived category of $\mathbf{S}$-semimodules. In other words, any $\mathcal{C}$-coacyclic quite semiflat complex of $\mathcal{C}$-coflat $\mathbf{S}$-semimodules can be obtained from the complexes of $\mathbf{S}$-semimodules induced from the total complexes of exact triples of complexes of coflat $\mathcal{C}$-comodules using the operations of cone and infinite direct sum.

It follows (by Lemmas 2.2 and 1.2.2) that the restriction of the functor of semitensor product $\mathsf{Hot}(\mathrm{simod}\text{-}\mathbf{S}) \times \mathsf{Hot}(\mathbf{S}\text{-}\mathrm{simod}\text{-}\mathbf{\mathcal{T}}) \dashrightarrow \mathsf{D}^{\mathsf{si}}(\mathrm{simod}\text{-}\mathbf{\mathcal{T}})$ to the Cartesian product of the homotopy category of quite semiflat complexes of $\mathcal{C}$-coflat right $\mathbf{S}$-semimodules and the homotopy category of complexes of $\mathbf{S}$-$\mathbf{\mathcal{T}}$-bisemimodules factorizes through the Cartesian product of semiderived categories of right $\mathbf{S}$-semimodules and $\mathbf{S}$-$\mathbf{\mathcal{T}}$-bisemimodules. So the desired derived functors are defined; and the associativity isomorphism follows from Proposition 1.4.4. Notice that this definition of a double-sided derived functor is *not* a particular case of the construction of Lemma 2.7.

Question. Can one use arbitrary semiflat complexes of $\mathcal{C}$-coflat $\mathbf{S}$-semimodules or, at least, semiflat complexes of semiflat $\mathbf{S}$-semimodules instead of quite semiflat complexes in this construction? In other words, assume that $\mathbf{N}^\bullet$ is a $\mathcal{C}$-coacyclic semiflat complex of semiflat right $\mathbf{S}$-semimodules and $\mathbf{\mathcal{K}}$ is an $\mathbf{S}$-$\mathbf{\mathcal{T}}$-bisemimodule. Is the complex $\mathbf{N}^\bullet \lozenge_\mathbf{S} \mathbf{\mathcal{K}}$ necessarily $\mathcal{D}$-coacyclic? (Cf. 4.9.)

Now suppose that we want to have derived semitensor products of any number of factors. Let $\mathbf{S}$ be a semialgebra over a coring $\mathcal{C}$ over a k-algebra A, $\mathbf{\mathcal{T}}$ be a semialgebra over a coring $\mathcal{D}$ over a k-algebra B, and $\mathbf{\mathcal{R}}$ be a semialgebra over a coring $\mathcal{E}$ over a k-algebra F, all three satisfying the conditions of 2.6. We would like to define the derived functor of semitensor product

$$\lozenge_\mathbf{\mathcal{R}}^\mathbb{D} : \mathsf{D}^{\mathsf{si}}(\mathbf{S}\text{-}\mathrm{simod}\text{-}\mathbf{\mathcal{R}}) \times \mathsf{D}^{\mathsf{si}}(\mathbf{\mathcal{R}}\text{-}\mathrm{simod}\text{-}\mathbf{\mathcal{T}}) \longrightarrow \mathsf{D}^{\mathsf{si}}(\mathbf{S}\text{-}\mathrm{simod}\text{-}\mathbf{\mathcal{T}}).$$

This can be done, assuming that the k-algebras A, B, and F are flat k-modules.

Let us call a complex of F-flat $\mathbf{S}$-$\mathbf{\mathcal{R}}$-bisemimodules *strongly $\mathbf{\mathcal{R}}$-semiflat* if its semitensor product over $\mathbf{\mathcal{R}}$ with any $\mathcal{E}$-$\mathcal{D}$-coacyclic complex of $\mathbf{\mathcal{R}}$-$\mathbf{\mathcal{T}}$-bisemimodules is a $\mathcal{C}$-$\mathcal{D}$-coacyclic complex of $\mathbf{S}$-$\mathbf{\mathcal{T}}$-bisemimodules for any semialgebra $\mathbf{\mathcal{T}}$. Using bimodule versions of the constructions of Lemmas 1.3.2 and 1.3.3, one can prove that the quotient category of the homotopy category of strongly $\mathbf{\mathcal{R}}$-semiflat complexes of F-flat $\mathbf{S}$-$\mathbf{\mathcal{R}}$-bisemimodules by its intersection with the thick subcategory of $\mathcal{C}$-$\mathcal{E}$-coacyclic bisemimodules is equivalent to the semiderived category of $\mathbf{S}$-$\mathbf{\mathcal{R}}$-bisemimodules, and the analogous result holds for the homotopy category of strongly $\mathbf{S}$-semiflat and strongly $\mathbf{\mathcal{R}}$-semiflat complexes of A-flat and F-flat $\mathbf{S}$-$\mathbf{\mathcal{R}}$-bisemimodules. One just uses the functor $G(M) = \bigoplus_{m \in M} A \otimes_k F$ in the construction of Lemma 1.1.3, considers the bicoaction and bisemiaction morphisms in place of the coaction and semiaction morphisms, etc. (As we only want our A-F-bimodules to be A-flat and F-flat, no assumption about the homological dimension of $A \otimes_k F$ is needed.) So Lemma 2.7 is applicable to the functor of semitensor product $\mathsf{Hot}(\mathcal{S}\text{-simod-}\mathcal{R}) \times \mathsf{Hot}(\mathcal{R}\text{-simod-}\mathcal{T}) \dashrightarrow \mathsf{D}^{\mathsf{si}}(\mathcal{S}\text{-simod-}\mathcal{T})$ and we obtain the desired double-sided derived functor. There is an associativity isomorphism

$$(\mathbf{N}^{\bullet} \lozenge_{\mathcal{S}}^{\mathbb{D}} \mathcal{K}^{\bullet}) \lozenge_{\mathcal{T}}^{\mathbb{D}} \mathcal{M}^{\bullet} \simeq \mathbf{N}^{\bullet} \lozenge_{\mathcal{S}}^{\mathbb{D}} (\mathcal{K}^{\bullet} \lozenge_{\mathcal{T}}^{\mathbb{D}} \mathcal{M}^{\bullet}).$$

In the case of derived cotensor product of bicomodules, one does not need to introduce quite coflat or strongly coflat complexes. It suffices to consider complexes of $\mathcal{C}$-coflat $\mathcal{C}$-comodules or complexes of ($\mathcal{C}$-coflat and) $\mathcal{E}$-coflat $\mathcal{C}$-$\mathcal{E}$-bicomodules. One can define double-sided derived functors

$$\square_{\mathcal{C}}^{\mathbb{D}} \colon \mathsf{D}^{\mathsf{co}}(\mathsf{comod}\text{-}\mathcal{C}) \times \mathsf{D}^{\mathsf{co}}(\mathcal{C}\text{-comod-}\mathcal{D}) \longrightarrow \mathsf{D}^{\mathsf{co}}(\mathsf{comod}\text{-}\mathcal{D})$$
$$\square_{\mathcal{D}}^{\mathbb{D}} \colon \mathsf{D}^{\mathsf{co}}(\mathcal{C}\text{-comod-}\mathcal{D}) \times \mathsf{D}^{\mathsf{co}}(\mathcal{D}\text{-comod}) \longrightarrow \mathsf{D}^{\mathsf{co}}(\mathcal{C}\text{-comod})$$

and prove the associativity isomorphism

$$\operatorname{Cotor}^{\mathcal{D}}(\mathbf{N}^{\bullet} \square_{\mathcal{C}}^{\mathbb{D}} \mathcal{K}^{\bullet}, \mathcal{M}^{\bullet}) \simeq \operatorname{Cotor}^{\mathcal{C}}(\mathbf{N}^{\bullet}, \mathcal{K}^{\bullet} \square_{\mathcal{D}}^{\mathbb{D}} \mathcal{M}^{\bullet})$$

by replacing the complex of right $\mathcal{C}$-comodules $\mathbf{N}^{\bullet}$ with a complex of coflat right $\mathcal{C}$-comodules and the complex of left $\mathcal{D}$-comodules $\mathcal{M}^{\bullet}$ by a complex of coflat left $\mathcal{D}$-comodules representing the same object in the coderived category of comodules. The derived functors $\square_{\mathcal{C}}^{\mathbb{D}}$ and $\square_{\mathcal{D}}^{\mathbb{D}}$ are well defined, since any coacyclic complex of coflat comodules is coacyclic with respect to the exact category of coflat comodules (see 7.2.2). If the k-modules A and F are flat, one can prove that the quotient category of the homotopy category of $\mathcal{E}$-coflat $\mathcal{C}$-$\mathcal{E}$-bicomodules by its intersection with the thick subcategory of coacyclic complexes of $\mathcal{C}$-$\mathcal{E}$-bicomodules is equivalent to the coderived category of bicomodules, and the same applies to the homotopy category of $\mathcal{C}$-coflat and $\mathcal{E}$-coflat $\mathcal{C}$-$\mathcal{E}$-bicomodules. Then one can apply Lemma 2.7 in order to define the double-sided derived functor

$$\square_{\mathcal{E}}^{\mathbb{D}} \colon \mathsf{D}^{\mathsf{co}}(\mathcal{C}\text{-comod-}\mathcal{E}) \times \mathsf{D}^{\mathsf{co}}(\mathcal{E}\text{-comod-}\mathcal{D}) \longrightarrow \mathsf{D}^{\mathsf{co}}(\mathcal{C}\text{-comod-}\mathcal{D}),$$

and there is an associativity isomorphism

$$(\mathbf{N}^{\bullet} \square_{\mathcal{C}}^{\mathbb{D}} \mathcal{K}^{\bullet}) \square_{\mathcal{D}}^{\mathbb{D}} \mathcal{M}^{\bullet} \simeq \mathbf{N}^{\bullet} \square_{\mathcal{C}}^{\mathbb{D}} (\mathcal{K}^{\bullet} \square_{\mathcal{D}}^{\mathbb{D}} \mathcal{M}^{\bullet}).$$

3 Semicontramodules and Semihomomorphisms

Throughout Chapters 3–11, $k^\vee$ is an injective cogenerator of the category of k-modules. One can always take $k^\vee = \operatorname{Hom}_{\mathbb{Z}}(k, \mathbb{Q}/\mathbb{Z})$.

3.1 Contramodules

For two k-algebras A and B, we will denote by $A\text{-mod-}B$ the category of k-modules with an A-B-bimodule structure.

3.1.1 The identity $\operatorname{Hom}_A(K \otimes_A M, P) \simeq \operatorname{Hom}_A(M, \operatorname{Hom}_A(K, P))$ for left A-modules M, P and an A-A-bimodule K means that the category opposite to the category of left A-modules is a right module category over the tensor category of A-A-bimodules with the functor of right action $(P^{\mathrm{op}}, N) \longmapsto \operatorname{Hom}(N, P)^{\mathrm{op}}$. Therefore, one can consider module objects in this module category over ring objects in $A\text{-mod-}A$ and comodule objects in this module category over coring objects in $A\text{-mod-}A$.

Clearly, a ring object B in $A\text{-mod-}A$ is just a k-algebra endowed with a k-algebra morphism $A \longrightarrow B$. A B-module in $A\text{-mod}^{\mathrm{op}}$ is an A-module P endowed with a map $P \longrightarrow \operatorname{Hom}_A(B, P)$; so one can easily see that B-modules in $A\text{-mod}^{\mathrm{op}}$ are just (objects of the category opposite to the category of) usual left B-modules.

Let $\mathcal{C}$ be a coring over A. The category of *left contramodules* over $\mathcal{C}$ is the opposite category to the category of comodule objects in the right module category $A\text{-mod}^{\mathrm{op}}$ over the coring object $\mathcal{C}$ in the tensor category $A\text{-mod-}A$. In other words, a left $\mathcal{C}$-contramodule $\mathfrak{P}$ is a left A-module endowed with a *left contraaction* map $\operatorname{Hom}_A(\mathcal{C}, \mathfrak{P}) \longrightarrow \mathfrak{P}$, which should be a morphism of left A-modules satisfying the following *contraassociativity* and *counity* equations. First, two maps

$$\operatorname{Hom}_A(\mathcal{C} \otimes_A \mathcal{C},\, \mathfrak{P}) = \operatorname{Hom}_A(\mathcal{C}, \operatorname{Hom}_A(\mathcal{C}, \mathfrak{P})) \rightrightarrows \operatorname{Hom}_A(\mathcal{C}, \mathfrak{P}),$$

one of which is induced by the comultiplication map of $\mathcal{C}$ and the other by the contraaction map, should have equal compositions with the contraaction map $\operatorname{Hom}_A(\mathcal{C}, \mathfrak{P}) \longrightarrow \mathfrak{P}$, and second, the composition

$$\mathfrak{P} = \operatorname{Hom}_A(A, \mathfrak{P}) \longrightarrow \operatorname{Hom}_A(\mathcal{C}, \mathfrak{P}) \longrightarrow \mathfrak{P}$$

of the map induced by the counit map of $\mathcal{C}$ with the contraaction map should be equal to the identity map of $\mathfrak{P}$. A *right contramodule* $\mathfrak{R}$ over $\mathcal{C}$ is a right A-module endowed with a *right contraaction* map $\operatorname{Hom}_{A^{\mathrm{op}}}(\mathcal{C}, \mathfrak{R}) \longrightarrow \mathfrak{R}$, which should be a map of right A-modules satisfying the analogous equations.

3.1.2   The standard example of a $\mathcal{C}$-contramodule: for any right $\mathcal{C}$-comodule $\mathcal{N}$ endowed with a left action of a k-algebra B by $\mathcal{C}$-comodule endomorphisms and any left B-module V, the left A-module $\mathrm{Hom}_B(\mathcal{N}, V)$ has a natural left $\mathcal{C}$-contramodule structure. The left $\mathcal{C}$-contramodule $\mathrm{Hom}_A(\mathcal{C}, V)$ is called the $\mathcal{C}$-contramodule *induced* from a left A-module V. According to Lemma 1.1.2, the k-module of contramodule homomorphisms from the induced $\mathcal{C}$-contramodule to an arbitrary $\mathcal{C}$-contramodule is described by the formula

$$\mathrm{Hom}^{\mathcal{C}}(\mathrm{Hom}_A(\mathcal{C}, V), \mathfrak{P}) \simeq \mathrm{Hom}_A(V, \mathfrak{P}).$$

We will denote the category of left $\mathcal{C}$-contramodules by $\mathcal{C}$-contra and the category of right $\mathcal{C}$-contramodules by contra-$\mathcal{C}$. The category of left $\mathcal{C}$-contramodules is abelian whenever $\mathcal{C}$ is a projective left A-module. Moreover, the left A-module $\mathcal{C}$ is projective if and only if the category $\mathcal{C}$-contra is abelian and the forgetful functor $\mathcal{C}$-contra $\longrightarrow$ A-mod is exact. This can be proven by the same adjoint functor argument as the analogous result for $\mathcal{C}$-comodules.

For any coring $\mathcal{C}$, there are two natural exact categories of left contramodules: the exact category of A-injective $\mathcal{C}$-contramodules and the exact category of arbitrary $\mathcal{C}$-contramodules with A-split exact triples. Besides, any morphism of $\mathcal{C}$-contramodules has a kernel and the forgetful functor $\mathcal{C}$-contra $\longrightarrow$ A-mod preserves kernels. When a morphism of $\mathcal{C}$-contramodules has the property that its cokernel in the category of A-modules is preserved by the functors of homomorphisms from $\mathcal{C}$ and $\mathcal{C} \otimes_A \mathcal{C}$ over A, this cokernel has a natural $\mathcal{C}$-contramodule structure, which makes it the cokernel of that morphism in the category of $\mathcal{C}$-contramodules.

Infinite products always exist in the category of $\mathcal{C}$-contramodules and the forgetful functor $\mathcal{C}$-contra $\longrightarrow$ A-mod preserves them. The induction functor A-mod $\longrightarrow$ $\mathcal{C}$-contra preserves both infinite direct sums and infinite products. To construct direct sums of $\mathcal{C}$-contramodules, one can present them as cokernels of morphisms of induced contramodules. All cokernels exist in the category of $\mathcal{C}$-contramodules [6], so the category of $\mathcal{C}$-contramodules has infinite direct sums.

Question. If $\mathcal{C}$ is a flat right A-module, then subcomodules of finite direct sums of copies of $\mathcal{C}$ constitute a set of generators of the category of left $\mathcal{C}$-comodules [23, 3.13]. Does the category of $\mathcal{C}$-contramodules have a set of cogenerators?

3.1.3   Assume that the coring $\mathcal{C}$ is a projective left and a flat right A-module and the ring A has a finite left homological dimension (homological dimension of the category of left A-modules).

Lemma.

(a) *There exists a (not always additive) functor assigning to any left $\mathcal{C}$-comodule a surjective map onto it from an A-projective $\mathcal{C}$-comodule. Moreover, the kernel of this map is an iterated extension of coinduced $\mathcal{C}$-comodules.*

(b) *There exists a (not always additive) functor assigning to any left $\mathcal{C}$-contra-module an injective map from it into an A-injective $\mathcal{C}$-contramodule. More-over, the cokernel of this map is an iterated extension of induced $\mathcal{C}$-contra-modules.*

Proof. The proof of part (a) is completely analogous to the proof of Lemma 1.1.3 and part (b) is proven in the following way. Let $P \longrightarrow G(P)$ be an injective map from an A-module P into an injective A-module $G(P)$ functorially depending on P. For example, one can take $G(P)$ to be the direct product of copies of the A-module $\operatorname{Hom}_A(A, k^\vee)$ numbered by all k-module homomorphisms $P \longrightarrow k^\vee$. Let $\mathfrak{P}$ be a left $\mathcal{C}$-contramodule. Consider the contraaction map $\operatorname{Hom}_A(\mathcal{C}, \mathfrak{P}) \longrightarrow \mathfrak{P}$; it is a surjective morphism of $\mathcal{C}$-contramodules; let $\mathfrak{K}(\mathfrak{P})$ denote its kernel. Let $\mathfrak{Q}(\mathfrak{P})$ be the cokernel of the composition

$$\mathfrak{K}(\mathfrak{P}) \longrightarrow \operatorname{Hom}_A(\mathcal{C}, \mathfrak{P}) \longrightarrow \operatorname{Hom}_A(\mathcal{C}, G(\mathfrak{P})).$$

Then the composition of maps $\operatorname{Hom}_A(\mathcal{C}, \mathfrak{P}) \longrightarrow \operatorname{Hom}_A(\mathcal{C}, G(\mathfrak{P})) \longrightarrow \mathfrak{Q}(\mathfrak{P})$ fac-torizes through the surjection $\operatorname{Hom}_A(\mathcal{C}, \mathfrak{P}) \longrightarrow \mathfrak{P}$, so there is a natural injective morphism of $\mathcal{C}$-contramodules $\mathfrak{P} \longrightarrow \mathfrak{Q}(\mathfrak{P})$. Let us show that the injective di-mension $\operatorname{di}_A \mathfrak{Q}(\mathfrak{P})$ of the A-module $\mathfrak{Q}(\mathfrak{P})$ is smaller than that of $\mathfrak{P}$. Indeed, the A-module $\operatorname{Hom}_A(\mathcal{C}, G(\mathfrak{P}))$ is injective, hence

$$\operatorname{di}_A \mathfrak{Q}(\mathfrak{P}) = \operatorname{di}_A \mathfrak{K}(\mathfrak{P}) - 1 \leqslant \operatorname{di}_A \operatorname{Hom}_A(\mathcal{C}, \mathfrak{P}) - 1 \leqslant \operatorname{di}_A(\mathfrak{P}) - 1,$$

because the A-module $\mathfrak{K}(\mathfrak{P})$ is a direct summand of the A-module $\operatorname{Hom}_A(\mathcal{C}, \mathfrak{P})$ and an injective resolution of the A-module $\operatorname{Hom}_A(\mathcal{C}, \mathfrak{P})$ can be constructed by ap-plying the functor $\operatorname{Hom}_A(\mathcal{C}, -)$ to an injective resolution of $\mathfrak{P}$. Notice that the cok-ernel of the map $\mathfrak{P} \longrightarrow \mathfrak{Q}(\mathfrak{P})$ is an induced $\mathcal{C}$-contramodule $\operatorname{Hom}_A(\mathcal{C}, G(\mathfrak{P})/\mathfrak{P})$. It remains to iterate the functor $\mathfrak{P} \longmapsto \mathfrak{Q}(\mathfrak{P})$ sufficiently many times. $\square$

3.2 Cohomomorphisms

3.2.1 The k-module of *cohomomorphisms* $\operatorname{Cohom}_\mathcal{C}(\mathcal{M}, \mathfrak{P})$ from a left $\mathcal{C}$-comodule $\mathcal{M}$ to a left $\mathcal{C}$-contramodule $\mathfrak{P}$ is defined as the cokernel of the pair of maps

$$\operatorname{Hom}_A(\mathcal{C} \otimes_A \mathcal{M}, \, \mathfrak{P}) = \operatorname{Hom}_A(\mathcal{M}, \operatorname{Hom}_A(\mathcal{C}, \mathfrak{P})) \rightrightarrows \operatorname{Hom}_A(\mathcal{M}, \mathfrak{P})$$

one of which is induced by the $\mathcal{C}$-coaction in $\mathcal{M}$ and the other by the $\mathcal{C}$-contraaction in $\mathfrak{P}$. The functor of cohomomorphisms is neither left nor right exact in general; it is right exact if the ring A is semisimple. For any left A-module U and any left $\mathcal{C}$-contramodule $\mathfrak{P}$ there is a natural isomorphism

$$\operatorname{Cohom}_\mathcal{C}(\mathcal{C} \otimes_A U, \, \mathfrak{P}) \simeq \operatorname{Hom}_A(U, \mathfrak{P}),$$

and for any left $\mathcal{C}$-comodule $\mathcal{M}$ and any left A-module V there is a natural iso-morphism

$$\operatorname{Cohom}_\mathcal{C}(\mathcal{M}, \operatorname{Hom}_A(\mathcal{C}, V)) \simeq \operatorname{Hom}_A(\mathcal{M}, V).$$

These assertions follow from Lemma 1.2.1. Explicitly, the first isomorphism can be obtained by applying the functor $\mathrm{Hom}_A(U, -)$ to the split exact sequence of A-modules $\mathrm{Hom}_A(\mathcal{C} \otimes_A \mathcal{C}, \mathfrak{P}) \longrightarrow \mathrm{Hom}_A(\mathcal{C}, \mathfrak{P}) \longrightarrow \mathfrak{P}$ and the second one can be obtained by applying the functor $\mathrm{Hom}_A(-, V)$ to the split exact sequence of A-modules $\mathcal{M} \longrightarrow \mathcal{C} \otimes_A \mathcal{M} \longrightarrow \mathcal{C} \otimes_A \mathcal{C} \otimes_A \mathcal{M}$.

3.2.2   Assuming that $\mathcal{C}$ is a projective left A-module, a left comodule $\mathcal{M}$ over $\mathcal{C}$ is called *coprojective* if the functor of cohomomorphisms from $\mathcal{M}$ is exact on the category of left $\mathcal{C}$-contramodules. It is easy to see that any coprojective $\mathcal{C}$-comodule is a projective A-module. The $\mathcal{C}$-comodule coinduced from a projective A-module is coprojective. Assuming that $\mathcal{C}$ is a flat right A-module, a left contramodule $\mathfrak{P}$ over $\mathcal{C}$ is called *coinjective* if the functor of cohomomorphisms into $\mathfrak{P}$ is exact on the category of left $\mathcal{C}$-comodules. Any coinjective $\mathcal{C}$-contramodule is an injective A-module. The $\mathcal{C}$-contramodule induced from an injective A-module is coinjective.

A left comodule $\mathcal{M}$ over $\mathcal{C}$ is called *coprojective relative to A* ($\mathcal{C}/A$-coprojective) if the functor of cohomomorphisms from $\mathcal{M}$ maps exact triples of A-injective $\mathcal{C}$-contramodules to exact triples. A left contramodule $\mathfrak{P}$ over $\mathcal{C}$ is called *coinjective relative to A* ($\mathcal{C}/A$-coinjective) if the functor of cohomomorphisms into $\mathfrak{P}$ maps exact triples of A-projective $\mathcal{C}$-comodules to exact triples. Any coinduced $\mathcal{C}$-comodule is $\mathcal{C}/A$-coprojective and any induced $\mathcal{C}$-contramodule is $\mathcal{C}/A$-coinjective.

For any right $\mathcal{C}$-comodule $\mathcal{N}$ and any left $\mathcal{C}$-comodule $\mathcal{M}$ there is a natural isomorphism

$$\mathrm{Hom}_k(\mathcal{N} \,\square_{\mathcal{C}}\, \mathcal{M}, \ k^\vee) \simeq \mathrm{Cohom}_{\mathcal{C}}(\mathcal{M}, \mathrm{Hom}_k(\mathcal{N}, k^\vee)).$$

Therefore, any coprojective $\mathcal{C}$-comodule $\mathcal{M}$ is coflat and any $\mathcal{C}/A$-coprojective $\mathcal{C}$-comodule $\mathcal{M}$ is $\mathcal{C}/A$-coflat. Besides, a right $\mathcal{C}$-comodule $\mathcal{N}$ is coflat if and only if the left $\mathcal{C}$-contramodule $\mathrm{Hom}_k(\mathcal{N}, k^\vee)$ is coinjective; if a right $\mathcal{C}$-comodule $\mathcal{N}$ is $\mathcal{C}/A$-coflat, then the left $\mathcal{C}$-contramodule $\mathrm{Hom}_k(\mathcal{N}, k^\vee)$ is $\mathcal{C}/A$-coinjective (and the converse can be deduced from Lemma 3.1.3(a) and the proof of the lemma below in the assumptions of 3.1.3).

It appears that the notion of a relatively coprojective left $\mathcal{C}$-comodule is useful when $\mathcal{C}$ is a flat right A-module, and the notion of a relatively coinjective left $\mathcal{C}$-contramodule is useful when $\mathcal{C}$ is a projective left A-module.

Lemma.

(a) *Assume that $\mathcal{C}$ is a flat right A-module. Then the class of $\mathcal{C}/A$-coprojective left $\mathcal{C}$-comodules is closed under extensions and cokernels of injective morphisms. The functor of cohomomorphisms into an A-injective left $\mathcal{C}$-contramodule maps exact triples of $\mathcal{C}/A$-coprojective left $\mathcal{C}$-comodules to exact triples.*

(b) *Assume that $\mathcal{C}$ is a projective left A-module. Then the class of $\mathcal{C}/A$-coinjective left $\mathcal{C}$-contramodules is closed under extensions and kernels of surjective* *morphisms. The functor of cohomomorphisms from an A-projective left C-co-module maps exact triples of C/A-coinjective left C-contramodules to exact triples.*

Proof. Part (a): these results follow from the standard properties of the left derived functor of the right exact functor of cohomomorphisms on the Cartesian product of the abelian category of left C-comodules and the exact category of A-injective left C-contramodules. One can define the k-modules $\operatorname{Coext}^i_{\mathfrak{C}}(\mathfrak{M}, \mathfrak{P})$, $i = 0, -1, \ldots$ as the cohomology of the bar complex

$$\cdots \longrightarrow \operatorname{Hom}_A(\mathfrak{C} \otimes_A \mathfrak{C} \otimes_A \mathfrak{M}, \mathfrak{P}) \longrightarrow \operatorname{Hom}_A(\mathfrak{C} \otimes_A \mathfrak{M}, \mathfrak{P}) \longrightarrow \operatorname{Hom}_A(\mathfrak{M}, \mathfrak{P})$$

for any left C-comodule $\mathfrak{M}$ and any A-injective left C-contramodule $\mathfrak{P}$. Then $\operatorname{Coext}^0_{\mathfrak{C}}(\mathfrak{M}, \mathfrak{P}) \simeq \operatorname{Cohom}_{\mathfrak{C}}(\mathfrak{M}, \mathfrak{P})$ and there are long exact sequences of $\operatorname{Coext}^*_{\mathfrak{C}}$ associated with exact triples of comodules and contramodules. Now a left C-co-module $\mathfrak{M}$ is C/A-coprojective if and only if $\operatorname{Coext}^i_{\mathfrak{C}}(\mathfrak{M}, \mathfrak{P}) = 0$ for any A-injective left C-contramodule $\mathfrak{P}$ and all $i < 0$. Indeed, the "if" assertion follows from the homological exact sequence, and "only if" holds since the bar complex is isomorphic to the complex of cohomomorphisms from the C-comodule $\mathfrak{M}$ into the bar resolution

$$\cdots \longrightarrow \operatorname{Hom}_A(\mathfrak{C}, \operatorname{Hom}_A(\mathfrak{C}, \mathfrak{P})) \longrightarrow \operatorname{Hom}_A(\mathfrak{C}, \mathfrak{P})$$

of the C-contramodule $\mathfrak{P}$, which is a complex of A-injective C-contramodules, exact except in degree 0 and split over A. The proof of part (b) is completely analogous; it uses the left derived functor of the functor of cohomomorphisms on the Cartesian product of the exact category of A-projective left C-comodules and the abelian category of left C-contramodules.   □

Remark. It follows from Lemma 5.2 that any extension of an A-projective C-co-module by a coprojective C-comodule splits, and any extension of a coinjective C-contramodule by an A-injective C-contramodule splits. The analogues of the results of Remark 1.2.2 also hold for (relatively) coprojective comodules and coin-jective contramodules in the assumptions of 3.1.3; see the proof of Lemma 5.3.2 for details.

Question. Are all relatively coflat C-comodules relatively coprojective? Are all A-projective coflat C-comodules coprojective?

3.2.3 Let C be an arbitrary coring. Let us call a left C-comodule $\mathfrak{M}$ *quasicoprojec-tive* if the functor of cohomomorphisms from $\mathfrak{M}$ is left exact on the category of left C-contramodules, i.e., this functor preserves kernels. Any coinduced C-comodule is quasicoprojective. Any quasicoprojective comodule is quasicoflat. Let us call a left C-contramodule $\mathfrak{P}$ *quasicoinjective* if the functor of cohomomorphisms into $\mathfrak{P}$ is left exact on the category of left C-comodules, i.e., this functor maps cokernels to kernels. Any induced C-contramodule is quasicoinjective. (Cf. Lemma 5.2.)

Proposition 1. *Let $\mathcal{M}$ be a left $\mathcal{C}$-comodule, $\mathcal{K}$ be a right $\mathcal{C}$-comodule endowed with a left action of a k-algebra B by comodule endomorphisms, and P be a left B-module. Then there is a natural k-module map*

$$\mathrm{Cohom}_{\mathcal{C}}(\mathcal{M}, \mathrm{Hom}_B(\mathcal{K}, P)) \longrightarrow \mathrm{Hom}_B(\mathcal{K} \,\square_{\mathcal{C}}\, \mathcal{M}, \, P),$$

which is an isomorphism, at least, in the following cases:

(a) *P is an injective left B-module;*

(b) *$\mathcal{M}$ is a quasicoprojective left $\mathcal{C}$-comodule;*

(c) *$\mathcal{C}$ is a projective left A-module, $\mathcal{M}$ is a projective left A-module, $\mathcal{K}$ is a $\mathcal{C}/A$-coflat right $\mathcal{C}$-comodule, $\mathcal{K}$ is a projective left B-module, and the ring B has a finite left homological dimension;*

(d) *$\mathcal{K}$ as a right $\mathcal{C}$-comodule with a left B-module structure is coinduced from a B-A-bimodule.*

Besides, in the case (c) the left B-module $\mathcal{K} \,\square_{\mathcal{C}}\, \mathcal{M}$ is projective.

Proof. The map $\mathrm{Hom}_B(\mathcal{K} \otimes_A \mathcal{M}, \, P) \longrightarrow \mathrm{Hom}_B(\mathcal{K} \,\square_{\mathcal{C}}\, \mathcal{M}, \, P)$ annihilates the difference of two maps $\mathrm{Hom}_B(\mathcal{K} \otimes_A \mathcal{C} \otimes_A \mathcal{M}, \, P) \rightrightarrows \mathrm{Hom}_B(\mathcal{K} \otimes_A \mathcal{M}, \, P)$ and this pair of maps can be identified with the pair of maps $\mathrm{Hom}_A(\mathcal{C} \otimes_A \mathcal{M}, \mathrm{Hom}_B(\mathcal{K}, P)) \rightrightarrows \mathrm{Hom}_A(\mathcal{M}, \mathrm{Hom}_B(\mathcal{K}, P))$ whose cokernel is, by the definition, the cohomomorphism module $\mathrm{Cohom}_{\mathcal{C}}(\mathcal{M}, \mathrm{Hom}_B(\mathcal{K}, P))$. Hence there is a natural map $\mathrm{Cohom}_{\mathcal{C}}(\mathcal{M}, \mathrm{Hom}_B(\mathcal{K}, P)) \longrightarrow \mathrm{Hom}_B(\mathcal{K} \,\square_{\mathcal{C}}\, \mathcal{M}, \, P)$. The case (a) is obvious. In the case (b), it suffices to present P as the kernel of a map of injective B-modules. The rest of the proof is completely analogous to the proof of Proposition 1.2.3 (with flat modules replaced by projective ones and the left and right sides switched). $\square$

Proposition 2. *Let $\mathfrak{P}$ be a left $\mathcal{C}$-contramodule, $\mathcal{K}$ be a left $\mathcal{C}$-comodule endowed with a right action of a k-algebra B by comodule endomorphisms, and M be a left B-module. Then there is a natural k-module map*

$$\mathrm{Cohom}_{\mathcal{C}}(\mathcal{K} \otimes_B M, \, \mathfrak{P}) \longrightarrow \mathrm{Hom}_B(M, \mathrm{Cohom}_{\mathcal{C}}(\mathcal{K}, \mathfrak{P})),$$

which is an isomorphism, at least, in the following cases:

(a) *M is a projective left B-module;*

(b) *$\mathfrak{P}$ is a quasicoinjective left $\mathcal{C}$-contramodule;*

(c) *$\mathcal{C}$ is a flat right A-module, $\mathfrak{P}$ is an injective left A-module, $\mathcal{K}$ is a $\mathcal{C}/A$-coprojective left $\mathcal{C}$-comodule, $\mathcal{K}$ is a flat right B-module, and the ring B has a finite left homological dimension;*

(d) *$\mathcal{K}$ as a left $\mathcal{C}$-comodule with a right B-module structure is coinduced from an A-B-bimodule.*

Besides, in the case (c) the left B-module $\mathrm{Cohom}_{\mathcal{C}}(\mathcal{K}, \mathfrak{P})$ is injective.

Proof. The map $\mathrm{Hom}_B(M, \mathrm{Hom}_A(\mathfrak{K}, \mathfrak{P})) \longrightarrow \mathrm{Hom}_B(M, \mathrm{Cohom}_\mathbb{C}(\mathfrak{K}, \mathfrak{P}))$ annihilates the difference of two maps

$$\mathrm{Hom}_B(M, \ \mathrm{Hom}_A(\mathbb{C} \otimes_A \mathfrak{K}, \ \mathfrak{P})) \rightrightarrows \mathrm{Hom}_B(M, \mathrm{Hom}_A(\mathfrak{K}, \mathfrak{P}))$$

and this pair of maps can be identified with the pair of maps $\mathrm{Hom}_A(\mathbb{C} \otimes_A \mathfrak{K} \otimes_B M,$ $\mathfrak{P}) \rightrightarrows \mathrm{Hom}_A(\mathfrak{K} \otimes_B M, \ \mathfrak{P})$ whose cokernel is, by the definition, the cohomomorphism module $\mathrm{Cohom}_\mathbb{C}(\mathfrak{K} \otimes_B M, \ \mathfrak{P})$. Hence there is a natural map $\mathrm{Cohom}_\mathbb{C}(\mathfrak{K} \otimes_B M, \ \mathfrak{P}) \longrightarrow \mathrm{Hom}_B(M, \mathrm{Cohom}_\mathbb{C}(\mathfrak{K}, \mathfrak{P}))$. The case (a) is obvious. In the case (b), it suffices to present M as the cokernel of a map of projective B-modules. To prove (c) and (d), consider the bar complex

$$\cdots \longrightarrow \mathrm{Hom}_A(\mathbb{C} \otimes_A \mathbb{C} \otimes_A \mathfrak{K}, \ \mathfrak{P}) \longrightarrow \mathrm{Hom}_A(\mathbb{C} \otimes_A \mathfrak{K}, \ \mathfrak{P})$$
$$\longrightarrow \mathrm{Hom}_A(\mathfrak{K}, \mathfrak{P}) \longrightarrow \mathrm{Cohom}_\mathbb{C}(\mathfrak{K}, \mathfrak{P}). \quad (3.1)$$

In the case (c) this complex is exact, since it is the complex of cohomomorphisms from a $\mathbb{C}/A$-coprojective $\mathbb{C}$-comodule $\mathfrak{K}$ into an A-split exact complex of A-injective $\mathbb{C}$-contramodules $\cdots \longrightarrow \mathrm{Hom}_A(\mathbb{C} \otimes_A \mathbb{C}, \ \mathfrak{P}) \longrightarrow \mathrm{Hom}_A(\mathbb{C}, \mathfrak{P}) \longrightarrow \mathfrak{P}$. Since all the terms of the complex (3.1), except possibly the rightmost one, are injective left B-modules and the left homological dimension of the ring B is finite, the rightmost term $\mathrm{Cohom}_\mathbb{C}(\mathfrak{K}, \mathfrak{P})$ is also an injective B-module, the complex of left B-modules (3.1) is contractible, and the complex of B-module homomorphisms from the left B-module M into (3.1) is exact. In the case (d), the complex (3.1) is also a split exact complex of left B-modules. $\quad\square$

3.2.4 Let $\mathbb{C}$ be a coring over a k-algebra A and $\mathcal{D}$ be a coring over a k-algebra B. Assume that $\mathcal{D}$ is a projective left B-module. Let $\mathfrak{K}$ be a $\mathbb{C}$-$\mathcal{D}$-bicomodule and $\mathfrak{P}$ be a left $\mathbb{C}$-contramodule. Then the module of cohomomorphisms $\mathrm{Cohom}_\mathbb{C}(\mathfrak{K}, \mathfrak{P})$ is endowed with a left $\mathcal{D}$-contramodule structure as the cokernel of a pair of contramodule morphisms $\mathrm{Hom}_A(\mathbb{C} \otimes_A \mathfrak{K}, \ \mathfrak{P}) \rightrightarrows \mathrm{Hom}_A(\mathfrak{K}, \mathfrak{P})$.

More generally, let $\mathbb{C}$ and $\mathcal{D}$ be arbitrary corings. Assume that the functor of homomorphisms from $\mathcal{D}$ over B preserves the cokernel of the pair of maps $\mathrm{Hom}_A(\mathbb{C} \otimes_A \mathfrak{K}, \ \mathfrak{P}) \rightrightarrows \mathrm{Hom}_A(\mathfrak{K}, \mathfrak{P})$, that is the natural map $\mathrm{Cohom}_\mathbb{C}(\mathfrak{K} \otimes_B \mathcal{D}, \ \mathfrak{P}) \longrightarrow \mathrm{Hom}_B(\mathcal{D}, \mathrm{Cohom}_\mathbb{C}(\mathfrak{K}, \mathfrak{P}))$ is an isomorphism. Then one can define a left contraaction map $\mathrm{Hom}_B(\mathcal{D}, \mathrm{Cohom}_\mathbb{C}(\mathfrak{K}, \mathfrak{P})) \longrightarrow \mathrm{Cohom}_\mathbb{C}(\mathfrak{K}, \mathfrak{P})$ by taking the cohomomorphisms over $\mathbb{C}$ from the right $\mathcal{D}$-coaction map $\mathfrak{K} \longrightarrow \mathfrak{K} \otimes_B \mathcal{D}$ into the contramodule $\mathfrak{P}$. This contraaction is counital and contraassociative, at least, if the natural map $\mathrm{Cohom}_\mathbb{C}(\mathfrak{K} \otimes_B \mathcal{D} \otimes_B \mathcal{D}, \ \mathfrak{P}) \longrightarrow \mathrm{Hom}_B(\mathcal{D} \otimes_B \mathcal{D}, \mathrm{Cohom}_\mathbb{C}(\mathfrak{K}, \mathfrak{P}))$ is also an isomorphism.

In particular, if one of the conditions of Proposition 3.2.3.2 is satisfied (for $M = \mathcal{D}$), then the left B-module $\mathrm{Cohom}_\mathbb{C}(\mathfrak{K}, \mathfrak{P})$ has a natural $\mathcal{D}$-contramodule structure.

3.2.5 Let $\mathbb{C}$ be a coring over a k-algebra A and $\mathcal{D}$ be a coring over a k-algebra B.

Proposition. *Let $\mathcal{M}$ be a left $\mathcal{D}$-comodule, $\mathcal{K}$ be a $\mathcal{C}$-$\mathcal{D}$-bicomodule, and $\mathfrak{P}$ be a left $\mathcal{C}$-contramodule. Then the iterated cohomomorphism modules*

$$\mathrm{Cohom}_{\mathcal{C}}(\mathcal{K} \,\square_{\mathcal{D}}\, \mathcal{M},\, \mathfrak{P}) \quad and \quad \mathrm{Cohom}_{\mathcal{D}}(\mathcal{M}, \mathrm{Cohom}_{\mathcal{C}}(\mathcal{K}, \mathfrak{P}))$$

are naturally isomorphic, at least, in the following cases:

(a) $\mathcal{D}$ *is a projective left B-module, $\mathcal{M}$ is a projective left B-module, $\mathcal{C}$ is a flat right, and $\mathfrak{P}$ is an injective left A-module;*

(b) $\mathcal{D}$ *is a projective left B-module and $\mathcal{M}$ is a coprojective left $\mathcal{D}$-comodule;*

(c) $\mathcal{C}$ *is a flat right A-module and $\mathfrak{P}$ is a coinjective left $\mathcal{C}$-contramodule;*

(d) $\mathcal{D}$ *is a projective left B-module, $\mathcal{M}$ is a projective left B-module, $\mathcal{K}$ is a $\mathcal{D}/B$-coflat right $\mathcal{D}$-comodule, $\mathcal{K}$ is a projective left A-module, and the ring A has a finite left homological dimension;*

(e) $\mathcal{C}$ *is a flat right A-module, $\mathfrak{P}$ is an injective left A-module, $\mathcal{K}$ is a $\mathcal{C}/A$-coprojective left $\mathcal{C}$-comodule, $\mathcal{K}$ is a flat right B-module, and the ring B has a finite left homological dimension;*

(f) $\mathcal{D}$ *is a projective left B-module, $\mathcal{M}$ is a projective left B-module, and $\mathcal{K}$ as a right $\mathcal{D}$-comodule with a left A-module structure is coinduced from an A-B-bimodule;*

(g) $\mathcal{C}$ *is a flat right A-module, $\mathfrak{P}$ is an injective left A-module, and $\mathcal{K}$ as a left $\mathcal{C}$-comodule with a right B-module structure is coinduced from an A-B-bimodule;*

(h) $\mathcal{M}$ *is a quasicoprojective left $\mathcal{D}$-comodule and $\mathcal{K}$ as a left $\mathcal{C}$-comodule with a right B-module structure is coinduced from an A-B-bimodule;*

(i) $\mathfrak{P}$ *is a quasicoinjective left $\mathcal{C}$-contramodule and $\mathcal{K}$ as a right $\mathcal{D}$-comodule with a left A-module structure is coinduced from an A-B-bimodule;*

(j) $\mathcal{K}$ *as a left $\mathcal{C}$-comodule with a right B-module structure is coinduced from an A-B-bimodule and $\mathcal{K}$ as a right $\mathcal{D}$-comodule with a left A-module structure is coinduced from an A-B-bimodule.*

More precisely, in all cases in this list the natural maps from the k-module $\mathrm{Hom}_A(\mathcal{K} \otimes_B \mathcal{M},\, \mathfrak{P}) = \mathrm{Hom}_B(\mathcal{M}, \mathrm{Hom}_A(\mathcal{K}, \mathfrak{P}))$ into both iterated cohomomorphism modules under consideration are surjective, their kernels coincide and are equal to the sum of the kernels of two maps from this module onto its quotient modules $\mathrm{Cohom}_{\mathcal{C}}(\mathcal{K} \otimes_B \mathcal{M},\, \mathfrak{P})$ and $\mathrm{Cohom}_{\mathcal{D}}(\mathcal{M}, \mathrm{Hom}_A(\mathcal{K}, \mathfrak{P}))$.

Proof. One can easily see that whenever both maps

$$\mathrm{Cohom}_{\mathcal{D}}(\mathcal{M}, \mathrm{Hom}_A(\mathcal{K}, \mathfrak{P})) \longrightarrow \mathrm{Hom}_A(\mathcal{K} \,\square_{\mathcal{D}}\, \mathcal{M},\, \mathfrak{P})$$

and

$$\mathrm{Cohom}_{\mathcal{D}}(\mathcal{M}, \mathrm{Hom}_A(\mathcal{K}, \mathrm{Hom}_A(\mathcal{C}, \mathfrak{P}))) \longrightarrow \mathrm{Hom}_A(\mathcal{K} \,\square_{\mathcal{D}}\, \mathcal{M}, \mathrm{Hom}_A(\mathcal{C}, \mathfrak{P}))$$

are isomorphisms, the natural map $\mathrm{Hom}_A(\mathcal{K} \otimes_B \mathcal{M}, \mathfrak{P}) \longrightarrow \mathrm{Cohom}_\mathcal{C}(\mathcal{K} \square_\mathcal{D} \mathcal{M}, \mathfrak{P})$ is surjective and its kernel coincides with the desired sum of two kernels of maps from $\mathrm{Hom}_A(\mathcal{K} \otimes_B \mathcal{M}, \mathfrak{P})$ onto its quotient modules. Analogously, whenever both maps $\mathrm{Cohom}_\mathcal{C}(\mathcal{K} \otimes_B \mathcal{M}, \mathfrak{P}) \longrightarrow \mathrm{Hom}_B(\mathcal{M}, \mathrm{Cohom}_\mathcal{C}(\mathcal{K}, \mathfrak{P}))$ and $\mathrm{Cohom}_\mathcal{C}(\mathcal{K} \otimes_B \mathcal{D} \otimes_B \mathcal{M}, \mathfrak{P}) \longrightarrow \mathrm{Hom}_B(\mathcal{D} \otimes_B \mathcal{M}, \mathrm{Cohom}_\mathcal{C}(\mathcal{K}, \mathfrak{P}))$ are isomorphisms, the natural map $\mathrm{Hom}_B(\mathcal{M}, \mathrm{Hom}_A(\mathcal{K}, \mathfrak{P})) \longrightarrow \mathrm{Cohom}_\mathcal{D}(\mathcal{M}, \mathrm{Cohom}_\mathcal{C}(\mathcal{K}, \mathfrak{P}))$ is surjective and it kernel coincides with the desired sum of two kernels in $\mathrm{Hom}_B(\mathcal{M}, \mathrm{Hom}_A(\mathcal{K}, \mathfrak{P}))$. Thus it remains to apply Propositions 3.2.3.1 and 3.2.3.2. $\qquad\square$

Commutativity of pentagonal diagrams of associativity isomorphisms between iterated cohomomorphism modules can be established in a way analogous to the case of iterated cotensor products. Namely, each of the five iterated cohomomorphism modules $\mathrm{Cohom}_\mathcal{C}((\mathcal{K}\square_\mathcal{E}\mathcal{L})\square_\mathcal{D}\mathcal{M}, \mathfrak{P})$, $\mathrm{Cohom}_\mathcal{C}(\mathcal{K}\square_\mathcal{E}(\mathcal{L}\square_\mathcal{D}\mathcal{M}), \mathfrak{P})$, $\mathrm{Cohom}_\mathcal{E}(\mathcal{L}\square_\mathcal{D}\mathcal{M}, \mathrm{Cohom}_\mathcal{C}(\mathcal{K}, \mathfrak{P}))$, $\mathrm{Cohom}_\mathcal{D}(\mathcal{M}, \mathrm{Cohom}_\mathcal{E}(\mathcal{L}, \mathrm{Cohom}_\mathcal{C}(\mathcal{K}, \mathfrak{P})))$, and $\mathrm{Cohom}_\mathcal{D}(\mathcal{M}, \mathrm{Cohom}_\mathcal{C}(\mathcal{K}\square_\mathcal{E}\mathcal{L}, \mathfrak{P}))$ is endowed with a natural map into it from the homomorphism module $\mathrm{Hom}_A(\mathcal{K}\otimes_F\mathcal{L}\otimes_B\mathcal{M}, \mathfrak{P})$, and since the associativity isomorphisms are, presumably, compatible with these maps, it suffices to check that at least one of these five maps is surjective in order to show that the pentagonal diagram commutes. In particular, if Proposition 3.2.5 together with Proposition 1.2.5 provide all the five isomorphisms constituting the pentagonal diagram and either $\mathcal{M}$ is a projective left B-module, or $\mathfrak{P}$ is an injective left A-module, or both $\mathcal{K}$ and $\mathcal{L}$ as left (right) comodules with right (left) module structures are coinduced from bimodules, then the pentagonal diagram is commutative.

We will say that multiple cohomomorphisms between several bicomodules and a contramodule $\mathrm{Cohom}_\mathcal{C}(\mathcal{K} \square_\mathcal{E} \cdots \square_\mathcal{D} \mathcal{M}, \mathfrak{P})$ are associative if the multiple cotensor product $\mathcal{K}\square_\mathcal{E}\cdots\square_\mathcal{D}\mathcal{M}$ is associative and for any possible way of representing this multiple cohomomorphism module in terms of iterated cotensor product and cohomomorphism operations all the intermediate cohomomorphism modules can be endowed with contramodule structures via the construction of 3.2.4, all possible associativity isomorphisms between iterated cohomomorphism modules exist in the sense of the last assertion of Proposition 3.2.5 and preserve contramodule structures, and all the pentagonal diagrams commute. Associativity isomorphisms and contramodule structures on associative multiple cohomomorphisms are preserved by the morphisms between them induced by any bicomodule and contramodule morphisms of the factors.

3.3 Semicontramodules

3.3.1 Depending on the (co)flatness, (co)projectivity, and/or (co)injectivity conditions imposed, there are several ways to make the category opposite to a category of left $\mathcal{C}$-contramodules into a right module category over a tensor category of $\mathcal{C}$-$\mathcal{C}$-bicomodules with respect to the functor $\mathrm{Cohom}_\mathcal{C}$. Moreover, a category of left $\mathcal{C}$-comodules typically can be made into a left module category over the same tensor category, so that the functor $\mathrm{Cohom}_{\mathcal{C}}$ would provide also a pairing between these left and right module categories taking values in the category $k\text{-mod}^{\mathrm{op}}$.

It follows from Proposition 3.2.5(b) that whenever $\mathcal{C}$ is a projective left A-module, the category opposite to the category of left $\mathcal{C}$-contramodules is a right module category over the tensor category of $\mathcal{C}$-$\mathcal{C}$-bicomodules that are coprojective left $\mathcal{C}$-comodules; the category of coprojective left $\mathcal{C}$-comodules is a left module category over this tensor category. If follows from Proposition 3.2.5(c) that whenever $\mathcal{C}$ is a flat right A-module, the category opposite to the category of coinjective left $\mathcal{C}$-contramodules is a right module category over the tensor category of $\mathcal{C}$-$\mathcal{C}$-bico-modules that are coflat right $\mathcal{C}$-comodules; the category of left $\mathcal{C}$-comodules is a left module category over this tensor category. It follows from Proposition 3.2.5(d) that whenever $\mathcal{C}$ is a projective left A-module and the ring A has a finite left homological dimension, the category opposite to the category of left $\mathcal{C}$-contramodules is a right module category over the tensor category of $\mathcal{C}$-$\mathcal{C}$-bicomodules that are projective left A-modules and $\mathcal{C}/A$-coflat right $\mathcal{C}$-comodules; the category of A-projective left $\mathcal{C}$-comodules is a left module category over this tensor category. It follows from Proposition 3.2.5(e) that whenever $\mathcal{C}$ is a flat right A-module and the ring A has a finite left homological dimension, the category opposite to the category of A-in-jective left $\mathcal{C}$-contramodules is a right module category over the tensor category of $\mathcal{C}$-$\mathcal{C}$-bicomodules that are flat right A-modules and $\mathcal{C}/A$-coprojective left $\mathcal{C}$-comod-ules; the category of left $\mathcal{C}$-comodules is a left module category over this tensor category. Finally, it follows from Proposition 3.2.5(a) that whenever the ring A is semisimple, the category opposite to the category of left $\mathcal{C}$-contramodules is a right module category over the tensor category of $\mathcal{C}$-$\mathcal{C}$-bicomodules; the category of left $\mathcal{C}$-comodules is a left module category over this tensor category. In each case, there is a pairing between these left and right module categories compatible with their module category structures and taking values in the category opposite to the category of k-modules.

A *left semicontramodule* over a semialgebra $\mathbf{S}$ is an object of the category opposite to the category of module objects in one of the right module categories of the above kind (opposite to a category of left $\mathcal{C}$-contramodules) over the ring object $\mathbf{S}$ in the corresponding tensor category of $\mathcal{C}$-$\mathcal{C}$-bicomodules. In other words, a left $\mathbf{S}$-semicontramodule $\mathfrak{P}$ is a left $\mathcal{C}$-contramodule endowed with a left $\mathcal{C}$-con-tramodule morphism of *left semicontraaction* $\mathfrak{P} \longrightarrow \mathrm{Cohom}_{\mathcal{C}}(\mathbf{S}, \mathfrak{P})$ satisfying the associativity and unity equations. Namely, two compositions

$$\mathfrak{P} \longrightarrow \mathrm{Cohom}_{\mathcal{C}}(\mathbf{S}, \mathfrak{P}) \rightrightarrows \mathrm{Cohom}_{\mathcal{C}}(\mathbf{S} \,\square_{\mathcal{C}}\, \mathbf{S}, \mathfrak{P})$$

of the semicontraaction morphism $\mathfrak{P} \longrightarrow \mathrm{Cohom}_{\mathcal{C}}(\mathbf{S}, \mathfrak{P})$ with the morphisms

$$\mathrm{Cohom}_{\mathcal{C}}(\mathbf{S}, \mathfrak{P}) \rightrightarrows \mathrm{Cohom}_{\mathcal{C}}(\mathbf{S} \,\square_{\mathcal{C}}\, \mathbf{S}, \mathfrak{P}) = \mathrm{Cohom}_{\mathcal{C}}(\mathbf{S}, \mathrm{Cohom}_{\mathcal{C}}(\mathbf{S}, \mathfrak{P}))$$

induced by the semimultiplication morphism of $\mathbf{S}$ and the semicontraaction mor-phism should coincide with each other, and the composition

$$\mathfrak{P} \longrightarrow \mathrm{Cohom}_{\mathcal{C}}(\mathbf{S}, \mathfrak{P}) \longrightarrow \mathrm{Cohom}_{\mathcal{C}}(\mathcal{C}, \mathfrak{P}) = \mathfrak{P}$$

of the semicontraaction morphism with the morphism induced by the semiunit morphism of S should coincide with the identity morphism of $\mathfrak{P}$. For this definition to make sense, (co)flatness, (co)projectivity, and/or (co)injectivity conditions imposed on S and/or $\mathfrak{P}$ must guarantee associativity of multiple cohomomorphism modules of the form $\mathrm{Cohom}_{\mathcal{C}}(S \,\square_{\mathcal{C}} \cdots \square_{\mathcal{C}} S,\, \mathfrak{P})$. *Right semicontramodules* over S are defined in the analogous way.

If $\mathfrak{Q}$ is a left $\mathcal{C}$-contramodule for which multiple cohomomorphisms

$$\mathrm{Cohom}_{\mathcal{C}}(S \,\square_{\mathcal{C}} \cdots \square_{\mathcal{C}} S,\, \mathfrak{Q})$$

are associative, then there is a natural left S-semicontramodule structure on the cohomomorphism module $\mathrm{Cohom}_{\mathcal{C}}(S, \mathfrak{Q})$. The semicontramodule $\mathrm{Cohom}_{\mathcal{C}}(S, \mathfrak{Q})$ is called the S-semicontramodule *coinduced* from a $\mathcal{C}$-contramodule $\mathfrak{Q}$. According to Lemma 1.1.2, the k-module of semicontramodule homomorphisms from an arbitrary S-semicontramodule into the coinduced S-semicontramodule is described by the formula

$$\mathrm{Hom}^{S}(\mathfrak{P}, \mathrm{Cohom}_{\mathcal{C}}(S, \mathfrak{Q})) \simeq \mathrm{Hom}^{\mathcal{C}}(\mathfrak{P}, \mathfrak{Q}).$$

We will denote the category of left S-semicontramodules by S-sicntr and the category of right S-semicontramodules by sicntr-S. This notation presumes that one can speak of (left or right) S-semicontramodules with no (co)injectivity conditions imposed on them. If $\mathcal{C}$ is a projective left A-module and S is a coprojective left $\mathcal{C}$-comodule, then the category of left semicontramodules over S is abelian and the forgetful functor S-sicntr $\longrightarrow$ $\mathcal{C}$-contra is exact.

If $\mathcal{C}$ is a projective left A-module and either S is a coprojective left $\mathcal{C}$-comodule, or S is a projective left A-module and a $\mathcal{C}/A$-coflat right $\mathcal{C}$-comodule and A has a finite left homological dimension, or A is semisimple, then both infinite direct sums and infinite products exist in the category of left S-semicontramodules and both are preserved by the forgetful functor S-sicntr $\longrightarrow$ $\mathcal{C}$-contra, even though only infinite products are preserved by the full forgetful functor S-sicntr $\longrightarrow$ A-mod.

If $\mathcal{C}$ is a flat right A-module, S is a flat right A-module and a $\mathcal{C}/A$-coprojective left $\mathcal{C}$-comodule, and A has a finite left homological dimension, then the category of A-injective left S-semicontramodules is exact. If $\mathcal{C}$ is a projective left A-module, S is a projective left A-module and a $\mathcal{C}/A$-coflat right $\mathcal{C}$-comodule, and A has a finite left homological dimension, then the category of $\mathcal{C}/A$-coinjective left S-semicontramodules is exact. If $\mathcal{C}$ is a flat right A-module and S is a coflat right $\mathcal{C}$-comodule, then the category of $\mathcal{C}$-coinjective left S-semicontramodules is exact. If A is semisimple, the category of $\mathcal{C}$-coinjective S-semicontramodules is exact. Infinite products exist in all of these exact categories, and the forgetful functors preserve them.

Question. When $\mathcal{C}$ is a flat right A-module and S is a coflat right $\mathcal{C}$-comodule, a right adjoint functor to the forgetful functor S-simod $\longrightarrow$ $\mathcal{C}$-comod exists according to the abstract adjoint functor existence theorem [67, V.8]. Indeed, the forgetful functor preserves colimits and the category of left S-semimodules has a set of generators (since the category of left $\mathcal{C}$-comodules does; see Question 3.1.2). Does a left adjoint functor to the forgetful functor $\mathbf{S}$-sicntr $\longrightarrow$ $\mathcal{C}$-contra exist? Can one describe these functors more explicitly?

3.3.2   Assume that the coring $\mathcal{C}$ is a projective left and a flat right A-module and the ring A has a finite left homological dimension.

Lemma.

(a) *If the semialgebra $\mathbf{S}$ is a coflat right $\mathcal{C}$-comodule and a projective left A-module, then there exists a (not always additive) functor assigning to any left $\mathbf{S}$-semimodule a surjective map onto it from an A-projective $\mathbf{S}$-semimodule.*

(b) *If the semialgebra $\mathbf{S}$ is a coprojective left $\mathcal{C}$-comodule and a flat right A-module, then there exists a (not always additive) functor assigning to any left $\mathbf{S}$-semicontramodule an injective map from it into an A-injective $\mathbf{S}$-semicontramodule.*

Proof. The proof of part (a) is completely analogous to the proof of Lemma 1.3.2 (with the last assertion of Proposition 3.2.3.1 used as needed); and part (b) is proven in the following way. Let $\mathfrak{P} \longrightarrow \mathfrak{J}(\mathfrak{P})$ denote the functorial injective morphism from a $\mathcal{C}$-contramodule $\mathfrak{P}$ into an A-injective $\mathcal{C}$-contramodule $\mathfrak{J}(\mathfrak{P})$ constructed in Lemma 3.1.3. Then for any $\mathbf{S}$-semicontramodule $\mathfrak{P}$ the composition of maps

$$\mathfrak{P} \longrightarrow \mathrm{Cohom}_{\mathcal{C}}(\mathbf{S}, \mathfrak{P}) \longrightarrow \mathrm{Cohom}_{\mathcal{C}}(\mathbf{S}, \mathfrak{J}(\mathfrak{P}))$$

provides the desired injective morphism of $\mathbf{S}$-semicontramodules. According to the last assertion of Proposition 3.2.3.2, the A-module $\mathfrak{J}(\mathfrak{P}) = \mathrm{Cohom}_{\mathcal{C}}(\mathbf{S}, \mathfrak{J}(\mathfrak{P}))$ is injective.   ☐

Remark. The analogues of the result of Remark 1.3.2 hold for $\mathcal{C}/A$-coprojective/semiprojective $\mathbf{S}$-semimodules and $\mathcal{C}/A$-coinjective/semiinjective $\mathbf{S}$-semicontramodules; see the proof of Lemma 9.2.1 for details.

3.3.3   Let $\mathbf{S}$ be a semialgebra over a coring $\mathcal{C}$ over a k-algebra A.

Lemma.

(a) *Assume that $\mathcal{C}$ is a projective left A-module, $\mathbf{S}$ is a coprojective left $\mathcal{C}$-comodule and a $\mathcal{C}/A$-coflat right $\mathcal{C}$-comodule, and the ring A has a finite left homological dimension. Then there exist*

   - *an exact functor assigning to any A-projective left $\mathbf{S}$-semimodule an A-split injective morphism from it into a $\mathcal{C}$-coprojective $\mathbf{S}$-semimodule, and*

   - *an exact functor assigning to any left $\mathbf{S}$-semicontramodule a surjective morphism onto it from a $\mathcal{C}/A$-coinjective $\mathbf{S}$-semicontramodule.*

(b) *Assume that $\mathcal{C}$ is a flat right A-module, $\mathbf{S}$ is a coflat right $\mathcal{C}$-comodule and a $\mathcal{C}/A$-coprojective left $\mathcal{C}$-comodule, and the ring A has a finite left homological dimension. Then there exist*

- *an exact functor assigning to any A-injective left $\mathbf{S}$-semicontramodule an A-split surjective morphism onto it from a $\mathcal{C}$-coinjective $\mathbf{S}$-semicontramodule, and*

- *an exact functor assigning to any left $\mathbf{S}$-semimodule an injective morphism from it into a $\mathcal{C}/A$-coprojective $\mathbf{S}$-semimodule.*

(c) *When both the assumptions of (a) and (b) are satisfied, the two functors acting in categories of semimodules (can be made to) agree and the two functors acting in categories of semicontramodules (can be made to) agree.*

Proof. The proof of the first assertion of part (a) and the second assertion of part (b) is based on the construction completely analogous to that of the proof of Lemma 1.3.3, with (co)flat (co)modules replaced by (co)projective ones, and the left and right sides switches as needed. The only difference is that the inductive limit of a sequence of coprojective comodules does not have to be coprojective, because even the inductive limit of a sequence of projective modules does not have to be projective. This obstacle is dealt with in the following way.

Sublemma A. *Assume that $\mathcal{C}$ is a projective left A-module. Let $\mathcal{U}_1 \longrightarrow \mathcal{U}_2 \longrightarrow \mathcal{U}_3 \longrightarrow \mathcal{U}_4 \longrightarrow \cdots$ be an inductive system of left $\mathcal{C}$-comodules, where the comodules $\mathcal{U}_{2i}$ are coprojective, while the morphisms of comodules $\mathcal{U}_{2i-1} \longrightarrow \mathcal{U}_{2i+1}$ are injective and split over A. Then the inductive limit $\varinjlim \mathcal{U}_j$ is a coprojective $\mathcal{C}$-comodule.*

Proof. Let us first show that for any $\mathcal{C}$-contramodule $\mathfrak{P}$ there is an isomorphism $\mathrm{Cohom}_{\mathcal{C}}(\varinjlim \mathcal{U}_j, \mathfrak{P}) = \varprojlim \mathrm{Cohom}_{\mathcal{C}}(\mathcal{U}_j, \mathfrak{P})$. Denote by $G_j^{\bullet}$ the bar complex

$$\cdots \longrightarrow \mathrm{Hom}_A(\mathcal{C}\otimes_A\mathcal{C}\otimes_A\mathcal{U}_j, \mathfrak{P}) \longrightarrow \mathrm{Hom}_A(\mathcal{C}\otimes_A\mathcal{U}_j, \mathfrak{P}) \longrightarrow \mathrm{Hom}_A(\mathcal{U}_j, \mathfrak{P});$$

we will denote the terms of this complex by upper indices, so that $G_j^n = 0$ for $n > 0$ and $H^0(G_j^{\bullet}) = \mathrm{Cohom}_{\mathcal{C}}(\mathcal{U}_j, \mathfrak{P})$. Clearly, we have $H^0(\varinjlim G_j^{\bullet}) = \mathrm{Cohom}_{\mathcal{C}}(\varinjlim \mathcal{U}_j, \mathfrak{P})$. Since the comodules $\mathcal{U}_{2i}$ are coprojective, $H^n(G_{2i}^{\bullet}) = 0$ for $n \neq 0$, as the complex $G_{2i}^{\bullet}$ can be obtained by applying the functor $\mathrm{Cohom}_{\mathcal{C}}(\mathcal{U}_{2i}, -)$ to the complex of $\mathcal{C}$-contramodules $\cdots \longrightarrow \mathrm{Hom}_A(\mathcal{C}\otimes_A\mathcal{C}, \mathfrak{P}) \longrightarrow \mathrm{Hom}_A(\mathcal{C}, \mathfrak{P})$, which is exact except at degree 0. Since the maps of A-modules $\mathcal{U}_{2i-1} \longrightarrow \mathcal{U}_{2i+1}$ are split injective, the morphisms of complexes $G_{2i+1}^{\bullet} \longrightarrow G_{2i-1}^{\bullet}$ are surjective. Therefore, $\varprojlim{}^1 G_j^{\bullet} = \varprojlim{}^1 G_{2i-1}^{\bullet} = 0$, hence there is a "universal coefficients" sequence [86, Theorem 3.5.8]

$$0 \longrightarrow \varprojlim{}^1 H^{n-1}(G_j^{\bullet}) \longrightarrow H^n(\varprojlim G_j^{\bullet}) \longrightarrow \varprojlim H^n(G_j^{\bullet}) \longrightarrow 0.$$

In particular, for $n = 0$ we obtain the desired isomorphism $H^0(\varprojlim G_j^{\bullet}) = \varprojlim H^0(G_j^{\bullet})$, because $\varprojlim{}^1 H^{-1}(G_j^{\bullet}) = \varprojlim{}^1 H^{-1}(G_{2i}^{\bullet}) = 0$.

Now for any exact triple of $\mathcal{C}$-contramodules $\mathfrak{P}' \to \mathfrak{P} \to \mathfrak{P}''$ we have an exact triple of projective systems $\mathrm{Cohom}_{\mathcal{C}}(\mathcal{U}_{2i}, \mathfrak{P}') \longrightarrow \mathrm{Cohom}_{\mathcal{C}}(\mathcal{U}_{2i}, \mathfrak{P}) \longrightarrow \mathrm{Cohom}_{\mathcal{C}}(\mathcal{U}_{2i}, \mathfrak{P}'')$ and $\varprojlim{}^1 \mathrm{Cohom}_{\mathcal{C}}(\mathcal{U}_{2i}, \mathfrak{P}') = \varprojlim{}^1 \mathrm{Cohom}_{\mathcal{C}}(\mathcal{U}_{2i-1}, \mathfrak{P}') = 0$, hence the triple remains exact after passing to the projective limit. $\square$

Sublemma B. *Assume that $\mathcal{C}$ is a flat right A-module. Let $\mathcal{U}_1 \longrightarrow \mathcal{U}_2 \longrightarrow \mathcal{U}_3 \longrightarrow$ $\mathcal{U}_4 \longrightarrow \cdots$ be an inductive system of left $\mathcal{C}$-comodules, where the comodules $\mathcal{U}_{2i}$ are $\mathcal{C}/A$-coprojective, while the morphisms of comodules $\mathcal{U}_{2i-1} \longrightarrow \mathcal{U}_{2i+1}$ are injective. Then the inductive limit $\varinjlim \mathcal{U}_j$ is a $\mathcal{C}/A$-coprojective $\mathcal{C}$-comodule.*

Proof. Analogous to the proof of Sublemma A, the only changes being that $\mathfrak{P}$, $\mathfrak{P}'$, $\mathfrak{P}''$ are now A-injective $\mathcal{C}$-contramodules and the complex $\cdots \longrightarrow \operatorname{Hom}_A(\mathcal{C} \otimes_A \mathcal{C}, \mathfrak{P}) \longrightarrow \operatorname{Hom}_A(\mathcal{C}, \mathfrak{P}) \longrightarrow \mathfrak{P}$ is an A-split exact sequence of A-injective $\mathcal{C}$-contramodules. $\square$

Proof of the first assertion of part (b): for any A-injective $\mathcal{C}$-contramodule $\mathfrak{P}$, set $\mathfrak{G}(\mathfrak{P}) = \operatorname{Hom}_A(\mathcal{C}, \mathfrak{P})$. Then the contraaction map $\mathfrak{G}(\mathfrak{P}) \longrightarrow \mathfrak{P}$ is a surjective morphism of $\mathcal{C}$-contramodules, the contramodule $\mathfrak{G}(\mathfrak{P})$ is coinjective, and the kernel of this morphism is A-injective. Now let $\mathfrak{P}$ be an A-injective left $\mathcal{S}$-semicontramodule. The semicontraaction map $\mathfrak{P} \longrightarrow \operatorname{Cohom}_\mathcal{C}(\mathcal{S}, \mathfrak{P})$ is an injective morphism of A-injective $\mathcal{S}$-semicontramodules; let $\mathfrak{K}(\mathfrak{P})$ denote its cokernel. The map $\operatorname{Cohom}_\mathcal{C}(\mathcal{S}, \mathfrak{G}(\mathfrak{P})) \longrightarrow \operatorname{Cohom}_\mathcal{C}(\mathcal{S}, \mathfrak{P}))$ is a surjective morphism of $\mathcal{S}$-semicontramodules with an A-injective kernel $\operatorname{Cohom}_\mathcal{C}(\mathcal{S}, \ker(\mathfrak{G}(\mathfrak{P}) \to \mathfrak{P}))$. Let $\mathfrak{Q}(\mathfrak{P})$ be the kernel of the composition

$$\operatorname{Cohom}_\mathcal{C}(\mathcal{S}, \mathfrak{G}(\mathfrak{P})) \longrightarrow \operatorname{Cohom}_\mathcal{C}(\mathcal{S}, \mathfrak{P}) \longrightarrow \mathfrak{K}(\mathfrak{P}).$$

Then the composition of maps $\mathfrak{Q}(\mathfrak{P}) \longrightarrow \operatorname{Cohom}_\mathcal{C}(\mathcal{S}, \mathfrak{G}(\mathfrak{P})) \longrightarrow \operatorname{Cohom}_\mathcal{C}(\mathcal{S}, \mathfrak{P})$ factorizes through the injection $\mathfrak{P} \longrightarrow \operatorname{Cohom}_\mathcal{C}(\mathcal{S}, \mathfrak{P})$, so there is a natural surjective morphism of $\mathcal{S}$-semicontramodules $\mathfrak{Q}(\mathfrak{P}) \longrightarrow \mathfrak{P}$. The kernel of the map $\mathfrak{Q}(\mathfrak{P}) \longrightarrow \mathfrak{P}$ is isomorphic to the kernel of the map $\operatorname{Cohom}_\mathcal{C}(\mathcal{S}, \mathfrak{G}(\mathfrak{P})) \longrightarrow \operatorname{Cohom}_\mathcal{C}(\mathcal{S}, \mathfrak{P})$, hence both $\ker(\mathfrak{Q}(\mathfrak{P}) \to \mathfrak{P})$ and $\mathfrak{Q}(\mathfrak{P})$ are injective A-modules.

Notice that the semicontramodule morphism $\mathfrak{Q}(\mathfrak{P}) \longrightarrow \mathfrak{P}$ can be extended to a contramodule morphism $\operatorname{Cohom}_\mathcal{C}(\mathcal{S}, \mathfrak{G}(\mathfrak{P})) \longrightarrow \mathfrak{P}$. Indeed, the map $\mathfrak{Q}(\mathfrak{P}) \longrightarrow \mathfrak{P}$ can be presented as the composition

$$\mathfrak{Q}(\mathfrak{P}) \longrightarrow \operatorname{Cohom}_\mathcal{C}(\mathcal{S}, \mathfrak{G}(\mathfrak{P})) \longrightarrow \operatorname{Cohom}_\mathcal{C}(\mathcal{S}, \mathfrak{P}) \longrightarrow \mathfrak{P},$$

where the map $\operatorname{Cohom}_\mathcal{C}(\mathcal{S}, \mathfrak{P}) \longrightarrow \mathfrak{P}$ is induced by the semiunit morphism $\mathcal{C} \longrightarrow \mathcal{S}$ of the semialgebra $\mathcal{S}$.

Iterating this construction, we obtain a projective system of $\mathcal{C}$-contramodule morphisms

$$\mathfrak{P} \longleftarrow \operatorname{Cohom}_\mathcal{C}(\mathcal{S}, \mathfrak{G}(\mathfrak{P})) \longleftarrow \mathfrak{Q}(\mathfrak{P})$$
$$\longleftarrow \operatorname{Cohom}_\mathcal{C}(\mathcal{S}, \mathfrak{G}(\mathfrak{Q}(\mathfrak{P}))) \longleftarrow \mathfrak{Q}(\mathfrak{Q}(\mathfrak{P})) \longleftarrow \cdots,$$

where the maps

$$\mathfrak{P} \longleftarrow \mathfrak{Q}(\mathfrak{P}) \longleftarrow \mathfrak{Q}(\mathfrak{Q}(\mathfrak{P})) \longleftarrow \cdots$$

are A-split surjective morphisms of A-injective $\mathcal{S}$-semicontramodules, while the $\mathcal{C}$-contramodules $\operatorname{Cohom}_\mathcal{C}(\mathcal{S}, \mathfrak{G}(\mathfrak{P}))$, $\operatorname{Cohom}_\mathcal{C}(\mathcal{S}, \mathfrak{G}(\mathfrak{Q}(\mathfrak{P})))$, ... are coinjective.

Denote by $\mathfrak{F}(\mathfrak{P})$ the projective limit of this system; then $\mathfrak{F}(\mathfrak{P}) \longrightarrow \mathfrak{P}$ is an A-split surjective morphism of $\mathbb{S}$-semicontramodules, while coinjectivity of the $\mathbb{C}$-contramodule $\mathfrak{F}(\mathfrak{P})$ follows from the next sublemma.

Sublemma C. *Assume that $\mathbb{C}$ is a flat right A-module. Let $\mathfrak{U}_1 \longleftarrow \mathfrak{U}_2 \longleftarrow \mathfrak{U}_3 \longleftarrow \mathfrak{U}_4 \longleftarrow \cdots$ be a projective system of left $\mathbb{C}$-contramodules, where the contramodules $\mathfrak{U}_{2i}$ are coinjective, while the morphisms of contramodules $\mathfrak{U}_{2i+1} \longrightarrow \mathfrak{U}_{2i-1}$ are surjective and split over A. Then the projective limit $\varprojlim \mathfrak{U}_j$ is a coinjective $\mathbb{C}$-contramodule.*

Proof. Completely analogous to the proof of Sublemma A. One considers the projective system of bar-complexes $\cdots \longrightarrow \mathrm{Hom}_A(\mathbb{C} \otimes_A \mathbb{C} \otimes_A \mathcal{M}, \, \mathfrak{U}_j) \longrightarrow \mathrm{Hom}_A(\mathbb{C} \otimes_A \mathcal{M}, \, \mathfrak{U}_j) \longrightarrow \mathrm{Hom}_A(\mathcal{M}, \mathfrak{U}_j)$, etc. $\qquad\square$

The *proof of the second assertion of part* (a) is based on the same construction; the only changes are that A-modules are no longer injective, for any left $\mathbb{C}$-contramodule $\mathfrak{P}$ the $\mathbb{C}$-contramodule $\mathfrak{G}(\mathfrak{P}) = \mathrm{Hom}_A(\mathbb{C}, \mathfrak{P})$ is $\mathbb{C}/A$-coinjective, and therefore the $\mathbb{S}$-semicontramodule $\mathrm{Cohom}_{\mathbb{C}}(\mathbb{S}, \mathfrak{G}(\mathfrak{P}))$ is $\mathbb{C}/A$-coinjective. The projective limit $\mathfrak{F}(\mathfrak{P})$ is $\mathbb{C}/A$-coinjective according to the following sublemma.

Sublemma D. *Assume that $\mathbb{C}$ is a projective left A-module. Let $\mathfrak{U}_1 \longleftarrow \mathfrak{U}_2 \longleftarrow \mathfrak{U}_3 \longleftarrow \mathfrak{U}_4 \longleftarrow \cdots$ be a projective system of left $\mathbb{C}$-contramodules, where the contramodules $\mathfrak{U}_{2i}$ are $\mathbb{C}/A$-coinjective, while the morphisms of contramodules $\mathfrak{U}_{2i+1} \longrightarrow \mathfrak{U}_{2i-1}$ are surjective. Then the projective limit $\varprojlim \mathfrak{U}_j$ is a $\mathbb{C}/A$-coinjective $\mathbb{C}$-contramodule.* $\qquad\square$

Both functors $\mathfrak{F}$ are exact, since the cokernels of injective maps, the kernels of surjective maps, and the projective limits of Mittag-Leffler sequences of k-modules preserve exact triples. Part (c) is clear from the constructions. $\qquad\square$

3.4 Semihomomorphisms

3.4.1 Assume that the coring $\mathbb{C}$ is a projective left A-module, the semialgebra $\mathbb{S}$ is a projective left A-module and a $\mathbb{C}/A$-coflat right A-module, and the ring A has a finite left homological dimension. Let $\mathcal{M}$ be an A-projective left $\mathbb{S}$-semimodule and $\mathfrak{P}$ be a left $\mathbb{S}$-semicontramodule. The k-module of *semihomomorphisms* $\mathrm{SemiHom}_{\mathbb{S}}(\mathcal{M}, \mathfrak{P})$ is defined as the kernel of the pair of maps

$$\mathrm{Cohom}_{\mathbb{C}}(\mathcal{M}, \mathfrak{P}) \rightrightarrows \mathrm{Cohom}_{\mathbb{C}}(\mathbb{S} \,\square_{\mathbb{C}}\, \mathcal{M}, \, \mathfrak{P}) = \mathrm{Cohom}_{\mathbb{C}}(\mathcal{M}, \mathrm{Cohom}_{\mathbb{C}}(\mathbb{S}, \mathfrak{P}))$$

one of which is induced by the $\mathbb{S}$-semiaction in $\mathcal{M}$ and the other by the $\mathbb{S}$-semicontraaction in $\mathfrak{P}$.

For any A-projective left $\mathbb{C}$-comodule $\mathcal{L}$ and any left $\mathbb{S}$-semicontramodule $\mathfrak{P}$ there is a natural isomorphism

$$\mathrm{SemiHom}_{\mathbb{S}}(\mathbb{S} \,\square_{\mathbb{C}}\, \mathcal{L}, \, \mathfrak{P}) \simeq \mathrm{Cohom}_{\mathbb{C}}(\mathcal{L}, \mathfrak{P}).$$

Analogously, for any A-projective left $\mathbf{S}$-semimodule $\mathcal{M}$ and any left $\mathcal{C}$-contramodule $\mathfrak{Q}$ there is a natural isomorphism

$$\mathrm{SemiHom}_{\mathbf{S}}(\mathcal{M}, \mathrm{Cohom}_{\mathcal{C}}(\mathbf{S}, \mathfrak{Q})) \simeq \mathrm{Cohom}_{\mathcal{C}}(\mathcal{M}, \mathfrak{Q}).$$

These assertions follow from Lemma 1.2.1.

3.4.2 Assume that the coring $\mathcal{C}$ is a flat right A-module, the semialgebra $\mathbf{S}$ is a flat right A-module and a $\mathcal{C}/A$-coprojective left A-module, and the ring A has a finite left homological dimension. Let $\mathcal{M}$ be a left $\mathbf{S}$-semimodule and $\mathfrak{P}$ be an A-injective left $\mathbf{S}$-semicontramodule. As above, the k-module of *semihomomorphisms* $\mathrm{SemiHom}_{\mathbf{S}}(\mathcal{M}, \mathfrak{P})$ is defined as the kernel of the pair of maps $\mathrm{Cohom}_{\mathcal{C}}(\mathcal{M}, \mathfrak{P}) \rightrightarrows$ $\mathrm{Cohom}_{\mathcal{C}}(\mathbf{S} \,\square_{\mathcal{C}}\, \mathcal{M},\ \mathfrak{P}) = \mathrm{Cohom}_{\mathcal{C}}(\mathcal{M}, \mathrm{Cohom}_{\mathcal{C}}(\mathbf{S}, \mathfrak{P}))$ one of which is induced by the $\mathbf{S}$-semiaction in $\mathcal{M}$ and the other by the $\mathbf{S}$-semicontraaction in $\mathfrak{P}$.

For any left $\mathcal{C}$-comodule $\mathcal{L}$ and any A-injective left $\mathbf{S}$-semicontramodule $\mathfrak{P}$ there is a natural isomorphism $\mathrm{SemiHom}_{\mathbf{S}}(\mathbf{S} \,\square_{\mathcal{C}}\, \mathcal{L},\ \mathfrak{P}) \simeq \mathrm{Cohom}_{\mathcal{C}}(\mathcal{L}, \mathfrak{P})$. Analogously, for any left $\mathbf{S}$-semimodule $\mathcal{M}$ and any A-injective left $\mathcal{C}$-contramodule $\mathfrak{Q}$ there is a natural isomorphism $\mathrm{SemiHom}_{\mathbf{S}}(\mathcal{M}, \mathrm{Cohom}_{\mathcal{C}}(\mathbf{S}, \mathfrak{Q})) \simeq \mathrm{Cohom}_{\mathcal{C}}(\mathcal{M}, \mathfrak{Q})$.

Notice that even under the strongest of our assumptions on A, $\mathcal{C}$ and $\mathbf{S}$, the A-projectivity of $\mathcal{M}$ or the A-injectivity of $\mathfrak{P}$ is still needed to guarantee that the triple cohomomorphisms $\mathrm{Cohom}_{\mathcal{C}}(\mathbf{S} \,\square_{\mathcal{C}}\, \mathcal{M},\ \mathfrak{P})$ are associative.

3.4.3 If the coring $\mathcal{C}$ is a projective left A-module and the semialgebra $\mathbf{S}$ is a coprojective left $\mathcal{C}$-comodule, one can define the module of semihomomorphisms from a $\mathcal{C}$-coprojective left $\mathbf{S}$-semimodule into an arbitrary left $\mathbf{S}$-semicontramodule. In these assumptions, a $\mathcal{C}$-coprojective left $\mathbf{S}$-semimodule $\mathcal{M}$ is called *semiprojective* if the functor of semihomomorphisms from $\mathcal{M}$ is exact on the abelian category of left $\mathbf{S}$-semicontramodules. The $\mathbf{S}$-semimodule induced from a coprojective $\mathcal{C}$-comodule is semiprojective. Any semiprojective $\mathbf{S}$-semimodule is semiflat.

If the coring $\mathcal{C}$ is a flat right A-module and the semialgebra $\mathbf{S}$ is a coflat right $\mathcal{C}$-comodule, one can define the module of semihomomorphisms from an arbitrary left $\mathbf{S}$-semimodule into a $\mathcal{C}$-coinjective left $\mathbf{S}$-semicontramodule. In these assumptions, a $\mathcal{C}$-coinjective left $\mathbf{S}$-semicontramodule $\mathfrak{P}$ is called *semiinjective* if the functor of semihomomorphisms into $\mathfrak{P}$ is exact on the abelian category of left $\mathbf{S}$-semimodules. The $\mathbf{S}$-semicontramodule coinduced from a coinjective $\mathcal{C}$-contramodule is semiinjective.

When the ring A is semisimple, the module of semihomomorphisms from an arbitrary $\mathbf{S}$-semimodule into an arbitrary $\mathbf{S}$-semicontramodule is defined without any conditions on the coring $\mathcal{C}$ and the semialgebra $\mathbf{S}$.

3.4.4 Let $\mathbf{S}$ be a semialgebra over a coring $\mathcal{C}$ over a k-algebra A and $\mathcal{T}$ be a semialgebra over a coring $\mathcal{D}$ over a k-algebra B. Let $\mathcal{K}$ be an $\mathbf{S}$-$\mathcal{T}$-bisemimodule and $\mathfrak{P}$ be a left $\mathbf{S}$-semicontramodule. We would like to define a left $\mathcal{T}$-semicontramodule structure on the module of semihomomorphisms $\mathrm{SemiHom}_{\mathbf{S}}(\mathcal{K}, \mathfrak{P})$.

Assume that multiple cohomomorphisms of the form $\mathrm{Cohom}_{\mathcal{C}}(\mathcal{S}\square_{\mathcal{C}}\mathcal{K}\square_{\mathcal{D}}\mathcal{J}\square_{\mathcal{D}}$ $\cdots\square_{\mathcal{D}}\mathcal{J},\,\mathfrak{P})$ are associative. Then, in particular, the k-modules of semihomomor-phisms $\mathrm{SemiHom}_{\mathcal{S}}(\mathcal{K}\square_{\mathcal{D}}\mathcal{J}\square_{\mathcal{D}}\cdots\square_{\mathcal{D}}\mathcal{J},\,\mathfrak{P})$ can be defined. Assume in addition that multiple cohomomorphisms of the form $\mathrm{Cohom}_{\mathcal{C}}(\mathcal{K}\square_{\mathcal{D}}\mathcal{J}\square_{\mathcal{D}}\cdots\square_{\mathcal{D}}\mathcal{J},\,\mathfrak{P})$ are asso-ciative. Then the semihomomorphism modules $\mathrm{SemiHom}_{\mathcal{S}}(\mathcal{K}\square_{\mathcal{D}}\mathcal{J}\square_{\mathcal{D}}\cdots\square_{\mathcal{D}}\mathcal{J},\,\mathfrak{P})$ have natural left $\mathcal{D}$-contramodule structures as kernels of $\mathcal{D}$-contramodule mor-phisms. Assume that multiple cohomomorphisms of the form $\mathrm{Cohom}_{\mathcal{D}}(\mathcal{J}\square_{\mathcal{D}}\cdots\square_{\mathcal{D}}$ $\mathcal{J},\,\mathrm{SemiHom}_{\mathcal{S}}(\mathcal{K},\,\mathfrak{P}))$ are also associative. Finally, assume that the cohomomor-phisms from $\mathcal{J}^{\square m}$ preserve the kernel of the pair of morphisms $\mathrm{Cohom}_{\mathcal{C}}(\mathcal{K},\,\mathfrak{P})\rightrightarrows$ $\mathrm{Cohom}_{\mathcal{C}}(\mathcal{S}\,\square_{\mathcal{C}}\,\mathcal{K},\,\mathfrak{P})$ for $m=1$ and 2, that is the contramodule morphisms $\mathrm{Cohom}_{\mathcal{D}}(\mathcal{J}^{\square m},\,\mathrm{SemiHom}_{\mathcal{S}}(\mathcal{K},\,\mathfrak{P}))\longrightarrow\mathrm{SemiHom}_{\mathcal{S}}(\mathcal{K}\square_{\mathcal{D}}\,\mathcal{J}^{\square m},\,\mathfrak{P})$ are isomor-phisms. Then one can define an associative and unital semicontraaction morphism $\mathrm{SemiHom}_{\mathcal{S}}(\mathcal{K},\,\mathfrak{P})\longrightarrow\mathrm{Cohom}_{\mathcal{D}}(\mathcal{J},\,\mathrm{SemiHom}_{\mathcal{S}}(\mathcal{K},\,\mathfrak{P}))$ by taking the semihomo-morphisms over $\mathcal{S}$ from the right $\mathcal{J}$-semiaction morphism $\mathcal{K}\,\square_{\mathcal{D}}\,\mathcal{J}\longrightarrow\mathcal{K}$ into the semicontramodule $\mathfrak{P}$.

For example, if $\mathcal{D}$ is a projective left B-module, $\mathcal{J}$ is a coprojective left $\mathcal{D}$-co-module, A has a finite left homological dimension, and either $\mathcal{C}$ is a projective left A-module, $\mathcal{S}$ is a projective left A-module and a $\mathcal{C}/A$-coflat right $\mathcal{C}$-comodule, and $\mathcal{K}$ is a projective left A-module, or $\mathcal{C}$ is a flat right A-module, $\mathcal{S}$ is a flat right A-module and a $\mathcal{C}/A$-coprojective left $\mathcal{C}$-comodule, and $\mathfrak{P}$ is an injective left A-module, then the module of semihomomorphisms $\mathrm{SemiHom}_{\mathcal{S}}(\mathcal{K},\,\mathfrak{P})$ has a natural left $\mathcal{J}$-semicontramodule structure. Since the category of left $\mathcal{J}$-semi-contramodules is abelian in this case, the $\mathcal{J}$-semicontramodule $\mathrm{SemiHom}_{\mathcal{S}}(\mathcal{K},\,\mathfrak{P})$ can be simply defined as the kernel of the pair of semicontramodule morphisms $\mathrm{Cohom}_{\mathcal{C}}(\mathcal{K},\,\mathfrak{P})\rightrightarrows\mathrm{Cohom}_{\mathcal{C}}(\mathcal{S}\,\square_{\mathcal{C}}\,\mathcal{K},\,\mathfrak{P})$.

Proposition. *Let* $\mathcal{M}$ *be a left* $\mathcal{J}$-*semimodule,* $\mathcal{K}$ *be an* $\mathcal{S}$-$\mathcal{J}$-*bisemimodule, and* $\mathfrak{P}$ *be a left* $\mathcal{S}$-*semicontramodule. Then the iterated semihomomorphism modules*

$$\mathrm{SemiHom}_{\mathcal{S}}(\mathcal{K}\,\lozenge_{\mathcal{J}}\,\mathcal{M},\,\mathfrak{P}) \quad \text{and} \quad \mathrm{SemiHom}_{\mathcal{J}}(\mathcal{M},\,\mathrm{SemiHom}_{\mathcal{S}}(\mathcal{K},\,\mathfrak{P}))$$

are well defined and naturally isomorphic, at least, in the following cases:

(a) $\mathcal{D}$ *is a projective left* B-*module,* $\mathcal{J}$ *is a coprojective left* $\mathcal{D}$-*comodule,* $\mathcal{M}$ *is a coprojective left* $\mathcal{D}$-*comodule,* $\mathcal{C}$ *is a flat right* A-*module,* $\mathcal{S}$ *is a coflat right* $\mathcal{C}$-*comodule, and* $\mathfrak{P}$ *is a coinjective left* $\mathcal{C}$-*contramodule;*

(b) $\mathcal{D}$ *is a projective left* B-*module,* $\mathcal{J}$ *is a coprojective left* $\mathcal{D}$-*comodule,* $\mathcal{M}$ *is a semiprojective left* $\mathcal{J}$-*semimodule, and either*

  - $\mathcal{C}$ *is a projective left* A-*module,* $\mathcal{S}$ *is a coprojective left* $\mathcal{C}$-*comodule, and* $\mathcal{K}$ *is a coprojective left* $\mathcal{C}$-*comodule, or*

  - $\mathcal{C}$ *is a projective left* A-*module,* $\mathcal{S}$ *is a projective left* A-*module and a* $\mathcal{C}/A$-*coflat right* $\mathcal{C}$-*comodule, the ring* A *has a finite left homological dimension, and* $\mathcal{K}$ *is a projective left* A-*module, or*

- $\mathcal{C}$ *is a flat right A-module,* $\mathbf{S}$ *is a flat right A-module and a* $\mathcal{C}/A$*-coprojective left* $\mathcal{C}$*-comodule, and* $\mathfrak{P}$ *is an injective left A-module, or*
- *the ring A is semisimple;*

(c) $\mathcal{C}$ *is a flat right A-module,* $\mathbf{S}$ *is a coflat right* $\mathcal{C}$*-comodule,* $\mathfrak{P}$ *is a semiinjective left* $\mathbf{S}$*-semicontramodule, and either*

- $\mathcal{D}$ *is a flat right B-module,* $\mathbf{T}$ *is a coflat right* $\mathcal{D}$*-comodule, and* $\mathcal{K}$ *is a coflat right* $\mathcal{D}$*-comodule, or*
- $\mathcal{D}$ *is a flat right B-module,* $\mathbf{T}$ *is a flat right B-module and a* $\mathcal{D}/B$*-coprojective left* $\mathcal{D}$*-comodule, the ring B has a finite left homological dimension, and* $\mathcal{K}$ *is a flat right B-module, or*
- $\mathcal{D}$ *is a projective left B-module,* $\mathbf{T}$ *is a projective left B-module and a* $\mathcal{D}/B$*-coflat right* $\mathcal{D}$*-comodule, the ring B has a finite left homological dimension, and* $\mathbf{M}$ *is a projective left B-module, or*
- *the ring B is semisimple;*

(d) $\mathcal{D}$ *is a projective left B-module,* $\mathbf{T}$ *is a coprojective left* $\mathcal{D}$*-comodule,* $\mathbf{M}$ *is a coprojective left* $\mathcal{D}$*-comodule, and either*

- $\mathcal{C}$ *is a projective left A-module,* $\mathbf{S}$ *is a coprojective left* $\mathcal{C}$*-comodule, and* $\mathcal{K}$ *as a right* $\mathbf{T}$*-semimodule with a left* $\mathcal{C}$*-comodule structure is induced from a* $\mathcal{C}$*-coprojective* $\mathcal{C}$*-$\mathcal{D}$-bicomodule, or*
- $\mathcal{C}$ *is a projective left A-module,* $\mathbf{S}$ *is a projective left A-module and a* $\mathcal{C}/A$*-coflat right* $\mathcal{C}$*-comodule, the ring A has a finite left homological dimension, and* $\mathcal{K}$ *as a right* $\mathbf{T}$*-semimodule with a left* $\mathcal{C}$*-comodule structure is induced from an A-projective* $\mathcal{C}$*-$\mathcal{D}$-bicomodule, or*
- $\mathcal{C}$ *is a flat right A-module,* $\mathbf{S}$ *is a flat right A-module and a* $\mathcal{C}/A$*-coprojective left* $\mathcal{C}$*-comodule, the ring A has a finite left homological dimension,* $\mathcal{K}$ *as a right* $\mathbf{T}$*-semimodule with a left* $\mathcal{C}$*-comodule structure is induced from a* $\mathcal{C}$*-$\mathcal{D}$-bicomodule, and* $\mathfrak{P}$ *is an injective left A-module, or*
- *the ring A is semisimple and* $\mathcal{K}$ *as a right* $\mathbf{T}$*-semimodule with a left* $\mathcal{C}$*-comodule structure is induced from a* $\mathcal{C}$*-$\mathcal{D}$-bicomodule;*

(e) $\mathcal{C}$ *is a flat right A-module,* $\mathbf{S}$ *is a coflat right* $\mathcal{C}$*-comodule,* $\mathfrak{P}$ *is a coinjective left* $\mathcal{C}$*-contramodule, and either*

- $\mathcal{D}$ *is a flat right B-module,* $\mathbf{T}$ *is a coflat right* $\mathcal{D}$*-comodule, and* $\mathcal{K}$ *as a left* $\mathbf{S}$*-semimodule with a right* $\mathcal{D}$*-comodule structure is induced from a* $\mathcal{D}$*-coflat* $\mathcal{C}$*-$\mathcal{D}$-bicomodule, or*
- $\mathcal{D}$ *is a flat right B-module,* $\mathbf{T}$ *is a flat right B-module and a* $\mathcal{D}/B$*-coprojective left* $\mathcal{D}$*-comodule, the ring B has a finite left homological dimension, and* $\mathcal{K}$ *as a left* $\mathbf{S}$*-semimodule with a right* $\mathcal{D}$*-comodule structure is induced from a B-flat* $\mathcal{C}$*-$\mathcal{D}$-bicomodule, or*
- $\mathcal{D}$ *is a projective left B-module,* $\mathbf{T}$ *is a projective left B-module and a* $\mathcal{D}/B$*-coflat right* $\mathcal{D}$*-comodule, the ring B has a finite left homological*

> *dimension, $\mathcal{K}$ as a left $\mathbf{S}$-semimodule with a right $\mathcal{D}$-comodule structure is induced from a $\mathcal{C}$-$\mathcal{D}$-bicomodule, and $\mathfrak{M}$ is a projective left B-module, or*

- *the ring B is semisimple and $\mathcal{K}$ as a left $\mathbf{S}$-semimodule with a right $\mathcal{D}$-comodule structure is induced from a $\mathcal{C}$-$\mathcal{D}$-bicomodule.*

More precisely, in all cases in this list the natural maps from both iterated semihomomorphism modules under consideration into the iterated cohomomorphism module $\mathrm{Cohom}_{\mathcal{C}}(\mathcal{K}\square_{\mathcal{D}}\mathfrak{M}, \mathfrak{P}) \simeq \mathrm{Cohom}_{\mathcal{D}}(\mathfrak{M}, \mathrm{Cohom}_{\mathcal{C}}(\mathcal{K}, \mathfrak{P}))$ are injective, their images coincide and are equal to the intersection of two submodules $\mathrm{SemiHom}_{\mathbf{S}}(\mathcal{K}\square_{\mathcal{D}}\mathfrak{M}, \mathfrak{P})$ and $\mathrm{SemiHom}_{\mathcal{T}}(\mathfrak{M}, \mathrm{Cohom}_{\mathcal{C}}(\mathcal{K}, \mathfrak{P}))$ in this k-module.

Proof. Analogous to the proof of Proposition 1.4.4 (see also the proof of Proposition 3.2.5). $\qquad\square$

4 Derived Functor SemiExt

4.1 Contraderived categories

Let A be an exact category in which all infinite products exist and the functors of infinite product are exact. A complex $C^\bullet$ over A is called *contraacyclic* if it belongs to the minimal triangulated subcategory $\mathsf{Acycl}^{\mathrm{ctr}}(\mathsf{A})$ of the homotopy category $\mathsf{Hot}(\mathsf{A})$ containing all the total complexes of exact triples $'K^\bullet \to K^\bullet \to ''K^\bullet$ of complexes over A and closed under infinite products. Any contraacyclic complex is acyclic. It follows from the next lemma that any acyclic complex bounded from above is contraacyclic.

Lemma. *Let* $\cdots \to P^{-1,\bullet} \to P^{0,\bullet} \to 0$ *be an exact sequence, bounded from above, of arbitrary complexes over* A. *Then the total complex* $T^\bullet$ *of the bicomplex* $P^{\bullet,\bullet}$ *constructed by taking infinite products along the diagonals is contraacyclic.*

Proof. See the proof of Lemma 2.1. □

The category of contraacyclic complexes $\mathsf{Acycl}^{\mathrm{ctr}}(\mathsf{A})$ is a thick subcategory of the homotopy category $\mathsf{Hot}(\mathsf{A})$, since it is a triangulated subcategory with infinite products. The *contraderived category* $\mathsf{D}^{\mathrm{ctr}}(\mathsf{A})$ of an exact category A is defined as the quotient category $\mathsf{Hot}(\mathsf{A})/\mathsf{Acycl}^{\mathrm{ctr}}(\mathsf{A})$.

Remark. One can check that for any exact category A and any thick subcategory T in $\mathsf{Hot}(\mathsf{A})$ contained in the thick subcategory of acyclic complexes, containing all bounded acyclic complexes, and containing with every exact complex its subcomplexes and quotient complexes of canonical filtration, the groups of homomorphisms $\mathrm{Hom}_{\mathsf{Hot}(\mathsf{A})/\mathsf{T}}(X, Y[i])$ between complexes with a single nonzero term coincide with the Yoneda extension groups $\mathrm{Ext}^i_\mathsf{A}(X, Y)$. Moreover, the natural functors $\mathsf{Hot}^{+/-/\mathrm{b}}(\mathsf{A})/(\mathsf{T} \cap \mathsf{Hot}^{+/-/\mathrm{b}}(\mathsf{A})) \longrightarrow \mathsf{Hot}(\mathsf{A})/\mathsf{T}$ between the "T-derived categories" with various bounding conditions are all fully faithful. Besides, whenever A is an abelian category there is a (degenerate) t-structure on $\mathsf{Hot}(\mathsf{A})/\mathsf{T}$ formed by the full subcategories of complexes concentrated in nonpositive and nonnegative degrees. The core of this t-structure coincides with A. In particular, all these assertions hold if $\mathsf{T} \subset \mathsf{Hot}(\mathsf{A})$ consists of acyclic complexes and contains either all exact complexes bounded from above or all exact complexes bounded from below.

4.2 Coprojective and coinjective complexes

Let $\mathcal{C}$ be a coring over a k-algebra A. The complex of cohomomorphisms $\mathrm{Cohom}_\mathcal{C}(\mathcal{M}^\bullet, \mathfrak{P}^\bullet)$ from a complex of left $\mathcal{C}$-comodules $\mathcal{M}^\bullet$ into a complex of

L. Positselski, *Homological Algebra of Semimodules and Semicontramodules*, Monografie Matematyczne 70, DOI 10.1007/978-3-0346-0436-9_5, © Springer Basel AG 2010

left $\mathcal{C}$-contramodules $\mathfrak{P}^\bullet$ is defined as the total complex of the bicomplex $\mathrm{Cohom}_{\mathcal{C}}(\mathcal{M}^i, \mathfrak{P}^j)$, constructed by taking infinite products along the diagonals.

If $\mathcal{C}$ is a projective left A-module, the category of left $\mathcal{C}$-contramodules is an abelian category with exact functors of infinite products, so the contraderived category $\mathsf{D}^{\mathrm{ctr}}(\mathcal{C}\text{-contra})$ is defined. When speaking about *contraacyclic complexes* of $\mathcal{C}$-contramodules, we will always mean contraacyclic complexes with respect to the abelian category of $\mathcal{C}$-contramodules, unless another exact category of $\mathcal{C}$-contramodules is explicitly mentioned.

Assuming that $\mathcal{C}$ is a projective left A-module, a complex of left $\mathcal{C}$-comodules $\mathcal{M}^\bullet$ is called *coprojective* if the complex $\mathrm{Cohom}_{\mathcal{C}}(\mathcal{M}^\bullet, \mathfrak{P}^\bullet)$ is acyclic whenever a complex of left $\mathcal{C}$-contramodules $\mathfrak{P}^\bullet$ is contraacyclic. Assuming that $\mathcal{C}$ is a flat right A-module, a complex of left $\mathcal{C}$-contramodules $\mathfrak{P}^\bullet$ is called *coinjective* if the complex $\mathrm{Cohom}_{\mathcal{C}}(\mathcal{M}^\bullet, \mathfrak{P}^\bullet)$ is acyclic whenever a complex of left $\mathcal{C}$-comodules $\mathcal{M}^\bullet$ is coacyclic.

Lemma.

(a) *Any complex of coprojective $\mathcal{C}$-comodules is coprojective.*

(b) *Any complex of coinjective $\mathcal{C}$-contramodules is coinjective.*

Proof. Argue as in the proof of Lemma 2.2, using the fact that the functor of cohomomorphisms of complexes maps infinite direct sums in the first argument into infinite products and preserves infinite products in the second argument. $\square$

If the ring A has a finite left homological dimension, then any coprojective complex of left $\mathcal{C}$-comodules is a projective complex of A-modules in the sense of 0.1.2 and any coinjective complex of left $\mathcal{C}$-contramodules is an injective complex of A-modules. The complex of $\mathcal{C}$-comodules $\mathcal{C} \otimes_A U^\bullet$ coinduced from a projective complex of A-modules $U^\bullet$ is coprojective and the complex of $\mathcal{C}$-contramodules $\mathrm{Hom}_A(\mathcal{C}, V^\bullet)$ induced from an injective complex of A-modules is coinjective.

4.3 Semiderived categories

Let $\mathbf{S}$ be a semialgebra over a coring $\mathcal{C}$. Assume that $\mathcal{C}$ is a projective left A-module and the semialgebra $\mathbf{S}$ is a coprojective left $\mathcal{C}$-comodule, so that the category of left $\mathbf{S}$-semicontramodules is abelian. The *semiderived category* of left $\mathbf{S}$-semicontramodules $\mathsf{D}^{\mathrm{si}}(\mathbf{S}\text{-sicntr})$ is defined as the quotient category of the homotopy category $\mathsf{Hot}(\mathbf{S}\text{-sicntr})$ by the thick subcategory $\mathsf{Acycl}^{\mathrm{ctr}\text{-}\mathcal{C}}(\mathbf{S}\text{-sicntr})$ of complexes of $\mathbf{S}$-semicontramodules that are *contraacyclic as complexes of $\mathcal{C}$-contramodules*.

4.4 Semiprojective and semiinjective complexes

Let $\mathbf{S}$ be a semialgebra. The complex of semihomomorphisms $\mathrm{SemiHom}_{\mathbf{S}}(\mathcal{M}^\bullet, \mathfrak{P}^\bullet)$ from a complex of left $\mathbf{S}$-semimodules $\mathcal{M}^\bullet$ to a complex of left $\mathbf{S}$-semicontramodules $\mathfrak{P}^\bullet$ is defined as the total complex of the bicomplex $\mathrm{SemiHom}_{\mathbf{S}}(\mathcal{M}^i, \mathfrak{P}^j)$, constructed by taking infinite products along the diagonals. Of course, appropriate conditions must be imposed on S, $\mathcal{M}^\bullet$, and $\mathfrak{P}^\bullet$ for this definition to make sense.

Assume that the coring $\mathcal{C}$ is a projective left A-module and a flat right A-module, the semialgebra S is a coprojective left S-semimodule and a coflat right S-semimodule, and the ring A has a finite left homological dimension.

A complex of A-projective left S-semimodules $\mathcal{M}^\bullet$ is called *semiprojective* if the complex $\mathrm{SemiHom}_S(\mathcal{M}^\bullet, \mathfrak{P}^\bullet)$ is acyclic whenever a complex of left S-semicontramodules $\mathfrak{P}^\bullet$ is $\mathcal{C}$-contraacyclic. Any semiprojective complex of S-semimodules is a coprojective complex of $\mathcal{C}$-comodules. The complex of S-semimodules $S \,\square_{\mathcal{C}}\, \mathcal{L}^\bullet$ induced from a coprojective complex of A-flat $\mathcal{C}$-comodules is semiprojective. Any semiprojective complex of S-semimodules is semiflat. Analogously, a complex of A-injective left S-semicontramodules $\mathfrak{P}^\bullet$ is called *semiinjective* if the complex $\mathrm{SemiHom}_S(\mathcal{M}^\bullet, \mathfrak{P}^\bullet)$ is acyclic whenever a complex of left S-semimodules $\mathcal{M}^\bullet$ is $\mathcal{C}$-coacyclic. Any semiinjective complex of S-semicontramodules is a coinjective complex of $\mathcal{C}$-contramodules. The complex of S-semicontramodules $\mathrm{Cohom}_{\mathcal{C}}(S, \mathfrak{Q}^\bullet)$ coinduced from a coinjective complex of A-injective $\mathcal{C}$-contramodules is semiinjective.

Notice that not every complex of semiprojective semimodules is semiprojective and not every complex of semiinjective semicontramodules is semiinjective. On the other hand, any complex of semiprojective semimodules bounded from above is semiprojective. Moreover, if $\cdots \to \mathcal{M}^{-1,\bullet} \to \mathcal{M}^{0,\bullet} \to 0$ is a complex, bounded from above, of semiprojective complexes of S-semimodules, then the total complex $\mathcal{E}^\bullet$ of the bicomplex $\mathcal{M}^{\bullet,\bullet}$ constructed by taking infinite direct sums along the diagonals is semiprojective. Indeed, the category of semiprojective complexes is closed under shifts, cones, and infinite direct sums, so one can apply Lemma 2.4. Analogously, any complex of semiinjective semicontramodules bounded from below is semiinjective. Moreover, if $0 \to \mathfrak{P}^{0,\bullet} \to \mathfrak{P}^{1,\bullet} \to \cdots$ is a complex, bounded from below, of semiinjective complexes of S-semicontramodules, then the total complex $\mathfrak{E}^\bullet$ of the bicomplex $\mathfrak{P}^{\bullet,\bullet}$ constructed by taking infinite products along the diagonals is semiinjective. Indeed, the category of semiinjective complexes is closed under shifts, cones, and infinite products, so one can apply the following lemma.

Lemma. *Let* $0 \to P^{0,\bullet} \to P^{1,\bullet} \to \cdots$ *be a complex, bounded from below, of arbitrary complexes over an additive category* A *where infinite products exist. Then the total complex* $E^\bullet$ *of the bicomplex* $P^{\bullet,\bullet}$ *up to the homotopy equivalence can be obtained from the complexes* $P^{i,\bullet}$ *using the operations of shift, cone, and infinite product.*

Proof. See the proof of Lemma 2.4. $\square$

4.5 Main theorem for comodules and contramodules

Assume that the coring $\mathcal{C}$ is a projective left and a flat right A-module and the ring A has a finite left homological dimension.

Theorem.

(a) *The functor mapping the quotient category of the homotopy category of complexes of coprojective left $\mathcal{C}$-comodules (coprojective complexes of left $\mathcal{C}$-comodules) by its intersection with the thick subcategory of coacyclic complexes of $\mathcal{C}$-comodules into the coderived category of left $\mathcal{C}$-comodules is an equivalence of triangulated categories.*

(b) *The functor mapping the quotient category of the homotopy category of complexes of coinjective left $\mathcal{C}$-contramodules (coinjective complexes of left $\mathcal{C}$-contramodules) by its intersection with the thick subcategory of contraacyclic complexes of $\mathcal{C}$-contramodules into the contraderived category of left $\mathcal{C}$-contramodules is an equivalence of triangulated categories.*

Proof. The proof of part (a) is completely analogous to the proof of Theorem 2.5. It is based on the same constructions of resolutions $\mathbb{L}_1$ and $\mathbb{R}_2$, and uses the result of Lemma 3.1.3(a) instead of Lemma 1.1.3.

To prove part (b), we will show that any complex of left $\mathcal{C}$-contramodules $\mathfrak{K}^\bullet$ can be connected with a complex of coinjective $\mathcal{C}$-contramodules in a functorial way by a chain of two morphisms $\mathfrak{K}^\bullet \longrightarrow \mathbb{L}_2(\mathfrak{K}^\bullet) \longleftarrow \mathbb{L}_2\mathbb{R}_1(\mathfrak{K}^\bullet)$ with contraacyclic cones. Moreover, if the complex $\mathfrak{K}^\bullet$ is a complex of coinjective $\mathcal{C}$-contramodules (coinjective complex of $\mathcal{C}$-contramodules), then the intermediate complex $\mathbb{L}_2(\mathfrak{K}^\bullet)$ is also a complex of coinjective $\mathcal{C}$-contramodules (coinjective complex of $\mathcal{C}$-contramodules). Then we will apply Lemma 2.5 in the way explained in the end of the proof of Theorem 2.5.

Let $\mathfrak{K}^\bullet$ be a complex of left $\mathcal{C}$-contramodules. Let $\mathfrak{P} \longrightarrow \mathfrak{I}(\mathfrak{P})$ denote the functorial injective morphism from an arbitrary left $\mathcal{C}$-contramodule $\mathfrak{P}$ into an A-injective $\mathcal{C}$-contramodule $\mathfrak{I}(\mathfrak{P})$ constructed in Lemma 3.1.3(b). The functor $\mathfrak{I}$ is the direct sum of a constant functor $\mathfrak{P} \longmapsto \mathfrak{I}(0)$ and a functor $\mathfrak{I}^+$ sending zero morphisms to zero morphisms. For any $\mathcal{C}$-contramodule $\mathfrak{P}$, the contramodule $\mathfrak{I}^+(\mathfrak{P})$ is A-injective and the morphism $\mathfrak{P} \longrightarrow \mathfrak{I}^+(\mathfrak{P})$ is injective. Set $\mathfrak{I}^0(\mathfrak{K}^\bullet) = \mathfrak{I}^+(\mathfrak{K}^\bullet)$, $\mathfrak{I}^1(\mathfrak{K}^\bullet) = \mathfrak{I}^+(\mathrm{coker}(\mathfrak{K}^\bullet \to \mathfrak{I}^0(\mathfrak{K}^\bullet)))$, etc. For d large enough, the cokernel $\mathfrak{Z}(\mathfrak{K}^\bullet)$ of the morphism $\mathfrak{I}^{d-2}(\mathfrak{K}^\bullet) \longrightarrow \mathfrak{I}^{d-1}(\mathfrak{K}^\bullet)$ will be a complex of A-injective $\mathcal{C}$-contramodules. Let $\mathbb{R}_1(\mathfrak{K}^\bullet)$ be the total complex of the bicomplex

$$\mathfrak{I}^0(\mathfrak{K}^\bullet) \longrightarrow \mathfrak{I}^1(\mathfrak{K}^\bullet) \longrightarrow \cdots \longrightarrow \mathfrak{I}^{d-1}(\mathfrak{K}^\bullet) \longrightarrow \mathfrak{Z}(\mathfrak{K}^\bullet).$$

Then $\mathbb{R}_1(\mathfrak{K}^\bullet)$ is a complex of A-injective $\mathcal{C}$-contramodules and the cone of the morphism $\mathfrak{K}^\bullet \longrightarrow \mathbb{R}_1(\mathfrak{K}^\bullet)$ is the total complex of a finite exact sequence of complexes of $\mathcal{C}$-contramodules, and therefore, a contraacyclic complex.

Now let $\mathfrak{R}^\bullet$ be a complex of A-injective left $\mathcal{C}$-contramodules. Consider the bar construction

$$\cdots \longrightarrow \mathrm{Hom}_A(\mathcal{C}, \mathrm{Hom}_A(\mathcal{C}, \mathfrak{R}^\bullet)) \longrightarrow \mathrm{Hom}_A(\mathcal{C}, \mathfrak{R}^\bullet).$$

Let $\mathbb{L}_2(\mathfrak{R}^\bullet)$ be the total complex of this bicomplex, constructed by taking infinite products along the diagonals. Then $\mathbb{L}_2(\mathfrak{R}^\bullet)$ is a complex of coinjective $\mathcal{C}$-contramodules. The functor $\mathbb{L}_2$ can be extended to arbitrary complexes of $\mathcal{C}$-contramodules; for any complex $\mathfrak{K}^\bullet$, the cone of the morphism $\mathbb{L}_2(\mathfrak{K}^\bullet) \longrightarrow \mathfrak{K}^\bullet$ is contraacyclic by Lemma 4.1.

Finally, if $\mathfrak{K}^\bullet$ is a coinjective complex of $\mathcal{C}$-contramodules, then $\mathbb{L}_2(\mathfrak{K}^\bullet)$ is also a coinjective complex of $\mathcal{C}$-contramodules, since the complex of cohomomorphisms from a complex of left $\mathcal{C}$-comodules $\mathcal{M}^\bullet$ into $\mathbb{L}_2(\mathfrak{K}^\bullet)$ coincides with the complex of cohomomorphisms into $\mathfrak{K}^\bullet$ from the total cobar complex $\mathbb{R}_2(\mathcal{M}^\bullet)$, and the latter is coacyclic whenever $\mathcal{M}^\bullet$ is coacyclic. $\square$

Remark. Another proof of the theorem (for complexes of coprojective comodules and complexes of coinjective contramodules) can be deduced from the results of Chapter 5. In addition, it will follow that any coacyclic complex of coprojective left $\mathcal{C}$-comodules is contractible and any contraacyclic complex of coinjective left $\mathcal{C}$-contramodules is contractible (see Remark 5.5).

4.6 Main theorem for semimodules and semicontramodules

Assume that the coring $\mathcal{C}$ is a projective left and a flat right A-module, the semialgebra $\mathbf{S}$ is a coprojective left and a coflat right $\mathcal{C}$-comodule, and the ring A has a finite left homological dimension.

Theorem.

(a) *The functor mapping the quotient category of the homotopy category of semiprojective complexes of A-projective ($\mathcal{C}$-coprojective, semiprojective) left $\mathbf{S}$-semimodules by its intersection with the thick subcategory of $\mathcal{C}$-coacyclic complexes of $\mathbf{S}$-semimodules into the semiderived category of left $\mathbf{S}$-semimodules is an equivalence of triangulated categories.*

(b) *The functor mapping the quotient category of the homotopy category of semiinjective complexes of A-injective ($\mathcal{C}$-coinjective, semiinjective) left $\mathbf{S}$-semicontramodules by its intersection with the thick subcategory of $\mathcal{C}$-contraacyclic complexes of $\mathbf{S}$-semicontramodules into the semiderived category of left $\mathbf{S}$-semicontramodules is an equivalence of triangulated categories.*

Proof. There are two approaches: one can argue as in 2.5 or as in 2.6. Either way, the proof is based on the constructions of intermediate resolutions $\mathbb{L}_i$ and $\mathbb{R}_j$. For part (a), it is the same constructions that were presented in the proof of Theorem 2.6. One just has to use the results of Lemmas 3.3.2(a) and 3.3.3(a) instead of Lemmas 1.3.2 and 1.3.3. Let us introduce the analogous constructions for part (b).

Let $\mathfrak{K}^\bullet$ be a complex of left $\mathbf{S}$-semicontramodules. Let $\mathfrak{P} \longrightarrow \mathfrak{J}(\mathfrak{P})$ denote the functorial injective morphism from an arbitrary left $\mathbf{S}$-semicontramodule $\mathfrak{P}$ into an A-injective $\mathbf{S}$-semicontramodule $\mathfrak{J}(\mathfrak{P})$ constructed in Lemma 3.3.2(b). The functor $\mathfrak{J}$ is the direct sum of a constant functor $\mathfrak{P} \longmapsto \mathfrak{J}(0)$ and a functor $\mathfrak{J}^+$ sending zero morphisms to zero morphisms. For any $\mathbf{S}$-semicontramodule $\mathfrak{P}$, the semicontramodule $\mathfrak{J}^+(\mathfrak{P})$ is A-injective and the morphism $\mathfrak{P} \longrightarrow \mathfrak{J}^+(\mathfrak{P})$ is injective. Set $\mathfrak{J}^0(\mathfrak{K}^\bullet) = \mathfrak{J}^+(\mathfrak{K}^\bullet)$, $\mathfrak{J}^1(\mathfrak{K}^\bullet) = \mathfrak{J}^+(\mathrm{coker}(\mathfrak{K}^\bullet \to \mathfrak{J}^0(\mathfrak{K}^\bullet)))$, etc. For d large enough, the cokernel $\mathfrak{Z}(\mathfrak{K}^\bullet)$ of the morphism $\mathfrak{J}^{d-2}(\mathfrak{K}^\bullet) \longrightarrow \mathfrak{J}^{d-1}(\mathfrak{K}^\bullet)$ will be a complex of A-injective S-semicontramodules. Let $\mathbb{R}_1(\mathfrak{K}^\bullet)$ be the total complex of the bicomplex

$$\mathfrak{J}^0(\mathfrak{K}^\bullet) \longrightarrow \mathfrak{J}^1(\mathfrak{K}^\bullet) \longrightarrow \cdots \longrightarrow \mathfrak{J}^{d-1}(\mathfrak{K}^\bullet) \longrightarrow \mathfrak{Z}(\mathfrak{K}^\bullet).$$

Then $\mathbb{R}_1(\mathfrak{K}^\bullet)$ is a complex of A-injective S-semicontramodules and the cone of the morphism $\mathfrak{K}^\bullet \longrightarrow \mathbb{R}_1(\mathfrak{K}^\bullet)$ is the total complex of a finite exact sequence of complexes of S-semicontramodules, and therefore, a $\mathcal{C}$-contraacyclic complex (and even an S-contraacyclic complex).

Now let $\mathfrak{R}^\bullet$ be a complex of A-injective left S-semicontramodules. Let $\mathfrak{F}(\mathfrak{P}) \longrightarrow \mathfrak{P}$ denote the functorial surjective morphism onto an arbitrary A-injective S-semicontramodule $\mathfrak{P}$ from a $\mathcal{C}$-coinjective S-semicontramodule $\mathfrak{F}(\mathfrak{P})$ with an A-injective kernel $\mathrm{ker}(\mathfrak{F}(\mathfrak{P}) \to \mathfrak{P})$ constructed in Lemma 3.3.3(b). Set $\mathfrak{F}_0(\mathfrak{R}^\bullet) = \mathfrak{F}(\mathfrak{R}^\bullet)$, $\mathfrak{F}_1(\mathfrak{R}^\bullet) = \mathfrak{F}(\mathrm{ker}(\mathfrak{F}_0(\mathfrak{R}^\bullet) \to \mathfrak{R}^\bullet))$, etc. Let $\mathbb{L}_2(\mathfrak{R}^\bullet)$ be the total complex of the bicomplex

$$\cdots \longrightarrow \mathfrak{F}_2(\mathfrak{R}^\bullet) \longrightarrow \mathfrak{F}_1(\mathfrak{R}^\bullet) \longrightarrow \mathfrak{F}_0(\mathfrak{R}^\bullet),$$

constructed by taking infinite products along the diagonals. Then $\mathbb{L}_2(\mathfrak{R}^\bullet)$ is a complex of $\mathcal{C}$-coinjective S-semicontramodules. Since the surjection $\mathfrak{F}(\mathfrak{P}) \longrightarrow \mathfrak{P}$ can be defined for arbitrary left S-semicontramodules, the functor $\mathbb{L}_2$ can be extended to arbitrary complexes of S-semicontramodules. For any complex $\mathfrak{K}^\bullet$, the cone of the morphism $\mathbb{L}_2(\mathfrak{K}^\bullet) \longrightarrow \mathfrak{K}^\bullet$ is a $\mathcal{C}$-contraacyclic complex (and even an S-contraacyclic complex) by Lemma 4.1.

Finally, let $\mathfrak{P}^\bullet$ be a $\mathcal{C}$-coinjective complex of A-injective left S-semicontramodules. Then the complex $\mathrm{Cohom}_{\mathcal{C}}(\mathsf{S}, \mathfrak{P}^\bullet)$ is a semiinjective complex of A-injective left S-semicontramodules. Moreover, if $\mathfrak{P}^\bullet$ is a complex of $\mathcal{C}$-coinjective S-semicontramodules, then $\mathrm{Cohom}_{\mathcal{C}}(\mathsf{S}, \mathfrak{P}^\bullet)$ is a semiinjective complex of semiinjective S-semicontramodules. Consider the cobar construction

$$\mathrm{Cohom}_{\mathcal{C}}(\mathsf{S}, \mathfrak{P}^\bullet) \longrightarrow \mathrm{Cohom}_{\mathcal{C}}(\mathsf{S}, \mathrm{Cohom}_{\mathcal{C}}(\mathsf{S}, \mathfrak{P}^\bullet)) \longrightarrow \cdots$$

Let $\mathbb{R}_3(\mathfrak{P}^\bullet)$ be the total complex of this bicomplex, constructed by taking infinite products along the diagonals. Then complex $\mathbb{R}_3(\mathfrak{P}^\bullet)$ is semiinjective by Lemma 4.4. The functor $\mathbb{R}_3$ can be extended to arbitrary complexes of S-semicontramodules; for any complex $\mathfrak{K}^\bullet$, the cone of the morphism $\mathfrak{K}^\bullet \longrightarrow \mathbb{R}_3(\mathfrak{K}^\bullet)$ is not only $\mathcal{C}$-contraacyclic, but even $\mathcal{C}$-contractible (the contracting homotopy being induced by the semiunit morphism $\mathcal{C} \longrightarrow \mathsf{S}$.)

It follows that the natural functors between the quotient categories of the homotopy categories of semiinjective complexes of semiinjective S-semicontramodules, semiinjective complexes of $\mathcal{C}$-coinjective S-semicontramodules, complexes of $\mathcal{C}$-coinjective $\mathbf{S}$-semicontramodules, semiinjective complexes of A-injective $\mathbf{S}$-semicontramodules, $\mathcal{C}$-coinjective complexes of A-injective $\mathbf{S}$-semicontramodules, complexes of A-injective $\mathbf{S}$-semicontramodules by their intersections with the thick subcategory of $\mathcal{C}$-contraacycliccomplexes and the semiderived category of left $\mathbf{S}$-semicontramodules are all equivalences of triangulated categories. Moreover, any complex of left $\mathbf{S}$-semicontramodules $\mathfrak{K}^{\bullet}$ can be connected with a semiinjective complex of semiinjective $\mathbf{S}$-semicontramodules in a functorial way by a chain of three morphisms $\mathfrak{K}^{\bullet} \longrightarrow \mathbb{R}_3(\mathfrak{K}^{\bullet}) \longleftarrow \mathbb{R}_3\mathbb{L}_2(\mathfrak{K}^{\bullet}) \longrightarrow \mathbb{R}_3\mathbb{L}_2\mathbb{R}_1(\mathfrak{K}^{\bullet})$ with $\mathcal{C}$-contraacyclic cones, and when $\mathfrak{K}^{\bullet}$ is a semiinjective complex of (A-injective, $\mathcal{C}$-coinjective, or semiinjective) $\mathbf{S}$-semicontramodules, all complexes in this chain are also semiinjective complexes of (A-injective, $\mathcal{C}$-coinjective, or semiinjective) $\mathbf{S}$-semicontramodules. $\square$

Remark. One can show using the methods developed in Chapter 6 that any $\mathcal{C}$-coacyclic semiprojective complex of $\mathcal{C}$-coprojective left $\mathbf{S}$-semimodules is contractible, and analogously, any $\mathcal{C}$-contraacyclic semiinjective complex of $\mathcal{C}$-coinjective left $\mathbf{S}$-semicontramodules is contractible (see Remark 6.4).

4.7 Derived functor SemiExt

Assume that the coring $\mathcal{C}$ is a projective left and a flat right A-module, the semialgebra $\mathbf{S}$ is a coprojective left and a coflat right $\mathcal{C}$-comodule, and the ring A has a finite left homological dimension.

The double-sided derived functor

$$\mathrm{SemiExt}_{\mathbf{S}} : \mathsf{D}^{\mathsf{si}}(\mathbf{S}\text{-simod}) \times \mathsf{D}^{\mathsf{si}}(\mathbf{S}\text{-sicntr}) \longrightarrow \mathsf{D}(k\text{-mod})$$

is defined as follows. Consider the partially defined functor of semihomomorphisms of complexes

$$\mathrm{SemiHom}_{\mathbf{S}} : \mathsf{Hot}(\mathbf{S}\text{-simod})^{\mathsf{op}} \times \mathsf{Hot}(\mathbf{S}\text{-sicntr}) \dashrightarrow \mathsf{Hot}(k\text{-mod}).$$

This functor is defined on the full subcategory of the Cartesian product of homotopy categories that consists of pairs of complexes $(\mathcal{M}^{\bullet}, \mathfrak{P}^{\bullet})$ such that either $\mathcal{M}^{\bullet}$ is a complex of A-projective $\mathbf{S}$-semimodules, or $\mathfrak{P}^{\bullet}$ is a complex of A-injective $\mathbf{S}$-semicontramodules. Compose it with the functor of localization $\mathsf{Hot}(k\text{-mod}) \longrightarrow \mathsf{D}(k\text{-mod})$ and restrict either to the Cartesian product of the homotopy category of semiprojective complexes of A-projective $\mathbf{S}$-semimodules and the homotopy category of $\mathbf{S}$-semicontramodules, or to the Cartesian product of the homotopy category of $\mathbf{S}$-semimodules and the homotopy category of semiinjective complexes of A-injective $\mathbf{S}$-semicontramodules.

By Theorem 4.6 and Lemma 2.7, both functors so obtained factorize through the Cartesian product of semiderived categories of left semimodules and left semicontramodules and the derived functors so defined are naturally isomorphic. The same derived functor is obtained by restricting the functor of semihomomorphisms to the Cartesian product of the homotopy categories of semiprojective complexes of A-projective $\mathbf{S}$-semimodules and semiinjective complexes of A-injective $\mathbf{S}$-semicontramodules. One can also use semiprojective complexes of $\mathcal{C}$-coprojective $\mathbf{S}$-semimodules or semiinjective complexes of $\mathcal{C}$-coinjective $\mathbf{S}$-semicontramodules, etc.

In particular, when the coring $\mathcal{C}$ is a projective left and a flat right A-module and the ring A has a finite left homological dimension, one defines the double-sided derived functor

$$\mathrm{Coext}_{\mathcal{C}} \colon \mathsf{D}^{\mathrm{co}}(\mathcal{C}\text{-comod})^{\mathrm{op}} \times \mathsf{D}^{\mathrm{ctr}}(\mathcal{C}\text{-contra}) \longrightarrow \mathsf{D}(k\text{-mod})$$

by composing the functor of cohomomorphisms

$$\mathrm{Cohom}_{\mathcal{C}} \colon \mathsf{Hot}(\mathcal{C}\text{-comod})^{\mathrm{op}} \times \mathsf{Hot}(\mathcal{C}\text{-contra}) \longrightarrow \mathsf{Hot}(k\text{-mod})$$

with the functor of localization $\mathsf{Hot}(k\text{-mod}) \longrightarrow \mathsf{D}(k\text{-mod})$ and restricting it to either the Cartesian product of the homotopy category of complexes of coprojective $\mathcal{C}$-comodules and the homotopy category of arbitrary complexes of $\mathcal{C}$-contramodules, or the Cartesian product of the homotopy category of arbitrary complexes of $\mathcal{C}$-comodules and the homotopy category of complexes of coinjective $\mathcal{C}$-contramodules. The same derived functor is obtained by restricting the functor of cohomomorphisms to the Cartesian product of the homotopy categories of coprojective $\mathcal{C}$-comodules and coinjective $\mathcal{C}$-contramodules. One can also use coprojective complexes of $\mathcal{C}$-comodules or coinjective complexes of $\mathcal{C}$-contramodules.

Question. Assuming only that $\mathcal{C}$ is a flat left and right A-module, one can define the double-sided derived functor $\mathrm{Cotor}^{\mathcal{C}}$ on the Cartesian product of coderived categories of the exact categories of right and left $\mathcal{C}$-comodules of flat dimension over A not exceeding d, for any given d, using Lemma 2.7 and the corresponding version of Lemma 1.1.3. Analogously, assuming that $\mathcal{C}$ is a projective left and a flat right A-module, one can define the double-sided derived functor $\mathrm{Coext}_{\mathcal{C}}$ on the Cartesian product of the coderived category of left $\mathcal{C}$-comodules of projective dimension over A not exceeding d and the contraderived category of $\mathcal{C}$-contramodules of injective dimension over A not exceeding d. One can even do with the homological dimension assumption on only one of the arguments of $\mathrm{Cotor}^{\mathcal{C}}$ and $\mathrm{Coext}_{\mathcal{C}}$, using the corresponding versions of the results of Theorem 7.2.2. Can one define, at least, a derived functor $\mathrm{SemiTor}^{\mathbf{S}}$ for complexes of A-flat $\mathbf{S}$-semimodules and a derived functor $\mathrm{SemiExt}_{\mathbf{S}}$ for complexes of A-projective $\mathbf{S}$-semimodules and A-injective $\mathbf{S}$-semicontramodules without the homological dimension assumptions on A? The only problem one encounters attempting to do so comes from the homological dimension conditions in Propositions 1.2.3(c) and 3.2.3.1-2(c) and consequently in Lemmas 1.3.3 and 3.3.3; when $\mathbf{S}$ satisfies the conditions of Proposition 1.2.5(f) there is no problem.

Remark. In a way completely analogous to Remark 2.7, without any homological dimension assumptions one can define the double-sided derived functor $\mathrm{IndCoext}_{\mathcal{C}}$

for complexes of left $\mathcal{C}$-comodules in k-mod$^\omega$ and complexes of left $\mathcal{C}$-contramodules in the category k-mod$_\omega$ of ind-objects over k-mod representable by countable filtered inductive systems of k-modules. Here the category opposite to k-mod$_\omega$ is considered as a module category over the tensor category k-mod$^\omega$ and $\mathcal{C}$ is a coring over a ring $\mathcal{A}$ in k-mod$^\omega$. Appropriate coflatness and "contraprojectivity" conditions have to be imposed on $\mathcal{C}$. The countability assumption can be dropped.

4.8 Relatively semiprojective and semiinjective complexes

We keep the assumptions and notation of 4.5, 4.6, and 4.7.

One can compute the derived functor Coext$_\mathcal{C}$ using resolutions of other kinds. Namely, the complex of cohomomorphisms Cohom$_\mathcal{C}(\mathcal{M}^\bullet, \mathfrak{P}^\bullet)$ from a complex of $\mathcal{C}/\mathcal{A}$-coprojective left $\mathcal{C}$-comodules $\mathcal{M}^\bullet$ into a complex of $\mathcal{A}$-injective left $\mathcal{C}$-contramodules $\mathfrak{P}^\bullet$ represents an object naturally isomorphic to Coext$_\mathcal{C}(\mathcal{M}^\bullet, \mathfrak{P}^\bullet)$ in the derived category of k-modules. Indeed, the complex $\mathbb{L}_2(\mathfrak{P}^\bullet)$ is a complex of coinjective $\mathcal{C}$-contramodules and the cone of the morphism $\mathbb{L}_2(\mathfrak{P}^\bullet) \longrightarrow \mathfrak{P}^\bullet$ is contraacyclic with respect to the exact category of $\mathcal{A}$-injective $\mathcal{C}$-contramodules, hence the morphism Cohom$_\mathcal{C}(\mathcal{M}^\bullet, \mathbb{L}_2(\mathfrak{P}^\bullet)) \longrightarrow$ Cohom$_\mathcal{C}(\mathcal{M}^\bullet, \mathfrak{P}^\bullet)$ is an isomorphism. Analogously, the complex of cohomomorphisms Cohom$_\mathcal{C}(\mathcal{M}^\bullet, \mathfrak{P}^\bullet)$ from a complex of $\mathcal{A}$-projective left $\mathcal{C}$-comodules $\mathcal{M}^\bullet$ into a complex of $\mathcal{C}/\mathcal{A}$-coinjective left $\mathcal{C}$-contramodules $\mathfrak{P}^\bullet$ represents an object naturally isomorphic to Coext$_\mathcal{C}(\mathcal{M}^\bullet, \mathfrak{P}^\bullet)$ in the derived category of k-modules.

One can also compute the derived functor SemiExt$_\mathcal{C}$ using resolutions of other kinds. Namely, a complex of left $\mathcal{S}$-semimodules is called *semiprojective relative to $\mathcal{A}$* if the complex of semihomomorphisms from it into any complex of $\mathcal{A}$-injective left $\mathcal{S}$-semicontramodules that as a complex of $\mathcal{C}$-contramodules is contraacyclic with respect to the exact category of $\mathcal{A}$-injective $\mathcal{C}$-contramodules is acyclic (cf. Theorem 7.2.2(c)). The complex of semihomomorphisms SemiHom$_\mathcal{C}(\mathcal{M}^\bullet, \mathfrak{P}^\bullet)$ from a complex of left $\mathcal{S}$-semimodules $\mathcal{M}^\bullet$ semiprojective relative to $\mathcal{A}$ into a complex of $\mathcal{A}$-injective left $\mathcal{S}$-semicontramodules $\mathfrak{P}^\bullet$ represents an object naturally isomorphic to SemiExt$_\mathcal{S}(\mathcal{M}^\bullet, \mathfrak{P}^\bullet)$ in the derived category of k-modules. Indeed, $\mathbb{R}_3\mathbb{L}_2(\mathfrak{P}^\bullet)$ is a semiinjective complex of $\mathcal{S}$-semicontramodules connected with $\mathfrak{P}^\bullet$ by a chain of morphisms $\mathfrak{P}^\bullet \longleftarrow \mathbb{L}_2(\mathfrak{P}^\bullet) \longrightarrow \mathbb{R}_3\mathbb{L}_2(\mathfrak{P}^\bullet)$ whose cones are contraacyclic with respect to the exact category of $\mathcal{A}$-injective $\mathcal{C}$-contramodules and contractible over $\mathcal{C}$, respectively. Analogously, a complex of left $\mathcal{S}$-semicontramodules is called *semiinjective relative to $\mathcal{A}$* if the complex of semihomomorphisms into it from any complex of $\mathcal{A}$-projective left $\mathcal{S}$-semimodules that as a complex of $\mathcal{C}$-comodules is coacyclic with respect to the exact category of $\mathcal{A}$-projective $\mathcal{C}$-comodules is acyclic (cf. Theorem 7.2.2(b)). The complex of semihomomorphisms SemiHom$_\mathcal{C}(\mathcal{M}^\bullet, \mathfrak{P}^\bullet)$ from a complex of $\mathcal{A}$-projective left $\mathcal{S}$-semimodules to a complex of left $\mathcal{S}$-semicontramodules semiinjective relative to $\mathcal{A}$ represents an object naturally isomorphic to SemiExt$_\mathcal{S}(\mathcal{M}^\bullet, \mathfrak{P}^\bullet)$ in the derived category of k-modules. For example, the complex of $\mathcal{S}$-semimodules induced from a complex of $\mathcal{C}/\mathcal{A}$-coprojective left $\mathcal{C}$-comodules is semiprojective relative to A and the complex of S-semicontramodules coinduced from a complex of $\mathcal{C}/A$-coinjective left $\mathcal{C}$-contramodules is semiinjective relative to A.

A complex of left S-semimodules is called *semiprojective relative to* $\mathcal{C}$ if the complex of semihomomorphisms from it into any $\mathcal{C}$-contractible complex of $\mathcal{C}$-coinjective left S-semicontramodules is acyclic. The complex of semihomomorphisms $\mathrm{SemiHom}_{\mathcal{C}}(\mathcal{M}^{\bullet}, \mathfrak{P}^{\bullet})$ from a complex of left S-semimodules $\mathcal{M}^{\bullet}$ semiprojective relative to $\mathcal{C}$ into a complex of $\mathcal{C}$-coinjective left S-semicontramodules $\mathfrak{P}^{\bullet}$ represents an object naturally isomorphic to $\mathrm{SemiExt}_{\mathsf{S}}(\mathcal{M}^{\bullet}, \mathfrak{P}^{\bullet})$ in the derived category of k-modules. Indeed, $\mathbb{R}_3(\mathfrak{P}^{\bullet})$ is a semiinjective complex of S-semicontramodules and the cone of the morphism $\mathfrak{P}^{\bullet} \longrightarrow \mathbb{R}_3(\mathfrak{P}^{\bullet})$ is a $\mathcal{C}$-contractible complex of $\mathcal{C}$-coinjective S-semicontramodules. Analogously, a complex of left S-semicontramodules is called *semiinjective relative to* $\mathcal{C}$ if the complex of semihomomorphisms into it from any $\mathcal{C}$-contractible complex of $\mathcal{C}$-coprojective left S-semimodules is acyclic. The complex of semihomomorphisms $\mathrm{SemiHom}_{\mathcal{C}}(\mathcal{M}^{\bullet}, \mathfrak{P}^{\bullet})$ from a complex of $\mathcal{C}$-coprojective left S-semimodules $\mathcal{M}^{\bullet}$ into a complex of left S-semicontramodules $\mathfrak{P}^{\bullet}$ semiinjective relative to $\mathcal{C}$ represents an object naturally isomorphic to $\mathrm{SemiExt}_{\mathsf{S}}(\mathcal{M}^{\bullet}, \mathfrak{P}^{\bullet})$ in the derived category of k-modules. It follows that the complex of semihomomorphisms from a complex of left S-semimodules semiprojective relative to $\mathcal{C}$ into a $\mathcal{C}$-contraacyclic complex of $\mathcal{C}$-coinjective left S-semicontramodules is acyclic, and the complex of semihomomorphisms into a complex of left S-semicontramodules semiinjective relative to $\mathcal{C}$ from a $\mathcal{C}$-coacyclic complex of $\mathcal{C}$-coprojective left S-semimodules is acyclic. For example, the complex of S-semimodules induced from a complex of left $\mathcal{C}$-comodules is semiprojective relative to $\mathcal{C}$ and the complex of S-semicontramodules coinduced from a complex of left $\mathcal{C}$-contramodules is semiinjective relative to $\mathcal{C}$.

At last, a complex of A-projective left S-semimodules is called *semiprojective relative to* $\mathcal{C}$ *relative to* A ($\mathsf{S}/\mathcal{C}/A$-semiprojective) if the complex of semihomomorphisms from it into any $\mathcal{C}$-contractible complex of $\mathcal{C}/A$-coinjective left S-semicontramodules is acyclic. The complex of semihomomorphisms $\mathrm{SemiHom}_{\mathcal{C}}(\mathcal{M}^{\bullet}, \mathfrak{P}^{\bullet})$ from an $\mathsf{S}/\mathcal{C}/A$-semiprojective complex of A-projective left S-semimodules $\mathcal{M}^{\bullet}$ into a complex of $\mathcal{C}/A$-coinjective left S-semicontramodules $\mathfrak{P}^{\bullet}$ represents an object naturally isomorphic to $\mathrm{SemiExt}_{\mathsf{S}}(\mathcal{M}^{\bullet}, \mathfrak{P}^{\bullet})$ in the derived category of k-modules. Indeed, $\mathbb{R}_3(\mathfrak{P}^{\bullet})$ is a complex of left S-semicontramodules semiinjective relative to A and the cone of the morphism $\mathfrak{P}^{\bullet} \longrightarrow \mathbb{R}_3(\mathfrak{P}^{\bullet})$ is a $\mathcal{C}$-contractible complex of $\mathcal{C}/A$-coinjective S-semicontramodules.

Analogously, a complex of A-injective left S-semicontramodules is called *semiinjective relative to* $\mathcal{C}$ *relative to* A ($\mathsf{S}/\mathcal{C}/A$-semiinjective) if the complex of semihomomorphisms into it from any $\mathcal{C}$-contractible complex of $\mathcal{C}/A$-coprojective left S-semimodules is acyclic. The complex of semihomomorphisms $\mathrm{SemiHom}_{\mathcal{C}}(\mathcal{M}^{\bullet}, \mathfrak{P}^{\bullet})$ from a complex of $\mathcal{C}/A$-coprojective left S-semimodules $\mathcal{M}^{\bullet}$ into an $\mathsf{S}/\mathcal{C}/A$-semiinjective complex of A-injective left S-semicontramodules $\mathfrak{P}^{\bullet}$ represents an object naturally isomorphic to $\mathrm{SemiExt}_{\mathsf{S}}(\mathcal{M}^{\bullet}, \mathfrak{P}^{\bullet})$ in the derived category of k-modules. It follows that the complex of semihomomorphisms from an $S/C/A$-semiprojective complex of A-projective left S-semimodules into a C-contraacyclic complex of C/A-coinjective left S-semicontramodules is acyclic, and the complex of semihomomorphisms into an $S/C/A$-semiinjective complex of A-injective left S-semicontramodules from a C-coacyclic complex of C/A-coprojective left S-semimodules is acyclic. For example, the complex of S-semimodules induced from a complex of A-projective left C-comodules is $S/C/A$-semiprojective and the complex of S-semicontramodules coinduced from a complex of A-injective left C-contramodules is $S/C/A$-semiinjective.

The functors mapping the quotient categories of the homotopy categories of complexes of S-semimodules semiprojective relative to A, complexes of S-semimodules semiprojective relative to C, and $S/C/A$-semiprojective complexes of S-semimodules by their intersections with the thick subcategory of C-coacyclic complexes into the semiderived category of left S-semimodules are equivalences of triangulated categories. Analogously, the functors mapping the quotient categories of the homotopy categories of complexes of S-semicontramodules semiinjective relative to A, complexes of S-semicontramodules semiinjective relative to C, and $S/C/A$-semiinjective complexes of S-semicontramodules by their intersections with the thick subcategory of C-contraacyclic complexes into the semiderived category of left S-semicontramodules are equivalences of triangulated categories. The same applies to complexes of A-projective, C-coprojective, and C/A-coprojective S-semimodules and complexes of A-injective, C-coinjective, and C/A-coinjective S-semicontramodules. These results follow easily from either of Lemmas 2.5 or 2.6. So one can define the derived functor SemiExt$_S$ by restricting the functor of semihomomorphisms to these categories of complexes as explained above.

Remark. One can define the double-sided or right derived functor SemiExt$_S$ in the assumptions analogous to those of Remark 2.8 in completely analogous ways.

4.9 Remarks on derived semihomomorphisms from bisemimodules

Let S be a semialgebra over a coring C and $\mathcal{T}$ be a semialgebra over a coring $\mathcal{D}$, both satisfying the conditions of 4.6. One can define the double-sided derived functor

$$\mathbb{D}\,\mathrm{SemiHom}_S : \mathsf{D}^{\mathsf{si}}(S\text{-simod-}\mathcal{T})^{\mathrm{op}} \times \mathsf{D}^{\mathsf{si}}(S\text{-sicntr}) \longrightarrow \mathsf{D}^{\mathsf{si}}(\mathcal{T}\text{-sicntr})$$

by restricting the functor of semihomomorphisms SemiHom$_S :$ Hot$(S\text{-simod-}\mathcal{T})^{\mathrm{op}} \times$ Hot$(S\text{-sicntr}) \dashrightarrow \mathsf{D}^{\mathsf{si}}(\mathcal{T}\text{-sicntr})$ to the Cartesian product of the homotopy category of complexes of S-$\mathcal{T}$-bisemimodules and the homotopy category of semiinjective complexes of C-coinjective left S-semicontramodules (using the result of Remark 6.4). There is an associativity isomorphism

$$\mathrm{SemiExt}_S(\mathcal{K}^{\bullet} \lozenge_{\mathcal{T}}^{\mathcal{D}} \mathcal{M}^{\bullet}, \mathfrak{P}^{\bullet}) \simeq \mathrm{SemiExt}_{\mathcal{T}}(\mathcal{M}^{\bullet}, \mathbb{D}\,\mathrm{SemiHom}_S(\mathcal{K}^{\bullet}, \mathfrak{P}^{\bullet})).$$

Let $\mathfrak{R}$ be a semialgebra over a coring $\mathcal{E}$ satisfying the conditions of 4.6. If the k-algebra A is a flat k-module and the k-algebras B and F are projective k-modules, then the derived functor $\mathbb{D}\operatorname{SemiHom}$ can be defined using Lemma 2.7 in terms of *strongly $\mathcal{S}$-semiprojective* complexes of A-projective $\mathcal{S}$-$\mathcal{T}$-bisemimodules and semiinjective complexes of $\mathcal{C}$-coinjective left $\mathcal{S}$-semicontramodules (or *strongly semiinjective* complexes of A-injective left $\mathcal{S}$-semicontramodules). Here a complex of A-projective $\mathcal{S}$-$\mathcal{T}$-bisemimodules $\mathcal{K}^\bullet$ is called strongly $\mathcal{S}$-semiprojective if for any $\mathcal{C}$-contraacyclic complex of left $\mathcal{S}$-semicontramodules $\mathfrak{P}^\bullet$ the complex of left $\mathcal{T}$-semicontramodules $\operatorname{SemiHom}_{\mathcal{S}}(\mathcal{K}^\bullet, \mathfrak{P}^\bullet)$ is $\mathcal{D}$-contraacyclic; strongly semi-injective complexes are defined in an analogous way. In this case, there is an associativity isomorphism

$$\mathbb{D}\operatorname{SemiHom}_{\mathcal{S}}(\mathcal{K}^\bullet \lozenge_{\mathcal{T}}^{\mathcal{D}} \mathcal{M}^\bullet,\ \mathfrak{P}^\bullet) \simeq \mathbb{D}\operatorname{SemiHom}_{\mathcal{T}}(\mathcal{M}^\bullet, \mathbb{D}\operatorname{SemiHom}_{\mathcal{S}}(\mathcal{K}^\bullet, \mathfrak{P}^\bullet))$$

for any complex of $\mathcal{T}$-$\mathfrak{R}$-bisemimodules $\mathcal{M}^\bullet$, any complex of $\mathcal{S}$-$\mathcal{T}$-bisemimodules $\mathcal{K}^\bullet$, and any complex of left $\mathcal{S}$-semicontramodules $\mathfrak{P}^\bullet$.

In particular, even without any conditions on the k-module A, for any complex of right $\mathcal{S}$-semimodules $\mathcal{N}^\bullet$ and any complex of left $\mathcal{S}$-semimodules $\mathcal{M}^\bullet$ there is a natural isomorphism

$$\operatorname{Hom}_k(\operatorname{SemiTor}^{\mathcal{S}}(\mathcal{N}^\bullet, \mathcal{M}^\bullet), k^\vee) \simeq \operatorname{SemiExt}_{\mathcal{S}}(\mathcal{M}^\bullet, \operatorname{Hom}_k(\mathcal{N}^\bullet, k^\vee)).$$

5 Comodule-Contramodule Correspondence

5.1 Contratensor product and comodule/contramodule homomorphisms

Let $\mathcal{C}$ be a coring over a k-algebra A.

5.1.1 The *contratensor product* $\mathcal{N} \odot_{\mathcal{C}} \mathfrak{P}$ of a right $\mathcal{C}$-comodule $\mathcal{N}$ and a left $\mathcal{C}$-contramodule $\mathfrak{P}$ is a k-module defined as the cokernel of the pair of maps

$$\mathcal{N} \otimes_A \operatorname{Hom}_A(\mathcal{C}, \mathfrak{P}) \rightrightarrows \mathcal{N} \otimes_A \mathfrak{P}$$

one of which is induced by the $\mathcal{C}$-contraaction in $\mathfrak{P}$, while the other is the composition of the map induced by the $\mathcal{C}$-coaction in $\mathcal{N}$ and the map induced by the evaluation map $\mathcal{C} \otimes_A \operatorname{Hom}_A(\mathcal{C}, \mathfrak{P}) \longrightarrow \mathfrak{P}$.

The contratensor product operation is dual to homomorphisms in the category of contramodules: for any right $\mathcal{C}$-comodule $\mathcal{N}$ with a left action of a k-algebra B by $\mathcal{C}$-comodule endomorphisms, any left $\mathcal{C}$-contramodule $\mathfrak{P}$, and any left B-module U there is a natural isomorphism

$$\operatorname{Hom}_B(\mathcal{N} \odot_{\mathcal{C}} \mathfrak{P},\, U) \simeq \operatorname{Hom}^{\mathcal{C}}(\mathfrak{P}, \operatorname{Hom}_B(\mathcal{N}, U)).$$

Indeed, both k-modules are isomorphic to the kernel of the same pair of maps $\operatorname{Hom}_A(\mathfrak{P}, \operatorname{Hom}_B(\mathcal{N}, U)) \rightrightarrows \operatorname{Hom}_A(\operatorname{Hom}_A(\mathcal{C}, \mathfrak{P}), \operatorname{Hom}_B(\mathcal{N}, U))$. Taking $B = k$, one can conclude that for any right $\mathcal{C}$-comodule $\mathcal{N}$ and any left A-module V there is a natural isomorphism

$$\mathcal{N} \odot_{\mathcal{C}} \operatorname{Hom}_A(\mathcal{C}, V) \simeq \mathcal{N} \otimes_A V.$$

When $\mathcal{C}$ is a projective left A-module, the functor of contratensor product over $\mathcal{C}$ is right exact in both its arguments.

5.1.2 Let $\mathcal{D}$ be a coring over a k-algebra B. For any $\mathcal{C}$-$\mathcal{D}$-bicomodule $\mathcal{K}$ and any left $\mathcal{C}$-comodule $\mathcal{M}$, the k-module $\operatorname{Hom}_{\mathcal{C}}(\mathcal{K}, \mathcal{M})$ has a natural left $\mathcal{D}$-contramodule structure as the kernel of a pair of $\mathcal{D}$-contramodule morphisms $\operatorname{Hom}_A(\mathcal{K}, \mathcal{M}) \rightrightarrows \operatorname{Hom}_A(\mathcal{K}, \mathcal{M} \otimes_B \mathcal{D})$. Analogously, for any $\mathcal{D}$-$\mathcal{C}$-bicomodule $\mathcal{K}$ and any left $\mathcal{C}$-contramodule $\mathfrak{P}$, the k-module $\mathcal{K} \odot_{\mathcal{C}} \mathfrak{P}$ has a natural left $\mathcal{D}$-comodule structure as the cokernel of a pair of $\mathcal{D}$-comodule morphisms $\mathcal{K} \otimes_A \operatorname{Hom}_A(\mathcal{C}, \mathfrak{P}) \rightrightarrows \mathcal{K} \otimes_A \mathfrak{P}$.

For any left $\mathcal{D}$-comodule $\mathcal{M}$, any $\mathcal{D}$-$\mathcal{C}$-bicomodule $\mathcal{K}$, and any left $\mathcal{C}$-contramodule $\mathfrak{P}$ there is a natural isomorphism

$$\operatorname{Hom}_{\mathcal{D}}(\mathcal{K} \odot_{\mathcal{C}} \mathfrak{P},\, \mathcal{M}) \simeq \operatorname{Hom}^{\mathcal{C}}(\mathfrak{P}, \operatorname{Hom}_{\mathcal{D}}(\mathcal{K}, \mathcal{M})).$$

L. Positselski, *Homological Algebra of Semimodules and Semicontramodules*, Monografie Matematyczne 70, DOI 10.1007/978-3-0346-0436-9_6, © Springer Basel AG 2010

Indeed, a B-module map $\mathcal{K} \otimes_A \mathfrak{P} \longrightarrow \mathcal{M}$ factorizes through $\mathcal{K} \odot_{\mathcal{C}} \mathcal{M}$ if and only if the corresponding A-module map $\mathfrak{P} \longrightarrow \mathrm{Hom}_B(\mathcal{K}, \mathcal{M})$ is a $\mathcal{C}$-contramodule morphism, and a B-module map $\mathcal{K} \otimes_A \mathfrak{P} \longrightarrow \mathcal{M}$ is a $\mathcal{D}$-comodule morphism if and only if the corresponding A-module map $\mathfrak{P} \longrightarrow \mathrm{Hom}_B(\mathcal{K}, \mathcal{M})$ factorizes through $\mathrm{Hom}_{\mathcal{D}}(\mathcal{K}, \mathcal{M})$.

In particular, there is a pair of adjoint functors $\Psi_{\mathcal{C}} \colon \mathcal{C}$-comod $\longrightarrow \mathcal{C}$-contra and $\Phi_{\mathcal{C}} \colon \mathcal{C}$-contra $\longrightarrow \mathcal{C}$-comod between the categories of left $\mathcal{C}$-comodules and left $\mathcal{C}$-contramodules defined by the rules

$$\Psi_{\mathcal{C}}(\mathcal{M}) = \mathrm{Hom}_{\mathcal{C}}(\mathcal{C}, \mathcal{M}) \quad \text{and} \quad \Phi_{\mathcal{C}}(\mathfrak{P}) = \mathcal{C} \odot_{\mathcal{C}} \mathfrak{P}.$$

5.1.3   A left $\mathcal{C}$-comodule $\mathcal{M}$ is called *quite injective relative to A* (quite $\mathcal{C}/A$-injective) if the functor of $\mathcal{C}$-comodule homomorphisms into $\mathcal{M}$ maps A-split exact triples of left $\mathcal{C}$-comodules to exact triples. It is easy to see that a $\mathcal{C}$-comodule is quite $\mathcal{C}/A$-injective if and only if it is a direct summand of a coinduced $\mathcal{C}$-comodule. Analogously, a left $\mathcal{C}$-contramodule $\mathfrak{P}$ is called *quite projective relative to A* (quite $\mathcal{C}/A$-projective) if the functor of $\mathcal{C}$-contramodule homomorphisms from $\mathfrak{P}$ maps A-split exact triples of left $\mathcal{C}$-contramodules to exact triples. A $\mathcal{C}$-contramodule is quite $\mathcal{C}/A$-projective if and only if it is a direct summand of an induced $\mathcal{C}$-contramodule.

The restrictions of the functors $\Psi_{\mathcal{C}}$ and $\Phi_{\mathcal{C}}$ on the subcategories of quite $\mathcal{C}/A$-injective left $\mathcal{C}$-comodules and quite $\mathcal{C}/A$-projective left $\mathcal{C}$-contramodules are mutually inverse equivalences between these subcategories. Indeed, one has

$$\mathrm{Hom}_{\mathcal{C}}(\mathcal{C}, \ \mathcal{C} \otimes_A V) = \mathrm{Hom}_A(\mathcal{C}, V) \quad \text{and} \quad \mathcal{C} \odot \mathrm{Hom}_A(\mathcal{C}, V) = \mathcal{C} \otimes_A V,$$

so the functors $\Psi_{\mathcal{C}}$ and $\Phi_{\mathcal{C}}$ transform the coinduced comodule $\mathcal{C} \otimes_A V$ into the induced contramodule $\mathrm{Hom}_A(\mathcal{C}, V)$ and back. This equivalence between the categories of coinduced left $\mathcal{C}$-comodules and induced left $\mathcal{C}$-contramodules is a particular case of the isomorphism of Kleisli categories [26].

5.1.4   A left $\mathcal{C}$-comodule $\mathcal{M}$ is called *injective relative to A* ($\mathcal{C}/A$-injective) if the functor of homomorphisms into $\mathcal{M}$ maps exact triples of A-projective left $\mathcal{C}$-comodules to exact triples. A left $\mathcal{C}$-contramodule $\mathfrak{P}$ is called *projective relative to A* ($\mathcal{C}/A$-projective) if the functor of homomorphisms from $\mathfrak{P}$ maps exact triples of A-injective left $\mathcal{C}$-contramodules to exact triples. (Cf. Lemma 5.3.2.)

Remark.   What we call quite relatively injective comodules are usually called relatively injective comodules [23] (cf. [47]). We chose this nontraditional terminology for coherence with our definitions of relative coflatness, relative coprojectivity, etc., and also because what we call relatively injective comodules is a more important notion from our point of view.

Question.   One can compute modules Ext in the exact category of left $\mathcal{C}$-comodules with A-split exact triples in terms of the cobar resolution. When $\mathcal{C}$ is a projective left A-module, this resolution can be also used to compute modules Ext in the exact category of A-projective left $\mathcal{C}$-comodules, which therefore turn out to be the same. How can one compute modules Ext in the exact category of A-projective $\mathcal{C}$-comodules without making any projectivity assumptions on $\mathcal{C}$?

5.1.5 When $\mathcal{C}$ is a flat right A-module, the coinduction functor A-mod $\longrightarrow$ $\mathcal{C}$-comod preserves injective objects. It follows easily that any left $\mathcal{C}$-comodule is a subcomodule of an injective $\mathcal{C}$-comodule; a $\mathcal{C}$-comodule is injective if and only if it is a direct summand of a $\mathcal{C}$-comodule coinduced from an injective A-module. Analogously, when $\mathcal{C}$ is a projective left A-module, the induction functor A-mod $\longrightarrow$ $\mathcal{C}$-contra preserves projective objects. Hence any left $\mathcal{C}$-contramodule is a quotient contramodule of a projective $\mathcal{C}$-contramodule; a $\mathcal{C}$-contramodule is projective if and only if it is a direct summand of a $\mathcal{C}$-contramodule induced from a projective A-module.

5.1.6 When $\mathcal{C}$ is a flat left A-module, a left $\mathcal{C}$-contramodule $\mathfrak{P}$ is called *contraflat* if the functor of contratensor product with $\mathfrak{P}$ is exact on the category of right $\mathcal{C}$-comodules. The $\mathcal{C}$-contramodule induced from a flat A-module is contraflat. Any projective $\mathcal{C}$-contramodule is contraflat.

A left $\mathcal{C}$-contramodule $\mathfrak{P}$ is called *quite $\mathcal{C}/A$-contraflat* if the functor of contratensor product with $\mathfrak{P}$ maps those exact triples of right $\mathcal{C}$-comodules which as exact triples of A-modules remain exact after the tensor product with any left A-module to exact triples. Any quite $\mathcal{C}/A$-projective $\mathcal{C}$-contramodule is quite $\mathcal{C}/A$-contraflat. A left $\mathcal{C}$-contramodule $\mathfrak{P}$ is called $\mathcal{C}/A$-*contraflat* if the functor of contratensor product with $\mathfrak{P}$ maps exact triples of A-flat right $\mathcal{C}$-comodules to exact triples. Using the dualization functor $\mathrm{Hom}_k(-, k^\vee)$, one can easily check that any $\mathcal{C}/A$-projective $\mathcal{C}$-comodule is $\mathcal{C}/A$-contraflat.

5.2 Associativity isomorphisms

Let $\mathcal{C}$ be a coring over a k-algebra A and $\mathcal{D}$ be a coring over a k-algebra B. The following three propositions will be mostly applied to the case of $\mathcal{K} = \mathcal{D} = \mathcal{C}$ in the sequel.

Proposition 1. *Let $\mathcal{N}$ be a right $\mathcal{D}$-comodule, $\mathcal{K}$ be a $\mathcal{D}$-$\mathcal{C}$-bicomodule, and $\mathfrak{P}$ be a left $\mathcal{C}$-contramodule. Then there is a natural map*

$$(\mathcal{N} \,\square_\mathcal{D}\, \mathcal{K}) \odot_\mathcal{C} \mathfrak{P} \longrightarrow \mathcal{N} \,\square_\mathcal{D}\, (\mathcal{K} \odot_\mathcal{C} \mathfrak{P})$$

whenever the cotensor product $\mathcal{N} \square_\mathcal{D} \mathcal{K}$ is endowed with a right $\mathcal{C}$-comodule structure such that the map $\mathcal{N} \square_\mathcal{D} \mathcal{K} \longrightarrow \mathcal{N} \otimes_B \mathcal{K}$ is a $\mathcal{C}$-comodule morphism. This natural map is an isomorphism, at least, in the following cases:

(a) *$\mathcal{C}$ is a flat left A-module and $\mathfrak{P}$ is a contraflat left $\mathcal{C}$-contramodule,*

(b) *$\mathfrak{P}$ is a quite $\mathcal{C}/A$-contraflat left $\mathcal{C}$-contramodule and $\mathcal{K}$ as a left $\mathcal{D}$-comodule with a right A-module structure is coinduced from a B-A-bimodule;*

(c) $\mathfrak{P}$ is a $\mathcal{C}/A$-contraflat left $\mathcal{C}$-contramodule, $\mathcal{D}$ is a flat right B-module, $\mathcal{N}$ is a flat right B-module, and $\mathcal{K}$ as a left $\mathcal{D}$-comodule with a right A-module structure is coinduced from an A-flat B-A-bimodule;

(d) $\mathfrak{P}$ is a $\mathcal{C}/A$-contraflat left $\mathcal{C}$-contramodule, $\mathcal{D}$ is a flat right B-module, $\mathcal{N}$ is a flat right B-module, $\mathcal{K}$ is a flat right A-module, $\mathcal{K}$ is a $\mathcal{D}/B$-coflat left $\mathcal{D}$-comodule, and the ring A has a finite weak homological dimension;

(e) $\mathcal{N}$ is a quasicoflat right $\mathcal{D}$-comodule.

Proof. The map $(\mathcal{N} \,\square_{\mathcal{D}}\, \mathcal{K}) \odot_{\mathcal{C}} \mathfrak{P} \longrightarrow \mathcal{N} \otimes_B \mathcal{K} \odot_{\mathcal{C}} \mathfrak{P}$ has equal compositions with two maps $\mathcal{N} \otimes_B \mathcal{K} \odot_{\mathcal{C}} \mathfrak{P} \rightrightarrows \mathcal{N} \otimes_B \mathcal{D} \otimes_B \mathcal{K} \odot_{\mathcal{C}} \mathfrak{P}$, so there is a natural map $(\mathcal{N} \,\square_{\mathcal{D}}\, \mathcal{K}) \odot_{\mathcal{C}} \mathfrak{P} \longrightarrow \mathcal{N} \,\square_{\mathcal{D}}\, (\mathcal{K} \odot_{\mathcal{C}} \mathfrak{P})$. Besides, the composition

$$(\mathcal{N} \,\square_{\mathcal{D}}\, \mathcal{K}) \otimes_A \mathfrak{P} \longrightarrow \mathcal{N} \,\square_{\mathcal{D}}\, (\mathcal{K} \otimes_A \mathfrak{P}) \longrightarrow \mathcal{N} \,\square_{\mathcal{D}}\, (\mathcal{K} \odot_{\mathcal{C}} \mathfrak{P})$$

annihilates the difference between two maps $(\mathcal{N} \,\square_{\mathcal{D}}\, \mathcal{K}) \otimes_A \mathrm{Hom}_A(\mathcal{C}, \mathfrak{P}) \rightrightarrows (\mathcal{N} \,\square_{\mathcal{D}}\, \mathcal{K}) \otimes_A \mathfrak{P}$, which leads to the same natural map $(\mathcal{N}\square_{\mathcal{D}}\mathcal{K}) \odot_{\mathcal{C}} \mathfrak{P} \longrightarrow \mathcal{N}\square_{\mathcal{D}}(\mathcal{K}\odot_{\mathcal{C}}\mathfrak{P})$. To prove cases (a-d), one shows that the sequence $0 \longrightarrow \mathcal{N}\square_{\mathcal{D}}\mathcal{K} \longrightarrow \mathcal{N}\otimes_B\mathcal{K} \longrightarrow \mathcal{N}\otimes_B\mathcal{D}\otimes_B\mathcal{K}$ remains exact after taking the contratensor product with $\mathfrak{P}$. Indeed, the case (a) is obvious, in the cases (b-c) this exact sequence of right A-modules splits, and in the cases (c-d) this sequence of right A-modules is exact with respect to the exact category of flat A-modules (see the proof of Proposition 1.2.3). To prove (e), one notices that the sequence $\mathcal{K} \otimes_A \mathrm{Hom}_A(\mathcal{C}, \mathfrak{P}) \longrightarrow \mathcal{K} \otimes_A \mathfrak{P} \longrightarrow \mathcal{K} \odot_{\mathcal{C}} \mathfrak{P} \longrightarrow 0$ remains exact after taking the cotensor product with $\mathcal{N}$ and uses Proposition 1.2.3(b). $\square$

Proposition 2. *Let $\mathcal{L}$ be a left $\mathcal{D}$-comodule, $\mathcal{K}$ be a $\mathcal{C}$-$\mathcal{D}$-bicomodule, and $\mathcal{M}$ be a left $\mathcal{C}$-comodule. Then there is a natural map*

$$\mathrm{Cohom}_{\mathcal{D}}(\mathcal{L}, \mathrm{Hom}_{\mathcal{C}}(\mathcal{K}, \mathcal{M})) \longrightarrow \mathrm{Hom}_{\mathcal{C}}(\mathcal{K} \,\square_{\mathcal{D}}\, \mathcal{L}, \mathcal{M})$$

whenever the cotensor product $\mathcal{K}\square_{\mathcal{D}}\mathcal{L}$ is endowed with a left $\mathcal{C}$-comodule structure such that the map $\mathcal{K} \,\square_{\mathcal{D}}\, \mathcal{L} \longrightarrow \mathcal{K} \otimes_B \mathcal{L}$ is a $\mathcal{C}$-comodule morphism. This natural map is an isomorphism, at least, in the following cases:

(a) $\mathcal{C}$ is a flat right A-module and $\mathcal{M}$ is an injective left $\mathcal{C}$-comodule;

(b) $\mathcal{M}$ is a quite $\mathcal{C}/A$-injective left $\mathcal{C}$-comodule and $\mathcal{K}$ as a right $\mathcal{D}$-comodule with a left A-module structure is coinduced from an A-B-bimodule;

(c) $\mathcal{M}$ is a $\mathcal{C}/A$-injective left $\mathcal{C}$-comodule, $\mathcal{D}$ is a projective left B-module, $\mathcal{L}$ is a projective left B-module, and $\mathcal{K}$ as a right $\mathcal{D}$-comodule with a left A-module structure is coinduced from an A-projective A-B-bimodule;

(d) $\mathcal{M}$ is a $\mathcal{C}/A$-injective left $\mathcal{C}$-comodule, $\mathcal{D}$ is a projective left B-module, $\mathcal{L}$ is a projective left B-module, $\mathcal{K}$ is a projective left A-module, $\mathcal{K}$ is a $\mathcal{D}/B$-coflat right $\mathcal{D}$-comodule, and the ring A has a finite left homological dimension;

(e) $\mathcal{L}$ is a quasiprojective left $\mathcal{D}$-comodule.

Proof. Analogous to the proof of Proposition 1 and Proposition 3 below (see also the proof of Proposition 3.2.3.1). In particular, to prove (e) one notices that the sequence $0 \longrightarrow \mathrm{Hom}_{\mathcal{C}}(\mathcal{K}, \mathcal{M}) \longrightarrow \mathrm{Hom}_A(\mathcal{K}, \mathcal{M}) \longrightarrow \mathrm{Hom}_A(\mathcal{K}, \mathcal{C} \otimes_A \mathcal{M})$ remains exact after taking the cohomomorphisms from $\mathcal{L}$. □

Proposition 3. *Let $\mathfrak{P}$ be a left $\mathcal{C}$-contramodule, $\mathcal{K}$ be a $\mathcal{D}$-$\mathcal{C}$-bicomodule, and $\mathfrak{Q}$ be a left $\mathcal{D}$-contramodule. Then there is a natural map*

$$\mathrm{Cohom}_{\mathcal{D}}(\mathcal{K} \odot_{\mathcal{C}} \mathfrak{P}, \, \mathfrak{Q}) \longrightarrow \mathrm{Hom}^{\mathcal{C}}(\mathfrak{P}, \mathrm{Cohom}_{\mathcal{D}}(\mathcal{K}, \mathfrak{Q}))$$

whenever the cohomomorphism module $\mathrm{Cohom}_{\mathcal{D}}(\mathcal{K}, \mathfrak{Q})$ is endowed with a left $\mathcal{C}$-contramodule structure such that the map $\mathrm{Hom}_B(\mathcal{K}, \mathfrak{Q}) \longrightarrow \mathrm{Cohom}_{\mathcal{D}}(\mathcal{K}, \mathfrak{Q})$ is a $\mathcal{C}$-contramodule morphism. This natural map is an isomorphism, at least, in the following cases:

(a) *$\mathcal{C}$ is a projective left A-module and $\mathfrak{P}$ is a projective left $\mathcal{C}$-contramodule;*

(b) *$\mathfrak{P}$ is a quite $\mathcal{C}/A$-projective left $\mathcal{C}$-contramodule and $\mathcal{K}$ as a left $\mathcal{D}$-comodule with a right A-module structure is coinduced from a B-A-bimodule,*

(c) *$\mathfrak{P}$ is a $\mathcal{C}/A$-projective left $\mathcal{C}$-contramodule, $\mathcal{D}$ is a flat right B-module, $\mathfrak{Q}$ is an injective left B-module, and $\mathcal{K}$ as a left $\mathcal{D}$-comodule with a right A-module structure is coinduced from an A-flat B-A-bimodule;*

(d) *$\mathfrak{P}$ is a $\mathcal{C}/A$-projective left $\mathcal{C}$-contramodule, $\mathcal{D}$ is a flat right B-module, $\mathfrak{Q}$ is an injective left B-module, $\mathcal{K}$ is a flat right A-module, $\mathcal{K}$ is a $\mathcal{D}/B$-projective left $\mathcal{D}$-comodule, and the ring A has a finite left homological dimension;*

(e) *$\mathfrak{Q}$ is a quasicoinjective left $\mathcal{D}$-contramodule.*

Proof. The map $\mathrm{Hom}^{\mathcal{C}}(\mathfrak{P}, \mathrm{Hom}_B(\mathcal{K}, \mathfrak{Q})) \longrightarrow \mathrm{Hom}^{\mathcal{C}}(\mathfrak{P}, \mathrm{Cohom}_{\mathcal{D}}(K, \mathfrak{Q}))$ annihilates the difference of two maps

$$\mathrm{Hom}^{\mathcal{C}}(\mathfrak{P}, \mathrm{Hom}_B(\mathcal{D} \otimes_B \mathcal{K}, \mathfrak{Q})) \rightrightarrows \mathrm{Hom}^{\mathcal{C}}(\mathfrak{P}, \mathrm{Hom}_B(\mathcal{K}, \mathfrak{Q}))$$

and this pair of maps can be identified with the pair of maps $\mathrm{Hom}_B(\mathcal{D} \otimes_B \mathcal{K} \odot_{\mathcal{C}} \mathfrak{P}, \mathfrak{Q}) \rightrightarrows \mathrm{Hom}_B(\mathcal{K} \odot_{\mathcal{C}} \mathfrak{P}, \mathfrak{Q})$ whose cokernel is, by the definition, the cohomomorphism module $\mathrm{Cohom}_{\mathcal{D}}(\mathcal{K} \odot_{\mathcal{C}} \mathfrak{P}, \mathfrak{Q})$. Hence there is a natural map $\mathrm{Cohom}_{\mathcal{D}}(\mathcal{K} \odot_{\mathcal{C}} \mathfrak{P}, \mathfrak{Q}) \longrightarrow \mathrm{Hom}^{\mathcal{C}}(\mathfrak{P}, \mathrm{Cohom}_{\mathcal{D}}(\mathcal{K}, \mathfrak{Q}))$. Besides, the composition

$$\mathrm{Cohom}_{\mathcal{D}}(\mathcal{K} \odot_{\mathcal{C}} \mathfrak{P}, \, \mathfrak{Q}) \longrightarrow \mathrm{Cohom}_{\mathcal{D}}(\mathcal{K} \otimes_A \mathfrak{P}, \, \mathfrak{Q}) \longrightarrow \mathrm{Hom}_A(\mathfrak{P}, \mathrm{Cohom}_{\mathcal{D}}(\mathcal{K}, \mathfrak{Q}))$$

has equal compositions with the two maps

$$\mathrm{Hom}_A(\mathfrak{P}, \mathrm{Cohom}_{\mathcal{D}}(\mathcal{K}, \mathfrak{Q})) \rightrightarrows \mathrm{Hom}_A(\mathrm{Hom}_A(\mathcal{C}, \mathfrak{P}), \mathrm{Cohom}_{\mathcal{D}}(\mathcal{K}, \mathfrak{Q})),$$

which leads to the same natural map

$$\mathrm{Cohom}_{\mathcal{D}}(\mathcal{K} \odot_{\mathcal{C}} \mathfrak{P}, \, \mathfrak{Q}) \longrightarrow \mathrm{Hom}^{\mathcal{C}}(\mathfrak{P}, \mathrm{Cohom}_{\mathcal{D}}(\mathcal{K}, \mathfrak{Q})).$$

To prove cases (a–d), one shows that the sequence $\mathrm{Hom}_B(\mathcal{D} \otimes_B \mathcal{K}, \, \mathfrak{Q}) \longrightarrow \mathrm{Hom}_B(\mathcal{K}, \mathfrak{Q}) \longrightarrow \mathrm{Cohom}_{\mathcal{D}}(\mathcal{K}, \mathfrak{Q}) \longrightarrow 0$ remains exact after applying the functor $\mathrm{Hom}^{\mathcal{C}}(\mathfrak{P}, -)$. Indeed, the case (a) is obvious, in the cases (b–d) this sequence of left A-modules splits, and in the cases (c–d) it is also an exact sequence of injective A-modules (see the proof of Proposition 3.2.3.2). To prove (e), one notices that the sequence $\mathcal{K} \otimes_A \mathrm{Hom}_A(\mathcal{C}, \mathfrak{P}) \longrightarrow \mathcal{K} \otimes_A \mathfrak{P} \longrightarrow \mathcal{K} \odot_{\mathcal{C}} \mathfrak{P} \longrightarrow 0$ remains exact after taking the cohomomorphisms into $\mathfrak{Q}$ and uses Proposition 3.2.3.2(b).   □

In the case of $\mathcal{K} = \mathcal{D} = \mathcal{C}$, the natural maps defined in Propositions 2–3 have the following property of compatibility with the adjoint functors $\Psi_{\mathcal{C}}$ and $\Phi_{\mathcal{C}}$: for any left $\mathcal{C}$-comodule $\mathcal{M}$ and any left $\mathcal{C}$-contramodule $\mathfrak{P}$ the maps

$$\mathrm{Cohom}_{\mathcal{C}}(\Phi_{\mathcal{C}}(\mathfrak{P}), \Psi_{\mathcal{C}}(\mathcal{M})) \longrightarrow \mathrm{Hom}_{\mathcal{C}}(\Phi_{\mathcal{C}}(\mathfrak{P}), \mathcal{M})$$

and

$$\mathrm{Cohom}_{\mathcal{C}}(\Phi_{\mathcal{C}}(\mathfrak{P}), \Psi_{\mathcal{C}}(\mathcal{M})) \longrightarrow \mathrm{Hom}^{\mathcal{C}}(\mathfrak{P}, \Psi_{\mathcal{C}}(\mathcal{M}))$$

form a commutative diagram with the isomorphism

$$\mathrm{Hom}_{\mathcal{C}}(\Phi_{\mathcal{C}}(\mathfrak{P}), \mathcal{M}) \simeq \mathrm{Hom}^{\mathcal{C}}(\mathfrak{P}, \Psi_{\mathcal{C}}(\mathcal{M})).$$

The following important lemma is deduced as a corollary of Propositions 2–3.

Lemma.

(a) *A $\mathcal{C}$-comodule is quasicoprojective if and only if it is quite $\mathcal{C}/A$-injective. If $\mathcal{C}$ is a projective left A-module, then a left $\mathcal{C}$-comodule is coprojective if and only if it is a direct summand of a comodule coinduced from a projective A-module.*

(b) *A $\mathcal{C}$-contramodule is quasicoinjective if and only if it is quite $\mathcal{C}/A$-projective. If $\mathcal{C}$ is a flat right A-module, then a left $\mathcal{C}$-contramodule is coinjective if and only if it is a direct summand of a contramodule induced from an injective A-module.*

Proof. Part (a): let $\mathcal{M}$ be a quasicoprojective left $\mathcal{C}$-comodule. Denote by l the coaction map $\mathcal{M} \longrightarrow \mathcal{C} \otimes_A \mathcal{M}$. It is an A-split injective morphism of quasicoprojective $\mathcal{C}$-comodules. According to Proposition 2(e), we have an isomorphism of morphisms $\mathrm{Hom}_{\mathcal{C}}(l, \mathcal{M}) \simeq \mathrm{Cohom}_{\mathcal{C}}(l, \mathrm{Hom}_{\mathcal{C}}(\mathcal{C}, \mathcal{M}))$. But the map $\mathrm{Cohom}_{\mathcal{C}}(l, \mathfrak{P})$ is surjective for any left $\mathcal{C}$-contramodule $\mathfrak{P}$. Therefore, the map $\mathrm{Hom}_{\mathcal{C}}(l, \mathcal{M})$ is also surjective, hence the morphism l splits and the comodule $\mathcal{M}$ is quite $\mathcal{C}/A$-injective. Now suppose that $\mathcal{M}$ is coprojective; then we already know that $\mathcal{M}$ is quite $\mathcal{C}/A$-injective. Set $\mathfrak{P} = \Psi_{\mathcal{C}}(\mathcal{M})$. It follows from Proposition 3(b) that there is an isomorphism of functors $\mathrm{Hom}^{\mathcal{C}}(\mathfrak{P}, -) \simeq \mathrm{Cohom}_{\mathcal{C}}(\mathcal{M}, -)$ on the category of left $\mathcal{C}$-contramodules. Therefore, the $\mathcal{C}$-contramodule $\mathfrak{P}$ is projective, hence it is a direct summand of a $\mathcal{C}$-contramodule induced from a projective A-module and $\mathcal{M}$ is a direct summand of the $\mathcal{C}$-comodule coinduced from the same projective A-module. The proof of part (b) is completely analogous; it uses Propositions 3(e) and 2(b).   □

Question. Are there any analogues of the results of the lemma for (quasi)coflat comodules and (quite relatively) contraflat contramodules?

5.3 Relatively injective comodules and relatively projective contramodules

Assume that $\mathcal{C}$ is a projective left A-module. For any right $\mathcal{C}$-comodule $\mathcal{N}$ and any left $\mathcal{C}$-contramodule $\mathfrak{P}$ denote by $\mathrm{Ctrtor}_i^{\mathcal{C}}(\mathcal{N}, \mathfrak{P})$ the sequence of left derived functors in the second argument of the right exact functor of contratensor product $\mathcal{N} \odot_{\mathcal{C}} \mathfrak{P}$. By the definition, the k-modules $\mathrm{Ctrtor}_i^{\mathcal{C}}(\mathcal{N}, \mathfrak{P})$ are computed using a left projective resolution of the $\mathcal{C}$-contramodule $\mathfrak{P}$. Since projective contramodules are contraflat, the functor $\mathrm{Ctrtor}_*^{\mathcal{C}}(\mathcal{N}, \mathfrak{P})$ assigns long exact sequences to exact triples in either of its arguments.

Question. Can one compute the derived functor Ctrtor using contraflat resolutions of the second argument? In other words, is it true that $\mathrm{Ctrtor}_i^{\mathcal{C}}(\mathcal{N}, \mathfrak{P}) = 0$ for any right $\mathcal{C}$-comodule $\mathcal{N}$, any contraflat left $\mathcal{C}$-contramodule $\mathfrak{P}$, and all $i > 0$? Also, is it true that $\mathrm{Ctrtor}_i^{\mathcal{C}}(\mathcal{N}, \mathfrak{P}) = 0$ for any A-flat right $\mathcal{C}$-comodule $\mathcal{N}$, any (quite) $\mathcal{C}/A$-contraflat left $\mathcal{C}$-contramodule $\mathfrak{P}$, and all $i > 0$? A related question: is $\mathrm{Ctrtor}_{>0}^{\mathcal{C}}(\mathcal{N}, \mathfrak{P})$ an effaceable functor of its first argument?

Now assume that $\mathcal{C}$ is a projective left and a flat right A-module and the ring A has a finite left homological dimension.

Lemma 1.

(a) *A left $\mathcal{C}$-comodule $\mathcal{M}$ is $\mathcal{C}/A$-injective if and only if for any A-projective left $\mathcal{C}$-comodule $\mathcal{L}$ the k-modules $\mathrm{Ext}_{\mathcal{C}}^i(\mathcal{L}, \mathcal{M})$ of Yoneda extensions in the abelian category of left $\mathcal{C}$-comodules vanish for all $i > 0$. In particular, the functor of $\mathcal{C}$-comodule homomorphisms from an A-projective left $\mathcal{C}$-comodule $\mathcal{L}$ maps exact triples of $\mathcal{C}/A$-injective left $\mathcal{C}$-comodules to exact triples. Besides, the class of $\mathcal{C}/A$-injective left $\mathcal{C}$-comodules is closed under extensions and cokernels of injective morphisms.*

(b) *A left $\mathcal{C}$-contramodule $\mathfrak{P}$ is $\mathcal{C}/A$-projective if and only if for any A-injective left $\mathcal{C}$-contramodule $\mathfrak{Q}$ the k-modules $\mathrm{Ext}^{\mathcal{C},i}(\mathfrak{P}, \mathfrak{Q})$ of Yoneda extensions in the abelian category of left $\mathcal{C}$-contramodules vanish for all $i > 0$. In particular, the functor of $\mathcal{C}$-contramodule homomorphisms into an A-injective left $\mathcal{C}$-contramodule $\mathfrak{Q}$ maps exact triples of $\mathcal{C}/A$-projective left $\mathcal{C}$-contramodules to exact triples. Besides, the class of $\mathcal{C}/A$-projective left $\mathcal{C}$-contramodules is closed under extensions and kernels of surjective morphisms.*

(c) *For any $\mathcal{C}/A$-projective left $\mathcal{C}$-contramodule $\mathfrak{P}$ and any A-flat right $\mathcal{C}$-comodule $\mathcal{N}$ the k-modules $\mathrm{Ctrtor}_i^{\mathcal{C}}(\mathcal{N}, \mathfrak{P})$ vanish for all $i > 0$. In particular, the functor of contratensor product with an A-flat right $\mathcal{C}$-comodule maps exact triples of $\mathcal{C}/A$-projective left $\mathcal{C}$-contramodules to exact triples.*

Proof. Part (a): the "if" part of the first assertion is obvious; let us prove the "only if" part. An arbitrary element of $\mathrm{Ext}_{\mathcal{C}}^i(\mathcal{L}, \mathcal{M})$ can be represented by a morphism of degree i from an exact complex $\cdots \to \mathcal{L}_i \to \cdots \to \mathcal{L}_0 \to \mathcal{L} \to 0$ to the comodule $\mathcal{M}$. According to Lemma 3.1.3(a), any left $\mathcal{C}$-comodule is a surjective image of an A-projective $\mathcal{C}$-comodule. Therefore, one can assume that the comodules $\mathcal{L}_i$ are A-projective. Now if $\mathcal{L}$ is also A-projective, then our exact complex of $\mathcal{C}$-comodules is composed of exact triples of A-projective $\mathcal{C}$-comodules, so if $\mathcal{M}$ is $\mathcal{C}/A$-injective, then the complex of homomorphisms into $\mathcal{M}$ from this complex of $\mathcal{C}$-comodules is acyclic. The remaining two assertions follow from the first one. The proof of part (b) is completely analogous. To prove (c), notice the isomorphism $\mathrm{Hom}_k(\mathrm{Ctrtor}_i^{\mathcal{C}}(\mathcal{N}, \mathfrak{P}), k^\vee) \simeq \mathrm{Ext}^{\mathcal{C},i}(\mathfrak{P}, \mathrm{Hom}_k(\mathcal{N}, k^\vee))$. $\qquad\square$

Remark. Analogues of the third assertion of Lemma 1(a) and the third assertion of Lemma 1(b) are *not* true for quite relatively injective comodules and quite relatively projective contramodules (see Remark 7.4.3; cf. Remark 9.1).

Theorem. *For any $\mathcal{C}/A$-injective left $\mathcal{C}$-comodule $\mathcal{M}$ the left $\mathcal{C}$-contramodule $\Psi_{\mathcal{C}}(\mathcal{M})$ is $\mathcal{C}/A$-projective and for any $\mathcal{C}/A$-projective left $\mathcal{C}$-contramodule $\mathfrak{P}$ the left $\mathcal{C}$-comodule $\Phi_{\mathcal{C}}(\mathcal{M})$ is $\mathcal{C}/A$-injective. The restrictions of the functors $\Psi_{\mathcal{C}}$ and $\Phi_{\mathcal{C}}$ to the full subcategories of $\mathcal{C}/A$-injective $\mathcal{C}$-comodules and $\mathcal{C}/A$-projective $\mathcal{C}$-contramodules are mutually inverse equivalences between these subcategories.*

Proof. Let us first show that the injective dimension of a $\mathcal{C}/A$-injective left $\mathcal{C}$-comodule $\mathcal{M}$ in the abelian category of $\mathcal{C}$-comodules does not exceed the left homological dimension d of the ring A. Indeed, it follows from Lemma 3.1.3(a) that any left $\mathcal{C}$-comodule $\mathcal{L}$ has a finite resolution $0 \to \mathcal{L}_d \to \cdots \to \mathcal{L}_0 \to \mathcal{L} \to 0$ with A-projective $\mathcal{C}$-comodules $\mathcal{L}_j$; and since $\mathrm{Ext}^i(\mathcal{L}_j, \mathcal{M}) = 0$ for all j and all $i > 0$, the complex $\mathrm{Hom}_{\mathcal{C}}(\mathcal{L}_\bullet, \mathcal{M})$ computes $\mathrm{Ext}_{\mathcal{C}}^*(\mathcal{L}, \mathcal{M})$. So the $\mathcal{C}$-comodule $\mathcal{M}$ has a finite injective resolution, and consequently it has a finite resolution $0 \to \mathcal{M} \to \mathcal{K}^0 \to \cdots \to \mathcal{K}^d \to 0$ consisting of quite $\mathcal{C}/A$-injective $\mathcal{C}$-comodules $\mathcal{K}^j$. According to Lemma 1(a), this exact sequence is composed of exact triples of $\mathcal{C}/A$-injective $\mathcal{C}$-comodules, which the functor $\Psi_{\mathcal{C}}$ maps to exact triples; so the sequence $0 \longrightarrow \Psi_{\mathcal{C}}(\mathcal{M}) \longrightarrow \Psi_{\mathcal{C}}(\mathcal{K}^0) \longrightarrow \cdots \longrightarrow \Psi_{\mathcal{C}}(\mathcal{K}^d) \longrightarrow 0$ is also exact. Since the $\mathcal{C}$-contramodules $\Psi_{\mathcal{C}}(\mathcal{K}^j)$ are quite $\mathcal{C}/A$-projective, it follows from Lemma 1(b) that the $\mathcal{C}$-contramodule $\Psi_{\mathcal{C}}(\mathcal{M})$ is $\mathcal{C}/A$-projective and the latter exact sequence is composed of exact triples of $\mathcal{C}/A$-projective $\mathcal{C}$-contramodules. Thus it follows from Lemma 1(c) that the sequence $0 \longrightarrow \Phi_{\mathcal{C}}\Psi_{\mathcal{C}}(\mathcal{M}) \longrightarrow \Phi_{\mathcal{C}}\Psi_{\mathcal{C}}(\mathcal{K}^0) \longrightarrow \cdots \longrightarrow \Phi_{\mathcal{C}}\Psi_{\mathcal{C}}(\mathcal{K}^d) \longrightarrow 0$ is also exact. Now since the adjunction maps $\Phi_{\mathcal{C}}\Psi_{\mathcal{C}}(\mathcal{K}^j) \longrightarrow \mathcal{K}^j$ are isomorphisms, the adjunction map $\Phi_{\mathcal{C}}\Psi_{\mathcal{C}}(\mathcal{M}) \longrightarrow \mathcal{M}$ is also an isomorphism. The remaining assertions are proven in a completely analogous way. $\qquad\square$

Lemma 2.

(a) *In the above assumptions, a left $\mathcal{C}$-comodule is $\mathcal{C}/A$-coprojective if and only if it is $\mathcal{C}/A$-injective.*

(b) *In the above assumptions, a left $\mathcal{C}$-contramodule is $\mathcal{C}/A$-coinjective if and only if it is $\mathcal{C}/A$-projective.*

Proof. Part (a) in the "if" direction: it follows from Proposition 5.2.3(c) that whenever a left $\mathcal{C}$-contramodule $\mathfrak{P}$ is $\mathcal{C}/A$-projective, the $\mathcal{C}$-comodule $\Phi_{\mathcal{C}}(\mathfrak{P})$ is

$\mathcal{C}/A$-coprojective. Now if a left $\mathcal{C}$-comodule $\mathcal{M}$ is $\mathcal{C}/A$-injective, then the $\mathcal{C}$-contramodule $\mathfrak{P} = \Psi_{\mathcal{C}}(\mathcal{M})$ is $\mathcal{C}/A$-projective and $\mathcal{M} = \Phi_{\mathcal{C}}(\mathfrak{P})$ by the above theorem. Part (a) in the "only if" direction: in view of Lemma 1(a), the construction of Lemmas 1.1.3 and 3.1.3(a) represents any left $\mathcal{C}$-comodule $\mathcal{M}$ as the quotient comodule of an A-projective $\mathcal{C}$-comodule $\mathcal{P}(\mathcal{M})$ by a $\mathcal{C}/A$-injective $\mathcal{C}$-comodule. We will show that whenever $\mathcal{M}$ is a $\mathcal{C}/A$-coprojective $\mathcal{C}$-comodule, $\mathcal{P}(\mathcal{M})$ is a coprojective $\mathcal{C}$-comodule; then it will follow that $\mathcal{M}$ is a $\mathcal{C}/A$-injective $\mathcal{C}$-comodule by Lemma 5.2(a) and Lemma 1(a). Indeed, an extension of $\mathcal{C}/A$-coprojective left $\mathcal{C}$-comodules is $\mathcal{C}/A$-coprojective by Lemma 3.2.2(a); let us check that an A-projective $\mathcal{C}/A$-coprojective $\mathcal{C}$-comodule is coprojective. For any left $\mathcal{C}$-comodule $\mathcal{M}$ and any left $\mathcal{C}$-contramodule $\mathfrak{P}$ denote by $\operatorname{Coext}^i_{\mathcal{C}}(\mathcal{M}, \mathfrak{P})$ the cohomology of the object $\operatorname{Coext}_{\mathcal{C}}(\mathcal{M}, \mathfrak{P})$ of the derived category $\mathsf{D}(k\text{-mod})$ that was constructed in 4.7. This definition agrees with the definition of $\operatorname{Coext}^*_{\mathcal{C}}(\mathcal{M}, \mathfrak{P})$ for an A-projective $\mathcal{C}$-comodule $\mathcal{M}$ or an A-injective $\mathcal{C}$-contramodule $\mathfrak{P}$ given in the proof of Lemma 3.2.2. The functor $\operatorname{Coext}^*_{\mathcal{C}}(\mathcal{M}, \mathfrak{P})$ assigns long exact sequences to exact triples in either of its arguments. For any A-projective left $\mathcal{C}$-comodule $\mathcal{M}$ and any left $\mathcal{C}$-contramodule $\mathfrak{P}$ one has $\operatorname{Coext}^i_{\mathcal{C}}(\mathcal{M}, \mathfrak{P}) = 0$ for all $i > 0$ and $\operatorname{Coext}^0_{\mathcal{C}}(\mathcal{M}, \mathfrak{P}) \simeq \operatorname{Cohom}_{\mathcal{C}}(\mathcal{M}, \mathfrak{P})$. Therefore, an A-projective left $\mathcal{C}$-comodule $\mathcal{M}$ is coprojective if and only if $\operatorname{Cohom}^i_{\mathcal{C}}(\mathcal{M}, \mathfrak{P}) = 0$ for any left $\mathcal{C}$-contramodule $\mathfrak{P}$ and all $i \neq 0$. For any $\mathcal{C}/A$-coprojective left $\mathcal{C}$-comodule $\mathcal{M}$ and any left $\mathcal{C}$-comodule $\mathfrak{P}$ one has $\operatorname{Coext}^i_{\mathcal{C}}(\mathcal{M}, \mathfrak{P}) = 0$ for all $i < 0$, since one can compute $\operatorname{Coext}_{\mathcal{C}}(\mathcal{M}, \mathfrak{P})$ using a finite A-injective right resolution of $\mathfrak{P}$ by the result of 4.8. Thus an A-projective $\mathcal{C}/A$-coprojective left $\mathcal{C}$-comodule is coprojective. The proof of part (b) is completely analogous; it uses Proposition 5.2.2(c) and Lemma 3.1.3(b). □

Question. It follows from Proposition 1(c) that if $\mathcal{C}$ is a flat right A-module, then whenever a left $\mathcal{C}$-contramodule $\mathfrak{P}$ is $\mathcal{C}/A$-contraflat the $\mathcal{C}$-comodule $\Phi_{\mathcal{C}}(\mathfrak{P})$ is $\mathcal{C}/A$-coflat. Does the converse hold?

5.4 Comodule-contramodule correspondence

Assume that the coring $\mathcal{C}$ is a projective left and a flat right A-module and the ring A has a finite left homological dimension.

The categories of $\mathcal{C}/A$-injective left $\mathcal{C}$-comodules and $\mathcal{C}/A$-projective left $\mathcal{C}$-contramodules have natural exact category structures as full subcategories, closed under extensions, of the abelian categories of left $\mathcal{C}$-comodules and left $\mathcal{C}$-contramodules.

Theorem.

(a) *The functor mapping the quotient category of the homotopy category of complexes of $\mathcal{C}/A$-injective left $\mathcal{C}$-comodules by its minimal triangulated subcategory containing the total complexes of exact triples of complexes of $\mathcal{C}/A$-injective $\mathcal{C}$-comodules into the coderived category of left $\mathcal{C}$-comodules is an equivalence of triangulated categories.*

(b) *The functor mapping the quotient category of the homotopy category of complexes of $\mathcal{C}/A$-projective left $\mathcal{C}$-contramodules by its minimal triangulated subcategory containing the total complexes of $\mathcal{C}/A$-projective $\mathcal{C}$-contramodules into the contraderived category of left $\mathcal{C}$-contramodules is an equivalence of triangulated categories.*

Proof. Part (a): let $\mathcal{M}^\bullet$ be a complex of left $\mathcal{C}$-comodules. Then the total complex of the cobar bicomplex $\mathcal{C} \otimes_A \mathcal{M}^\bullet \longrightarrow \mathcal{C} \otimes_A \mathcal{C} \otimes_A \mathcal{M}^\bullet \longrightarrow \cdots$ is a complex of (quite) $\mathcal{C}/A$-injective $\mathcal{C}$-comodules, the complex $\mathcal{M}^\bullet$ maps into this total complex, and the cone of this map is coacyclic. Hence it follows from Lemma 2.6 that the coderived category of left $\mathcal{C}$-comodules is equivalent to the quotient category of the homotopy category of complexes of $\mathcal{C}/A$-injective $\mathcal{C}$-comodules by its intersection with the thick subcategory of coacyclic complexes of $\mathcal{C}$-comodules. It remains to show that this intersection of subcategories coincides with the minimal triangulated subcategory containing the total complexes of exact triples of complexes of $\mathcal{C}/A$-injective $\mathcal{C}$-comodules.

Lemma.

(a) *For any exact category A where infinite direct sums exist and preserve exact triples, the complex of homomorphisms from a coacyclic complex over A into a complex of injective objects with respect to A is acyclic.*

(b) *For any exact category A where infinite products exist and preserve exact triples, the complex of homomorphisms from a complex of projective objects with respect to A into a contraacyclic complex over A is acyclic.*

Proof. Analogous to the proofs of Lemmas 2.2 and 4.2. Part (a): let $M^\bullet$ be a complex of injective objects with respect to A. Since the functor of homomorphisms into $M^\bullet$ maps distinguished triangles in the homotopy category to distinguished triangles and infinite direct sums to infinite products, it suffices to check that the complex $\mathrm{Hom}_{\mathsf{A}}(L^\bullet, M^\bullet)$ is acyclic whenever $L^\bullet$ is the total complex of an exact triple ${}'K^\bullet \to K^\bullet \to {}''K^\bullet$ of complexes over A. But the complex $\mathrm{Hom}_{\mathsf{A}}(L^\bullet, M^\bullet)$ is the total complex of an exact triple of complexes of abelian groups $\mathrm{Hom}_{\mathsf{A}}({}''K^\bullet, M^\bullet) \longrightarrow \mathrm{Hom}_{\mathsf{A}}(K^\bullet, M^\bullet) \longrightarrow \mathrm{Hom}_{\mathsf{A}}({}'K^\bullet, M^\bullet)$ in this case. The proof of part (b) is dual.   □

In the remaining part of the proof we will show that the following two triangulated subcategories:

(i) the minimal triangulated subcategory containing the total complexes of exact triples of complexes of $\mathcal{C}/A$-injective $\mathcal{C}$-comodules, and

(ii) the homotopy category of complexes of injective $\mathcal{C}$-comodules

form a semiorthogonal decomposition [18] of the homotopy category of complexes of $\mathcal{C}/A$-injective left $\mathcal{C}$-comodules. This means, in addition to the subcategory (i) being left orthogonal to the subcategory (ii), that for any complex $\mathcal{K}^\bullet$ of $\mathcal{C}/A$-injective $\mathcal{C}$-comodules there exists a (unique and functorial) distinguished triangle $\mathcal{L}^\bullet \to$

$\mathcal{K}^\bullet \longrightarrow \mathcal{M}^\bullet \longrightarrow \mathcal{L}^\bullet[1]$ in the homotopy category of $\mathcal{C}$-comodules, where $\mathcal{L}^\bullet$ belongs to the subcategory (i) and $\mathcal{M}^\bullet$ belongs to the subcategory (ii). Then it will follow that the subcategory (i) is the maximal subcategory of the homotopy category of complexes of $\mathcal{C}/A$-injective $\mathcal{C}$-comodules left orthogonal to the subcategory (ii), hence the subcategory (i) contains the intersection of the homotopy category of complexes of $\mathcal{C}/A$-injective $\mathcal{C}$-comodules with the thick subcategory of coacyclic complexes of $\mathcal{C}$-comodules.

Indeed, let $\mathcal{K}^\bullet$ be a complex of $\mathcal{C}/A$-injective left $\mathcal{C}$-comodules. Choose for every n an injection j^n of the $\mathcal{C}$-comodule $\mathcal{K}^n$ into an injective $\mathcal{C}$-comodule $\mathcal{J}^n$. Consider the complex $\mathcal{E}^\bullet = \mathcal{E}(\mathcal{K}^\bullet)$ whose terms are the $\mathcal{C}$-comodules $\mathcal{E}^n = \mathcal{J}^n \oplus \mathcal{J}^{n+1}$ and the differential $d_{\mathcal{E}}^n \colon \mathcal{E}^n \longrightarrow \mathcal{E}^{n+1}$ maps $\mathcal{J}^{n+1}$ into itself by the identity map and vanishes in the restriction to $\mathcal{J}^n$ and in the projection to $\mathcal{J}^{n+2}$. There is a natural injective morphism of complexes $\mathcal{K}^\bullet \longrightarrow \mathcal{E}^\bullet$ formed by the $\mathcal{C}$-comodule maps $\mathcal{K}^n \longrightarrow \mathcal{E}^n$ whose components are $j^n \colon \mathcal{K}^n \longrightarrow \mathcal{J}^n$ and $j^{n+1}d_{\mathcal{K}}^n \colon \mathcal{K}^n \longrightarrow \mathcal{J}^{n+1}$. Set $^0\mathcal{E}^\bullet = \mathcal{E}(\mathcal{K}^\bullet)$, $^1\mathcal{E}^\bullet = \mathcal{E}(^0\mathcal{E}^\bullet/\mathcal{K}^\bullet)$, etc. As it was shown in the proof of Theorem 5.3, the injective dimension of a $\mathcal{C}/A$-injective left $\mathcal{C}$-comodule does not exceed the left homological dimension d of the ring A. Therefore, the complex $\mathcal{Z}^\bullet = \mathrm{coker}(^{d-2}\mathcal{E}^\bullet \to {}^{d-1}\mathcal{E}^\bullet)$ is a complex of injective $\mathcal{C}$-comodules. Now it is clear that the total complex $\mathcal{M}^\bullet$ of the bicomplex $^0\mathcal{E}^\bullet \longrightarrow {}^1\mathcal{E}^\bullet \longrightarrow \cdots \longrightarrow {}^{d-1}\mathcal{E}^\bullet \longrightarrow \mathcal{Z}^\bullet$ is a complex of injective $\mathcal{C}$-comodules and the cone $\mathcal{L}^\bullet$ of the morphism $\mathcal{K}^\bullet \longrightarrow \mathcal{M}^\bullet$ belongs to the minimal triangulated subcategory containing the total complexes of exact triples of complexes of $\mathcal{C}/A$-injective $\mathcal{C}$-comodules by Lemma 5.3.1(a).

Part (a) is proven; the proof of part (b) is completely analogous and uses Lemma 5.3.1(b).   □

Remark. Let A be an exact category where infinite direct sums exist and preserve exact triples, every object admits an admissible monomorphism into an object injective relative to A, and the class of such injective objects is closed under infinite direct sums. Then the thick subcategory of coacyclic complexes with respect to A and the triangulated subcategory of complexes of injective objects form a semiorthogonal decomposition of the homotopy category $\mathsf{Hot}(A)$, so the coderived category $\mathsf{D}^{co}(A)$ is equivalent to the homotopy category of complexes of injectives in A. Indeed, orthogonality is already proven in the lemma, so it remains to construct a morphism from any complex $C^\bullet$ over A into a complex of injectives $M^\bullet$ with a coacyclic cone. To do so, one proceeds as in the proof of the theorem, constructing a morphism from $C^\bullet$ into a complex of injectives $^0E^\bullet$ that is an admissible monomorphism in every degree, taking the quotient complex, constructing an analogous morphism from it into a complex of injectives $^1E^\bullet$, etc. Finally, one constructs the total complex $M^\bullet$ of the bicomplex $^\bullet E^\bullet$ by taking infinite direct sums along the diagonals; then $M^\bullet$ is a complex of injectives and the cone of the morphism $C^\bullet \longrightarrow M^\bullet$ is coacyclic by Lemma 2.1. Consequently, the homotopy category of acyclic complexes of injectives in A is equivalent to the quotient category $\mathsf{Acycl}(A)/\mathsf{Acycl}^{co}(A)$ and to the kernel of the localization functor $\mathsf{D}^{co}(A) \longrightarrow \mathsf{D}(A)$ (cf. [64]). When A has a finite homological dimension, the condition that the class of injectives is closed under infinite direct sums is not needed in this argument. This is a somewhat trivial situation, though; see Remark 2.1. Moreover, let A be an exact category where infinite direct sums exist and preserve exact triples and every object admits an admissible monomorphism into an injective. Let $F \subset A$ be a class of objects closed under cokernels of admissible monomorphisms, containing the injectives, and consisting of objects of finite injective dimension. Then every complex over F that is coacyclic as a complex over A belongs to the minimal triangulated subcategory of $Hot(A)$ containing the total complexes of exact triples of complexes over F. When there is a class $F \subset A$ closed under infinite direct sums, consisting of objects of finite injective dimension, and such that every object of A admits an admissible monomorphism into an object of F, the coderived category $D^{co}(A)$ is equivalent to the homotopy category of complexes of injectives in A (cf. [55], where in the dual situation the role of the class F is played by flat modules). To show this, one has to repeat twice the above construction of a resolution $^\bullet E^\bullet$, taking infinite direct sums along the diagonals for the first time and finite direct sums along the diagonals of the canonical truncation for the second time. When A is the abelian category of C-comodules, one can take F to be the class of C/A-injective C-comodules or quite C/A-injective C-comodules. The related results for comodules and contramodules are obtained in Theorem 5.5 and Remark 5.5.

Corollary. *The restrictions of the functors Ψ_C and Φ_C (applied to complexes termwise) to the homotopy category of complexes of C/A-injective C-comodules and the homotopy category of complexes of C/A-projective C-contramodules define mutually inverse equivalences $\mathbb{R}\Psi_C$ and $\mathbb{L}\Phi_C$ between the coderived category of left C-comodules and the contraderived category of left C-contramodules.*

Proof. By Theorem 5.3, the functors Ψ_C and Φ_C induce mutually inverse equivalences between the homotopy categories of C/A-injective left C-comodules and C/A-projective left C-contramodules. According to Lemma 5.3.1(a) and (c), the total complexes of exact triples of complexes of C/A-injective C-comodules correspond to the total complexes of exact triples of complexes of C/A-projective C-contramodules under this equivalence. So it remains to apply the above theorem. $\square$

Question. Can one obtain a version of the derived comodule-contramodule correspondence (an equivalence between appropriately defined exotic derived categories of left C-comodules and left C-contramodules) not depending on any assumptions about the homological dimension of the ring A? It is not difficult to see that one can weaken the assumption that A has a finite left homological dimension to the assumption that A is left Gorenstein, i.e., the classes of left A-modules of finite projective and injective dimensions coincide. In this case, the coderived category of left C-comodules and the contraderived category of left C-contramodules are naturally equivalent whenever the coring C is a projective left and a flat right A-module. Indeed, arguing as in Theorem 5.5 below, one can show that the coderived category of left C-comodules is equivalent to the quotient category of the homotopy category of complexes of C-comodules coinduced from left A-modules of finite projective (injective) dimension by its minimal triangulated subcategory containing the total complexes of exact triples of complexes of C-comodules that at every term are exact triples of C-comodules coinduced from exact triples of A-modules of finite projective (injective) dimension. Analogously, the contraderived category of left C-contramodules is equivalent to the quotient category of the homotopy category of complexes of C-contramodules induced from left A-modules of finite projective (injective) dimension by its minimal triangulated subcategory containing the total complexes of exact triples of complexes of C-contramodules that at every term are exact triples of C-contramodules induced from exact triples of A-modules of finite projective (injective) dimension. The key step is to notice that the class of left A-modules of finite projective (injective) dimension is closed under infinite direct sums and products.

5.5 Derived functor Ctrtor

The following analogue of Theorem 5.4 holds under slightly weaker conditions.

Theorem.

(a) *Assume that the coring C is a flat right A-module and the ring A has a finite left homological dimension. Then the functor mapping the homotopy category of complexes of injective left C-comodules into the coderived category of left C-comodules is an equivalence of triangulated categories. In addition, the functor mapping the quotient category of the homotopy category of complexes of quite C/A-injective left C-comodules by the minimal triangulated subcategory containing the total complexes of exact triples of complexes of coinduced C-comodules that at every term are exact triples of C-comodules coinduced from exact triples of A-modules into the coderived category of left C-comodules is an equivalence of triangulated categories.*

(b) *Assume that the coring C is a projective left A-module and the ring A has a finite left homological dimension. Then the functor mapping the homotopy category of complexes of projective left C-contramodules into the contraderived category of left C-contramodules is an equivalence of triangulated categories. In addition, the functor mapping the quotient category of the homotopy category of complexes of quite C/A-projective left C-contramodules by the minimal triangulated subcategory containing the total complexes of exact triples of complexes of induced C-contramodules that at every term are exact triples of C-contramodules induced from exact triples of A-modules into the contraderived category of left C-contramodules is an equivalence of triangulated categories.*

Proof. Part (a): when C is also a projective left A-module, the first assertion follows from the proof of Theorem 5.4. To prove both assertions in the general case, we will show that

(i) the minimal triangulated subcategory containing the total complexes of exact triples of complexes of coinduced $\mathcal{C}$-comodules that at every term are exact triples of $\mathcal{C}$-comodules coinduced from exact triples of A-modules, and

(ii) the homotopy category of complexes of injective $\mathcal{C}$-comodules

form a semiorthogonal decomposition of the homotopy category of complexes of quite $\mathcal{C}/A$-injective left $\mathcal{C}$-comodules. Then we will argue as in the proof of Theorem 5.4.

Any complex of quite $\mathcal{C}/A$-injective $\mathcal{C}$-comodules is homotopy equivalent to a complex of coinduced $\mathcal{C}$-comodules. Let $\mathcal{K}^\bullet$ be a complex of coinduced left $\mathcal{C}$-comodules; then $\mathcal{K}^n \simeq \mathcal{C} \otimes_A V^n$ for certain A-modules V^n. Let $V^i \longrightarrow I^i$ be injective maps of the A-modules V^n into injective A-modules I^n; set $\mathcal{J}^n = \mathcal{C} \otimes_A I^n$. Then $\mathcal{J}^n$ are injective $\mathcal{C}$-comodules endowed with injective $\mathcal{C}$-comodule morphisms $\mathcal{K}^n \longrightarrow \mathcal{J}^n$. As in the proof of Theorem 5.4, we construct the complex of injective $\mathcal{C}$-comodules $\mathcal{E}^\bullet$ with $\mathcal{E}^n = \mathcal{J}^n \oplus \mathcal{J}^{n+1}$ and the injective morphism of complexes $\mathcal{K}^\bullet \longrightarrow \mathcal{E}^\bullet$. Let us show that there exists an automorphism of the $\mathcal{C}$-comodule $\mathcal{E}^n$ such that its composition with the injection $\mathcal{K}^n \longrightarrow \mathcal{E}^n$ is the injection whose components are $j_n \colon \mathcal{K}^n \longrightarrow \mathcal{J}^n$ and the zero map $\mathcal{K}^n \longrightarrow \mathcal{J}^{n+1}$. Since the comodule $\mathcal{J}^{n+1}$ is injective, the component $\mathcal{K}^n \longrightarrow \mathcal{J}^{n+1}$ of the morphism $\mathcal{K}^n \longrightarrow \mathcal{E}^n$ can be extended from the comodule $\mathcal{K}^n$ to comodule $\mathcal{J}^n$ containing it. Denote the morphism so obtained by $h^n \colon \mathcal{J}^n \longrightarrow \mathcal{J}^{n+1}$; then the automorphism of the comodule $\mathcal{E}^n$ whose components are $-h^n$, the identity automorphisms of $\mathcal{J}^n$ and $\mathcal{J}^{n+1}$, and zero has the desired property. Now it is clear that the triple $\mathcal{K}^\bullet \longrightarrow \mathcal{E}^\bullet \longrightarrow \mathcal{E}^\bullet/\mathcal{K}^\bullet$ is an exact triple of complexes of coinduced $\mathcal{C}$-comodules which at every term is an exact triple of $\mathcal{C}$-comodules coinduced from an exact triple of A-modules. Moreover, $\mathcal{E}^n/\mathcal{K}^n \simeq \mathcal{C} \otimes_A W^n$, where the injective dimension $\mathrm{di}_A W^n$ is equal to $\mathrm{di}_A V^n - 1$. So we can iterate this (nonfunctorial) construction, setting ${}^0\mathcal{E}^\bullet = \mathcal{E}(\mathcal{K}^\bullet) = \mathcal{E}^\bullet$, ${}^1\mathcal{E}^\bullet = \mathcal{E}({}^0\mathcal{E}^\bullet/\mathcal{K}^\bullet)$, etc., and $\mathcal{Z}^\bullet = \mathrm{coker}({}^{d-2}\mathcal{E}^\bullet \to {}^{d-1}\mathcal{E}^\bullet)$. Then the total complex $\mathcal{M}^\bullet$ of the bicomplex ${}^0\mathcal{E}^\bullet \longrightarrow {}^1\mathcal{E}^\bullet \longrightarrow \cdots \longrightarrow {}^{d-1}\mathcal{E}^\bullet \longrightarrow \mathcal{Z}^\bullet$ is a complex of injective $\mathcal{C}$-comodules and the cone $\mathcal{L}^\bullet$ of the morphism $\mathcal{K}^\bullet \longrightarrow \mathcal{M}^\bullet$ belongs to the minimal triangulated subcategory containing the total complexes of exact triples of complexes of coinduced $\mathcal{C}$-comodules that at every term is an exact triple of $\mathcal{C}$-comodules coinduced from an exact triple of A-modules. $\square$

Remark. Theorem 5.5 provides an alternative way of proving Corollary 5.4. Besides, it follows from 5.1.3, Lemma 5.2, and Theorem 5.5 that in the assumptions of 5.4 the functor mapping the homotopy category of complexes of coprojective left $\mathcal{C}$-comodules into the coderived category of left $\mathcal{C}$-comodules and the functor mapping the homotopy category of complexes of coinjective left $\mathcal{C}$-contramodules into the contraderived category of left $\mathcal{C}$-contramodules are equivalences of triangulated categories. This is a stronger result than Theorem 4.5.

The contratensor product $\mathcal{N}^\bullet \odot_\mathcal{C} \mathfrak{P}^\bullet$ of a complex of right $\mathcal{C}$-comodules $\mathcal{N}^\bullet$ and a complex of left $\mathcal{C}$-contramodules $\mathfrak{P}^\bullet$ is defined as the total complex of the bicomplex $\mathcal{N}^i \odot_\mathcal{C} \mathfrak{P}^j$, constructed by taking infinite direct sums along the diagonals.

Assume that the coring $\mathcal{C}$ is a projective left A-module and the ring A has a finite left homological dimension. One can prove in a way completely analogous to the proof of Lemma 2.2 that the contratensor product of a coacyclic complex of right $\mathcal{C}$-comodules and a complex of contraflat (and in particular, projective) left $\mathcal{C}$-contramodules is acyclic. The left derived functor of contratensor product

$$\mathrm{Ctrtor}^{\mathcal{C}} \colon \mathsf{D}^{\mathrm{co}}(\mathrm{comod}\text{-}\mathcal{C}) \times \mathsf{D}^{\mathrm{ctr}}(\mathcal{C}\text{-contra}) \longrightarrow \mathsf{D}(k\text{-mod})$$

is defined by restricting the functor of contratensor product to the Cartesian product of the homotopy category of right $\mathcal{C}$-comodules and the homotopy category of complexes of projective left $\mathcal{C}$-contramodules.

The same derived functor can be obtained by restricting the functor of contratensor product to the Cartesian product of the homotopy category of complexes of A-flat right $\mathcal{C}$-comodules and the homotopy category of complexes of quite $\mathcal{C}/A$-projective left $\mathcal{C}$-contramodules. Indeed, it follows from part (b) of the theorem that the contratensor product of a complex of A-flat right $\mathcal{C}$-comodules and a contraacyclic complex of quite $\mathcal{C}/A$-projective left $\mathcal{C}$-contramodules is acyclic. Now if $\mathcal{N}^{\bullet}$ is a complex of A-flat right $\mathcal{C}$-comodules, $\mathfrak{P}^{\bullet}$ is a complex of quite $\mathcal{C}/A$-projective left $\mathcal{C}$-contramodules, and $'\mathfrak{P}^{\bullet} \longrightarrow \mathfrak{P}^{\bullet}$ is a morphism from a complex of projective $\mathcal{C}$-contramodules $'\mathfrak{P}^{\bullet}$ into $\mathfrak{P}^{\bullet}$ with a contraacyclic cone, then the map $\mathcal{N}^{\bullet} \odot_{\mathcal{C}} \mathfrak{P}^{\bullet} \longrightarrow \mathcal{N}^{\bullet} \odot_{\mathcal{C}} {'\mathfrak{P}^{\bullet}}$ is a quasi-isomorphism. In particular, if the complex $\mathcal{N}^{\bullet}$ is coacyclic, then the complex $\mathcal{N}^{\bullet} \odot_{\mathcal{C}} \mathfrak{P}^{\bullet}$ is acyclic, since the complex $\mathcal{N}^{\bullet} \odot_{\mathcal{C}} {'\mathfrak{P}^{\bullet}}$ is. When $\mathcal{C}$ is also a flat right A-module, one can use complexes of $\mathcal{C}/A$-projective $\mathcal{C}$-contramodules instead of complexes of quite $\mathcal{C}/A$-projective $\mathcal{C}$-contramodules, because the contratensor product of a complex of A-flat right $\mathcal{C}$-comodules and a contraacyclic complex of $\mathcal{C}/A$-projective left $\mathcal{C}$-contramodules is acyclic by Theorem 5.4(b) and Lemma 5.3.1(c). Notice that this definition of the derived functor $\mathrm{Ctrtor}^{\mathcal{C}}$ is *not* a particular case of Lemma 2.7 (instead, it is a particular case of Lemma 6.5.2 below).

Analogously, assume that the coring $\mathcal{C}$ is a flat right A-module and the ring A has a finite left homological dimension. According to Lemma 5.4, the complex of homomorphisms from a coacyclic complex of left $\mathcal{C}$-comodules into a complex of injective left $\mathcal{C}$-comodules is acyclic. Therefore, the natural map

$$\mathrm{Hom}_{\mathsf{Hot}(\mathcal{C}\text{-comod})}(\mathcal{L}^{\bullet}, \mathcal{M}^{\bullet}) \longrightarrow \mathrm{Hom}_{\mathsf{D}^{\mathrm{co}}(\mathcal{C}\text{-comod})}(\mathcal{L}^{\bullet}, \mathcal{M}^{\bullet})$$

is an isomorphism whenever $\mathcal{M}^{\bullet}$ is a complex of injective $\mathcal{C}$-comodules. So the functor of homomorphisms in the coderived category of left $\mathcal{C}$-comodules can be lifted to a functor

$$\mathrm{Ext}_{\mathcal{C}} \colon \mathsf{D}^{\mathrm{co}}(\mathcal{C}\text{-comod})^{\mathrm{op}} \times \mathsf{D}^{\mathrm{co}}(\mathcal{C}\text{-comod}) \longrightarrow \mathsf{D}(k\text{-mod}),$$

which is defined by restricting the functor of homomorphisms of complexes of $\mathcal{C}$-comodules to the Cartesian product of the homotopy category of left $\mathcal{C}$-comodules and the homotopy category of complexes of injective left $\mathcal{C}$-comodules.

The same functor $\mathrm{Ext}_{\mathcal{C}}$ can be obtained by restricting the functor of homomorphisms to the Cartesian product of the homotopy category of complexes of A-projective left $\mathcal{C}$-comodules and the homotopy category of complexes of quite $\mathcal{C}/A$-injective left $\mathcal{C}$-comodules. Indeed, it follows from part (a) of Theorem 5.5 that the complex of homomorphisms from a complex of A-projective left $\mathcal{C}$-comodules into a coacyclic complex of quite $\mathcal{C}/A$-injective left $\mathcal{C}$-comodules is acyclic. Now if $\mathcal{L}^{\bullet}$ is a complex of A-projective left $\mathcal{C}$-comodules, $\mathcal{M}^{\bullet}$ is a complex of quite $\mathcal{C}/A$-injective left $\mathcal{C}$-comodules, and $\mathcal{M}^{\bullet} \longrightarrow {}'\mathcal{M}^{\bullet}$ is a morphism from $\mathcal{M}^{\bullet}$ into a complex of injective $\mathcal{C}$-comodules ${}'\mathcal{M}^{\bullet}$ with a coacyclic cone, then the map $\mathrm{Hom}_{\mathcal{C}}(\mathcal{L}^{\bullet}, \mathcal{M}^{\bullet}) \longrightarrow \mathrm{Hom}_{\mathcal{C}}(\mathcal{L}^{\bullet}, {}'\mathcal{M}^{\bullet})$ is a quasi-isomorphism. When $\mathcal{C}$ is also a projective left A-module, one can use complexes of $\mathcal{C}/A$-injective $\mathcal{C}$-comodules instead of complexes of quite $\mathcal{C}/A$-injective $\mathcal{C}$-comodules.

Finally, assume that the coring $\mathcal{C}$ is a projective left A-module and the ring A has a finite left homological dimension. By Lemma 5.4, the natural map

$$\mathrm{Hom}_{\mathsf{Hot}(\mathcal{C}\text{-contra})}(\mathfrak{P}^{\bullet}, \mathfrak{Q}^{\bullet}) \longrightarrow \mathrm{Hom}_{\mathsf{D}^{\mathsf{ctr}}(\mathcal{C}\text{-contra})}(\mathfrak{P}^{\bullet}, \mathfrak{Q}^{\bullet})$$

is an isomorphism whenever $\mathfrak{P}^{\bullet}$ is a complex of projective $\mathcal{C}$-contramodules. So the functor of homomorphisms in the contraderived category of left $\mathcal{C}$-contramodules can be lifted to a functor

$$\mathrm{Ext}^{\mathcal{C}} \colon \mathsf{D}^{\mathsf{ctr}}(\mathcal{C}\text{-contra})^{\mathsf{op}} \times \mathsf{D}^{\mathsf{ctr}}(\mathcal{C}\text{-contra}) \longrightarrow \mathsf{D}(k\text{-mod}),$$

which is defined by restricting the functor of homomorphisms of complexes of $\mathcal{C}$-contramodules to the Cartesian product of the homotopy category of complexes of projective left $\mathcal{C}$-contramodules and the homotopy category of left $\mathcal{C}$-contramodules.

The same functor $\mathrm{Ext}^{\mathcal{C}}$ can be obtained by restricting the functor of homomorphisms to the Cartesian product of the homotopy category of complexes of quite $\mathcal{C}/A$-projective left $\mathcal{C}$-contramodules and the homotopy category of complexes of A-injective left $\mathcal{C}$-contramodules. When $\mathcal{C}$ is also a flat right A-module, one can use complexes of $\mathcal{C}/A$-projective $\mathcal{C}$-contramodules instead of complexes of quite $\mathcal{C}/A$-projective $\mathcal{C}$-contramodules.

5.6 Coext and Ext, Cotor and Ctrtor

Assume that the coring $\mathcal{C}$ is a projective left and a flat right A-module and the ring A has a finite left homological dimension.

Corollary.

(a) *There are natural isomorphisms of functors*

$$\mathrm{Coext}_{\mathcal{C}}(\mathcal{M}^{\bullet}, \mathfrak{P}^{\bullet}) \simeq \mathrm{Ext}_{\mathcal{C}}(\mathcal{M}^{\bullet}, \mathbb{L}\Phi_{\mathcal{C}}(\mathfrak{P}^{\bullet})) \simeq \mathrm{Ext}^{\mathcal{C}}(\mathbb{R}\Psi_{\mathcal{C}}(\mathcal{M}^{\bullet}), \mathfrak{P}^{\bullet})$$

on the Cartesian product of the category opposite to the coderived category of left $\mathcal{C}$-comodules and the contraderived category of left $\mathcal{C}$-contramodules.

(b) *There is a natural isomorphism of functors*

$$\mathrm{Cotor}^{\mathcal{C}}(\mathcal{N}^{\bullet}, \mathcal{M}^{\bullet}) \simeq \mathrm{Ctrtor}^{\mathcal{C}}(\mathcal{N}^{\bullet}, \mathbb{R}\Psi_{\mathcal{C}}(\mathcal{M}^{\bullet}))$$

on the Cartesian product of the coderived category of right $\mathcal{C}$-comodules and the coderived category of left $\mathcal{C}$-comodules.

Proof. Clearly, it suffices to construct natural isomorphisms

$$\mathrm{Coext}_{\mathcal{C}}(\mathcal{L}^{\bullet}, \mathbb{R}\Psi_{\mathcal{C}}(\mathcal{M}^{\bullet})) \simeq \mathrm{Ext}_{\mathcal{C}}(\mathcal{L}^{\bullet}, \mathcal{M}^{\bullet}), \quad \mathrm{Coext}_{\mathcal{C}}(\mathbb{L}\Phi_{\mathcal{C}}(\mathfrak{P}^{\bullet}), \mathfrak{Q}^{\bullet}) \simeq \mathrm{Ext}^{\mathcal{C}}(\mathfrak{P}^{\bullet}, \mathfrak{Q}^{\bullet}),$$

$$\text{and} \quad \mathrm{Cotor}^{\mathcal{C}}(\mathcal{N}^{\bullet}, \mathbb{L}\Phi_{\mathcal{C}}(\mathfrak{P}^{\bullet})) \simeq \mathrm{Ctrtor}^{\mathcal{C}}(\mathcal{N}^{\bullet}, \mathfrak{P}^{\bullet}).$$

In the first case, represent the image of $\mathcal{M}^{\bullet}$ in $\mathsf{D}^{\mathrm{co}}(\mathcal{C}\text{-comod})$ by a complex of injective $\mathcal{C}$-comodules, notice that the functor $\Psi_{\mathcal{C}}$ maps injective comodules to coinjective contramodules, and use Proposition 5.2.2(a). Alternatively, represent the image of $\mathcal{M}^{\bullet}$ in $\mathsf{D}^{\mathrm{co}}(\mathcal{C}\text{-comod})$ by a complex of $\mathcal{C}/A$-injective $\mathcal{C}$-comodules and the image of $\mathcal{L}^{\bullet}$ in $\mathsf{D}^{\mathrm{co}}(\mathcal{C}\text{-comod})$ by a complex of coprojective $\mathcal{C}$-comodules, and use Proposition 5.2.2(e); or represent the image of $\mathcal{M}^{\bullet}$ in $\mathsf{D}^{\mathrm{co}}(\mathcal{C}\text{-comod})$ by a complex of $\mathcal{C}/A$-injective $\mathcal{C}$-comodules and the image of $\mathcal{L}^{\bullet}$ in $\mathsf{D}^{\mathrm{co}}(\mathcal{C}\text{-comod})$ by a complex of A-projective $\mathcal{C}$-comodules, and use Proposition 5.2.2(c), Lemma 5.3.2(b), and the result of 4.8.

In the second case, represent the image of $\mathfrak{P}^{\bullet}$ in $\mathsf{D}^{\mathrm{ctr}}(\mathcal{C}\text{-contra})$ by a complex of projective $\mathcal{C}$-contramodules, notice that the functor $\Phi_{\mathcal{C}}$ maps projective contramodules to coprojective comodules, and use Proposition 5.2.3(a). Alternatively, represent the image of $\mathfrak{P}^{\bullet}$ in $\mathsf{D}^{\mathrm{ctr}}(\mathcal{C}\text{-contra})$ by a complex of $\mathcal{C}/A$-projective $\mathcal{C}$-contramodules and the image of $\mathfrak{Q}^{\bullet}$ in $\mathsf{D}^{\mathrm{ctr}}(\mathcal{C}\text{-contra})$ by a complex of coinjective $\mathcal{C}$-contramodules, and use Proposition 5.2.3(e); or represent the image of $\mathfrak{P}^{\bullet}$ in $\mathsf{D}^{\mathrm{ctr}}(\mathcal{C}\text{-contra})$ by a complex of $\mathcal{C}/A$-projective $\mathcal{C}$-contramodules and the image of $\mathfrak{Q}^{\bullet}$ in $\mathsf{D}^{\mathrm{ctr}}(\mathcal{C}\text{-contra})$ by a complex of A-injective $\mathcal{C}$-contramodules, and use Proposition 5.2.3(c), Lemma 5.3.2(a), and the result of 4.8.

In the third case, represent the image of $\mathfrak{P}^{\bullet}$ in $\mathsf{D}^{\mathrm{ctr}}(\mathcal{C}\text{-contra})$ by a complex of projective $\mathcal{C}$-contramodules, notice that the functor $\Phi_{\mathcal{C}}$ maps projective contramodules to coprojective comodules, and use Proposition 5.2.1(a). Alternatively, represent the image of $\mathfrak{P}^{\bullet}$ in $\mathsf{D}^{\mathrm{ctr}}(\mathcal{C}\text{-contra})$ by a complex of $\mathcal{C}/A$-projective $\mathcal{C}$-contramodules and the image of $\mathcal{N}^{\bullet}$ in $\mathsf{D}^{\mathrm{co}}(\mathrm{comod}\text{-}\mathcal{C})$ by a complex of coflat $\mathcal{C}$-comodules, and use Proposition 5.2.1(e); or represent the image of $\mathfrak{P}^{\bullet}$ in $\mathsf{D}^{\mathrm{ctr}}(\mathcal{C}\text{-contra})$ by a complex of $\mathcal{C}/A$-projective $\mathcal{C}$-contramodules and the image of $\mathcal{N}^{\bullet}$ in $\mathsf{D}^{\mathrm{co}}(\mathrm{comod}\text{-}\mathcal{C})$ by a complex of A-flat $\mathcal{C}$-comodules, and use Proposition 5.2.1(c), Lemma 5.3.2(a), and the result of 2.8.

Finally, to show that the three pairwise isomorphisms between the functors $\mathrm{Coext}_{\mathcal{C}}(\mathcal{M}^{\bullet}, \mathfrak{P}^{\bullet})$, $\mathrm{Ext}_{\mathcal{C}}(\mathcal{M}^{\bullet}, \mathbb{L}\Phi_{\mathcal{C}}(\mathfrak{P}^{\bullet}))$, and $\mathrm{Ext}^{\mathcal{C}}(\mathbb{R}\Psi_{\mathcal{C}}(\mathcal{M}^{\bullet}), \mathfrak{P}^{\bullet})$ form a commutative diagram, one can represent the image of $\mathcal{M}^{\bullet}$ in $\mathsf{D}^{\mathrm{co}}(\mathcal{C}\text{-comod})$ by a complex of coprojective $\mathcal{C}$-comodules and the image of $\mathfrak{P}^{\bullet}$ in $\mathsf{D}^{\mathrm{ctr}}(\mathcal{C}\text{-contra})$ by a complex of coinjective $\mathcal{C}$-contramodules (having in mind Lemma 5.2), and use a result of 5.2. $\qquad\square$

6 Semimodule-Semicontramodule Correspondence

6.1 Contratensor product and semimodule/semicontramodule homomorphisms

Let S be a semialgebra over a coring $\mathcal{C}$.

6.1.1 We would like to define the operation of contratensor product of a right S-semimodule and a left S-semicontramodule. Depending on the (co)flatness and/or (co)projectivity conditions on $\mathcal{C}$ and S, one can speak of S-semimodules and S-semicontramodules with various (co)flatness and (co)injectivity conditions imposed on them. In particular, when $\mathcal{C}$ is a projective left A-module and either S is a coprojective left $\mathcal{C}$-comodule, or S is a projective left A-module and a $\mathcal{C}/A$-coflat right $\mathcal{C}$-comodule and A has a finite left homological dimension, or A is semisimple, one can consider right S-semimodules and left S-semicontramodules with no (co)flatness or (co)injectivity conditions imposed. When $\mathcal{C}$ is a flat right A-module, S is a flat right A-module and a $\mathcal{C}/A$-projective left $\mathcal{C}$-comodule, and A has a finite left homological dimension, one can consider A-flat right S-semimodules and A-injective left S-semicontramodules. When $\mathcal{C}$ is a flat right A-module and S is a coflat right $\mathcal{C}$-comodule, one can consider $\mathcal{C}$-coflat right S-semimodules and $\mathcal{C}$-coinjective left S-semicontramodules.

The *contratensor product* $\mathcal{N} \odot_S \mathfrak{P}$ of a right S-semimodule $\mathcal{N}$ and a left S-semicontramodule $\mathfrak{P}$ is a k-module defined as the cokernel of the following pair of maps

$$(\mathcal{N} \,\square_{\mathcal{C}}\, S) \odot_{\mathcal{C}} \mathfrak{P} \rightrightarrows \mathcal{N} \odot_{\mathcal{C}} \mathfrak{P}.$$

The first map is induced by the right S-semiaction morphism $\mathcal{N} \,\square_{\mathcal{C}}\, S \longrightarrow \mathcal{N}$. The second map is the composition of the map induced by the left S-semicontraaction morphism $\mathfrak{P} \longrightarrow \mathrm{Cohom}_{\mathcal{C}}(S, \mathfrak{P})$ and the natural "evaluation" map

$$\eta_S \colon (\mathcal{N} \,\square_{\mathcal{C}}\, S) \odot_{\mathcal{C}} \mathrm{Cohom}_{\mathcal{C}}(S, \mathfrak{P}) \longrightarrow \mathcal{N} \odot_{\mathcal{C}} \mathfrak{P}.$$

The latter is defined in the following generality. Let $\mathcal{C}$ be a coring over a k-algebra A and $\mathcal{D}$ be a coring over a k-algebra B. Let $\mathcal{K}$ be a $\mathcal{C}$-$\mathcal{D}$-bicomodule, $\mathcal{N}$ be a right $\mathcal{C}$-comodule, and $\mathfrak{P}$ be a left $\mathcal{C}$-contramodule. Suppose that the cotensor product $\mathcal{N}\square_{\mathcal{C}}\mathcal{K}$ is endowed with a right $\mathcal{D}$-comodule structure via the construction of 1.2.4 and the cohomomorphism module $\mathrm{Cohom}_{\mathcal{C}}(\mathcal{K}, \mathfrak{P})$ is endowed with a left $\mathcal{D}$-contramodule structure via the construction of 3.2.4. Then the composition of maps

$$(\mathcal{N} \,\square_{\mathcal{C}}\, \mathcal{K}) \otimes_B \mathrm{Hom}_A(\mathcal{K}, \mathfrak{P}) \longrightarrow \mathcal{N} \otimes_A \mathcal{K} \otimes_B \mathrm{Hom}_A(\mathcal{K}, \mathfrak{P}) \longrightarrow \mathcal{N} \otimes_A \mathfrak{P} \longrightarrow \mathcal{N} \odot_{\mathcal{C}} \mathfrak{P}$$

L. Positselski, *Homological Algebra of Semimodules and Semicontramodules*, Monografie Matematyczne 70, DOI 10.1007/978-3-0346-0436-9_7, © Springer Basel AG 2010

factorizes through the surjection

$$(\mathcal{N} \,\square_{\mathcal{C}}\, \mathcal{K}) \otimes_B \operatorname{Hom}_A(\mathcal{K}, \mathfrak{P}) \longrightarrow (\mathcal{N} \,\square_{\mathcal{C}}\, \mathcal{K}) \odot_{\mathcal{D}} \operatorname{Cohom}_{\mathcal{C}}(\mathcal{K}, \mathfrak{P}),$$

so there is a natural map

$$\eta_{\mathcal{K}} \colon (\mathcal{N} \,\square_{\mathcal{C}}\, \mathcal{K}) \odot_{\mathcal{D}} \operatorname{Cohom}_{\mathcal{C}}(\mathcal{K}, \mathfrak{P}) \longrightarrow \mathcal{N} \odot_{\mathcal{C}} \mathfrak{P}.$$

Indeed, the kernel of this surjection is equal to the sum of the images of the difference of two maps $(\mathcal{N}\square_{\mathcal{C}}\mathcal{K})\otimes_B\operatorname{Hom}_A(\mathcal{K}\otimes_B\mathcal{D}, \mathfrak{P}) \rightrightarrows (\mathcal{N}\square_{\mathcal{C}}\mathcal{K})\otimes_B\operatorname{Hom}_A(\mathcal{K}, \mathfrak{P})$ and the difference of two maps $(\mathcal{N} \,\square_{\mathcal{C}}\, \mathcal{K}) \otimes_B \operatorname{Hom}_A(\mathcal{C} \otimes_A \mathcal{K},\, \mathfrak{P}) \rightrightarrows (\mathcal{N} \,\square_{\mathcal{C}}\, \mathcal{K}) \otimes_B \operatorname{Hom}_A(\mathcal{K}, \mathfrak{P})$. The difference of the first pair of maps vanishes already in the composition with the map $(\mathcal{N} \,\square_{\mathcal{C}}\, \mathcal{K}) \otimes_B \operatorname{Hom}_A(\mathcal{K}, \mathfrak{P}) \longrightarrow \mathcal{N} \otimes_A \mathfrak{P}$, while the second pair of maps can be presented as the composition of the map $(\mathcal{N}\square_{\mathcal{C}}\mathcal{K}) \otimes_B \operatorname{Hom}_A(\mathcal{C}\otimes_A\mathcal{K}, \mathfrak{P}) \longrightarrow \mathcal{N}\otimes_A\operatorname{Hom}_A(\mathcal{C}, \mathfrak{P})$ and the pair of maps $\mathcal{N}\otimes_A\operatorname{Hom}_A(\mathcal{C}, \mathfrak{P}) \rightrightarrows \mathcal{N} \otimes_A \mathfrak{P}$ whose cokernel is, by the definition, $\mathcal{N} \odot_{\mathcal{C}} \mathfrak{P}$. The "evaluation" map $\eta_{\mathcal{K}}$ is dual to the map

$\operatorname{Hom}_k(\eta_{\mathcal{K}}, k^{\vee}) = \operatorname{Cohom}_{\mathcal{C}}(\mathcal{K}, -)\colon$

$\operatorname{Hom}^{\mathcal{C}}(\mathfrak{P}, \operatorname{Hom}_k(\mathcal{N}, k^{\vee})) \longrightarrow \operatorname{Hom}^{\mathcal{D}}(\operatorname{Cohom}_{\mathcal{C}}(\mathcal{K}, \mathfrak{P}), \operatorname{Cohom}_{\mathcal{C}}(\mathcal{K}, \operatorname{Hom}_k(\mathcal{N}, k^{\vee}))).$

6.1.2 The operation of contratensor product over $\mathcal{S}$ is dual to homomorphisms in the category of left $\mathcal{S}$-semicontramodules: for any right $\mathcal{S}$-semimodule $\mathcal{N}$ and any left $\mathcal{S}$-semicontramodule $\mathfrak{P}$ there is a natural isomorphism

$$\operatorname{Hom}_k(\mathcal{N} \odot_{\mathcal{S}} \mathfrak{P},\ k^{\vee}) \simeq \operatorname{Hom}^{\mathcal{S}}(\mathfrak{P}, \operatorname{Hom}_k(\mathcal{N}, k^{\vee})).$$

Indeed, both k-modules are isomorphic to the kernel of the same pair of maps $\operatorname{Hom}^{\mathcal{C}}(\mathfrak{P}, \operatorname{Hom}_k(\mathcal{N}, k^{\vee})) \rightrightarrows \operatorname{Hom}^{\mathcal{C}}(\mathfrak{P}, \operatorname{Cohom}_{\mathcal{C}}(\mathcal{S}, \operatorname{Hom}_k(\mathcal{N}, k^{\vee})))$. It follows that for any right $\mathcal{C}$-comodule $\mathcal{R}$ for which the induced right $\mathcal{S}$-semimodule $\mathcal{R} \,\square_{\mathcal{C}}\, \mathcal{S}$ is defined and any left $\mathcal{S}$-semicontramodule $\mathfrak{P}$ the composition of the map $(\mathcal{R} \,\square_{\mathcal{C}}\, \mathcal{S}) \odot_{\mathcal{C}} \mathfrak{P} \longrightarrow (\mathcal{R}\,\square_{\mathcal{C}}\, \mathcal{S}) \odot_{\mathcal{C}} \operatorname{Cohom}_{\mathcal{C}}(\mathcal{S}, \mathfrak{P})$ induced by the $\mathcal{S}$-semicontraaction in $\mathfrak{P}$ with the "evaluation" map $(\mathcal{R} \,\square_{\mathcal{C}}\, \mathcal{S}) \odot_{\mathcal{C}} \operatorname{Cohom}_{\mathcal{C}}(\mathcal{S}, \mathfrak{P}) \longrightarrow \mathcal{R} \odot_{\mathcal{C}} \mathfrak{P}$ induces a natural isomorphism

$$(\mathcal{R} \,\square_{\mathcal{C}}\, \mathcal{S}) \odot_{\mathcal{S}} \mathfrak{P} \simeq \mathcal{R} \odot_{\mathcal{C}} \mathfrak{P}.$$

When $\mathcal{C}$ is a projective left A-module and $\mathcal{S}$ is a coprojective left $\mathcal{C}$-comodule, the functor of contratensor product over $\mathcal{S}$ is right exact in both its arguments.

6.1.3 Let $\mathcal{S}$ be a semialgebra over a coring $\mathcal{C}$ and $\mathcal{T}$ be a semialgebra over a coring $\mathcal{D}$. We would like to define a $\mathcal{T}$-semimodule structure on the contratensor product of a $\mathcal{T}$-$\mathcal{S}$-bisemimodule and an $\mathcal{S}$-semicontramodule, and an $\mathcal{S}$-semicontramodule structure on semimodule homomorphisms from a $\mathcal{T}$-$\mathcal{S}$-bisemimodule to a $\mathcal{T}$-semimodule.

Let $\mathcal{N}$ be a right $\mathcal{D}$-comodule, $\mathcal{K}$ be $\mathcal{D}$-$\mathcal{C}$-bicomodule with a right $\mathcal{S}$-semimodule structure such that the multiple cotensor products $\mathcal{N} \,\square_{\mathcal{D}}\, \mathcal{K} \,\square_{\mathcal{C}}\, \mathcal{S} \,\square_{\mathcal{C}} \cdots \square_{\mathcal{C}}\, \mathcal{S}$

are associative and the semiaction map $\mathcal{K} \,\square_{\mathcal{C}}\, \mathcal{S} \longrightarrow \mathcal{K}$ is a left $\mathcal{D}$-comodule morphism, and $\mathfrak{P}$ be a left $\mathcal{S}$-semicontramodule. Then the contratensor product $\mathcal{K} \odot_{\mathcal{S}} \mathfrak{P}$ has a natural left $\mathcal{D}$-comodule structure as the cokernel of a pair of $\mathcal{D}$-comodule morphisms $(\mathcal{K} \,\square_{\mathcal{C}}\, \mathcal{S}) \odot_{\mathcal{C}} \mathfrak{P} \rightrightarrows \mathcal{K} \odot_{\mathcal{C}} \mathfrak{P}$. The composition of maps $(\mathcal{N} \,\square_{\mathcal{D}}\, \mathcal{K}) \odot_{\mathcal{C}} \mathfrak{P} \longrightarrow \mathcal{N} \,\square_{\mathcal{D}}\, (\mathcal{K} \odot_{\mathcal{C}} \mathfrak{P}) \longrightarrow \mathcal{N} \,\square_{\mathcal{D}}\, (\mathcal{K} \odot_{\mathcal{S}} \mathfrak{P})$ factorizes through the surjection $(\mathcal{N} \,\square_{\mathcal{D}}\, \mathcal{K}) \odot_{\mathcal{C}} \mathfrak{P} \longrightarrow (\mathcal{N} \,\square_{\mathcal{D}}\, \mathcal{K}) \odot_{\mathcal{S}} \mathfrak{P}$, so there is a natural map $(\mathcal{N} \,\square_{\mathcal{D}}\, \mathcal{K}) \odot_{\mathcal{S}} \mathfrak{P} \longrightarrow \mathcal{N} \,\square_{\mathcal{D}}\, (\mathcal{K} \odot_{\mathcal{S}} \mathfrak{P})$.

Indeed, the composition of the pair of maps $(\mathcal{N} \,\square_{\mathcal{D}}\, \mathcal{K} \,\square_{\mathcal{C}}\, \mathcal{S}) \odot_{\mathcal{C}} \mathfrak{P} \rightrightarrows (\mathcal{N} \,\square_{\mathcal{D}}\, \mathcal{K}) \odot_{\mathcal{C}} \mathfrak{P}$ whose cokernel is, by the definition, $(\mathcal{N} \,\square_{\mathcal{D}}\, \mathcal{K}) \odot_{\mathcal{S}} \mathfrak{P}$, with the map $(\mathcal{N} \,\square_{\mathcal{D}}\, \mathcal{K}) \odot_{\mathcal{C}} \mathfrak{P} \longrightarrow \mathcal{N} \,\square_{\mathcal{D}}\, (\mathcal{K} \odot_{\mathcal{C}} \mathfrak{P})$ is equal to the composition of the map $(\mathcal{N} \,\square_{\mathcal{D}}\, \mathcal{K} \,\square_{\mathcal{C}}\, \mathcal{S}) \odot_{\mathcal{C}} \mathfrak{P} \longrightarrow \mathcal{N} \,\square_{\mathcal{D}}\, ((\mathcal{K} \,\square_{\mathcal{C}}\, \mathcal{S}) \odot_{\mathcal{C}} \mathfrak{P})$ with the pair of maps $\mathcal{N} \,\square_{\mathcal{D}}\, ((\mathcal{K} \,\square_{\mathcal{C}}\, \mathcal{S}) \odot_{\mathcal{C}} \mathfrak{P}) \rightrightarrows \mathcal{N} \,\square_{\mathcal{D}}\, (\mathcal{K} \odot_{\mathcal{C}} \mathfrak{P})$.

Now let $\mathcal{K}$ be a $\mathcal{T}$-$\mathcal{S}$-bisemimodule and $\mathfrak{P}$ be a left $\mathcal{S}$-semicontramodule. Assume that the multiple cotensor products $\mathcal{T} \,\square_{\mathcal{D}}\, \cdots \,\square_{\mathcal{D}}\, \mathcal{T} \,\square_{\mathcal{D}}\, (\mathcal{K} \odot_{\mathcal{S}} \mathfrak{P})$ are associative and the $\mathcal{D}$-comodule morphisms $(\mathcal{T}^{\square m} \,\square_{\mathcal{D}}\, \mathcal{K}) \odot_{\mathcal{S}} \mathfrak{P} \longrightarrow \mathcal{T}^{\square m} \,\square_{\mathcal{D}}\, (\mathcal{K} \odot_{\mathcal{S}} \mathfrak{P})$ are isomorphisms for $m \leqslant 2$. Then one can define an associative and unital semiaction morphism $\mathcal{T} \,\square_{\mathcal{D}}\, (\mathcal{K} \odot_{\mathcal{S}} \mathfrak{P}) \longrightarrow \mathcal{K} \odot_{\mathcal{S}} \mathfrak{P}$ by taking the contratensor product over $\mathcal{S}$ of the semiaction morphism $\mathcal{T} \,\square_{\mathcal{D}}\, \mathcal{K} \longrightarrow \mathcal{K}$ with the semicontramodule $\mathfrak{P}$.

Analogously, let $\mathcal{L}$ be a left $\mathcal{C}$-comodule, $\mathcal{K}$ be a $\mathcal{D}$-$\mathcal{C}$-bicomodule with a left $\mathcal{T}$-semimodule structure such that the multiple cotensor products $\mathcal{T} \,\square_{\mathcal{D}}\, \cdots \,\square_{\mathcal{D}}\, \mathcal{T} \,\square_{\mathcal{D}}\, \mathcal{K} \,\square_{\mathcal{C}}\, \mathcal{L}$ are associative and the semiaction map $\mathcal{T} \,\square_{\mathcal{D}}\, \mathcal{K} \longrightarrow \mathcal{K}$ is a right $\mathcal{C}$-comodule morphism, and $\mathcal{M}$ be a left $\mathcal{T}$-semimodule. Then the module of homomorphisms $\mathrm{Hom}_{\mathcal{T}}(\mathcal{K}, \mathcal{M})$ has a natural left $\mathcal{C}$-contramodule structure as the kernel of a pair of $\mathcal{C}$-contramodule morphisms $\mathrm{Hom}_{\mathcal{D}}(\mathcal{K}, \mathcal{M}) \rightrightarrows \mathrm{Hom}_{\mathcal{D}}(\mathcal{T} \,\square_{\mathcal{D}}\, \mathcal{K}, \mathcal{M})$. The composition of maps $\mathrm{Cohom}_{\mathcal{C}}(\mathcal{L}, \mathrm{Hom}_{\mathcal{T}}(\mathcal{K}, \mathcal{M})) \longrightarrow \mathrm{Cohom}_{\mathcal{C}}(\mathcal{L}, \mathrm{Hom}_{\mathcal{D}}(\mathcal{K}, \mathcal{M})) \longrightarrow \mathrm{Hom}_{\mathcal{D}}(\mathcal{K} \,\square_{\mathcal{C}}\, \mathcal{L}, \mathcal{M})$ factorizes through the injection $\mathrm{Hom}_{\mathcal{T}}(\mathcal{K} \,\square_{\mathcal{C}}\, \mathcal{L}, \mathcal{M}) \longrightarrow \mathrm{Hom}_{\mathcal{D}}(\mathcal{K} \,\square_{\mathcal{C}}\, \mathcal{L}, \mathcal{M})$, so there is a natural map $\mathrm{Cohom}_{\mathcal{C}}(\mathcal{L}, \mathrm{Hom}_{\mathcal{T}}(\mathcal{K}, \mathcal{M})) \longrightarrow \mathrm{Hom}_{\mathcal{T}}(\mathcal{K} \,\square_{\mathcal{C}}\, \mathcal{L}, \mathcal{M})$.

Now let $\mathcal{K}$ be a $\mathcal{T}$-$\mathcal{S}$-bisemimodule and $\mathcal{M}$ be a left $\mathcal{T}$-semimodule. Assume that the multiple cohomomorphisms $\mathrm{Cohom}_{\mathcal{C}}(\mathcal{S} \,\square_{\mathcal{C}}\, \cdots \,\square_{\mathcal{C}}\, \mathcal{S}, \mathrm{Hom}_{\mathcal{T}}(\mathcal{K}, \mathcal{M}))$ are associative and the $\mathcal{C}$-contramodule morphisms $\mathrm{Cohom}_{\mathcal{C}}(\mathcal{S}^{\square n}, \mathrm{Hom}_{\mathcal{T}}(\mathcal{K}, \mathcal{M})) \longrightarrow \mathrm{Hom}_{\mathcal{T}}(\mathcal{K} \,\square_{\mathcal{C}}\, \mathcal{S}^{\square n}, \mathcal{M})$ are isomorphisms for $n \leqslant 2$.

Then one can define an associative and unital semicontraaction morphism $\mathrm{Hom}_{\mathcal{T}}(\mathcal{K}, \mathcal{M}) \longrightarrow \mathrm{Cohom}_{\mathcal{C}}(\mathcal{S}, \mathrm{Hom}_{\mathcal{T}}(\mathcal{K}, \mathcal{M}))$ by taking the $\mathcal{T}$-semimodule homomorphisms from the semiaction morphism $\mathcal{K} \,\square_{\mathcal{C}}\, \mathcal{S} \longrightarrow \mathcal{K}$ into the semimodule $\mathcal{M}$.

6.1.4 Let $\mathcal{M}$ be a left $\mathcal{T}$-semimodule, $\mathcal{K}$ be a $\mathcal{T}$-$\mathcal{S}$-bisemimodule, and $\mathfrak{P}$ be a left $\mathcal{S}$-semicontramodule. Assume that a left $\mathcal{T}$-semimodule structure on $\mathcal{K} \odot_{\mathcal{S}} \mathfrak{P}$ and a left $\mathcal{S}$-semicontramodule structure on $\mathrm{Hom}_{\mathcal{T}}(\mathcal{K}, \mathcal{M})$ are defined via the constructions of 6.1.3. Then there is a natural adjunction isomorphism

$$\mathrm{Hom}_{\mathcal{T}}(\mathcal{K} \odot_{\mathcal{S}} \mathfrak{P},\, \mathcal{M}) \simeq \mathrm{Hom}^{\mathcal{S}}(\mathfrak{P}, \mathrm{Hom}_{\mathcal{T}}(\mathcal{K}, \mathcal{M})).$$

Indeed, the module $\mathrm{Hom}_{\mathcal{T}}(\mathcal{K} \odot_{\mathcal{S}} \mathfrak{P},\, \mathcal{M})$ is the kernel of the pair of maps $\mathrm{Hom}_{\mathcal{D}}(\mathcal{K} \odot_{\mathcal{S}} \mathfrak{P},\, \mathcal{M}) \rightrightarrows \mathrm{Hom}_{\mathcal{D}}(\mathcal{T} \,\square_{\mathcal{D}}\, \mathcal{K} \odot_{\mathcal{S}} \mathfrak{P},\, \mathcal{M})$ and there is an injection

$\mathrm{Hom}_{\mathcal{D}}(\mathfrak{T}\square_{\mathcal{D}}\mathcal{K}\odot_{\mathbf{S}}\mathfrak{P}, \mathcal{M}) \longrightarrow \mathrm{Hom}_{\mathcal{D}}((\mathfrak{T}\square_{\mathcal{D}}\mathcal{K})\odot_{\mathcal{C}}\mathfrak{P}, \mathcal{M})$. The module $\mathrm{Hom}_{\mathcal{D}}(\mathcal{K}\odot_{\mathbf{S}}$ $\mathfrak{P}, \mathcal{M})$ is the kernel of the pair of maps $\mathrm{Hom}_{\mathcal{D}}(\mathcal{K}\odot_{\mathcal{C}}\mathfrak{P}, \mathcal{M}) \rightrightarrows \mathrm{Hom}_{\mathcal{D}}((\mathcal{K}\square_{\mathcal{C}}$ $\mathbf{S})\odot_{\mathcal{C}}\mathfrak{P}, \mathcal{M})$. There is a pair of natural maps $\mathrm{Hom}_{\mathcal{D}}(\mathcal{K}\odot_{\mathcal{C}}\mathfrak{P}, \mathcal{M}) \rightrightarrows \mathrm{Hom}_{\mathcal{D}}((\mathfrak{T}\square_{\mathcal{D}}$ $\mathcal{K})\odot_{\mathcal{C}}\mathfrak{P}, \mathcal{M})$ (one of which goes through $\mathrm{Hom}_{\mathcal{D}}(\mathfrak{T}\square_{\mathcal{D}}(\mathcal{K}\odot_{\mathcal{C}}\mathfrak{P}), \mathcal{M}))$ extending the pair of maps $\mathrm{Hom}_{\mathcal{D}}(\mathcal{K}\odot_{\mathbf{S}}\mathfrak{P}, \mathcal{M}) \rightrightarrows \mathrm{Hom}_{\mathcal{D}}(\mathfrak{T}\square_{\mathcal{D}}\mathcal{K}\odot_{\mathbf{S}}\mathfrak{P}, \mathcal{M})$. Therefore, the module $\mathrm{Hom}_{\mathfrak{T}}(\mathcal{K}\odot_{\mathbf{S}}\mathfrak{P}, \mathcal{M})$ is isomorphic to the intersection of the kernels of two pairs of maps $\mathrm{Hom}_{\mathcal{D}}(\mathcal{K}\odot_{\mathcal{C}}\mathfrak{P}, \mathcal{M}) \rightrightarrows \mathrm{Hom}_{\mathcal{D}}((\mathcal{K}\square_{\mathcal{C}}\mathbf{S})\odot_{\mathcal{C}}\mathfrak{P}, \mathcal{M})$ and $\mathrm{Hom}_{\mathcal{D}}(\mathcal{K}\odot_{\mathcal{C}}\mathfrak{P}, \mathcal{M}) \rightrightarrows \mathrm{Hom}_{\mathcal{D}}((\mathfrak{T}\square_{\mathcal{D}}\mathcal{K})\odot_{\mathcal{C}}\mathfrak{P}, \mathcal{M})$. Analogously, the module $\mathrm{Hom}^{\mathbf{S}}(\mathfrak{P}, \mathrm{Hom}_{\mathfrak{T}}(\mathcal{K}, \mathcal{M}))$ is embedded into $\mathrm{Hom}^{\mathcal{C}}(\mathfrak{P}, \mathrm{Hom}_{\mathcal{D}}(\mathcal{K}, \mathcal{M}))$ by the composition of maps $\mathrm{Hom}^{\mathbf{S}}(\mathfrak{P}, \mathrm{Hom}_{\mathfrak{T}}(\mathcal{K}, \mathcal{M})) \longrightarrow \mathrm{Hom}^{\mathcal{C}}(\mathfrak{P}, \mathrm{Hom}_{\mathfrak{T}}(\mathcal{K}, \mathcal{M})) \longrightarrow \mathrm{Hom}^{\mathcal{C}}(\mathfrak{P}, \mathrm{Hom}_{\mathcal{D}}(\mathcal{K}, \mathcal{M}))$ and its image coincides with the intersection of the kernels of two pairs of maps $\mathrm{Hom}^{\mathcal{C}}(\mathfrak{P}, \mathrm{Hom}_{\mathcal{D}}(\mathcal{K}, \mathcal{M})) \rightrightarrows \mathrm{Hom}^{\mathcal{C}}(\mathfrak{P}, \mathrm{Hom}_{\mathcal{D}}(\mathfrak{T}\square_{\mathcal{D}}\mathcal{K}, \mathcal{M}))$ and $\mathrm{Hom}^{\mathcal{C}}(\mathfrak{P}, \mathrm{Hom}_{\mathcal{D}}(\mathcal{K}, \mathcal{M})) \rightrightarrows \mathrm{Hom}^{\mathcal{C}}(\mathfrak{P}, \mathrm{Hom}_{\mathcal{D}}(\mathcal{K}\square_{\mathcal{C}}\mathbf{S}, \mathcal{M}))$. These are the same two pairs of maps.

In order to obtain adjoint functors and equivalences between specific categories of left semimodules and left semicontramodules, we will have to prove associativity isomorphisms needed for the constructions of 6.1.3 to work.

6.2   Associativity isomorphisms

Let $\mathbf{S}$ be a semialgebra over a coring $\mathcal{C}$ over a k-algebra A and $\mathfrak{T}$ be a semialgebra over a coring $\mathcal{D}$ over a k-algebra B. The following three propositions will be mostly applied to the cases of $\mathcal{K} = \mathfrak{T} = \mathbf{S}$ or $\mathfrak{T} - \mathcal{D} - \mathcal{C}$, $\mathcal{K} - \mathbf{S}$ in the sequel.

Proposition 1. *Let* $\mathcal{N}$ *be a right* $\mathfrak{T}$-*semimodule,* $\mathcal{K}$ *be a* $\mathfrak{T}$-$\mathbf{S}$-*bisemimodule, and* $\mathfrak{P}$ *be a left* $\mathbf{S}$-*semicontramodule. Then there is a natural map*

$$(\mathcal{N}\lozenge_{\mathfrak{T}}\mathcal{K})\odot_{\mathbf{S}}\mathfrak{P} \longrightarrow \mathcal{N}\lozenge_{\mathbf{S}}(\mathcal{K}\odot_{\mathbf{S}}\mathfrak{P})$$

whenever both modules are defined via the constructions of 1.4.4 and 6.1.3. This map is an isomorphism, at least, in the following cases:

(a) $\mathcal{D}$ *is a flat left* B-*module,* $\mathcal{C}$ *is a projective left* A-*module,* $\mathfrak{P}$ *is a contraflat left* $\mathcal{C}$-*contramodule, and either*

- $\mathfrak{T}$ *is a coflat left* $\mathcal{D}$-*comodule,* $\mathbf{S}$ *is a coprojective left* $\mathcal{C}$-*comodule, and* $\mathcal{K}$ *as a right* $\mathbf{S}$-*semimodule with a left* $\mathcal{D}$-*comodule structure is induced from a* $\mathcal{D}$-*coflat* $\mathcal{D}$-$\mathcal{C}$-*bicomodule, or*

- $\mathfrak{T}$ *is a flat left* B-*module and a* $\mathcal{D}/B$-*coflat right* $\mathcal{D}$-*comodule,* $\mathbf{S}$ *is a projective left* A-*module and a* $\mathcal{C}/A$-*coflat right* $\mathcal{C}$-*comodule, the ring* A *(resp.,* B) *has a finite left (resp., weak) homological dimension,* $\mathcal{K}$ *as a right* $\mathbf{S}$-*semimodule with a left* $\mathcal{D}$-*comodule structure is induced from a* B-*flat and* $\mathcal{C}/A$-*coflat* $\mathcal{D}$-$\mathcal{C}$-*bicomodule, and* $\mathcal{K}$ *as a left* $\mathfrak{T}$-*semimodule with a right* $\mathcal{C}$-*comodule structure is induced from a* B-*flat* $\mathcal{D}$-$\mathcal{C}$-*bicomodule, or*

- *the ring A is semisimple, the ring B is absolutely flat, $\mathcal{K}$ as a right $\mathcal{S}$-semimodule with a left $\mathcal{D}$-comodule structure is induced from a $\mathcal{D}$-$\mathcal{C}$-bicomodule, and $\mathcal{K}$ as a left $\mathcal{T}$-semimodule with a right $\mathcal{C}$-comodule structure is induced from a $\mathcal{D}$-$\mathcal{C}$-bicomodule;*

(b) *$\mathcal{N}$ is a flat right B-module, $\mathcal{D}$ is a flat right B-module, $\mathcal{T}$ is a flat right B-module and a $\mathcal{D}/B$-coflat left $\mathcal{D}$-comodule, $\mathcal{C}$ is a flat right A-module, $\mathcal{S}$ is a flat right A-module and a $\mathcal{C}/A$-coprojective left $\mathcal{C}$-comodule, the ring A (resp., B) has a finite left (resp., weak) homological dimension, $\mathcal{K}$ as a right $\mathcal{S}$-semimodule with a left $\mathcal{D}$-comodule structure is induced from an A-flat and $\mathcal{D}/B$-coflat $\mathcal{D}$-$\mathcal{C}$-bicomodule, $\mathcal{K}$ as a left $\mathcal{T}$-semimodule with a right $\mathcal{C}$-comodule structure is induced from an A-flat and $\mathcal{D}/B$-coflat $\mathcal{D}$-$\mathcal{C}$-bicomodule, and $\mathfrak{P}$ is an A-injective and $\mathcal{C}/A$-contraflat left $\mathcal{C}$-contramodule;*

(c) *$\mathcal{N}$ is a flat right B-module, $\mathcal{D}$ is a flat right B-module, $\mathcal{T}$ is a flat right B-module and a $\mathcal{D}/B$-coflat left $\mathcal{D}$-comodule, the ring B has a finite weak homological dimension, $\mathcal{K}$ as a right $\mathcal{S}$-semimodule with a left $\mathcal{D}$-comodule structure is induced from an A-flat $\mathcal{D}$-$\mathcal{C}$-bicomodule, $\mathcal{C}$ is a projective left A-module, $\mathfrak{P}$ is a $\mathcal{C}/A$-contraflat left $\mathcal{C}$-contramodule, and either*

  - *$\mathcal{S}$ is a coprojective left $\mathcal{C}$-comodule and the ring A has a finite weak homological dimension, or*

  - *$\mathcal{S}$ is a projective left A-module and a $\mathcal{C}/A$-coflat right $\mathcal{C}$-comodule, the ring A has a finite left homological dimension, and $\mathcal{K}$ as a left $\mathcal{T}$-semimodule with a right $\mathcal{C}$-comodule structure is induced from a $\mathcal{D}$-$\mathcal{C}$-bicomodule;*

(d) *$\mathcal{D}$ is a flat right B-module, $\mathcal{T}$ is a coflat right $\mathcal{D}$-comodule, $\mathcal{N}$ is a coflat right $\mathcal{D}$-comodule, and either*

  - *$\mathcal{C}$ is a projective left A-module and $\mathcal{S}$ is a coprojective left $\mathcal{C}$-comodule, or*

  - *$\mathcal{C}$ is a projective left A-module, $\mathcal{S}$ is a projective left A-module and a $\mathcal{C}/A$-coflat right $\mathcal{C}$-comodule, the right A has a finite left homological dimension, and $\mathcal{K}$ as a left $\mathcal{T}$-semimodule with a right $\mathcal{C}$-comodule structure is induced from a $\mathcal{D}$-$\mathcal{C}$-bicomodule, or*

  - *$\mathcal{C}$ is a flat right A-module, $\mathcal{S}$ is a flat right A-module and a $\mathcal{C}/A$-coprojective left $\mathcal{C}$-comodule, the ring A has a finite left homological dimension, $\mathcal{K}$ as a left $\mathcal{T}$-semimodule with a right $\mathcal{C}$-comodule structure is induced from an A-flat $\mathcal{D}$-$\mathcal{C}$-bicomodule, and $\mathfrak{P}$ is an injective left A-module, or*

  - *$\mathcal{C}$ is a flat right A-module, $\mathcal{S}$ is a coflat right $\mathcal{C}$-comodule, $\mathcal{K}$ as a left $\mathcal{T}$-semimodule with a right $\mathcal{C}$-comodule structure is induced from a $\mathcal{C}$-coflat $\mathcal{D}$-$\mathcal{C}$-bicomodule, and $\mathfrak{P}$ is a coinjective left $\mathcal{C}$-contramodule.*

Proof. If $\mathcal{N}''' \to \mathcal{N}'' \to \mathcal{N}' \to 0$ is a sequence of right $\mathcal{S}$-semimodule morphisms which is exact in the category of A-modules and remains exact after taking the cotensor product with S over C, then for any left S-semicontramodule $\mathfrak{P}$ there is an exact sequence $\mathcal{N}''' \odot_S \mathfrak{P} \longrightarrow \mathcal{N}'' \odot_S \mathfrak{P} \longrightarrow \mathcal{N}' \odot_S \mathfrak{P} \longrightarrow 0$. Hence whenever a right S-semimodule structure on $\mathcal{N} \Diamond_T \mathcal{K}$ is defined via the construction of 1.4.4, the k-module $(\mathcal{N} \Diamond_T \mathcal{K}) \odot_S \mathfrak{P}$ is the cokernel of the pair of maps $(\mathcal{N} \square_D \mathcal{T} \square_D \mathcal{K}) \odot_S \mathfrak{P} \rightrightarrows (\mathcal{N} \square_D \mathcal{K}) \odot_S \mathfrak{P}$. By the definition, the semitensor product $\mathcal{N} \Diamond_T (\mathcal{K} \odot_S \mathfrak{P})$ is the cokernel of the pair of maps $\mathcal{N} \square_D \mathcal{T} \square_D (\mathcal{K} \odot_S \mathfrak{P}) \rightrightarrows \mathcal{N} \square_D (\mathcal{K} \odot_S \mathfrak{P})$. There are natural maps $(\mathcal{N} \square_D \mathcal{K}) \odot_S \mathfrak{P} \longrightarrow \mathcal{N} \square_D (\mathcal{K} \odot_S \mathfrak{P})$ and $(\mathcal{N} \square_D \mathcal{T} \square_D \mathcal{K}) \odot_S \mathfrak{P} \longrightarrow \mathcal{N} \square_D \mathcal{T} \square_D (\mathcal{K} \odot_S \mathfrak{P})$ constructed in 6.1.3. Whenever the left $\mathcal{T}$-semimodule structure on $\mathcal{K} \odot_S \mathfrak{P}$ is defined via the construction of 6.1.3, the corresponding (two) square diagrams commute. So there is a natural map $(\mathcal{N} \Diamond_T \mathcal{K}) \odot_S \mathfrak{P} \longrightarrow \mathcal{N} \Diamond_T (\mathcal{K} \odot_S \mathfrak{P})$, which is an isomorphism provided that the map $(\mathcal{N} \square_D \mathcal{K}) \odot_S \mathfrak{P} \longrightarrow \mathcal{N} \square_D (\mathcal{K} \odot_S \mathfrak{P})$ and the analogous map for $\mathcal{N} \square_D \mathcal{T}$ in place of $\mathcal{N}$ are isomorphisms; and the left $\mathcal{T}$-semimodule structure on $\mathcal{K} \odot_S \mathfrak{P}$ is defined provided that the analogous map for $\mathcal{T}$ in place of $\mathcal{N}$ is an isomorphism. It is straightforward to check that in each case (a-d) a right S-semimodule structure on $\mathcal{N} \Diamond_T \mathcal{K}$ is defined via the construction of 1.4.4 (that is where the conditions that $\mathcal{K}$ as a left $\mathcal{T}$-semimodule with a right C-comodule structure is induced from a D-C-bimodule are used). It is also easy to verify the (co)flatness conditions on $\mathcal{K} \odot_S \mathfrak{P}$ that are needed to guarantee that the semitensor product $\mathcal{N} \Diamond_T (\mathcal{K} \odot_S \mathfrak{P})$ is defined in the case (a). Thus it remains to show that the map $(\mathcal{N} \square_D \mathcal{K}) \odot_S \mathfrak{P} \longrightarrow \mathcal{N} \square_D (\mathcal{K} \odot_S \mathfrak{P})$ is an isomorphism.

In the case (d), the map $(\mathcal{N} \square_D \mathcal{K}) \odot_C \mathfrak{P} \longrightarrow \mathcal{N} \square_D (\mathcal{K} \odot_C \mathfrak{P})$ and the analogous map for $\mathcal{K} \square_C S$ in place of $\mathcal{K}$ are isomorphisms by Proposition 5.2.1(e) and the module $\mathcal{N} \sqcup_D (\mathcal{K} \odot_S \mathfrak{P})$ is the cokernel of the pair of maps $\mathcal{N} \square_D ((\mathcal{K} \square_C S) \odot_C \mathfrak{P}) \rightrightarrows \mathcal{N} \square_D (\mathcal{K} \odot_C \mathfrak{P})$, so it is clear from the construction of the map $(\mathcal{N} \square_D \mathcal{K}) \odot_S \mathfrak{P} \longrightarrow \mathcal{N} \square_D (\mathcal{K} \odot_S \mathfrak{P})$ that it is an isomorphism. In the cases (a-c), one has $\mathcal{K} \simeq \mathcal{K} \square_C S$ and the multiple cotensor products $\mathcal{N} \square_D \mathcal{K} \square_C S \square_C \cdots \square_C S$ are associative. So the map $(\mathcal{N} \square_D \mathcal{K}) \odot_S \mathfrak{P} \longrightarrow \mathcal{N} \square_D (\mathcal{K} \odot_S \mathfrak{P})$ is naturally isomorphic to the map $(\mathcal{N} \square_D \mathcal{K}) \odot_C \mathfrak{P} \longrightarrow \mathcal{N} \square_D (\mathcal{K} \odot_C \mathfrak{P})$. The latter is an isomorphism by Proposition 5.2.1(a) in the case (a) and by Proposition 5.2.1(d) in the cases (b-c). □

Proposition 2. *Let $\mathcal{L}$ be a left C-semimodule, $\mathcal{K}$ be a $\mathcal{T}$-S-bisemimodule, and $\mathcal{M}$ be a left $\mathcal{T}$-semimodule. Then there is a natural map*

$$\mathrm{SemiHom}_S(\mathcal{L}, \mathrm{Hom}_T(\mathcal{K}, \mathcal{M})) \longrightarrow \mathrm{Hom}_T(\mathcal{K} \Diamond_S \mathcal{L}, \mathcal{M})$$

whenever both modules are defined via the constructions of 1.4.4 and 6.1.3. This map is an isomorphism, at least, in the following cases:

(a) *C is a flat right A-module, D is a flat right B-module, $\mathcal{M}$ is an injective left D-comodule, and either*

   • *S is a coflat right C-comodule, $\mathcal{T}$ is a coflat right D-comodule, and $\mathcal{K}$ as a left $\mathcal{T}$-semimodule with a right C-comodule structure is induced from a C-coflat D-C-bicomodule, or*

- S *is a flat right A-module and a C/A-coprojective left C-comodule, T is a flat right B-module and a D/B-coflat left D-comodule, the ring A (resp., B) has a finite left (resp., weak) homological dimension, K as a left T-semimodule with a right C-comodule structure is induced from an A-flat and D/B-coflat D-C-bicomodule, and K as a right S-semimodule with a left D-comodule structure is induced from an A-flat D-C-bicomodule, or*

- *the ring A is semisimple, the ring B is absolutely flat, K as a left T-semimodule with a right C-comodule structure is induced from a D-C-bicomodule, and K as a right S-semimodule with a left D-comodule structure is induced from a D-C-bicomodule;*

(b) L *is a projective left A-module, C is a projective left A-module, S is a projective left A-module and a C/A-coflat right C-comodule, D is a flat left B-module, T is a flat left B-module and a D/B-coflat right D-comodule, the rings A and B have finite left homological dimensions, K as a left T-semimodule with a right C-comodule structure is induced from a B-projective and C/A-coflat D-C-bicomodule, K as a right S-semimodule with a left D-comodule structure is induced from a B-flat and C/A-coflat D-C-bicomodule, and M is a B-flat and D/B-injective left D-comodule;*

(c) L *is a projective left A-module, C is a projective left A-module, S is a projective left A-module and a C/A-coflat right C-comodule, the rings A and B have finite left homological dimensions, K as a left T-semimodule with a right C-comodule structure is induced from a B-projective D-C-bicomodule, D is a flat right B-module, M is a D/B-injective left D-comodule, and either*

  - T *is a coflat right D-comodule, or*

  - T *is a flat right B-module and a D/B-coflat left D-comodule, and K as a right S-semimodule with a left D-comodule structure is induced from a D-C-bicomodule;*

(d) C *is a projective left A-module, S is a coprojective left C-comodule, L is a coprojective left C-comodule, and either*

  - D *is a flat right B-module and T is a coflat right D-comodule, or*

  - D *is a flat right B-module, T is a flat right B-module and a D/B-coflat left D-comodule, and K as a right S-semimodule with a left D-comodule structure is induced from a D-C-bicomodule, or*

  - D *is a flat left B-module, T is a flat left B-module and a D/B-coflat right D-comodule, K as a right S-semimodule with a left D-comodule structure is induced from a B-flat D-C-bicomodule, and M is a flat left B-module, or*

  - D *is a flat left B-module, T is a coflat left D-comodule, K as a right S-semimodule with a left D-comodule structure is induced from a D-coflat D-C-bicomodule, and M is a coflat left D-comodule.*

Proof. Any sequence $\mathcal{L}'' \to \mathcal{L}'' \to \mathcal{L}' \to 0$ of $\mathcal{T}$-semimodule morphisms which is exact in the category of B-modules and remains exact after taking the cotensor product with $\mathcal{T}$ over $\mathcal{D}$ is exact in the category of $\mathcal{T}$-semimodules, i.e., for any $\mathcal{T}$-semimodule $\mathcal{M}$ there is an exact sequence

$$0 \longrightarrow \operatorname{Hom}_{\mathcal{T}}(\mathcal{L}', \mathcal{M}) \longrightarrow \operatorname{Hom}_{\mathcal{T}}(\mathcal{L}'', \mathcal{M}) \longrightarrow \operatorname{Hom}_{\mathcal{T}}(\mathcal{L}''', \mathcal{M}).$$

Hence whenever a left $\mathcal{T}$-semimodule structure is defined on $\mathcal{K} \lozenge_{\mathbf{S}} \mathcal{L}$ via the construction of 1.4.4, the k-module $\operatorname{Hom}_{\mathcal{T}}(\mathcal{K} \lozenge_{\mathbf{S}} \mathcal{L}, \mathcal{M})$ is the kernel of the pair of maps

$$\operatorname{Hom}_{\mathcal{T}}(\mathcal{K} \,\square_{\mathcal{C}}\, \mathcal{L}, \mathcal{M}) \rightrightarrows \operatorname{Hom}_{\mathcal{T}}(\mathcal{K} \,\square_{\mathcal{C}}\, \mathbf{S} \,\square_{\mathcal{C}}\, \mathcal{L}, \mathcal{M}).$$

By the definition, the k-module $\operatorname{SemiHom}_{\mathbf{S}}(\mathcal{L}, \operatorname{Hom}_{\mathcal{T}}(\mathcal{K}, \mathcal{M}))$ is the kernel of the pair of maps $\operatorname{Cohom}_{\mathcal{C}}(\mathcal{L}, \operatorname{Hom}_{\mathcal{T}}(\mathcal{K}, \mathcal{M})) \rightrightarrows \operatorname{Cohom}_{\mathcal{C}}(\mathbf{S} \,\square_{\mathcal{C}}\, \mathcal{L}, \operatorname{Hom}_{\mathcal{T}}(\mathcal{K}, \mathcal{M})) = \operatorname{Cohom}_{\mathcal{C}}(\mathcal{L}, \operatorname{Cohom}_{\mathcal{C}}(\mathbf{S}, \operatorname{Hom}_{\mathcal{T}}(\mathcal{K}, \mathcal{M})))$. There are natural maps

$$\operatorname{Cohom}_{\mathcal{C}}(\mathcal{L}, \operatorname{Hom}_{\mathcal{T}}(\mathcal{K}, \mathcal{M})) \longrightarrow \operatorname{Hom}_{\mathcal{T}}(\mathcal{K} \,\square_{\mathcal{C}}\, \mathcal{L}, \mathcal{M})$$

and

$$\operatorname{Cohom}_{\mathcal{C}}(\mathbf{S} \,\square_{\mathcal{C}}\, \mathcal{L}, \operatorname{Hom}_{\mathcal{T}}(\mathcal{K}, \mathcal{M})) \longrightarrow \operatorname{Hom}_{\mathcal{T}}(\mathcal{K} \,\square_{\mathcal{C}}\, \mathbf{S} \,\square_{\mathcal{C}}\, \mathcal{L}, \mathcal{M})$$

constructed in 6.1.3.

Whenever the left $\mathbf{S}$-semicontramodule structure on $\operatorname{Hom}_{\mathcal{T}}(\mathcal{K}, \mathcal{M})$ is defined via the construction of 6.1.3, the corresponding (two) square diagrams commute. So there is a natural map

$$\operatorname{SemiHom}_{\mathbf{S}}(\mathcal{L}, \operatorname{Hom}_{\mathcal{T}}(\mathcal{K}, \mathcal{M})) \longrightarrow \operatorname{Hom}_{\mathcal{T}}(\mathcal{K} \lozenge_{\mathbf{S}} \mathcal{L}, \mathcal{M}),$$

which is an isomorphism provided that the map $\operatorname{Cohom}_{\mathcal{C}}(\mathcal{L}, \operatorname{Hom}_{\mathcal{T}}(\mathcal{K}, \mathcal{M})) \longrightarrow \operatorname{Hom}_{\mathcal{T}}(\mathcal{K} \,\square_{\mathcal{C}}\, \mathcal{L}, \mathcal{M})$ and the analogous map for $\mathbf{S} \,\square_{\mathcal{C}}\, \mathcal{L}$ in place of $\mathcal{L}$ are isomorphisms; and the left $\mathbf{S}$-semicontramodule structure on $\operatorname{Hom}_{\mathcal{T}}(\mathcal{K}, \mathcal{M})$ is defined provided that the analogous map for $\mathbf{S}$ in place of $\mathcal{L}$ is an isomorphism.

It is straightforward to check that in each case (a-d) a left $\mathcal{T}$-semimodule structure on $\mathcal{K} \lozenge_{\mathbf{S}} \mathcal{L}$ is defined via the construction of 1.4.4. It is also easy to verify (using Proposition 5.2.2(a)) the (co)injectivity conditions on $\operatorname{Hom}_{\mathcal{T}}(\mathcal{K}, \mathcal{M})$ that are needed to guarantee that the semihomomorphism module $\operatorname{SemiHom}_{\mathbf{S}}(\mathcal{L}, \operatorname{Hom}_{\mathcal{T}}(\mathcal{K}, \mathcal{M}))$ is defined in the case (a). Thus it remains to show that the map $\operatorname{Cohom}_{\mathcal{C}}(\mathcal{L}, \operatorname{Hom}_{\mathcal{T}}(\mathcal{K}, \mathcal{M})) \longrightarrow \operatorname{Hom}_{\mathcal{T}}(\mathcal{K} \,\square_{\mathcal{C}}\, \mathcal{L}, \mathcal{M})$ is an isomorphism.

In the case (d), the map $\operatorname{Cohom}_{\mathcal{C}}(\mathcal{L}, \operatorname{Hom}_{\mathcal{D}}(\mathcal{K}, \mathcal{M})) \longrightarrow \operatorname{Hom}_{\mathcal{D}}(\mathcal{K} \square_{\mathcal{C}} \mathcal{L}, \mathcal{M})$ and the analogous map for $\mathcal{T} \square_{\mathcal{D}} \mathcal{K}$ in place of $\mathcal{K}$ are isomorphisms by Proposition 5.2.2(e) and the module $\operatorname{Cohom}_{\mathcal{C}}(\mathcal{L}, \operatorname{Hom}_{\mathcal{T}}(\mathcal{K}, \mathcal{M}))$ is the kernel of the pair of maps $\operatorname{Cohom}_{\mathcal{C}}(\mathcal{L}, \operatorname{Hom}_{\mathcal{T}}(\mathcal{K}, \mathcal{M})) \rightrightarrows \operatorname{Cohom}_{\mathcal{C}}(\mathcal{L}, \operatorname{Hom}_{\mathcal{T}}(\mathcal{T} \square_{\mathcal{D}} \mathcal{K}, \mathcal{M}))$, so it is clear from the construction of the map $\operatorname{Cohom}_{\mathcal{C}}(\mathcal{L}, \operatorname{Hom}_{\mathcal{T}}(\mathcal{K}, \mathcal{M})) \longrightarrow \operatorname{Hom}_{\mathcal{T}}(\mathcal{K} \,\square_{\mathcal{C}}\, \mathcal{L}, \mathcal{M})$ that it is an isomorphism. In the cases (a-c), one has $\mathcal{K} = \mathcal{T} \square_{\mathcal{D}} \mathcal{K}$ and multiple cotensor products $\mathcal{T} \,\square_{\mathcal{D}} \cdots \square_{\mathcal{D}}\, \mathcal{T} \,\square_{\mathcal{D}}\, \mathcal{K} \,\square_{\mathcal{C}}\, \mathcal{L}$ are associative. So the map $\mathrm{Cohom}_{\mathcal{C}}(\mathcal{L}, \mathrm{Hom}_{\mathcal{T}}(\mathcal{K}, \mathcal{M})) \longrightarrow \mathrm{Hom}_{\mathcal{T}}(\mathcal{K} \,\square_{\mathcal{C}}\, \mathcal{L},\, \mathcal{M})$ is naturally isomorphic to the map $\mathrm{Cohom}_{\mathcal{C}}(\mathcal{L}, \mathrm{Hom}_{\mathcal{D}}(\mathcal{K}, \mathcal{M})) \longrightarrow \mathrm{Hom}_{\mathcal{D}}(\mathcal{K} \,\square_{\mathcal{C}}\, \mathcal{L},\, \mathcal{M})$. The latter is an isomorphism by Proposition 5.2.2(a) in the case (a) and by 5.2.2(d) in the cases (b-c). $\hspace{2cm}\square$

Proposition 3. *Let $\mathfrak{P}$ be a left $\mathbf{S}$-semicontramodule, $\mathcal{K}$ be a $\mathcal{T}$-$\mathbf{S}$-bisemimodule, and $\mathfrak{Q}$ be a left $\mathcal{T}$-semicontramodule. Then there is a natural map*

$$\mathrm{SemiHom}_{\mathcal{T}}(\mathcal{K} \odot_{\mathbf{S}} \mathfrak{P},\, \mathfrak{Q}) \longrightarrow \mathrm{Hom}^{\mathbf{S}}(\mathfrak{P}, \mathrm{SemiHom}_{\mathcal{T}}(\mathcal{K}, \mathfrak{Q}))$$

whenever both modules are defined via the constructions of 3.4.4 and 6.1.3. This map is an isomorphism, at least, in the following cases:

(a) *$\mathcal{D}$ is a projective left B-module, $\mathcal{C}$ is a projective left A-module, $\mathfrak{P}$ is a projective left $\mathcal{C}$-contramodule, and either*

- *$\mathcal{T}$ is a coprojective left $\mathcal{D}$-comodule, $\mathbf{S}$ is a coprojective left $\mathcal{C}$-comodule, and $\mathcal{K}$ as a right $\mathbf{S}$-semimodule with a left $\mathcal{D}$-comodule structure is induced from a $\mathcal{D}$-coprojective $\mathcal{D}$-$\mathcal{C}$-bicomodule, or*

- *$\mathcal{T}$ is a projective left B-module and a $\mathcal{D}/B$-coflat right $\mathcal{D}$-comodule, $\mathbf{S}$ is a projective left A-module and a $\mathcal{C}/A$-coflat right $\mathcal{C}$-comodule, the rings A and B have finite left homological dimensions, $\mathcal{K}$ as a right $\mathbf{S}$-semimodule with a left $\mathcal{D}$-comodule structure is induced from a B-projective and $\mathcal{C}/A$-coflat $\mathcal{D}$-$\mathcal{C}$-bicomodule, and $\mathcal{K}$ as a left $\mathcal{T}$-semimodule with a right $\mathcal{C}$-comodule structure is induced from a B-projective $\mathcal{D}$-$\mathcal{C}$-bicomodule, or*

- *the rings A and B are semisimple, $\mathcal{K}$ as a right $\mathbf{S}$-semimodule with a left $\mathcal{D}$-comodule structure is induced from a $\mathcal{D}$-$\mathcal{C}$-bicomodule, and $\mathcal{K}$ as a left $\mathcal{T}$-semimodule with a right $\mathcal{C}$-comodule structure is induced from a $\mathcal{D}$-$\mathcal{C}$-bicomodule;*

(b) *$\mathcal{D}$ is a flat right B-module, $\mathcal{T}$ is a flat right B-module and a $\mathcal{D}/B$-coflat left $\mathcal{D}$-comodule, $\mathfrak{Q}$ is an injective left B-module, $\mathcal{C}$ is a flat right A-module, $\mathbf{S}$ is a flat right A-module and a $\mathcal{C}/A$-coprojective left $\mathcal{C}$-comodule, the rings A and B have finite left homological dimensions, $\mathcal{K}$ as a right $\mathbf{S}$-semimodule with a left $\mathcal{D}$-comodule structure is induced from an A-flat and $\mathcal{D}/B$-coprojective $\mathcal{D}$-$\mathcal{C}$-bicomodule, $\mathcal{K}$ as a left $\mathcal{T}$-semimodule with a right $\mathcal{C}$-comodule structure is induced from an A-flat and $\mathcal{D}/B$-coprojective $\mathcal{D}$-$\mathcal{C}$-bicomodule, and $\mathfrak{P}$ is a coinjective left $\mathcal{C}$-contramodule;*

(c) *$\mathcal{D}$ is a flat right B-module, $\mathcal{T}$ is a flat right B-module and a $\mathcal{D}/B$-coflat left $\mathcal{D}$-comodule, $\mathfrak{Q}$ is an injective left B-module, the rings A and B have finite left homological dimensions, $\mathcal{K}$ as a right $\mathbf{S}$-semimodule with a left $\mathcal{D}$-comodule structure is induced from an A-flat $\mathcal{D}$-$\mathcal{C}$-bicomodule, $\mathcal{C}$ is a projective left A-module, $\mathfrak{P}$ is a $\mathcal{C}/A$-projective left $\mathcal{C}$-contramodule, and either*

- $\mathcal{S}$ *is a coprojective left* $\mathcal{C}$*-comodule, or*
- $\mathcal{S}$ *is a projective left* A*-module and a* $\mathcal{C}/A$*-coflat right* $\mathcal{C}$*-comodule, and* $\mathcal{K}$ *as a left* $\mathcal{T}$*-semimodule with a right* $\mathcal{C}$*-comodule structure is induced from a* $\mathcal{D}$*-*$\mathcal{C}$*-bicomodule;*

(d) $\mathcal{D}$ *is a flat right* B*-module,* $\mathcal{T}$ *is a coflat right* $\mathcal{D}$*-comodule,* $\mathfrak{Q}$ *is a coinjective left* $\mathcal{D}$*-comodule, and one of the conditions of the list of Proposition* 1(d) *holds.*

Proof. Let $\mathfrak{Q}$ be a left $\mathcal{D}$-contramodule, $\mathcal{K}$ be a $\mathcal{D}$-$\mathcal{C}$-bicomodule with a right $\mathcal{S}$-semimodule structure such that multiple cohomomorphisms $\mathrm{Cohom}_{\mathcal{D}}(\mathcal{K}\square_{\mathcal{C}}\mathcal{S}\square_{\mathcal{C}} \cdots \square_{\mathcal{C}}\mathcal{S},\, \mathfrak{Q})$ are associative and the semiaction map $\mathcal{K}\square_{\mathcal{C}}\mathcal{S} \longrightarrow \mathcal{K}$ is a left $\mathcal{D}$-comodule morphism, and $\mathfrak{P}$ be a left $\mathcal{S}$-semicontramodule.

Then there is a natural left $\mathcal{S}$-semicontramodule structure on the module $\mathrm{Cohom}_{\mathcal{D}}(\mathcal{K}, \mathfrak{Q})$. The composition of maps $\mathrm{Cohom}_{\mathcal{D}}(\mathcal{K}\odot_{\mathcal{S}}\mathfrak{P},\, \mathfrak{Q}) \longrightarrow \mathrm{Cohom}_{\mathcal{D}}(\mathcal{K}\odot_{\mathcal{C}}\mathfrak{P},\, \mathfrak{Q}) \longrightarrow \mathrm{Hom}^{\mathcal{C}}(\mathfrak{P}, \mathrm{Cohom}_{\mathcal{D}}(\mathcal{K}, \mathfrak{Q}))$ factorizes through the injection $\mathrm{Hom}^{\mathcal{S}}(\mathfrak{P}, \mathrm{Cohom}_{\mathcal{D}}(\mathcal{K}, \mathfrak{Q})) \longrightarrow \mathrm{Hom}^{\mathcal{C}}(\mathfrak{P}, \mathrm{Cohom}_{\mathcal{D}}(\mathcal{K}, \mathfrak{Q}))$, so there is a natural map $\mathrm{Cohom}_{\mathcal{D}}(\mathcal{K}\odot_{\mathcal{S}}\mathfrak{P},\, \mathfrak{Q}) \longrightarrow \mathrm{Hom}^{\mathcal{S}}(\mathfrak{P}, \mathrm{Cohom}_{\mathcal{D}}(\mathcal{K}, \mathfrak{Q}))$. The rest of the proof is analogous to the proofs of Propositions 1 and 2. $\square$

Assume that $\mathcal{C}$ is a projective left A-module, $\mathcal{S}$ is a coprojective left $\mathcal{C}$-comodule, $\mathcal{D}$ is a flat right B-module, and $\mathcal{T}$ is a coflat right $\mathcal{D}$-comodule. Then it follows from 6.1.4 together with Propositions 1(d) and 2(d) that for any left $\mathcal{T}$ semimodule $\mathfrak{P}$, and $\mathcal{T}$-$\mathcal{S}$-bisemimodule $\mathcal{K}$, and any left $\mathcal{S}$-semicontramodule $\mathfrak{P}$ there is a natural isomorphism $\mathrm{Hom}_{\mathcal{T}}(\mathcal{K}\odot_{\mathcal{S}}\mathfrak{P},\, \mathcal{M}) \simeq \mathrm{Hom}^{\mathcal{S}}(\mathfrak{P}, \mathrm{Hom}_{\mathcal{T}}(\mathcal{K}, \mathcal{M}))$.

In particular, when $\mathcal{C}$ is a projective left and a flat right A-module and $\mathcal{S}$ is a coprojective left and a coflat right $\mathcal{C}$-comodule, there is a pair of adjoint functors $\Psi_{\mathcal{S}}\colon \mathcal{S}$-simod $\longrightarrow \mathcal{S}$-sicntr and $\Phi_{\mathcal{S}}\colon \mathcal{S}$-sicntr $\longrightarrow \mathcal{S}$-simod compatible with the functors $\Psi_{\mathcal{C}}\colon \mathcal{C}$-comod $\longrightarrow \mathcal{C}$-contra and $\Phi_{\mathcal{C}}\colon \mathcal{C}$-contra $\longrightarrow \mathcal{C}$-comod. In other words, the $\mathcal{S}$-semimodule $\Psi_{\mathcal{S}}(\mathcal{M})$ as a $\mathcal{C}$-comodule is naturally isomorphic to $\Psi_{\mathcal{C}}(\mathcal{M})$ and the $\mathcal{S}$-semicontramodule $\Phi_{\mathcal{S}}(\mathfrak{P})$ as a $\mathcal{C}$-contramodule is naturally isomorphic to $\Phi_{\mathcal{C}}(\mathfrak{P})$.

Assume that $\mathcal{C}$ is a projective left A-module and either $\mathcal{S}$ is a coprojective left $\mathcal{C}$-comodule, or $\mathcal{S}$ is a projective left A-module and a $\mathcal{C}/A$-coflat right $\mathcal{C}$-comodule and A has a finite left homological dimension. Then it follows from Propositions 1(a) and 2(b,d) that the categories of $\mathcal{C}$-coprojective left $\mathcal{S}$-semimodules and $\mathcal{C}$-projective left $\mathcal{S}$-semicontramodules are naturally equivalent.

Assume that $\mathcal{C}$ is a flat right A-module and either $\mathcal{S}$ is a coflat right $\mathcal{C}$-comodule, or $\mathcal{S}$ is a flat right A-module and a $\mathcal{C}/A$-coprojective left $\mathcal{C}$-comodule and A has a finite left homological dimension. Then if follows from Propositions 1(b,d) and 2(a) that the categories of $\mathcal{C}$-injective left $\mathcal{S}$-semimodules and $\mathcal{C}$-coinjective left $\mathcal{S}$-semicontramodules are naturally equivalent.

Assume that $\mathcal{C}$ is a projective left A-module and a flat right A-module, A has a finite left homological dimension, and either $\mathcal{S}$ is a coprojective left $\mathcal{C}$-comodule and a flat right A-module, or S is a projective left A-module and a coflat right $\mathcal{C}$-co-module. Then it follows from Propositions 1(c,d) and 2(c,d) that the categories of $\mathcal{C}/A$-injective left S-semimodules and $\mathcal{C}/A$-projective left S-semicontramodules are naturally equivalent.

Finally, assume that the ring A is semisimple. Then it follows from Propositions 1(a) and 2(a) that the categories of $\mathcal{C}$-injective left S-semimodules and $\mathcal{C}$-projective left S-semicontramodules are naturally equivalent.

In each of these cases, the natural maps defined in Propositions 2–3 in the case of $\mathcal{K} = \mathcal{T} = S$ have the following property of compatibility with the adjoint functors between categories of S-semimodules and S-semicontramodules. For any left S-semimodule $\mathcal{M}$ and any left S-semicontramodule $\mathfrak{P}$ such that the S-semi-module $\Phi_S(\mathfrak{P}) = S \odot_S \mathfrak{P}$, the S-semicontramodule $\Psi_S(\mathcal{M}) = \mathrm{Hom}_S(S, \mathcal{M})$, and the k-module of semihomomorphisms $\mathrm{SemiHom}_S(\Phi_S(\mathfrak{P}), \Psi_S(\mathcal{M}))$ are defined via the constructions of 6.1.3 and 3.4.4, the maps $\mathrm{SemiHom}_S(\Phi_S(\mathfrak{P}), \Psi_S(\mathcal{M})) \longrightarrow$ $\mathrm{Hom}_S(\Phi_S(\mathfrak{P}), \mathcal{M})$ and $\mathrm{SemiHom}_S(\Phi_S(\mathfrak{P}), \Psi_S(\mathcal{M})) \longrightarrow \mathrm{Hom}^S(\mathfrak{P}, \Psi_S(\mathcal{M}))$ form a commutative diagram with the adjunction isomorphism $\mathrm{Hom}_S(\Phi_S(\mathfrak{P}), \mathcal{M}) \simeq$ $\mathrm{Hom}^S(\mathfrak{P}, \Psi_S(\mathcal{M}))$.

6.3 Semimodule-semicontramodule correspondence

Assume that the coring $\mathcal{C}$ is a projective left and a flat right A-module, the semi-algebra S is a coprojective left and a coflat right $\mathcal{C}$-comodule, and the ring A has a finite left homological dimension.

Theorem.

(a) *The functor mapping the quotient category of complexes of $\mathcal{C}/A$-injective left S-semimodules by the thick subcategory of $\mathcal{C}$-coacyclic complexes of $\mathcal{C}/A$-in-jective S-semimodules into the semiderived category of left S-semimodules is an equivalence of triangulated categories.*

(b) *The functor mapping the quotient category of complexes of $\mathcal{C}/A$-projective left S-semicontramodules by the thick subcategory of $\mathcal{C}$-contraacyclic complexes of $\mathcal{C}/A$-projective S-semicontramodules into the semiderived category of left S-semicontramodules is an equivalence of triangulated categories.*

Proof. Part (b) follows from Lemma 5.3.2(b) and Lemma 2.6 applied to the con-struction of the morphism of complexes $\mathbb{L}_2(\mathfrak{P}^\bullet) \longrightarrow \mathfrak{P}^\bullet$ from the proof of The-orem 4.6(b). As an alternative to using Lemma 5.3.2, one can show that $\mathbb{L}_2(\mathfrak{P}^\bullet)$ is a complex of $\mathcal{C}/A$-projective S-semicontramodules in the following way. Use Lemma 3.3.2(b) to construct a finite right A-injective resolution of every term of the complex of left S-semicontramodules $\mathfrak{P}^\bullet$, then apply the functor $\mathbb{L}_2$, which maps exact triples of complexes to exact triples, and use Lemmas 3.3.3(c), 5.2(b), and 5.3.1(b). The proof of part (a) is completely analogous. $\square$

Remark. The analogue of the above theorem holds for complexes of quite $\mathcal{C}/A$-in-jective S-semimodules and quite $\mathcal{C}/A$-projective S-semicontramodules. Moreover, for any complex of left S-semimodules $\mathcal{M}^\bullet$ there exists a morphism from $\mathcal{M}^\bullet$ into a complex of $\mathcal{C}$-injective S-semimodules with a $\mathcal{C}$-coacyclic cone, and for any complex of left S-semicontramodules $\mathfrak{P}^\bullet$ there exists a morphism into $\mathfrak{P}^\bullet$ from a complex of $\mathcal{C}$-projective S-semicontramodules with a $\mathcal{C}$-contraacyclic cone. Indeed, consider the complex of $\mathcal{C}/A$-injective S-semimodules $\Phi_\mathsf{S}\mathbb{L}_2(\mathfrak{P}^\bullet)$ and apply to it the construction of the morphism of complexes $\mathbb{L}_1(\mathcal{K}^\bullet) \longrightarrow \mathcal{K}^\bullet$ from the proof of Theorems 2.6 and 4.6(a). For any complex of $\mathcal{C}/A$-injective S-semimodules $\mathcal{K}^\bullet$, the complex $\mathbb{L}_1(\mathcal{K}^\bullet)$ is a complex of $\mathcal{C}$-coprojective S-semimodules by Remark 3.2.2 and Lemma 5.3.2(a) (or simply because the class of $\mathcal{C}/A$-injective left $\mathcal{C}$-comodules is closed under extensions and any A-projective $\mathcal{C}/A$-injective left $\mathcal{C}$-comodule is coprojective, which is easy to check). So the complex of $\mathcal{C}$-coprojective S-semimodules $\mathbb{L}_1(\Phi_\mathsf{S}\mathbb{L}_2(\mathfrak{P}^\bullet))$ maps into $\Phi_\mathsf{S}\mathbb{L}_2(\mathfrak{P}^\bullet)$ with a $\mathcal{C}$-coacyclic cone, hence the complex of $\mathcal{C}$-projective S-semicontramodules $\Psi_\mathsf{S}\mathbb{L}_1(\Phi_\mathsf{S}\mathbb{L}_2(\mathfrak{P}^\bullet))$ maps into $\mathbb{L}_2\mathfrak{P}^\bullet$ and $\mathfrak{P}^\bullet$ with $\mathcal{C}$-contraacyclic cones.

Corollary. *The restrictions of the functors* Ψ_S *and* Φ_S *(applied to complexes termwise) to the homotopy category of complexes of* $\mathcal{C}/A$-*injective* S-*semimodules and* $\mathcal{C}/A$-*projective* S-*semicontramodules define mutually inverse equivalences* $\mathbb{R}\Psi_\mathsf{S}$ *and* $\mathbb{L}\Phi_\mathsf{S}$ *between the semiderived category of left* S-*semimodules and the semiderived category of left* S-*semicontramodules.*

Proof. By Corollary 5.4, the restrictions of functors Ψ_S and Φ_S induce mutually inverse equivalences between the quotient category of the homotopy category of $\mathcal{C}/A$-injective S-semimodules by its intersection with the thick subcategory of $\mathcal{C}$-coacyclic complexes and the quotient category of the homotopy category of $\mathcal{C}/A$-projective S-semicontramodules by its intersection with the thick subcategory of $\mathcal{C}$-contraacyclic complexes. Thus it remains to take in account Theorem 6.3.   □

6.4 Birelatively contraflat, projective, and injective complexes

We keep the assumptions of 6.3.

A complex of left S-semimodules $\mathcal{M}^\bullet$ is called *projective relative to* $\mathcal{C}$ *relative to* A ($\mathsf{S}/\mathcal{C}/A$-*projective*) if the complex of homomorphisms over S from $\mathcal{M}^\bullet$ into any $\mathcal{C}$-coacyclic complex of $\mathcal{C}/A$-injective S-semimodules is acyclic. A complex of left S-semicontramodules $\mathfrak{P}^\bullet$ is called *injective relative to* $\mathcal{C}$ *relative to* A ($\mathsf{S}/\mathcal{C}/A$-*injective*) if the complex of homomorphisms over S into $\mathfrak{P}^\bullet$ from any $\mathcal{C}$-contraacyclic complex of $\mathcal{C}/A$-projective S-semicontramodules is acyclic.

The contratensor product $\mathcal{N}^\bullet \odot_\mathsf{S} \mathfrak{P}^\bullet$ of a complex $\mathcal{N}^\bullet$ of right S-semimodules and a complex $\mathfrak{P}^\bullet$ of left S-semicontramodules is defined as the total complex of the bicomplex $\mathcal{N}^i \odot_\mathsf{S} \mathfrak{P}^j$, constructed by taking infinite direct sums along the diagonals. A complex of right S-semimodules $\mathcal{N}^\bullet$ is called *contraflat relative to* $\mathcal{C}$ *relative to* A ($\mathsf{S}/\mathcal{C}/A$-*contraflat*) if the contratensor product over S of the complex $\mathcal{N}^\bullet$ and any $\mathcal{C}$-contraacyclic complex of $\mathcal{C}/A$-projective left S-semicontramodules is acyclic.

It follows from Theorem 5.4 and Lemma 5.3.1 that the complex of left $\mathbf{S}$-semi-modules induced from a complex of A-projective $\mathcal{C}$-comodules is $\mathbf{S}/\mathcal{C}/A$-projective, the complex of left $\mathbf{S}$-semicontramodules coinduced from a complex of A-injective $\mathcal{C}$-contramodules is $\mathbf{S}/\mathcal{C}/A$-injective, and the complex of right $\mathbf{S}$-semimodules induced from a complex of A-flat $\mathcal{C}$-comodules is $\mathbf{S}/\mathcal{C}/A$-contraflat.

Lemma.

(a) *Any $\mathbf{S}/\mathcal{C}/A$-semiflat complex of A-flat right $\mathbf{S}$-semimodules (in the sense of 2.8) is $\mathbf{S}/\mathcal{C}/A$-contraflat.*

(b) *A complex of A-projective left $\mathbf{S}$-semimodules is $\mathbf{S}/\mathcal{C}/A$-projective if and only if it is $\mathbf{S}/\mathcal{C}/A$-semiprojective (in the sense of 4.8).*

(c) *A complex of A-injective left $\mathbf{S}$-semicontramodules is $\mathbf{S}/\mathcal{C}/A$-injective if and only if it is $\mathbf{S}/\mathcal{C}/A$-semiinjective (in the sense of 4.8).*

Proof. The functors $\Psi_{\mathbf{S}}$ and $\Phi_{\mathbf{S}}$ define an equivalence between the category of $\mathcal{C}$-coacyclic complexes of $\mathcal{C}/A$-injective left $\mathbf{S}$-semimodules and the category of $\mathcal{C}$-contraacyclic complexes of $\mathcal{C}/A$-projective left $\mathbf{S}$-semicontramodules. Therefore, part (a) follows from Proposition 6.2.1(c) (applied to $\mathcal{K} = \mathcal{T} = \mathbf{S}$) and Lemma 5.3.2(a), part (b) follows from Proposition 6.2.2(c) and Lemma 5.3.2(b), and part (c) follows from Proposition 6.2.3(c) and Lemma 5.3.2(a). $\square$

In view of the relevant results of 4.8, it is also clear that a complex of A-projective left $\mathbf{S}$-semimodules is $\mathbf{S}/\mathcal{C}/A$-projective if the complex of $\mathbf{S}$-semimodule homomorphisms from it into any $\mathcal{C}$-contractible complex of $\mathcal{C}/A$-injective $\mathbf{S}$-semimodules is acyclic. Analogously, a complex of A-injective left $\mathbf{S}$-semicontramodules is $\mathbf{S}/\mathcal{C}/A$-injective if the complex of $\mathbf{S}$-semicontramodule homomorphisms into it from any $\mathcal{C}$-contractible complex of $\mathcal{C}/A$-projective $\mathbf{S}$-semicontramodules is acyclic.

Question. One can show using the construction of the morphism of complexes of left $\mathbf{S}$-semimodules $\mathcal{L}^{\bullet} \longrightarrow \mathbb{R}_2(\mathcal{L}^{\bullet})$ and Lemma 1.2.2 that any $\mathbf{S}/\mathcal{C}/A$-contraflat complex of (appropriately defined) $\mathbf{S}/\mathcal{C}/A$-semiflat right $\mathbf{S}$-semimodules is $\mathbf{S}/\mathcal{C}/A$-semiflat. One can also show using the functor $\mathrm{SemiTor}^{\mathbf{S}}$ that any A-flat $\mathbf{S}/\mathcal{C}/A$-contraflat right $\mathbf{S}$-semimodule (defined in terms of exact triples of $\mathcal{C}/A$-projective or $\mathcal{C}/A$-contraflat left $\mathbf{S}$-semicontramodules) is $\mathbf{S}/\mathcal{C}/A$-semiflat; the converse is clear (cf. 9.2). Are all $\mathbf{S}/\mathcal{C}/A$-contraflat (in either definition) right $\mathbf{S}$-semimodules A-flat? Are all $\mathbf{S}/\mathcal{C}/A$-contraflat complexes of A-flat right $\mathbf{S}$-semimodules $\mathbf{S}/\mathcal{C}/A$-semiflat?

The functor mapping the quotient category of $\mathbf{S}/\mathcal{C}/A$-contraflat complexes of right $\mathbf{S}$-semimodules by its intersection with the thick subcategory of $\mathcal{C}$-coacyclic complexes into the semiderived category of right $\mathbf{S}$-semimodules is an equivalence of triangulated categories, since the complex $\mathbb{L}_3\mathbb{L}_1(\mathcal{K}^{\bullet})$ is $\mathbf{S}/\mathcal{C}/A$-contraflat for any complex of right $\mathbf{S}$-semimodules $\mathcal{K}^{\bullet}$. The analogous results for $\mathbf{S}/\mathcal{C}/A$-projective complexes of left $\mathbf{S}$-semimodules and $\mathbf{S}/\mathcal{C}/A$-injective complexes of left $\mathbf{S}$-semicontramodules follow from the corresponding results of 4.8.

Remark. It follows from the above lemma and Lemma 5.2 that any $\mathcal{C}$-coacyclic semiprojective complex of $\mathcal{C}$-coprojective left $\mathcal{S}$-semimodules is contractible. Indeed, such a complex is simultaneously an $\mathcal{S}/\mathcal{C}/A$-projective complex and a $\mathcal{C}$-coacyclic complex of $\mathcal{C}/A$-injective $\mathcal{S}$-semimodules.

Analogously, any $\mathcal{C}$-contraacyclic semiinjective complex of $\mathcal{C}$-coinjective left $\mathcal{S}$-semicontramodules is contractible. Hence the homotopy category of semiprojective complexes of $\mathcal{C}$-coprojective $\mathcal{S}$-semimodules is equivalent to the semiderived category of left $\mathcal{S}$-semimodules and the homotopy category of semiinjective complexes of $\mathcal{C}$-coinjective $\mathcal{S}$-semicontramodules is equivalent to the semiderived category of left $\mathcal{S}$-semicontramodules (see Theorem 4.6).

Furthermore, it follows that the homotopy category of semiprojective complexes of $\mathcal{C}$-coprojective $\mathcal{S}$-semimodules is the minimal triangulated subcategory containing the complexes of left $\mathcal{S}$-semimodules induced from complexes of $\mathcal{C}$-coprojective $\mathcal{C}$-comodules and closed under infinite direct sums. Analogously, the homotopy category of semiinjective complexes of $\mathcal{C}$-coinjective $\mathcal{S}$-semicontramodules is the minimal triangulated subcategory containing the complexes of left $\mathcal{S}$-semicontramodules coinduced from complexes of $\mathcal{C}$-coinjective $\mathcal{C}$-contramodules and closed under infinite products. (Cf. 2.9.)

6.5 Derived functor CtrTor

The following lemmas provide a general approach to one-sided derived functors of any number of arguments. They are essentially due to P. Deligne [29, 1.2.1–2].

Lemma 1. *Let* H *be a category and* S *be a localizing class of morphisms in* H. *Let* P *and* J *be full subcategories of* H *such that either*

(a) *the map* $\mathrm{Hom}_{\mathsf{H}}(Q, j)$ *is bijective for any object* $Q \in \mathsf{P}$ *and any morphism* $j \in \mathsf{S} \cap \mathsf{J}$, *and for any object* $Y \in \mathsf{H}$ *there is an object* $J \in \mathsf{J}$ *together with a morphism* $Y \longrightarrow J$ *belonging to* S, *or*

(b) *the map* $\mathrm{Hom}_{\mathsf{H}}(q, J)$ *is bijective for any morphism* $q \in \mathsf{S} \cap \mathsf{P}$ *and any object* $J \in \mathsf{J}$, *and for any object* $X \in \mathsf{H}$ *there is an object* $Q \in \mathsf{P}$ *together with a morphism* $Q \longrightarrow X$ *belonging to* S.

Then for any objects $P \in \mathsf{P}$ *and* $I \in \mathsf{J}$ *the natural map* $\mathrm{Hom}_{\mathsf{H}}(P, I) \longrightarrow \mathrm{Hom}_{\mathsf{H}[\mathsf{S}^{-1}]}(P, I)$ *is bijective.*

Proof. Part (b): any element of $\mathrm{Hom}_{\mathsf{H}[\mathsf{S}^{-1}]}(P, I)$ can be represented by a fraction of morphisms $P \longleftarrow X \longrightarrow I$ in H, where the morphism $X \longrightarrow P$ belongs to S. Choose an object $Q \in \mathsf{P}$ together with a morphism $Q \longrightarrow X$ belonging to S. Then the composition $Q \longrightarrow X \longrightarrow P$ belongs to $\mathsf{S} \cap \mathsf{P}$, hence the map $\mathrm{Hom}_{\mathsf{H}}(P, I) \longrightarrow \mathrm{Hom}_{\mathsf{H}}(Q, I)$ is bijective and there exists a morphism $P \longrightarrow I$ that forms a commutative triangle with the morphisms $Q \longrightarrow X \longrightarrow P$ and $Q \longrightarrow X \longrightarrow I$. Obviously, this morphism $P \longrightarrow I$ represents the same morphism in $\mathsf{H}[\mathsf{S}^{-1}]$ that the fraction $P \longleftarrow X \longrightarrow I$. Now suppose that there are two morphisms $P \rightrightarrows I$ in H whose images in $\mathsf{H}[\mathsf{S}^{-1}]$ coincide. Then there exists a morphism $X \longrightarrow P$ belonging to H which has equal compositions with the morphisms $P \rightrightarrows I$. Choose an object $Q \in \mathsf{P}$ together with a morphism $Q \longrightarrow X$ belonging to H again. The composition $Q \longrightarrow X \longrightarrow P$ has equal compositions with the morphisms $P \rightrightarrows I$, and since the map $\mathrm{Hom}_\mathsf{H}(P, I) \longrightarrow \mathrm{Hom}_\mathsf{H}(Q, I)$ is bijective, our two morphisms $P \rightrightarrows I$ are equal. Proof of part (a) is dual. $\square$

Lemma 2. *Let* H_i, $i = 1, \ldots, n$ *be several categories,* S_i *be localizing classes of morphisms in* H_i, *and* F_1 *be full subcategories of* H_i. *Assume that for any object* X *in* H_i *there is an object* U *in* F_i *together with a morphism* $U \longrightarrow X$ *from* S_i. *Let* K *be a category and* $\Theta \colon \mathsf{H}_1 \times \cdots \times \mathsf{H}_n \longrightarrow \mathsf{K}$ *be a functor such that the morphism* $\Theta(U_1, \ldots, U_{i-1}, t, U_{i+1}, \ldots, U_n)$ *is an isomorphism for any objects* $U_j \in \mathsf{F}_j$ *and any morphism* $t \in \mathsf{S}_i \cap \mathsf{F}_i$. *Then the left derived functor* $\mathbb{L}\Theta \colon \mathsf{H}_1[\mathsf{S}_1^{-1}] \times \cdots \times \mathsf{H}_n[\mathsf{S}_n^{-1}] \longrightarrow \mathsf{K}$ *obtained by restricting* Θ *to* $\mathsf{F}_1 \times \cdots \times \mathsf{F}_n$ *is a universal final object in the category of all functors* $\Xi \colon \mathsf{H}_1 \times \cdots \times \mathsf{H}_n \longrightarrow \mathsf{K}$ *factorizable through* $\mathsf{H}_1[\mathsf{S}_1^{-1}] \times \cdots \times \mathsf{H}_n[\mathsf{S}_n^{-1}]$ *and endowed with a morphism of functors* $\Xi \longrightarrow \Theta$.

Proof. It suffices to consider a single category $\mathsf{H} = \mathsf{H}_1 \times \cdots \times \mathsf{H}_n$ with the class of morphisms $\mathsf{S} = \mathsf{S}_1 \times \cdots \times \mathsf{S}_n$, the full subcategory $\mathsf{F} = \mathsf{F}_1 \times \cdots \times \mathsf{F}_n$, and the functor of one argument $\Theta \colon \mathsf{H} \longrightarrow \mathsf{K}$. The functor $\mathsf{F}[(\mathsf{S} \cap \mathsf{F})^{-1}] \longrightarrow \mathsf{H}[\mathsf{S}^{-1}]$ is an equivalence of categories by Lemma 2.6, so the derived functor $\mathbb{L}\Theta$ can be defined. For any object $X \in \mathsf{H}$, choose an object $U_X \in \mathsf{F}$ together with a morphism $U_X \longrightarrow X$ from S; then we have the induced morphism $\mathbb{L}\Theta(X) = \Theta(U_X) \longrightarrow \Theta(X)$. For any morphism $X \longrightarrow Y$ in H there exists an object V in F together with a morphism $V \longrightarrow U_X$ belonging to S and a morphism $V \longrightarrow U_Y$ in H forming a commutative diagram with the morphisms $U_X \longrightarrow X \longrightarrow Y$ and $V_X \longrightarrow Y$. So we have constructed a morphism of functors $\mathbb{L}\Theta \longrightarrow \Theta$. Now if a functor $\Xi \colon \mathsf{H} \longrightarrow \mathsf{K}$ factorizable through $\mathsf{H}[\mathsf{S}^{-1}]$ is endowed with a morphism of functors $\Xi \longrightarrow \Theta$, then the desired morphism of functors $\Xi \longrightarrow \mathbb{L}\Theta$ can be obtained by restricting the morphism of functors $\Xi \longrightarrow \Theta$ to the subcategory $\mathsf{F} \subset \mathsf{H}$. $\square$

Notice the difference between the construction of a double-sided derived functor of two arguments in Lemma 2.7 and the construction of a left derived functor of any number of arguments in Lemma 2. While in the former construction only *one* of the two arguments is resolved, and the conditions imposed on the resolutions guarantee that the two derived functors obtained in this way coincide, in the latter construction *all* of the arguments are resolved at once and it would not suffice to resolve only some of them. In other words, the construction of Lemma 2.7 only works to define *balanced* double-sided derived functors, while construction of Lemma 2 allows us to define *nonbalanced* one-sided derived functors.

Assume that the semialgebra $\mathcal{S}$ satisfies the conditions of 6.3. According to Lemma 1(a) and (the proof of) Theorem 6.3(a), the natural map

$$\mathrm{Hom}_{\mathsf{Hot}(\mathcal{S}\text{-simod})}(\mathcal{L}^\bullet, \mathcal{M}^\bullet) \longrightarrow \mathrm{Hom}_{\mathsf{D}^{\mathrm{si}}(\mathcal{S}\text{-simod})}(\mathcal{L}^\bullet, \mathcal{M}^\bullet)$$

is an isomorphism whenever $\mathcal{L}^{\bullet}$ is a complex of $\mathcal{S}/\mathcal{C}/A$-projective $\mathcal{S}$-semimodules and $\mathcal{M}^{\bullet}$ is a complex of $\mathcal{C}/A$-injective $\mathcal{S}$-semimodules. So the functor of homomorphisms in the semiderived category of left $\mathcal{S}$-semimodules can be lifted to a functor

$$\operatorname{Ext}_{\mathcal{S}} \colon \mathsf{D}^{\mathsf{si}}(\mathcal{S}\text{-simod})^{\mathrm{op}} \times \mathsf{D}^{\mathsf{si}}(\mathcal{S}\text{-simod}) \longrightarrow \mathsf{D}(k\text{-mod}),$$

which is defined by restricting the functor of homomorphisms of complexes of left $\mathcal{S}$-semimodules to the Cartesian product of the homotopy category of $\mathcal{S}/\mathcal{C}/A$-projective complexes of $\mathcal{S}$-semimodules and the homotopy category of complexes of $\mathcal{C}/A$-injective $\mathcal{S}$-semimodules. By Lemma 2, this construction of the right derived functor $\operatorname{Ext}_{\mathcal{S}}$ does not depend on the choice of subcategories of adjusted complexes.

Analogously, according to Lemma 1(b) and (the proof of) Theorem 6.3(b), the natural map

$$\operatorname{Hom}_{\mathsf{Hot}(\mathcal{S}\text{-sicntr})}(\mathfrak{P}^{\bullet}, \mathfrak{Q}^{\bullet}) \longrightarrow \operatorname{Hom}_{\mathsf{D}^{\mathsf{si}}(\mathcal{S}\text{-sicntr})}(\mathfrak{P}^{\bullet}, \mathfrak{Q}^{\bullet})$$

is an isomorphism whenever $\mathfrak{P}^{\bullet}$ is a complex of $\mathcal{C}/A$-injective $\mathcal{S}$-semicontramodules and $\mathfrak{Q}^{\bullet}$ is a complex of $\mathcal{S}/\mathcal{C}/A$-injective $\mathcal{S}$-semicontramodules. So the functor of homomorphisms in the semiderived category of left $\mathcal{S}$-semicontramodules can be lifted to a functor

$$\operatorname{Ext}^{\mathcal{S}} \colon \mathsf{D}^{\mathsf{si}}(\mathcal{S}\text{-sicntr})^{\mathrm{op}} \times \mathsf{D}^{\mathsf{si}}(\mathcal{S}\text{-sicntr}) \longrightarrow \mathsf{D}(k\text{-mod}),$$

which is defined by restricting the functor of homomorphisms of complexes of left $\mathcal{S}$-semicontramodules to the Cartesian product of the homotopy category of complexes of $\mathcal{C}/A$-projective $\mathcal{S}$-semicontramodules and the homotopy category of $\mathcal{S}/\mathcal{C}/A$-injective complexes of $\mathcal{S}$-semicontramodules.

Finally, the left derived functor of contratensor product

$$\operatorname{CtrTor}^{\mathcal{S}} \colon \mathsf{D}^{\mathsf{si}}(\text{simod-}\mathcal{S}) \times \mathsf{D}^{\mathsf{si}}(\mathcal{S}\text{-sicntr}) \longrightarrow \mathsf{D}(k\text{-mod})$$

is defined by restricting the functor of contratensor product over $\mathcal{S}$ to the Cartesian product of the homotopy category of $\mathcal{S}/\mathcal{C}/A$-contraflat complexes of right $\mathcal{S}$-semimodules and the homotopy category of complexes of $\mathcal{C}/A$-projective left $\mathcal{S}$-semicontramodules. By the definition, this restriction factorizes through the semiderived category of left $\mathcal{S}$-semicontramodules in the second argument; let us show that it also factorizes through the semiderived category of right $\mathcal{S}$-semimodules in the first argument. The complex of left $\mathcal{S}$-semicontramodules $\operatorname{Hom}_k(\mathcal{N}^{\bullet}, k^{\vee})$ is $\mathcal{S}/\mathcal{C}/A$-injective whenever a complex of right $\mathcal{S}$-semimodules $\mathcal{N}^{\bullet}$ is $\mathcal{S}/\mathcal{C}/A$-contraflat; and the complex $\operatorname{Hom}_k(\mathcal{N}^{\bullet}, k^{\vee})$ is $\mathcal{C}$-contraacyclic whenever the complex $\mathcal{N}^{\bullet}$ is $\mathcal{C}$-coacyclic. Hence if $\mathcal{N}^{\bullet}$ is a $\mathcal{C}$-coacyclic $\mathcal{S}/\mathcal{C}/A$-contraflat complex of right $\mathcal{S}$-semimodules and $\mathfrak{P}^{\bullet}$ is a complex of $\mathcal{C}/A$-projective left $\mathcal{S}$-semicontramodules, then the complex $\operatorname{Hom}^{\mathcal{S}}(\mathfrak{P}^{\bullet}, \operatorname{Hom}_k(\mathcal{N}^{\bullet}, k^{\vee}))$ is acyclic, so the complex $\mathcal{N}^{\bullet} \odot_{\mathcal{S}} \mathfrak{P}^{\bullet}$ is also acyclic. By Lemma 2, this construction of the left derived functor $\operatorname{CtrTor}^{\mathcal{S}}$ does not depend on the choice of subcategories of adjusted complexes.

Notice that the constructions of derived functors $\mathbb{R}\Psi_{\mathcal{S}}$ and $\mathbb{L}\Phi_{\mathcal{S}}$ in Corollary 6.3 are also particular cases of Lemma 2.

Remark. To define/compute the composition multiplication $\mathrm{Ext}_{\mathbf{S}}(\mathcal{L}^{\bullet}, \mathcal{M}^{\bullet}) \otimes_{k}^{\mathbb{L}} \mathrm{Ext}_{\mathbf{S}}(\mathcal{K}^{\bullet}, \mathcal{L}^{\bullet}) \longrightarrow \mathrm{Ext}_{\mathbf{S}}(\mathcal{K}^{\bullet}, \mathcal{M}^{\bullet})$ it suffices to represent the images of $\mathcal{K}^{\bullet}$, $\mathcal{L}^{\bullet}$, and $\mathcal{M}^{\bullet}$ in the semiderived category of left $\mathbf{S}$-semimodules by semiprojective complexes of $\mathcal{C}$-coprojective $\mathbf{S}$-semimodules. The same applies to the functor $\mathrm{Ext}^{\mathbf{S}}$ and semiinjective complexes of $\mathcal{C}$-coinjective $\mathbf{S}$-semicontramodules. Besides, one can compute the functors $\mathrm{Ext}_{\mathbf{S}}$, $\mathrm{Ext}^{\mathbf{S}}$, and $\mathrm{CtrTor}^{\mathbf{S}}$ using resolutions of other kinds. In particular, one can use complexes of $\mathcal{C}$-injective $\mathbf{S}$-semimodules and complexes of $\mathcal{C}$-projective $\mathbf{S}$-semicontramodules (see Remark 6.3) together with (appropriately defined) $\mathbf{S}/\mathcal{C}$-projective complexes of left $\mathbf{S}$-semimodules, $\mathbf{S}/\mathcal{C}$-injective complexes of left $\mathbf{S}$-semicontramodules, and $\mathbf{S}/\mathcal{C}$-contraflat complexes of right $\mathbf{S}$-semimodules. One can also compute the functor $\mathrm{Ext}_{\mathbf{S}}$ in terms of injective complexes of $\mathbf{S}$-semimodules (defined as complexes right orthogonal to $\mathcal{C}$-coacyclic complexes in $\mathrm{Hot}(\mathbf{S}\text{-simod})$) and the functor $\mathrm{Ext}^{\mathbf{S}}$ in terms of projective complexes of $\mathbf{S}$-semicontramodules. These can be obtained by applying the functor $\Phi_{\mathbf{S}}$ to semiinjective complexes of $\mathcal{C}$-coinjective $\mathbf{S}$-semicontramodules and the functor $\Psi_{\mathbf{S}}$ to semiprojective complexes of $\mathcal{C}$-coprojective $\mathbf{S}$-semimodules, and using Propositions 6.2.2(a) and 6.2.3(a). Injective complexes of $\mathbf{S}$-semimodules can be also constructed using the functor right adjoint to the forgetful functor $\mathbf{S}\text{-simod} \longrightarrow \mathcal{C}\text{-comod}$ (see Question 3.3.1) and infinite products of complexes of $\mathbf{S}$-semimodules; this approach works assuming only that $\mathcal{C}$ is a flat right A-module, $\mathbf{S}$ is a coflat right $\mathcal{C}$-comodule, and A has a finite left homological dimension.

6.6 SemiExt and Ext, SemiTor and CtrTor

We keep the assumptions of 6.3.

Corollary.

(a) *There are natural isomorphisms of functors*

$$\mathrm{SemiExt}_{\mathbf{S}}(\mathcal{M}^{\bullet}, \mathfrak{P}^{\bullet}) \simeq \mathrm{Ext}_{\mathbf{S}}(\mathcal{M}^{\bullet}, \mathbb{L}\Phi_{\mathbf{S}}(\mathfrak{P}^{\bullet})) \simeq \mathrm{Ext}^{\mathbf{S}}(\mathbb{R}\Psi_{\mathbf{S}}(\mathcal{M}^{\bullet}), \mathfrak{P}^{\bullet})$$

on the Cartesian product of the category opposite to the semiderived category of left $\mathbf{S}$-semimodules and the semiderived category of left $\mathbf{S}$-semicontramodules.

(b) *There is a natural isomorphism of functors*

$$\mathrm{SemiTor}^{\mathbf{S}}(\mathcal{N}^{\bullet}, \mathcal{M}^{\bullet}) \simeq \mathrm{CtrTor}^{\mathbf{S}}(\mathcal{N}^{\bullet}, \mathbb{R}\Psi_{\mathbf{S}}(\mathcal{M}^{\bullet}))$$

on the Cartesian product of the semiderived category of right $\mathbf{S}$-semimodules and the semiderived category of left $\mathbf{S}$-semimodules.

Proof. It suffices to construct natural isomorphisms

$$\mathrm{SemiExt}_{\mathbf{S}}(\mathcal{L}^{\bullet}, \mathbb{R}\Psi_{\mathbf{S}}(\mathcal{M}^{\bullet})) \simeq \mathrm{Ext}_{\mathbf{S}}(\mathcal{L}^{\bullet}, \mathcal{M}^{\bullet}),$$
$$\mathrm{SemiExt}_{\mathbf{S}}(\mathbb{L}\Phi_{\mathbf{S}}(\mathfrak{P}^{\bullet}), \mathfrak{Q}^{\bullet}) \simeq \mathrm{Ext}^{\mathbf{S}}(\mathfrak{P}^{\bullet}, \mathfrak{Q}^{\bullet}),$$
$$\mathrm{SemiTor}^{\mathbf{S}}(\mathcal{N}^{\bullet}, \mathbb{L}\Phi_{\mathbf{S}}(\mathfrak{P}^{\bullet})) \simeq \mathrm{CtrTor}^{\mathbf{S}}(\mathcal{N}^{\bullet}, \mathfrak{P}^{\bullet}).$$

In the first case, represent the image of $\mathcal{M}^\bullet$ in $\mathsf{D}^{\mathsf{si}}(\mathsf{S}\text{-simod})$ by a complex of $\mathcal{C}/A$-injective S-semimodules and the image of $\mathcal{L}^\bullet$ in $\mathsf{D}^{\mathsf{si}}(\mathsf{S}\text{-simod})$ by a semi-projective complex of $\mathcal{C}$-coprojective S-semimodules, and use Proposition 6.2.2(d) and Lemma 6.4(b). Alternatively, represent the image of $\mathcal{M}^\bullet$ in $\mathsf{D}^{\mathsf{si}}(\mathsf{S}\text{-simod})$ by a complex of $\mathcal{C}/A$-injective S-semimodules and the image of $\mathcal{L}^\bullet$ in $\mathsf{D}^{\mathsf{si}}(\mathsf{S}\text{-simod})$ by an $\mathsf{S}/\mathcal{C}/A$-semiprojective complex of A-projective S-semimodules (see 4.8), and use Proposition 6.2.2(c), Lemma 6.4(b), and Lemma 5.3.2(b).

In the second case, represent the image of $\mathfrak{P}^\bullet$ in $\mathsf{D}^{\mathsf{si}}(\mathsf{S}\text{-sicntr})$ by a complex of $\mathcal{C}/A$-projective S-semicontramodules and the image of $\mathfrak{Q}^\bullet$ in $\mathsf{D}^{\mathsf{si}}(\mathsf{S}\text{-sicntr})$ by a semiinjective complex of $\mathcal{C}$-coinjective S-semimodules, and use Proposition 6.2.3(d) and Lemma 6.4(c). Alternatively, represent the image of $\mathfrak{P}^\bullet$ in $\mathsf{D}^{\mathsf{si}}(\mathsf{S}\text{-sicntr})$ by a complex of $\mathcal{C}/A$-projective S-semicontramodules and the image of $\mathfrak{Q}^\bullet$ in $\mathsf{D}^{\mathsf{si}}(\mathsf{S}\text{-sicntr})$ by an $\mathsf{S}/\mathcal{C}/A$-semiinjective complex of A-injective S-semicontramodules (see 4.8), and use Proposition 6.2.3(c), Lemma 6.4(c), and Lemma 5.3.2(a).

In the third case, represent the image of $\mathfrak{P}^\bullet$ in $\mathsf{D}^{\mathsf{si}}(\mathsf{S}\text{-sicntr})$ by a complex of $\mathcal{C}/A$-projective S-semicontramodules and the image of $\mathcal{N}^\bullet$ in $\mathsf{D}^{\mathsf{si}}(\mathsf{simod}\text{-}\mathsf{S})$ by a semiflat complex of $\mathcal{C}$-coflat S-semimodules, and use Proposition 6.2.1(d) and Lemma 6.4(a). Alternatively, represent the image of $\mathfrak{P}^\bullet$ in $\mathsf{D}^{\mathsf{si}}(\mathsf{S}\text{-sicntr})$ by a complex of $\mathcal{C}/A$-projective S-semicontramodules and the image of $\mathcal{N}^\bullet$ in $\mathsf{D}^{\mathsf{si}}(\mathsf{simod}\text{-}\mathsf{S})$ by an $\mathsf{S}/\mathcal{C}/A$-semiflat complex of A-flat S-semimodules (see 2.8), and use Proposition 6.2.1(c), Lemma 6.4(a), and Lemma 5.3.2(a).

Finally, to show that the three pairwise isomorphisms between the functors $\mathrm{SemiExt}_\mathsf{S}(\mathcal{M}^\bullet, \mathfrak{P}^\bullet)$, $\mathrm{Ext}_\mathsf{S}(\mathcal{M}^\bullet, \mathbb{L}\Phi_\mathsf{S}(\mathfrak{P}^\bullet))$, and $\mathrm{Ext}^\mathsf{S}(\mathbb{R}\Psi_\mathsf{S}(\mathcal{M}^\bullet), \mathfrak{P}^\bullet)$ form a commutative diagram, one can represent the image of $\mathcal{M}^\bullet$ in $\mathsf{D}^{\mathsf{si}}(\mathsf{S}\text{-simod})$ by a semiprojective complex of $\mathcal{C}$-coprojective S-semimodules and the image of $\mathfrak{P}^\bullet$ in $\mathsf{D}^{\mathsf{si}}(\mathsf{S}\text{-sicntr})$ by a semiinjective complex of $\mathcal{C}$-coinjective S-semicontramodules (having in mind Lemmas 6.4 and 5.2), and use a result of 6.2. $\square$

7 Functoriality in the Coring

7.1 Compatible morphisms

Let $\mathcal{C}$ be a coring over a k-algebra A and $\mathcal{D}$ be a coring over a k-algebra B.

7.1.1 We will say that a map $\mathcal{C} \longrightarrow \mathcal{D}$ is compatible with a k-algebra morphism $A \longrightarrow B$ if the biaction maps $A \otimes_k \mathcal{C} \otimes_k A \longrightarrow \mathcal{C}$ and $B \otimes_k \mathcal{D} \otimes_k B \longrightarrow \mathcal{D}$ form a commutative diagram with the maps $\mathcal{C} \longrightarrow \mathcal{D}$ and $A \otimes_k \mathcal{C} \otimes_k A \longrightarrow B \otimes_k \mathcal{D} \otimes_k B$ (in other words, the map $\mathcal{C} \longrightarrow \mathcal{D}$ is an A-A-bimodule morphism) and the comultiplication maps $\mathcal{C} \longrightarrow \mathcal{C} \otimes_A \mathcal{C}$ and $\mathcal{D} \longrightarrow \mathcal{D} \otimes_B \mathcal{D}$, as well as the counit maps $\mathcal{C} \longrightarrow A$ and $\mathcal{D} \longrightarrow B$, form commutative diagrams with the maps $A \longrightarrow B$, $\mathcal{C} \longrightarrow \mathcal{D}$, and $\mathcal{C} \otimes_A \mathcal{C} \longrightarrow \mathcal{D} \otimes_B \mathcal{D}$.

Let $\mathcal{C} \longrightarrow \mathcal{D}$ be a map of corings compatible with a k-algebra map $A \longrightarrow B$. Let $\mathcal{M}$ be a left comodule over $\mathcal{C}$ and $\mathcal{N}$ be a left comodule over B. We will say that a map $\mathcal{M} \longrightarrow \mathcal{N}$ is compatible with the maps $A \longrightarrow B$ and $\mathcal{C} \longrightarrow \mathcal{D}$ if the action maps $A \otimes_k \mathcal{M} \longrightarrow \mathcal{M}$ and $B \otimes_k \mathcal{N} \longrightarrow \mathcal{N}$ form a commutative diagram with the maps $\mathcal{M} \longrightarrow \mathcal{N}$ and $A \otimes_k \mathcal{M} \longrightarrow B \otimes_k \mathcal{N}$ (that is the map $\mathcal{M} \longrightarrow \mathcal{N}$ is an A-module morphism) and the coaction maps $\mathcal{M} \longrightarrow \mathcal{C} \otimes_A \mathcal{M}$ and $\mathcal{N} \longrightarrow \mathcal{D} \otimes_B \mathcal{N}$ form a commutative diagram with the maps $\mathcal{M} \longrightarrow \mathcal{N}$ and $\mathcal{C} \otimes_A \mathcal{M} \longrightarrow \mathcal{D} \otimes_B \mathcal{M}$. Analogously, let $\mathfrak{P}$ be a left contramodule over $\mathcal{C}$ and $\mathfrak{Q}$ be a left contramodule over $\mathcal{D}$. We will say that a map $\mathfrak{Q} \longrightarrow \mathfrak{P}$ is compatible with the maps $A \longrightarrow B$ and $\mathcal{C} \longrightarrow \mathcal{D}$ if the action maps $\mathfrak{P} \longrightarrow \operatorname{Hom}_k(A, \mathfrak{P})$ and $\mathfrak{Q} \longrightarrow \operatorname{Hom}_k(B, \mathfrak{Q})$ form a commutative diagram with the maps $\mathfrak{Q} \longrightarrow \mathfrak{P}$ and $\operatorname{Hom}_k(B, \mathfrak{Q}) \longrightarrow \operatorname{Hom}_k(A, \mathfrak{P})$ (that is the map $\mathfrak{Q} \longrightarrow \mathfrak{P}$ is an A-module morphism) and the contraaction maps $\operatorname{Hom}_A(\mathcal{C}, \mathfrak{P}) \longrightarrow \mathfrak{P}$ and $\operatorname{Hom}_B(\mathcal{D}, \mathfrak{Q}) \longrightarrow \mathfrak{Q}$ form a commutative diagram with the maps $\mathfrak{Q} \longrightarrow \mathfrak{P}$ and $\operatorname{Hom}_B(\mathcal{D}, \mathfrak{Q}) \longrightarrow \operatorname{Hom}_A(\mathcal{C}, \mathfrak{P})$.

Let $\mathcal{M}' \longrightarrow \mathcal{N}'$ be a map from a right $\mathcal{C}$-comodule $\mathcal{M}'$ to a right $\mathcal{D}$-comodule $\mathcal{N}'$ compatible with the maps $A \longrightarrow B$ and $\mathcal{C} \longrightarrow \mathcal{D}$, and $\mathcal{M}'' \longrightarrow \mathcal{N}''$ be a map from a left $\mathcal{C}$-comodule $\mathcal{M}''$ to a left $\mathcal{D}$-comodule $\mathcal{N}''$ compatible with the maps $A \longrightarrow B$ and $\mathcal{C} \longrightarrow \mathcal{D}$. Then there is a natural map $\mathcal{M}' \square_{\mathcal{C}} \mathcal{M}'' \longrightarrow \mathcal{N}' \square_{\mathcal{D}} \mathcal{N}''$. Analogously, let $\mathcal{M} \longrightarrow \mathcal{N}$ be a map from a left $\mathcal{C}$-comodule $\mathcal{M}$ to a left $\mathcal{D}$-comodule $\mathcal{N}$ compatible with the maps $A \longrightarrow B$ and $\mathcal{C} \longrightarrow \mathcal{D}$, and $\mathfrak{Q} \longrightarrow \mathfrak{P}$ be a map from a left $\mathcal{D}$-contramodule $\mathfrak{Q}$ to a left $\mathcal{C}$-contramodule $\mathfrak{P}$ compatible with the maps $A \longrightarrow B$ and $\mathcal{C} \longrightarrow \mathcal{D}$. Then there is a natural map $\operatorname{Cohom}_{\mathcal{D}}(\mathcal{N}, \mathfrak{Q}) \longrightarrow \operatorname{Cohom}_{\mathcal{C}}(\mathcal{M}, \mathfrak{P})$.

7.1.2 Let $\mathcal{C} \longrightarrow \mathcal{D}$ be a map of corings compatible with a k-algebra map $A \longrightarrow B$. Then there is a functor from the category of left $\mathcal{C}$-comodules to the category of left $\mathcal{D}$-comodules assigning to a $\mathcal{C}$-comodule $\mathcal{M}$ the $\mathcal{D}$-comodule

$$_B\mathcal{M} = B \otimes_A \mathcal{M}$$

with the $\mathcal{D}$-coaction map defined as the composition $B \otimes_A \mathcal{M} \longrightarrow B \otimes_A \mathcal{C} \otimes_A \mathcal{M} \longrightarrow \mathcal{D} \otimes_A \mathcal{M} = \mathcal{D} \otimes_B (B \otimes_A \mathcal{M})$ of the map induced by the $\mathcal{C}$-coaction in $\mathcal{M}$ and the map induced by the map $\mathcal{C} \longrightarrow \mathcal{D}$ and the left B-action in $\mathcal{D}$. The functor $\mathcal{M} \longmapsto \mathcal{M}_B$ from the category of right $\mathcal{C}$-comodules to the category of right $\mathcal{D}$-comodules is defined in an analogous way. Furthermore, there is a functor from the category of left $\mathcal{C}$-contramodules to the category of left $\mathcal{D}$-contramodules assigning to a $\mathcal{C}$-contramodule $\mathfrak{P}$ the $\mathcal{D}$-contramodule

$$^B\mathfrak{P} = \operatorname{Hom}_A(B, \mathfrak{P})$$

with the contraaction map defined as the composition $\operatorname{Hom}_B(\mathcal{D}, \operatorname{Hom}_A(B, \mathfrak{P})) = \operatorname{Hom}_A(\mathcal{D}, \mathfrak{P}) \longrightarrow \operatorname{Hom}_A(\mathcal{C} \otimes_A B, \mathfrak{P}) = \operatorname{Hom}_A(B, \operatorname{Hom}_A(\mathcal{C}, \mathfrak{P})) \longrightarrow \operatorname{Hom}_A(B, \mathfrak{P})$ of the map induced by the map $\mathcal{C} \longrightarrow \mathcal{D}$ and the right B-action in $\mathcal{D}$ with the map induced by the $\mathcal{C}$-contraaction in $\mathfrak{P}$.

  If $\mathcal{C}$ is a flat right A-module, then the functor $\mathcal{M} \longmapsto {}_B\mathcal{M}$ has a right adjoint functor assigning to a left $\mathcal{D}$-comodule $\mathcal{N}$ the left $\mathcal{D}$-comodule

$$_{\mathcal{C}}\mathcal{N} = \mathcal{C}_B \,\square_{\mathcal{D}}\, \mathcal{N},$$

where $\mathcal{C}_B = \mathcal{C} \otimes_A B$ is a $\mathcal{C}$-$\mathcal{D}$-bicomodule with the right $\mathcal{D}$-comodule structure provided by the above construction. These functors are adjoint since both k-modules $\operatorname{Hom}_{\mathcal{D}}({}_B\mathcal{M}, \mathcal{N})$ and $\operatorname{Hom}_{\mathcal{C}}(\mathcal{M}, {}_{\mathcal{C}}\mathcal{N})$ are isomorphic to the k-module of all maps of comodules $\mathcal{M} \longrightarrow \mathcal{N}$ compatible with the maps $A \longrightarrow B$ and $\mathcal{C} \longrightarrow \mathcal{D}$, so

$$\operatorname{Hom}_{\mathcal{D}}({}_B\mathcal{M}, \mathcal{N}) \simeq \operatorname{Hom}_{\mathcal{C}}(\mathcal{M}, {}_{\mathcal{C}}\mathcal{N}).$$

Without any assumptions on the coring $\mathcal{C}$, the functor $\mathcal{N} \longmapsto {}_{\mathcal{C}}\mathcal{N}$ is defined on the full subcategory of left $\mathcal{D}$-comodules such that the cotensor product $\mathcal{C}_B \,\square_{\mathcal{D}}\, \mathcal{N}$ can be endowed with a left $\mathcal{C}$-comodule structure via the construction of 1.2.4; this includes, in particular, quasicoflat $\mathcal{D}$-comodules. Analogously, if $\mathcal{C}$ is a flat left A-module, then the functor $\mathcal{M} \longmapsto \mathcal{M}_B$ has a right adjoint functor assigning to a right $\mathcal{D}$-comodule $\mathcal{N}$ the right $\mathcal{C}$-comodule $\mathcal{N}_{\mathcal{C}} = \mathcal{N} \,\square_{\mathcal{D}}\, {}_B\mathcal{C}$, where $_B\mathcal{C} = B \otimes_A \mathcal{C}$ is a $\mathcal{D}$-$\mathcal{C}$-bicomodule with the left $\mathcal{D}$-comodule structure provided by the above construction.

  Furthermore, if $\mathcal{C}$ is a projective left A-module, then the functor $\mathfrak{P} \longmapsto {}^B\mathfrak{P}$ has a left adjoint functor assigning to a left $\mathcal{D}$-contramodule $\mathfrak{Q}$ the left $\mathcal{C}$-contra-module

$$^{\mathcal{C}}\mathfrak{Q} = \operatorname{Cohom}_{\mathcal{D}}({}_B\mathcal{C}, \mathfrak{Q}).$$

These functors are adjoint since both k-modules $\operatorname{Hom}^{\mathcal{D}}(\mathfrak{Q}, {}^B\mathfrak{P})$ and $\operatorname{Hom}^{\mathcal{C}}({}^{\mathcal{C}}\mathfrak{Q}, \mathfrak{P})$ are isomorphic to the k-module of all maps of contramodules $\mathfrak{Q} \longrightarrow \mathfrak{P}$ compatible with the maps $A \longrightarrow B$ and $\mathcal{C} \longrightarrow \mathcal{D}$, so

$$\operatorname{Hom}^{\mathcal{D}}(\mathfrak{Q}, {}^B\mathfrak{P}) \simeq \operatorname{Hom}^{\mathcal{C}}({}^{\mathcal{C}}\mathfrak{Q}, \mathfrak{P}).$$

Without any assumptions on the coring $\mathcal{C}$, the functor $\mathfrak{Q} \longmapsto {}^{\mathcal{C}}\mathfrak{Q}$ is defined on the full subcategory of left $\mathcal{D}$-contramodules such that the cohomomorphism module

$\mathrm{Cohom}_{\mathcal{D}}({}_B\mathcal{C},\mathfrak{Q})$ can be endowed with a left $\mathcal{C}$-contramodule structure via the construction of 3.2.4; this includes, in particular, quasicoinjective $\mathcal{D}$-contramodules.

If $\mathcal{C}$ is a projective left A-module, then for any right $\mathcal{C}$-comodule $\mathcal{M}$ and any left $\mathcal{D}$-contramodule $\mathfrak{Q}$ there is a natural isomorphism

$$\mathcal{M}_B \odot_{\mathcal{D}} \mathfrak{Q} \simeq \mathcal{M} \odot_{\mathcal{C}} {}^{\mathcal{C}}\mathfrak{Q}.$$

Indeed, both k-modules are isomorphic to the cokernel of the pair of maps $\mathcal{M} \otimes_A \mathrm{Hom}_B(\mathcal{D},\mathfrak{Q}) \rightrightarrows \mathcal{M} \otimes_A \mathfrak{Q}$, one of which is induced by the $\mathcal{D}$-contraaction in $\mathfrak{Q}$ and the other is the composition of the map induced by the $\mathcal{C}$-coaction in $\mathcal{M}$ and the map induced by the evaluation map $\mathcal{C}_B \otimes_B \mathrm{Hom}_B(\mathcal{D},\mathfrak{Q}) \longrightarrow \mathfrak{Q}$. This is obvious for $\mathcal{M}_B \odot_{\mathcal{D}} \mathfrak{Q}$, and in order to show this for $\mathcal{M} \odot_{\mathcal{C}} {}^{\mathcal{C}}\mathfrak{Q}$ it suffices to represent ${}^{\mathcal{C}}\mathfrak{Q}$ as the cokernel of the pair of $\mathcal{C}$-contramodule morphisms $\mathrm{Hom}_B({}_B\mathcal{C}, \mathrm{Hom}_B(\mathcal{D},\mathfrak{Q})) \rightrightarrows \mathrm{Hom}_B({}_B\mathcal{C},\mathfrak{Q})$. Without any assumptions on the coring $\mathcal{C}$, there is a natural isomorphism $\mathcal{M}_B \odot_{\mathcal{D}} \mathfrak{Q} \simeq \mathcal{M} \odot_{\mathcal{C}} {}^{\mathcal{C}}\mathfrak{Q}$ for any right $\mathcal{C}$-comodule $\mathcal{M}$ and any left $\mathcal{D}$-contramodule $\mathfrak{Q}$ for which the $\mathcal{C}$-contramodule ${}^{\mathcal{C}}\mathfrak{Q} = \mathrm{Cohom}_{\mathcal{D}}({}_B\mathcal{C},\mathfrak{Q})$ is defined via the construction of 3.2.4.

7.1.3 Let $\mathcal{C} \longrightarrow \mathcal{D}$ be a map of corings compatible with a k-algebra map $A \longrightarrow B$.

Proposition.

(a) *For any left $\mathcal{C}$-comodule $\mathcal{M}$ and any right $\mathcal{D}$-comodule $\mathcal{N}$ for which the right $\mathcal{C}$-comodule $\mathcal{N}_{\mathcal{C}}$ is defined there is a natural map*

$$\mathcal{N}_{\mathcal{C}} \,\square_{\mathcal{C}}\, \mathcal{M} \longrightarrow \mathcal{N} \square_{\mathcal{D}} {}_B\mathcal{M},$$

which is an isomorphism, at least, when $\mathcal{C}$ and $\mathcal{M}$ are flat left A-modules or $\mathcal{N}$ is a quasicoflat right $\mathcal{D}$-comodule.

(b) *For any left $\mathcal{C}$-contramodule $\mathfrak{P}$ and any left $\mathcal{D}$-comodule $\mathcal{N}$ for which the left $\mathcal{C}$-comodule ${}_{\mathcal{C}}\mathcal{N}$ is defined there is a natural map*

$$\mathrm{Cohom}_{\mathcal{D}}(\mathcal{N}, {}^B\mathfrak{P}) \longrightarrow \mathrm{Cohom}_{\mathcal{C}}({}_{\mathcal{C}}\mathcal{N}, \mathfrak{P}),$$

which is an isomorphism, at least, when either $\mathcal{C}$ is a flat right A-module and $\mathfrak{P}$ is an injective left A-module, or $\mathcal{N}$ is a quasicoprojective left $\mathcal{D}$-comodule.

(c) *For any left $\mathcal{C}$-comodule $\mathcal{M}$ and any left $\mathcal{D}$-contramodule $\mathfrak{Q}$ for which the left $\mathcal{C}$-contramodule ${}^{\mathcal{C}}\mathfrak{Q}$ is defined there is a natural map*

$$\mathrm{Cohom}_{\mathcal{D}}({}_B\mathcal{M}, \mathfrak{Q}) \longrightarrow \mathrm{Cohom}_{\mathcal{C}}(\mathcal{M}, {}^{\mathcal{C}}\mathfrak{Q}),$$

which is an isomorphism, at least, when $\mathcal{C}$ and $\mathcal{M}$ are projective left A-modules or $\mathfrak{Q}$ is a quasicoinjective left $\mathcal{D}$-contramodule.

Proof. Part (a): for any left $\mathcal{C}$-comodule $\mathcal{M}$ and any right $\mathcal{D}$-comodule $\mathcal{N}$ there are maps of comodules $\mathcal{M} \longrightarrow {}_B\mathcal{M}$ and $\mathcal{N}_{\mathcal{C}} \longrightarrow \mathcal{N}$ compatible with the maps

$A \longrightarrow B$ and $\mathcal{C} \longrightarrow \mathcal{D}$. So there is the induced map $\mathcal{N}_\mathcal{C} \,\square_\mathcal{C}\, \mathcal{M} \longrightarrow \mathcal{N} \,\square_\mathcal{D}\, _B\mathcal{M}$. On the other hand, for any left $\mathcal{C}$-comodule $\mathcal{M}$ there is a natural isomorphism of left $\mathcal{D}$-comodules $_B\mathcal{M} \simeq {_B}\mathcal{C} \,\square_\mathcal{C}\, \mathcal{M}$, hence $\mathcal{N}_\mathcal{C} \,\square_\mathcal{C}\, \mathcal{M} = (\mathcal{N} \,\square_\mathcal{D}\, {_B}\mathcal{C}) \,\square_\mathcal{C}\, \mathcal{M}$ and $\mathcal{N}\square_\mathcal{D}\, {_B}\mathcal{M} = \mathcal{N}\square_\mathcal{D} ({_B}\mathcal{C}\square_\mathcal{C} \mathcal{M})$. Let us check that the maps $\mathcal{N}_\mathcal{C}\,\square_\mathcal{C}\, \mathcal{M} \longrightarrow \mathcal{N}\square_\mathcal{D}\, {_B}\mathcal{M}$, $\mathcal{N}_\mathcal{C}\square_\mathcal{C}\mathcal{M} \longrightarrow \mathcal{N}\otimes_B {_B}\mathcal{C}\otimes_A\mathcal{M}$ and $\mathcal{N}\square_\mathcal{D}\,{_B}\mathcal{M} \longrightarrow \mathcal{N}\otimes_B {_B}\mathcal{C}\otimes_A\mathcal{M}$ form a commutative diagram. Indeed, the map $(\mathcal{N}\,\square_\mathcal{D}\, {_B}\mathcal{C}) \otimes_A \mathcal{M} \longrightarrow \mathcal{N} \otimes_B {_B}\mathcal{C} \otimes_A \mathcal{M}$ is equal to the composition of the map $(\mathcal{N}\,\square_\mathcal{D}\, {_B}\mathcal{C}) \otimes_A \mathcal{M} \longrightarrow (\mathcal{N}\,\square_\mathcal{D}\, {_B}\mathcal{C}) \otimes_A \mathcal{C} \otimes_A \mathcal{M}$ induced by the $\mathcal{C}$-coaction in $\mathcal{N}\square_\mathcal{D}\, {_B}\mathcal{C}$ with the map $(\mathcal{N}\square_\mathcal{D}\, {_B}\mathcal{C})\otimes_A \mathcal{C}\otimes_A\mathcal{M} \longrightarrow \mathcal{N}\otimes_B {_B}\mathcal{C}\otimes_A\mathcal{M}$ induced by the maps $\mathcal{N}\,\square_\mathcal{D}\, {_B}\mathcal{C} \longrightarrow \mathcal{N}$ and $\mathcal{C} \longrightarrow {_B}\mathcal{C}$; while the composition of maps $(\mathcal{N}\,\square_\mathcal{D}\, {_B}\mathcal{C}) \otimes_A \mathcal{M} \longrightarrow \mathcal{N} \otimes_B ({_B}\mathcal{C}\,\square_\mathcal{C}\, \mathcal{M}) \longrightarrow \mathcal{N} \otimes_B {_B}\mathcal{C} \otimes_A \mathcal{M}$ is equal to the composition of the map $(\mathcal{N}\square_\mathcal{D}\, {_B}\mathcal{C})\otimes_A\mathcal{M} \longrightarrow (\mathcal{N}\square_\mathcal{D}\, {_B}\mathcal{C})\otimes_A \mathcal{C}\otimes_A\mathcal{M} \longrightarrow \mathcal{N}\otimes_B {_B}\mathcal{C}\otimes_A\mathcal{M}$ induced by the $\mathcal{C}$-coaction in $\mathcal{M}$ with the same map $(\mathcal{N}\square_\mathcal{D}\, {_B}\mathcal{C})\otimes_A \mathcal{C}\otimes_A\mathcal{M} \longrightarrow \mathcal{N}\otimes_B {_B}\mathcal{C}\otimes_A \mathcal{M}$. It remains to apply Proposition 1.2.5(d) and (e) with the left and right sides switched. The proofs of parts (b) and (c) are completely analogous; the proof of (b) uses Proposition 3.2.5(g,h) and the proof of (c) uses Proposition 3.2.5(f,i).   □

7.1.4   Let $\mathcal{C} \longrightarrow \mathcal{D}$ be a map of corings compatible with a k-algebra map $A \longrightarrow B$. Assume that $\mathcal{C}$ is a projective left and a flat right A-module.

Then for any left $\mathcal{D}$-contramodule $\mathfrak{Q}$ there is a natural morphism of $\mathcal{C}$-co-modules

$$\Phi_\mathcal{C}({^\mathcal{C}}\mathfrak{Q}) \longrightarrow {_\mathcal{C}}(\Phi_\mathcal{D}\mathfrak{Q}),$$

which is an isomorphism, at least, when $\mathcal{D}$ is a flat right B-module and $\mathfrak{Q}$ is a $\mathcal{D}/B$-contraflat left $\mathcal{D}$-contramodule. Indeed, $\Phi_\mathcal{C}({^\mathcal{C}}\mathfrak{Q}) = \mathcal{C} \odot_\mathcal{C} {^\mathcal{C}}\mathfrak{Q} \simeq \mathcal{C}_B \odot_\mathcal{D} \mathfrak{Q}$ as a left $\mathcal{C}$-comodule and $_\mathcal{C}(\Phi_\mathcal{D}\mathfrak{Q}) = \mathcal{C}_B \,\square_\mathcal{D}\, (\mathcal{D} \odot_\mathcal{D} \mathfrak{Q})$, so it remains to apply Proposition 5.2.1(c). Analogously, for any left $\mathcal{D}$-comodule $\mathcal{N}$ there is a natural morphism of $\mathcal{C}$-contramodules

$$^\mathcal{C}(\Psi_\mathcal{D}\mathcal{N}) \longrightarrow \Psi_\mathcal{C}({_\mathcal{C}}\mathcal{N}),$$

which is an isomorphism, at least, when $\mathcal{D}$ is a projective left B-module and $\mathcal{N}$ is a $\mathcal{D}/B$-injective left $\mathcal{D}$-comodule. Indeed, $\Psi_\mathcal{C}({_\mathcal{C}}\mathcal{N}) = \mathrm{Hom}_\mathcal{C}(\mathcal{C}, {_\mathcal{C}}\mathcal{N}) \simeq \mathrm{Hom}_\mathcal{D}({_B}\mathcal{C}, \mathcal{N})$ as a left $\mathcal{C}$-contramodule and $^\mathcal{C}(\Psi_\mathcal{D}\mathcal{N}) = \mathrm{Cohom}_\mathcal{D}({_B}\mathcal{C}, \mathrm{Hom}_\mathcal{D}(\mathcal{D}, \mathcal{N}))$, so it remains to apply Proposition 5.2.2(c).

Without any assumptions on the corings $\mathcal{C}$ and $\mathcal{D}$, there is a natural isomorphism $\Phi_\mathcal{C}({^\mathcal{C}}\mathfrak{Q}) \simeq {_\mathcal{C}}(\Phi_\mathcal{D}\mathfrak{Q})$ for any quite $\mathcal{D}/B$-projective $\mathcal{D}$-contramodule $\mathfrak{Q}$ and a natural isomorphism $^\mathcal{C}(\Psi_\mathcal{D}\mathcal{N}) \simeq \Psi_\mathcal{C}({_\mathcal{C}}\mathcal{N})$ for any quite $\mathcal{D}/B$-injective $\mathcal{D}$-comodule $\mathcal{N}$.

The natural morphisms $\Phi_\mathcal{C}({^\mathcal{C}}\mathfrak{Q}) \longrightarrow {_\mathcal{C}}(\Phi_\mathcal{D}\mathfrak{Q})$ and $^\mathcal{C}(\Psi_\mathcal{D}\mathcal{N}) \longrightarrow \Psi_\mathcal{C}({_\mathcal{C}}\mathcal{N})$ have the following compatibility property. For any left $\mathcal{D}$-comodule $\mathcal{N}$ and left $\mathcal{D}$-contramodule $\mathfrak{Q}$ for which the $\mathcal{C}$-comodule $_\mathcal{C}\mathcal{N}$ and the $\mathcal{C}$-contramodule $^\mathcal{C}\mathfrak{Q}$ are defined via the constructions of 1.2.4 and 3.2.4, for any pair of morphisms $\Phi_\mathcal{D}\mathfrak{Q} \longrightarrow \mathcal{N}$ and $\mathfrak{Q} \longrightarrow \Psi_\mathcal{D}\mathcal{N}$ corresponding to each other under the adjunction of functors $\Psi_\mathcal{D}$ and $\Phi_\mathcal{D}$, the compositions $\Phi_\mathcal{C}({^\mathcal{C}}\mathfrak{Q}) \longrightarrow {_\mathcal{C}}(\Phi_\mathcal{D}\mathfrak{Q}) \longrightarrow {_\mathcal{C}}\mathcal{N}$ and $^\mathcal{C}\mathfrak{Q} \longrightarrow {^\mathcal{C}}(\Psi_\mathcal{D}\mathcal{N}) \longrightarrow \Psi_\mathcal{C}({_\mathcal{C}}\mathcal{N})$ correspond to each other under the adjunction of functors $\Psi_\mathcal{C}$ and $\Phi_\mathcal{C}$.

7.2 Properties of the pull-back and push-forward functors

7.2.1 Let $\mathcal{C} \longrightarrow \mathcal{D}$ be a map of corings compatible with a k-algebra map $A \longrightarrow B$.

Theorem.

(a) *Assume that $\mathcal{C}$ is a flat right A-module. Then the functor $\mathcal{N} \longmapsto {}_{\mathcal{C}}\mathcal{N}$ maps $\mathcal{D}/B$-coflat ($\mathcal{D}/B$-coprojective) left $\mathcal{D}$-comodules to $\mathcal{C}/A$-coflat ($\mathcal{C}/A$-coprojective) left $\mathcal{C}$-comodules. Assume additionally that $\mathcal{D}$ is a flat right B-module. Then the same functor applied to complexes maps coacyclic complexes of $\mathcal{D}/B$-coflat $\mathcal{D}$-comodules to coacyclic complexes of $\mathcal{C}$-comodules.*

(b) *Assume that $\mathcal{C}$ is a projective left A-module. Then the functor $\mathfrak{Q} \longmapsto {}^{\mathcal{C}}\mathfrak{Q}$ maps $\mathcal{D}/B$-coinjective left $\mathcal{D}$-contramodules to $\mathcal{C}/A$-coinjective left $\mathcal{C}$-contramodules. Assume additionally that $\mathcal{D}$ is a projective left B-module. Then the same functor applied to complexes maps contraacyclic complexes of $\mathcal{D}/B$-coinjective $\mathcal{D}$-contramodules to contraacyclic complexes of $\mathcal{C}$-contramodules.*

Proof. Part (a): the first assertion follows from parts (a) (with the left and right sides switched) and (b) of Proposition 7.1.3. To prove the second assertion, denote by $\mathcal{K}^\bullet$ the cobar resolution $\mathcal{C}_B \otimes_B \mathcal{D} \longrightarrow \mathcal{C}_B \otimes_B \mathcal{D} \otimes_B \mathcal{D} \longrightarrow \cdots$ of the right $\mathcal{D}$-comodule $\mathcal{C}_B$. Then $\mathcal{K}^\bullet$ is a complex of $\mathcal{D}$-coflat $\mathcal{C}$-$\mathcal{D}$-bicomodules and the cone of the morphism $\mathcal{C}_B \longrightarrow \mathcal{K}^\bullet$ is coacyclic with respect to the exact category of B-flat $\mathcal{C}$-$\mathcal{D}$-bicomodules. Thus if $\mathcal{N}^\bullet$ is a coacyclic complex of left $\mathcal{D}$-comodules, then the complex of left $\mathcal{C}$-comodules $\mathcal{K}^\bullet \square_{\mathcal{D}} \mathcal{N}^\bullet$ is coacyclic and if $\mathcal{N}^\bullet$ is a complex of $\mathcal{D}/B$-coflat left $\mathcal{D}$-comodules, then the cone of the morphism $\mathcal{C}_B \square_{\mathcal{D}} \mathcal{N}^\bullet \longrightarrow \mathcal{K}^\bullet \square_{\mathcal{D}} \mathcal{N}^\bullet$ is coacyclic. The proof of part (b) is completely analogous. $\square$

7.2.2 It is obvious that the functor $\mathcal{M} \longmapsto {}_B\mathcal{M}$ maps complexes of A-flat $\mathcal{C}$-comodules to complexes of B-flat $\mathcal{D}$-comodules. It will follow from the next theorem that it maps coacyclic complexes of A-flat $\mathcal{C}$-comodules to coacyclic complexes of $\mathcal{D}$-comodules.

Theorem.

(a) *Assume that the coring $\mathcal{C}$ is a flat left and right A-module and the ring A has a finite weak homological dimension. Then any complex of A-flat $\mathcal{C}$-comodules that is coacyclic as a complex of $\mathcal{C}$-comodules is coacyclic with respect to the exact category of A-flat $\mathcal{C}$-comodules.*

(b) *Assume that the coring $\mathcal{C}$ is a projective left and a flat right A-module and the ring A has a finite left homological dimension. Then any complex of A-projective left $\mathcal{C}$-comodules that is coacyclic as a complex of $\mathcal{C}$-comodules is coacyclic with respect to the exact category of A-projective left $\mathcal{C}$-comodules.*

(c) *In the assumptions of part (b), any complex of A-injective left $\mathcal{C}$-contramodules that is contraacyclic as a complex of $\mathcal{C}$-contramodules is contraacyclic with respect to the exact category of A-injective left $\mathcal{C}$-contramodules.*

Proof. The proof is not difficult when k is a field, as in this case the functors of Lemmas 1.1.3 and 3.1.3 can be made additive and exact. Then it follows that for any coacyclic complex of $\mathcal{C}$-comodules $\mathcal{M}^\bullet$ the complex $\mathbb{L}_1(\mathcal{M}^\bullet)$ is coacyclic with respect to the exact category of A-flat $\mathcal{C}$-comodules, while it is clear that for any complex of A-flat $\mathcal{C}$-comodules $\mathcal{M}^\bullet$ the cone of the morphism $\mathbb{L}_1(\mathcal{M}^\bullet) \longrightarrow \mathcal{M}^\bullet$ is coacyclic with respect to the exact category of A-flat $\mathcal{C}$-comodules. Besides, parts (b) and (c) can be derived from the result of Remark 5.5 using the cobar and bar constructions for $\mathcal{C}$-comodules and $\mathcal{C}$-contramodules. Finally, part (a) can be deduced from part (b) using Lemma 3.1.3(a), but this argument requires stronger assumptions on $\mathcal{C}$ and A.

Here is a direct proof of part (a). Let us call a complex of $\mathcal{C}$-comodules m-flat if its terms considered as A-modules have weak homological dimensions not exceeding m, and let us call an m-flat complex of $\mathcal{C}$-comodules m-coacyclic if it is coacyclic with respect to the exact category of $\mathcal{C}$-comodules whose weak homological dimension over A does not exceed m. We will show that for any m-coacyclic complex of $\mathcal{C}$-comodules $\mathcal{M}^\bullet$ there exists an $(m-1)$-coacyclic complex of $\mathcal{C}$-comodules $\mathcal{L}^\bullet$ together with a surjective morphism of complexes $\mathcal{L}^\bullet \longrightarrow \mathcal{M}^\bullet$ whose kernel $\mathcal{K}^\bullet$ is also $(m-1)$-coacyclic. It will follow that any $(m-1)$-flat m-coacyclic complex of $\mathcal{C}$-comodules $\mathcal{M}$ is $(m-1)$-coacyclic, since the total complex of the exact triple $\mathcal{K}^\bullet \to \mathcal{L}^\bullet \to \mathcal{M}^\bullet$ is $(m-1)$-coacyclic, as is the cone of the morphism $\mathcal{K}^\bullet \longrightarrow \mathcal{L}^\bullet$. By induction we will deduce that any 0-flat d-coacyclic complex of $\mathcal{C}$-comodules is 0-coacyclic, where d denotes the weak homological dimension of the ring A; that is a reformulation of the assertion (a).

Let $\mathcal{M}^\bullet$ be the total complex of an exact triple of m-flat complexes of $\mathcal{C}$-comodules $'\mathcal{M}^\bullet \to ''\mathcal{M}^\bullet \to '''\mathcal{M}^\bullet$. Let us choose for each degree n projective A-modules $'G^n$ and $'''G^n$ endowed with surjective A-module maps $'G^n \longrightarrow '\mathcal{M}^n$ and $'''G^n \longrightarrow '''\mathcal{M}^n$. The latter map can be lifted to an A-module map $'''G^n \longrightarrow ''\mathcal{M}^n$, leading to a surjective map from the exact triple of A-modules $'G^n \to 'G^n \oplus '''G^n \to '''G^n$ to the exact triple of $\mathcal{C}$-comodules $'\mathcal{M}^n \to ''\mathcal{M}^n \to '''\mathcal{M}^n$. Applying the construction of Lemma 1.1.3, one can obtain a surjective map from an exact triple of A-flat $\mathcal{C}$-comodules $'\mathcal{P}^n \to ''\mathcal{P}^n \to '''\mathcal{P}^n$ to the exact triple of $\mathcal{C}$-comodules $'\mathcal{M}^n \to ''\mathcal{M}^n \to '''\mathcal{M}^n$. Now consider three complexes of $\mathcal{C}$-comodules $'\mathcal{L}^\bullet$, $''\mathcal{L}^\bullet$, and $'''\mathcal{L}^\bullet$ whose terms are $^{(i)}\mathcal{L}^n = {}^{(i)}\mathcal{P}^{n-1} \oplus {}^{(i)}\mathcal{P}^n$ and the differential $d^n_{(i)\mathcal{L}} : {}^{(i)}\mathcal{L}^n \longrightarrow {}^{(i)}\mathcal{L}^{n+1}$ maps $^{(i)}\mathcal{P}^n$ into itself by the identity map and vanishes in the restriction to $^{(i)}\mathcal{P}^{n-1}$ and in the projection to $^{(i)}\mathcal{P}^{n+1}$. There are natural surjective morphisms of complexes $^{(i)}\mathcal{L}^\bullet \longrightarrow {}^{(i)}\mathcal{M}^\bullet$ constructed as in the proof of Theorem 5.4. Taken together, they form a surjective map from the exact triple of complexes $'\mathcal{L}^\bullet \to ''\mathcal{L}^\bullet \to '''\mathcal{L}^\bullet$ onto the exact triple of complexes $'\mathcal{M}^\bullet \to ''\mathcal{M}^\bullet \to '''\mathcal{M}^\bullet$. Let $'\mathcal{K}^\bullet \to ''\mathcal{K}^\bullet \to '''\mathcal{K}^\bullet$ be the kernel of this map of exact triples of complexes; then the complexes $^{(i)}\mathcal{L}^\bullet$ are 0-flat, while the complexes $^{(i)}\mathcal{K}^\bullet$ are $(m-1)$-flat. Therefore, the total complex $\mathcal{L}^\bullet$ of the exact triple $'\mathcal{L}^\bullet \to ''\mathcal{L}^\bullet \to '''\mathcal{L}^\bullet$ is 0-coacyclic, while the total complex $\mathcal{K}^\bullet$ of the exact triple $'\mathcal{K}^\bullet \to ''\mathcal{K}^\bullet \to '''\mathcal{K}^\bullet$ is $(m-1)$-coacyclic. There is a surjective morphism of complexes $\mathcal{L}^\bullet \longrightarrow \mathcal{M}^\bullet$ with the kernel $\mathcal{K}^\bullet$.

Now let $'\mathcal{K}^\bullet \to '\mathcal{L}^\bullet \to '\mathcal{M}^\bullet$ and $''\mathcal{K}^\bullet \to ''\mathcal{L}^\bullet \to ''\mathcal{M}^\bullet$ be exact triples of complexes of $\mathcal{C}$-comodules where the complexes $'\mathcal{K}^\bullet$, $'\mathcal{L}^\bullet$, $''\mathcal{K}^\bullet$, and $''\mathcal{L}^\bullet$ are $(m-1)$-coacyclic, and suppose that there is a morphism of complexes $'\mathcal{M}^\bullet \longrightarrow ''\mathcal{M}^\bullet$. Let us construct for the complex $\mathcal{M}^\bullet = \mathrm{cone}('\mathcal{M}^\bullet \to ''\mathcal{M}^\bullet)$ an exact triple of complexes $\mathcal{K}^\bullet \to \mathcal{L}^\bullet \to \mathcal{M}^\bullet$ with $(m-1)$-coacyclic complexes $\mathcal{K}^\bullet$ and $\mathcal{L}^\bullet$. Denote by $'''\mathcal{L}^\bullet$ the complex $'\mathcal{L}^\bullet \oplus ''\mathcal{L}^\bullet$; there is the embedding of a direct summand $'\mathcal{L}^\bullet \longrightarrow '''\mathcal{L}^\bullet$ and the surjective morphism of complexes $'''\mathcal{L}^\bullet \longrightarrow ''\mathcal{M}^\bullet$ whose components are the composition $'\mathcal{L}^\bullet \longrightarrow '\mathcal{M}^\bullet \longrightarrow ''\mathcal{M}^\bullet$ and the surjective morphism $''\mathcal{L}^\bullet \longrightarrow ''\mathcal{M}^\bullet$. These two morphisms form a commutative square with the morphisms $'\mathcal{L}^\bullet \longrightarrow '\mathcal{M}^\bullet$ and $'\mathcal{M}^\bullet \longrightarrow ''\mathcal{M}^\bullet$. The kernel $'''\mathcal{K}^\bullet$ of the morphism $'''\mathcal{L}^\bullet \longrightarrow ''\mathcal{M}^\bullet$ is the middle term of an exact triple of complexes $''\mathcal{K}^\bullet \longrightarrow '''\mathcal{K}^\bullet \longrightarrow '\mathcal{L}^\bullet$. Since the complexes $''\mathcal{K}^\bullet$ and $'\mathcal{L}^\bullet$ are $(m-1)$-coacyclic, the complex $'''\mathcal{K}^\bullet$ is also $(m-1)$-coacyclic. Set $\mathcal{L}^\bullet = \mathrm{cone}('\mathcal{L}^\bullet \longrightarrow '''\mathcal{L}^\bullet)$ and $\mathcal{K}^\bullet = \mathrm{cone}('\mathcal{K}^\bullet \longrightarrow '''\mathcal{K}^\bullet)$; then there is an exact triple of complexes $\mathcal{K}^\bullet \to \mathcal{L}^\bullet \to \mathcal{M}^\bullet$ with the desired properties.

Obviously, if certain complexes of $\mathcal{C}$-comodules $\mathcal{M}_\alpha^\bullet$ can be presented as quotient complexes of $(m-1)$-coacyclic complexes by $(m-1)$-coacyclic subcomplexes, then their direct sum $\bigoplus \mathcal{M}_\alpha^\bullet$ can be also presented in this way.

Finally, let $\mathcal{M}^\bullet \longrightarrow '\mathcal{M}^\bullet$ be a homotopy equivalence of m-flat complexes of $\mathcal{C}$-comodules, and suppose that there is an exact triple $'\mathcal{K}^\bullet \to '\mathcal{L}^\bullet \to '\mathcal{M}^\bullet$ with $(m-1)$-coacyclic complexes $'\mathcal{K}^\bullet$ and $'\mathcal{L}^\bullet$. Let us construct an exact triple of complexes $\mathcal{K}^\bullet \to \mathcal{L}^\bullet \to \mathcal{M}^\bullet$ with $(m-1)$-coacyclic complexes $\mathcal{K}^\bullet$ and $\mathcal{L}^\bullet$. Consider the cone of the morphism $\mathcal{M}^\bullet \longrightarrow '\mathcal{M}^\bullet$; it is contractible, and therefore isomorphic to the cone of the identity endomorphism of a complex of $\mathcal{C}$-comodules $\mathcal{N}^\bullet$ with zero differential. The complex $\mathcal{N}^\bullet$ is m-flat, so it can be presented as the quotient complex of a complex of A-flat $\mathcal{C}$-comodules $\mathcal{P}^\bullet$ by its $(m-1)$-flat subcomplex $\mathcal{Q}^\bullet$. Hence the complex $\mathrm{cone}(\mathcal{M}^\bullet \to '\mathcal{M}^\bullet)$ is isomorphic to the quotient complex of a 0-flat contractible complex $\mathrm{cone}(\mathrm{id}_{\mathcal{P}^\bullet})$ by an $(m-1)$-flat contractible subcomplex $\mathrm{cone}(\mathrm{id}_{\mathcal{Q}^\bullet})$.

As we have proven, for the cocone $''\mathcal{M}^\bullet$ of the morphism $'\mathcal{M}^\bullet \longrightarrow \mathrm{cone}(\mathcal{M}^\bullet \to '\mathcal{M}^\bullet)$ there exists an exact triple $''\mathcal{K}^\bullet \to ''\mathcal{L}^\bullet \to ''\mathcal{M}^\bullet$ with $(m-1)$-coacyclic complexes $''\mathcal{K}^\bullet$ and $''\mathcal{L}^\bullet$. The complex $''\mathcal{M}^\bullet$ is isomorphic to the direct sum of the complex $\mathcal{M}^\bullet$ and the cocone of the identity endomorphism of the complex $'\mathcal{M}^\bullet$. (Indeed, there is a term-wise split exact triple of complexes $\mathrm{cone}(\mathrm{id}_{'\mathcal{M}^\bullet})[-1] \longrightarrow ''\mathcal{M}^\bullet \longrightarrow \mathcal{M}^\bullet$ and the complex $\mathrm{cone}(\mathrm{id}_{'\mathcal{M}^\bullet})[-1]$ is contractible.) The latter cocone can be presented as the quotient complex of an $(m-1)$-flat contractible complex $'\mathcal{P}^\bullet$ by an $(m-1)$-flat contractible subcomplex $'\mathcal{Q}^\bullet$, e.g., by taking $'\mathcal{P}^\bullet = \mathrm{cone}(\mathrm{id}_{'\mathcal{L}^\bullet})[-1]$ and $'\mathcal{Q}^\bullet = \mathrm{cone}(\mathrm{id}_{'\mathcal{K}^\bullet})[-1]$.

Now suppose that there are exact triples $''\mathcal{K}^\bullet \to ''\mathcal{L}^\bullet \to ''\mathcal{M}^\bullet$ and $'\mathcal{Q}^\bullet \to '\mathcal{P}^\bullet \to '\mathcal{N}^\bullet$ with $(m-1)$-coacyclic complexes $''\mathcal{K}^\bullet$, $''\mathcal{L}^\bullet$, $'\mathcal{Q}^\bullet$, and $'\mathcal{P}^\bullet$ for certain complexes $''\mathcal{M}^\bullet = \mathcal{M}^\bullet \oplus '\mathcal{N}^\bullet$ and $'\mathcal{N}^\bullet$. Let us construct an exact triple $\mathcal{K}^\bullet \to \mathcal{L}^\bullet \to \mathcal{M}^\bullet$ with $(m-1)$-coacyclic complexes $\mathcal{K}^\bullet$ and $\mathcal{L}^\bullet$ (in fact, we will have $\mathcal{K}^\bullet = ''\mathcal{K}^\bullet$ and our construction with obvious modifications will work for the kernel $\mathcal{M}^\bullet$ of a surjective morphism of complexes $''\mathcal{M}^\bullet \longrightarrow '\mathcal{N}^\bullet$). Set $'''\mathcal{M}^\bullet = \mathcal{M}^\bullet \oplus '\mathcal{P}^\bullet$; then there is a surjective morphism of complexes $'''\mathcal{M}^\bullet \longrightarrow ''\mathcal{M}^\bullet$ with the kernel $'\mathcal{Q}^\bullet$. Let $'''\mathcal{L}^\bullet$ be the fibered product of the complexes $'''\mathcal{M}^\bullet$ and $''\mathcal{L}^\bullet$ over $''\mathcal{M}^\bullet$; then there are exact triples of complexes $''\mathcal{K}^\bullet \longrightarrow '''\mathcal{L}^\bullet \longrightarrow '''\mathcal{M}^\bullet$ and $'\mathcal{Q}^\bullet \longrightarrow '''\mathcal{L}^\bullet \longrightarrow ''\mathcal{L}^\bullet$. It follows from the latter exact triple that the complex $'''\mathcal{L}^\bullet$ is $(m-1)$-coacyclic. Furthermore, there is an injective morphism of complexes $\mathcal{M}^\bullet \longrightarrow '''\mathcal{M}^\bullet$ with the cokernel $'\mathcal{P}^\bullet$. Let $\mathcal{L}^\bullet$ be the fibered product of the complexes $\mathcal{M}^\bullet$ and $'''\mathcal{L}^\bullet$ over $'''\mathcal{M}^\bullet$; then there are exact triples of complexes $''\mathcal{K}^\bullet \longrightarrow \mathcal{L}^\bullet \longrightarrow \mathcal{M}^\bullet$ and $\mathcal{L}^\bullet \longrightarrow '''\mathcal{L}^\bullet \longrightarrow '\mathcal{P}^\bullet$. It follows from the latter exact triple that the complex $\mathcal{L}^\bullet$ is $(m-1)$-coacyclic.

Part (a) is proven; the proofs of parts (b) and (c) are completely analogous.   □

Remark. It follows from part (a) of the theorem that (in the same assumptions) any coacyclic complex of coflat $\mathcal{C}$-comodules is coacyclic with respect to the exact category of coflat $\mathcal{C}$-comodules. Indeed, for any complex of $\mathcal{C}$-comodules $\mathcal{M}^\bullet$ coacyclic with respect to the exact category of A-flat $\mathcal{C}$-comodules the complex $\mathbb{R}_2(\mathcal{M}^\bullet)$ is coacyclic with respect to the exact category of coflat $\mathcal{C}$-comodules, and for any complex of coflat $\mathcal{C}$-comodules $\mathcal{M}^\bullet$ the cone of the morphism $\mathcal{M}^\bullet \longrightarrow \mathbb{R}_2(\mathcal{M}^\bullet)$ is coacyclic with respect to the exact category of coflat $\mathcal{C}$-comodules (by Lemma 1.2.2). Analogously, if $\mathcal{C}$ is a flat right A-module then any coacyclic complex of $\mathcal{C}/A$-coflat left $\mathcal{C}$-comodules is coacyclic with respect to the exact category of $\mathcal{C}/A$-coflat left $\mathcal{C}$-comodules. For coprojective $\mathcal{C}$-comodules, coinjective $\mathcal{C}$-contramodules, (quite) $\mathcal{C}/A$-injective $\mathcal{C}$-comodules, and (quite) $\mathcal{C}/A$-projective $\mathcal{C}$-contramodules even stronger results are provided by Remark 5.5, Theorem 5.4, and Theorem 5.5.

7.3   Derived functors of pull-back and push-forward

Let $\mathcal{C} \longrightarrow \mathcal{D}$ be a map of corings compatible with a k-algebra map $A \longrightarrow B$.

Assume that $\mathcal{C}$ is a flat right A-module and $\mathcal{D}$ is a flat right B-module. Then the functor mapping the quotient category of the homotopy category of complexes of $\mathcal{D}/B$-coflat left $\mathcal{D}$-comodules by its intersection with the thick subcategory of coacyclic complexes to the coderived category of left $\mathcal{D}$-comodules is an equivalence of triangulated categories by Lemma 2.6. Indeed, for any complex of left $\mathcal{D}$-comodules $\mathcal{N}^\bullet$ there is a morphism from $\mathcal{N}^\bullet$ into a complex of $\mathcal{D}/B$-coflat $\mathcal{D}$-comodules $\mathbb{R}_2(\mathcal{N}^\bullet)$ with a coacyclic cone, which was constructed in 2.5. Compose the functor $\mathcal{N}^\bullet \longmapsto {}_{\mathcal{C}}\mathcal{N}^\bullet$ acting from the homotopy category of left $\mathcal{D}$-comodules to the homotopy category of left $\mathcal{C}$-comodules with the localization functor $\mathsf{Hot}(\mathcal{C}\text{-comod}) \longrightarrow \mathsf{D}^{\mathrm{co}}(\mathcal{C}\text{-comod})$ and restrict it to the full subcategory of complexes of $\mathcal{D}/B$-coflat $\mathcal{D}$-comodules. By Theorem 7.2.1(a), this restriction factorizes through the coderived category of left $\mathcal{D}$-comodules. Let us denote the right derived functor so obtained by

$$\mathcal{N}^\bullet \longmapsto {}_{\mathcal{C}}^{\mathbb{R}}\mathcal{N}^\bullet \colon \mathsf{D}^{\mathrm{co}}(\mathcal{D}\text{-comod}) \longrightarrow \mathsf{D}^{\mathrm{co}}(\mathcal{C}\text{-comod}).$$

According to Lemma 6.5.2, this definition of a right derived functor does not depend on the choice of a subcategory of adjusted complexes.

Assume that $\mathcal{C}$ is a flat left and right A-module, A has a finite weak homological dimension, and $\mathcal{D}$ is a flat right B-module. Then the functor mapping the quotient category of the homotopy category of complexes of A-flat $\mathcal{C}$-comodules by its intersection with the thick subcategory of coacyclic complexes to the coderived category of $\mathcal{C}$-comodules is an equivalence of triangulated categories by Lemma 2.6. Indeed, for any complex of $\mathcal{C}$-comodules $\mathcal{M}^\bullet$ there is a morphism into $\mathcal{M}^\bullet$ from a complex of A-flat $\mathcal{C}$-comodules $\mathbb{L}_1(\mathcal{M}^\bullet)$ with a coacyclic cone, which was constructed in 2.5. Compose the functor $\mathcal{M}^\bullet \longmapsto {}_B\mathcal{M}^\bullet$ acting from the homotopy category of left $\mathcal{C}$-comodules to the homotopy category of left $\mathcal{D}$-comodules with the localization functor $\mathsf{Hot}(\mathcal{D}\text{-comod}) \longrightarrow \mathsf{D}^{\mathrm{co}}(\mathcal{D}\text{-comod})$ and restrict it to the full subcategory of complexes of A-flat $\mathcal{C}$-comodules. It follows from Theorem 7.2.2(a) that this restriction factorizes through the coderived category of left $\mathcal{C}$-comodules. Let us denote the left derived functor so obtained by

$$\mathcal{M}^\bullet \longmapsto {}_B^{\mathbb{L}}\mathcal{M}^\bullet \colon \mathsf{D}^{\mathrm{co}}(\mathcal{C}\text{-comod}) \longrightarrow \mathsf{D}^{\mathrm{co}}(\mathcal{D}\text{-comod}).$$

According to Lemma 6.5.2, this definition of a left derived functor does not depend on the choice of a subcategory of adjusted complexes.

Analogously, assume that $\mathcal{C}$ is a projective left A-module and $\mathcal{D}$ is a projective left B-module. Then the left derived functor

$$\mathfrak{Q}^\bullet \longmapsto {}_{\mathbb{L}}^{\mathcal{C}}\mathfrak{Q}^\bullet \colon \mathsf{D}^{\mathrm{ctr}}(\mathcal{D}\text{-contra}) \longrightarrow \mathsf{D}^{\mathrm{ctr}}(\mathcal{C}\text{-contra})$$

is defined by restricting the functor $\mathfrak{Q}^\bullet \longmapsto {}^{\mathcal{C}}\mathfrak{Q}^\bullet$ to the full subcategory of complexes of $\mathcal{D}/B$-coinjective left $\mathcal{D}$-contramodules.

Assume that $\mathcal{C}$ is a projective left and a flat right A-module, A has a finite left homological dimension, and $\mathcal{D}$ is a projective left B-module. Then the right derived functor

$$\mathfrak{P}^\bullet \longmapsto {}_{\mathbb{R}}^{B}\mathfrak{P}^\bullet \colon \mathsf{D}^{\mathrm{ctr}}(\mathcal{C}\text{-contra}) \longrightarrow \mathsf{D}^{\mathrm{ctr}}(\mathcal{D}\text{-contra})$$

is defined by restricting the functor $\mathfrak{P}^\bullet \longmapsto {}^{B}\mathfrak{P}^\bullet$ to the full subcategory of complexes of A-injective left $\mathcal{C}$-contramodules.

Properties of the above-defined derived functors will be studied (in the greater generality of semimodules and semicontramodules) in Chapter 8. In particular, the functor $\mathcal{N}^\bullet \longmapsto {}_{\mathcal{C}}^{\mathbb{R}}\mathcal{N}^\bullet$ is right adjoint to the functor $\mathcal{M}^\bullet \longmapsto {}_B^{\mathbb{L}}\mathcal{M}^\bullet$ when the latter is defined; the functor $\mathfrak{Q}^\bullet \longmapsto {}_{\mathbb{L}}^{\mathcal{C}}\mathfrak{Q}^\bullet$ is left adjoint to the functor $\mathfrak{P}^\bullet \longmapsto {}_{\mathbb{R}}^{B}\mathfrak{P}^\bullet$ when the latter is defined; the equivalences of categories $\mathsf{D}^{\mathrm{co}}(\mathcal{C}\text{-comod}) \simeq \mathsf{D}^{\mathrm{ctr}}(\mathcal{C}\text{-contra})$ and $\mathsf{D}^{\mathrm{co}}(\mathcal{D}\text{-comod}) \simeq \mathsf{D}^{\mathrm{ctr}}(\mathcal{D}\text{-contra})$, when they are defined, transform the functor $\mathcal{N}^\bullet \longmapsto {}_{\mathcal{C}}^{\mathbb{R}}\mathcal{N}^\bullet$ into the functor $\mathfrak{Q}^\bullet \longmapsto {}_{\mathbb{L}}^{\mathcal{C}}\mathfrak{Q}^\bullet$; and there are formulas connecting our derived functors with the derived functors Ctrtor, Cotor and Coext.

7.4   Faithfully flat/projective base ring change

7.4.1   The main ideas of the following section are due to Kontsevich and Rosenberg [62, 63].

Let $\mathcal{C}$ be a coring over a k-algebra A and $A \longrightarrow B$ be a k-algebra morphism. The coring $_B\mathcal{C}_B$ over the k-algebra B is constructed in the following way. As a B-B-bimodule, $_B\mathcal{C}_B$ is equal to $B \otimes_A \mathcal{C} \otimes_A B$. The comultiplication in $_B\mathcal{C}_B$ is defined as the composition $B \otimes_A \mathcal{C} \otimes_A B \longrightarrow B \otimes_A \mathcal{C} \otimes_A \mathcal{C} \otimes_A B \longrightarrow B \otimes_A \mathcal{C} \otimes_A B \otimes_A \mathcal{C} \otimes_A B = (B \otimes_A \mathcal{C} \otimes_A B) \otimes_B (B \otimes_A \mathcal{C} \otimes_A B)$ of the map induced by the comultiplication in $\mathcal{C}$ and the map induced by the map $A \longrightarrow B$. The counit in $_B\mathcal{C}_B$ is defined as the composition $B \otimes_A \mathcal{C} \otimes_A B \longrightarrow B \otimes_A B \longrightarrow B$ of the map induced by the counit in $\mathcal{C}$ and the map induced by the multiplication in B.

The coring $_B\mathcal{C}_B$ is a universal initial object in the category of corings $\mathcal{D}$ over B endowed with a map $\mathcal{C} \longrightarrow \mathcal{D}$ compatible with the map $A \longrightarrow B$.

As always, B is called a faithfully flat right A-module if it is a flat right A-module and for any nonzero left A-module M the tensor product $B \otimes_A M$ is nonzero. Assuming the former condition, the latter one holds if and only if the map $M = A \otimes_A M \longrightarrow B \otimes_A M$ is injective for any left A-module M. Therefore, B is a faithfully flat right A-module if and only if the map $A \longrightarrow B$ is injective and its cokernel A/B is a flat right A-module. Analogously, the ring B is called a faithfully projective left A-module if it is a projective generator of the category of left A-modules, i.e., it is a projective left A-module and for any nonzero left A-module P the module $\mathrm{Hom}_A(B, P)$ is nonzero. Assuming the former condition, the latter one holds if and only if the map $\mathrm{Hom}_A(B, P) \longrightarrow \mathrm{Hom}_A(A, P) = P$ is surjective for any left A-module P. Therefore, B is a faithfully projective left A-module if and only if the map $A \longrightarrow B$ is injective and its cokernel A/B is a projective left A-module.

If the coring $\mathcal{C}$ is a flat right A-module and the ring B is a faithfully flat right A-module, then the functors $\mathcal{M} \longmapsto {}_B\mathcal{M}$ and $\mathcal{N} \longmapsto {}_\mathcal{C}\mathcal{N}$ are mutually inverse equivalences between the abelian categories of left $\mathcal{C}$-comodules and left $_B\mathcal{C}_B$-comodules,

$$\mathcal{C}\text{-comod} \simeq {}_B\mathcal{C}_B\text{-comod.}$$

Analogously, if $\mathcal{C}$ is a projective left A-module and B is a faithfully projective left A-module, then the functors $\mathfrak{P} \longmapsto {}^B\mathfrak{P}$ and $\mathfrak{Q} \longmapsto {}^\mathcal{C}\mathfrak{Q}$ are mutually inverse equivalences between the abelian categories of left $\mathcal{C}$-contramodules and left $_B\mathcal{C}_B$-contramodules,

$$\mathcal{C}\text{-contra} \simeq {}_B\mathcal{C}_B\text{-contra.}$$

Both assertions follow from the next general theorem, which is the particular case of Barr–Beck Theorem [67, VI.7] for abelian categories and exact functors.

Theorem. *If $\Delta\colon \mathsf{B} \longrightarrow \mathsf{A}$ is an exact functor between abelian categories mapping nonzero objects to nonzero objects and $\Gamma\colon \mathsf{A} \longrightarrow \mathsf{B}$ is a functor left (resp., right) adjoint to Δ, then the natural functor from the category B to the category of* *modules over the monad $\Delta\Gamma$ (resp., comodules over the comonad $\Delta\Gamma$) over the category* A *is an equivalence of abelian categories.* $\square$

To prove the first assertion, it suffices to apply the theorem to the functor $\Delta\colon \mathcal{C}$-comod $\longrightarrow$ B-mod mapping a $\mathcal{C}$-comodule $\mathcal{M}$ to the B-module $B\otimes_A\mathcal{M}$ and the functor $\Gamma\colon B$-mod $\longrightarrow$ $\mathcal{C}$-comod right adjoint to Δ mapping a B-module U to the $\mathcal{C}$-comodule $\mathcal{C}\otimes_A U$. To prove the second assertion, apply the theorem to the functor $\Delta\colon \mathcal{C}$-contra $\longrightarrow$ B-mod mapping a $\mathcal{C}$-contramodule $\mathfrak{P}$ to the B-module $\mathrm{Hom}_A(B,\mathfrak{P})$ and the functor $\Gamma\colon B$-mod $\longrightarrow$ $\mathcal{C}$-contra left adjoint to Δ mapping a B-module V to the $\mathcal{C}$-contramodule $\mathrm{Hom}_A(\mathcal{C},V)$.

7.4.2 Let $\mathcal{C}$ be a coring over a k-algebra A and $A\longrightarrow B$ be a k-algebra morphism.

Assume that $\mathcal{C}$ is a flat left and right A-module and B is a faithfully flat left and right A-module. Then it follows from Proposition 7.1.3(a) that for any right $\mathcal{C}$-comodule $\mathcal{N}$ and any left $\mathcal{C}$-comodule $\mathcal{M}$ there is a natural map $\mathcal{N}\,\square_{\mathcal{C}}\,\mathcal{M}\longrightarrow \mathcal{N}_B\,\square_{B\mathcal{C}_B}\,{}_B\mathcal{M}$, which is an isomorphism, at least, when one of the A-modules $\mathcal{N}$ and $\mathcal{M}$ is flat or one of the ${}_B\mathcal{C}_B$-comodules $\mathcal{N}_B$ and ${}_B\mathcal{M}$ is quasicoflat. Analogously, assume that $\mathcal{C}$ is a projective left and a flat right A-module and B is a faithfully projective left and a faithfully flat right A-module. Then it follows from Proposition 7.1.3(b-c) that for any left $\mathcal{C}$-comodule $\mathcal{M}$ and any left $\mathcal{C}$-contra-module $\mathfrak{P}$ there is a natural map $\mathrm{Cohom}_{B\mathcal{C}_B}({}_B\mathcal{M},\,{}^B\mathfrak{P})\longrightarrow \mathrm{Cohom}_{\mathcal{C}}(\mathcal{M},\mathfrak{P})$, which is an isomorphism, at least, when the A-module $\mathcal{M}$ is projective, the A-module $\mathfrak{P}$ is injective, the ${}_B\mathcal{C}_B$-comodule ${}_B\mathcal{M}$ is quasiprojective, or the ${}_B\mathcal{C}_B$-contramodule ${}^B\mathfrak{P}$ is quasicoinjective.

Remark. In general the map $\mathcal{N}\square_{\mathcal{C}}\mathcal{M}\longrightarrow \mathcal{N}_B\square_{B\mathcal{C}_B}{}_B\mathcal{M}$ is not an isomorphism, even under the strongest of our assumptions on A, B, and $\mathcal{C}$. For example, let $\mathcal{C}=A$ and ${}_B\mathcal{C}_B=B\otimes_A B$; then $\mathcal{N}\square_{\mathcal{C}}\mathcal{M}=\mathcal{N}\otimes_A\mathcal{M}$, while $\mathcal{N}_B\square_{B\mathcal{C}_B}{}_B\mathcal{M}$ is the kernel of the pair of maps $\mathcal{N}\otimes_A B\otimes_A\mathcal{M}\rightrightarrows \mathcal{N}\otimes_A B\otimes_A B\otimes_A\mathcal{M}$ induced by the map $A\longrightarrow B$. The sequence $0\longrightarrow \mathcal{N}\otimes_A\mathcal{M}\longrightarrow \mathcal{N}\otimes_A B\otimes_A\mathcal{M}\longrightarrow \mathcal{N}\otimes_A B\otimes_A B\otimes_A\mathcal{M}$ is exact if one of two A-modules $\mathcal{M}$ and $\mathcal{N}$ is flat or admits a B-module structure, but in general the map $\mathcal{N}\otimes_A\mathcal{M}\longrightarrow \mathcal{N}\otimes_A B\otimes_A\mathcal{M}$ is not injective. Indeed, let k be a field, $A=k[x]$ be the algebra of polynomials in one variable, and $B=k[x,\partial_x]$ be the algebra of differential operators in the affine line. Let $\mathcal{M}=k=\mathcal{N}$ be one-dimensional A-modules with the trivial action of x. Then the map $\mathcal{N}\otimes_A\mathcal{M}\longrightarrow \mathcal{N}\otimes_A B\otimes_A\mathcal{M}$ is zero, since $m\otimes 1\otimes n=m\otimes(\partial_x x-x\partial_x)\otimes n=0$ in $\mathcal{N}\otimes_A B\otimes_A\mathcal{M}$.

Assume that $\mathcal{C}$ is a projective left and a flat right A-module and B is a faithfully projective left and a faithfully flat right A-module. Then the equivalences between the categories $\mathcal{C}$-comod and ${}_B\mathcal{C}_B$-comod and between the categories $\mathcal{C}$-contra and ${}_B\mathcal{C}_B$-contra transform the functors $\Psi_{\mathcal{C}}$ and $\Phi_{\mathcal{C}}$ into the functors $\Psi_{B\mathcal{C}_B}$ and $\Phi_{B\mathcal{C}_B}$. Indeed, one has $\mathrm{Hom}_{B\mathcal{C}_B}({}_B\mathcal{C}_B,{}_B\mathcal{M})=\mathrm{Hom}_{\mathcal{C}}(\mathcal{C}_B,\mathcal{M})=\mathrm{Hom}_A(B,\mathrm{Hom}_{\mathcal{C}}(\mathcal{C},\mathcal{M}))$ and ${}_B\mathcal{C}_B\odot_{B\mathcal{C}_B}{}^B\mathfrak{P}={}_B\mathcal{C}\odot_{\mathcal{C}}\mathfrak{P}=B\otimes_A(\mathcal{C}\odot_{\mathcal{C}}\mathfrak{P})$. Alternatively, the same isomorphisms can be constructed as in 7.1.4 using Propositions 5.2.1(e) and 5.2.2(e).

7.4.3   Let $\mathcal{C}$ be a coring over a k-algebra A and $A \longrightarrow B$ be a k-algebra morphism.

Obviously, if $\mathcal{C}$ is a flat right A-module and B is a faithfully flat right A-module, then a complex of left $\mathcal{C}$-comodules $\mathcal{M}^\bullet$ is coacyclic if any only if the complex of left ${}_B\mathcal{C}_B$-comodules ${}_B\mathcal{M}^\bullet$ is coacyclic. So the functor $\mathcal{M}^\bullet \longmapsto {}_B\mathcal{M}^\bullet$ induces an equivalence of the coderived categories of left $\mathcal{C}$-comodules and left ${}_B\mathcal{C}_B$-comodules. If $\mathcal{C}$ is a projective left A-module and B is a faithfully projective left A-module, then a complex of left $\mathcal{C}$-contramodules $\mathfrak{P}^\bullet$ is contraacyclic if and only if the complex of ${}_B\mathcal{C}_B$-contramodules ${}^B\mathfrak{P}^\bullet$ is contraacyclic. So the functor $\mathfrak{P}^\bullet \longmapsto {}^B\mathfrak{P}^\bullet$ induces an equivalence of the contraderived categories of left $\mathcal{C}$-contramodules and left ${}_B\mathcal{C}_B$-contramodules.

If $\mathcal{C}$ is a flat left and right A-module, B is a faithfully flat left and right A-module, and A and B have finite weak homological dimensions, then the equivalences of categories

$$\mathsf{D}^{\mathrm{co}}(\text{comod-}\mathcal{C}) \simeq \mathsf{D}^{\mathrm{co}}(\text{comod-}{}_B\mathcal{C}_B) \quad \text{and} \quad \mathsf{D}^{\mathrm{co}}(\mathcal{C}\text{-comod}) \simeq \mathsf{D}^{\mathrm{co}}({}_B\mathcal{C}_B\text{-comod})$$

transform the derived functor $\mathrm{Cotor}^\mathcal{C}$ into the derived functor $\mathrm{Cotor}^{{}^B\mathcal{C}_B}$. If $\mathcal{C}$ is a projective left and a flat right A-module, B is a faithfully projective left and a faithfully flat right A-module, and A and B have finite left homological dimensions, then the equivalences of categories $\mathsf{D}^{\mathrm{co}}(\mathcal{C}\text{-comod}) \simeq \mathsf{D}^{\mathrm{co}}({}_B\mathcal{C}_B\text{-comod})$ and $\mathsf{D}^{\mathrm{ctr}}(\mathcal{C}\text{-contra}) \simeq \mathsf{D}^{\mathrm{ctr}}({}_B\mathcal{C}_B\text{-contra})$ transform the derived functor $\mathrm{Coext}_\mathcal{C}$ into the derived functor $\mathrm{Coext}_{{}_B\mathcal{C}_B}$. In the same assumptions, the same equivalences of categories transform the mutually inverse functors $\mathbb{R}\Psi_\mathcal{C}$ and $\mathbb{L}\Phi_\mathcal{C}$ into the mutually inverse functors $\mathbb{R}\Psi_{{}_B\mathcal{C}_B}$ and $\mathbb{L}\Phi_{{}_B\mathcal{C}_B}$. If $\mathcal{C}$ is a flat right A-module, B is a faithfully flat right A-module, and A and B have finite left homological dimensions, then the above equivalence of categories transforms the functor $\mathrm{Ext}_\mathcal{C}$ into the functor $\mathrm{Ext}_{{}_B\mathcal{C}_B}$. If $\mathcal{C}$ is a projective left A-module, B is a faithfully projective left A-module, and A and B have finite left homological dimensions, then the above equivalences of categories transform the functors $\mathrm{Ext}^\mathcal{C}$ and $\mathrm{Ctrtor}^\mathcal{C}$ into the functors $\mathrm{Ext}^{{}^B\mathcal{C}_B}$ and $\mathrm{Ctrtor}^{{}^B\mathcal{C}_B}$.

These isomorphisms of functors can be deduced from the uniqueness/universality assertions of Lemmas 2.7 and 6.5.2 or derived from the preservation/reflection results of the next remark. Besides, they are particular cases of the much more general isomorphisms constructed in Chapter 8.

Remark. In the strongest of the above flatness/projectivity and homological dimension assumptions, almost all the properties of comodules and contramodules over corings considered in this book are preserved by the passages from a coring $\mathcal{C}$ to the coring ${}_B\mathcal{C}_B$ and back. This applies to the properties of coflatness, coprojectivity, coinjectivity, relative coflatness, relative coprojectivity, relative coinjectivity, injectivity, projectivity, contraflatness, relative injectivity, relative projectivity, relative contraflatness. All of this follows from the facts that an A-module M is flat if and only if the B-module $B \otimes_A M$ is flat, an A-module M is projective if and only if the B-module $B \otimes_A M$ is projective, and an A-module P is injective if and only if the B-module $\mathrm{Hom}_A(B, P)$ is injective. Indeed, suppose that the left B-module $B \otimes_A M$ is flat. Any flat left B-module is a flat left A-module, since the ring B is a flat left A-module. Consider the tensor product of complexes $(A \to B) \otimes_A \cdots \otimes_A (A \to B) \otimes_A M$, where the number of factors $A \longrightarrow B$ is at least equal to the weak homological dimension of A. This complex is exact everywhere except its rightmost term, since the map $A \longrightarrow B$ is injective and B/A is a flat right A-module. Since all terms of this complex, except possibly the leftmost one, are flat left A-modules, the leftmost term A is also a flat left A-module. Alternatively, one can consider the Amitsur complex $M \longrightarrow B \otimes_A M \longrightarrow B \otimes_A B \otimes_A M \longrightarrow \cdots$ with the alternating sums of the maps induced by the map $A \longrightarrow B$ as the differentials; this complex of left A-modules is acyclic, since the induced complex of left B-modules is contractible. Notice that the assumption of finite weak homological dimension of the ring A is necessary for this argument, since otherwise the ring B can be absolutely flat while the ring A is not (see Remark 8.4.3). Assuming only that $\mathcal{C}$ is a flat right A-module and B is a faithfully flat right A-module, the right $_B\mathcal{C}_B$-comodule $\mathcal{N}_B$ is coflat if a right $\mathcal{C}$-comodule $\mathcal{N}$ is coflat, etc.

On the other hand, even under the strongest of the above assumptions there are more quite $\mathcal{C}/A$-injective $\mathcal{C}$-comodules than quite $_B\mathcal{C}_B/B$-injective $_B\mathcal{C}_B$-comodules and there are more quite $\mathcal{C}/A$-projective $\mathcal{C}$-contramodules than quite $_B\mathcal{C}_B/B$-projective $_B\mathcal{C}_B$-contramodules; i.e., quite relative injectivity and quite relative projectivity is not preserved by the equivalences of categories $\mathcal{M} \longmapsto {}_B\mathcal{M}$ and $\mathfrak{P} \longmapsto {}^B\mathfrak{P}$ in general. Analogously, there are more quasicoflat $\mathcal{C}$-comodules than quasicoflat $_B\mathcal{C}_B$-comodules.

Indeed, consider the case when $\mathcal{C} = A$ and $_B\mathcal{C}_B = B \otimes_A B$ is a "Galois coring" [23]. Then all $\mathcal{C}$-comodules are coinduced and all $\mathcal{C}$-contramodules are induced, while a $_B\mathcal{C}_B$-comodule is quite $_B\mathcal{C}_B/B$-injective, or a $_B\mathcal{C}_B$-contramodule is quite $_B\mathcal{C}_B/B$-projective, if and only if the corresponding A-module is a direct summand of an A-module admitting a B-module structure. For example, if $A = k[x]$ and $B = k[x, \partial_x]$ as in Remark 7.4.2, then the one-dimensional A-module M with the trivial action of x is not the direct summand of any A-module admitting a B-module structure, since the equation $xm = 0$ would imply $m = -x\partial_x m$. At the same time, any projective left A-module is a direct summand of a projective left B-module and any injective left A-module is a direct summand of an injective left B-module. It follows, in particular, that the cokernel of an injective morphism of quite $\mathcal{C}/A$-injective $\mathcal{C}$-comodules is not always quite $\mathcal{C}/A$-injective and the kernel of a surjective morphism of quite $\mathcal{C}/A$-projective $\mathcal{C}$-contramodules is not always quite $\mathcal{C}/A$-projective.

7.5 Remarks on Morita morphisms

7.5.1 A *Morita morphism* from a k-algebra A to a k-algebra B is an A-B-bimodule E such that E is a finitely generated projective right B-module. For any Morita morphism E from A to B, set $E^\vee = \mathrm{Hom}_{B^{op}}(E, B)$; then $E^\vee$ is a B-A-bimodule and a finitely generated projective left B-module. To any k-algebra morphism $A \longrightarrow B$, one can assign a Morita morphism $E = B = E^{\vee}$ from A to B.

Equivalently, a Morita morphism from A to B can be defined as a pair consisting of an A-B-bimodule E and a B-A-bimodule $E^{\vee}$ endowed with an A-A-bimodule morphism $A \longrightarrow E \otimes_B E^{\vee}$ and a B-B-bimodule morphism $E^{\vee} \otimes_A E \longrightarrow B$ such that the two compositions

$$E \longrightarrow E \otimes_B E^{\vee} \otimes_A E \longrightarrow E \quad \text{and} \quad E^{\vee} \longrightarrow E^{\vee} \otimes_A E \otimes_B E^{\vee} \longrightarrow E^{\vee}$$

are equal to the identity endomorphisms of E and $E^{\vee}$.

For any Morita morphism E from A to B the functor

$$N \longmapsto {}_A N = E \otimes_B N = \operatorname{Hom}_B(E^{\vee}, N)$$

from the category of left B-modules to the category of left A-modules has a left adjoint functor

$$M \longmapsto {}_B M = E^{\vee} \otimes_A M$$

and a right adjoint functor

$$P \longmapsto {}^B P = \operatorname{Hom}_A(E, P).$$

Analogously, the functor $N \longmapsto N_A = N \otimes_B E^{\vee} = \operatorname{Hom}_{B^{op}}(E, N)$ from the category of right B-modules to the category of right A-modules has a left adjoint functor $M \longmapsto M_B = M \otimes_A E$ and a right adjoint functor $P \longmapsto P^B = \operatorname{Hom}_{B^{op}}(E^{\vee}, P)$.

Let $\mathcal{C}$ be a coring over a k-algebra A and E be a Morita morphism from A to B. Then there is a coring structure on the B-B-bimodule

$$_B\mathcal{C}_B = E^{\vee} \otimes_A \mathcal{C} \otimes_A E$$

defined in the following way [24]. The comultiplication in $_B\mathcal{C}_B$ is the composition $E^{\vee} \otimes_A \mathcal{C} \otimes_A E \longrightarrow E^{\vee} \otimes_A \mathcal{C} \otimes_A \mathcal{C} \otimes_A E \longrightarrow E^{\vee} \otimes_A \mathcal{C} \otimes_A E \otimes_B E^{\vee} \otimes_A \mathcal{C} \otimes_A E$ of the map induced by the comultiplication in $\mathcal{C}$ and the map induced by the map $A \longrightarrow E \otimes_B E^{\vee}$. The counit in $_B\mathcal{C}_B$ is the composition $E^{\vee} \otimes_A \mathcal{C} \otimes_A E \longrightarrow E^{\vee} \otimes_A E \longrightarrow B$, where the first map is induced by the counit in $\mathcal{C}$.

All the results of 7.1–7.3 can be generalized to the situation of a Morita morphism E from a k-algebra A to a k-algebra B and a morphism $_B\mathcal{C}_B \longrightarrow \mathcal{D}$ of corings over B. In particular, for any left $\mathcal{C}$-comodule $\mathcal{M}$ there is a natural $\mathcal{D}$-comodule structure on the B-module $_B\mathcal{M} = E^{\vee} \otimes_A \mathcal{M}$, and analogously for right comodules and left contramodules. For any right $\mathcal{C}$-comodule $\mathcal{M}'$ and any left $\mathcal{C}$-comodule $\mathcal{M}''$ there is a natural map $\mathcal{M}' \square_{\mathcal{C}} \mathcal{M}'' \longrightarrow \mathcal{M}'_B \square_{\mathcal{D}} {}_B\mathcal{M}''$ compatible with the map $\mathcal{M}' \otimes_A \mathcal{M}'' \longrightarrow \mathcal{M}'_B \otimes_B {}_B\mathcal{M}''$, etc. All the results of 7.4 can be generalized to the case of a Morita morphism E from a k-algebra A to a k-algebra B. In particular, $E^{\vee}$ is a (faithfully) flat right A-module if and only if $E \otimes_B E^{\vee}$ is a (faithfully) flat right A-module, etc.

7.5.2 One would like to define a Morita morphism from a coring $\mathcal{C}$ to a coring $\mathcal{D}$ as a pair consisting of a $\mathcal{C}$-$\mathcal{D}$-bicomodule $\mathcal{E}$ and a $\mathcal{D}$-$\mathcal{C}$-bicomodule $\mathcal{E}^\vee$ endowed with maps $\mathcal{C} \longrightarrow \mathcal{E} \,\square_\mathcal{D}\, \mathcal{E}^\vee$ and $\mathcal{E}^\vee \,\square_\mathcal{C}\, \mathcal{E} \longrightarrow \mathcal{D}$ satisfying appropriate conditions. This works fine for coalgebras over fields, but in the coring situation it is not clear how to deal with the problems of nonassociativity of the cotensor product. That is why we restrict ourselves to the special case of coflat/coprojective Morita morphisms.

Notice that, assuming $\mathcal{D}$ to be a flat right B-module, a k-linear functor

$$\Lambda \colon \mathcal{C}\text{-comod} \longrightarrow \mathcal{D}\text{-comod}$$

is isomorphic to a functor of the form $\mathcal{M} \longmapsto \mathcal{K}\,\square_\mathcal{C}\,\mathcal{M}$ for a certain $\mathcal{D}$-$\mathcal{C}$-bicomodule $\mathcal{K}$ if and only if it preserves cokernels of the morphisms coinduced from morphisms of A-modules, kernels of A-split morphisms, and infinite direct sums. Analogously, assuming $\mathcal{D}$ to be a projective left B-module, a k-linear functor

$$\Lambda \colon \mathcal{C}\text{-contra} \longrightarrow \mathcal{D}\text{-contra}$$

is isomorphic to a functor of the form $\mathfrak{P} \longmapsto \mathrm{Cohom}_\mathcal{C}(\mathcal{K}, \mathfrak{P})$ for a certain $\mathcal{C}$-$\mathcal{D}$-bicomodule $\mathcal{K}$ if and only if it preserves kernels of the morphisms induced from morphisms of A-modules, cokernels of A-split morphisms, and infinite direct products. Indeed, let us compose our functor Λ with the induction functor A-mod $\longrightarrow$ $\mathcal{C}$-contra and with the forgetful functor $\mathcal{D}$-contra $\longrightarrow$ B-mod; then the functor A-mod $\longrightarrow$ B-mod so obtained has the form $U \longmapsto \mathrm{Hom}_A(\mathcal{K}, U)$ for an A-B-bimodule $\mathcal{K}$. This follows from a theorem of Watts [85] about representability of left exact product-preserving covariant functors on the category of modules over a ring, which is a particular case of the abstract adjoint functor existence theorem [67, V.8]. The morphism of functors $\mathrm{Hom}_A(\mathcal{C}, \mathrm{Hom}_A(\mathcal{C}, U)) \longrightarrow \mathrm{Hom}_A(\mathcal{C}, U)$ induces a left $\mathcal{C}$-coaction in $\mathcal{K}$, while the functorial $\mathcal{D}$-contramodule structures on the B-modules $\mathrm{Hom}_A(\mathcal{K}, U)$ induce a right $\mathcal{D}$-coaction in $\mathcal{K}$. Since the functor Λ sends the exact sequences $\mathrm{Hom}_A(\mathcal{C}, \mathrm{Hom}_A(\mathcal{C}, \mathfrak{P})) \longrightarrow \mathrm{Hom}_A(\mathcal{C}, \mathfrak{P}) \longrightarrow \mathfrak{P} \longrightarrow 0$ to exact sequences, it is isomorphic to the functor $\mathfrak{P} \longmapsto \mathrm{Cohom}_\mathcal{C}(\mathcal{K}, \mathfrak{P})$.

Let $\mathcal{C}$ be a coring over a k-algebra A and $\mathcal{D}$ be a coring over a k-algebra B. Assume that $\mathcal{C}$ is a flat right A-module and $\mathcal{D}$ is a flat right B-module. A *right coflat Morita morphism* from $\mathcal{C}$ to $\mathcal{D}$ is a pair consisting of a $\mathcal{D}$-coflat $\mathcal{C}$-$\mathcal{D}$-bicomodule $\mathcal{E}$ and a $\mathcal{C}$-coflat $\mathcal{D}$-$\mathcal{C}$-bicomodule $\mathcal{E}^\vee$ endowed with a $\mathcal{C}$-$\mathcal{C}$-bicomodule morphism $\mathcal{C} \longrightarrow \mathcal{E} \,\square_\mathcal{D}\, \mathcal{E}^\vee$ and a $\mathcal{D}$-$\mathcal{D}$-bicomodule morphism $\mathcal{E}^\vee \,\square_\mathcal{C}\, \mathcal{E} \longrightarrow \mathcal{D}$ such that the two compositions

$$\mathcal{E} \longrightarrow \mathcal{E} \,\square_\mathcal{D}\, \mathcal{E}^\vee \,\square_\mathcal{C}\, \mathcal{E} \longrightarrow \mathcal{E} \quad \text{and} \quad \mathcal{E}^\vee \longrightarrow \mathcal{E}^\vee \,\square_\mathcal{C}\, \mathcal{E} \,\square_\mathcal{D}\, \mathcal{E}^\vee \longrightarrow \mathcal{E}^\vee$$

are equal to the identity endomorphisms of $\mathcal{E}$ and $\mathcal{E}^\vee$. A right coflat Morita morphism $(\mathcal{E}, \mathcal{E}^\vee)$ from $\mathcal{C}$ to $\mathcal{D}$ induces an exact functor $\mathcal{M} \longmapsto {}_\mathcal{D}\mathcal{M} = \mathcal{E}^\vee \,\square_\mathcal{C}\, \mathcal{M}$ from the category of left $\mathcal{C}$-comodules to the category of left $\mathcal{D}$-comodules and an exact functor $\mathcal{N} \longmapsto {}_\mathcal{C}\mathcal{N} = \mathcal{E}\,\square_\mathcal{D}\,\mathcal{N}$ from the category of left $\mathcal{D}$-comodules to the category of left $\mathcal{C}$-comodules; the former functor is left adjoint to the latter one. Conversely, any pair of adjoint exact k-linear functors preserving infinite direct sums between the categories of left $\mathcal{C}$-comodules and left $\mathcal{D}$-comodules is induced by a right coflat Morita morphism.

Analogously, assume that $\mathcal{C}$ is a projective left A-module and $\mathcal{D}$ is a projective left B-module. A *left coprojective Morita morphism* from $\mathcal{C}$ to $\mathcal{D}$ is defined as a pair consisting of a $\mathcal{C}$-coprojective $\mathcal{C}$-$\mathcal{D}$-bicomodule $\mathcal{E}$ and a $\mathcal{D}$-coprojective $\mathcal{D}$-$\mathcal{C}$-bicomodule $\mathcal{E}^\vee$ endowed with a $\mathcal{C}$-$\mathcal{C}$-bicomodule morphism $\mathcal{C} \longrightarrow \mathcal{E} \,\square_\mathcal{D}\, \mathcal{E}^\vee$ and a $\mathcal{D}$-$\mathcal{D}$-bicomodule morphism $\mathcal{E}^\vee \square_\mathcal{C} \mathcal{E} \longrightarrow \mathcal{D}$ satisfying the same conditions as above. A left coprojective Morita morphism $(\mathcal{E}, \mathcal{E}^\vee)$ from $\mathcal{C}$ to $\mathcal{D}$ induces an exact functor $\mathfrak{P} \longmapsto {}^\mathcal{D}\mathfrak{P} = \mathrm{Cohom}_\mathcal{C}(\mathcal{E}, \mathfrak{P})$ from the category of left $\mathcal{C}$-contramodules to the category of left $\mathcal{D}$-contramodules and an exact functor $\mathfrak{Q} \longmapsto {}^\mathcal{C}\mathfrak{Q} = \mathrm{Cohom}_\mathcal{D}(\mathcal{E}^\vee, \mathfrak{Q})$ from the category of left $\mathcal{D}$-contramodules to the category of left $\mathcal{C}$-contramodules; the former functor is right adjoint to the latter one. Conversely, any pair of adjoint exact k-linear functors preserving infinite products between the categories of left $\mathcal{C}$-contramodules and left $\mathcal{D}$-contramodules is induced by a left coprojective Morita morphism.

All the results of 7.1–7.3 can be extended to the situation of a left coprojective and right coflat Morita morphism from a coring $\mathcal{C}$ to a coring $\mathcal{D}$. In particular, for any right $\mathcal{C}$-comodule $\mathcal{M}$ and any left $\mathcal{D}$-contramodule $\mathfrak{Q}$ the compositions $(\mathcal{M}\square_\mathcal{C} \mathcal{E})\odot_\mathcal{D}\mathfrak{Q} \longrightarrow (\mathcal{M}\square_\mathcal{C}\mathcal{E})\odot_\mathcal{D}\mathrm{Cohom}_\mathcal{C}(\mathcal{E}, \mathrm{Cohom}_\mathcal{D}(\mathcal{E}^\vee, \mathfrak{Q})) \longrightarrow \mathcal{M}\odot_\mathcal{C}\mathrm{Cohom}_\mathcal{D}(\mathcal{E}^\vee, \mathfrak{Q})$ and $\mathcal{M}\odot_\mathcal{C}\mathrm{Cohom}_\mathcal{D}(\mathcal{E}^\vee, \mathfrak{Q}) \longrightarrow (\mathcal{M}\,\square_\mathcal{C}\,\mathcal{E}\,\square_\mathcal{D}\,\mathcal{E}^\vee)\odot_\mathcal{C}\mathrm{Cohom}_\mathcal{D}(\mathcal{E}^\vee, \mathfrak{Q}) \longrightarrow (\mathcal{M}\,\square_\mathcal{C}\,\mathcal{E})\odot_\mathcal{D}\mathfrak{Q}$ of the maps induced by the morphisms $\mathcal{E}^\vee\square_\mathcal{C}\mathcal{E} \longrightarrow \mathcal{D}$ and $\mathcal{C} \longrightarrow \mathcal{E}\square_\mathcal{D}\mathcal{E}^\vee$ and the natural "evaluation" maps are mutually inverse isomorphisms between the k-modules $\mathcal{M}_\mathcal{D}\odot_\mathcal{D}\mathfrak{Q}$ and $\mathcal{M}\odot_\mathcal{C}{}^\mathcal{C}\mathfrak{Q}$. For any left $\mathcal{D}$-contramodule $\mathfrak{Q}$ there are natural isomorphisms of left $\mathcal{C}$-comodules $\Phi_\mathcal{C}({}^\mathcal{C}\mathfrak{Q}) = \mathcal{C}\odot_\mathcal{C}{}^\mathcal{C}\mathfrak{Q} \simeq \mathcal{C}_\mathcal{D}\odot_\mathcal{D}\mathfrak{Q} \simeq \mathcal{E}\odot_\mathcal{D}\mathfrak{Q} \simeq \mathcal{E}\,\square_\mathcal{D}\,(\mathcal{D}\odot_\mathcal{D}\mathfrak{Q}) = \mathcal{C}(\Phi_\mathcal{D}\mathfrak{Q})$ by Proposition 5.2.1(e), etc. However, one sometimes has to impose the homological dimension conditions on A and B where they were not previously needed and strengthen the quasicoflatness (quasicoprojectivity, quasicoinjectivity) conditions to coflatness (coprojectivity, coinjectivity) conditions.

7.5.3   A *right coflat Morita equivalence* between corings $\mathcal{C}$ and $\mathcal{D}$ is a right coflat Morita morphism $(\mathcal{E}, \mathcal{E}^\vee)$ from $\mathcal{C}$ to $\mathcal{D}$ such that the bicomodule morphisms $\mathcal{C} \longrightarrow \mathcal{E}\,\square_\mathcal{D}\,\mathcal{E}^\vee$ and $\mathcal{E}^\vee\,\square_\mathcal{C}\,\mathcal{E} \longrightarrow \mathcal{D}$ are isomorphisms; it can be also considered as a right coflat Morita morphism $(\mathcal{E}^\vee, \mathcal{E})$ from $\mathcal{D}$ to $\mathcal{C}$. *Left coflat Morita equivalences* and *left coprojective Morita equivalences* are defined in the analogous way. A right coflat Morita equivalence between corings $\mathcal{C}$ and $\mathcal{D}$ induces an equivalence of the categories of left $\mathcal{C}$-comodules and left $\mathcal{D}$-comodules, and, assuming that $\mathcal{C}$ is a flat right A-module and $\mathcal{D}$ is a flat right B-module, any equivalence between these two k-linear categories comes from a right coflat Morita equivalence. Analogously, a left coprojective Morita equivalence between corings $\mathcal{C}$ and $\mathcal{D}$ induces an equivalence of the categories of left $\mathcal{C}$-contramodules and left $\mathcal{D}$-contramodules, and, assuming that $\mathcal{C}$ is a projective left A-module and $\mathcal{D}$ is a projective left B-module, any equivalence between these two k-linear categories comes from a left coprojective Morita equivalence.

Let $\mathcal{C}$ be a coring over a k-algebra A and $(E, E^{\vee})$ be a Morita morphism from A to B. If $\mathcal{C}$ is a flat right A-module and $E^{\vee}$ is a faithfully flat right A-module, then the pair of bicomodules $\mathcal{E} = \mathcal{C}_B = \mathcal{C} \otimes_A E$ and $\mathcal{E}^{\vee} = {}_B\mathcal{C} = E^{\vee} \otimes_A \mathcal{C}$ is a right coflat Morita equivalence between the corings $\mathcal{C}$ and ${}_B\mathcal{C}_B$. Analogously, if $\mathcal{C}$ is a projective left A-module and E is a faithfully projective left A-module, then the same pair of bicomodules $\mathcal{E} = \mathcal{C}_B$ and $\mathcal{E}^{\vee} = {}_B\mathcal{C}$ is a left coprojective Morita equivalence between the corings $\mathcal{C}$ and ${}_B\mathcal{C}_B$. This is a reformulation of the results of 7.4.1 in the case of a Morita morphism of k-algebras.

All the results of 7.4.3 can be generalized to the situation of a Morita equivalence, satisfying appropriate coflatness/coprojectivity conditions, between corings $\mathcal{C}$ and $\mathcal{D}$. The same applies to the results of 7.4.2, with homological dimension conditions added when necessary and the quasicoflatness (quasicoprojectivity, quasicoinjectivity) conditions strengthened to coflatness (coprojectivity, coinjectivity) conditions.

Remark. When the rings A and B are semisimple, one can consider Morita morphisms from the coring $\mathcal{C}$ to the coring $\mathcal{D}$ without any coflatness/coprojectivity conditions imposed. Moreover, for any Morita morphism $(\mathcal{E}, \mathcal{E}^{\vee})$ from $\mathcal{C}$ to $\mathcal{D}$ the left $\mathcal{C}$-comodule $\mathcal{E}$ is coprojective and the right $\mathcal{C}$-comodule $\mathcal{E}^{\vee}$ is coprojective. In particular, any Morita equivalence between $\mathcal{C}$ and $\mathcal{D}$ is left and right coprojective. On the other hand, without such conditions on the rings A and B not every right coflat Morita equivalence between $\mathcal{C}$ and $\mathcal{D}$ is a left coflat Morita equivalence. For example, when $\mathcal{C}$ is a finite-dimensional coalgebra over a field k, B is the algebra over k dual to $\mathcal{C}$, and $\mathcal{D} = B$, the right coflat Morita equivalence between $\mathcal{C}$ and $\mathcal{D}$ inducing the equivalence of categories $\mathcal{C}$-comod $\simeq$ B-mod is not left coflat, since this equivalence of categories does not preserve coflatness of comodules.

8 Functoriality in the Semialgebra

8.1 Compatible morphisms

Let $\mathcal{C} \longrightarrow \mathcal{D}$ be a map of corings compatible with a k-algebra map $A \longrightarrow B$. Let $\mathbf{S}$ be a semialgebra over the coring $\mathcal{C}$ and $\mathbf{T}$ be a semialgebra over the coring $\mathcal{D}$.

8.1.1 A map $\mathbf{S} \longrightarrow \mathbf{T}$ is called compatible with the maps $A \longrightarrow B$ and $\mathcal{C} \longrightarrow \mathcal{D}$ if the biaction maps $A \otimes_k \mathbf{S} \otimes_k A \longrightarrow \mathbf{S}$ and $B \otimes_k \mathbf{T} \otimes_k B \longrightarrow \mathbf{T}$ form a commutative diagram with the maps $\mathbf{S} \longrightarrow \mathbf{T}$ and $A \otimes_k \mathbf{S} \otimes_k A \longrightarrow B \otimes_k \mathbf{T} \otimes_k B$ (that is the map $\mathbf{S} \longrightarrow \mathbf{T}$ is an A-A-bimodule morphism), the bicoaction maps $\mathbf{S} \longrightarrow \mathcal{C} \otimes_A \mathbf{S} \otimes_A \mathcal{C}$ and $\mathbf{T} \longrightarrow \mathcal{D} \otimes_B \mathbf{T} \otimes_B \mathcal{D}$ form a commutative diagram with the maps $\mathbf{S} \longrightarrow \mathbf{T}$ and $\mathcal{C} \otimes_A \mathbf{S} \otimes_A \mathcal{C} \longrightarrow \mathcal{D} \otimes_B \mathbf{T} \otimes_B \mathcal{D}$ (that it the induced map $B \otimes_A \mathbf{S} \otimes_A B \longrightarrow \mathbf{T}$ is a $\mathcal{D}$-$\mathcal{D}$-bicomodule morphism), and furthermore, the semimultiplication maps $\mathbf{S} \square_\mathcal{C} \mathbf{S} \longrightarrow \mathbf{S}$ and $\mathbf{T} \square_\mathcal{D} \mathbf{T} \longrightarrow \mathbf{T}$ and the semiunit maps $\mathcal{C} \longrightarrow \mathbf{S}$ and $\mathcal{D} \longrightarrow \mathbf{T}$ form commutative diagrams with the maps $\mathcal{C} \longrightarrow \mathcal{D}$, $\mathbf{S} \longrightarrow \mathbf{T}$, and $\mathbf{S} \square_\mathcal{C} \mathbf{S} \longrightarrow \mathbf{T} \square_\mathcal{D} \mathbf{T}$.

Let $\mathbf{S} \longrightarrow \mathbf{T}$ be a map of semialgebras compatible with a map of corings $\mathcal{C} \longrightarrow \mathcal{D}$ and a k-algebra map $A \longrightarrow B$. Let $\mathcal{M}$ be a left $\mathbf{S}$-semimodule and $\mathcal{N}$ be a left $\mathbf{T}$-semimodule. A map $\mathcal{M} \longrightarrow \mathcal{N}$ is called compatible with the maps $A \longrightarrow B$, $\mathcal{C} \longrightarrow \mathcal{D}$, and $\mathbf{S} \longrightarrow \mathbf{T}$ if it is compatible with the maps $A \longrightarrow B$ and $\mathcal{C} \longrightarrow \mathcal{D}$ as a map from a $\mathcal{C}$-comodule to a $\mathcal{D}$-comodule and the semiaction maps $\mathbf{S} \square_\mathcal{C} \mathcal{M} \longrightarrow \mathcal{M}$ and $\mathbf{T} \square_\mathcal{D} \mathcal{N} \longrightarrow \mathcal{N}$ form a commutative diagram with the maps $\mathcal{M} \longrightarrow \mathcal{N}$ and $\mathbf{S} \square_\mathcal{C} \mathcal{M} \longrightarrow \mathbf{T} \square_\mathcal{D} \mathcal{N}$. Analogously, let $\mathfrak{P}$ be a left $\mathbf{S}$-semicontramodule and $\mathfrak{Q}$ be a left $\mathbf{T}$-semicontramodule. A map $\mathfrak{Q} \longrightarrow \mathfrak{P}$ is called compatible with the maps $A \longrightarrow B$, $\mathcal{C} \longrightarrow \mathcal{D}$, and $\mathbf{S} \longrightarrow \mathbf{T}$ if it is compatible with the maps $A \longrightarrow B$ and $\mathcal{C} \longrightarrow \mathcal{D}$ as a map from a $\mathcal{D}$-contramodule to a $\mathcal{C}$-contramodule and the semicontraaction maps $\mathfrak{P} \longrightarrow \mathrm{Cohom}_\mathcal{C}(\mathbf{S}, \mathfrak{P})$ and $\mathfrak{Q} \longrightarrow \mathrm{Cohom}_\mathcal{D}(\mathbf{T}, \mathfrak{Q})$ form a commutative diagram with the maps $\mathfrak{Q} \longrightarrow \mathfrak{P}$ and $\mathrm{Cohom}_\mathcal{D}(\mathbf{T}, \mathfrak{Q}) \longrightarrow \mathrm{Cohom}_\mathcal{C}(\mathbf{S}, \mathfrak{P})$.

Let $\mathcal{M}' \longrightarrow \mathcal{N}'$ be a map from a right $\mathbf{S}$-semimodule to a right $\mathbf{T}$-semimodule compatible with the maps $A \longrightarrow B$, $\mathcal{C} \longrightarrow \mathcal{D}$, and $\mathbf{S} \longrightarrow \mathbf{T}$, and let $\mathcal{M}'' \longrightarrow \mathcal{N}''$ be a map from a left $\mathbf{S}$-semimodule to a left $\mathbf{T}$-semimodule compatible with the maps $A \longrightarrow B$, $\mathcal{C} \longrightarrow \mathcal{D}$, and $\mathbf{S} \longrightarrow \mathbf{T}$. Assume that the triple cotensor products $\mathcal{M}' \square_\mathcal{C} \mathbf{S} \square_\mathcal{C} \mathcal{M}''$ and $\mathcal{N}' \square_\mathcal{D} \mathbf{T} \square_\mathcal{D} \mathcal{N}''$ are associative. Then there is a natural map of k-modules $\mathcal{M}' \lozenge_\mathbf{S} \mathcal{M}'' \longrightarrow \mathcal{N}' \lozenge_\mathbf{S} \mathcal{N}''$. Analogously, let $\mathcal{M} \longrightarrow \mathcal{N}$ be a map from a left $\mathbf{S}$-semimodule to a left $\mathbf{T}$-semimodule compatible with the maps $A \longrightarrow B$, $\mathcal{C} \longrightarrow \mathcal{D}$, and $\mathbf{S} \longrightarrow \mathbf{T}$, and let $\mathfrak{Q} \longrightarrow \mathfrak{P}$ be a map from a left $\mathbf{T}$-semicontramodule to a left $\mathbf{S}$-semicontramodule compatible with the maps $A \longrightarrow B$, $\mathcal{C} \longrightarrow \mathcal{D}$, and $\mathbf{S} \longrightarrow \mathbf{T}$. Assume that the triple cohomomorphisms $\mathrm{Cohom}_\mathcal{C}(\mathbf{S} \square_\mathcal{C} \mathcal{M}, \mathfrak{P})$ and $\mathrm{Cohom}_\mathcal{D}(\mathbf{T} \square_\mathcal{D} \mathcal{N}, \mathfrak{Q})$ are associative. Then there is a natural map of k-modules $\mathrm{SemiHom}_\mathbf{T}(\mathcal{N}, \mathfrak{Q}) \longrightarrow \mathrm{SemiHom}_\mathbf{S}(\mathcal{M}, \mathfrak{P})$.

L. Positselski, *Homological Algebra of Semimodules and Semicontramodules*, Monografie Matematyczne 70, DOI 10.1007/978-3-0346-0436-9_9, © Springer Basel AG 2010

8.1.2 Let $\mathbf{S} \longrightarrow \mathbf{T}$ be a map of semialgebras compatible with a map of corings $\mathcal{C} \longrightarrow \mathcal{D}$ and a k-algebra map $A \longrightarrow B$.

Assume that $\mathcal{C}$ is a flat right A-module and either $\mathbf{S}$ is a coflat right $\mathcal{C}$-co-module, or $\mathbf{S}$ is a flat right A-module and a $\mathcal{C}/A$-coflat left $\mathcal{C}$-comodule and A has a finite weak homological dimension, or A is absolutely flat. Then for any left $\mathbf{T}$-semimodule $\mathcal{N}$ there is a natural $\mathbf{S}$-semimodule structure on the left $\mathcal{C}$-comodule ${}_{\mathcal{C}}\mathcal{N}$. It is constructed as follows: the composition $\mathbf{S} \square_{\mathcal{C}} {}_{\mathcal{C}}\mathcal{N} \longrightarrow \mathbf{T} \square_{\mathcal{D}} \mathcal{N} \longrightarrow \mathcal{N}$ of the map induced by the maps $\mathbf{S} \longrightarrow \mathbf{T}$ and ${}_{\mathcal{C}}\mathcal{N} \longrightarrow \mathcal{N}$ with the $\mathbf{T}$-semiaction in $\mathcal{N}$ is a map from a $\mathcal{C}$-comodule to a $\mathcal{D}$-comodule compatible with the maps $A \longrightarrow B$ and $\mathcal{C} \longrightarrow \mathcal{D}$, hence there is a $\mathcal{C}$-comodule map $\mathbf{S} \square_{\mathcal{C}} {}_{\mathcal{C}}\mathcal{N} \longrightarrow {}_{\mathcal{C}}\mathcal{N}$. Analogously, assume that $\mathcal{C}$ is a projective left A-module and either $\mathbf{S}$ is a coprojective left $\mathcal{C}$-comodule, or $\mathbf{S}$ is a projective left A-module and a $\mathcal{C}/A$-coflat right $\mathcal{C}$-comodule and A has a finite left homological dimension, or A is semisimple. Then for any left $\mathbf{T}$-semicon-tramodule $\mathfrak{Q}$ there is a natural $\mathbf{S}$-semicontramodule structure on the left $\mathcal{C}$-contra-module ${}^{\mathcal{C}}\mathfrak{Q}$. Indeed, the composition $\mathfrak{Q} \longrightarrow \mathrm{Cohom}_{\mathcal{D}}(\mathbf{T}, \mathfrak{Q}) \longrightarrow \mathrm{Cohom}_{\mathcal{C}}(\mathbf{S}, {}^{\mathcal{C}}\mathfrak{Q})$ is a map from a $\mathcal{D}$-contramodule to a $\mathcal{C}$-contramodule compatible with the maps $A \longrightarrow B$ and $\mathcal{C} \longrightarrow \mathcal{D}$, hence a $\mathcal{C}$-contramodule map ${}^{\mathcal{C}}\mathfrak{Q} \longrightarrow \mathrm{Cohom}_{\mathcal{C}}(\mathbf{S}, {}^{\mathcal{C}}\mathfrak{Q})$. Assuming that $\mathcal{D}$ is a flat right B-module, $\mathcal{C}$ is a flat right A-module, and $\mathbf{S}$ is a coflat right $\mathcal{C}$-comodule, for any $\mathcal{D}$-coflat right $\mathbf{T}$-semimodule $\mathcal{N}$ there is a natural $\mathbf{S}$-semimodule structure on the coflat right $\mathcal{C}$-comodule $\mathcal{N}_{\mathcal{C}}$ and for any $\mathcal{D}$-coinjective left $\mathbf{T}$-semicontramodule $\mathfrak{Q}$ there is a natural $\mathbf{S}$-semicontramodule structure on the coinjective left $\mathcal{C}$-contramodule ${}^{\mathcal{C}}\mathfrak{Q}$ provided that B is a flat right A-module.

Assume that $\mathcal{C}$ is a flat right A-module, $\mathbf{S}$ is a coflat right $\mathcal{C}$-comodule, $\mathcal{D}$ is a flat right B-module, and $\mathbf{T}$ is a coflat right $\mathcal{D}$-comodule. Then the functor $\mathcal{N} \longmapsto {}_{\mathcal{C}}\mathcal{N}$ from the category of left $\mathbf{T}$-semimodules to the category of left $\mathbf{S}$-semi-modules has a left adjoint functor $\mathcal{M} \longmapsto {}_{\mathbf{T}}\mathcal{M}$, which is constructed as follows. For induced left $\mathbf{S}$-semimodules, one has

$$\mathbf{T}(\mathbf{S} \square_{\mathcal{C}} \mathcal{L}) = \mathbf{T} \square_{\mathcal{D}} {}_{B}\mathcal{L};$$

to compute the $\mathbf{T}$-semimodule ${}_{\mathbf{T}}\mathcal{M}$ for an arbitrary left $\mathbf{S}$-semimodule $\mathcal{M}$, one can represent $\mathcal{M}$ as the cokernel of a morphism of induced $\mathbf{S}$-semimodules. Both k-modules $\mathrm{Hom}_{\mathbf{S}}(\mathcal{M}, {}_{\mathcal{C}}\mathcal{N})$ and $\mathrm{Hom}_{\mathbf{T}}({}_{\mathbf{T}}\mathcal{M}, \mathcal{N})$ are isomorphic to the k-module of all maps of semimodules $\mathcal{M} \longrightarrow \mathcal{N}$ compatible with the maps $A \longrightarrow B$, $\mathcal{C} \longrightarrow \mathcal{D}$, and $\mathbf{S} \longrightarrow \mathbf{T}$, hence

$$\mathrm{Hom}_{\mathbf{S}}(\mathcal{M}, {}_{\mathcal{C}}\mathcal{N}) \simeq \mathrm{Hom}_{\mathbf{T}}({}_{\mathbf{T}}\mathcal{M}, \mathcal{N}).$$

There are also a few situations when the functor $\mathcal{M} \longmapsto {}_{\mathbf{T}}\mathcal{M}$ is defined on the full subcategory of induced $\mathbf{S}$-semimodules. Under the analogous assumptions, the functor $\mathcal{M} \longmapsto \mathcal{M}_{\mathbf{T}}$ left adjoint to the functor $\mathcal{N} \longmapsto \mathcal{N}_{\mathcal{C}}$ acts from the category of right $\mathbf{S}$-semimodules to the category of right $\mathbf{T}$-semimodules.

Now assume that $\mathcal{C}$ is a flat left and right A-module, $\mathbf{S}$ is a flat left A-mod-ule and a coflat right $\mathcal{C}$-comodule, A has a finite weak homological dimension, $\mathcal{D}$

is a flat right B-module, and $\mathcal{T}$ is a coflat right $\mathcal{D}$-comodule. Then the functor $\mathcal{N} \longmapsto {}_{\mathcal{C}}\mathcal{N}$ can be constructed in a different way: when $\mathcal{M}$ is a flat left A-module, one has

$$\mathcal{T}\mathcal{M} = \mathcal{T}_{\mathcal{C}} \lozenge_{\mathcal{S}} \mathcal{M},$$

where $\mathcal{T}_{\mathcal{C}} = \mathcal{T}\square_{\mathcal{D}}{}_{B}\mathcal{C}$ is a $\mathcal{T}$-$\mathcal{S}$-bisemimodule with the right $\mathcal{S}$-semimodule structure provided by the above construction. To compute the $\mathcal{T}$-semimodule $\mathcal{T}\mathcal{M}$ for an arbitrary left $\mathcal{S}$-semimodule $\mathcal{M}$, one can represent $\mathcal{M}$ as the cokernel of a morphism of A-flat $\mathcal{S}$-semimodules. Assuming only that $\mathcal{C}$ is a flat right A-module, $\mathcal{S}$ is a coflat right $\mathcal{C}$-comodule, $\mathcal{D}$ is a flat right B-module, and $\mathcal{T}$ is a coflat right $\mathcal{D}$-comodule, the functor $\mathcal{M} \longmapsto \mathcal{T}\mathcal{M}$ can be defined by the formula $\mathcal{T}\mathcal{M} = \mathcal{T}_{\mathcal{C}} \lozenge_{\mathcal{S}} \mathcal{M}$ for any $\mathcal{M}$ whenever B is a flat right A-module. If $\mathcal{C}$ is a flat left and right A-module, $\mathcal{S}$ is a coflat left and right $\mathcal{C}$-comodule, $\mathcal{D}$ is a flat right B-module, and $\mathcal{T}$ is a coflat right $\mathcal{D}$-comodule, the functor $\mathcal{M} \longmapsto \mathcal{T}\mathcal{M}$ is given by the formula $\mathcal{T}\mathcal{M} = \mathcal{T}_{\mathcal{C}} \lozenge_{\mathcal{S}} \mathcal{M}$ on the full subcategory of $\mathcal{C}$-coflat $\mathcal{S}$-semimodules $\mathcal{M}$.

Furthermore, assume that $\mathcal{C}$ is a projective left A-module, $\mathcal{S}$ is a coprojective left $\mathcal{C}$-comodule, $\mathcal{D}$ is a projective left B-module, and $\mathcal{T}$ is a coprojective left $\mathcal{D}$-comodule. Then the functor $\mathfrak{Q} \longmapsto {}^{\mathcal{C}}\mathfrak{Q}$ from the category of left $\mathcal{T}$-semicontramodules to the category of left $\mathcal{S}$-semicontramodules has a right adjoint functor $\mathfrak{P} \longmapsto {}^{\mathcal{T}}\mathfrak{P}$, which is constructed as follows. For coinduced left $\mathcal{S}$-semicontramodules, one has

$$\mathcal{T}\mathrm{Cohom}_{\mathcal{C}}(\mathcal{S}, \mathfrak{R}) = \mathrm{Cohom}_{\mathcal{D}}(\mathcal{T}, {}^{B}\mathfrak{R});$$

to compute the $\mathcal{T}$-semicontramodule ${}^{\mathcal{T}}\mathfrak{P}$ for an arbitrary left $\mathcal{S}$-semicontramodule $\mathfrak{P}$, one can represent $\mathfrak{P}$ as the kernel of a morphism of coinduced $\mathcal{S}$-semicontramodules. Both k-modules $\mathrm{Hom}^{\mathcal{S}}({}^{\mathcal{C}}\mathfrak{Q}, \mathfrak{P})$ and $\mathrm{Hom}^{\mathcal{T}}(\mathfrak{Q}, {}^{\mathcal{T}}\mathfrak{P})$ are isomorphic to the k-module of all maps of semicontramodules $\mathfrak{Q} \longrightarrow \mathfrak{P}$ compatible with the maps $A \longrightarrow B$, $\mathcal{C} \longrightarrow \mathcal{D}$, and $\mathcal{S} \longrightarrow \mathcal{T}$, hence

$$\mathrm{Hom}^{\mathcal{S}}({}^{\mathcal{C}}\mathfrak{Q}, \mathfrak{P}) \simeq \mathrm{Hom}^{\mathcal{T}}(\mathfrak{Q}, {}^{\mathcal{T}}\mathfrak{P}).$$

There are also a few situations when the functor $\mathfrak{P} \longmapsto {}^{\mathcal{T}}\mathfrak{P}$ is defined on the full subcategory of coinduced $\mathcal{S}$-semicontramodules.

Now assume that $\mathcal{C}$ is a projective left and a flat right A-module, $\mathcal{S}$ is a coprojective left $\mathcal{C}$-comodule and a flat right A-module, A has a finite left homological dimension, $\mathcal{D}$ is a projective left B-module, and $\mathcal{T}$ is a coprojective left $\mathcal{D}$-comodule. Then the functor $\mathfrak{P} \longmapsto {}^{\mathcal{T}}\mathfrak{P}$ can be constructed in a different way: when $\mathfrak{P}$ is an injective left A-module,

$$\mathcal{T}\mathfrak{P} = \mathrm{SemiHom}_{\mathcal{S}}({}_{\mathcal{C}}\mathcal{T}, \mathfrak{P});$$

to compute the $\mathcal{T}$-semicontramodule ${}^{\mathcal{T}}\mathfrak{P}$ for an arbitrary left $\mathcal{S}$-semicontramodule $\mathfrak{P}$, one can represent $\mathfrak{P}$ as the kernel of a morphism of A-injective $\mathcal{S}$-semicontramodules. Assuming only that $\mathcal{C}$ is a projective left A-module, $\mathcal{S}$ is a coprojective left $\mathcal{C}$-comodule, $\mathcal{D}$ is a projective left B-module, and $\mathcal{T}$ is a coprojective left $\mathcal{D}$-comodule, the functor $\mathfrak{P} \longmapsto {}^{\mathcal{T}}\mathfrak{P}$ can be defined by the formula ${}^{\mathcal{T}}\mathfrak{P} =$

SemiHom$_S({}_C\mathcal{J}, \mathfrak{P})$ for any $\mathfrak{P}$ whenever B is a projective left A-module. If $\mathcal{C}$ is a projective left and a flat right A-module, $\mathcal{S}$ is a coprojective left and a coflat right $\mathcal{C}$-comodule, $\mathcal{D}$ is a projective left B-module, and $\mathcal{J}$ is a coprojective left $\mathcal{D}$-comodule, the functor $\mathfrak{P} \longmapsto {}^{\mathcal{J}}\mathfrak{P}$ is given by the formula ${}^{\mathcal{J}}\mathfrak{P} = \mathrm{SemiHom}_S({}_C\mathcal{J}, \mathfrak{P})$ on the full subcategory of $\mathcal{C}$-coinjective $\mathcal{S}$-semicontramodules $\mathfrak{P}$.

Assume that $\mathcal{C}$ is a projective left A-module, $\mathcal{S}$ is a coprojective left $\mathcal{C}$-comodule, $\mathcal{D}$ is a projective left B-module, and $\mathcal{J}$ is a coprojective left $\mathcal{D}$-comodule. Then for any right $\mathcal{S}$-semimodule $\mathcal{M}$ and any left $\mathcal{J}$-semicontramodule $\mathfrak{Q}$ there is a natural isomorphism

$$\mathcal{M}_{\mathcal{J}} \odot_{\mathcal{J}} \mathfrak{Q} \simeq \mathcal{M} \odot_S {}^{\mathcal{C}}\mathfrak{Q}.$$

Moreover, both k-modules are isomorphic to the cokernel of the pair of maps $(\mathcal{M} \square_C \mathcal{S})_B \odot_{\mathcal{D}} \mathfrak{Q} \rightrightarrows \mathcal{M}_B \odot_{\mathcal{D}} \mathfrak{Q}$ one of which is induced by the $\mathcal{S}$-semiaction in $\mathcal{M}$ and the other is defined in terms of the morphism $(\mathcal{M} \square_C \mathcal{S})_B \longrightarrow \mathcal{M}_B \square_{\mathcal{D}} \mathcal{J}$, the $\mathcal{J}$-semicontraaction in $\mathfrak{Q}$, and the natural "evaluation" map $(\mathcal{M}_B \square_{\mathcal{D}} \mathcal{J}) \odot_{\mathcal{D}} \mathrm{Cohom}_{\mathcal{D}}(\mathcal{J}, \mathfrak{Q}) \longrightarrow \mathcal{M}_B \odot_{\mathcal{D}} \mathfrak{Q}$. This is clear for $\mathcal{M} \odot_S {}^{\mathcal{C}}\mathfrak{Q}$, and to construct this isomorphism for $\mathcal{M}_{\mathcal{J}} \odot_{\mathcal{J}} \mathfrak{Q}$ it suffices to represent $\mathcal{M}$ as the cokernel of the pair of morphisms of induced $\mathcal{S}$-semimodules $\mathcal{M} \square_C \mathcal{S} \square_C \mathcal{S} \rightrightarrows \mathcal{M} \square_C \mathcal{S}$. In the above situations when $\mathcal{M}_{\mathcal{J}} = \mathcal{M} \lozenge_S {}_C\mathcal{J}$, this isomorphism can be also constructed by representing $\mathcal{M}_{\mathcal{J}}$ as the cokernel of the pair of $\mathcal{J}$-semimodule morphisms $\mathcal{M} \square_C \mathcal{S} \square_C {}_C\mathcal{J} \rightrightarrows \mathcal{M} \square_C {}_C\mathcal{J}$ and using the isomorphisms $\mathcal{M} \square_C {}_C\mathcal{J} \simeq \mathcal{M}_B \square_{\mathcal{D}} \mathcal{J}$.

8.1.3   Let $\mathcal{S} \longrightarrow \mathcal{J}$ be a map of semialgebras compatible with a map of corings $\mathcal{C} \longrightarrow \mathcal{D}$ and a k-algebra map $A \longrightarrow B$.

Proposition.

(a) *Let $\mathcal{M}$ be a left $\mathcal{S}$-semimodule and $\mathcal{N}$ be a right $\mathcal{J}$-semimodule. Then the semitensor product ${}_{\mathcal{J}}\mathcal{M} = \mathcal{J}_C \lozenge_S \mathcal{M}$ can be endowed with a left $\mathcal{J}$-semimodule structure via the construction of 1.4.4 and the map of semitensor products*

$$\mathcal{N}_C \lozenge_S \mathcal{M} \longrightarrow \mathcal{N} \lozenge_{\mathcal{J}} {}_{\mathcal{J}}\mathcal{M}$$

induced by the maps of semimodules $\mathcal{N}_C \longrightarrow \mathcal{N}$ and $\mathcal{M} \longrightarrow {}_{\mathcal{J}}\mathcal{M}$ is an isomorphism, at least, in the following cases:

- *$\mathcal{D}$ is a flat right B-module, $\mathcal{J}$ is a coflat right $\mathcal{D}$-comodule, $\mathcal{N}$ is a coflat right $\mathcal{D}$-comodule, $\mathcal{C}$ is a flat left A-module, $\mathcal{S}$ is a flat left A-module and a $\mathcal{C}/A$-coflat right $\mathcal{C}$-comodule, the ring A has a finite weak homological dimension, and $\mathcal{M}$ is a flat left A-module, or*

- *$\mathcal{D}$ is a flat right B-module, $\mathcal{J}$ is a coflat right $\mathcal{D}$-comodule, $\mathcal{N}$ is a coflat right $\mathcal{D}$-comodule, $\mathcal{C}$ is a flat left A-module, $\mathcal{S}$ is a coflat left $\mathcal{C}$-comodule, and $\mathcal{M}$ is a coflat left $\mathcal{C}$-comodule, or*

- *$\mathcal{D}$ is a flat right B-module, $\mathcal{J}$ is a coflat right $\mathcal{D}$-comodule, $\mathcal{N}$ is a coflat right $\mathcal{D}$-comodule, $\mathcal{C}$ is a flat right A-module, $\mathcal{S}$ is a coflat right $\mathcal{C}$-comodule, and B is a flat right A-module, or*

- $\mathcal{D}$ *is a flat left B-module, $\mathcal{T}$ is a flat left B-module and a $\mathcal{D}/B$-coflat right $\mathcal{D}$-comodule, the ring B has a finite weak homological dimension, $\mathcal{C}$ is a flat left A-module, $\mathcal{S}$ is a coflat left $\mathcal{C}$-comodule, and $\mathcal{M}$ is a semiflat left $\mathcal{S}$-semimodule, or*

- $\mathcal{D}$ *is a flat left B-module, $\mathcal{T}$ is a coflat left $\mathcal{D}$-comodule, $_B\mathcal{C}$ is a coflat left $\mathcal{D}$-comodule, $\mathcal{C}$ is a flat left A-module, $\mathcal{S}$ is a coflat left $\mathcal{C}$-comodule, and $\mathcal{M}$ is a semiflat left $\mathcal{S}$-semimodule.*

When the ring A (resp., B) is absolutely flat, the $\mathcal{C}/A$-coflatness (resp., $\mathcal{D}/B$-coflatness) assumption can be dropped.

(b) *Let $\mathfrak{P}$ be a left $\mathcal{S}$-semicontramodule and $\mathcal{N}$ be a left $\mathcal{T}$-semimodule. Then the module of semihomomorphisms $^{\mathcal{T}}\mathfrak{P} = \mathrm{SemiHom}_{\mathcal{S}}(_{\mathcal{C}}\mathcal{T}, \mathfrak{P})$ can be endowed with a left $\mathcal{T}$-semicontramodule structure via the construction of 3.4.4 and the map of the semihomomorphism modules*

$$\mathrm{SemiHom}_{\mathcal{T}}(\mathcal{N}, {}^{\mathcal{T}}\mathfrak{P}) \longrightarrow \mathrm{SemiHom}_{\mathcal{S}}(_{\mathcal{C}}\mathcal{N}, \mathfrak{P})$$

induced by the maps of semimodules and semicontramodules $_{\mathcal{C}}\mathcal{N} \longrightarrow \mathcal{N}$ and $^{\mathcal{T}}\mathfrak{P} \longrightarrow \mathfrak{P}$ is an isomorphism, at least, in the following cases:

- $\mathcal{D}$ *is a projective left B-module, $\mathcal{T}$ is a coprojective left $\mathcal{D}$-comodule, $\mathcal{N}$ is a coprojective left $\mathcal{D}$-comodule, $\mathcal{C}$ is a flat right A-module, $\mathcal{S}$ is a flat right A-module and a $\mathcal{C}/A$-coprojective left $\mathcal{C}$-comodule, the ring A has a finite left homological dimension, and $\mathfrak{P}$ is an injective left A-module, or*

- $\mathcal{D}$ *is a projective left B-module, $\mathcal{T}$ is a coprojective left $\mathcal{D}$-comodule, $\mathcal{N}$ is a coprojective left $\mathcal{D}$-comodule, $\mathcal{C}$ is a flat right A-module, $\mathcal{S}$ is a coflat right $\mathcal{C}$-comodule, and $\mathfrak{P}$ is a coinjective left $\mathcal{C}$-comodule, or*

- $\mathcal{D}$ *is a projective left B-module, $\mathcal{T}$ is a coprojective left $\mathcal{D}$-comodule, $\mathcal{N}$ is a coprojective left $\mathcal{D}$-comodule, $\mathcal{C}$ is a projective left A-module, $\mathcal{S}$ is a coprojective left $\mathcal{C}$-comodule, and B is a projective left A-module, or*

- $\mathcal{D}$ *is a flat right B-module, $\mathcal{T}$ is a flat right B-module and a $\mathcal{D}/B$-coprojective left $\mathcal{D}$-comodule, the ring B has a finite left homological dimension, $\mathcal{C}$ is a flat right A-module, $\mathcal{S}$ is a coflat right $\mathcal{C}$-comodule, and $\mathfrak{P}$ is a semiinjective left $\mathcal{S}$-semicontramodule, or*

- $\mathcal{D}$ *is a flat right B-module, $\mathcal{T}$ is a coflat right $\mathcal{D}$-comodule, $\mathcal{C}_B$ is a coflat right $\mathcal{D}$-comodule, $\mathcal{C}$ is a flat right A-module, $\mathcal{S}$ is a coflat right $\mathcal{C}$-comodule, and $\mathfrak{P}$ is a semiinjective left $\mathcal{S}$-semicontramodule.*

When the ring A (resp., B) is semisimple, the $\mathcal{C}/A$-coprojectivity (resp., $\mathcal{D}/B$-coprojectivity) assumption can be dropped.

(c) *Let $\mathcal{M}$ be a left $\mathcal{S}$-semimodule and $\mathfrak{Q}$ be a left $\mathcal{T}$-semicontramodule. Then the map of semihomomorphism modules*

$$\mathrm{SemiHom}_{\mathcal{T}}(_{\mathcal{T}}\mathcal{M}, \mathfrak{Q}) \longrightarrow \mathrm{SemiHom}_{\mathcal{S}}(\mathcal{M}, {}^{\mathcal{C}}\mathfrak{Q})$$

induced by the map of semimodules $\mathcal{M} \longrightarrow {}_{\mathcal{T}}\mathcal{M}$ *and the map of semicontra-modules* $\mathfrak{Q} \longrightarrow {}^{\mathcal{C}}\mathfrak{Q}$ *is an isomorphism, at least, in the following cases:*

- $\mathcal{D}$ *is a flat right B-module,* $\mathcal{T}$ *is a coflat right $\mathcal{D}$-comodule,* $\mathfrak{Q}$ *is a coinjective left $\mathcal{D}$-contramodule,* $\mathcal{C}$ *is a projective left A-module,* $\mathcal{S}$ *is a projective left A-module and a $\mathcal{C}/A$-coflat right $\mathcal{C}$-comodule, the ring A has a finite left homological dimension, and* $\mathcal{M}$ *is a projective left A-module, or*

- $\mathcal{D}$ *is a flat right B-module,* $\mathcal{T}$ *is a coflat right $\mathcal{D}$-comodule,* $\mathfrak{Q}$ *is a coinjective left $\mathcal{D}$-contramodule,* $\mathcal{C}$ *is a projective left A-module,* $\mathcal{S}$ *is a coprojective left $\mathcal{C}$-comodule, and* $\mathcal{M}$ *is a coprojective left $\mathcal{C}$-comodule, or*

- $\mathcal{D}$ *is a flat right B-module,* $\mathcal{T}$ *is a coflat right $\mathcal{D}$-comodule,* $\mathfrak{Q}$ *is a coinjective left $\mathcal{D}$-contramodule,* $\mathcal{C}$ *is a flat right A-module,* $\mathcal{S}$ *is a coflat right $\mathcal{C}$-comodule, and B is a flat right A-module, or*

- $\mathcal{D}$ *is a projective left B-module,* $\mathcal{T}$ *is a projective left B-module and a $\mathcal{D}/B$-coflat right $\mathcal{D}$-comodule, the ring B has a finite left homological dimension,* $\mathcal{C}$ *is a projective left A-module,* $\mathcal{S}$ *is a coprojective left $\mathcal{C}$-co-module, and* $\mathcal{M}$ *is a semiprojective left $\mathcal{S}$-semimodule, or*

- $\mathcal{D}$ *is a projective left B-module,* $\mathcal{T}$ *is a coprojective left $\mathcal{D}$-comodule,* ${}_B\mathcal{C}$ *is a coprojective left $\mathcal{D}$-comodule,* $\mathcal{C}$ *is a projective left A-module,* $\mathcal{S}$ *is a coprojective left $\mathcal{C}$-comodule, and* $\mathcal{M}$ *is a semiprojective left $\mathcal{S}$-semi-module.*

When the ring A (resp., B) is semisimple, the $\mathcal{C}/A$-coflatness (resp., $\mathcal{D}/B$-co-flatness) assumption can be dropped.

Proof. Part (a): under our assumptions, there is a natural isomorphism of right $\mathcal{S}$-semimodules $\mathcal{N}_{\mathcal{C}} \simeq \mathcal{N} \lozenge_{\mathcal{T}} \mathcal{T}_{\mathcal{C}}$. For any left $\mathcal{S}$-semimodule $\mathcal{M}$ and right $\mathcal{T}$-semi-module $\mathcal{N}$ for which the iterated semitensor products $(\mathcal{N} \lozenge_{\mathcal{T}} \mathcal{T}_{\mathcal{C}}) \lozenge_{\mathcal{S}} \mathcal{M}$ and $\mathcal{N} \lozenge_{\mathcal{T}} (\mathcal{T}_{\mathcal{C}} \lozenge_{\mathcal{S}} \mathcal{M})$ are defined and the triple cotensor product $\mathcal{N} \square_{\mathcal{D}} \mathcal{T}_{\mathcal{C}} \square_{\mathcal{C}} \mathcal{M}$ is associative, the map $(\mathcal{N} \lozenge_{\mathcal{T}} \mathcal{T}_{\mathcal{C}}) \lozenge_{\mathcal{S}} \mathcal{M} \longrightarrow \mathcal{N} \lozenge_{\mathcal{T}} (\mathcal{T}_{\mathcal{C}} \lozenge_{\mathcal{S}} \mathcal{M})$ induced by the bisemimodule maps $\mathcal{S} \longrightarrow \mathcal{T}_{\mathcal{C}} \longrightarrow \mathcal{T}$ compatible with the maps $A \longrightarrow B$, $\mathcal{C} \longrightarrow \mathcal{D}$, and $\mathcal{S} \longrightarrow \mathcal{T}$ forms a commutative diagram with the maps $\mathcal{N} \square_{\mathcal{D}} \mathcal{T}_{\mathcal{C}} \square_{\mathcal{C}} \mathcal{M} \longrightarrow (\mathcal{N} \lozenge_{\mathcal{T}} \mathcal{T}_{\mathcal{C}}) \lozenge_{\mathcal{S}} \mathcal{M}$ and $\mathcal{N} \square_{\mathcal{D}} \mathcal{T}_{\mathcal{C}} \square_{\mathcal{C}} \mathcal{M} \longrightarrow \mathcal{N} \lozenge_{\mathcal{T}} (\mathcal{T}_{\mathcal{C}} \lozenge_{\mathcal{S}} \mathcal{M})$. Indeed, the map $\mathcal{N} \square_{\mathcal{D}} \mathcal{T}_{\mathcal{C}} \square_{\mathcal{C}} \mathcal{M} \longrightarrow \mathcal{N} \square_{\mathcal{D}} (\mathcal{T}_{\mathcal{C}} \lozenge_{\mathcal{S}} \mathcal{M})$ is equal to the composition of the map $\mathcal{N} \square_{\mathcal{D}} \mathcal{T}_{\mathcal{C}} \square_{\mathcal{C}} \mathcal{M} \longrightarrow \mathcal{N} \square_{\mathcal{D}} \mathcal{T} \square_{\mathcal{D}} (\mathcal{T}_{\mathcal{C}} \lozenge_{\mathcal{S}} \mathcal{M})$ induced by the maps $\mathcal{T}_{\mathcal{C}} \longrightarrow \mathcal{T}$ and $\mathcal{M} \longrightarrow \mathcal{T}_{\mathcal{C}} \lozenge_{\mathcal{S}} \mathcal{M}$ and the map $\mathcal{N} \square_{\mathcal{D}} \mathcal{T} \square_{\mathcal{D}} (\mathcal{T}_{\mathcal{C}} \lozenge_{\mathcal{S}} \mathcal{M}) \longrightarrow \mathcal{N} \square_{\mathcal{D}} (\mathcal{T}_{\mathcal{C}} \lozenge_{\mathcal{S}} \mathcal{M})$ induced by the left $\mathcal{T}$-semiaction in $\mathcal{T}_{\mathcal{C}} \lozenge_{\mathcal{S}} \mathcal{M}$. To check this, one can notice that the diagram in question is obtained by taking the cotensor product with $\mathcal{N}$ of the diagram of maps $\mathcal{T}_{\mathcal{C}} \square_{\mathcal{C}} \mathcal{M} \longrightarrow \mathcal{T} \square_{\mathcal{D}} (\mathcal{T}_{\mathcal{C}} \lozenge_{\mathcal{S}} \mathcal{M}) \longrightarrow \mathcal{T}_{\mathcal{C}} \lozenge_{\mathcal{S}} \mathcal{M}$ and compose the latter diagram with the surjective map $\mathcal{T}_{\mathcal{C}} \square_{\mathcal{C}} \mathcal{S} \square_{\mathcal{C}} \mathcal{M} \longrightarrow \mathcal{T}_{\mathcal{C}} \square_{\mathcal{C}} \mathcal{M}$ induced by the left $\mathcal{S}$-semiaction in $\mathcal{M}$. On the other hand, the composition of maps $\mathcal{N} \square_{\mathcal{D}} \mathcal{T}_{\mathcal{C}} \square_{\mathcal{C}} \mathcal{M} \longrightarrow (\mathcal{N} \lozenge_{\mathcal{T}} \mathcal{T}_{\mathcal{C}}) \lozenge_{\mathcal{S}} \mathcal{M} \longrightarrow \mathcal{N} \square_{\mathcal{D}} (\mathcal{T}_{\mathcal{C}} \lozenge_{\mathcal{S}} \mathcal{M})$ is equal to the composition of the same map $\mathcal{N} \square_{\mathcal{D}} \mathcal{T}_{\mathcal{C}} \square_{\mathcal{C}} \mathcal{M} \longrightarrow \mathcal{N} \square_{\mathcal{D}} \mathcal{T} \square_{\mathcal{D}} (\mathcal{T}_{\mathcal{C}} \lozenge_{\mathcal{S}} \mathcal{M})$ and the map $\mathcal{N} \square_{\mathcal{D}} \mathcal{T} \square_{\mathcal{D}} (\mathcal{T}_{\mathcal{C}} \lozenge_{\mathcal{S}} \mathcal{M}) \longrightarrow \mathcal{N} \square_{\mathcal{D}} (\mathcal{T}_{\mathcal{C}} \lozenge_{\mathcal{S}} \mathcal{M})$

induced by the right $\mathcal{T}$-semiaction in $\mathcal{N}$, since both compositions are equal to the composition of the map $\mathcal{N} \square_{\mathcal{D}} \mathcal{T}_{\mathfrak{C}} \square_{\mathfrak{C}} \mathcal{M} \longrightarrow \mathcal{N} \square_{\mathfrak{C}} \mathcal{M}$ induced by the composition $\mathcal{N} \square_{\mathcal{D}} \mathcal{T}_{\mathfrak{C}} \longrightarrow \mathcal{N} \lozenge_{\mathcal{T}} \mathcal{T}_{\mathfrak{C}} \longrightarrow \mathcal{N}$ with the map $\mathcal{N} \square_{\mathfrak{C}} \mathcal{M} \longrightarrow \mathcal{N} \square_{\mathcal{D}} (\mathcal{T}_{\mathfrak{C}} \lozenge_{\mathbf{S}} \mathcal{M})$ induced by the map $\mathcal{M} \longrightarrow \mathcal{T}_{\mathfrak{C}} \lozenge_{\mathbf{S}} \mathcal{M}$. It remains to apply Proposition 1.4.4. The proofs of parts (b) and (c) are completely analogous. $\qquad\square$

8.1.4 Let $\mathbf{S} \longrightarrow \mathcal{T}$ be a map of semialgebras compatible with a map of corings $\mathfrak{C} \longrightarrow \mathcal{D}$ and a k-algebra map $A \longrightarrow B$.

Assume that $\mathfrak{C}$ is a projective left and a flat right A-module, $\mathbf{S}$ is a coprojective left and a coflat right $\mathfrak{C}$-comodule, $\mathcal{D}$ is a projective left and a flat right B-module, and $\mathcal{T}$ is a coprojective left and a coflat right $\mathfrak{C}$-comodule. Then for any left $\mathcal{T}$-semicontramodule $\mathfrak{Q}$ the natural map of $\mathfrak{C}$-comodules $\Phi_{\mathfrak{C}}(^{\mathfrak{C}}\mathfrak{Q}) \longrightarrow {}_{\mathfrak{C}}(\Phi_{\mathcal{D}}\mathfrak{Q})$ is an $\mathbf{S}$-semimodule morphism

$$\Phi_{\mathbf{S}}(^{\mathfrak{C}}\mathfrak{Q}) \longrightarrow {}_{\mathfrak{C}}(\Phi_{\mathcal{T}}\mathfrak{Q}).$$

Indeed, $\Phi_{\mathbf{S}}(^{\mathfrak{C}}\mathfrak{Q}) = \mathbf{S} \odot_{\mathbf{S}} {}^{\mathfrak{C}}\mathfrak{Q} \simeq \mathbf{S}_{\mathcal{T}} \odot_{\mathcal{T}} \mathfrak{Q} \simeq {}_{\mathfrak{C}}\mathcal{T} \odot_{\mathcal{T}} \mathfrak{Q}$ as a left $\mathbf{S}$-semimodule and ${}_{\mathfrak{C}}(\Phi_{\mathcal{T}}\mathfrak{Q}) = {}_{\mathfrak{C}}(\mathcal{T} \odot_{\mathcal{T}} \mathfrak{Q})$, so there is an $\mathbf{S}$-semimodule morphism $\Phi_{\mathbf{S}}(^{\mathfrak{C}}\mathfrak{Q}) \longrightarrow {}_{\mathfrak{C}}(\Phi_{\mathcal{T}}\mathfrak{Q})$; it coincides with the $\mathfrak{C}$-comodule morphism $\Phi_{\mathfrak{C}}(^{\mathfrak{C}}\mathfrak{Q}) \longrightarrow {}_{\mathfrak{C}}(\Phi_{\mathcal{D}}\mathfrak{Q})$ defined in 7.1.4. Analogously, for any left $\mathcal{T}$-semimodule $\mathcal{N}$ the natural map of $\mathfrak{C}$-contramodules ${}^{\mathfrak{C}}(\Psi_{\mathcal{D}}\mathcal{N}) \longrightarrow \Psi_{\mathfrak{C}}({}_{\mathfrak{C}}\mathcal{N})$ is an $\mathbf{S}$-semicontramodule morphism

$${}^{\mathfrak{C}}(\Psi_{\mathcal{T}}\mathcal{N}) \longrightarrow \Psi_{\mathbf{S}}({}_{\mathfrak{C}}\mathcal{N}).$$

Indeed, $\Psi_{\mathbf{S}}({}_{\mathfrak{C}}\mathcal{N}) = \operatorname{Hom}_{\mathbf{S}}(\mathbf{S}, {}_{\mathfrak{C}}\mathcal{N}) \simeq \operatorname{Hom}_{\mathcal{T}}(_{\mathcal{T}}\mathbf{S}, \mathcal{N}) \simeq \operatorname{Hom}_{\mathcal{T}}(\mathcal{T}_{\mathfrak{C}}, \mathcal{N})$ as a left $\mathbf{S}$-semicontramodule and ${}^{\mathfrak{C}}(\Psi_{\mathcal{T}}\mathcal{N}) = {}^{\mathfrak{C}}\operatorname{Hom}_{\mathcal{T}}(\mathcal{T}, \mathcal{N})$.

Assume that $\mathfrak{C}$ is a projective left A-module, $\mathbf{S}$ is a coprojective left $\mathfrak{C}$-comodule, $\mathcal{D}$ is a projective left B-module, $\mathcal{T}$ is a coprojective left $\mathcal{D}$-comodule, and B is a projective left A-module. Then the equivalence of categories of $\mathfrak{C}$-coprojective left $\mathbf{S}$-semimodules and $\mathfrak{C}$-projective left $\mathbf{S}$-semicontramodules and the equivalence of categories of $\mathcal{D}$-coprojective left $\mathcal{T}$-semimodules and $\mathcal{D}$-projective left $\mathbf{S}$-semicontramodules transform the functor $\mathcal{N} \longmapsto {}_{\mathfrak{C}}\mathcal{N}$ into the functor $\mathfrak{Q} \longmapsto {}^{\mathfrak{C}}\mathfrak{Q}$. Indeed, the above argument shows that for any $\mathcal{D}$-projective left $\mathcal{T}$-semicontramodule $\mathfrak{Q}$ the isomorphism $\Phi_{\mathfrak{C}}(^{\mathfrak{C}}\mathfrak{Q}) \simeq {}_{\mathfrak{C}}(\Phi_{\mathcal{D}}\mathfrak{Q})$ preserves the $\mathbf{S}$-semimodule structures.

Assume that $\mathfrak{C}$ is a flat right A-module, $\mathbf{S}$ is a coflat right $\mathfrak{C}$-comodule, $\mathcal{D}$ is a flat right B-module, $\mathcal{T}$ is a coflat right $\mathcal{D}$-comodule, and B is a flat right A-module. Then the equivalence of categories of $\mathfrak{C}$-injective left $\mathbf{S}$-semimodules and $\mathfrak{C}$-coinjective left $\mathbf{S}$-semicontramodules and the equivalence of categories of $\mathcal{D}$-injective left $\mathcal{T}$-semimodules and $\mathcal{D}$-coinjective left $\mathbf{S}$-semicontramodules transform the functor $\mathcal{N} \longmapsto {}_{\mathfrak{C}}\mathcal{N}$ into the functor $\mathfrak{Q} \longmapsto {}^{\mathfrak{C}}\mathfrak{Q}$. Indeed, the above argument shows that for any $\mathcal{D}$-injective left $\mathcal{T}$-semimodule $\mathcal{N}$ the isomorphism ${}^{\mathfrak{C}}(\Psi_{\mathcal{D}}\mathcal{N}) \simeq \Psi_{\mathfrak{C}}({}_{\mathfrak{C}}\mathcal{N})$ preserves the $\mathbf{S}$-semicontramodule structures.

Assume that $\mathfrak{C}$ is a projective left and a flat right A-module, $\mathbf{S}$ is a coprojective left $\mathfrak{C}$-comodule and a flat right A-module, $\mathcal{D}$ is a projective left and a flat right B-module, $\mathcal{T}$ is a coprojective left $\mathcal{D}$-comodule and a flat right B-module, and the rings A and B have finite left homological dimensions. Then the equivalence of categories of $\mathcal{C}/A$-injective left $\mathcal{S}$-semimodules and $\mathcal{C}/A$-projective left $\mathcal{S}$-semicontramodules and the equivalence of categories of $\mathcal{D}/B$-injective left $\mathcal{T}$-semimodules and $\mathcal{D}/B$-projective left $\mathcal{T}$-semicontramodules transform the functor $\mathcal{N} \longmapsto {}_{\mathbb{e}}\mathcal{N}$ into the functor $\mathfrak{Q} \longmapsto {}^{\mathbb{e}}\mathfrak{Q}$. Indeed, the above argument shows that for any $\mathcal{D}/B$-projective left $\mathcal{T}$-semicontramodule $\mathfrak{Q}$ the isomorphism $\Phi_{\mathbb{e}}({}^{\mathbb{e}}\mathfrak{Q}) \simeq {}_{\mathbb{e}}(\Phi_{\mathcal{D}}\mathfrak{Q})$ preserves the $\mathcal{S}$-semimodule structures. The analogous result holds when $\mathcal{S}$ is a projective left A-module and a coflat right $\mathcal{C}$-comodule and $\mathcal{T}$ is a projective left B-module and a coflat right $\mathcal{D}$-comodule; it can be proven by applying the above argument to the isomorphism ${}^{\mathbb{e}}(\Psi_{\mathcal{D}}\mathcal{N}) \simeq \Psi_{\mathbb{e}}({}_{\mathbb{e}}\mathcal{N})$ for a $\mathcal{D}/B$-injective left $\mathcal{T}$-semimodule $\mathcal{N}$.

Finally, assume that the rings A and B are semisimple. Then the equivalence of categories of $\mathcal{C}$-injective left $\mathcal{S}$-semimodules and $\mathcal{C}$-projective left $\mathcal{S}$-semicontramodules and the equivalence of categories of $\mathcal{D}$-injective left $\mathcal{T}$-semimodules and $\mathcal{D}$-projective left $\mathcal{T}$-semicontramodules transform the functor $\mathcal{N} \longmapsto {}_{\mathbb{e}}\mathcal{N}$ into the functor $\mathfrak{Q} \longmapsto {}^{\mathbb{e}}\mathfrak{Q}$. One can show this using the semialgebra analogues of the assertions of 7.1.2 related to quasicoflat comodules and quasicoinjective contramodules.

8.2   Complexes, adjusted to pull-backs and push-forwards

Let $\mathcal{S} \longrightarrow \mathcal{T}$ be a map of semialgebras compatible with a map of corings $\mathcal{C} \longrightarrow \mathcal{D}$ and a k-algebra map $A \longrightarrow B$. The following result generalizes Theorem 6.3.

Theorem 1.

(a) *Assume that $\mathcal{D}$ is a flat right B-module, $\mathcal{T}$ is a coflat right $\mathcal{D}$-comodule and a $\mathcal{D}/B$-coflat ($\mathcal{D}/B$-coprojective) left $\mathcal{D}$-comodule, and the ring B has a finite weak (left) homological dimension. Then the functor mapping the quotient category of the homotopy category of complexes of $\mathcal{D}/B$-coflat ($\mathcal{D}/B$-coprojective) left $\mathcal{T}$-semimodules by its intersection with the thick subcategory of $\mathcal{D}$-coacyclic complexes into the semiderived category of left $\mathcal{T}$-semimodules is an equivalence of triangulated categories.*

(b) *Assume that $\mathcal{D}$ is a projective left B-module, $\mathcal{T}$ is a coprojective left and a $\mathcal{D}/B$-coflat right $\mathcal{D}$-comodule, and the ring B has a finite left homological dimension. Then the functor mapping the quotient category of the homotopy category of complexes of $\mathcal{D}/B$-coinjective left $\mathcal{T}$-semicontramodules by its intersection with the thick subcategory of $\mathcal{D}$-contraacyclic complexes into the semiderived category of left $\mathcal{T}$-semicontramodules is an equivalence of triangulated categories.*

Proof. To prove part (a) for $\mathcal{D}/B$-coflat $\mathcal{T}$-semimodules, use Lemma 1.3.3, the construction of the morphism of complexes $\mathcal{L}^{\bullet} \longrightarrow \mathbb{R}_2(\mathcal{L}^{\bullet})$ from the proof of Theorem 2.6, and Lemma 2.6. To prove part (a) for $\mathcal{D}/B$-coprojective $\mathcal{T}$-semimodules, use Lemma 3.3.3(b). To prove part (b), use Lemma 3.3.3(a) and the construction of the morphism of complexes $\mathbb{L}_2(\mathfrak{R}^\bullet) \longrightarrow \mathfrak{R}^\bullet$ from the proof of Theorem 4.6. □

A complex of $\mathbf{S}$-semimodules is called *quite $\mathbf{S}/\mathcal{C}/A$-semiflat (quite $\mathbf{S}/\mathcal{C}/A$-semiprojective)* if it belongs to the minimal triangulated subcategory of the homotopy category of complexes of $\mathbf{S}$-semimodules containing the complexes induced from complexes of A-flat (A-projective) $\mathcal{C}$-comodules and closed under infinite direct sums. Analogously, a complex of $\mathbf{S}$-semicontramodules is called *quite $\mathbf{S}/\mathcal{C}/A$-semiinjective* if it belongs to the minimal triangulated subcategory of the homotopy category of complexes of $\mathbf{S}$-semicontramodules containing the complexes coinduced from complexes of A-injective $\mathcal{C}$-contramodules and closed under infinite products. Under appropriate assumptions on $\mathbf{S}$, $\mathcal{C}$, and A, any quite $\mathbf{S}/\mathcal{C}/A$-semiflat complex of A-flat $\mathbf{S}$-semimodules is $\mathbf{S}/\mathcal{C}/A$-semiflat in the sense of 2.8, and analogously for birelative semiprojectivity and semiinjectivity in the sense of 4.8. Any quite $\mathbf{S}/\mathcal{C}/A$-semiflat complex of right $\mathbf{S}$-semimodules is $\mathbf{S}/\mathcal{C}/A$-contraflat, any quite $\mathbf{S}/\mathcal{C}/A$-semiprojective complex of left $\mathbf{S}$-semimodules is $\mathbf{S}/\mathcal{C}/A$-projective, and any quite $\mathbf{S}/\mathcal{C}/A$-semiinjective complex of left $\mathbf{S}$-semicontramodules is $\mathbf{S}/\mathcal{C}/A$-injective in the sense of 6.4.

Theorem 2.

(a) *Assume that $\mathcal{C}$ is a flat (projective) left and a flat right A-module, $\mathcal{C}$ is a flat (projective) left A-module and a coflat right $\mathcal{C}$-comodule, and the ring A has a finite weak (left) homological dimension. Then the functor mapping the quotient category of the homotopy category of quite $\mathbf{S}/\mathcal{C}/A$-semiflat (quite $\mathbf{S}/\mathcal{C}/A$-semiprojective) complexes of left $\mathbf{S}$-semimodules by its minimal triangulated subcategory containing complexes induced from coacyclic complexes of A-flat (A-projective) $\mathbf{S}$-semimodules and closed under infinite direct sums into the semiderived category of left $\mathbf{S}$-semimodules is an equivalence of triangulated categories.*

(b) *Assume that $\mathcal{C}$ is a projective left and a flat right A-module, $\mathcal{C}$ is a coprojective left $\mathcal{C}$-comodule and a flat right A-module, and the ring A has a finite left homological dimension. Then the functor mapping the quotient category of the homotopy category of quite $\mathbf{S}/\mathcal{C}/A$-semiinjective complexes of left $\mathbf{S}$-semicontramodules by its minimal triangulated subcategory containing complexes of coinduced from contraacyclic complexes of A-injective $\mathcal{C}$-contramodules and closed under infinite products into the semiderived category of left $\mathbf{S}$-semicontramodules is an equivalence of triangulated categories.*

Proof. Proof of part (a): for any complex of $\mathbf{S}$-semimodules $\mathcal{K}^\bullet$ there is a natural morphism into $\mathcal{K}^\bullet$ from a quite $\mathbf{S}/\mathcal{C}/A$-semiflat complex of $\mathbf{S}$-semimodules $\mathbb{L}_3\mathbb{L}_1(\mathcal{K}^\bullet)$ with a $\mathcal{C}$-coacyclic cone. Hence it follows from Lemma 2.6 that the semiderived category of $\mathbf{S}$-semimodules is equivalent to the quotient category of the homotopy category of quite $\mathbf{S}/\mathcal{C}/A$-semiflat complexes of $\mathbf{S}$-semimodules by its intersection with the thick subcategory of $\mathcal{C}$-coacyclic complexes. It remains

to show that any C-coacyclic quite $S/C/A$-semiflat complex of S-semimodules belongs to the minimal triangulated subcategory containing the complexes induced
from coacyclic complexes of A-flat S-semimodules and closed under infinite direct
sums. Indeed, if a complex of A-flat left S-semimodules $\mathcal{M}^\bullet$ is C-coacyclic, then
the total complex $\mathbb{L}_3(\mathcal{M}^\bullet)$ of the bar bicomplex $\cdots \longrightarrow S\square_C S\square_C \mathcal{M}^\bullet \longrightarrow S\square_C \mathcal{M}^\bullet$
up to the homotopy equivalence can be obtained from complexes of S-semimod-
ules induced from coacyclic complexes of A-flat C-comodules using the operations
of cone and infinite direct sum. So the same applies to a C-coacyclic complex of
S-semimodules $\mathcal{M}^\bullet$ homotopy equivalent to a complex of A-flat S-semimodules.
On the other hand, if a complex of S-semimodules $\mathcal{M}^\bullet$ is induced from a complex
of C-comodules, then the cone of the morphism of complexes $\mathbb{L}_3(\mathcal{M}^\bullet) \longrightarrow \mathcal{M}^\bullet$ is
a contractible complex of S-semimodules, since it is isomorphic to the cotensor
product over C of the bar complex $\cdots \longrightarrow S\square_C S\square_C S \longrightarrow S\square_C S \longrightarrow S$, which is
contractible as a complex of left S-semimodules with right C-comodule structures,
and a certain complex of left C-comodules. So the same applies to any complex
of S-semimodules $\mathcal{M}^\bullet$ that up to the homotopy equivalence can be obtained from
complexes of S-semimodules induced from complexes of C-comodules using the
operations of cone and infinite direct sum. Part (a) for quite $S/C/A$-semiflat complexes is proven; the proofs of part (a) for quite $S/C/A$-semiprojective complexes
and part (b) are completely analogous. $\square$

Theorem 3.

(a) *Assume that C is a flat right A-module, S is a coflat right C-comodule, $\mathcal{D}$ is
a flat right B-module, and $\mathcal{T}$ is a coflat right $\mathcal{D}$-comodule. Then the functor
$\mathcal{M}^\bullet \longmapsto {}_{\mathcal{T}}\mathcal{M}^\bullet$ maps quite $S/C/A$-semiflat (quite $S/C/A$-semiprojective) complexes of left S-semimodules to quite $\mathcal{T}/\mathcal{D}/B$-semiflat (quite $\mathcal{T}/\mathcal{D}/B$-semi-
projective) complexes of left $\mathcal{T}$-semimodules. Assume additionally that C and
S are flat left A-modules and the ring A has a finite weak homological dimension. Then the same functor maps C-coacyclic quite $S/C/A$-semiflat com-
plexes of left S-semimodules to $\mathcal{D}$-coacyclic complexes of left $\mathcal{T}$-semimodules.*

(b) *Assume that C is a projective left A-module, S is a coprojective left C-comod-
ule, $\mathcal{D}$ is a projective left B-module, and $\mathcal{T}$ is a coprojective left $\mathcal{D}$-comod-
ule. Then the functor $\mathfrak{P}^\bullet \longmapsto {}^{\mathcal{T}}\mathfrak{P}^\bullet$ maps quite $S/C/A$-semiinjective complexes of left S-semicontramodules to quite $\mathcal{T}/\mathcal{D}/B$-semiinjective complexes
of left $\mathcal{T}$-semicontramodules. Assume additionally that C and S are flat right
A-modules and the ring A has a finite left homological dimension. Then the
same functor maps C-contraacyclic quite $S/C/A$-semiinjective complexes of
left S-semicontramodules to $\mathcal{D}$-contraacyclic complexes of left $\mathcal{T}$-semicontra-
modules.*

(c) *Assume that C is a projective left and a flat right A-module, S is a coprojective
left and a coflat right C-comodule, A has a finite left homological dimension,
$\mathcal{D}$ is a projective left and a flat right B-module, $\mathcal{T}$ is a coprojective left and
a coflat right $\mathcal{D}$-comodule, and B has a finite left homological dimension.
Then the functor $\mathcal{M}^\bullet \longmapsto {}_{\mathcal{T}}\mathcal{M}^\bullet$ maps $S/C/A$-projective complexes of left*

S-semimodules to $\mathfrak{T}/\mathcal{D}/B$-projective complexes left $\mathfrak{T}$-semimodules and the functor $\mathfrak{P}^\bullet \longmapsto {}^{\mathfrak{T}}\mathfrak{P}^\bullet$ maps $S/\mathcal{C}/A$-injective complexes of left S-semicontramodules to $\mathfrak{T}/\mathcal{D}/B$-injective complexes of left $\mathfrak{T}$-semicontramodules. The same functors map $\mathcal{C}$-coacyclic $S/\mathcal{C}/A$-projective complexes of left S-semimodules to $\mathcal{D}$-coacyclic complexes of left $\mathfrak{T}$-semimodules and $\mathcal{C}$-contraacyclic $S/\mathcal{C}/A$-injective complexes of left S-semicontramodules to $\mathcal{D}$-contraacyclic complexes of left $\mathfrak{T}$-semicontramodules.

Proof. Part (a): the functor $\mathcal{M} \longmapsto {}_{\mathfrak{T}}\mathcal{M}$ maps the S-semimodule induced from a $\mathcal{C}$-comodule $\mathcal{L}$ to the $\mathfrak{T}$-semimodule induced from the $\mathcal{D}$-comodule ${}_B\mathcal{L}$. The first assertion follows immediately; to prove the second one, use Theorem 7.2.2(a) and Theorem 2(a). The proof of part (b) is completely analogous. Part (c): the first assertion follows from the adjointness of functors $\mathcal{M}^\bullet \longmapsto {}_{\mathfrak{T}}\mathcal{M}^\bullet$ and $\mathcal{N}^\bullet \longmapsto {}_{\mathcal{C}}\mathcal{N}^\bullet$, the adjointness of functors $\mathfrak{P}^\bullet \longmapsto {}^{\mathfrak{T}}\mathfrak{P}^\bullet$ and $\mathfrak{Q}^\bullet \longmapsto {}^{\mathcal{C}}\mathfrak{Q}^\bullet$, and the second assertions of Theorem 7.2.1(a) and (b). The second assertion follows from the first assertions of Theorem 7.2.1(a-b), because a complex of left S-semimodules $\mathcal{M}^\bullet$ is $S/\mathcal{C}/A$-projective and $\mathcal{C}$-coacyclic if and only if the complex $\mathrm{Hom}_S(\mathcal{M}^\bullet, \mathcal{L}^\bullet)$ is acyclic for all complexes of $\mathcal{C}/A$-injective left S-semimodules $\mathcal{L}^\bullet$, and a complex of left S-semicontramodules $\mathfrak{P}^\bullet$ is $S/\mathcal{C}/A$-injective and $\mathcal{C}$-contraacyclic if and only if the complex $\mathrm{Hom}^S(\mathcal{R}^\bullet, \mathfrak{P}^\bullet)$ is acyclic for all complexes of $\mathcal{C}/A$-projective left S-semicontramodules $\mathcal{R}^\bullet$ (and analogously for complexes of $\mathfrak{T}$-semimodules and $\mathfrak{T}$-semicontramodules). This follows from Theorem 6.3 and the results of 6.5, since a complex of S-semimodules is $\mathcal{C}$-coacyclic iff it represents a zero object of the semiderived category of S-semimodules, and a complex of S-semicontramodules is $\mathcal{C}$-contraacyclic iff it represents a zero object of the semiderived category of S-semicontramodules. $\qquad\square$

8.3 Derived functors of pull-back and push-forward

Let $S \longrightarrow \mathfrak{T}$ be a map of semialgebras compatible with a map of corings $\mathcal{C} \longrightarrow \mathcal{D}$ and a k-algebra map $A \longrightarrow B$.

Assume that $\mathcal{C}$ is a flat right A-module, S is a coflat right $\mathcal{C}$-comodule, $\mathcal{D}$ is a flat right B-module, $\mathfrak{T}$ is a coflat right $\mathcal{D}$-comodule and a $\mathcal{D}/B$-coflat left $\mathcal{D}$-comodule, and B has a finite weak homological dimension. The right derived functor

$$\mathcal{N}^\bullet \longmapsto {}^{\mathbb{R}}_{\mathcal{C}}\mathcal{N}^\bullet \colon \mathsf{D}^{\mathsf{si}}(\mathfrak{T}\text{-simod}) \longrightarrow \mathsf{D}^{\mathsf{si}}(S\text{-simod})$$

is defined by composing the functor $\mathcal{N}^\bullet \longmapsto {}_{\mathcal{C}}\mathcal{N}^\bullet$ acting from the homotopy category of left $\mathfrak{T}$-semimodules to the homotopy category of left S-semimodules with the localization functor $\mathsf{Hot}(S\text{-simod}) \longrightarrow \mathsf{D}^{\mathsf{si}}(S\text{-simod})$ and restricting it to the full subcategory of complexes of $\mathcal{D}/B$-coflat $\mathfrak{T}$-semimodules. By Theorems 8.2.1(a) and 7.2.1(a), this restriction factorizes through the semiderived category of left $\mathfrak{T}$-semimodules.

Assume that $\mathcal{C}$ is a flat left and right A-module, $\mathcal{S}$ is a flat left A-module and a coflat right $\mathcal{C}$-comodule, A has a finite weak homological dimension, $\mathcal{D}$ is a flat right B-module, and $\mathcal{T}$ is a coflat right $\mathcal{D}$-comodule. The left derived functor

$$\mathcal{M}^\bullet \longmapsto {}_{\mathcal{T}}^{\mathbb{L}}\mathcal{M}^\bullet : \mathsf{D}^{\mathrm{si}}(\mathcal{S}\text{-simod}) \longrightarrow \mathsf{D}^{\mathrm{si}}(\mathcal{T}\text{-simod})$$

is defined by composing the functor $\mathcal{M}^\bullet \longmapsto {}_{\mathcal{T}}\mathcal{M}^\bullet$ acting from the homotopy category of left $\mathcal{S}$-semimodules to the homotopy category of left $\mathcal{T}$-semimodules with the localization functor $\mathsf{Hot}(\mathcal{T}\text{-simod}) \longrightarrow \mathsf{D}^{\mathrm{si}}(\mathcal{T}\text{-simod})$ and restricting it to the full subcategory of quite $\mathcal{S}/\mathcal{C}/A$-semiflat complexes of $\mathcal{S}$-semimodules. By Theorems 8.2.2(a) and 8.2.3(a), this restriction factorizes through the semiderived category of left $\mathcal{S}$-semimodules.

Analogously, assume that $\mathcal{C}$ is a projective left A-module, $\mathcal{S}$ is a coprojective left $\mathcal{C}$-comodule, $\mathcal{D}$ is a projective left B-module, $\mathcal{T}$ is a coprojective left $\mathcal{D}$-comodule and a $\mathcal{D}/B$-coflat right $\mathcal{D}$-comodule, and B has a finite left homological dimension. The left derived functor

$$\mathfrak{Q}^\bullet \longmapsto {}_{\mathbb{L}}^{\mathcal{C}}\mathfrak{Q}^\bullet : \mathsf{D}^{\mathrm{si}}(\mathcal{T}\text{-sicntr}) \longrightarrow \mathsf{D}^{\mathrm{si}}(\mathcal{S}\text{-sicntr})$$

is defined by composing the functor $\mathfrak{Q}^\bullet \longmapsto {}^{\mathcal{C}}\mathfrak{Q}^\bullet$ with the localization functor $\mathsf{Hot}(\mathcal{S}\text{-sicntr}) \longrightarrow \mathsf{D}^{\mathrm{si}}(\mathcal{S}\text{-sicntr})$ and restricting it to the full subcategory of complexes of $\mathcal{D}/B$-coinjective $\mathcal{T}$-semicontramodules. By Theorems 8.2.1(b) and 7.2.1(b), this restriction factorizes through the semiderived category of left $\mathcal{T}$-semicontramodules. According to Lemma 6.5.2, this definition of a left derived functor does not depend on the choice of a subcategory of adjusted complexes.

Assume that $\mathcal{C}$ is a projective left and a flat right A-module, $\mathcal{S}$ is a coprojective left $\mathcal{C}$-comodule and a flat right A-module, A has a finite left homological dimension, $\mathcal{D}$ is a projective left B-module, and $\mathcal{T}$ is a coprojective left $\mathcal{D}$-comodule.

$$\mathfrak{P}^\bullet \longmapsto {}_{\mathcal{T}}^{\mathbb{R}}\mathfrak{P}^\bullet : \mathsf{D}^{\mathrm{si}}(\mathcal{S}\text{-sicntr}) \longrightarrow \mathsf{D}^{\mathrm{si}}(\mathcal{T}\text{-sicntr})$$

is defined by composing the functor $\mathfrak{P}^\bullet \longmapsto {}^{\mathcal{T}}\mathfrak{P}^\bullet$ with the localization functor $\mathsf{Hot}(\mathcal{T}\text{-simod}) \longrightarrow \mathsf{D}^{\mathrm{si}}(\mathcal{T}\text{-simod})$ and restricting it to the full subcategory of quite $\mathcal{S}/\mathcal{C}/A$-semiflatcomplexes of $\mathcal{S}$-semicontramodules. By Theorems 8.2.2(b) and 8.2.3(b), this restriction factorizes through the semiderived category of left $\mathcal{S}$-semicontramodules. According to Lemma 6.5.2, this definition of a right derived functor does not depend on the choice of a subcategory of adjusted complexes.

Notice that in the assumptions of Theorem 8.2.3(c) above and Corollary 1(c) below one can also define the left derived functor $\mathcal{M}^\bullet \longmapsto {}_{\mathcal{T}}^{\mathbb{L}}\mathcal{M}^\bullet$ in terms of $\mathcal{S}/\mathcal{C}/A$-projective complexes of left $\mathcal{S}$-semimodules and the right derived functor $\mathfrak{P}^\bullet \longmapsto {}_{\mathcal{T}}^{\mathbb{R}}\mathfrak{P}^\bullet$ in terms of $\mathcal{S}/\mathcal{C}/A$-injective complexes of left $\mathcal{S}$-semicontramodules.

The derived functors $\mathcal{N}^\bullet \longmapsto {}_{\mathcal{C}}^{\mathbb{R}}\mathcal{N}^\bullet$ and $\mathfrak{Q}^\bullet \longmapsto {}_{\mathbb{L}}^{\mathcal{C}}\mathfrak{Q}^\bullet$ in the categories of semimodules and semicontramodules agree with the derived functors $\mathcal{N}^\bullet \longmapsto {}_{\mathcal{C}}^{\mathbb{R}}\mathcal{N}^\bullet$ and $\mathfrak{Q}^\bullet \longmapsto {}_{\mathbb{L}}^{\mathcal{C}}\mathfrak{Q}^\bullet$ in the categories of comodules and contramodules, so our notation is not ambiguous.

Remark 1. Under the assumptions that $\mathcal{C}$ is a flat right A-module, $\mathcal{S}$ is a coflat right $\mathcal{C}$-comodule, $\mathcal{D}$ is a flat right B-module, $\mathcal{T}$ is a coflat right $\mathcal{D}$-comodule, and B has a finite left homological dimension, one can define the derived functor $\mathcal{N}^{\bullet} \longmapsto {}^{\mathbb{R}}_{\mathcal{C}}\mathcal{N}^{\bullet}$ in terms of injective complexes of left $\mathcal{T}$-semimodules (see Remark 6.5).

Corollary 1.

(a) *The derived functor* $\mathcal{M}^{\bullet} \longmapsto {}^{\mathbb{L}}_{\mathcal{T}}\mathcal{M}^{\bullet}$ *is left adjoint to the derived functor* $\mathcal{N}^{\bullet} \longmapsto {}^{\mathbb{R}}_{\mathcal{C}}\mathcal{N}^{\bullet}$ *whenever both functors are defined by the above construction.*

(b) *The derived functor* $\mathfrak{P}^{\bullet} \longmapsto {}^{\mathcal{T}}_{\mathbb{R}}\mathfrak{P}^{\bullet}$ *is right adjoint to the derived functor* $\mathfrak{Q}^{\bullet} \longmapsto {}^{\mathcal{C}}_{\mathbb{L}}\mathfrak{Q}^{\bullet}$ *whenever both functors are defined by the above construction.*

(c) *Assume that* $\mathcal{C}$ *is a projective left and a flat right A-module, $\mathcal{S}$ is a coprojective left and a coflat right $\mathcal{C}$-comodule, A has a finite left homological dimension, $\mathcal{D}$ is a projective left and a flat right B-module, $\mathcal{T}$ is a coprojective left and a coflat right $\mathcal{D}$-comodule, and B has a finite left homological dimension. Then for any objects $\mathcal{M}^{\bullet}$ in $\mathsf{D}^{\mathrm{si}}(\mathrm{simod}\text{-}\mathcal{S})$ and $\mathfrak{Q}^{\bullet}$ in $\mathsf{D}^{\mathrm{si}}(\mathcal{T}\text{-}\mathrm{sicntr})$ there is a natural isomorphism*

$$\mathrm{CtrTor}^{\mathcal{T}}(\mathcal{M}{}^{\bullet}_{\mathcal{T}}{}^{\mathbb{L}}, \mathfrak{Q}^{\bullet}) \simeq \mathrm{CtrTor}^{\mathcal{S}}(\mathcal{M}^{\bullet}, {}^{\mathcal{C}}_{\mathbb{L}}\mathfrak{Q}^{\bullet})$$

in the derived category of k-modules.

Proof. In the assumptions of part (c), one can prove somewhat stronger versions of the assertions (a) and (b): for any $\mathcal{M}^{\bullet}$ in $\mathsf{D}^{\mathrm{si}}(\mathcal{S}\text{-}\mathrm{simod})$ and $\mathcal{N}^{\bullet}$ in $\mathsf{D}^{\mathrm{si}}(\mathcal{T}\text{-}\mathrm{simod})$, there is a natural isomorphism $\mathrm{Ext}_{\mathcal{T}}({}^{\mathbb{L}}_{\mathcal{T}}\mathcal{M}, \mathcal{N}) \simeq \mathrm{Ext}_{\mathcal{S}}(\mathcal{M}, {}^{\mathbb{R}}_{\mathcal{C}}\mathcal{N})$ and for any $\mathfrak{P}^{\bullet}$ in $\mathsf{D}^{\mathrm{si}}(\mathcal{S}\text{-}\mathrm{sicntr})$ and $\mathfrak{Q}^{\bullet}$ in $\mathsf{D}^{\mathrm{si}}(\mathcal{T}\text{-}\mathrm{sicntr})$ there is a natural isomorphism $\mathrm{Ext}^{\mathcal{T}}(\mathfrak{Q}^{\bullet}, {}^{\mathcal{T}}_{\mathbb{R}}\mathfrak{P}^{\bullet}) \simeq \mathrm{Ext}^{\mathcal{S}}({}^{\mathcal{C}}_{\mathbb{L}}\mathfrak{Q}^{\bullet}, \mathfrak{P}^{\bullet})$ in the derived category of k-modules. To obtain the first isomorphism, it suffices to represent the object $\mathcal{M}^{\bullet}$ by an $\mathcal{S}/\mathcal{C}/A$-projective complex of left $\mathcal{S}$-semimodules and the object $\mathcal{N}^{\bullet}$ by a complex of $\mathcal{D}/B$-injective left $\mathcal{T}$-semimodules, and use Lemma 5.3.2(a), Theorem 7.2.1(a), and Theorem 8.2.3(c). In the second case, one can represent the object $\mathfrak{P}^{\bullet}$ by an $\mathcal{S}/\mathcal{C}/A$-injective complex of left $\mathcal{S}$-semicontramodules and the object $\mathfrak{Q}^{\bullet}$ by a complex of $\mathcal{D}/B$-projective left $\mathcal{T}$-semicontramodules, and use Lemma 5.3.2(b), Theorem 7.2.1(b), and Theorem 8.2.3(c). To verify part (c), it suffices to represent the object $\mathcal{M}^{\bullet}$ by a quite $\mathcal{S}/\mathcal{C}/A$-semiflat complex of right $\mathcal{S}$-semimodules and the object $\mathfrak{Q}^{\bullet}$ by a complex of $\mathcal{D}/B$-projective left $\mathcal{S}$-semicontramodules, and use Lemma 5.3.2(b), Theorem 7.2.1(b), and Theorem 8.2.3(a). Finally, parts (a) and (b) in their weaker assumptions follow from the next lemma. $\square$

Lemma. *Let* H_1 *and* H_2 *be categories,* S_1 *and* S_2 *be localizing classes of morphisms in* H_1 *and* H_2, *and* F_1 *and* F_2 *be full subcategories in* H_1 *and* H_2. *Assume that for any object* $X \in \mathsf{H}_1$ *there exists an object* $U \in \mathsf{F}_1$ *together with a morphism* $U \longrightarrow X$ *from* S_1 *and for any object* $Y \in \mathsf{H}_2$ *there exists an object* $V \in \mathsf{F}_2$ *together with a morphism* $Y \longrightarrow V$ *from* S_2. *Let* $\Sigma \colon \mathsf{H}_1 \longrightarrow \mathsf{H}_2$ *be a functor and* $\Pi \colon \mathsf{H}_2 \longrightarrow \mathsf{H}_1$ *be a functor right adjoint to* Σ. *Assume that the morphism* $\Sigma(t)$ *belongs to* S_2 *for any morphism* $t \in \mathsf{F}_1 \cap \mathsf{S}_1$ *and the morphism* $\Pi(s)$ *belongs to* S_1 *for any morphism* $s \in \mathsf{F}_2 \cap \mathsf{S}_2$. *Then the right derived functor* $\mathbb{R}\Pi \colon \mathsf{H}_2[\mathsf{S}_2^{-1}] \longrightarrow$

$H_1[S_1^{-1}]$ *defined by restricting* Π *to* F_2 *is right adjoint to the left derived functor* $\mathbb{L}\Sigma\colon H_1[S_1^{-1}] \longrightarrow H_2[S_2^{-1}]$ *defined by restricting* Σ *to* F_1.

Proof. The functors $F_i[(F_i \cap S_i)^{-1}] \longrightarrow H_i[S_i^{-1}]$ are equivalences of categories by Lemma 2.6, so the derived functors $\mathbb{L}\Sigma$ and $\mathbb{R}\Pi$ can be defined. For any objects $U \in F_1$ and $V \in F_2$ we have to construct a bijection between the sets $\mathrm{Hom}_{H_1[S_1^{-1}]}(U, \Pi V)$ and $\mathrm{Hom}_{H_2[S_2^{-1}]}(\Sigma U, V)$, functorial in U and V. Any element of the first set can be represented by a fraction $U \leftarrow U' \to \Pi V$ in H_1 with the morphism $U' \longrightarrow U$ belonging to S_1. By assumption, one can choose U' to be an object of F_1. Assign to this fraction the element of the second set represented by the fraction $\Sigma U \leftarrow \Sigma U' \to V$. By assumption, the morphism $\Sigma U' \longrightarrow \Sigma U$ belongs to S_2. Analogously, any element of the second set can be represented by a fraction $\Sigma U \to V' \leftarrow V$ in H_2 with the morphism $V \longrightarrow V'$ belonging to S_2, and one can choose V' to be an object of F_2. Assign to this fraction the element of the first set represented by the fraction $U \to \Pi V' \leftarrow \Pi V$. The compositions of these two maps between sets of morphisms are identities, since the square formed by the morphisms $U' \longrightarrow U$, $U \longrightarrow \Pi V'$, $U' \longrightarrow \Pi V$, and $\Pi V \longrightarrow \Pi V'$ and the square formed by the morphisms $\Sigma U' \longrightarrow \Sigma U$, $\Sigma U \longrightarrow V'$, $\Sigma U' \longrightarrow V$, and $V \longrightarrow V'$ are commutative simultaneously. $\square$

Let $\mathcal{R}$ be a semialgebra over a coring $\mathcal{E}$ over a k-algebra F, and $\mathcal{T} \longrightarrow \mathcal{R}$ be a map of semialgebras compatible with a map of corings $\mathcal{D} \longrightarrow \mathcal{E}$ and a k-algebra map $B \quad \rangle \ F$. Then the composition provides a map of semialgebras $\mathcal{S} \longrightarrow \mathcal{R}$ compatible with a map of corings $\mathcal{C} \longrightarrow \mathcal{E}$ and a k-algebra map $A \longrightarrow B$.

Corollary 2.

(a) *There is a natural isomorphism* ${}_{\mathcal{C}}^{\mathcal{R}}({}_{\mathcal{D}}^{\mathcal{R}}\mathcal{L}^{\bullet}) \simeq {}_{\mathcal{C}}^{\mathcal{R}}\mathcal{L}^{\bullet}$ *for any object* $\mathcal{L}^{\bullet}$ *in* $\mathsf{D}^{\mathrm{si}}(\mathcal{R}\text{-simod})$ *whenever both functors* $\mathcal{L}^{\bullet} \longmapsto {}_{\mathcal{D}}^{\mathcal{R}}\mathcal{L}^{\bullet}$ *and* $\mathcal{N}^{\bullet} \longmapsto {}_{\mathcal{C}}^{\mathcal{R}}\mathcal{N}^{\bullet}$ *are defined by the above construction.*

(b) *There is a natural isomorphism* ${}_{\mathcal{R}}^{\mathrm{L}}({}_{\mathcal{T}}^{\mathrm{L}}\mathcal{M}^{\bullet}) \simeq {}_{\mathcal{R}}^{\mathrm{L}}\mathcal{M}^{\bullet}$ *for any object* $\mathcal{M}^{\bullet}$ *in* $\mathsf{D}^{\mathrm{si}}(\mathcal{S}\text{-simod})$ *whenever both functors* $\mathcal{M}^{\bullet} \longmapsto {}_{\mathcal{T}}^{\mathrm{L}}\mathcal{M}^{\bullet}$ *and* $\mathcal{N}^{\bullet} \longmapsto {}_{\mathcal{R}}^{\mathrm{L}}\mathcal{N}^{\bullet}$ *are defined by the above construction.*

(c) *There is a natural isomorphism* ${}_{\mathrm{L}}^{\mathcal{C}}({}_{\mathrm{L}}^{\mathcal{D}}\mathcal{K}^{\bullet}) \simeq {}_{\mathrm{L}}^{\mathcal{C}}\mathcal{K}^{\bullet}$ *for any object* $\mathcal{K}^{\bullet}$ *in* $\mathsf{D}^{\mathrm{si}}(\mathcal{R}\text{-sicntr})$ *whenever both functors* $\mathcal{K}^{\bullet} \longmapsto {}_{\mathrm{L}}^{\mathcal{D}}\mathcal{K}^{\bullet}$ *and* $\mathcal{Q}^{\bullet} \longmapsto {}_{\mathrm{L}}^{\mathcal{C}}\mathcal{Q}^{\bullet}$ *are defined by the above construction.*

(d) *There is a natural isomorphism* ${}_{\mathcal{R}}^{\mathcal{R}}({}_{\mathcal{R}}^{\mathcal{T}}\mathcal{P}^{\bullet}) \simeq {}_{\mathcal{R}}^{\mathcal{R}}\mathcal{P}^{\bullet}$ *for any object* $\mathcal{P}^{\bullet}$ *in* $\mathsf{D}^{\mathrm{si}}(\mathcal{S}\text{-sicntr})$ *whenever both functors* $\mathcal{P}^{\bullet} \longmapsto {}_{\mathcal{R}}^{\mathcal{T}}\mathcal{P}^{\bullet}$ *and* $\mathcal{Q}^{\bullet} \longmapsto {}_{\mathcal{R}}^{\mathcal{R}}\mathcal{Q}^{\bullet}$ *are defined by the above construction.*

Proof. Part (a) follows from the first assertion of Theorem 7.2.1(a), part (b) follows from the first assertion of Theorem 8.2.3(a), part (c) follows from the first assertion of Theorem 7.2.1(b), part (d) follows from the first assertion of Theorem 8.2.3(b). $\square$

Recall that a complex of $\mathcal{C}$-coflat right $\mathcal{S}$-semimodules is called quite semi-flat if it belongs to the minimal triangulated subcategory of the homotopy category of right $\mathcal{S}$-semimodules containing the complexes of $\mathcal{S}$-semimodules induced from complexes of coflat right $\mathcal{C}$-comodules and closed under infinite direct sums (see 2.9). This definition presumes that $\mathcal{C}$ is a flat right A-module and $\mathcal{S}$ is a coflat right $\mathcal{C}$-comodule.

Corollary 3.

(a) *Assume that $\mathcal{C}$ is a flat left and right A-module, $\mathcal{S}$ is a coflat left and right $\mathcal{C}$-comodule, A has a finite weak homological dimension, $\mathcal{D}$ is a flat left and right B-module, $\mathcal{T}$ is a coflat left and right $\mathcal{D}$-comodule, and B has a finite weak homological dimension. Then for any objects $\mathcal{M}^\bullet$ in $\mathsf{D}^{\mathrm{si}}(\mathcal{S}\text{-simod})$ and $\mathcal{N}^\bullet$ in $\mathsf{D}^{\mathrm{si}}(\mathrm{simod}\text{-}\mathcal{T})$ there is a natural isomorphism*

$$\mathrm{SemiTor}^{\mathcal{T}}(\mathcal{N}^\bullet, \tfrac{\mathsf{L}}{\mathcal{T}}\mathcal{M}^\bullet) \simeq \mathrm{SemiTor}^{\mathcal{S}}(\mathcal{N}^{\mathbb{R}}_{\mathcal{C}}, \mathcal{M}^\bullet)$$

in the derived category of k-modules.

(b) *Under the assumptions of Corollary 1(c), for any objects $\mathfrak{P}^\bullet$ in $\mathsf{D}^{\mathrm{si}}(\mathcal{S}\text{-sicntr})$ and $\mathcal{N}^\bullet$ in $\mathsf{D}^{\mathrm{si}}(\mathcal{T}\text{-simod})$ there is a natural isomorphism*

$$\mathrm{SemiExt}_{\mathcal{T}}(\mathcal{N}^\bullet, \tfrac{\mathcal{T}}{\mathbb{R}}\mathfrak{P}^\bullet) \simeq \mathrm{SemiExt}_{\mathcal{S}}(\tfrac{\mathbb{R}}{\mathcal{C}}\mathcal{N}^\bullet, \mathfrak{P}^\bullet)$$

in the derived category of k-modules.

(c) *Under the assumptions of Corollary 1(c), for any objects $\mathcal{M}^\bullet$ in $\mathsf{D}^{\mathrm{si}}(\mathcal{S}\text{-simod})$ and $\mathfrak{Q}^\bullet$ in $\mathsf{D}^{\mathrm{si}}(\mathcal{T}\text{-sicntr})$ there is a natural isomorphism*

$$\mathrm{SemiExt}_{\mathcal{T}}(\tfrac{\mathsf{L}}{\mathcal{T}}\mathcal{M}^\bullet, \mathfrak{Q}^\bullet) \simeq \mathrm{SemiExt}_{\mathcal{S}}(\mathcal{M}^\bullet, \tfrac{\mathcal{C}}{\mathsf{L}}\mathfrak{Q}^\bullet)$$

in the derived category of k-modules.

Proof. Part (a): represent the object $\mathcal{M}^\bullet$ by a quite semiflat complex of $\mathcal{S}$-semimodules and the object $\mathcal{N}^\bullet$ by a semiflat complex of $\mathcal{D}$-coflat $\mathcal{T}$-semimodules, and use the second case of Proposition 8.1.3(a). Alternatively, represent $\mathcal{M}^\bullet$ by a quite $\mathcal{S}/\mathcal{C}/A$-semiflat complex of A-flat $\mathcal{S}$-semimodules and $\mathcal{N}^\bullet$ by a complex of $\mathcal{D}$-coflat $\mathcal{T}$-semimodules, and use Theorem 7.2.1(a), Theorem 8.2.3(a), the result of 2.8, and the first case of Proposition 8.1.3(a); or represent $\mathcal{M}^\bullet$ by a quite semiflat complex of semiflat $\mathcal{S}$-semimodules and $\mathcal{N}^\bullet$ by a complex of $\mathcal{D}/B$-coflat $\mathcal{T}$-semimodules, and use the same theorems, the result of 2.8, and the fourth case of Proposition 8.1.3(a).

Part (b): represent the object $\mathfrak{P}^\bullet$ by a semiinjective complex of $\mathcal{C}$-coinjective $\mathcal{S}$-semicontramodules (having in mind Lemma 6.4(c) or Remark 6.4) and the object $\mathcal{N}^\bullet$ by a semiprojective complex of $\mathcal{D}$-coprojective $\mathcal{T}$-semimodules, and use the second case of Proposition 8.1.3(b). Alternatively, represent $\mathfrak{P}^\bullet$ by a quite $\mathcal{S}/\mathcal{C}/A$-semiinjective complex of A-injective $\mathcal{S}$-semicontramodules and $\mathcal{N}^\bullet$

by a complex of $\mathcal{D}$-coprojective $\mathcal{T}$-semimodules, and use Theorem 7.2.1(a), Theorem 8.2.3(b), the result of 4.8, and the first case of Proposition 8.1.3(b); or represent $\mathfrak{P}^{\bullet}$ by a semiinjective complex of semiinjective $\mathcal{S}$-semicontramodules and $\mathcal{N}^{\bullet}$ by a complex of $\mathcal{D}/B$-coprojective $\mathcal{T}$-semimodules, and use the same theorems, the result of 4.8, and the fourth case of Proposition 8.1.3(b).

Part (c): represent the object $\mathcal{M}^{\bullet}$ by a semiprojective complex of $\mathcal{C}$-coprojective $\mathcal{S}$-semicontramodules (having in mind Lemma 6.4(b) or Remark 6.4) and the object $\mathfrak{Q}^{\bullet}$ by a semiinjective complex of $\mathcal{D}$-coinjective $\mathcal{T}$-semicontramodules, and use the second case of Proposition 8.1.3(c). Alternatively, represent $\mathcal{M}^{\bullet}$ by a quite $\mathcal{S}/\mathcal{C}/A$-semiprojective complex of A-projective $\mathcal{S}$-semimodules and $\mathfrak{Q}^{\bullet}$ by a complex of $\mathcal{D}$-coinjective $\mathcal{T}$-semicontramodules, and use Theorem 7.2.1(b), Theorem 8.2.3(a), the result of 4.8, and the first case of Proposition 8.1.3(c); or represent $\mathcal{M}^{\bullet}$ by a semiprojective complex of semiprojective $\mathcal{S}$-semimodules and $\mathfrak{Q}^{\bullet}$ by a complex of $\mathcal{D}/B$-coinjective $\mathcal{T}$-semicontramodules, and use the same theorems, the result of 4.8, and the fourth case of Proposition 8.1.3(c). $\square$

Remark 2. Suppose that two objects $'\mathcal{M}^{\bullet}$ in $\mathsf{D}^{\mathsf{si}}(\mathsf{simod}\text{-}\mathcal{S})$ and $'\mathcal{N}^{\bullet}$ in $\mathsf{D}^{\mathsf{si}}(\mathsf{simod}\text{-}\mathcal{T})$ are endowed with a morphism $'\mathcal{M}^{\bullet L}_{\mathcal{T}} \longrightarrow '\mathcal{N}^{\bullet}$, or, which is the same, a morphism $'\mathcal{M}^{\bullet} \longrightarrow '\mathcal{N}^{\bullet R}_{\mathcal{C}}$, and two objects $''\mathcal{M}^{\bullet}$ in $\mathsf{D}^{\mathsf{si}}(\mathcal{S}\text{-}\mathsf{simod})$ and $''\mathcal{N}^{\bullet}$ in $\mathsf{D}^{\mathsf{si}}(\mathcal{T}\text{-}\mathsf{simod})$ are endowed with a morphism $^{L}_{\mathcal{T}}''\mathcal{M}^{\bullet} \longrightarrow ''\mathcal{N}^{\bullet}$, or, which is the same, a morphism $''\mathcal{M}^{\bullet} \longrightarrow {}^{R}_{\mathcal{C}}''\mathcal{N}^{\bullet}$. Then the two morphisms $\mathrm{SemiTor}^{\mathcal{S}}('\mathcal{M}^{\bullet}, ''\mathcal{M}^{\bullet}) \longrightarrow \mathrm{SemiTor}^{\mathcal{T}}('\mathcal{N}^{\bullet}, ''\mathcal{N}^{\bullet})$ in $\mathsf{D}(k\text{-}\mathsf{mod})$ provided by the compositions

$$\mathrm{SemiTor}^{\mathcal{S}}('\mathcal{M}^{\bullet}, ''\mathcal{M}^{\bullet}) \longrightarrow \mathrm{SemiTor}^{\mathcal{S}}('\mathcal{N}^{\bullet R}_{\mathcal{C}}, ''\mathcal{M}^{\bullet})$$
$$\simeq \mathrm{SemiTor}^{\mathcal{T}}('\mathcal{N}^{\bullet}, {}^{L}_{\mathcal{T}}''\mathcal{M}^{\bullet}) \longrightarrow \mathrm{SemiTor}^{\mathcal{T}}('\mathcal{N}^{\bullet}, ''\mathcal{N}^{\bullet})$$

and

$$\mathrm{SemiTor}^{\mathcal{S}}('\mathcal{M}^{\bullet}, ''\mathcal{M}^{\bullet}) \longrightarrow \mathrm{SemiTor}^{\mathcal{S}}('\mathcal{M}^{\bullet}, {}^{R}_{\mathcal{C}}''\mathcal{N}^{\bullet})$$
$$\simeq \mathrm{SemiTor}^{\mathcal{T}}('\mathcal{M}^{\bullet L}_{\mathcal{T}}, ''\mathcal{N}^{\bullet}) \longrightarrow \mathrm{SemiTor}^{\mathcal{T}}('\mathcal{N}^{\bullet}, ''\mathcal{N}^{\bullet})$$

coincide with each other. Indeed, let us represent the objects $'\mathcal{M}^{\bullet}$ and $'\mathcal{N}^{\bullet}$ by complexes of right $\mathcal{S}$-semimodules and $\mathcal{T}$-semimodules in such a way that the adjoint morphisms $'\mathcal{M}^{\bullet L}_{\mathcal{T}} \longrightarrow '\mathcal{N}^{\bullet}$ and $'\mathcal{M}^{\bullet} \longrightarrow '\mathcal{N}^{\bullet R}_{\mathcal{C}}$ could be represented by a map of complexes of semimodules $'\mathcal{M}^{\bullet} \longrightarrow '\mathcal{N}^{\bullet}$ compatible with the maps $A \longrightarrow B$, $\mathcal{C} \longrightarrow \mathcal{D}$, and $\mathcal{S} \longrightarrow \mathcal{T}$. Applying to the complexes of $'\mathcal{M}^{\bullet}$ and $'\mathcal{N}^{\bullet}$ simultaneously the constructions from the proof of Theorem 2.6, one can construct a map of quite semiflat complexes of right semimodules $\mathbb{L}_3\mathbb{R}_2\mathbb{L}_1('\mathcal{M}^{\bullet}) \longrightarrow \mathbb{L}_3\mathbb{R}_2\mathbb{L}_1('\mathcal{N}^{\bullet})$ representing the same adjoint morphisms in the semiderived categories of left semimodules. So one can assume $'\mathcal{M}_{\bullet}$ and $'\mathcal{N}^{\bullet}$ to be quite semiflat complexes. Analogously, represent the morphisms $^{L}_{\mathcal{T}}''\mathcal{M}^{\bullet} \longrightarrow ''\mathcal{N}^{\bullet}$ and $''\mathcal{M}^{\bullet} \longrightarrow {}^{R}_{\mathcal{C}}''\mathcal{N}^{\bullet}$ in the semiderived categories of left semimodules by a map of quite semiflat complexes of left semimodules $''\mathcal{M}^{\bullet} \longrightarrow ''\mathcal{N}^{\bullet}$ compatible with the maps $A \longrightarrow B$, $\mathcal{C} \longrightarrow \mathcal{D}$, and $\mathcal{S} \longrightarrow \mathcal{T}$. Then both compositions in question are represented by the same map of complexes of k-modules $'\mathcal{M}^{\bullet} \lozenge_{\mathcal{S}} ''\mathcal{M}^{\bullet} \longrightarrow '\mathcal{N}^{\bullet} \lozenge_{\mathcal{T}} ''\mathcal{N}^{\bullet}$.

Furthermore, suppose that two objects $\mathcal{M}^\bullet$ in $\mathsf{D}^{si}(\mathcal{S}\text{-simod})$ and $\mathcal{N}^\bullet$ in $\mathsf{D}^{si}(\mathcal{T}\text{-simod})$ are endowed with a morphism $^{\mathbb{L}}_\mathcal{T}\mathcal{M}^\bullet \longrightarrow \mathcal{N}^\bullet$, or, which is the same, a morphism $\mathcal{M}^\bullet \longrightarrow {}^{\mathbb{R}}_{\mathcal{C}}\mathcal{N}^\bullet$, and two objects $\mathfrak{P}^\bullet$ in $\mathsf{D}^{si}(\mathcal{S}\text{-sicntr})$ and $\mathfrak{Q}^\bullet$ in $\mathsf{D}^{si}(\mathcal{T}\text{-sicntr})$ are endowed with a morphism $\mathfrak{Q}^\bullet \longrightarrow {}^{\mathcal{T}}_{\mathbb{R}}\mathfrak{P}^\bullet$, or, which is the same, a morphism $^{\mathcal{C}}_{\mathbb{L}}\mathfrak{Q}^\bullet \longrightarrow \mathfrak{P}^\bullet$. Then the two morphisms $\mathrm{SemiExt}_\mathcal{T}(\mathcal{N}^\bullet, \mathfrak{Q}^\bullet) \longrightarrow \mathrm{SemiExt}_\mathcal{S}(\mathcal{M}^\bullet, \mathfrak{P}^\bullet)$ in $\mathsf{D}(k\text{-mod})$ provided by the compositions

$$\mathrm{SemiExt}_\mathcal{T}(\mathcal{N}^\bullet, \mathfrak{Q}^\bullet) \longrightarrow \mathrm{SemiExt}_\mathcal{T}(\mathcal{N}^\bullet, {}^{\mathbb{R}}_\mathcal{T}\mathfrak{P}^\bullet)$$
$$\simeq \mathrm{SemiExt}_\mathcal{S}({}^{\mathbb{R}}_\mathcal{C}\mathcal{N}^\bullet, \mathfrak{P}^\bullet) \longrightarrow \mathrm{SemiExt}_\mathcal{S}(\mathcal{M}^\bullet, \mathfrak{P}^\bullet)$$

and

$$\mathrm{SemiExt}_\mathcal{T}(\mathcal{N}^\bullet, \mathfrak{Q}^\bullet) \longrightarrow \mathrm{SemiExt}_\mathcal{T}({}^{\mathbb{L}}_\mathcal{T}\mathcal{M}^\bullet, \mathfrak{Q}^\bullet)$$
$$\simeq \mathrm{SemiExt}_\mathcal{S}(\mathcal{M}^\bullet, {}^{\mathcal{C}}_{\mathbb{L}}\mathfrak{Q}^\bullet) \longrightarrow \mathrm{SemiExt}_\mathcal{S}(\mathcal{M}^\bullet, \mathfrak{P}^\bullet)$$

coincide with each other.

Corollary 4. *Under the assumptions of Corollary 1(c), the mutually inverse equivalences of categories*

$$\mathbb{R}\Psi_\mathcal{S} \colon \mathsf{D}^{si}(\mathcal{S}\text{-simod}) \longrightarrow \mathsf{D}^{si}(\mathcal{S}\text{-sicntr}) \quad and \quad \mathbb{L}\Phi_\mathcal{S} \colon \mathsf{D}^{si}(\mathcal{S}\text{-sicntr}) \longrightarrow \mathsf{D}^{si}(\mathcal{S}\text{-simod})$$

and the mutually inverse equivalences of categories

$$\mathbb{R}\Psi_\mathcal{T} \colon \mathsf{D}^{si}(\mathcal{T}\text{-simod}) \longrightarrow \mathsf{D}^{si}(\mathcal{T}\text{-sicntr}) \quad and \quad \mathbb{L}\Phi_\mathcal{T} \colon \mathsf{D}^{si}(\mathcal{T}\text{-sicntr}) \longrightarrow \mathsf{D}^{si}(\mathcal{T}\text{-simod})$$

transform the derived functor $\mathcal{N}^\bullet \longmapsto {}^{\mathbb{R}}_\mathcal{C}\mathcal{N}^\bullet$ *into the derived functor* $\mathfrak{Q}^\bullet \longmapsto {}^{\mathcal{C}}_{\mathbb{L}}\mathfrak{Q}^\bullet$.

Proof. To construct the isomorphism $\mathbb{L}\Phi_\mathcal{S}({}^{\mathcal{C}}_{\mathbb{L}}\mathfrak{Q}^\bullet) \simeq {}^{\mathbb{R}}_\mathcal{C}(\mathbb{L}\Phi_\mathcal{T}\mathfrak{Q}^\bullet)$, represent the object $\mathfrak{Q}^\bullet$ by a complex of $\mathcal{D}/B$-projective $\mathcal{C}$-contramodules, and use Lemma 5.3.2, Theorem 7.2.1(b), and the results of 7.1.4 and 8.1.4. To construct the isomorphism ${}^{\mathcal{C}}_{\mathbb{L}}(\mathbb{R}\Psi_\mathcal{T}\mathcal{N}^\bullet) \simeq \mathbb{R}\Psi_\mathcal{S}({}^{\mathbb{R}}_\mathcal{C}\mathcal{N}^\bullet)$, represent the object $\mathcal{N}^\bullet$ by a complex of $\mathcal{D}/B$-injective $\mathcal{C}$-comodules, and use Lemma 5.3.2, Theorem 7.2.1(a), and the results of 7.1.4 and 8.1.4. To show that these isomorphisms agree, it suffices to check that for any adjoint morphisms $\mathbb{L}\Phi_\mathcal{T}\mathfrak{Q}^\bullet \longrightarrow \mathcal{N}^\bullet$ and $\mathfrak{Q}^\bullet \longrightarrow \mathbb{R}\Psi_\mathcal{T}\mathcal{N}^\bullet$ in the semiderived categories of $\mathcal{T}$-semimodules and $\mathcal{T}$-semicontramodules the compositions $\mathbb{L}\Phi_\mathcal{S}({}^{\mathcal{C}}_{\mathbb{L}}\mathfrak{Q}^\bullet) \longrightarrow {}^{\mathbb{R}}_\mathcal{C}(\mathbb{L}\Phi_\mathcal{T}\mathfrak{Q}^\bullet) \longrightarrow {}^{\mathbb{R}}_\mathcal{C}\mathcal{N}^\bullet$ and ${}^{\mathcal{C}}_{\mathbb{L}}\mathfrak{Q}^\bullet \longrightarrow {}^{\mathcal{C}}_{\mathbb{L}}(\mathbb{R}\Psi_\mathcal{T}\mathcal{N}^\bullet) \longrightarrow \mathbb{R}\Psi_\mathcal{S}({}^{\mathbb{R}}_\mathcal{C}\mathcal{N}^\bullet)$ are adjoint morphisms in the semiderived categories of $\mathcal{S}$-semimodules and $\mathcal{S}$-semicontramodules. Here one can represent $\mathcal{N}^\bullet$ by a semiprojective complex of $\mathcal{D}$-coprojective left $\mathcal{T}$-semimodules and $\mathfrak{Q}^\bullet$ by a semiinjective complex of $\mathcal{D}$-coinjective left $\mathcal{T}$-semicontramodules (having in mind Lemmas 5.2 and 6.4), and use a result of 7.1.4. □

Thus we have constructed three functors between the semiderived categories $\mathsf{D}^{si}(\mathcal{S}\text{-simod}) \simeq \mathsf{D}^{si}(\mathcal{S}\text{-sicntr})$ and $\mathsf{D}^{si}(\mathcal{T}\text{-simod}) \simeq \mathsf{D}^{si}(\mathcal{T}\text{-sicntr})$: the functor described in Corollary 4, and two functors adjoint to it from the left and from the right, described in Corollary 1.

Remark 3. One can show that the isomorphisms of derived functors from Corollary 6.6 are compatible with the change-of-semialgebra isomorphisms from Corollaries 1, 3, and 4 in the following way. To check that the compositions of isomorphisms

$$\operatorname{SemiExt}_{\mathcal{T}}(\tfrac{\mathbb{L}}{\mathcal{T}}\mathcal{M}^{\bullet}, \mathbb{R}\Psi_{\mathcal{T}}(\mathbf{N}^{\bullet})) \longrightarrow \operatorname{Ext}_{\mathcal{T}}(\tfrac{\mathbb{L}}{\mathcal{T}}\mathcal{M}^{\bullet}, \mathbf{N}^{\bullet}) \longrightarrow \operatorname{Ext}_{\mathcal{S}}(\mathcal{M}^{\bullet}, {}^{\mathbb{R}}_{\mathcal{C}}\mathbf{N}^{\bullet})$$

and

$$\operatorname{SemiExt}_{\mathcal{T}}(\tfrac{\mathbb{L}}{\mathcal{T}}\mathcal{M}^{\bullet}, \mathbb{R}\Psi_{\mathcal{T}}(\mathbf{N}^{\bullet})) \longrightarrow \operatorname{SemiExt}_{\mathcal{S}}(\mathcal{M}^{\bullet}, {}^{\mathcal{C}}_{\mathbb{L}}(\mathbb{R}\Psi_{\mathcal{T}}\mathbf{N}^{\bullet}))$$
$$\longrightarrow \operatorname{SemiExt}_{\mathcal{S}}(\mathcal{M}^{\bullet}, \mathbb{R}\Psi_{\mathcal{S}}({}^{\mathbb{R}}_{\mathcal{C}}\mathbf{N}^{\bullet})) \longrightarrow \operatorname{Ext}_{\mathcal{S}}(\mathcal{M}^{\bullet}, {}^{\mathbb{R}}_{\mathcal{C}}\mathbf{N}^{\bullet})$$

coincide, represent the object $\mathcal{M}^{\bullet}$ by a semiprojective complex of semiprojective left $\mathcal{S}$-semimodules and the object $\mathbf{N}^{\bullet}$ by a complex of $\mathcal{D}/B$-injective left $\mathcal{T}$-semimodules, and use the result of 4.8. To check that the compositions of isomorphisms

$$\operatorname{CtrTor}^{\mathcal{S}}(\mathcal{M}^{\bullet}, {}^{\mathcal{C}}_{\mathbb{L}}\mathfrak{Q}^{\bullet}) \longrightarrow \operatorname{CtrTor}^{\mathcal{T}}(\mathcal{M}^{\bullet\mathbb{L}}_{\mathcal{T}}, \mathfrak{Q}^{\bullet}) \longrightarrow \operatorname{SemiTor}^{\mathcal{T}}(\mathcal{M}^{\bullet\mathbb{L}}_{\mathcal{T}}, \mathbb{L}\Phi_{\mathcal{T}}(\mathfrak{Q}^{\bullet}))$$

and

$$\operatorname{CtrTor}^{\mathcal{S}}(\mathcal{M}^{\bullet}, {}^{\mathcal{C}}_{\mathbb{L}}\mathfrak{Q}^{\bullet}) \longrightarrow \operatorname{SemiTor}^{\mathcal{S}}(\mathcal{M}^{\bullet}, \mathbb{L}\Phi_{\mathcal{S}}({}^{\mathcal{C}}_{\mathbb{L}}\mathfrak{Q}^{\bullet}))$$
$$\longrightarrow \operatorname{SemiTor}^{\mathcal{S}}(\mathcal{M}^{\bullet}, {}^{\mathbb{R}}_{\mathcal{C}}(\mathbb{L}\Phi_{\mathcal{T}}\mathfrak{Q}^{\bullet})) \longrightarrow \operatorname{SemiTor}^{\mathcal{T}}(\mathcal{M}^{\bullet\mathbb{L}}_{\mathcal{T}}, \mathbb{L}\Phi_{\mathcal{T}}(\mathfrak{Q}^{\bullet}))$$

coincide, represent the object $\mathcal{M}^{\bullet}$ by a quite semiflat complex of semiflat right $\mathcal{S}$-semimodules and the object $\mathfrak{Q}^{\bullet}$ by a complex of $\mathcal{D}/B$-projective left $\mathcal{T}$-semicontramodules, and use the result of 2.8. Commutativity of the respective diagrams on the level of abelian categories is straightforward to verify under our assumptions on the terms of the complexes representing the objects $\mathcal{M}^{\bullet}$.

8.4 Remarks on Morita morphisms

8.4.1 Let $\mathcal{C}$ be a coring over a k-algebra A and $\mathcal{D}$ be a coring over a k-algebra B such that $\mathcal{C}$ is a flat right A-module and $\mathcal{D}$ is a flat right B-module. Let $(\mathcal{E}, \mathcal{E}^{\vee})$ be a right coflat Morita morphism from $\mathcal{C}$ to $\mathcal{D}$ and $\mathcal{T}$ be a semialgebra over the coring $\mathcal{D}$ such that $\mathcal{T}$ is a coflat right $\mathcal{D}$-comodule. In this case, there is a natural semialgebra structure on the $\mathcal{C}$-$\mathcal{C}$-bicomodule

$$_{\mathcal{C}}\mathcal{T}_{\mathcal{C}} = \mathcal{E} \,\square_{\mathcal{D}}\, \mathcal{T} \,\square_{\mathcal{D}}\, \mathcal{E}^{\vee}.$$

The semimultiplication in $_{\mathcal{C}}\mathcal{T}_{\mathcal{C}}$ is defined as the composition $\mathcal{E}\square_{\mathcal{D}}\mathcal{T}\square_{\mathcal{D}}\mathcal{E}^{\vee}\square_{\mathcal{C}}\mathcal{E}\square_{\mathcal{D}}$ $\mathcal{T}\square_{\mathcal{D}}\mathcal{E}^{\vee} \longrightarrow \mathcal{E}\square_{\mathcal{D}}\mathcal{T}\square_{\mathcal{D}}\mathcal{T}\square_{\mathcal{D}}\mathcal{E}^{\vee} \longrightarrow \mathcal{E}\square_{\mathcal{D}}\mathcal{T}\square_{\mathcal{D}}\mathcal{E}^{\vee}$ of the morphism induced by the morphism $\mathcal{E}^{\vee} \square_{\mathcal{C}} \mathcal{E} \longrightarrow \mathcal{D}$ and the morphism induced by the semimultiplication in $\mathcal{T}$. The semiunit in $_{\mathcal{C}}\mathcal{T}_{\mathcal{C}}$ is defined as the composition $\mathcal{C} \longrightarrow \mathcal{E} \square_{\mathcal{D}} \mathcal{E}^{\vee} \longrightarrow$ $\mathcal{E} \square_{\mathcal{D}} \mathcal{T} \square_{\mathcal{D}} \mathcal{E}^{\vee}$ of the morphism induced by the morphism $\mathcal{C} \longrightarrow \mathcal{E} \square_{\mathcal{D}} \mathcal{E}^{\vee}$ and the morphism induced by the semiunit in $\mathcal{T}$.

For example, if $\mathcal{C} \longrightarrow \mathcal{D}$ is a map of corings compatible with a k-algebra map $A \longrightarrow B$ such that B is a flat right A-module and $\mathcal{C}_B$ is a coflat right $\mathcal{D}$-comodule, one can take $\mathcal{E} = \mathcal{C}_B$ and $\mathcal{E}^\vee = {}_B\mathcal{C}$. Then the algebra ${}_\mathcal{C}\mathcal{T}_\mathcal{C}$ is a universal final object in the category of semialgebras $\mathbf{S}$ over $\mathcal{C}$ endowed with a map $\mathbf{S} \longrightarrow \mathcal{T}$ compatible with the maps $A \longrightarrow B$ and $\mathcal{C} \longrightarrow \mathcal{D}$. The semialgebra ${}_\mathcal{C}\mathcal{T}_\mathcal{C} = \mathcal{C}_B \,\square_\mathcal{D}\, \mathcal{T} \,\square_\mathcal{D}\, {}_B\mathcal{C}$ can be also defined, e.g., when $(E, E^\vee)$ is a Morita morphism from a k-algebra A to a k-algebra B and ${}_B\mathcal{C}_B = E^\vee \otimes_A \mathcal{C} \otimes_A E \longrightarrow \mathcal{D}$ is a morphism of corings over B such that $E^\vee$ is a flat right A-module, ${}_B\mathcal{C} = E^\vee \otimes_A \mathcal{C}$ is a $\mathcal{D}/B$-coflat left $\mathcal{D}$-comodule, $\mathcal{T}$ is a flat right B-module and a $\mathcal{D}/B$-coflat left $\mathcal{D}$-comodule, and the rings A and B have finite weak homological dimensions.

All the results of 8.1–8.3 can be extended to the situation of a left coprojective and right coflat Morita morphism $(\mathcal{E}, \mathcal{E}^\vee)$ from a coring $\mathcal{C}$ to a coring $\mathcal{D}$ and a morphism $\mathbf{S} \longrightarrow {}_\mathcal{C}\mathcal{T}_\mathcal{C}$ of semialgebras over $\mathcal{C}$. In particular, when $\mathcal{C}$ is a flat right A-module, $\mathcal{D}$ a flat right B-module, $\mathbf{S}$ is a coflat right $\mathcal{C}$-comodule, $\mathcal{T}$ is a coflat right $\mathcal{D}$-comodule, and $(\mathcal{E}, \mathcal{E}^\vee)$ is a right coflat Morita morphism, the functor

$$\mathbf{N} \longmapsto {}_\mathcal{C}\mathbf{N} = \mathcal{E} \,\square_\mathcal{D}\, \mathbf{N}$$

from the category of left $\mathcal{T}$-semimodules to the category of left $\mathbf{S}$-semimodules has a left adjoint functor

$$\mathbf{M} \longmapsto {}_\mathcal{T}\mathbf{M} = \mathcal{T}_\mathcal{C} \,\lozenge_\mathbf{S}\, \mathbf{M}.$$

Analogously, when $\mathcal{C}$ is a projective left A-module, $\mathcal{D}$ is a projective left B-module, $\mathbf{S}$ is a coprojective left $\mathcal{C}$-comodule, $\mathcal{T}$ is a coprojective left $\mathcal{D}$-comodule, and $(\mathcal{E}, \mathcal{E}^\vee)$ is a left coprojective Morita morphism, the functor

$$\mathfrak{Q} \longmapsto {}^\mathcal{C}\mathfrak{Q} = \mathrm{Cohom}_\mathcal{D}(\mathcal{E}^\vee, \mathfrak{Q})$$

from the category of left $\mathcal{T}$-semicontramodules to the category of left $\mathbf{S}$-semicontramodules has a right adjoint functor

$$\mathfrak{P} \longmapsto {}^\mathcal{T}\mathfrak{P} = \mathrm{SemiHom}_\mathbf{S}({}_\mathcal{C}\mathcal{T}, \mathfrak{P}),$$

etc. However, one sometimes has to impose the homological dimension conditions on A and B where they were not previously needed.

8.4.2 Assume that $\mathcal{C}$ is a flat right A-module and $\mathcal{D}$ is a flat right B-module. A right $\mathcal{D}$-comodule $\mathcal{K}$ is called *faithfully coflat* if it is a coflat $\mathcal{D}$-comodule and for any nonzero left $\mathcal{D}$-comodule $\mathcal{M}$ the cotensor product $\mathcal{K} \,\square_\mathcal{D}\, \mathcal{M}$ is nonzero. A right coflat Morita morphism $(\mathcal{E}, \mathcal{E}^\vee)$ from $\mathcal{C}$ to $\mathcal{D}$ is called *right faithfully coflat* if the right $\mathcal{D}$-comodule $\mathcal{E}$ is faithfully coflat. A right coflat Morita morphism $(\mathcal{E}, \mathcal{E}^\vee)$ is right faithfully coflat if and only if the right $\mathcal{D}$-comodule $\mathcal{E}^\vee \,\square_\mathcal{C}\, \mathcal{E}$ is faithfully coflat and if and only if the morphism $\mathcal{E}^\vee \,\square_\mathcal{C}\, \mathcal{E} \longrightarrow \mathcal{D}$ is surjective and its kernel is a coflat right $\mathcal{D}$-comodule. Indeed, the cotensor product $\mathcal{E} \,\square_\mathcal{D}\, \mathcal{M}$ is nonzero if and only if the morphism $\mathcal{E}^\vee \,\square_\mathcal{C}\, \mathcal{E} \,\square_\mathcal{D}\, \mathcal{M} \longrightarrow \mathcal{M}$ is nonzero; this holds for any nonzero left $\mathcal{D}$-comodule $\mathcal{M}$ if and only if the morphism $\mathcal{E}^\vee \,\square_\mathcal{C}\, \mathcal{E} \,\square_\mathcal{D}\, \mathcal{M} \longrightarrow \mathcal{M}$ is surjective for any left $\mathcal{D}$-comodule $\mathcal{M}$, and it remains to use the results of (the proof of) Lemma 1.2.2.

Let $(\mathcal{E}, \mathcal{E}^\vee)$ be a right faithfully coflat Morita morphism from $\mathcal{C}$ to $\mathcal{D}$ and $\mathcal{T}$ be a semialgebra over the coring $\mathcal{D}$ such that $\mathcal{T}$ is a coflat right $\mathcal{D}$-comodule. Then the functor $\mathbf{N} \longmapsto {}_{\mathcal{C}}\mathbf{N}$ is an equivalence of the abelian categories of left $\mathcal{T}$-semimodules and left ${}_{\mathcal{C}}\mathcal{T}_{\mathcal{C}}$-semimodules. This follows from Theorem 7.4.1 applied to the functor $\Delta \colon \mathcal{T}\text{-simod} \longrightarrow \mathcal{C}\text{-comod}$ mapping a $\mathcal{T}$-semimodule $\mathbf{N}$ to the $\mathcal{C}$-comodule ${}_{\mathcal{C}}\mathbf{N}$ and the functor $\Gamma \colon \mathcal{C}\text{-comod} \longrightarrow \mathcal{T}\text{-simod}$ left adjoint to Δ mapping a $\mathcal{C}$-comodule $\mathcal{M}$ to the $\mathcal{T}$-semimodule $\mathcal{T} \,\square_{\mathcal{D}}\, {}_{\mathcal{D}}\mathcal{M}$.

Now assume that $\mathcal{C}$ is a projective left A-module and $\mathcal{D}$ is a projective left B-module. A left $\mathcal{D}$-comodule $\mathcal{K}$ is called *faithfully coprojective* if it is a coprojective $\mathcal{D}$-comodule and for any nonzero left $\mathcal{D}$-contramodule $\mathfrak{P}$ the cohomomorphism module $\mathrm{Cohom}_{\mathcal{D}}(\mathcal{K}, \mathfrak{P})$ is nonzero. A faithfully coprojective $\mathcal{D}$-comodule is faithfully coflat. A left coprojective Morita morphism $(\mathcal{E}, \mathcal{E}^\vee)$ from $\mathcal{C}$ to $\mathcal{D}$ is called *left faithfully coprojective* if the left $\mathcal{D}$-comodule $\mathcal{E}^\vee$ is faithfully coprojective. A left coprojective Morita morphism $(\mathcal{E}, \mathcal{E}^\vee)$ is left faithfully coprojective if and only if the left $\mathcal{D}$-comodule $\mathcal{E}^\vee \,\square_{\mathcal{C}}\, \mathcal{E}$ is faithfully coprojective and if and only if the morphism $\mathcal{E}^\vee \,\square_{\mathcal{C}}\, \mathcal{E} \longrightarrow \mathcal{D}$ is surjective and its kernel is a coprojective left $\mathcal{D}$-comodule.

Let $(\mathcal{E}, \mathcal{E}^\vee)$ be a left faithfully coprojective Morita morphism from $\mathcal{C}$ to $\mathcal{D}$ and $\mathcal{T}$ be a semialgebra over the coring $\mathcal{D}$ such that $\mathcal{T}$ is a coprojective left $\mathcal{D}$-comodule. Then the functor $\mathfrak{Q} \longmapsto {}^{\mathcal{C}}\mathfrak{Q}$ is an equivalence of the abelian categories of left $\mathcal{T}$-semicontramodules and left ${}_{\mathcal{C}}\mathcal{T}_{\mathcal{C}}$-semicontramodules. This follows from Theorem 7.4.1 applied to the functor $\Delta \colon \mathcal{T}\text{-sicntr} \longrightarrow \mathcal{C}\text{-contra}$ mapping a $\mathcal{T}$-semicontramodule $\mathfrak{Q}$ to the $\mathcal{C}$-contramodule ${}^{\mathcal{C}}\mathfrak{Q}$ and the functor $\Gamma \colon \mathcal{C}\text{-contra} \longrightarrow \mathcal{T}\text{-sicntr}$ right adjoint to Δ mapping a $\mathcal{C}$-contramodule $\mathfrak{P}$ to the $\mathcal{T}$-semicontramodule $\mathrm{Cohom}_{\mathcal{D}}(\mathcal{T}, {}^{\mathcal{D}}\mathfrak{P})$.

8.4.3 Assume that $\mathcal{C}$ is a flat right A-module and $\mathcal{D}$ is a flat right B-module. Let $(\mathcal{E}, \mathcal{E}^\vee)$ be a right coflat Morita morphism from $\mathcal{C}$ to $\mathcal{D}$ and $\mathcal{T}$ be a semialgebra over the coring $\mathcal{D}$ such that $\mathcal{T}$ is a coflat right $\mathcal{D}$-comodule. Then the functor $\mathbf{N}^\bullet \longmapsto {}_{\mathcal{C}}\mathbf{N}^\bullet$ maps $\mathcal{D}$-coacyclic complexes of $\mathcal{T}$-semimodules to $\mathcal{C}$-coacyclic complexes of ${}_{\mathcal{C}}\mathcal{T}_{\mathcal{C}}$-semimodules and the semiderived category of left ${}_{\mathcal{C}}\mathcal{T}_{\mathcal{C}}$-semimodules is a localization of the semiderived category of left $\mathcal{T}$-semimodules by the kernel of the functor induced by $\mathbf{N}^\bullet \longmapsto {}_{\mathcal{C}}\mathbf{N}^\bullet$ (as one can check by computing the functor $\mathcal{M}^\bullet \longmapsto {}_{\mathcal{C}}(\tfrac{\mathbb{L}}{\mathcal{T}}\mathcal{M}^\bullet)$ on the semiderived category of left ${}_{\mathcal{C}}\mathcal{T}_{\mathcal{C}}$-semimodules). The triangulated categories $\mathsf{D}^{\mathsf{si}}(\mathcal{T}\text{-simod})$ and $\mathsf{D}^{\mathsf{si}}({}_{\mathcal{C}}\mathcal{T}_{\mathcal{C}}\text{-simod})$ are equivalent when $(\mathcal{E}, \mathcal{E}^\vee)$ is a right coflat Morita equivalence, or more generally when the morphism $\mathcal{E}^\vee \,\square_{\mathcal{C}}\, \mathcal{E} \longrightarrow \mathcal{D}$ is an isomorphism.

Analogously, assume that $\mathcal{C}$ is a flat right A-module and $\mathcal{D}$ is a projective left B-module. Let $(\mathcal{E}, \mathcal{E}^\vee)$ be a left coprojective Morita morphism from $\mathcal{C}$ to $\mathcal{D}$ and $\mathcal{T}$ be a semialgebra over the coring $\mathcal{D}$ such that $\mathcal{T}$ is a coprojective left $\mathcal{D}$-comodule. Then the functor $\mathfrak{Q}^\bullet \longmapsto {}^{\mathcal{C}}\mathfrak{Q}^\bullet$ maps $\mathcal{D}$-contraacyclic complexes of $\mathcal{T}$-semicontramodules to $\mathcal{C}$-contraacyclic complexes of ${}_{\mathcal{C}}\mathcal{T}_{\mathcal{C}}$-semicontramodules and the semiderived category of left ${}_{\mathcal{C}}\mathcal{T}_{\mathcal{C}}$-semicontramodules is a localization of the semiderived category of left $\mathcal{T}$-semicontramodules by the kernel of the functor induced by $\mathfrak{Q}^\bullet \longmapsto {}^{\mathcal{C}}\mathfrak{Q}^\bullet$. The triangulated categories $\mathsf{D}^{\mathsf{si}}(\mathcal{T}\text{-sicntr})$ and $\mathsf{D}^{\mathsf{si}}({}_{\mathcal{C}}\mathcal{T}_{\mathcal{C}}\text{-sicntr})$ are equivalent when $(\mathcal{E}, \mathcal{E}^\vee)$ is a left coprojective Morita equivalence, or more generally when the morphism $\mathcal{E}^\vee \,\square_\mathcal{C}\, \mathcal{E} \longrightarrow \mathcal{D}$ is an isomorphism.

Remark. The semiderived categories of left $\mathcal{T}$-semimodules and left ${}_\mathcal{C}\mathcal{T}_\mathcal{C}$-semimodules can be different even when $(\mathcal{E}, \mathcal{E}^\vee)$ is a right faithfully coflat Morita morphism and the abelian categories of left $\mathcal{T}$-semimodules and left ${}_\mathcal{C}\mathcal{T}_\mathcal{C}$-semimodules are equivalent. Indeed, let $A = B = k$ be a field and F be a finite-dimensional algebra over k. Let $\mathcal{D} = F^*$ and $\mathcal{C} = \text{End}(F)^*$ be the coalgebras over k dual to the finite-dimensional k-algebras F and $\text{End}(F)$. Then there is a coalgebra morphism $\mathcal{C} \longrightarrow \mathcal{D}$ dual to the algebra embedding $F \longrightarrow \text{End}(F)$ related to the action of F in itself by left multiplications. Since $\text{End}(F)$ is a free left F-module, $\mathcal{C}$ is a cofree right $\mathcal{D}$-comodule. Set $\mathcal{E} = \mathcal{C} = \mathcal{E}^\vee$; this is a right faithfully coprojective Morita morphism from $\mathcal{C}$ to $\mathcal{D}$. Now put $\mathcal{T} = \mathcal{D}$; then the semiderived category of left $\mathcal{T}$-semimodules coincides with the coderived category of left $\mathcal{D}$-comodules. At the same time, the coalgebra $\mathcal{C}$ is semisimple and a complex of $\mathcal{C}$-comodules is coacyclic if and only if it is acyclic, so the semiderived category of left ${}_\mathcal{C}\mathcal{T}_\mathcal{C}$-comodules is equivalent to the conventional derived category of left $\mathcal{D}$-comodules. When F is a Frobenius algebra, $\text{End}(F)$ is a free left and right F-module, so $(\mathcal{E}, \mathcal{E}^\vee)$ is a left and right faithfully coprojective Morita morphism, but the categories $\mathsf{D}^{\mathsf{si}}(\mathcal{T}\text{-simod})$ and $\mathsf{D}^{\mathsf{si}}({}_\mathcal{C}\mathcal{T}_\mathcal{C}\text{-simod})$ are still not equivalent when the homological dimension of F is infinite. Alternatively, one can consider the right coprojective Morita morphism from the coalgebra $\mathcal{C} = k$ to the coalgebra $\mathcal{D} = F^*$ with $\mathcal{E} = F^*$ and $\mathcal{E}^\vee = F$ and the same semialgebra $\mathcal{T} = \mathcal{D}$ over $\mathcal{D}$; then the semialgebra ${}_\mathcal{C}\mathcal{T}_\mathcal{C}$ over $\mathcal{C}$ is isomorphic to the algebra F over k; the category $\mathsf{D}^{\mathsf{si}}(\mathcal{T}\text{-simod})$ is the coderived category of F^*-comodules and the category $\mathsf{D}^{\mathsf{si}}({}_\mathcal{C}\mathcal{T}_\mathcal{C}\text{-simod})$ is the derived category of F-modules.

Assume that $\mathcal{C}$ is a flat left and right A-module, $\mathcal{D}$ is a flat left and right B-module, the rings A and B have finite weak homological dimensions, $\mathcal{T}$ is a coflat left and right $\mathcal{D}$-comodule, and $(\mathcal{E}, \mathcal{E}^\vee)$ is a left and right coflat Morita morphism from $\mathcal{C}$ to $\mathcal{D}$. Then whenever the functor $\mathbf{N}^\bullet \longmapsto {}_\mathcal{C}\mathbf{N}^\bullet$ induces an equivalence of the semiderived categories of left $\mathcal{T}$-semimodules and left ${}_\mathcal{C}\mathcal{T}_\mathcal{C}$-semimodules and the functor $\mathbf{N}^\bullet \longmapsto \mathbf{N}^\bullet_\mathcal{C}$ induces an equivalence of the semiderived categories of right $\mathcal{T}$-semimodules and right ${}_\mathcal{C}\mathcal{T}_\mathcal{C}$-semimodules, these equivalences of categories transform the functor $\text{SemiTor}^\mathcal{T}$ into the functor $\text{SemiTor}^{{}_\mathcal{C}\mathcal{T}_\mathcal{C}}$.

Assume that $\mathcal{C}$ is a projective left and a flat right A-module, $\mathcal{D}$ is a projective left and a flat right B-module, the rings A and B have finite left homological dimensions, $\mathcal{T}$ is a coprojective left and a coflat right $\mathcal{D}$-comodule, and $(\mathcal{E}, \mathcal{E}^\vee)$ is a left coprojective and right coflat Morita morphism from $\mathcal{C}$ to $\mathcal{D}$. Then whenever the functor $\mathbf{N}^\bullet \longmapsto {}_\mathcal{C}\mathbf{N}^\bullet$ induces an equivalence of the semiderived categories of left $\mathcal{T}$-semimodules and left ${}_\mathcal{C}\mathcal{T}_\mathcal{C}$-semimodules and the functor $\mathfrak{Q}^\bullet \longmapsto {}^\mathcal{C}\mathfrak{Q}^\bullet$ induces an equivalence of the semiderived categories of left $\mathcal{T}$-semicontramodules and left ${}_\mathcal{C}\mathcal{T}_\mathcal{C}$-semicontramodules, these equivalences of categories transform the functor $\text{SemiExt}_\mathcal{T}$ into the functor $\text{SemiExt}_{{}_\mathcal{C}\mathcal{T}_\mathcal{C}}$ and the equivalences of categories

$\mathbb{R}\Psi_{\mathcal{T}}$ and $\mathbb{L}\Phi_{\mathcal{T}}$ into the equivalences of categories $\mathbb{R}\Psi_{e\mathcal{T}e}$ and $\mathbb{L}\Phi_{e\mathcal{T}e}$. The same applies to the functors $\operatorname{Ext}_{\mathcal{T}}$, $\operatorname{Ext}^{\mathcal{T}}$, and $\operatorname{CtrTor}^{\mathcal{T}}$, under the relevant assumptions.

8.4.4   Here are some further partial results about equivalence of the semiderived categories related to $\mathcal{T}$ and $_e\mathcal{T}_e$. The problem is, essentially, to find conditions under which a complex of left $\mathcal{D}$-comodules $\mathcal{N}^\bullet$ is coacyclic whenever the complex of $\mathcal{C}$-comodules $_e\mathcal{N}^\bullet$ is coacyclic, or a complex of left $\mathcal{D}$-contramodules $\mathfrak{Q}^\bullet$ is contraacyclic whenever the complex of $\mathcal{C}$-contramodules $^e\mathfrak{Q}^\bullet$ is contraacyclic.

Consider the following general setting. Let A and B be exact categories with exact functors of infinite direct sum, $\Delta\colon \mathsf{B} \longrightarrow \mathsf{A}$ be an exact functor preserving infinite direct sums and such that a complex $C^\bullet$ over B is acyclic if the complex $\Delta(C^\bullet)$ over A is contractible, and $\Gamma\colon \mathsf{A} \longrightarrow \mathsf{B}$ be an exact functor left adjoint to Δ. Clearly, if a complex $C^\bullet$ is coacyclic then the complex $\Delta(C^\bullet)$ is coacyclic; we would like to know when the converse holds.

First, if a complex $C^\bullet$ is coacyclic whenever the complex $\Delta(C^\bullet)$ is contractible, then a complex $C^\bullet$ is coacyclic if and only if the complex $\Delta(C^\bullet)$ is coacyclic. Indeed, consider the bar bicomplex

$$\cdots \longrightarrow \Gamma\Delta\Gamma\Delta(C^\bullet) \longrightarrow \Gamma\Delta(C^\bullet) \longrightarrow C^\bullet$$

whose differentials are the alternating sums of morphisms induced by the adjunction morphism $\Gamma\Delta \longrightarrow \operatorname{Id}$. The total complex of this bicomplex constructed by taking infinite direct sums along the diagonals becomes contractible after applying the functor Δ; the contracting homotopy is induced by the adjunction morphism $\operatorname{Id} \longrightarrow \Delta\Gamma$. By assumption, it follows that the total complex itself is coacyclic over B. On the other hand, if the complex $\Delta(C^\bullet)$ is coacyclic over A, then every complex $(\Gamma\Delta)^n(C^\bullet)$ is coacyclic over B, since the functors Δ and Γ are exact and preserve infinite direct sums. The total complex of the bicomplex $\cdots \longrightarrow \Gamma\Delta\Gamma\Delta(C^\bullet) \longrightarrow \Gamma\Delta(C^\bullet)$ is homotopy equivalent to a complex obtained from the complexes $(\Gamma\Delta)^n(C^\bullet)$ using the operations of shift, cone, and infinite direct sum; hence the complex $C^\bullet$ is coacyclic.

By the same argument, a complex $C^\bullet$ is acyclic if and only if the complex $\Delta(C^\bullet)$ is acyclic, so if the exact category B has a finite homological dimension, then a complex $C^\bullet$ is coacyclic if and only if the complex $\Delta(C^\bullet)$ is coacyclic. This is the trivial case.

Finally, let us say that an exact functor $\Delta\colon \mathsf{B} \longrightarrow \mathsf{A}$ has a finite relative homological dimension if the category B with the exact category structure formed by the exact triples in B that split after applying Δ has a finite homological dimension. We claim that when the functor Δ has a finite relative homological dimension, a complex $C^\bullet$ over B is coacyclic if and only if the complex $\Delta(C^\bullet)$ is coacyclic, in our assumptions. Indeed, consider again the bar bicomplex $\cdots \longrightarrow \Gamma\Delta\Gamma\Delta(C^\bullet) \longrightarrow \Gamma\Delta(C^\bullet) \longrightarrow C^\bullet$. One can assume that the category B contains images of idempotent endomorphisms, as passing to the Karoubian closure doesn't change coacyclicity. One can also assume that the complex $C^\bullet$ is bounded from above, as any acyclic complex bounded from below is coacyclic. The complex $\cdots \longrightarrow \Gamma\Delta\Gamma\Delta(X) \longrightarrow \Gamma\Delta(X)$ is split exact in high enough (negative) degrees for any object $X \in \mathsf{B}$, since it is exact and the complex of homomorphisms from it to an object $Y \in \mathsf{B}$ computes $\mathrm{Ext}(X, Y)$ in the relative exact category. Let d be an integer not smaller than the relative homological dimension; denote by $Z(X)$ the image of the morphism $(\Gamma\Delta)^{d+1}(X) \longrightarrow (\Gamma\Delta)^d(X)$. Then the total complex of the bicomplex

$$\cdots \longrightarrow (\Gamma\Delta)^{d+2}(C^\bullet) \longrightarrow (\Gamma\Delta)^{d+1}(C^\bullet) \longrightarrow Z(C^\bullet)$$

is contractible, while the total complex of the bicomplex

$$(\Gamma\Delta)^d(C^\bullet)/Z(C^\bullet) \longrightarrow (\Gamma\Delta)^{d-1}(C^\bullet) \longrightarrow \cdots \longrightarrow \Gamma\Delta(C^\bullet) \longrightarrow C^\bullet$$

is coacyclic. If the complex $\Delta(C^\bullet)$ is coacyclic, the total complex of the bicomplex $\cdots \longrightarrow \Gamma\Delta\Gamma\Delta(C^\bullet) \longrightarrow \Gamma\Delta(C^\bullet)$ is also coacyclic; thus the complex $C^\bullet$ is coacyclic.

8.4.5 Let S be a semialgebra over a coring $\mathcal{C}$ and $\mathcal{T}$ be a semialgebra over a coring $\mathcal{D}$. Assume that $\mathcal{C}$ is a flat right A-module, $\mathcal{D}$ is a flat right B-module, S is a coflat right $\mathcal{C}$-comodule, and $\mathcal{T}$ is a coflat right $\mathcal{D}$-comodule. A *right semiflat Morita morphism* from S to $\mathcal{T}$ is a pair consisting of a $\mathcal{T}$-semiflat S-$\mathcal{T}$-bisemimodule $\mathcal{E}$ and an S-semiflat $\mathcal{T}$-S-bisemimodule $\mathcal{E}^\vee$ endowed with an S-S-bisemimodule morphism $\mathsf{S} \longrightarrow \mathcal{E} \lozenge_\mathcal{T} \mathcal{E}^\vee$ and a $\mathcal{T}$-$\mathcal{T}$-bisemimodule morphism $\mathcal{E}^\vee \lozenge_\mathsf{S} \mathcal{E} \longrightarrow \mathcal{T}$ such that the two compositions

$$\mathcal{E} \longrightarrow \mathcal{E} \lozenge_\mathcal{T} \mathcal{E}^\vee \lozenge_\mathsf{S} \mathcal{E} \longrightarrow \mathcal{E} \quad \text{and} \quad \mathcal{E}^\vee \longrightarrow \mathcal{E}^\vee \lozenge_\mathsf{S} \mathcal{E} \lozenge_\mathcal{T} \mathcal{E}^\vee \longrightarrow \mathcal{E}^\vee$$

are equal to the identity endomorphisms of $\mathcal{E}$ and $\mathcal{E}^\vee$. A right semiflat Morita morphism $(\mathcal{E}, \mathcal{E}^\vee)$ from S to $\mathcal{T}$ induces an exact functor

$$\mathcal{M} \longmapsto \mathcal{T}\mathcal{M} = \mathcal{E}^\vee \lozenge_\mathsf{S} \mathcal{M}$$

from the category of left S-semimodules to the category of left $\mathcal{T}$-semimodules and an exact functor

$$\mathcal{N} \longmapsto \mathsf{S}\mathcal{N} = \mathcal{E} \lozenge_\mathcal{T} \mathcal{N}$$

from the category of left $\mathcal{T}$-semimodules to the category of left S-semimodules; the former functor is left adjoint to the latter one. Conversely, any pair of adjoint exact k-linear functors preserving infinite direct sums between the category of left S-semimodules and left $\mathcal{T}$-semimodules is induced by a right semiflat Morita morphism. Indeed, any exact k-linear functor S-simod $\longrightarrow$ $\mathcal{T}$-simod preserving infinite direct sums is the functor of semitensor product with an S-semiflat $\mathcal{T}$-S-bisemimodule; this can be proven as in 7.5.2.

Analogously, assume that $\mathcal{C}$ is a projective left A-module, $\mathcal{D}$ is a projective left B-module, S is a coprojective left $\mathcal{C}$-comodule, and $\mathcal{T}$ is a coprojective left $\mathcal{D}$-comodule. A *left semiprojective Morita morphism* from S to $\mathcal{T}$ is defined as a pair consisting of an S-semiprojective S-T-bisemimodule $\mathcal{E}$ and a T-semiprojective T-S-bisemimodule $\mathcal{E}^\vee$ endowed with an S-S-bisemimodule morphism $S \longrightarrow \mathcal{E} \lozenge_T \mathcal{E}^\vee$ and a T-T-bisemimodule morphism $\mathcal{E}^\vee \lozenge_S \mathcal{E} \longrightarrow T$ satisfying the same conditions as above. A left semiprojective Morita morphism $(\mathcal{E}, \mathcal{E}^\vee)$ from S to T induces an exact functor

$$\mathfrak{P} \longmapsto {}^T\mathfrak{P} = \mathrm{SemiHom}_S(\mathcal{E}, \mathfrak{P})$$

from the category of left S-semicontramodules to the category of left T-semicontramodules and an exact functor

$$\mathfrak{Q} \longmapsto {}^S\mathfrak{Q} = \mathrm{SemiHom}_T(\mathcal{E}^\vee, \mathfrak{Q})$$

from the category of left T-semicontramodules to the category of left S-semicontramodules; the former functor is right adjoint to the latter one. Conversely, any pair of adjoint exact k-linear functors preserving infinite products between the category of left S-semicontramodules and left T-semicontramodules is induced by a left semiprojective Morita morphism. Indeed, any exact k-linear functor S-sicntr $\longrightarrow$ T-sicntr preserving infinite products is the functor of semihomomorphisms from an S-semiprojective S-T-bisemimodule.

A right semiflat Morita morphism $(\mathcal{E}, \mathcal{E}^\vee)$ from S to T is called a *right semiflat Morita equivalence* if the bisemimodule morphisms $S \longrightarrow \mathcal{E} \lozenge_T \mathcal{E}^\vee$ and $\mathcal{E}^\vee \lozenge_S \mathcal{E} \longrightarrow T$ are isomorphisms; then it can be also considered as a right semiflat Morita morphism $(\mathcal{E}^\vee, \mathcal{E})$ from T to S. *Left semiprojective Morita equivalences* are defined in an analogous way. A right semiflat Morita equivalence between semialgebras S and T induces an equivalence of the abelian categories of left S-semimodules and left T-semimodules, and in the relevant above assumptions any equivalence between these two k-linear categories comes from a right semiflat Morita equivalence. Analogously, a left semiprojective Morita equivalence between S and T induces an equivalence of the abelian categories of left S-semicontramodules and left T-semicontramodules, and in the relevant above assumptions any equivalence between these two k-linear categories comes from a left semiprojective Morita equivalence.

Assume that the coring $\mathcal{C}$ is a flat right A-module and the coring $\mathcal{D}$ is a flat right B-module. Let T be a semialgebra over $\mathcal{D}$ such that T is a coflat right $\mathcal{D}$-comodule and $(\mathcal{E}, \mathcal{E}^\vee)$ be a right faithfully coflat Morita morphism from $\mathcal{C}$ to $\mathcal{D}$. Then the pair of bisemimodules $\mathcal{E} = {}_\mathcal{C}T$ and $\mathcal{E}^\vee = T_\mathcal{C}$ is a right semiflat Morita equivalence between the semialgebras T and ${}_\mathcal{C}T_\mathcal{C}$. Analogously, assume that $\mathcal{C}$ is a projective left A-module and $\mathcal{D}$ is a projective left B-module. Let T be a semialgebra over $\mathcal{D}$ such that T is a coprojective left $\mathcal{D}$-comodule and $(\mathcal{E}, \mathcal{E}^\vee)$ be a left faithfully coprojective Morita morphism from $\mathcal{C}$ to $\mathcal{D}$. Then the same pair of bisemimodules $\mathcal{E}$ and $\mathcal{E}^\vee$ is a left semiprojective Morita equivalence between T and ${}_\mathcal{C}T_\mathcal{C}$.

All the results of 8.1 can be extended to the case of a left semiprojective and right semiflat Morita morphism $(\mathcal{E}, \mathcal{E}^\vee)$ from a semialgebra S to a semialgebra $\mathcal{T}$. In particular, for any left $\mathcal{T}$-semimodule $\mathcal{N}$ there are natural isomorphisms of left $\mathcal{S}$-semicontramodules $\Psi_{\mathcal{S}}({}_{\mathcal{S}}\mathcal{N}) = \operatorname{Hom}_{\mathcal{S}}(\mathcal{S}, {}_{\mathcal{S}}\mathcal{N}) \simeq \operatorname{Hom}_{\mathcal{T}}({}_{\mathcal{T}}\mathcal{S}, \mathcal{N}) \simeq \operatorname{Hom}_{\mathcal{T}}(\mathcal{E}^{\vee}, \mathcal{N}) \simeq \operatorname{SemiHom}_{\mathcal{T}}(\mathcal{E}^{\vee}, \operatorname{Hom}_{\mathcal{T}}(\mathcal{T}, \mathcal{N})) = {}^{\mathcal{S}}(\Psi_{\mathcal{T}}\mathcal{N})$ by Proposition 6.2.2(d), etc. However, one sometimes has to strengthen the coflatness (coprojectivity, coinjectivity) conditions to the semiflatness (semiprojectivity, semiinjectivity) conditions.

The first assertions of Theorem 8.2.3(a), (b) and (c) do *not* hold for Morita morphisms of semialgebras, though. The derived functors $\mathcal{M}^{\bullet} \longmapsto {}^{L}_{\mathcal{T}}\mathcal{M}^{\bullet}$ and $\mathfrak{P}^{\bullet} \longmapsto {}^{\mathcal{T}}_{R}\mathfrak{P}^{\bullet}$ still can be defined in terms of $\mathcal{S}/\mathcal{C}$-projective ($=$ quite $\mathcal{S}/\mathcal{C}$-semiflat) complexes of $\mathcal{S}$-semimodules and $\mathcal{S}/\mathcal{C}$-injective ($=$ quite $\mathcal{S}/\mathcal{C}$-semiinjective) complexes of $\mathcal{S}$-semicontramodules. The right derived functor $\mathcal{N}^{\bullet} \longmapsto {}^{R}_{\mathcal{S}}\mathcal{N}^{\bullet}$ can be defined in terms of injective complexes of $\mathcal{T}$-semimodules and the left derived functor $\mathfrak{Q}^{\bullet} \longmapsto {}^{\mathcal{S}}_{L}\mathfrak{Q}^{\bullet}$ can be defined in terms of projective complexes of $\mathcal{T}$-semicontramodules (see Remark 6.5).

The results of Corollaries 8.3.2–8.3.4 do *not* hold for Morita morphisms of semialgebras, as one can see in the example of the Morita equivalence related to a Frobenius algebra from Remark 8.4.3 considered as a Morita morphism in the inverse direction. The mentioned results remain valid for left semiprojective and right semiflat Morita morphisms from $\mathcal{S}$ to $\mathcal{T}$ when the categories of $\mathcal{C}$-comodules and $\mathcal{C}$-contramodules have finite homological dimensions, or the Morita morphism of semialgebras is induced by a left coprojective and right coflat Morita morphism of corings, or more generally when the functors $\mathcal{N}^{\bullet} \longmapsto {}_{\mathcal{S}}\mathcal{N}^{\bullet}$, $\mathcal{N}^{\bullet} \longmapsto \mathcal{N}^{\bullet}_{\mathcal{S}}$, and $\mathfrak{Q}^{\bullet} \longmapsto {}^{\mathcal{S}}\mathfrak{Q}^{\bullet}$ map $\mathcal{D}$-coacyclic and $\mathcal{D}$-contraacyclic complexes to $\mathcal{C}$-coacyclic and $\mathcal{C}$-contraacyclic complexes.

Morita equivalences of semialgebras do *not* induce equivalences of the semiderived categories of semimodules and semicontramodules, except in rather special cases. A right semiflat Morita equivalence between $\mathcal{S}$ and $\mathcal{T}$ does induce an equivalence of the semiderived categories of left $\mathcal{S}$-semimodules and left $\mathcal{T}$-semimodules when the categories of left $\mathcal{C}$-comodules and left $\mathcal{D}$-comodules have finite homological dimensions, or when the Morita equivalence comes from a right faithfully coflat Morita morphism of corings and one of the conditions of 8.4.3–8.4.4 is satisfied.

8.4.6 A short summary: one encounters no problems generalizing the results of 7.1–7.4 and 8.1–8.3 to the case of a Morita morphism of k-algebras and related maps of corings and semialgebras. The problems are manageable when one considers Morita morphisms of corings. And there are severe problems with Morita morphisms/equivalences of semialgebras, which do not always respect the essential structure of "an object split in two halves" (see Introduction).

9 Closed Model Category Structures

By a *closed model category* we mean a model category in the sense of Hovey [49]. The closed model categories that we will construct are also *abelian model categories* in the sense of [51], so our results can be viewed as particular cases of the general framework developed in [50].

9.1 Complexes of comodules and contramodules

Let $\mathcal{C}$ be a coring over a k-algebra A. Assume that $\mathcal{C}$ is a projective left and a flat right A-module and the ring A has a finite left homological dimension.

Theorem.

(a) *The category of complexes of left $\mathcal{C}$-comodules has a closed model category structure with the following properties. A morphism is a weak equivalence if and only if its cone is coacyclic. A morphism is a cofibration if and only if it is injective and its cokernel is a complex of A-projective $\mathcal{C}$-comodules. A morphism is a fibration if and only if it is surjective and its kernel is a complex of $\mathcal{C}/A$-injective $\mathcal{C}$-comodules. An object is simultaneously fibrant and cofibrant if and only if it is a complex of coprojective left $\mathcal{C}$-comodules.*

(b) *The category of complexes of left $\mathcal{C}$-contramodules has a closed model category structure with the following properties. A morphism is a weak equivalence if and only if its cone is contraacyclic. A morphism is a cofibration if and only if it is injective and its cokernel is a complex of $\mathcal{C}/A$-projective $\mathcal{C}$-contramodules. A morphism is a fibration if and only if it is surjective and its kernel is a complex of A-injective $\mathcal{C}$-contramodules. An object is simultaneously fibrant and cofibrant if and only if it is a complex of coinjective left $\mathcal{C}$-contramodules.*

Proof. Part (a): the category of complexes of left $\mathcal{C}$-comodules has arbitrary limits and colimits, since it is an abelian category with infinite direct sums and products. The two-out-of-three property of weak equivalences follows from the octahedron axiom, since coacyclic complexes form a triangulated subcategory of the homotopy category of left $\mathcal{C}$-comodules. The retraction properties are clear, since the classes of projective A-modules, $\mathcal{C}/A$-injective $\mathcal{C}$-comodules, and coacyclic complexes of $\mathcal{C}$-comodules are closed under direct summands. It is also clear that a morphism is a trivial cofibration if and only if it is injective and its cokernel is a coacyclic complex of A-projective $\mathcal{C}$-comodules, and a morphism is a trivial fibration if and only if it is surjective and its kernel is a coacyclic complex of $\mathcal{C}/A$-injective $\mathcal{C}$-comodules. Now let us verify the lifting properties.

Lemma 1. *Let U, V, X, and Y be four objects of an abelian category A, $U \longrightarrow V$ be an injective morphism with the cokernel E, and $X \longrightarrow Y$ be a surjective mor-*

L. Positselski, *Homological Algebra of Semimodules and Semicontramodules*, Monografie Matematyczne 70, DOI 10.1007/978-3-0346-0436-9_10, © Springer Basel AG 2010

phism with the kernel K. Suppose that $\mathrm{Ext}^1_A(E, K) = 0$. Then for any two morphisms $U \longrightarrow X$ and $V \longrightarrow Y$ forming a commutative square with the above two morphisms there exists a morphism $V \longrightarrow X$ forming two commutative triangles with the given four morphisms.

Proof. Let us first find a morphism $V \longrightarrow X$ making a commutative triangle with the morphisms $U \longrightarrow X$ and $U \longrightarrow V$. The obstruction to extending the morphism $U \longrightarrow X$ from U to V lies in the group $\mathrm{Ext}^1_A(E, X)$. Since the composition $U \longrightarrow X \longrightarrow Y$ admits an extension to V, our element of $\mathrm{Ext}^1_A(E, X)$ becomes zero in $\mathrm{Ext}^1_A(E, Y)$ and therefore comes from the group $\mathrm{Ext}^1_A(E, K)$. Now let us modify the obtained morphism so that the new morphism $V \longrightarrow X$ forms also a commutative triangle with the morphisms $V \longrightarrow Y$ and $X \longrightarrow Y$. The difference between the given morphism $V \longrightarrow Y$ and the composition $V \longrightarrow X \longrightarrow Y$ is a morphism $V \longrightarrow Y$ annihilating U, so it comes from a morphism $E \longrightarrow Y$. We need to lift the latter to a morphism $E \longrightarrow X$. The obstruction to this lies in $\mathrm{Ext}^1_A(E, K)$.   □

To verify the condition of Lemma 1, consider an extension $\mathcal{E}^\bullet \longrightarrow \mathcal{M}^\bullet \longrightarrow \mathcal{K}^\bullet$ of a complex of A-projective left $\mathcal{C}$-comodules $\mathcal{K}^\bullet$ by a complex of $\mathcal{C}/A$-injective left $\mathcal{C}$-comodules $\mathcal{E}^\bullet$. By Lemma 5.3.1(a), this extension is term-wise split, so it comes from a morphism of complexes of $\mathcal{C}$-comodules $\mathcal{K}^\bullet \longrightarrow \mathcal{E}^\bullet[1]$. Now suppose that one of the complexes $\mathcal{K}^\bullet$ and $\mathcal{E}^\bullet$ is coacyclic. Then any morphism $\mathcal{K}^\bullet \longrightarrow \mathcal{E}^\bullet[1]$ is homotopic to zero by a result of 5.5, hence the extension of complexes $\mathcal{E}^\bullet \longrightarrow \mathcal{M}^\bullet \longrightarrow \mathcal{K}^\bullet$ is split. The lifting properties are proven.

It remains to construct the functorial factorizations. These constructions use two building blocks: one is Lemma 3.1.3(a), the other one is the following Lemma 2(a).

Lemma 2.

(a) *There exists a (not always additive) functor assigning to any left $\mathcal{C}$-comodule an injective morphism from it into a $\mathcal{C}/A$-injective left $\mathcal{C}$-comodule with an A-projective cokernel.*

(b) *There exists a (not always additive) functor assigning to any left $\mathcal{C}$-contramodule a surjective morphism onto it from a $\mathcal{C}/A$-projective left $\mathcal{C}$-contramodule with an A-injective kernel.*

Proof. Part (a): let $\mathcal{M}$ be a left $\mathcal{C}$-comodule and $\mathcal{P}(\mathcal{M}) \longrightarrow \mathcal{M}$ be the surjective morphism onto it from an A-projective $\mathcal{C}$-comodule $\mathcal{P}(\mathcal{M})$ constructed in Lemma 3.1.3(a). Let $\mathcal{K}$ be kernel of the map $\mathcal{P}(\mathcal{M}) \longrightarrow \mathcal{M}$ and let $\mathcal{P}(\mathcal{M}) \longrightarrow \mathcal{C} \otimes_A \mathcal{P}(\mathcal{M})$ be the $\mathcal{C}$-coaction map. Set $\mathcal{J}(\mathcal{M})$ to be the cokernel of the composition

$$\mathcal{K} \longrightarrow \mathcal{P}(\mathcal{M}) \longrightarrow \mathcal{C} \otimes_A \mathcal{P}(\mathcal{M}).$$

Then the composition of maps $\mathcal{P}(\mathcal{M}) \longrightarrow \mathcal{C} \otimes_A \mathcal{P}(\mathcal{M}) \longrightarrow \mathcal{J}(\mathcal{M})$ factorizes through the surjection $\mathcal{P}(\mathcal{M}) \longrightarrow \mathcal{M}$, so there is a natural injective morphism of $\mathcal{C}$-comodules $\mathcal{M} \longrightarrow \mathcal{J}(\mathcal{M})$. The $\mathcal{C}$-comodule $\mathcal{J}(\mathcal{M})$ is $\mathcal{C}/A$-injective as the cokernel of an injective map of $\mathcal{C}/A$-injective $\mathcal{C}$-comodules $\mathcal{K} \longrightarrow \mathcal{C} \otimes_A \mathcal{P}(\mathcal{M})$. The cokernel of the map $\mathcal{M} \longrightarrow \mathcal{J}(\mathcal{M})$ is isomorphic to the cokernel of the map $\mathcal{P}(\mathcal{M}) \longrightarrow \mathcal{C} \otimes_A \mathcal{P}(\mathcal{M})$ and hence A-projective. Part (a) is proven; the construction of the surjective morphism of $\mathcal{C}$-contramodules $\mathfrak{F}(\mathfrak{P}) \longrightarrow \mathfrak{P}$ in part (b) is completely analogous. $\qquad\square$

Let us first decompose functorially an arbitrary morphism of complexes of left $\mathcal{C}$-comodules $\mathcal{L}^\bullet \longrightarrow \mathcal{M}^\bullet$ into a cofibration followed by a fibration. This can be done in either of two dual ways. Let us start with a surjective morphism $\mathcal{P}^+(\mathcal{M}^\bullet) \longrightarrow \mathcal{M}^\bullet$ onto the complex $\mathcal{M}^\bullet$ from a complex of A-projective left $\mathcal{C}$-comodules $\mathcal{P}^+(\mathcal{M}^\bullet)$ constructed as in the proof of Theorem 2.5. Let $\mathcal{K}^\bullet$ be the kernel of the morphism $\mathcal{L}^\bullet \oplus \mathcal{P}^+(\mathcal{M}^\bullet) \longrightarrow \mathcal{M}^\bullet$ and let $\mathcal{K}^\bullet \longrightarrow \mathcal{J}^+(\mathcal{K}^\bullet)$ be an injective morphism from the complex $\mathcal{K}^\bullet$ into a complex of $\mathcal{C}/A$-injective left $\mathcal{C}$-comodules $\mathcal{J}^+(\mathcal{K}^\bullet)$ constructed in an analogous way using Lemma 2. The cokernel of this morphism is a complex of A-projective $\mathcal{C}$-comodules. Let $\mathcal{E}^\bullet$ denote the fibered coproduct of $\mathcal{L}^\bullet \oplus \mathcal{P}^+(\mathcal{M}^\bullet)$ and $\mathcal{J}^+(\mathcal{K}^\bullet)$ over $\mathcal{K}^\bullet$. There is a natural morphism of complexes $\mathcal{E}^\bullet \longrightarrow \mathcal{M}^\bullet$ whose composition with the morphism $\mathcal{J}^+(\mathcal{K}^\bullet) \longrightarrow \mathcal{E}^\bullet$ is zero and composition with the morphism $\mathcal{L}^\bullet \oplus \mathcal{P}^+(\mathcal{M}^\bullet) \longrightarrow \mathcal{E}^\bullet$ is equal to our morphism $\mathcal{L}^\bullet \oplus \mathcal{P}^+(\mathcal{M}^\bullet) \longrightarrow \mathcal{M}^\bullet$. The morphism $\mathcal{L}^\bullet \longrightarrow \mathcal{M}^\bullet$ is equal to the composition

$$\mathcal{L}^\bullet \longrightarrow \mathcal{E}^\bullet \longrightarrow \mathcal{M}^\bullet.$$

The cokernel of the morphism $\mathcal{L}^\bullet \longrightarrow \mathcal{E}^\bullet$ is an extension of the cokernel of the morphism $\mathcal{K}^\bullet \longrightarrow \mathcal{J}^+(\mathcal{K}^\bullet)$ and the complex $\mathcal{P}^+(\mathcal{M}^\bullet)$, hence a complex of A-projective $\mathcal{C}$-comodules. The kernel of the morphism $\mathcal{E}^\bullet \longrightarrow \mathcal{M}^\bullet$ is isomorphic to $\mathcal{J}^+(\mathcal{K}^\bullet)$, which is a complex of $\mathcal{C}/A$-injective $\mathcal{C}$-comodules. Another way is to start with an injective morphism $\mathcal{L}^\bullet \longrightarrow \mathcal{J}^+(\mathcal{L}^\bullet)$ and consider the cokernel of the morphism $\mathcal{L}^\bullet \longrightarrow \mathcal{M}^\bullet \oplus \mathcal{J}^+(\mathcal{L}^\bullet)$.

Now let us construct a factorization of the morphism $\mathcal{L}^\bullet \longrightarrow \mathcal{M}^\bullet$ into a cofibration followed by a trivial fibration. Represent the kernel of the morphism $\mathcal{E}^\bullet \longrightarrow \mathcal{M}^\bullet$ as the quotient complex of a complex of A-projective left $\mathcal{C}$-comodules $\mathcal{E}_1^\bullet$ by a complex of $\mathcal{C}/A$-injective $\mathcal{C}$-comodules; represent the latter complex as the quotient complex of a complex $\mathcal{E}_2^\bullet$ with the analogous properties, etc. The complexes $\mathcal{E}_i^\bullet$ are also complexes of $\mathcal{C}/A$-injective $\mathcal{C}$-comodules as extensions of complexes of $\mathcal{C}/A$-injective $\mathcal{C}$-comodules. For d large enough, the kernel $\mathcal{Z}^\bullet$ of the morphism $\mathcal{E}_d^\bullet \longrightarrow \mathcal{E}_{d-1}^\bullet$ will be a complex of A-projective $\mathcal{C}$-comodules. Actually, $\mathcal{E}_i^\bullet$ and $\mathcal{Z}^\bullet$ are complexes of coprojective $\mathcal{C}$-comodules, as a $\mathcal{C}/A$-injective A-projective left $\mathcal{C}$-comodule $\mathcal{Q}$ is coprojective (since the injection of $\mathcal{C}$-comodules $\mathcal{Q} \longrightarrow \mathcal{C} \otimes_A \mathcal{Q}$ splits, $\mathcal{Q} \longrightarrow \mathcal{C} \otimes_A \mathcal{Q} \longrightarrow \mathcal{C} \otimes_A \mathcal{Q}/\mathcal{Q}$ being an exact triple of A-projective $\mathcal{C}$-comodules). Let $\mathcal{K}^\bullet$ be the total complex of the bicomplex

$$\mathcal{Z}^\bullet \longrightarrow \mathcal{E}_d^\bullet \longrightarrow \cdots \longrightarrow \mathcal{E}_1^\bullet \longrightarrow \mathcal{E}^\bullet.$$

Then the morphism $\mathcal{L}^\bullet \longrightarrow \mathcal{M}^\bullet$ factorizes through $\mathcal{K}^\bullet$ in a natural way, the kernel of the morphism $\mathcal{K}^\bullet \longrightarrow \mathcal{M}^\bullet$ is a coacyclic complex of $\mathcal{C}/A$-injective $\mathcal{C}$-comodules, and the cokernel of the morphism $\mathcal{L}^\bullet \longrightarrow \mathcal{K}^\bullet$ is a complex of A-projective $\mathcal{C}$-comodules. Notice that the complex $\mathcal{K}^\bullet$ is the cone of the natural morphism $\mathbb{L}_1(\ker(\mathcal{E}^\bullet \to \mathcal{M}^\bullet)) \longrightarrow \mathcal{E}^\bullet$, where $\mathbb{L}_1$ denotes the functor from the proof of Theorem 4.5.

Finally, let us construct a factorization of the morphism $\mathcal{L}^\bullet \longrightarrow \mathcal{M}^\bullet$ into a trivial cofibration followed by a fibration. Represent the cokernel of the morphism $\mathcal{L}^\bullet \longrightarrow \mathcal{E}^\bullet$ as a subcomplex of a complex of $\mathcal{C}/A$-injective left $\mathcal{C}$-comodules $^1\mathcal{E}^\bullet$ such that the quotient complex is a complex of A-projective $\mathcal{C}$-comodules; represent this quotient complex as a subcomplex of a complex $^2\mathcal{E}^\bullet$ with the analogous properties, etc. The complexes $^i\mathcal{E}^\bullet$ are also complexes of A-projective $\mathcal{C}$-comodules as extensions of complexes of A-projective $\mathcal{C}$-comodules (so they are complexes of coprojective $\mathcal{C}$-comodules). Let $\mathcal{K}^\bullet$ be the total complex of the bicomplex

$$\mathcal{E}^\bullet \longrightarrow {}^1\mathcal{E}^\bullet \longrightarrow {}^2\mathcal{E}^\bullet \longrightarrow \cdots,$$

constructed by taking infinite direct sums along the diagonals. Then the morphism $\mathcal{L}^\bullet \longrightarrow \mathcal{M}^\bullet$ factorizes through $\mathcal{K}^\bullet$ in a natural way, the cokernel of the morphism $\mathcal{L}^\bullet \longrightarrow \mathcal{K}^\bullet$ is a coacyclic complex of A-projective $\mathcal{C}$-comodules, and the kernel of the morphism $\mathcal{K}^\bullet \longrightarrow \mathcal{M}^\bullet$ is a complex of $\mathcal{C}/A$-injective $\mathcal{C}$-comodules. The class of $\mathcal{C}/A$-injective $\mathcal{C}$-comodules is closed under infinite direct sums by Lemma 5.3.2(a).

Part (a) is proven; the proof of part (b) is completely analogous. $\square$

Remark. It follows from the proof of Lemma 2 that any $\mathcal{C}/A$-injective left $\mathcal{C}$-comodule can be obtained from coinduced $\mathcal{C}$-comodules by taking extensions, cokernels of injective morphisms, and direct summands. Analogously, any $\mathcal{C}/A$-projective left $\mathcal{C}$-contramodule can be obtained from induced $\mathcal{C}$-contramodules by taking extensions, kernels of surjective morphisms, and direct summands.

Let $\mathcal{C}$ and $\mathcal{D}$ be two corings satisfying the above assumptions and $\mathcal{C} \longrightarrow \mathcal{D}$ be a map of corings compatible with a k-algebra map $A \longrightarrow B$. Then the pair of adjoint functors $\mathcal{M}^\bullet \longmapsto {}_B\mathcal{M}^\bullet$ and $\mathcal{N}^\bullet \longmapsto {}_{\mathfrak{C}}\mathcal{N}^\bullet$ is a Quillen adjunction [49] from the category of complexes of left $\mathcal{C}$-comodules to the category of complexes of left $\mathcal{D}$-comodules; the pair of adjoint functors $\mathfrak{Q}^\bullet \longmapsto {}^{\mathcal{C}}\mathfrak{Q}^\bullet$ and $\mathfrak{P}^\bullet \longmapsto {}^{B}\mathfrak{P}^\bullet$ is a Quillen adjunction from the category of complexes of left $\mathcal{D}$-contramodules to the category of complexes of left $\mathcal{C}$-contramodules. The same applies to the case of a Morita morphism $(E, E^\vee)$ from A to B and a morphism ${}_B\mathcal{C}_B \longrightarrow \mathcal{D}$ of corings over B.

The pair of adjoint functors $\Phi_{\mathcal{C}}$ and $\Psi_{\mathcal{C}}$ applied to complexes term-wise is *not* a Quillen equivalence, and not even a Quillen adjunction, between the model category of complexes of left $\mathcal{C}$-contramodules and the model category of complexes of left $\mathcal{C}$-comodules. This pair of functors *is* a Quillen equivalence, however, when $\mathcal{C}$ is a coring over a semisimple ring A. In general, the model categories of complexes of left $\mathcal{C}$-comodules and left $\mathcal{C}$-contramodules can be connected by a chain of three Quillen equivalences (see Remark 9.2.2).

9.2 Complexes of semimodules and semicontramodules

Let S be a semialgebra over a coring $\mathcal{C}$ over a k-algebra A. Assume that $\mathcal{C}$ is a projective left and a flat right A-module, S is a coprojective left and a coflat right $\mathcal{C}$-comodule, and the ring A has a finite left homological dimension.

A left S-semimodule $\mathcal{L}$ is called $S/\mathcal{C}/A$-*projective* if the functor of S-semimodule homomorphisms from $\mathcal{L}$ maps exact triples of $\mathcal{C}/A$-injective left S-semimodules to exact triples. An A-projective left S-semimodule $\mathcal{L}$ is called $S/\mathcal{C}/A$-*semiprojective* if the functor of semihomomorphisms from $\mathcal{L}$ over S maps exact triples of $\mathcal{C}/A$-coinjective left S-semicontramodules to exact triples. Analogously, a left S-semicontramodule $\mathfrak{Q}$ is called $S/\mathcal{C}/A$-*injective* if the functor of S-semicontramodule homomorphisms into $\mathfrak{Q}$ maps exact triples of $\mathcal{C}/A$-projective left S-semicontramodules to exact triples. An A-injective left S-semicontramodule $\mathfrak{Q}$ is called $S/\mathcal{C}/A$-*semiinjective* if the functor of semihomomorphisms into $\mathfrak{Q}$ over S maps exact triples of $\mathcal{C}/A$-coprojective left S-semimodules to exact triples.

As in Lemma 6.4, it follows from Proposition 6.2.2(c) that an A-projective left S-semimodule is $S/\mathcal{C}/A$-projective if and only if it is $S/\mathcal{C}/A$-semiprojective. Analogously, it follows from Proposition 6.2.3(c) that an A-injective left S-semicontramodule is $S/\mathcal{C}/A$-injective if and only if it is $S/\mathcal{C}/A$-semiinjective. It will be shown below that any $S/\mathcal{C}/A$-projective left S-semimodule is A-projective and any $S/\mathcal{C}/A$-injective left S-semicontramodule is A-injective.

Theorem.

(a) *The category of complexes of left S-semimodules has a closed model category structure with the following properties. A morphism is a weak equivalence if and only if its cone is $\mathcal{C}$-coacyclic. A morphism is a cofibration if and only if it is injective and its cokernel is an $S/\mathcal{C}/A$-projective complex of $S/\mathcal{C}/A$-projective S-semimodules. A morphism is a fibration if and only if it is surjective and its kernel is a complex of $\mathcal{C}/A$-injective S-semimodules. An object is simultaneously fibrant and cofibrant if and only if it is a semiprojective complex of semiprojective left S-semimodules.*

(b) *The category of complexes of left S-semicontramodules has a closed model category structure with the following properties. A morphism is a weak equivalence if and only if its cone is $\mathcal{C}$-contraacyclic. A morphism is a cofibration if and only if it is injective and its cokernel is a complex of $\mathcal{C}/A$-projective S-semicontramodules. A morphism is a fibration if and only if it is injective and its cokernel is an $S/\mathcal{C}/A$-injective complex of $S/\mathcal{C}/A$-injective S-semicontramodules. An object is simultaneously fibrant and cofibrant if and only if it is a semiinjective complex of semiinjective left S-semicontramodules.*

Proof. Part (a): existence of limits and colimits, the two-out-of-three property of weak equivalences, and the retraction properties are verified as in the proof of Theorem 9.1. It is clear that a morphism is a trivial cofibration if and only if it is injective and its cokernel is a $\mathcal{C}$-coacyclic $S/\mathcal{C}/A$-projective complex of $S/\mathcal{C}/A$-projective S-semimodules, and a morphism is a trivial fibration if and only if it is surjective and its kernel is a $\mathcal{C}/A$-coacyclic complex of $\mathcal{C}/A$-injective $\mathcal{S}$-semi-modules. To check the lifting properties, use Lemma 9.1.1. Consider an extension $\mathcal{E}^\bullet \longrightarrow \mathcal{M}^\bullet \longrightarrow \mathcal{K}^\bullet$ of an $\mathcal{S}/\mathcal{C}/A$-projective complex of $\mathcal{S}/\mathcal{C}/A$-projective left $\mathcal{S}$-semimodules $\mathcal{K}^\bullet$ by a complex of $\mathcal{C}/A$-injective left $\mathcal{S}$-semimodules $\mathcal{E}^\bullet$. By the next Lemma 1(a), this extension is term-wise split, so it comes from a morphism of complexes of $\mathcal{S}$-semimodules $\mathcal{K}^\bullet \longrightarrow \mathcal{E}^\bullet[1]$. Now suppose that one of the complexes $\mathcal{K}^\bullet$ and $\mathcal{E}^\bullet$ is $\mathcal{C}$-coacyclic. Then any morphism $\mathcal{K}^\bullet \longrightarrow \mathcal{E}^\bullet[1]$ is homotopic to zero by a result of 6.5, hence the extension of complexes $\mathcal{E}^\bullet \longrightarrow \mathcal{M}^\bullet \longrightarrow \mathcal{K}^\bullet$ is split. So after Lemma 1 is proven it will remain to construct the functorial factorizations.

Lemma 1.

(a) *A left $\mathcal{S}$-semimodule $\mathcal{L}$ is $\mathcal{S}/\mathcal{C}/A$-projective if and only if for any $\mathcal{C}/A$-injective left $\mathcal{S}$-semimodule $\mathcal{M}$ the k-modules $\mathrm{Ext}_{\mathcal{S}}^i(\mathcal{L}, \mathcal{M})$ of Yoneda extensions in the abelian category of left $\mathcal{S}$-semimodules vanish for all $i > 0$. The functor of $\mathcal{S}$-semimodule homomorphisms into a $\mathcal{C}/A$-injective $\mathcal{S}$-semimodule maps exact triples of $\mathcal{S}/\mathcal{C}/A$-projective left $\mathcal{S}$-semimodules to exact triples. The classes of $\mathcal{S}/\mathcal{C}/A$-projective left $\mathcal{S}$-semimodules and $\mathcal{S}/\mathcal{C}/A$-projective complexes of $\mathcal{S}/\mathcal{C}/A$-projective left $\mathcal{S}$-semimodules are closed under extensions and kernels of surjective morphisms.*

(b) *A left $\mathcal{S}$-semicontramodule $\mathfrak{Q}$ is $\mathcal{S}/\mathcal{C}/A$-injective if and only if for any $\mathcal{C}/A$-projective left $\mathcal{S}$-semicontramodule $\mathfrak{P}$ the k-modules $\mathrm{Ext}^{\mathcal{S},i}(\mathfrak{P}, \mathfrak{Q})$ of Yoneda extensions in the abelian category of left $\mathcal{S}$-semicontramodules vanish for all $i > 0$. The functor of $\mathcal{S}$-semicontramodule homomorphisms from a $\mathcal{C}/A$-projective $\mathcal{S}$-semicontramodule maps exact triples of $\mathcal{S}/\mathcal{C}/A$-injective left $\mathcal{S}$-semicontramodules to exact triples. The classes of $\mathcal{S}/\mathcal{C}/A$-injective left $\mathcal{S}$-semicontramodules and $\mathcal{S}/\mathcal{C}/A$-injective complexes of $\mathcal{S}/\mathcal{C}/A$-injective left $\mathcal{S}$-semicontramodules are closed under extensions and cokernels of injective morphisms.*

Proof. Part (a): the forgetful functor $\mathcal{S}$-simod $\longrightarrow \mathcal{C}$-comod preserves injective objects, since it is right adjoint to the exact functor of induction. Let us show that any $\mathcal{C}/A$-injective left $\mathcal{S}$-semimodule $\mathcal{M}$ is a subsemimodule of an injective $\mathcal{S}$-semimodule (it will follow that the category of left $\mathcal{S}$-semimodules has enough injectives).

The construction of Lemma 3.3.2(b) assigns to a $\mathcal{C}/A$-coinjective left $\mathcal{S}$-semicontramodule $\mathfrak{P}$ an injective map from it into a semiinjective $\mathcal{S}$-semi-contramodule $\mathfrak{J}(\mathfrak{P})$ with a $\mathcal{C}/A$-coinjective cokernel. Indeed, the cokernel of the map $\mathfrak{P} \longrightarrow \mathfrak{J}(\mathfrak{P})$ is a $\mathcal{C}/A$-coinjective $\mathcal{C}$-contramodule by Lemma 3.1.3(b), so $\mathfrak{J}(\mathfrak{P})$ is a $\mathcal{C}/A$-coinjective $\mathcal{C}$-contramodule as an extension of two $\mathcal{C}/A$-coin-jective $\mathcal{C}$-contramodules and a coinjective $\mathcal{C}$-contramodule as an A-injective and $\mathcal{C}/A$-coinjective $\mathcal{C}$-contramodule. Hence $\mathfrak{J}(\mathfrak{P}) = \mathrm{Cohom}_{\mathcal{C}}(\mathcal{S}, \mathfrak{J}(\mathfrak{P}))$ is a semiinjective $\mathcal{S}$-semicontramodule. The cokernel of the composition of injective morphisms

$$\mathfrak{P} \longrightarrow \mathrm{Cohom}_{\mathcal{C}}(\mathcal{S}, \mathfrak{P}) \longrightarrow \mathrm{Cohom}_{\mathcal{C}}(\mathcal{S}, \mathfrak{J}(\mathfrak{P}))$$

is an extension of the cokernel of the morphism

$$\mathrm{Cohom}_{\mathcal{C}}(\mathcal{S},\mathfrak{P}) \longrightarrow \mathrm{Cohom}_{\mathcal{C}}(\mathcal{S},\mathfrak{I}(\mathfrak{P}))$$

and the cokernel of the morphism $\mathfrak{P} \longrightarrow \mathrm{Cohom}_{\mathcal{C}}(\mathcal{S},\mathfrak{P})$; the former is $\mathcal{C}/A$-coinjective since the cokernel of the morphism $\mathfrak{P} \longrightarrow \mathfrak{I}(\mathfrak{P})$ is, and the latter is $\mathcal{C}/A$-coinjective as a $\mathcal{C}$-contramodule direct summand of $\mathrm{Cohom}_{\mathcal{C}}(\mathcal{S},\mathfrak{P})$. Hence the cokernel of the morphism $\mathfrak{P} \longrightarrow \mathfrak{I}(\mathfrak{P})$ is $\mathcal{C}/A$-coinjective.

Applying these observations to the $\mathcal{S}$-semicontramodule $\mathfrak{P} = \Psi_{\mathcal{S}}(\mathcal{M})$ and using Lemmas 5.3.2(b) and 5.3.1(c), we conclude that

$$\mathcal{M} \simeq \Phi_{\mathcal{S}}\Psi_{\mathcal{S}}(\mathcal{M}) \longrightarrow \Phi_{\mathcal{S}}\mathfrak{I}(\Psi_{\mathcal{S}}\mathcal{M})$$

is an injective morphism of $\mathcal{S}$-semimodules for any $\mathcal{C}/A$-injective left $\mathcal{S}$-semimodule $\mathcal{M}$. Now the functor $\Phi_{\mathcal{S}}$ maps semiinjective $\mathcal{S}$-semicontramodules to injective $\mathcal{S}$-semimodules by Proposition 6.2.2(a).

So any $\mathcal{C}/A$-injective left $\mathcal{S}$-semimodule $\mathcal{M}$ has an injective right resolution; by the construction or by Lemma 5.3.1(a), this resolution is exact with respect to the exact category of $\mathcal{C}/A$-injective $\mathcal{S}$-semimodules. Applying to this resolution the functor of $\mathcal{S}$-semimodule homomorphisms from an $\mathcal{S}/\mathcal{C}/A$-projective left $\mathcal{S}$-semimodule $\mathcal{L}$, we obtain the desired vanishing $\mathrm{Ext}^i_{\mathcal{S}}(\mathcal{L},\mathcal{M}) = 0$ for all $i > 0$. The remaining assertions follow (to verify the assertions related to complexes, notice that the class of acyclic complexes of k-modules is closed under extensions and cokernels of injective morphisms). Part (a) is proven; the proof of part (b) is completely analogous and based on the construction of a semicontramodule projective resolution.

Alternatively, one can argue as in the proof of Lemma 5.3.1(a-b). □

The analogous results for $\mathcal{S}/\mathcal{C}/A$-semiprojective (complexes of) left $\mathcal{S}$-semimodules and $\mathcal{S}/\mathcal{C}/A$-semiinjective (complexes of) left $\mathcal{S}$-semicontramodules can be obtained by considering the derived functor $\mathrm{SemiExt}^*_{\mathcal{S}}(\mathcal{M},\mathfrak{P})$, defined as the cohomology of the object $\mathrm{SemiExt}_{\mathcal{S}}(\mathcal{M},\mathfrak{P})$ of $\mathsf{D}(k\text{-mod})$. For an A-projective $\mathcal{S}$-semimodule $\mathcal{M}$ and a $\mathcal{C}/A$-coinjective $\mathcal{S}$-semicontramodule $\mathfrak{P}$ or a $\mathcal{C}/A$-coprojective $\mathcal{S}$-semimodule $\mathcal{M}$ and an A-injective $\mathcal{S}$-semicontramodule $\mathfrak{P}$ it is computed by the cobar-complex $\mathrm{Cohom}_{\mathcal{C}}(\mathcal{M},\mathfrak{P}) \longrightarrow \mathrm{Cohom}_{\mathcal{C}}(\mathcal{S} \,\square_{\mathcal{C}}\, \mathcal{M},\, \mathfrak{P}) \longrightarrow \cdots$, hence $\mathrm{SemiExt}^i_{\mathcal{S}}(\mathcal{M},\mathfrak{P}) = 0$ for $i > 0$ and $\mathrm{SemiExt}^0_{\mathcal{S}}(\mathcal{M},\mathfrak{P}) = \mathrm{SemiHom}_{\mathcal{S}}(\mathcal{M},\mathfrak{P})$.

Lemma 2.

(a) *There exists a (not always additive) functor assigning to any left $\mathcal{S}$-semimodule an injective morphism from it into a $\mathcal{C}/A$-injective $\mathcal{S}$-semimodule with an $\mathcal{S}/\mathcal{C}/A$-projective cokernel. Furthermore, there exists a functor assigning to any complex of left $\mathcal{S}$-semimodules an injective morphism from it into a complex of $\mathcal{C}/A$-injective $\mathcal{S}$-semimodules such that the cokernel is a $\mathcal{C}$-coacyclic $\mathcal{S}/\mathcal{C}/A$-projective complex of $\mathcal{S}/\mathcal{C}/A$-projective $\mathcal{S}$-semimodules.*

(b) *There exists a (not always additive) functor assigning to any left $\mathbf{S}$-semicon-tramodule a surjective morphism onto it from a $\mathcal{C}/A$-projective $\mathbf{S}$-semicon-tramodule with an $\mathbf{S}/\mathcal{C}/A$-injective kernel. Furthermore, there exists a func-tor assigning to any complex of left $\mathbf{S}$-semicontramodules a surjective mor-phism onto it from a complex of $\mathcal{C}/A$-projective $\mathbf{S}$-semicontramodules such that the kernel is a $\mathcal{C}$-contraacyclic $\mathbf{S}/\mathcal{C}/A$-injective complex of $\mathbf{S}/\mathcal{C}/A$-in-jective $\mathbf{S}$-semicontramodules.*

Proof. Part (a): modify the construction of the second assertion of Lemma 1.3.3, replacing the injective morphism of $\mathcal{C}$-comodules $\mathcal{M} \longrightarrow \mathcal{G}(\mathcal{M}) = \mathcal{C} \otimes_A \mathcal{M}$ with the injective morphism of $\mathcal{C}$-comodules $\mathcal{M} \longrightarrow \mathcal{J}(\mathcal{M})$ constructed in Lemma 9.1.2(a). In other words, for any left $\mathbf{S}$-semimodule $\mathcal{M}$ denote by $\mathcal{K}(\mathcal{M})$ the kernel of the morphism $\mathbf{S}\square_{\mathcal{C}}\mathcal{M} \longrightarrow \mathcal{M}$ and by $\mathcal{Q}(\mathcal{M})$ the cokernel of the composition $\mathcal{K}(\mathcal{M}) \longrightarrow \mathbf{S}\square_{\mathcal{C}}\mathcal{M} \longrightarrow \mathbf{S}\square_{\mathcal{C}}\mathcal{J}(\mathcal{M})$. The composition of maps $\mathbf{S}\square_{\mathcal{C}}\mathcal{M} \longrightarrow \mathbf{S}\square_{\mathcal{C}}\mathcal{J}(\mathcal{M}) \longrightarrow \mathcal{Q}(\mathcal{M})$ factorizes through the surjection $\mathbf{S}\square_{\mathcal{C}}\mathcal{M} \longrightarrow \mathcal{M}$, so there is a natural injective morphism of $\mathbf{S}$-semimodules $\mathcal{M} \longrightarrow \mathcal{Q}(\mathcal{M})$. The cokernel of this morphism is iso-morphic to $\mathbf{S}\square_{\mathcal{C}}(\mathcal{J}(\mathcal{M})/\mathcal{M})$, which is an $\mathbf{S}/\mathcal{C}/A$-projective $\mathbf{S}$-semimodule because $\mathcal{J}(\mathcal{M})/\mathcal{M}$ is an A-projective $\mathcal{C}$-comodule. As in the proof of Lemma 1.3.3, the $\mathbf{S}$-semimodule morphism $\mathcal{M} \longrightarrow \mathcal{Q}(\mathcal{M})$ can be lifted to a $\mathcal{C}$-comodule morphism $\mathcal{M} \longrightarrow \mathbf{S}\square_{\mathcal{C}}\mathcal{J}(\mathcal{M})$. Let $\mathbf{J}(\mathcal{M})$ denote the inductive limit of the sequence

$$\mathcal{M} \longrightarrow \mathbf{S}\square_{\mathcal{C}}\mathcal{J}(\mathcal{M}) \longrightarrow \mathcal{Q}(\mathcal{M}) \longrightarrow \mathbf{S}\square_{\mathcal{C}}\mathcal{J}(\mathcal{Q}(\mathcal{M})) \longrightarrow \mathcal{Q}(\mathcal{Q}(\mathcal{M})) \longrightarrow \cdots;$$

it is the desired $\mathcal{C}/A$-injective $\mathbf{S}$-semimodule into which $\mathcal{M}$ maps injectively with an $\mathbf{S}/\mathcal{C}/A$-projective cokernel. Indeed, $\mathbf{J}(\mathcal{M})$ is $\mathcal{C}/A$-injective by Sublemma 3.3.3.B and the cokernel of the morphism $\mathcal{M} \longrightarrow \mathbf{J}(\mathcal{M})$ is $\mathbf{S}/\mathcal{C}/A$-projective by the next sublemma.

Sublemma.

(a) *Let $0 = \mathcal{U}_0^\bullet \longrightarrow \mathcal{U}_1^\bullet \longrightarrow \mathcal{U}_2^\bullet \longrightarrow \cdots$ be an inductive system of complexes of left $\mathbf{S}$-semimodules such that the successive cokernels $\mathrm{coker}(\mathcal{U}_{i-1}^\bullet \to \mathcal{U}_i^\bullet)$ are $\mathbf{S}/\mathcal{C}/A$-projective complexes of $\mathbf{S}/\mathcal{C}/A$-projective $\mathbf{S}$-semimodules. Then the inductive limit $\varinjlim \mathcal{U}_i^\bullet$ is an $\mathbf{S}/\mathcal{C}/A$-projective complex of $\mathbf{S}/\mathcal{C}/A$-projective $\mathbf{S}$-semimodules.*

(b) *Let $0 = \mathfrak{U}_0^\bullet \longleftarrow \mathfrak{U}_1^\bullet \longleftarrow \mathfrak{U}_2^\bullet \longleftarrow \cdots$ be a projective system of complexes of left $\mathbf{S}$-semicontramodules such that the successive kernels $\ker(\mathfrak{U}_i^\bullet \to \mathfrak{U}_{i-1}^\bullet)$ are $\mathbf{S}/\mathcal{C}/A$-injective complexes of $\mathbf{S}/\mathcal{C}/A$-injective $\mathbf{S}$-semimodules. Then the projective limit $\varprojlim \mathfrak{U}_i^\bullet$ is an $\mathbf{S}/\mathcal{C}/A$-injective complex of $\mathbf{S}/\mathcal{C}/A$-injective $\mathbf{S}$-semimodules.*

Proof. The forgetful functor $\mathbf{S}\text{-simod} \longrightarrow A\text{-mod}$ preserves inductive limits, since it preserves cokernels and infinite direct sums, so one has $\mathrm{Hom}_{\mathbf{S}}(\varinjlim \mathcal{U}_i^\bullet, \mathcal{M}^\bullet) = \varprojlim \mathrm{Hom}_{\mathbf{S}}(\mathcal{U}_i^\bullet, \mathcal{M}^\bullet)$ for any complex of left $\mathbf{S}$-semimodules $\mathcal{M}^\bullet$. Analogously, the forgetful functor $\mathbf{S}\text{-sicntr} \longrightarrow A\text{-mod}$ preserves projective limits, since it preserves kernels and infinite products, so one has $\mathrm{Hom}^{\mathbf{S}}(\mathfrak{P}^\bullet, \varprojlim \mathfrak{U}_i^\bullet) = \varprojlim \mathrm{Hom}^{\mathbf{S}}(\mathfrak{P}^\bullet, \mathfrak{U}_i^\bullet)$

for any complex of left S-semicontramodules $\mathfrak{P}^\bullet$. As the projective limits of sequences of surjective maps preserve exact triples and acyclic complexes, the assertions of the sublemma follow from Lemma 1.   □

The first statement of Lemma 2(a) is proven. To prove the second one, consider the functor assigning to a complex of left S-semimodules $\mathcal{M}^\bullet$ the injective map from it into the complex $\mathfrak{J}^+(\mathcal{M}^\bullet)$, which is constructed in terms of the functor $\mathcal{M} \longmapsto \mathfrak{J}(\mathcal{M})$ as in the proof of Theorem 2.5. By Sublemma 9.2, the cokernel of the morphism $\mathcal{M}^\bullet \longrightarrow \mathfrak{J}^+(\mathcal{M}^\bullet)$ is an $S/\mathcal{C}/A$-projective complex of $S/\mathcal{C}/A$-projective S-semimodules, since a complex of S-semimodules induced from a complex of A-projective $\mathcal{C}$-comodules belongs to this class. Set $^0\mathfrak{J}^\bullet = \mathfrak{J}^+(\mathcal{M}^\bullet)$, $^1\mathfrak{J}^\bullet = \mathfrak{J}^+(\mathrm{coker}(\mathcal{M}^\bullet \to {}^0\mathfrak{J}^\bullet))$, etc. The complexes $^i\mathfrak{J}^\bullet$ are complexes of $\mathcal{C}/A$-injective S-semimodules and the complexes $\mathrm{coker}(\mathcal{M}^\bullet \to {}^0\mathfrak{J}^\bullet)$, $\mathrm{coker}(^{i-1}\mathfrak{J}^\bullet \to {}^i\mathfrak{J}^\bullet)$ are $S/\mathcal{C}/A$-projective complexes of $S/\mathcal{C}/A$-projective S-semimodules. The complexes $^i\mathfrak{J}^\bullet$ for $i > 0$ are also $S/\mathcal{C}/A$-projective complexes of $S/\mathcal{C}/A$-projective S-semimodules as extensions of complexes with these properties. Let $\mathcal{K}^\bullet$ be the total complex of the bicomplex

$$^0\mathfrak{J}^\bullet \longrightarrow {}^1\mathfrak{J}^\bullet \longrightarrow {}^2\mathfrak{J}^\bullet \longrightarrow \cdots,$$

constructed by taking infinite direct sums along the diagonals. Then $\mathcal{K}^\bullet$ is a complex of $\mathcal{C}/A$-injective S-semimodules and the cokernel of the injective morphism $\mathcal{M}^\bullet \longrightarrow \mathcal{K}^\bullet$ is a $\mathcal{C}$-coacyclic (and even S-coacyclic) $S/\mathcal{C}/A$-projective complex of $S/\mathcal{C}/A$-projective S-semimodules. To check the latter properties, one can apply Sublemma 9.2 to the canonical filtration of the complex $^0\mathfrak{J}^\bullet/\mathcal{M}^\bullet \longrightarrow {}^1\mathfrak{J}^\bullet \longrightarrow {}^2\mathfrak{J}^\bullet \longrightarrow \cdots$

The proof of Lemma 2(b) is completely analogous and based on the modification of the construction of the second assertion of Lemma 3.3.3(a) using the surjective morphism of $\mathcal{C}$-contramodules $\mathfrak{F}(\mathfrak{P}) \longrightarrow \mathfrak{P}$ from Lemma 9.1.2(b) in place of the morphism $\mathfrak{G}(\mathfrak{P}) = \mathrm{Hom}_A(\mathcal{C}, \mathfrak{P}) \longrightarrow \mathfrak{P}$.   □

In the sequel we will denote by $\mathcal{M} \longmapsto \mathfrak{J}(\mathcal{M})$ the functor constructed in Lemma 2 rather than its more simplistic version from Lemmas 1.3.3 and 3.3.3.

Lemma 3.

(a) *There exists a (not always additive) functor assigning to any left S-semimodule a surjective map onto it from an $S/\mathcal{C}/A$-projective S-semimodule with a $\mathcal{C}/A$-injective kernel. Furthermore, there exists a functor assigning to any complex of left S-semimodules a surjective map onto it from an $S/\mathcal{C}/A$-projective complex of $S/\mathcal{C}/A$-projective S-semimodules such that the kernel is a $\mathcal{C}$-coacyclic complex of $\mathcal{C}/A$-injective S-semimodules.*

(b) *There exists a (not always additive) functor assigning to any left S-semicontramodule an injective map from it into an $S/\mathcal{C}/A$-injective S-semicontramodule with a $\mathcal{C}/A$-projective cokernel. Furthermore, there exists a functor assigning to any complex of left S-semicontramodules an injective map from it into an $S/\mathcal{C}/A$-injective complex of $S/\mathcal{C}/A$-injective S-semicontramodules* *such that the cokernel is a $\mathcal{C}$-contraacyclic complex of $\mathcal{C}/A$-projective $\mathbf{S}$-semi-contramodules.*

Proof. Part (a): for any left $\mathbf{S}$-semimodule $\mathbf{L}$, consider the injective morphism $\mathbf{L} \longrightarrow \mathbf{J}(\mathbf{L})$ from Lemma 2 and denote by $\mathbf{K}(\mathbf{L})$ its cokernel. The functor $\mathbf{M} \longmapsto \mathbf{P}(\mathbf{M})$ of Lemmas 1.3.2 and 3.3.2 assigns to a $\mathcal{C}/A$-injective left $\mathbf{S}$-semimodule $\mathbf{M}$ a surjective morphism onto it from the $\mathbf{S}$-semimodule $\mathbf{P}(\mathbf{M})$ induced from a coprojective $\mathcal{C}$-comodule $\mathcal{P}(\mathbf{M})$ such that the kernel is a $\mathcal{C}/A$-injective $\mathbf{S}$-semimod-ule (see the proof of Lemma 1). Denote by $\mathbf{F}(\mathbf{L})$ the kernel of the composition

$$\mathbf{P}(\mathbf{J}(\mathbf{L})) \longrightarrow \mathbf{J}(\mathbf{L}) \longrightarrow \mathbf{K}(\mathbf{L}).$$

The composition of maps $\mathbf{F}(\mathbf{L}) \longrightarrow \mathbf{P}(\mathbf{J}(\mathbf{L})) \longrightarrow \mathbf{J}(\mathbf{L})$ factorizes through the injection $\mathbf{L} \longrightarrow \mathbf{J}(\mathbf{L})$, so there is a natural surjective morphism of $\mathbf{S}$-semimodules $\mathbf{F}(\mathbf{L}) \longrightarrow \mathbf{L}$. The $\mathbf{S}$-semimodules $\mathbf{P}(\mathbf{J}(\mathbf{L}))$ and $\mathbf{K}(\mathbf{L})$ are $\mathbf{S}/\mathcal{C}/A$-projective, hence the $\mathbf{S}$-semimodule $\mathbf{F}(\mathbf{L})$ is also $\mathbf{S}/\mathcal{C}/A$-projective. The kernel of the morphism $\mathbf{F}(\mathbf{L}) \longrightarrow \mathbf{L}$ is $\mathcal{C}/A$-injective, since it is isomorphic to the kernel of the morphism $\mathbf{P}(\mathbf{J}(\mathbf{L})) \longrightarrow \mathbf{J}(\mathbf{L})$. This proves the first statement of part (a).

Now consider the functor assigning to any complex of left $\mathbf{S}$-semimodules $\mathbf{L}^\bullet$ the surjective map onto it from the complex $\mathbf{F}^+(\mathbf{L}^\bullet)$. The complex $\mathbf{F}^+(\mathbf{L}^\bullet)$ is $\mathbf{S}/\mathcal{C}/A$-projective as the kernel of a surjective morphism of $\mathbf{S}/\mathcal{C}/A$-projective complexes; it is also a complex of $\mathbf{S}/\mathcal{C}/A$-projective and A-projective $\mathbf{S}$-semimod-ules. Set $\mathbf{F}_0^\bullet = \mathbf{F}^+(\mathbf{L}^\bullet)$, $\mathbf{F}_1^\bullet = \mathbf{F}^+(\ker(\mathbf{F}_0^\bullet \to \mathbf{L}^\bullet))$, etc. The complexes $\ker(\mathbf{F}_0^\bullet \to \mathbf{L}^\bullet)$, $\ker(\mathbf{F}_i^\bullet \to \mathbf{F}_{i-1}^\bullet)$ are complexes of $\mathcal{C}/A$-injective $\mathbf{S}$-semimodules, hence the complexes $\mathbf{F}_i^\bullet$ for $i > 0$ are also complexes of $\mathcal{C}/A$-injective $\mathbf{S}$-semimodules as extensions of complexes of $\mathcal{C}/A$-injective $\mathbf{S}$-semimodules. For d large enough, the kernel $\mathbf{Z}^\bullet$ of the morphism $\mathbf{F}_{d-1}^\bullet \longrightarrow \mathbf{F}_{d-2}^\bullet$ will be a complex of A-projective $\mathbf{S}$-semimodules, and consequently a complex of $\mathcal{C}$-coprojective $\mathbf{S}$-semimodules. Let $\mathcal{E}^\bullet$ be the total complex of the bicomplex

$$\cdots \longrightarrow \mathbf{S} \square_\mathcal{C} \mathbf{S} \square_\mathcal{C} \mathbf{Z}^\bullet \longrightarrow \mathbf{S} \square_\mathcal{C} \mathbf{Z}^\bullet \longrightarrow \mathbf{F}_{d-1}^\bullet \longrightarrow \mathbf{F}_{d-2}^\bullet \longrightarrow \cdots \longrightarrow \mathbf{F}_0^\bullet,$$

constructed by taking infinite direct sums along the diagonals. Then the complex $\mathcal{E}^\bullet$ is a complex of $\mathbf{S}/\mathcal{C}/A$-projective $\mathbf{S}$-semimodules, and it is an $\mathbf{S}/\mathcal{C}/A$-projec-tive complex since it is homotopy equivalent to a complex obtained from such complexes using the operations of cone and infinite direct sum. The kernel of the morphism $\mathcal{E}^\bullet \longrightarrow \mathbf{L}^\bullet$ is a complex of $\mathcal{C}/A$-injective $\mathbf{S}$-semimodules, and it is $\mathcal{C}$-coacyclic since it contains a $\mathcal{C}$-contractible subcomplex of $\mathbf{S}$-semimodules such that the quotient complex is the total complex of a finite exact complex of com-plexes of $\mathbf{S}$-semimodules. Part (a) is proven; the proof of part (b) is completely analogous.   □

Let us show that any $\mathbf{S}/\mathcal{C}/A$-projective left $\mathbf{S}$-semimodule $\mathbf{L}$ is A-projective. Consider the surjective morphism $\mathbf{F}(\mathbf{L}) \longrightarrow \mathbf{L}$ from Lemma 3. Since its kernel is $\mathcal{C}/A$-injective, we have an extension of an $\mathbf{S}/\mathcal{C}/A$-projective left $\mathbf{S}$-semimodule by a $\mathcal{C}/A$-injective left $\mathbf{S}$-semimodule, which is always trivial by Lemma 1. Therefore, $\mathbf{\mathcal{L}}$ is a direct summand of $\mathbf{\mathcal{F}}(\mathbf{\mathcal{L}})$, while $\mathbf{\mathcal{F}}(\mathbf{\mathcal{L}})$ is A-projective by the construction. Analogously, any $\mathbf{S}/\mathcal{C}/A$-injective left $\mathbf{S}$-semicontramodule is A-injective.

Let us show that any $\mathbf{S}/\mathcal{C}/A$-projective $\mathcal{C}/A$-injective left $\mathbf{S}$-semimodule $\mathbf{\mathcal{M}}$ is a direct summand of the $\mathbf{S}$-semimodule induced from the $\mathcal{C}$-comodule coinduced from a projective A-module; in particular, a left $\mathbf{S}$-semimodule is simultaneously $\mathbf{S}/\mathcal{C}/A$-projective and $\mathcal{C}/A$-injective if and only if it is semiprojective. Consider the exact triple $\mathbf{\mathcal{K}} \longrightarrow \mathbf{S} \,\square_{\mathcal{C}}\, \mathbf{\mathcal{M}} \longrightarrow \mathbf{\mathcal{M}}$, where $\mathbf{\mathcal{K}} = \ker(\mathbf{S} \,\square_{\mathcal{C}}\, \mathbf{\mathcal{M}} \to \mathbf{\mathcal{M}})$. If an $\mathbf{S}$-semimodule $\mathbf{\mathcal{M}}$ is $\mathcal{C}/A$-injective, then so is the $\mathbf{S}$-semimodule $\mathbf{S} \,\square_{\mathcal{C}}\, \mathbf{\mathcal{M}}$, since $\mathcal{C}/A$-injectivity is equivalent to $\mathcal{C}/A$-coprojectivity; then the $\mathbf{S}$-semimodule $\mathbf{\mathcal{K}}$ is $\mathcal{C}/A$-injective as a $\mathcal{C}$-comodule direct summand of $\mathbf{S} \,\square_{\mathcal{C}}\, \mathbf{\mathcal{M}}$. If the $\mathbf{S}$-semimodule $\mathbf{\mathcal{M}}$ is also $\mathbf{S}/\mathcal{C}/A$-projective, then our exact triple splits over $\mathbf{S}$ and $\mathbf{\mathcal{M}}$ is a direct summand of the induced $\mathbf{S}$-semimodule $\mathbf{S}\square_{\mathcal{C}}\mathbf{\mathcal{M}}$. Since the $\mathcal{C}$-comodule $\mathbf{\mathcal{M}}$ is A-projective and $\mathcal{C}/A$-injective, it is a direct summand of the $\mathcal{C}$-comodule coinduced from a projective A-module. Analogously, any $\mathbf{S}/\mathcal{C}/A$-injective $\mathcal{C}/A$-projective left $\mathbf{S}$-semicontramodule $\mathbf{\mathfrak{P}}$ is a direct summand of the $\mathbf{S}$-semicontramodule coinduced from the $\mathcal{C}$-contramodule induced from an injective A-module; in particular, a left $\mathbf{S}$-semicontramodule is simultaneously $\mathbf{S}/\mathcal{C}/A$-injective and $\mathcal{C}/A$-projective if and only if it is semiinjective. In other words, $\mathbf{\mathcal{M}}$ is a direct summand of a direct sum of copies of the $\mathbf{S}$-semimodule $\mathbf{S}$ and $\mathbf{\mathfrak{P}}$ is a direct summand of a product of copies of the $\mathbf{S}$-semicontramodule $\mathrm{Hom}_k(\mathbf{S}, k^\vee)$.

An $\mathbf{S}/\mathcal{C}/A$-projective complex of $\mathcal{C}$-coprojective left $\mathbf{S}$-semimodules $\mathbf{\mathcal{M}}^\bullet$ is homotopy equivalent to a complex obtained from complexes of $\mathbf{S}$-semimodules induced from complexes of $\mathcal{C}$-coprojective $\mathcal{C}$-comodules using the operations of cone and infinite direct sum. In particular, the complex $\mathbf{\mathcal{M}}^\bullet$ is semiprojective. Indeed, the total complex of the bicomplex $\cdots \longrightarrow \mathbf{S} \,\square_{\mathcal{C}}\, \mathbf{S} \,\square_{\mathcal{C}}\, \mathbf{\mathcal{M}} \longrightarrow \mathbf{S} \,\square_{\mathcal{C}}\, \mathbf{\mathcal{M}} \longrightarrow \mathbf{\mathcal{M}}$ is contractible, being a $\mathcal{C}$-coacyclic $\mathbf{S}/\mathcal{C}/A$-projective complex of $\mathcal{C}/A$-injective left $\mathbf{S}$-semimodules. Analogously, an $\mathbf{S}/\mathcal{C}/A$-injective complex of $\mathcal{C}$-coinjective left $\mathbf{S}$-semicontramodules $\mathbf{\mathfrak{P}}^\bullet$ is homotopy equivalent to a complex obtained from complexes of $\mathbf{S}$-semicontramodules coinduced from complexes of $\mathcal{C}$-coinjective $\mathcal{C}$-contramodules using the operations of cone and infinite product. In particular, the complex $\mathbf{\mathfrak{P}}^\bullet$ is semiinjective.

Finally we turn to the construction of functorial factorizations. As in the proof of Theorem 9.1, let us first decompose an arbitrary morphism of complexes of left $\mathbf{S}$-semimodules $\mathbf{\mathcal{L}}^\bullet \longrightarrow \mathbf{\mathcal{M}}^\bullet$ into a cofibration followed by a fibration. This can be done in either of two dual ways. Let us start with an injective morphism from the complex $\mathbf{\mathcal{L}}^\bullet$ into the complex $\mathbf{\mathcal{J}}^+(\mathbf{\mathcal{L}}^\bullet)$ constructed in Lemma 2. Let $\mathbf{\mathcal{K}}^\bullet$ be the cokernel of the morphism $\mathbf{\mathcal{L}}^\bullet \longrightarrow \mathbf{\mathcal{M}}^\bullet \oplus \mathbf{\mathcal{J}}^+(\mathbf{\mathcal{L}}^\bullet)$ and let $\mathbf{\mathcal{F}}^+(\mathbf{\mathcal{K}}^\bullet) \longrightarrow \mathbf{\mathcal{K}}^\bullet$ be the surjective morphism onto the complex $\mathbf{\mathcal{K}}^\bullet$ from the complex $\mathbf{\mathcal{F}}^+(\mathbf{\mathcal{K}}^\bullet)$ constructed in Lemma 3. Let $\mathbf{\mathcal{L}}^\bullet \longrightarrow \mathbf{\mathcal{E}}^\bullet \longrightarrow \mathbf{\mathcal{F}}^+(\mathbf{\mathcal{K}}^\bullet)$ be the pull-back of the exact triple $\mathbf{\mathcal{L}}^\bullet \longrightarrow \mathbf{\mathcal{M}}^\bullet \oplus \mathbf{\mathcal{J}}^+(\mathbf{\mathcal{L}}^\bullet) \longrightarrow \mathbf{\mathcal{K}}^\bullet$ with respect to the morphism $\mathbf{\mathcal{F}}^+(\mathbf{\mathcal{K}}^\bullet) \longrightarrow \mathbf{\mathcal{K}}^\bullet$. Then the morphism $\mathbf{\mathcal{L}}^\bullet \longrightarrow \mathbf{\mathcal{M}}^\bullet$ is equal to the composition

$$\mathbf{\mathcal{L}}^\bullet \longrightarrow \mathbf{\mathcal{E}}^\bullet \longrightarrow \mathbf{\mathcal{M}}^\bullet.$$

The cokernel $\mathcal{F}^+(\mathcal{K}^\bullet)$ of the morphism $\mathcal{L}^\bullet \longrightarrow \mathcal{E}^\bullet$ is an $\mathcal{S}/\mathcal{C}/A$-projective complex of $\mathcal{S}/\mathcal{C}/A$-projective $\mathcal{S}$-semimodules. The kernel of the morphism $\mathcal{E}^\bullet \longrightarrow \mathcal{M}^\bullet$ is an extension of the complex $\mathcal{J}^+(\mathcal{L}^\bullet)$ and the kernel of the morphism $\mathcal{F}^+(\mathcal{K}^\bullet) \longrightarrow \mathcal{K}^\bullet$, hence a complex of $\mathcal{C}/A$-injective $\mathcal{S}$-semimodules. Another way is to start with the surjective morphism $\mathcal{F}^+(\mathcal{M}^\bullet) \longrightarrow \mathcal{M}^\bullet$ and consider the kernel of the morphism $\mathcal{L}^\bullet \oplus \mathcal{F}^+(\mathcal{M}^\bullet) \longrightarrow \mathcal{M}^\bullet$.

Now let us construct a factorization of the morphism $\mathcal{L}^\bullet \longrightarrow \mathcal{M}^\bullet$ into a cofibration followed by a trivial fibration. Represent the kernel of the morphism $\mathcal{E}^\bullet \longrightarrow \mathcal{M}^\bullet$ as the quotient complex of an $\mathcal{S}/\mathcal{C}/A$-projective complex of $\mathcal{S}/\mathcal{C}/A$-projective left $\mathcal{S}$-semimodules $\mathcal{P}^\bullet$ by a $\mathcal{C}$-coacyclic complex of $\mathcal{C}/A$-injective $\mathcal{S}$-semimodules. Then the complex $\mathcal{P}^\bullet$ is also a complex of $\mathcal{C}/A$-injective $\mathcal{S}$-semimodules (so it is even a semiprojective complex of semiprojective $\mathcal{S}$-semimodules). Let $\mathcal{K}^\bullet$ be the cone of the morphism $\mathcal{P}^\bullet \longrightarrow \mathcal{E}^\bullet$. Then the morphism $\mathcal{L}^\bullet \longrightarrow \mathcal{M}^\bullet$ factorizes through $\mathcal{K}^\bullet$ in a natural way, the kernel of the morphism $\mathcal{K}^\bullet \longrightarrow \mathcal{M}^\bullet$ is a $\mathcal{C}$-coacyclic complex of $\mathcal{C}/A$-injective $\mathcal{S}$-semimodules, and the cokernel of the morphism $\mathcal{L}^\bullet \longrightarrow \mathcal{K}^\bullet$ is an $\mathcal{S}/\mathcal{C}/A$-projective complex of $\mathcal{S}/\mathcal{C}/A$-projective $\mathcal{S}$-semimodules.

It remains to construct a factorization of the morphism $\mathcal{L}^\bullet \longrightarrow \mathcal{M}^\bullet$ into a trivial cofibration followed by a fibration. Represent the cokernel of the morphism $\mathcal{L}^\bullet \longrightarrow \mathcal{E}^\bullet$ as a subcomplex of a complex of $\mathcal{C}/A$-injective $\mathcal{S}$-semimodules $\mathcal{Q}^\bullet$ such that the quotient complex is a $\mathcal{C}$-coacyclic $\mathcal{S}/\mathcal{C}/A$-projective complex of $\mathcal{S}/\mathcal{C}/A$-projective $\mathcal{S}$-semimodules. Then the complex $\mathcal{Q}^\bullet$ is also an $\mathcal{S}/\mathcal{C}/A$-projective complex of $\mathcal{S}/\mathcal{C}/A$-projective $\mathcal{S}$-semimodules (hence a semiprojective complex of semiprojective $\mathcal{S}$-semimodules). Let $\mathcal{K}^\bullet$ be the cocone of the morphism $\mathcal{E}^\bullet \longrightarrow \mathcal{Q}^\bullet$. Then the morphism $\mathcal{L}^\bullet \longrightarrow \mathcal{M}^\bullet$ factorizes through $\mathcal{K}^\bullet$ in a natural way, the kernel of the morphism $\mathcal{K}^\bullet \longrightarrow \mathcal{M}^\bullet$ is a complex of $\mathcal{C}/A$-injective $\mathcal{S}$-semimodules, and the cokernel of the morphism $\mathcal{L}^\bullet \longrightarrow \mathcal{K}^\bullet$ is a $\mathcal{C}$-coacyclic $\mathcal{S}/\mathcal{C}/A$-projective complex of $\mathcal{S}/\mathcal{C}/A$-projective $\mathcal{S}$-semimodules.

Part (a) of the theorem is proven; the proof of part (b) is completely analogous. $\square$

Remark 1. One can obtain descriptions of $\mathcal{S}/\mathcal{C}/A$-projective complexes of $\mathcal{S}/\mathcal{C}/A$-projective $\mathcal{S}$-semimodules, $\mathcal{C}$-coacyclic $\mathcal{S}/\mathcal{C}/A$-projective complexes of $\mathcal{S}/\mathcal{C}/A$-projective $\mathcal{S}$-semimodules, etc., from the proof of the above theorem. Namely, let $\mathcal{M}^\bullet$ be an $\mathcal{S}/\mathcal{C}/A$-projective complex of $\mathcal{S}/\mathcal{C}/A$-projective left $\mathcal{S}$-semimodules; decompose the morphism $0 \longrightarrow \mathcal{M}^\bullet$ into a cofibration $0 \longrightarrow \mathcal{K}^\bullet$ followed by a trivial fibration $\mathcal{K}^\bullet \longrightarrow \mathcal{M}^\bullet$ by the above construction (this can be also obtained directly from Lemma 3). Then the complex $\mathcal{M}^\bullet$ is a direct summand of $\mathcal{K}^\bullet$ and therefore can be obtained from complexes of $\mathcal{S}$-semimodules induced from complexes of A-projective $\mathcal{C}$-comodules using the operations of cone, infinitely iterated extension in the sense of inductive limit, and kernel of surjective morphism. Let $\mathcal{M}^\bullet$ be a $\mathcal{C}$-coacyclic $\mathcal{S}/\mathcal{C}/A$-projective complex of $\mathcal{S}/\mathcal{C}/A$-projective left $\mathcal{S}$-semimodules; decompose the morphism $0 \longrightarrow \mathcal{M}^\bullet$ into a trivial cofibration $0 \longrightarrow \mathcal{K}^\bullet$ followed by a fibration $\mathcal{K}^\bullet \longrightarrow \mathcal{M}^\bullet$ by the above construction. Then the complex $\mathcal{M}^\bullet$ is a direct summand of $\mathcal{K}^\bullet$ and therefore up to the homotopy equivalence can be obtained from the total complexes of exact triples of $\mathcal{S}/\mathcal{C}/A$-projective complexes of $\mathcal{S}/\mathcal{C}/A$-projective $\mathcal{S}$-semimodules using the operations of cone and infinite direct sum. The analogous results hold for complexes of left $\mathcal{S}$-semicontramodules.

Let $\mathcal{S}$ and $\mathcal{T}$ be two semialgebras satisfying the above assumptions and $\mathcal{S} \longrightarrow \mathcal{T}$ be a map of semialgebras compatible with a map of corings $\mathcal{C} \longrightarrow \mathcal{D}$ and a k-algebra map $A \longrightarrow B$. Then the pair of adjoint functors $\mathcal{M}^\bullet \longmapsto {}_\mathcal{T}\mathcal{M}^\bullet$ and $\mathcal{N}^\bullet \longmapsto {}_\mathcal{C}\mathcal{N}^\bullet$ is a Quillen adjunction from the category of complexes of left $\mathcal{S}$-semimodules to the category of complexes of left $\mathcal{T}$-semimodules; the pair of adjoint functors $\mathfrak{Q}^\bullet \longmapsto {}^\mathcal{C}\mathfrak{Q}^\bullet$ and $\mathfrak{P}^\bullet \longmapsto {}^\mathcal{T}\mathfrak{P}^\bullet$ is a Quillen adjunction from the category of complexes of left $\mathcal{T}$-semicontramodules to the category of complexes of left $\mathcal{S}$-semicontramodules. These results follow from Theorems 7.2.1 and 8.2.3(c). They also hold in the case of a left coprojective and right coflat Morita morphism $(\mathcal{E}, \mathcal{E}^\vee)$ from $\mathcal{C}$ to $\mathcal{D}$ and a morphism $\mathcal{S} \longrightarrow {}_\mathcal{C}\mathcal{T}_\mathcal{C}$ of semialgebras over $\mathcal{C}$.

The pair of adjoint functors $\Phi_\mathcal{S}$ and $\Psi_\mathcal{S}$ applied to complexes term-wise is *not* a Quillen equivalence, and not even a Quillen adjunction, between the model category of complexes of left $\mathcal{S}$-semicontramodules and the model category of complexes of left $\mathcal{S}$-semimodules. Instead, this pair of adjoint functors between closed model categories has the following properties.

The functor $\Phi_\mathcal{S}$ maps trivial cofibrations of complexes of left $\mathcal{S}$-semicontramodules to weak equivalences of complexes of left $\mathcal{S}$-semimodules. The functor $\Psi_\mathcal{S}$ maps trivial fibrations of complexes of left $\mathcal{S}$-semimodules to weak equivalences of complexes of left $\mathcal{S}$-semicontramodules. For a cofibrant complex of left $\mathcal{S}$-semicontramodules $\mathfrak{P}^\bullet$ and a fibrant complex of left $\mathcal{S}$-semimodules $\mathcal{M}^\bullet$, a morphism $\Phi_\mathcal{S}(\mathfrak{P}^\bullet) \longrightarrow \mathcal{M}^\bullet$ is a weak equivalence if and only if the corresponding morphism $\mathfrak{P}^\bullet \longrightarrow \Psi_\mathcal{S}(\mathcal{M}^\bullet)$ is a weak equivalence. Furthermore, the functor $\Phi_\mathcal{S}$ maps cofibrant complexes to fibrant ones, while the functor $\Psi_\mathcal{S}$ maps fibrant complexes to cofibrant ones. The restrictions of the functors $\Phi_\mathcal{S}$ and $\Psi_\mathcal{S}$ are mutually inverse equivalences between the full subcategories formed by cofibrant complexes of left $\mathcal{S}$-semicontramodules and fibrant complexes of right $\mathcal{S}$-semimodules. These restrictions of functors also send weak equivalences to weak equivalences.

Remark 2. One can connect the above model categories of complexes of left $\mathcal{S}$-semimodules and left $\mathcal{S}$-semicontramodules by a chain of three Quillen adjunctions by considering other model category structures on these two categories. The above model category structure on the category of complexes of left $\mathcal{S}$-semimodules can be called the semiprojective model structure, and the model category structure on the category of complexes of left $\mathcal{S}$-semicontramodules can be called the semiinjective model structure. In addition to these, there is also the injective model structure on the category of complexes of left $\mathcal{S}$-semimodules, and the projective model structure on the category of complexes of left $\mathcal{S}$-semicontramodules. In these alternative model structures, weak equivalences are still morphisms with $\mathcal{C}$-coacyclic or $\mathcal{C}$-contraacyclic cones, respectively. A morphism of complexes of semimodules is a cofibration if and only if it is injective, and a morphism of complexes of semicontramodules is a fibration if and only if it is surjective. A morphism of complexes of semimodules is a fibration if and only if it is surjective and its kernel is an injective complex of injective semimodules in the sense of Remark 6.5 and the proof of Lemma 1 above; a morphism of complexes of semicontramodules is a cofibration if and only if it is injective and its cokernel is a projective complex of projective semicontramodules. One can check that these are model category structures in a way analogous to (and much simpler than) the proof of Theorem 9.2 above. The functors Φ_S and Ψ_S are a Quillen equivalence between the injective and the projective model category structures; the identity functors are Quillen equivalences between the semiprojective and the injective model structures, and between the semiinjective and the projective model structures.

10   A Construction of Semialgebras

10.1   Construction of comodules and contramodules

10.1.1   Let A and B be associative k-algebras.

For any projective finitely generated left A-module U and any left A-module V there is a natural isomorphism $\mathrm{Hom}_A(U, A) \otimes_A V \simeq \mathrm{Hom}_A(U, V)$ given by the formula $u^* \otimes v \longmapsto (u \mapsto \langle u, u^* \rangle v)$. In particular, for any A-B-bimodule C and any projective finitely generated left B-module D there are natural isomorphisms $\mathrm{Hom}_A(C \otimes_B D, A) \simeq \mathrm{Hom}_B(D, \mathrm{Hom}_A(C, A)) \simeq \mathrm{Hom}_B(D, B) \otimes_B \mathrm{Hom}_A(C, A)$.

It follows that there is a tensor anti-equivalence between the tensor category of A-A-bimodules that are projective and finitely generated as left A-modules and the tensor category of A-A-bimodules that are projective and finitely generated as right A-modules, given by the mutually-inverse functors $C \longmapsto \mathrm{Hom}_A(C, A)$ and $K \longmapsto \mathrm{Hom}_{A^{\mathrm{op}}}(K, A)$. Therefore, noncommutative ring structures on a right-projective and finitely generated A-A-bimodule K correspond bijectively to coring structures on the left-projective and finitely generated A-A-bimodule $\mathrm{Hom}_{A^{\mathrm{op}}}(K, A)$. On the other hand, for any coring $\mathcal{C}$ over A there is a natural structure of a k-algebra on $\mathrm{Hom}_A(\mathcal{C}, A)$ together with a morphism of k-algebras $A \longrightarrow \mathrm{Hom}_A(\mathcal{C}, A)$.

Furthermore, let K be a k-algebra endowed with a k-algebra map $A \longrightarrow K$ such that K is a finitely generated projective right A-module, and let $\mathcal{C} = \mathrm{Hom}_{A^{\mathrm{op}}}(K, A)$ be the corresponding coring over A. Then the natural isomorphism $N \otimes_A \mathcal{C} = \mathrm{Hom}_{A^{\mathrm{op}}}(K, N)$ for a right A-module N provides a bijective correspondence between the structures of right K-module and right $\mathcal{C}$-comodule on N. Analogously, the natural isomorphism $\mathrm{Hom}_A(\mathcal{C}, P) = K \otimes_A P$ for a left A-module P provides a bijective correspondence between the structures of left K-module and left $\mathcal{C}$-contramodule on P. In other words, there are isomorphisms of abelian categories comod-$\mathcal{C} \simeq$ mod-K and $\mathcal{C}$-contra $\simeq K$-mod.

10.1.2   Here is a generalization of the situation we just described. Let $\mathcal{C}$ be a coring over a k-algebra A and K be a k-algebra endowed with a k-algebra map $A \longrightarrow K$. Suppose that we are given a pairing

$$\phi \colon \mathcal{C} \otimes_A K \longrightarrow A$$

which is an A-A-bimodule map satisfying the following conditions of compatibility with the comultiplication in $\mathcal{C}$ and the multiplication in K and with the counit of $\mathcal{C}$ and the unit of K. First, the composition $\mathcal{C} \otimes_A K \otimes_A K \longrightarrow \mathcal{C} \otimes_A \mathcal{C} \otimes_A K \otimes_A K \longrightarrow \mathcal{C} \otimes_A K \longrightarrow A$ of the map induced by the comultiplication in $\mathcal{C}$, the map induced by the pairing ϕ, and the pairing ϕ itself should be equal to the composition $\mathcal{C} \otimes_A K \otimes_A K \longrightarrow \mathcal{C} \otimes_A K \longrightarrow A$ of the map induced by the multiplication in A

L. Positselski, *Homological Algebra of Semimodules and Semicontramodules*, Monografie Matematyczne 70, DOI 10.1007/978-3-0346-0436-9_11, © Springer Basel AG 2010

and the pairing ϕ. Second, the composition $\mathcal{C} = \mathcal{C} \otimes_A A \longrightarrow \mathcal{C} \otimes_A K \longrightarrow A$ of the map coming from the unit of K with the pairing ϕ should be equal to the counit of $\mathcal{C}$. Equivalently, the map $K \longrightarrow \operatorname{Hom}_A(\mathcal{C}, A)$ induced by ϕ should be a morphism of k-algebras.

Then for any right $\mathcal{C}$-comodule $\mathcal{N}$ the composition $\mathcal{N} \otimes_A K \longrightarrow \mathcal{N} \otimes_A \mathcal{C} \otimes_A K \longrightarrow \mathcal{N}$ of the map induced by the $\mathcal{C}$-coaction in $\mathcal{N}$ and the map induced by the pairing ϕ defines a structure of right K-module on $\mathcal{N}$. Analogously, for any left $\mathcal{C}$-contramodule $\mathfrak{P}$ the composition $K \otimes_A \mathfrak{P} \longrightarrow \operatorname{Hom}_A(\mathcal{C}, \mathfrak{P}) \longrightarrow \mathfrak{P}$ of the map given by the formula $k' \otimes p \longmapsto (c \mapsto \phi(c, k')p)$ and the $\mathcal{C}$-contraaction map defines a structure of left K-module on $\mathfrak{P}$. So the pairing ϕ induces faithful functors

$$\Delta_\phi \colon \mathsf{comod}\text{-}\mathcal{C} \longrightarrow \mathsf{mod}\text{-}K \quad \text{and} \quad \Delta^\phi \colon \mathcal{C}\text{-}\mathsf{contra} \longrightarrow K\text{-}\mathsf{mod}.$$

In particular, a pairing ϕ provides the coring $\mathcal{C}$ with a structure of left $\mathcal{C}$-comodule endowed with a right action of the k-algebra K by $\mathcal{C}$-comodule endomorphisms. Moreover, the data of a right action of K by endomorphisms of the left $\mathcal{C}$-comodule $\mathcal{C}$ agreeing with the right action of A in $\mathcal{C}$ is equivalent to the data of a pairing ϕ.

10.1.3 When $\mathcal{C}$ is a projective left A-module, the functor Δ^ϕ has a left adjoint functor

$$\Gamma^\phi \colon K\text{-}\mathsf{mod} \longrightarrow \mathcal{C}\text{-}\mathsf{contra}.$$

This functor sends the induced left K-module $K \otimes_A V$ to the induced left $\mathcal{C}$-contramodule $\operatorname{Hom}_A(\mathcal{C}, V)$ for any left A-module V; to compute $\Gamma^\phi(M)$ for an arbitrary left K-module M, one can represent M as the cokernel of a morphism of K-modules induced from A-modules. Analogously, when $\mathcal{C}$ is a flat right A-module, the functor Δ_ϕ has a right adjoint functor

$$\Gamma_\phi \colon \mathsf{mod}\text{-}K \longrightarrow \mathsf{comod}\text{-}\mathcal{C}.$$

This functor sends the coinduced right K-module $\operatorname{Hom}_{A^{\mathrm{op}}}(K, U)$ to the coinduced right $\mathcal{C}$-comodule $U \otimes_A \mathcal{C}$ for any right A-module U; to compute $\Gamma_\phi(N)$ for an arbitrary right K-module N, one can represent N as the kernel of a morphism of K-modules coinduced from A-modules.

Without any conditions on the coring $\mathcal{C}$, the composition of functors $\Psi_\mathcal{C} \colon \mathcal{C}\text{-}\mathsf{comod} \longrightarrow \mathcal{C}\text{-}\mathsf{contra}$ and $\Delta^\phi \colon \mathcal{C}\text{-}\mathsf{contra} \longrightarrow K\text{-}\mathsf{mod}$ has a left adjoint functor of the form

$$K\text{-}\mathsf{mod} \longrightarrow \mathcal{C}\text{-}\mathsf{comod}, \qquad M \longmapsto \mathcal{C} \otimes_K M.$$

Analogously, the composition of the functors $\Phi_{\mathcal{C}^{\mathrm{op}}} \colon \mathsf{contra}\text{-}\mathcal{C} \longrightarrow \mathsf{comod}\text{-}\mathcal{C}$ and $\Delta_\phi \colon \mathsf{comod}\text{-}\mathcal{C} \longrightarrow \mathsf{mod}\text{-}K$ has a right adjoint functor given by

$$\mathsf{mod}\text{-}K \longrightarrow \mathsf{contra}\text{-}\mathcal{C}, \qquad N \longmapsto \operatorname{Hom}_{K^{\mathrm{op}}}(\mathcal{C}, N).$$

So one can compute the compositions of functors $\Phi_\mathcal{C}\Gamma^\phi$ and $\Psi_{\mathcal{C}^{\mathrm{op}}}\Gamma_\phi$ in this way.

10.1.4 It is easy to see that the functor Δ_ϕ is fully faithful whenever for any right A-module N the map

$$N \otimes_A \mathcal{C} \longrightarrow \mathrm{Hom}_{A^{\mathrm{op}}}(K, N), \qquad n \otimes c \longmapsto (k' \mapsto n\phi(c, k'))$$

is injective (cf. [23, 19.2]). In particular, when A is a semisimple ring, the functor Δ_ϕ is fully faithful if the map $\mathcal{C} \longrightarrow \mathrm{Hom}_{A^{\mathrm{op}}}(K, A)$ induced by the pairing ϕ is injective, i.e., the pairing ϕ is nondegenerate in $\mathcal{C}$.

10.2 Construction of semialgebras

10.2.1 Assume that a coring $\mathcal{C}$ over a k-algebra A is a flat left A-module. Let K be a k-algebra endowed with a k-algebra map $A \longrightarrow K$ and a pairing $\phi: \mathcal{C} \otimes_A K \longrightarrow A$ satisfying the conditions of 10.1.2, and let R be a k-algebra endowed with a k-algebra map $f: K \longrightarrow R$ such that R is a flat left K-module. Then the tensor product $\mathcal{C} \otimes_K R$ is a coflat left $\mathcal{C}$-comodule endowed with a right action of the k-algebra K (and even of the k-algebra R) by left $\mathcal{C}$-comodule endomorphisms.

Suppose that there exists a structure of right $\mathcal{C}$-comodule on $\mathcal{C} \otimes_K R$ inducing the existing structure of right K-module and such that the following three maps are right $\mathcal{C}$-comodule morphisms:

(i) the left $\mathcal{C}$-coaction map $\mathcal{C} \otimes_K R \longrightarrow \mathcal{C} \otimes_A (\mathcal{C} \otimes_K R)$,

(ii) the semiunit map $\mathcal{C} = \mathcal{C} \otimes_K K \longrightarrow \mathcal{C} \otimes_K R$ induced by the k-algebra map $f: K \longrightarrow R$, and

(iii) the semimultiplication map $(\mathcal{C} \otimes_K R) \square_{\mathcal{C}} (\mathcal{C} \otimes_K R) \simeq \mathcal{C} \otimes_K R \otimes_K R \longrightarrow \mathcal{C} \otimes_K R$, where the isomorphism is the inverse of the natural isomorphism of Proposition 1.2.3(a) and the map $\mathcal{C} \otimes_K R \otimes_K R \longrightarrow \mathcal{C} \otimes_K R$ is induced by the multiplication in R.

Then the semiunit and semimultiplication maps (ii) and (iii) define a semialgebra structure on the $\mathcal{C}$-$\mathcal{C}$-bicomodule $\mathbf{S} = \mathcal{C} \otimes_K R$.

Notice that the maps (i-iii) always preserve the right K-module structures. If the functor Δ_ϕ is fully faithful, then a right $\mathcal{C}$-comodule structure inducing a given right K-module structure on $\mathcal{C} \otimes_K R$ is unique provided that it exists, and the maps (i-iii) preserve this structure automatically. If the functor Δ_ϕ is an equivalence of categories, then a unique right $\mathcal{C}$-comodule structure with the desired properties always exists on $\mathcal{C} \otimes_K R$.

The associativity of semimultiplication in $\mathbf{S}$ follows from the associativity of multiplication in R and the commutativity of diagrams of associativity isomorphisms of cotensor products.

10.2.2 By Proposition 1.2.3(a), for any right $\mathcal{C}$-comodule $\mathcal{N}$ there is a natural isomorphism

$$\mathcal{N} \square_{\mathcal{C}} \mathbf{S} \simeq \mathcal{N} \otimes_K R,$$

hence every right $\mathbf{S}$-semimodule has a natural structure of right R-module. So there is a faithful exact functor

$$\Delta_{\phi,f} \colon \mathsf{simod}\text{-}\mathbf{S} \longrightarrow \mathsf{mod}\text{-}R$$

which agrees with the functor $\Delta_\phi \colon \mathsf{comod}\text{-}\mathcal{C} \longrightarrow \mathsf{mod}\text{-}K$. Moreover, the category of right $\mathbf{S}$-semimodules is isomorphic to the category of k-modules $\mathcal{N}$ endowed with a right $\mathcal{C}$-comodule and right R-module structures satisfying the following compatibility conditions: first, the induced right K-module structures should coincide, and second, the action map $\mathcal{N} \otimes_K R \longrightarrow \mathcal{N}$ should be a morphism of right $\mathcal{C}$-comodules, where the right $\mathcal{C}$-comodule structure on $\mathcal{N} \otimes_K R$ is provided by the isomorphism $\mathcal{N} \otimes_K R = \mathcal{N} \,\square_\mathcal{C}\, \mathbf{S}$. When the functor Δ_ϕ is fully faithful, the category $\mathsf{simod}\text{-}\mathbf{S}$ is simply described as the full subcategory of the category of right R-modules consisting of those modules whose right K-module structure comes from a right $\mathcal{C}$-comodule structure.

Analogously, if $\mathcal{C}$ is a projective left A-module and R is a projective left K-module, then $\mathbf{S}$ is a coprojective left $\mathcal{C}$-comodule and by Proposition 3.2.3.2(a) for any left $\mathcal{C}$-contramodule $\mathfrak{P}$ there is a natural isomorphism

$$\mathrm{Cohom}_\mathcal{C}(\mathbf{S},\mathfrak{P}) \simeq \mathrm{Hom}_K(R,\mathfrak{P}),$$

hence any left $\mathbf{S}$-semicontramodule has a natural structure of left R-module. So there is a faithful exact functor

$$\Delta^{\phi,f} \colon \mathbf{S}\text{-}\mathsf{sicntr} \longrightarrow R\text{-}\mathsf{mod}$$

which agrees with the functor $\Delta^\phi \colon \mathcal{C}\text{-}\mathsf{comod} \longrightarrow K\text{-}\mathsf{mod}$. Moreover, the category of left $\mathbf{S}$-semicontramodules is isomorphic to the category of k-modules $\mathfrak{P}$ endowed with a left $\mathcal{C}$-contramodule and a left R-module structures satisfying the following compatibility conditions: first, the induced left K-module structures should coincide, and second, the action map $\mathfrak{P} \longrightarrow \mathrm{Hom}_K(R,\mathfrak{P})$ should be a morphism of $\mathcal{C}$-contramodules, where the left $\mathcal{C}$-contramodule structure on $\mathrm{Hom}_K(R,\mathfrak{P})$ is provided by the isomorphism $\mathrm{Hom}_K(R,\mathfrak{P}) = \mathrm{Cohom}_\mathcal{C}(\mathbf{S},\mathfrak{P})$.

10.2.3   When K is a projective finitely generated right A-module and the pairing ϕ corresponds to an isomorphism $\mathcal{C} \simeq \mathrm{Hom}_{A^{\mathrm{op}}}(K,A)$, the isomorphisms of categories $\Delta_\phi \colon \mathsf{comod}\text{-}\mathcal{C} \simeq \mathsf{mod}\text{-}K$ and $\Delta^\phi \colon \mathcal{C}\text{-}\mathsf{contra} \simeq K\text{-}\mathsf{mod}$ transform the functor of contratensor product over $\mathcal{C}$ into the functor of tensor product over K:

$$\mathcal{N} \odot_\mathcal{C} \mathfrak{P} \simeq \mathcal{N} \otimes_K \mathfrak{P}.$$

This follows from the isomorphism $\mathrm{Hom}_A(\mathcal{C},\mathfrak{P}) = K \otimes_A \mathfrak{P}$. When in addition R is a projective left K-module, the isomorphisms of categories $\Delta_{\phi,f} \colon \mathsf{simod}\text{-}\mathbf{S} \simeq \mathsf{mod}\text{-}R$ and $\Delta^{\phi,f} \colon \mathbf{S}\text{-}\mathsf{sicntr} \simeq R\text{-}\mathsf{mod}$ transform the functor of contratensor product over $\mathbf{S}$ into the functor of tensor product over R:

$$\mathcal{N} \odot_\mathbf{S} \mathfrak{P} \simeq \mathcal{N} \otimes_R \mathfrak{P}.$$

One checks this using the isomorphism $\mathcal{N} \,\square_\mathcal{C}\, \mathbf{S} = \mathcal{N} \otimes_K R$ and the above formula for the contratensor product over $\mathcal{C}$.

10.2.4 The functor $\Delta_{\phi,f}$ has a right adjoint functor

$$\Gamma_{\phi,f} : \text{mod-}R \longrightarrow \text{simod-}\mathbf{S},$$

which agrees with the functor $\Gamma_\phi : \text{mod-}K \longrightarrow \text{comod-}\mathcal{C}$. The functor $\Gamma_{\phi,f}$ is constructed as follows. Let N be a right R-module; it has an induced right K-module structure. Consider the composition $\Delta_\phi(\Gamma_\phi(N) \,\square_\mathcal{C}\, \mathbf{S}) = \Delta_\phi \Gamma_\phi(N) \otimes_K R \longrightarrow N \otimes_K R \longrightarrow N$ of the isomorphism of Proposition 1.2.3(a), the map induced by the adjunction map $\Delta_\phi \Gamma_\phi(N) \longrightarrow N$, and the right action map. By adjunction, this composition corresponds to a right $\mathcal{C}$-comodule morphism $\Gamma_\phi(N) \,\square_\mathcal{C}\, \mathbf{S} \longrightarrow \Gamma_\phi(N)$, which provides a right $\mathbf{S}$-semimodule structure on $\Gamma_\phi(N)$.

Analogously, if $\mathcal{C}$ is a projective left A-module and R is a projective left K-module, then the functor $\Delta^{\phi,f}$ has a left adjoint functor

$$\Gamma^{\phi,f} : R\text{-mod} \longrightarrow \mathbf{S}\text{-sicntr},$$

which agrees with the functor $\Gamma^\phi : K\text{-mod} \longrightarrow \mathcal{C}\text{-contra}$. The functor $\Gamma^{\phi,f}$ is constructed as follows. Let P be a left R-module; it has an induced left K-module structure. Consider the composition

$$P \longrightarrow \text{Hom}_K(R, P) \longrightarrow \text{Hom}_K(R, \Delta^\phi \Gamma^\phi(P)) = \Delta^\phi(\text{Cohom}_\mathcal{C}(\mathbf{S}, \Gamma^\phi(P)))$$

of the action map, the map induced by the adjunction map $P \longrightarrow \Delta^\phi \Gamma^\phi(P)$, and the isomorphism of Proposition 3.2.3.2(a). By adjunction, this composition corresponds to a left $\mathcal{C}$-contramodule morphism $\Gamma^\phi(P) \longrightarrow \text{Cohom}_\mathcal{C}(\mathbf{S}, \Gamma^\phi(P))$, which provides a left $\mathbf{S}$-semicontramodule structure on $\Gamma^\phi(P)$.

Notice that the semialgebra $\mathbf{S}$ has a structure of left $\mathbf{S}$-semimodule endowed with a right action of the k-algebra R by left $\mathbf{S}$-semimodule endomorphisms. So when $\mathcal{C}$ is a flat right A-module and $\mathbf{S}$ turns out to be a coflat right $\mathcal{C}$-comodule, there is the functor

$$\mathbf{S}\text{-simod} \longrightarrow R\text{-mod}, \qquad \mathcal{M} \longmapsto \text{Hom}_\mathbf{S}(\mathbf{S}, \mathcal{M}) = \text{Hom}_\mathcal{C}(\mathcal{C}, \mathcal{M}).$$

This functor has the left adjoint functor

$$R\text{-mod} \longrightarrow \mathbf{S}\text{-simod}, \qquad M \longmapsto \mathbf{S} \otimes_R M = \mathcal{C} \otimes_K M.$$

In the case when $\mathcal{C}$ is a projective left A-module and R is a projective left K-module, the former functor is isomorphic to $\Delta^{\phi,f} \Psi_\mathbf{S}$, and consequently the latter functor is isomorphic to $\Phi_\mathbf{S} \Gamma^{\phi,f}$. Analogously, when $\mathcal{C}$ is a projective right A-module and $\mathbf{S}$ turns out to be a coprojective right $\mathcal{C}$-comodule, the functor $\Psi_{\mathbf{S}^{\text{op}}} \Gamma_{\phi,f}$ maps a right R-module N to the right $\mathbf{S}$-semicontramodule $\text{Hom}_{R^{\text{op}}}(\mathbf{S}, N) = \text{Hom}_{K^{\text{op}}}(\mathcal{C}, N)$.

Let us point out that *no explicit description of the category of left $\mathbf{S}$-semimodules is in general available*. We only described the categories of right $\mathbf{S}$-semimodules and left $\mathbf{S}$-semicontramodules, and constructed certain functors acting to and from the category of left $\mathbf{S}$-semimodules.

10.2.5   The following observations were inspired by [3, Section 5].

Suppose that there is a commutative diagram of k-algebra maps $A \longrightarrow K$, $K \longrightarrow R$, $A' \longrightarrow K'$, $K' \longrightarrow R'$, $A \longrightarrow A'$, $K \longrightarrow K'$, $R \longrightarrow R'$. Let $\mathcal{C}$ be a coring over A and $\mathcal{C}'$ be a coring over A' endowed with a map of corings $\mathcal{C} \longrightarrow \mathcal{C}'$ compatible with the k-algebra map $A \longrightarrow A'$. Assume that $\mathcal{C}$ is a flat left A-module, $\mathcal{C}'$ is a flat left A'-module, R is a flat left K-module, and R' is a flat left K'-module. Let $\phi \colon \mathcal{C} \otimes_A K \longrightarrow A$ and $\phi' \colon \mathcal{C}' \otimes_{A'} K' \longrightarrow A'$ be two pairings satisfying the conditions of 10.1.2 and forming a commutative diagram with the maps $\mathcal{C} \otimes_A K \longrightarrow \mathcal{C}' \otimes_{A'} K'$ and $A \longrightarrow A'$. Furthermore, suppose that the natural map $K \otimes_A A' \longrightarrow K'$ is an isomorphism. Assume that there is a structure of right $\mathcal{C}$-comodule on $\mathcal{C} \otimes_K R$ and a structure of right $\mathcal{C}'$-comodule on $\mathcal{C}' \otimes_{K'} R'$ satisfying the conditions of 10.2.1, so that $\mathcal{C} \otimes_K R$ is a semialgebra over $\mathcal{C}$ and $\mathcal{C}' \otimes_{K'} R'$ is a semialgebra over $\mathcal{C}'$. Then the natural map from the right $\mathcal{C}'$-comodule $\mathcal{C} \otimes_K R \otimes_A A'$ to the right $\mathcal{C}'$-comodule $\mathcal{C}' \otimes_{K'} R'$ is a morphism of right K'-modules. If it is also a morphism of right $\mathcal{C}'$-comodules, then the map $\mathcal{C} \otimes_K R \longrightarrow \mathcal{C}' \otimes_{K'} R'$ is a map of semialgebras compatible with the map of corings $\mathcal{C} \longrightarrow \mathcal{C}'$ and the k-algebra map $A \longrightarrow A'$.

Suppose that there is a commutative diagram of k-algebra maps $A \longrightarrow K$, $K \longrightarrow R$, $A \longrightarrow K'$, $K' \longrightarrow R'$, $K \longrightarrow K'$, $R \longrightarrow R'$. Let $\mathcal{C}$ and $\mathcal{C}'$ be two corings over A and $\mathcal{C}' \longrightarrow \mathcal{C}$ be a morphism of corings over A. Assume that $\mathcal{C}$ and $\mathcal{C}'$ are flat left A-modules, R is a flat left K-module, and R' is a flat left K'-module. Let $\phi \colon \mathcal{C} \otimes_A K \longrightarrow A$ and $\phi' \colon \mathcal{C}' \otimes_A K' \longrightarrow A$ be two pairings satisfying the conditions of 10.1.2 and forming a commutative diagram with the maps $\mathcal{C}' \otimes_A K \longrightarrow \mathcal{C} \otimes_A K$ and $\mathcal{C}' \otimes_A K \longrightarrow \mathcal{C}' \otimes_A K'$. Furthermore, suppose that the natural map $K' \otimes_K R \longrightarrow R'$ is an isomorphism. Assume that there is a structure of right $\mathcal{C}$-comodule on $\mathcal{C} \otimes_K R$ and a structure of right $\mathcal{C}'$-comodule on $\mathcal{C}' \otimes_{K'} R'$ satisfying the conditions of 10.2.1, so that $\mathcal{C} \otimes_K R$ is a semialgebra over $\mathcal{C}$ and $\mathcal{C}' \otimes_{K'} R'$ is a semialgebra over $\mathcal{C}'$. In this case, if the right K-module map $\mathcal{C}' \otimes_{K'} R' = \mathcal{C}' \otimes_{K'} K' \otimes_K R \simeq \mathcal{C}' \otimes_K R \longrightarrow \mathcal{C} \otimes_K R$ is a morphism of right $\mathcal{C}$-comodules, then it is a map of semialgebras compatible with the morphism $\mathcal{C}' \longrightarrow \mathcal{C}$ of corings over A.

10.3   Entwining structures

An important particular case of the above construction of semialgebras was considered in [21]. Namely, it was noticed that there is a set of data from which one can construct *both* a coring and a semialgebra.

10.3.1   Let $\mathcal{C}$ be a coring over a k-algebra A and $A \longrightarrow B$ be a morphism of k-algebras. A *right entwining structure* for $\mathcal{C}$ and B over A is an A-A-bimodule map

$$\psi \colon \mathcal{C} \otimes_A B \longrightarrow B \otimes_A \mathcal{C}$$

satisfying the following equations:

(i) the composition $\mathcal{C} \otimes_A B \otimes_A B \longrightarrow B \otimes_A \mathcal{C} \otimes_A B \longrightarrow B \otimes_A B \otimes_A \mathcal{C} \longrightarrow B \otimes_A \mathcal{C}$ of two maps induced by the map ψ and the map induced by the multiplication in B is equal to the composition $\mathcal{C} \otimes_A B \otimes_A B \longrightarrow \mathcal{C} \otimes_A B \longrightarrow B \otimes_A \mathcal{C}$ of the map induced by the multiplication in B and the map ψ;

(ii) the map ψ forms a commutative triangle with the maps $\mathcal{C} \longrightarrow \mathcal{C} \otimes_A B$ and $\mathcal{C} \longrightarrow B \otimes_A \mathcal{C}$ coming from the unit of B;

(iii) the composition $\mathcal{C} \otimes_A B \longrightarrow \mathcal{C} \otimes_A \mathcal{C} \otimes_A B \longrightarrow \mathcal{C} \otimes_A B \otimes_A \mathcal{C} \longrightarrow B \otimes_A B \otimes_A \mathcal{C}$ of the map induced by the comultiplication in $\mathcal{C}$ and two maps induced by the map ψ is equal to the composition $\mathcal{C} \otimes_A B \longrightarrow B \otimes_A \mathcal{C} \longrightarrow B \otimes_A \mathcal{C} \otimes_A \mathcal{C}$ of the map ψ and the map induced by the comultiplication in $\mathcal{C}$;

(iv) the map ψ forms a commutative triangle with the maps $\mathcal{C} \otimes_A B \longrightarrow B$ and $B \otimes_A \mathcal{C} \longrightarrow B$ coming from the counit of $\mathcal{C}$.

A *left entwining structure* for $\mathcal{C}$ and B over A is defined as an A-A-bimodule map
$$\psi^{\#} : B \otimes_A \mathcal{C} \longrightarrow \mathcal{C} \otimes_A B$$
satisfying the opposite equations. Notice that whenever a map $\psi \colon \mathcal{C} \otimes_A B \longrightarrow B \otimes_A \mathcal{C}$ is invertible the map ψ is a right entwining structure if and only if the map $\psi^{\#} = \psi^{-1}$ is a left entwining structure.

10.3.2 A (right) *entwined module* over a right entwining structure $\psi \colon \mathcal{C} \otimes_A B \longrightarrow B \otimes_A \mathcal{C}$ is a k-module $\mathcal{N}$ endowed with a right $\mathcal{C}$-comodule and a right B-module structures such that the corresponding right A-module structures coincide and the following equation holds: the composition $\mathcal{N} \otimes_A B \longrightarrow \mathcal{N} \otimes_A \mathcal{C} \otimes_A B \longrightarrow \mathcal{N} \otimes_A B \otimes_A \mathcal{C} \longrightarrow \mathcal{N} \otimes_A \mathcal{C}$ of the map induced by the $\mathcal{C}$-coaction in $\mathcal{N}$, the map induced by the map ψ, and the map induced by the B-action in $\mathcal{N}$ is equal to the composition $\mathcal{N} \otimes_A B \longrightarrow \mathcal{N} \longrightarrow \mathcal{N} \otimes_A \mathcal{C}$ of the B-action map and the $\mathcal{C}$-coaction map.

A (left) *entwined contramodule* over a right entwining structure ψ is a k-module $\mathfrak{P}$ endowed with a left $\mathcal{C}$-contramodule and a left B-module structures such that the corresponding left A-module structures coincide and the following equation holds: the composition $\mathrm{Hom}_A(\mathcal{C}, \mathfrak{P}) \longrightarrow \mathrm{Hom}_A(\mathcal{C}, \mathrm{Hom}_A(B, \mathfrak{P})) = \mathrm{Hom}_A(B \otimes_A \mathcal{C}, \mathfrak{P}) = \mathrm{Hom}_A(\mathcal{C} \otimes_A B, \mathfrak{P}) = \mathrm{Hom}_A(B, \mathrm{Hom}_A(\mathcal{C}, \mathfrak{P})) \longrightarrow \mathrm{Hom}_A(B, \mathfrak{P})$ of the map induced by the B-action in $\mathfrak{P}$, the map induced by the map ψ, and the map induced by the $\mathcal{C}$-contraaction in $\mathfrak{P}$ is equal to the composition $\mathrm{Hom}_A(\mathcal{C}, \mathfrak{P}) \longrightarrow \mathfrak{P} \longrightarrow \mathrm{Hom}_A(B, \mathfrak{P})$ of the $\mathcal{C}$-contraaction map and the B-action map.

(Left) *entwined modules* and (right) *entwined contramodules* over a left entwining structure are defined in an analogous way.

10.3.3 Let $\psi \colon \mathcal{C} \otimes_A B \longrightarrow B \otimes_A \mathcal{C}$ be a right entwining structure. Define a coring $\mathcal{D}$ over B as the left B-module
$$\mathcal{D} = B \otimes_A \mathcal{C}$$

endowed with the following right action of B, comultiplication, and counit. The right B-action is the composition $(B \otimes_A \mathcal{C}) \otimes_A B \longrightarrow B \otimes_A B \otimes_A \mathcal{C} \longrightarrow B \otimes_A \mathcal{C}$ of the map induced by the map ψ and the multiplication in B. The comultiplication is the map $B \otimes_A \mathcal{C} \longrightarrow B \otimes_A \mathcal{C} \otimes_A \mathcal{C} = (B \otimes_A \mathcal{C}) \otimes_B (B \otimes_A \mathcal{C})$ induced by the comultiplication in $\mathcal{C}$. The counit is the map $B \otimes_A \mathcal{C} \longrightarrow B \otimes_A A = B$ coming from the counit of $\mathcal{C}$. One has to use the equation (i) on the entwining map ψ to check that the right action of B is associative, the equation (ii) to check that the right action of B agrees with the existing right action of A, and the equations (iii) and (iv) to check that the comultiplication and counit are right B-module maps.

Analogously, for a left entwining structure $\psi^{\#} : B \otimes_A \mathcal{C} \longrightarrow \mathcal{C} \otimes_A B$ one defines a coring $\mathcal{D}^{\#} = \mathcal{C} \otimes_A B$ over B. When $\psi^{\#} = \psi^{-1}$ are two inverse maps satisfying the entwining structure equations, the maps ψ and $\psi^{\#}$ themselves are mutually inverse isomorphisms $\mathcal{D}^{\#} \simeq \mathcal{D}$ between the corresponding corings over B.

10.3.4   Let $\psi : \mathcal{C} \otimes_A B \longrightarrow B \otimes_A \mathcal{C}$ be a right entwining structure. Define a semialgebra $\mathbf{S}$ over $\mathcal{C}$ as the left $\mathcal{C}$-comodule

$$\mathbf{S} = \mathcal{C} \otimes_A B$$

endowed with the following right coaction of $\mathcal{C}$, semimultiplication, and semiunit. The right $\mathcal{C}$-coaction is the composition $\mathcal{C} \otimes_A B \longrightarrow \mathcal{C} \otimes_A \mathcal{C} \otimes_A B \longrightarrow (\mathcal{C} \otimes_A B) \otimes_A \mathcal{C}$ of the map induced by the comultiplication in $\mathcal{C}$ and the map induced by the map ψ. The semimultiplication is the map $(\mathcal{C} \otimes_A B) \square_{\mathcal{C}} (\mathcal{C} \otimes_A B) = \mathcal{C} \otimes_A B \otimes_A B \longrightarrow \mathcal{C} \otimes_A B$ induced by the multiplication in B. The semiunit is the map $\mathcal{C} = \mathcal{C} \otimes_A A \longrightarrow \mathcal{C} \otimes_A B$ coming from the unit of B. The multiple cotensor products $\mathcal{N} \square_{\mathcal{C}} \mathbf{S} \square_{\mathcal{C}} \mathbf{S} \square_{\mathcal{C}} \cdots \square_{\mathcal{C}} \mathbf{S}$ and the multiple cohomomorphisms $\operatorname{Cohom}_{\mathcal{C}}(\mathbf{S} \square_{\mathcal{C}} \cdots \square_{\mathcal{C}} \mathbf{S}, \mathfrak{P})$ are associative for any right $\mathcal{C}$-comodule $\mathcal{N}$ and any left $\mathcal{C}$-contramodule $\mathfrak{P}$ by Propositions 1.2.5(e) and 3.2.5(h).

Analogously, for a left entwining structure $\psi^{\#} : B \otimes_A \mathcal{C} \longrightarrow \mathcal{C} \otimes_A B$ one defines a semialgebra $\mathbf{S}^{\#} = B \otimes_A \mathcal{C}$ over $\mathcal{C}$. When $\psi^{\#} = \psi^{-1}$ are two inverse maps satisfying the entwining structure equations, the maps ψ and $\psi^{\#}$ themselves are mutually inverse isomorphisms $\mathbf{S} \simeq \mathbf{S}^{\#}$ between the corresponding semialgebras over $\mathcal{C}$.

10.3.5   An entwined module over a right entwining structure ψ is the same thing as a right $\mathcal{D}$-comodule and the same thing as a right $\mathbf{S}$-semimodule; in other words, the corresponding categories are isomorphic. Analogously, an entwined module over a left entwining structure $\psi^{\#}$ is the same as a left $\mathcal{D}^{\#}$-comodule and the same as a left $\mathbf{S}^{\#}$-semimodule. Similar assertions apply to contramodules: an entwined contramodule over a right entwining structure ψ is the same thing as a left $\mathcal{D}$-contramodule and the same thing as a left $\mathbf{S}$-semicontramodule; analogously for a left entwining structure.

For any entwined module $\mathcal{N}$ over a right entwining structure ψ there is a natural injective morphism

$$\mathcal{N} \longrightarrow \mathcal{N} \otimes_B \mathcal{D} \simeq \mathcal{N} \otimes_A \mathcal{C}$$

from $\mathcal{N}$ into an entwined module which as a $\mathcal{C}$-comodule is coinduced from an A-module. Analogously, for any left entwined contramodule $\mathfrak{P}$ over ψ there is a natural surjective morphism

$$\mathrm{Hom}_A(\mathcal{C}, \mathfrak{P}) \simeq \mathrm{Hom}_B(\mathcal{D}, \mathfrak{P}) \longrightarrow \mathfrak{P}$$

onto $\mathfrak{P}$ from an entwined contramodule which as a $\mathcal{C}$-contramodule is induced from an A-module. So we obtain, *in the entwining structure case*, a functorial injection from an arbitrary $\mathbf{S}$-semimodule into a $\mathcal{C}/A$-injective $\mathbf{S}$-semimodule and a functorial surjection onto an arbitrary $\mathbf{S}$-semicontramodule from a $\mathcal{C}/A$-projective $\mathbf{S}$-semicontramodule constructed in a way much simpler than that of Lemmas 1.3.3 and 3.3.3 (cf. [2, 3, 17]).

When the ring A is semisimple, there is also a functorial surjection onto an arbitrary $\mathcal{D}$-comodule $\mathcal{N}$ from a B-projective $\mathcal{D}$-comodule $\mathcal{N}\square_\mathcal{C}\mathbf{S} \simeq \mathcal{N}\otimes_A B$ and a functorial injection from an arbitrary $\mathcal{D}$-contramodule $\mathfrak{P}$ into a B-injective $\mathcal{D}$-contramodule $\mathrm{Cohom}_\mathcal{C}(\mathbf{S}, \mathfrak{P}) \simeq \mathrm{Hom}_A(B, \mathfrak{P})$; these are much simpler constructions than those of Lemmas 1.1.3 and 3.1.3.

When B is a flat right A-module, the construction of the semialgebra $\mathbf{S} = \mathcal{C}\otimes_A B$ corresponding to an entwining structure ψ becomes a particular case of the construction of the semialgebra $\mathbf{S} = \mathcal{C}\otimes_K R$ corresponding to a pairing ϕ (take $K = A$, $R = B$, and the only possible ϕ).

10.3.6 When $\psi^\# = \psi^{-1}$ are two inverse entwining structures, there is an explicit description of *both* the categories of left and right comodules over $\mathcal{D}^\# \simeq \mathcal{D}$ and *both* the categories of left and right semimodules over $\mathbf{S} \simeq \mathbf{S}^\#$.

When ψ is invertible, the multiple cotensor products $\mathcal{N}\square_\mathcal{C}\mathbf{S}\square_\mathcal{C}\cdots\square_\mathcal{C}\mathbf{S}\square_\mathcal{C}\mathcal{M}$ and the multiple cohomomorphisms $\mathrm{Cohom}_\mathcal{C}(\mathbf{S}\square_\mathcal{C}\cdots\square_\mathcal{C}\mathbf{S}\square_\mathcal{C}\mathcal{M}, \mathfrak{P})$ are associative for any right $\mathcal{C}$-comodule $\mathcal{N}$, left $\mathcal{C}$-comodule $\mathcal{M}$, and left $\mathcal{C}$-contramodule $\mathfrak{P}$ by Propositions 1.2.5(f) and 3.2.5(j), so the functors of semitensor product and semihomomorphism over $\mathbf{S}$ are everywhere defined.

10.4 Semiproduct and semimorphisms

Let $\psi\colon \mathcal{C}\otimes_A B \longrightarrow B\otimes_A \mathcal{C}$ be a right entwining structure; suppose that ψ is an invertible map. Let $\mathbf{S} = \mathcal{C}\otimes_A B$ and $\mathcal{D} = B\otimes_A \mathcal{C}$ be the corresponding semialgebra over $\mathcal{C}$ and coring over B.

One defines [79] the *semiproduct* $\mathcal{N}\otimes_B^\mathcal{C}\mathcal{M}$ of a right entwined module $\mathcal{N}$ over ψ and a left entwined module $\mathcal{M}$ over ψ^{-1} as the image of the composition of maps

$$\mathcal{N}\square_\mathcal{C}\mathcal{M} \longrightarrow \mathcal{N}\otimes_A \mathcal{M} \longrightarrow \mathcal{N}\otimes_B \mathcal{M}.$$

Analogously, one defines the k-module of *semimorphisms* $\mathrm{Hom}_B^\mathcal{C}(\mathcal{M}, \mathfrak{P})$ from a left entwined module $\mathcal{M}$ over ψ^{-1} to a left entwined contramodule $\mathfrak{P}$ over ψ as the image of the composition of maps

$$\mathrm{Hom}_B(\mathcal{M}, \mathfrak{P}) \longrightarrow \mathrm{Hom}_A(\mathcal{M}, \mathfrak{P}) \longrightarrow \mathrm{Cohom}_\mathcal{C}(\mathcal{M}, \mathfrak{P}).$$

There is a natural map of semialgebras $\mathbf{S} \longrightarrow B$ compatible with the map $\mathcal{C} \longrightarrow A$ of corings over A. Hence for any entwined modules $\mathcal{N}$ over ψ and $\mathcal{M}$ over ψ^{-1} there is a natural injective map from the pair of morphisms $\mathcal{N}\square_{\mathcal{C}}\mathbf{S}\square_{\mathcal{C}}\mathcal{M} \rightrightarrows \mathcal{N}\square_{\mathcal{C}}\mathcal{M}$ to the pair of morphisms $\mathcal{N} \otimes_A B \otimes_A \mathcal{M} \rightrightarrows \mathcal{N} \otimes_A \mathcal{M}$. Therefore, we have a natural surjective map

$$\mathcal{N} \lozenge_{\mathbf{S}} \mathcal{M} \longrightarrow \mathcal{N} \otimes_B^{\mathcal{C}} \mathcal{M},$$

which is an isomorphism if and only if the map $\mathcal{N} \lozenge_{\mathbf{S}} \mathcal{M} \longrightarrow \mathcal{N} \otimes_B \mathcal{M}$ is injective. Analogously, for any entwined module $\mathcal{M}$ over ψ^{-1} and entwined contramodule $\mathfrak{P}$ over ψ there is a natural surjective map from the pair of morphisms $\mathrm{Hom}_A(\mathcal{M}, \mathfrak{P}) \rightrightarrows \mathrm{Hom}_A(B \otimes_A \mathcal{M}, \mathfrak{P})$ to the pair of morphisms $\mathrm{Cohom}_{\mathcal{C}}(\mathcal{M}, \mathfrak{P}) \rightrightarrows \mathrm{Cohom}_{\mathcal{C}}(\mathbf{S} \square_{\mathcal{C}} \mathcal{M}, \mathfrak{P})$. So we get a natural injective map

$$\mathrm{Hom}_B^{\mathcal{C}}(\mathcal{M}, \mathfrak{P}) \longrightarrow \mathrm{SemiHom}_{\mathbf{S}}(\mathcal{M}, \mathfrak{P}),$$

which is an isomorphism if and only if the map $\mathrm{Hom}_B(\mathcal{M}, \mathfrak{P}) \to \mathrm{SemiHom}_{\mathbf{S}}(\mathcal{M}, \mathfrak{P})$ is surjective.

Consider the natural injective morphism of entwined modules $\mathcal{N} \longrightarrow \mathcal{N} \otimes_B \mathcal{D} = \mathcal{N} \otimes_A \mathcal{C}$. Taking the semitensor product of this morphism with $\mathcal{M}$ over $\mathbf{S}$, we obtain the map $\mathcal{N} \lozenge_{\mathbf{S}} \mathcal{M} \longrightarrow (\mathcal{N} \otimes_A \mathcal{C}) \lozenge_{\mathbf{S}} \mathcal{M} \simeq \mathcal{N} \otimes_B \mathcal{M}$ that we are interested in. Hence the natural map $\mathcal{N} \lozenge_{\mathbf{S}} \mathcal{M} \longrightarrow \mathcal{N} \otimes_B^{\mathcal{C}} \mathcal{M}$ is an isomorphism whenever the semitensor product with $\mathcal{M}$ maps A-split injections of right $\mathbf{S}$-semimodules to injections or $\mathcal{N}$ has such property with respect to left $\mathbf{S}$-semimodules. This includes the cases when $\mathcal{N}$ or $\mathcal{M}$ is an $\mathbf{S}$-semimodule induced from a $\mathcal{C}$-comodule.

Analogously, consider the natural surjective morphism of entwined contramodules $\mathrm{Hom}_A(\mathcal{C}, \mathfrak{P}) = \mathrm{Hom}_B(\mathcal{D}, \mathfrak{P}) \longrightarrow \mathfrak{P}$. The map $\mathrm{Hom}_B(\mathcal{M}, \mathfrak{P}) \longrightarrow \mathrm{SemiHom}_{\mathbf{S}}(\mathcal{M}, \mathfrak{P})$ can be obtained by taking the semihomomorphisms over $\mathbf{S}$ from $\mathcal{M}$ to the morphism $\mathrm{Hom}_A(\mathcal{C}, \mathfrak{P}) \longrightarrow \mathfrak{P}$ or by taking the semihomomorphisms over $\mathbf{S}$ from the morphism $\mathcal{M} \longrightarrow \mathcal{C} \otimes_A \mathcal{M}$ to $\mathfrak{P}$. Thus the natural map $\mathrm{Hom}_B^{\mathcal{C}}(\mathcal{M}, \mathfrak{P}) \longrightarrow \mathrm{SemiHom}_{\mathbf{S}}(\mathcal{M}, \mathfrak{P})$ is an isomorphism whenever the functor of semihomomorphisms from $\mathcal{M}$ maps A-split surjections of left $\mathbf{S}$-semicontramodules to surjections or the functor of semihomomorphisms into $\mathfrak{P}$ maps A-split injections of left $\mathbf{S}$-semimodules to surjections. This includes the cases when $\mathcal{M}$ is an $\mathbf{S}$-semimodule induced from a $\mathcal{C}$-comodule or $\mathfrak{P}$ is an $\mathbf{S}$-semicontramodule coinduced from a $\mathcal{C}$-contramodule.

In the same way one constructs a natural injective map $\mathcal{N} \otimes_B^{\mathcal{C}} \mathcal{M} \longrightarrow \mathcal{N} \square_{\mathcal{D}} \mathcal{M}$ and shows that it is an isomorphism whenever the cotensor product with $\mathcal{N}$ or $\mathcal{M}$ over $\mathcal{D}$ maps surjections of $\mathcal{D}$-comodules to surjections, in particular, when one of the $\mathcal{D}$-comodules $\mathcal{M}$ or $\mathcal{N}$ is quasicoflat. Analogously, there is a natural surjective map $\mathrm{Cohom}_{\mathcal{D}}(\mathcal{M}, \mathfrak{P}) \longrightarrow \mathrm{Hom}_B^{\mathcal{C}}(\mathcal{M}, \mathfrak{P})$, which is an isomorphism whenever the functor of cohomomorphisms from $\mathcal{M}$ over $\mathcal{D}$ maps injections of left $\mathcal{D}$-contramodules to injections or the functor of cohomomorphisms into $\mathfrak{P}$ over $\mathcal{D}$ maps surjections of left $\mathcal{D}$-comodules to injections, in particular, when $\mathcal{M}$ is a quasicoprojective $\mathcal{D}$-comodule or $\mathfrak{P}$ is a quasicoinjective $\mathcal{D}$-contramodule.

11 Relative Nonhomogeneous Koszul Duality

11.1 Graded semialgebras

11.1.1 All the constructions of Chapters 1–10 can be carried out with the category of k-modules replaced by the category of graded k-modules.

So one would consider a graded k-algebra A, a coring object $\mathcal{C}$ in the tensor category of graded A-A-bimodules, a ring object $\mathbf{S}$ in a tensor category of graded $\mathcal{C}$-$\mathcal{C}$-bicomodules, assume A to have a finite graded homological dimension, consider graded $\mathbf{S}$-semimodules and graded $\mathbf{S}$-semicontramodules. All of our definitions and results can be transferred to the graded situation without any difficulties. All the functors so obtained commute with the shift of grading in modules.

Furthermore, there are *two* forgetful functors Σ and Π from the category of graded k-modules k-mod^{gr} to the category k-mod, the functor Σ sending a graded k-module to the infinite direct sum of its components and the functor Π sending it to their infinite product. For any graded semialgebra $\mathbf{S}$ over a graded coring $\mathcal{C}$ over a graded k-algebra A, there are natural structures of a k-algebra on ΣA, of a coring over ΣA on $\Sigma \mathcal{C}$, and of a semialgebra over $\Sigma \mathcal{C}$ on $\Sigma \mathbf{S}$. For any graded $\mathbf{S}$-semimodule $\mathcal{M}$ there is a natural structure of a $\Sigma \mathbf{S}$-semimodule on $\Sigma \mathcal{M}$ and for any graded $\mathbf{S}$-semicontramodule $\mathfrak{P}$ there is a natural structure of a $\Sigma \mathbf{S}$-semicontramodule on $\Pi \mathfrak{P}$.

The functors of semitensor product and semihomomorphism defined in the graded setting are related to their ungraded analogues by the formulas

$$\Sigma(\mathcal{N} \lozenge_{\mathbf{S}}^{\mathrm{gr}} \mathcal{M}) \simeq \Sigma \mathcal{N} \lozenge_{\Sigma \mathbf{S}} \Sigma \mathcal{M}$$

and

$$\Pi \operatorname{SemiHom}_{\mathbf{S}}^{\mathrm{gr}}(\mathcal{M}, \mathfrak{P}) \simeq \operatorname{SemiHom}_{\Sigma \mathbf{S}}(\Sigma \mathcal{M}, \Pi \mathfrak{P}).$$

The functors $\mathcal{N} \longmapsto {}_{\mathcal{C}}\mathcal{N}$ and $\mathcal{M} \longmapsto {}_{\mathcal{J}}\mathcal{M}$ commute with the forgetful functors Σ and the functors $\mathfrak{Q} \longmapsto {}^{\mathcal{C}}\mathfrak{Q}$ and $\mathfrak{P} \longmapsto {}^{\mathcal{J}}\mathfrak{P}$ commute with the forgetful functors Π. The corresponding derived functors SemiTor, SemiExt, etc., have the same properties. However, the functors $\operatorname{Hom}_{\mathbf{S}}$, $\operatorname{Hom}^{\mathbf{S}}$, $\operatorname{CtrTor}^{\mathbf{S}}$, $\Psi_{\mathbf{S}}$, and $\Phi_{\mathbf{S}}$ and their derived functors have *no* properties of compatibility with the functors of forgetting the grading. Thus one has to be aware of the distinction between $\operatorname{Hom}_{\mathbf{S}}$ and $\operatorname{Hom}_{\mathbf{S}}^{\mathrm{gr}}$, $\Phi_{\mathbf{S}}$ and $\Phi_{\mathbf{S}}^{\mathrm{gr}}$, etc.

11.1.2 Assume that A is a nonnegatively graded k-algebra, $\mathcal{C}$ is a nonnegatively graded coring over A, and $\mathbf{S}$ is a nonnegatively graded semialgebra over $\mathcal{C}$. Let $\mathbf{S}$-simod$^{\uparrow}$ and simod$^{\uparrow}$-$\mathbf{S}$ denote the categories of nonnegatively graded $\mathbf{S}$-semimod-

L. Positselski, *Homological Algebra of Semimodules and Semicontramodules*, Monografie Matematyczne 70, DOI 10.1007/978-3-0346-0436-9_12, © Springer Basel AG 2010

ules, and S-contra$^\downarrow$ denote the category of nonpositively graded S-semicontramodules.

All the constructions of Chapters 1–4 in their graded versions preserve the categories of comodules and semimodules graded by nonnegative integers and the categories of contramodules and semicontramodules graded by nonpositive integers. All the definitions and results of these chapters can be transferred to the described situation of bounded grading and no problems occur. In particular, one can apply Lemma 2.7 to define the functors SemiTor and SemiExt in the bounded grading case. Moreover, the functors so obtained agree with the functors SemiTor$^{\mathsf{S}}_{\mathrm{gr}}$ and SemiExt$^{\mathrm{gr}}_{\mathsf{S}}$ defined in terms of complexes with unbounded grading. This is so because the constructions of resolutions agree. For the same reasons, in the assumptions of 6.3 the functors $\mathsf{D}^{\mathsf{si}}(\mathsf{S}\text{-simod}^\uparrow) \longrightarrow \mathsf{D}^{\mathsf{si}}(\mathsf{S}\text{-simod}^{\mathrm{gr}})$ and $\mathsf{D}^{\mathsf{si}}(\mathsf{S}\text{-sicntr}^\downarrow) \longrightarrow \mathsf{D}^{\mathsf{si}}(\mathsf{S}\text{-sicntr}^{\mathrm{gr}})$ are fully faithful, and the functor CtrTor defined by applying Lemma 6.5.2 in the bounded grading case agrees with the functor CtrTor$^{\mathsf{S}}_{\mathrm{gr}}$. But the functors $\Psi^{\mathrm{gr}}_{\mathsf{S}}$ and $\Phi^{\mathrm{gr}}_{\mathsf{S}}$ do *not* preserve the bounded grading.

11.2   Differential semialgebras

11.2.1   Let B be a graded k-algebra endowed with an odd derivation d_B of degree 1 and $\mathcal{D}$ be a graded coring over B. A homogeneous map $d_{\mathcal{D}} \colon \mathcal{D} \longrightarrow \mathcal{D}$ of degree 1 is called a *coderivation of $\mathcal{D}$ with respect to d_B* if the biaction map $B \otimes_k \mathcal{D} \otimes_k B \longrightarrow \mathcal{D}$ and the comultiplication map $\mathcal{D} \longrightarrow \mathcal{D} \otimes_B \mathcal{D}$ are morphisms in the category of graded k-modules endowed with endomorphisms of degree 1, where the endomorphisms of the tensor products are defined by the usual super-Leibniz rule $d(xy) = d(x)y + (-1)^{|x|}xd(y)$ (the degree of a homogeneous element x being denoted by $|x|$). In this case, it follows that the counit map $\mathcal{D} \longrightarrow B$ satisfies the same condition. In the particular case when B is concentrated in the degree 0 and $d_B = 0$, the condition on the biaction map simply means that $d_{\mathcal{D}}$ is a B-B-bimodule morphism.

Now assume that B is a DG-algebra over k, i.e., $d_B^2 = 0$. A *DG-coring* over B is a graded coring $\mathcal{D}$ over the graded ring B endowed with a coderivation $d_{\mathcal{D}} \colon \mathcal{D} \longrightarrow \mathcal{D}$ with respect to d_B such that $d_{\mathcal{D}}^2 = 0$.

Let $\mathcal{D}$ be a DG-coring over a DG-algebra B. Then the cohomology $H(\mathcal{D})$ is endowed with a natural structure of a graded coring over the graded algebra $H(B)$ provided that the natural maps $H(\mathcal{D}) \otimes_{H(B)} H(\mathcal{D}) \longrightarrow H(\mathcal{D} \otimes_B \mathcal{D})$ and $H(\mathcal{D}) \otimes_{H(B)} H(\mathcal{D}) \otimes_{H(B)} H(\mathcal{D}) \longrightarrow H(\mathcal{D} \otimes_B \mathcal{D} \otimes_B \mathcal{D})$ are isomorphisms. A map of DG-corings $\mathcal{C} \longrightarrow \mathcal{D}$ compatible with a morphism of DG-algebras $A \longrightarrow B$ induces a map of graded corings $H(\mathcal{C}) \longrightarrow H(\mathcal{D})$ compatible with the morphism of graded algebras $H(A) \longrightarrow H(B)$ whenever both DG-corings $\mathcal{C}$ and $\mathcal{D}$ satisfy the above two conditions. Here a map $\mathcal{C} \longrightarrow \mathcal{D}$ from a DG-coring $\mathcal{C}$ over a DG-algebra A to a DG-coring $\mathcal{D}$ over a DG-algebra B is called compatible with a morphism of DG-algebras $A \longrightarrow B$ over k if the map of graded corings $\mathcal{C} \longrightarrow \mathcal{D}$ is compatible with the morphism of graded algebras $A \longrightarrow B$ and the map $\mathcal{C} \longrightarrow \mathcal{D}$ is a morphism of complexes.

11.2.2 Coderivations of a graded coring $\mathcal{D}$ of degree -1 with respect to coderivations of a graded k-algebra B of degree -1 are defined in the same way as above.

Now let A be an ungraded k-algebra. A *quasi-differential coring* $\mathcal{D}^{\sim}$ over A is a graded coring over A endowed with a coderivation ∂ of degree 1 (with respect to the zero derivation of the k-algebra A, which is considered as a graded k-algebra concentrated in degree 0) such that $\partial^2 = 0$ and the cohomology of ∂ vanishes. If $\mathcal{D}^{\sim}$ is a quasi-differential coring over a k-algebra A, then the cokernel $\mathcal{D}^{\sim}/\operatorname{im}\partial$ of the derivation ∂ has a natural structure of graded coring over A. A *quasi-differential structure* on a graded coring $\mathcal{D}$ is the data of a quasi-differential coring $\mathcal{D}^{\sim}$ together with an isomorphism of graded corings

$$\mathcal{D}^{\sim}/\operatorname{im}\partial \simeq \mathcal{D}.$$

We will denote the grading of a quasi-differential coring $\mathcal{D}^{\sim}$ by lower indices, even though the differential raises the degree. This terminology and notation is explained by the following construction (cf. 0.4.4).

We will use Sweedler's notation [82] $p \longmapsto p_{(1)} \otimes p_{(2)}$ for the comultiplication map of a coring $\mathcal{D}$ over A; here $p \in \mathcal{D}$ and $p_{(1)} \otimes p_{(2)} \in \mathcal{D} \otimes_A \mathcal{D}$. A *CDG-coring* $\mathcal{D}$ over a k-algebra A is a graded coring over A endowed with a coderivation d of degree -1 (with respect to the zero derivation of A) and an A-A-bimodule map $h \colon \mathcal{D}_2 \longrightarrow A$ satisfying the equations $d^2(p) = h(p_{(1)})p_{(2)} - p_{(1)}h(p_{(2)})$ and $h(d(p)) = 0$ for all $p \in \mathcal{D}$, where the map h is considered to be extended by zero to the components $\mathcal{D}_i$ with $i \neq 2$. Given a CDG-coring $\mathcal{D}$ over a k-algebra A and a CDG-coring $\mathcal{E}$ over a k-algebra B, a morphism of CDG-corings $\mathcal{D} \longrightarrow \mathcal{E}$ compatible with a morphism of k-algebras $A \longrightarrow B$ is a pair (g, a), where $g \colon \mathcal{D} \longrightarrow \mathcal{E}$ is a map of graded corings compatible with a morphism of k-algebras $A \longrightarrow B$ and $a \colon \mathcal{D}_1 \longrightarrow B$ is an A-A-bimodule map satisfying the equations $d(g(p)) = g(d(p)) + a(p_{(1)})g(p_{(2)}) + (-1)^{|p|}g(p_{(1)})a(p_{(2)})$ and $h(g(q)) = h(q) + a(d(q)) + a(q_{(1)})a(q_{(2)})$ hold for all $p \in \mathcal{D}$ and $q \in \mathcal{D}_2$ (where the map a is extended by zero to the components $\mathcal{D}_i$ with $i \neq 1$).

Composition of morphisms of CDG-corings is defined by the rule

$$(g', a')(g'', a'') = (g'g'',\ a'g'' + a'');$$

identity morphisms are the morphisms $(\mathrm{id}, 0)$. So the category of CDG-corings is defined. Notice that two CDG-corings of the form $(\mathcal{D}, d', h')$ and $(\mathcal{D}, d'', h'')$ over a k-algebra A with $d''(p) = d'(p) + a(p_{(1)})p_{(2)} + (-1)^{|p|}p_{(1)}a(p_{(2)})$ and $h''(q) = h'(q) + a(d'(q)) + a(q_{(1)})a(q_{(2)})$, where $a \colon \mathcal{D}_1 \longrightarrow A$ is an A-A-bimodule map, are always naturally isomorphic to each other, the isomorphism being given by $(\mathrm{id}, a) \colon (\mathcal{D}, d', h') \longrightarrow (\mathcal{D}, d'', h'')$.

The category of DG-corings (over ungraded k-algebras considered as DG-algebras concentrated in degree zero) has DG-corings $\mathcal{D}$ over k-algebras A as objects and maps of DG-corings $\mathcal{D} \longrightarrow \mathcal{E}$ compatible with morphisms of k-algebras

$A \longrightarrow B$ as morphisms. The category of quasi-differential corings can be defined as the full subcategory of the category of DG-corings whose objects are the DG-corings with acyclic differentials. One can also consider the category of DG-corings (over ungraded k-algebras) with coderivations of degree -1. There is an obvious faithful, but not fully faithful functor from the latter category to the category of CDG-corings, assigning the CDG-coring $(\mathcal{D}, d, h)$ with $h = 0$ to a DG-coring $(\mathcal{D}, d)$ and the morphism of CDG-corings $(g, 0)$ to a map of DG-corings $g \colon \mathcal{D} \longrightarrow \mathcal{E}$ compatible with a morphism of k-algebras $A \longrightarrow B$.

There is a natural fully faithful functor from the category of CDG-corings to the category of quasi-differential corings, whose image consists of the quasi-differential corings $\mathcal{D}^\sim$ over A for which the counit map $\mathcal{D}_0^\sim \longrightarrow A$ can be presented as the composition of the coderivation component $\partial_0 \colon \mathcal{D}_0^\sim \longrightarrow \mathcal{D}_1^\sim$ and some A-A-bimodule map $\delta \colon \mathcal{D}_1^\sim \longrightarrow A$. In other words, a quasi-differential coring comes from a CDG-coring if and only if the counit map $\mathcal{D}_0^\sim / \partial_{-1} \mathcal{D}_{-1}^\sim \longrightarrow A$ can be extended to an A-A-bimodule map $\mathcal{D}_1^\sim \longrightarrow A$, where the A-A-bimodule $\mathcal{D}_0^\sim / \partial_{-1} \mathcal{D}_{-1}^\sim$ is embedded into the A-A-bimodule $\mathcal{D}_1^\sim$ by the map ∂_0. In particular, the categories of quasi-differential corings and CDG-corings over a field $A = k$ (*quasi-differential coalgebras* and *CDG-coalgebras* over k) are naturally equivalent.

Let us first construct the inverse functor. Given a quasi-differential coring $(\mathcal{D}^\sim, \partial)$ and a map $\delta \colon \mathcal{D}_1^\sim \longrightarrow A$ as above, set $\mathcal{D} = \mathcal{D}^\sim / \operatorname{im} \partial$ and define d and h by the formulas $d(\overline{p}) = \delta(p_{(1)}) \overline{p_{(2)}} + (-1)^{|p|} \overline{p_{(1)}} \delta(p_{(2)})$ and $h(\overline{q}) = \delta(q_{(1)}) \delta(q_{(2)})$ for $p \in \mathcal{D}^\sim$ and $q \in \mathcal{D}_2^\sim$, where the map δ is extended by zero to the components $\mathcal{D}_i^\sim$ with $i \neq 1$ and $\overline{r} \in \mathcal{D}$ denotes the image of an element $r \in \mathcal{D}^\sim$. To a map of quasi-differential corings $g \colon \mathcal{D}^\sim \longrightarrow \mathcal{E}^\sim$ endowed with maps $\delta_\mathcal{D} \colon \mathcal{D}_1^\sim \longrightarrow A$ and $\delta_\mathcal{E} \colon \mathcal{E}_1^\sim \longrightarrow B$ with the above property, compatible with a morphism of k-algebras $f \colon A \longrightarrow B$, one assigns the morphism of CDG-corings $(\overline{g}, \, \delta_\mathcal{E} g - f \delta_\mathcal{D})$, where $\overline{g} \colon \mathcal{D} \longrightarrow \mathcal{E}$ denotes the induced morphism on the cokernels of the coderivations ∂.

Conversely, to a CDG-coring $(\mathcal{D}, d, h)$ over a k-algebra A one assigns the quasi-differential coring $(\mathcal{D}^\sim, \partial)$ over A whose graded components are the A-A-bimodules $\mathcal{D}_i^\sim = \mathcal{D}_i \oplus \mathcal{D}_{i-1}$, the coderivation ∂ is given by the formula $\partial(\tau p + \overline{\partial} q) = \overline{\partial} p$, and the comultiplication is given by the formula $\tau p + \overline{\partial} q \longmapsto \tau p_{(1)} \otimes \tau p_{(2)} + (-1)^{|p_{(1)}|} \tau d(p_{(1)}) \otimes \overline{\partial} p_{(2)} + (-1)^{|p_{(2)}|} h(p_{(1)}) \overline{\partial} p_{(2)} \otimes \overline{\partial} p_{(3)} + \overline{\partial} q_{(1)} \otimes \tau q_{(2)} + (-1)^{|q_{(1)}|} \tau q_{(1)} \otimes \overline{\partial} q_{(2)} + (-1)^{|q_{(1)}|} \overline{\partial} d(q_{(1)}) \otimes \overline{\partial} q_{(2)}$, where $\tau p + \overline{\partial} q = (p, q)$ is a formal notation for an element of $\bigoplus_i (\mathcal{D}_i \oplus \mathcal{D}_{i-1})$. To a morphism of CDG-corings $(g, a) \colon \mathcal{D} \longrightarrow \mathcal{E}$, the morphism of quasi-differential corings $\bigoplus_i (\mathcal{D}_i \oplus \mathcal{D}_{i-1}) \longrightarrow \bigoplus_i (\mathcal{E}_i \oplus \mathcal{E}_{i-1})$ given by the formula $\tau p + \overline{\partial} q \longmapsto \tau g(p) + a(p_{(1)}) \overline{\partial} g(p_{(2)}) + \overline{\partial} g(q)$ is assigned. For a quasi-differential coring $(\mathcal{D}^\sim, \partial)$ over a k-algebra A endowed with a map $\delta \colon \mathcal{D}_1^\sim \longrightarrow A$ with the above property and the corresponding CDG-coring $(\mathcal{D}, d, h)$, the natural morphism of quasi-differential corings $\mathcal{D}^\sim \longrightarrow \bigoplus_i (\mathcal{D}_i \oplus \mathcal{D}_{i-1})$ over A is given by the formula $p \longmapsto \tau \overline{p} + \delta(p_{(1)}) \overline{\partial p_{(2)}}$ for $p \in \mathcal{D}^\sim$. This morphism is an isomorphism, since the induced morphism of the cokernels of the coderivations ∂ is an isomorphism.

11.2.3 Let B be a graded k-algebra endowed with a derivation d_B of degree 1 and $\mathcal{D}$ be a graded coring over B endowed with a coderivation $\partial_{\mathcal{D}}$ with respect to d_B. Let $\mathcal{T}$ be a graded semialgebra over $\mathcal{D}$. A homogeneous map $d_{\mathcal{T}} \colon \mathcal{T} \longrightarrow \mathcal{T}$ of degree 1 is called a *semiderivation of $\mathcal{T}$ with respect to $d_{\mathcal{D}}$ and d_B* if the biaction map $B \otimes_k \mathcal{T} \otimes_k B \longrightarrow \mathcal{T}$, the bicoaction map $\mathcal{T} \longrightarrow \mathcal{D} \otimes_B \mathcal{T} \otimes_B \mathcal{D}$, and the semimultiplication map $\mathcal{T} \square_{\mathcal{D}} \mathcal{T} \longrightarrow \mathcal{T}$ are morphisms in the category of graded k-modules endowed with endomorphisms of degree 1. In this case, it follows that the semiunit map $\mathcal{D} \longrightarrow \mathcal{T}$ satisfies the same condition. In the particular case when B and $\mathcal{D}$ are concentrated in degree 0 and $d_B = 0 = d_{\mathcal{D}}$, the conditions on the biaction and bicoaction map simply mean that $d_{\mathcal{T}}$ is a $\mathcal{D}$-$\mathcal{D}$-bicomodule morphism.

Let B be a DG-algebra over k and $\mathcal{D}$ be a DG-coring over B. A *DG-semialgebra* over $\mathcal{D}$ is a graded semialgebra over the graded coring $\mathcal{D}$ endowed with a semiderivation $d_{\mathcal{T}}$ with respect to $d_{\mathcal{D}}$ and d_B such that $d_{\mathcal{T}}^2 = 0$.

Let $\mathcal{T}$ be a DG-semialgebra over a DG-coring $\mathcal{D}$. Then the cohomology $H(\mathcal{T})$ is endowed with a natural structure of graded semialgebra over the graded coring $H(\mathcal{D})$ provided that

(i) the natural maps from the tensor products of cohomology to the cohomology of the tensor products are isomorphisms for the tensor products $\mathcal{D} \otimes_B \mathcal{D}$, $\mathcal{D} \otimes_B \mathcal{D} \otimes_B \mathcal{D}$, $\mathcal{D} \otimes_B \mathcal{T}$, $\mathcal{T} \otimes_B \mathcal{D}$, $\mathcal{D} \otimes_B \mathcal{D} \otimes_B \mathcal{T}$, $\mathcal{T} \otimes_B \mathcal{D} \otimes_B \mathcal{D}$, $\mathcal{D} \otimes_B \mathcal{T} \otimes_B \mathcal{D}$, $\mathcal{T} \otimes_B \mathcal{T}$, $\mathcal{D} \otimes_B \mathcal{T} \otimes_B \mathcal{T}$, $\mathcal{T} \otimes_B \mathcal{T} \otimes_B \mathcal{D}$, $\mathcal{T} \otimes_B \mathcal{D} \otimes_B \mathcal{T}$, $\mathcal{D} \otimes_B \mathcal{T} \otimes_B \mathcal{D} \otimes_B \mathcal{T}$, $\mathcal{T} \otimes_B \mathcal{D} \otimes_B \mathcal{T} \otimes_B \mathcal{D}$, $\mathcal{T} \otimes_B \mathcal{T} \otimes_B \mathcal{T}$, $\mathcal{T} \otimes_B \mathcal{D} \otimes_B \mathcal{T} \otimes_B \mathcal{T}$, $\mathcal{T} \otimes_B \mathcal{T} \otimes_B \mathcal{D} \otimes_B \mathcal{T}$;

(ii) the multiple cotensor products $H(\mathcal{T}) \square_{H(\mathcal{D})} \cdots \square_{H(\mathcal{D})} H(\mathcal{T})$ are associative, where the graded $H(\mathcal{D})$-$H(\mathcal{D})$-bicomodule structure on $H(\mathcal{T})$ is well defined in view of (i); and

(iii) the natural maps $H(\mathcal{T} \square_{\mathcal{D}} \mathcal{T}) \longrightarrow H(\mathcal{T}) \square_{H(\mathcal{D})} H(\mathcal{T})$, $H(\mathcal{D} \otimes_B \mathcal{T} \square_{\mathcal{D}} \mathcal{T}) \longrightarrow H(\mathcal{D}) \otimes_{H(B)} H(\mathcal{T}) \square_{H(\mathcal{D})} H(\mathcal{T})$, $H(\mathcal{T} \square_{\mathcal{D}} \mathcal{T} \otimes_B \mathcal{D}) \longrightarrow H(\mathcal{T}) \square_{H(\mathcal{D})} H(\mathcal{T}) \otimes_{H(B)} H(\mathcal{D})$, and $H(\mathcal{T} \square_{\mathcal{D}} \mathcal{T} \square_{\mathcal{D}} \mathcal{T}) \longrightarrow H(\mathcal{T}) \square_{H(\mathcal{D})} H(\mathcal{T}) \square_{H(\mathcal{D})} H(\mathcal{T})$, which are well defined in view of (i) and (ii), are isomorphisms.

A map of DG-semialgebras $\mathcal{S} \longrightarrow \mathcal{T}$ compatible with a map of DG-corings $\mathcal{C} \longrightarrow \mathcal{D}$ and a morphism of DG-algebras $A \longrightarrow B$ induces a map of graded semialgebras $H(\mathcal{S}) \longrightarrow H(\mathcal{T})$ compatible with the map of graded corings $H(\mathcal{C}) \longrightarrow H(\mathcal{D})$ and the morphism of graded k-algebras $H(A) \longrightarrow H(B)$ whenever both DG-semialgebras $\mathcal{S}$ and $\mathcal{T}$ satisfy the above three conditions. Here a map $\mathcal{S} \longrightarrow \mathcal{T}$ from a DG-semialgebra $\mathcal{S}$ over a DG-coring $\mathcal{C}$ to a DG-semialgebra $\mathcal{T}$ over a DG-coring $\mathcal{D}$ is called compatible with a map of DG-corings $\mathcal{C} \longrightarrow \mathcal{D}$ and a morphism of DG-algebras $A \longrightarrow B$ if the map of graded semialgebras $\mathcal{S} \longrightarrow \mathcal{T}$ is compatible with the map of graded corings $\mathcal{C} \longrightarrow \mathcal{D}$ and the morphism of graded k-algebras $A \longrightarrow B$, and the maps $\mathcal{S} \longrightarrow \mathcal{T}$ and $\mathcal{C} \longrightarrow \mathcal{D}$ are morphisms of complexes.

11.3   One-sided SemiTor

The only purpose of this section is to weaken certain (co)flatness assumptions as far as possible. Let S be a semialgebra over a coring C over a k-algebra A. We will consider two situations separately.

11.3.1   Assume that C is a flat right A-module and S is a coflat right C-comodule.

Consider the functor of semitensor product over S on the Cartesian product of the homotopy category of complexes of C-coflat right S-semimodules and the homotopy category of complexes of left S-semimodules. The semiderived category of C-coflat right S-semimodules is defined as the quotient category of the homotopy category of C-coflat right S-semimodules by the thick subcategory of complexes of right S-semimodules that as complexes of C-comodules are coacyclic with respect to the exact category of coflat right C-comodules. A complex of left S-semimodules $\mathcal{M}^\bullet$ is called *semiflat relative to* C if the complex $\mathcal{N}^\bullet \lozenge_S \mathcal{M}^\bullet$ is acyclic for any C-contractible complex of C-coflat right S-semimodules $\mathcal{N}^\bullet$ (cf. 2.8).

The left derived functor SemiTor^S on the Cartesian product of the semiderived category of C-coflat right S-semimodules and the semiderived category of left S-semimodules is defined by restricting the functor of semitensor product to the Cartesian product of the homotopy category of C-coflat right S-semimodules and the homotopy category of complexes of left S-semimodules semiflat relative to C, or to the Cartesian product of the homotopy category of semiflat complexes of right S-semimodules and the homotopy category of left S-semimodules. This definition of a derived functor is a particular case of *both* Lemmas 2.7 and 6.5.2. If $\mathcal{N}^\bullet$ is a complex of C-coflat right S-semimodules and $\mathcal{M}^\bullet$ is a complex of left S-semimodules, then the total complex of the bar bicomplex $\cdots \longrightarrow \mathcal{N}^\bullet \square_C S \square_C S \square_C \mathcal{M}^\bullet \longrightarrow \mathcal{N}^\bullet \square_C S \square_C \mathcal{M}^\bullet \longrightarrow \mathcal{N}^\bullet \square_C \mathcal{M}^\bullet$, constructed by taking infinite direct sums along the diagonals, represents the object SemiTor$^S(\mathcal{M}^\bullet, \mathcal{N}^\bullet)$ in $\mathsf{D}(k\text{-mod})$. When the semiunit map $C \longrightarrow S$ is injective and its cokernel is a flat right A-module (and hence a coflat right C-comodule by Lemma 1.2.2), one can also use the reduced bar bicomplex $\cdots \longrightarrow \mathcal{N}^\bullet \square_C S/C \square_C S/C \square_C \mathcal{M}^\bullet \longrightarrow \mathcal{N}^\bullet \square_C S/C \square_C \mathcal{M}^\bullet \longrightarrow \mathcal{N}^\bullet \square_C \mathcal{M}^\bullet$.

In the case when S is a graded semialgebra one analogously defines the derived functor SemiTor$^S_{gr}$ acting from the Cartesian product of the semiderived category of C-coflat graded right S-semimodules and the semiderived category of graded left S-semimodules to the derived category of graded k-modules.

11.3.2   Assume that C is a flat right A-module, S is a flat right A-module and a C/A-coflat left C-comodule, and the ring A has a finite weak homological dimension.

Consider the functor of semitensor product over S on the Cartesian product of the homotopy category of complexes of A-flat right S-semimodules and the homotopy category of complexes of C/A-coflat left S-semimodules. The semiderived category of A-flat right S-semimodules (C/A-coflat left S-semimodules) is defined as the quotient category of the homotopy category of A-flat right S-semimodules

($\mathcal{C}/A$-coflat left $\mathbf{S}$-semimodules) by the thick subcategory of complexes of $\mathbf{S}$-semi-modules that as complexes of $\mathcal{C}$-comodules are coacyclic with respect to the exact category of A-flat right $\mathcal{C}$-comodules ($\mathcal{C}/A$-coflat left $\mathcal{C}$-comodules). A complex of $\mathcal{C}/A$-coflat left $\mathbf{S}$-semimodules $\mathbf{M}^\bullet$ is called *semiflat relative to A* if the complex of k-modules $\mathbf{N}^\bullet \lozenge_\mathbf{S} \mathbf{M}^\bullet$ is acyclic for any complex of right $\mathbf{S}$-semimodules $\mathbf{N}^\bullet$ that as a complex of right $\mathcal{C}$-comodules is coacyclic with respect to the exact cat-egory of A-flat right $\mathcal{C}$-comodules. A complex of A-flat right $\mathbf{S}$-semimodules $\mathbf{N}^\bullet$ is called $\mathbf{S}/\mathcal{C}/A$-*semiflat* if the complex of k-modules $\mathbf{N}^\bullet \lozenge_\mathbf{S} \mathbf{M}^\bullet$ is acyclic for any $\mathcal{C}$-contractible complex of $\mathcal{C}/A$-coflat left $\mathbf{S}$-semimodules $\mathbf{M}^\bullet$ (cf. 2.8).

The left derived functor $\mathrm{SemiTor}^\mathbf{S}$ on the Cartesian product of the semi-derived category of A-flat right $\mathbf{S}$-semimodules and the semiderived category of $\mathcal{C}/A$-coflat left $\mathbf{S}$-semimodules is defined by restricting the functor of semitensor product to the Cartesian product of the homotopy category of A-flat right $\mathbf{S}$-semi-modules and the homotopy category of complexes of $\mathcal{C}/A$-coflat left $\mathbf{S}$-semimodules semiflat relative to A, or to the Cartesian product of the homotopy category of $\mathbf{S}/\mathcal{C}/A$-semiflat complexes of A-flat right $\mathbf{S}$-semimodules and the homotopy cat-egory of $\mathcal{C}/A$-coflat left $\mathbf{S}$-semimodules. This definition of a derived functor is a particular case of *both* Lemmas 2.7 and 6.5.2. If $\mathbf{N}^\bullet$ is a complex of A-flat right $\mathbf{S}$-semimodules and $\mathbf{M}^\bullet$ is a complex of $\mathcal{C}/A$-coflat left $\mathbf{S}$-semimodules, then the to-tal complex of the bar bicomplex $\cdots \longrightarrow \mathbf{N}^\bullet \square_\mathcal{C} \mathbf{S} \square_\mathcal{C} \mathbf{S} \square_\mathcal{C} \mathbf{M}^\bullet \longrightarrow \mathbf{N}^\bullet \square_\mathcal{C} \mathbf{S} \square_\mathcal{C} \mathbf{M}^\bullet \longrightarrow \mathbf{N}^\bullet \square_\mathcal{C} \mathbf{M}^\bullet$, constructed by taking infinite direct sums along the diagonals, repre-sents the object $\mathrm{SemiTor}^\mathbf{S}(\mathbf{M}^\bullet, \mathbf{N}^\bullet)$ in $\mathsf{D}(k\text{-mod})$. When the semiunit map $\mathcal{C} \longrightarrow \mathbf{S}$ is injective and its cokernel is a flat right A-module (the cokernel is a $\mathcal{C}/A$-coflat left $\mathcal{C}$-comodule by Lemma 1.2.2), one can also use the reduced bar bicomplex $\cdots \longrightarrow \mathbf{N}^\bullet \square_\mathcal{C} \mathbf{S}/\mathcal{C} \square_\mathcal{C} \mathbf{S}/\mathcal{C} \square_\mathcal{C} \mathbf{M}^\bullet \longrightarrow \mathbf{N}^\bullet \square_\mathcal{C} \mathbf{S}/\mathcal{C} \square_\mathcal{C} \mathbf{M}^\bullet \longrightarrow \mathbf{N}^\bullet \square_\mathcal{C} \mathbf{M}^\bullet$.

In the case when $\mathbf{S}$ is a graded semialgebra one analogously defines the derived functor $\mathrm{SemiTor}^\mathbf{S}_{\mathrm{gr}}$ acting from the Cartesian product of the semiderived category of A-flat graded right $\mathbf{S}$-semimodules and the semiderived category of $\mathcal{C}/A$-coflat graded left $\mathbf{S}$-semimodules to the derived category of graded k-modules.

11.4 Koszul semialgebras and corings

11.4.1 Let $\mathbf{S}$ be a semialgebra over a coring $\mathcal{C}$ over a k-algebra A. Suppose that $\mathbf{S}$ is endowed with an augmentation, i.e., a morphism $\mathbf{S} \longrightarrow \mathcal{C}$ of semialgebras over $\mathcal{C}$; let $\mathbf{S}_+$ be the kernel of this map. We will denote by $\mathrm{Bar}^\bullet(\mathbf{S}, \mathcal{C})$ the reduced bar complex

$$\cdots \longrightarrow \mathbf{S}_+ \square_\mathcal{C} \mathbf{S}_+ \square_C \mathbf{S}_+ \longrightarrow \mathbf{S}_+ \square_\mathcal{C} \mathbf{S}_+ \longrightarrow \mathbf{S}_+ \longrightarrow \mathcal{C}.$$

It can be also defined as the coring $\bigoplus_{n=0}^\infty \mathbf{S}_+^{\square_\mathcal{C} n}$ over the k-algebra A (the "cotensor coring" of the $\mathcal{C}$-$\mathcal{C}$-bicomodule $\mathbf{S}$) endowed with the unique grading such that the component $\mathbf{S}_+$ is situated in degree -1 and the unique coderivation (with respect to the zero derivation of A) of degree 1 whose component mapping $\mathbf{S}_+ \square_\mathcal{C} \mathbf{S}_+$ to $\mathbf{S}_+$ is equal to the semimultiplication map $\mathbf{S}_+ \square_\mathcal{C} \mathbf{S}_+ \longrightarrow \mathbf{S}_+$. So $\mathrm{Bar}^\bullet(\mathbf{S}, \mathcal{C})$

is a DG-coring over the k-algebra A considered as a DG-algebra concentrated in degree 0.

Now let $\mathbf{S}$ be a graded semialgebra over a coring $\mathcal{C}$ over a k-algebra A, where A and $\mathcal{C}$ are considered as a graded k-algebra and a graded coring concentrated in degree 0; assume additionally that $\mathbf{S}$ is concentrated in nonnegative degrees, $\mathcal{C}$ is the component of degree 0 in $\mathbf{S}$, and the augmentation map $\mathbf{S} \longrightarrow \mathcal{C}$ is simply the projection of $\mathbf{S}$ to its component of degree 0. In this case there is a graded version $\mathrm{Bar}_{\mathrm{gr}}^{\bullet}(\mathbf{S}, \mathcal{C})$ of the above bar complex, which is a bigraded object with the grading denoted by upper indices coming from the cotensor powers of $\mathbf{S}_+$ and the grading denoted by lower indices coming from the grading of $\mathbf{S}_+$ itself. Notice that the component $\mathrm{Bar}_n^i(\mathbf{S}, \mathcal{C})$ can be only nonzero when $0 \leqslant -i \leqslant n$.

Let $\mathcal{C}$ and $\mathcal{D}$ be corings over a k-algebra A. Suppose that we are given two maps $\mathcal{C} \longrightarrow \mathcal{D}$ and $\mathcal{D} \longrightarrow \mathcal{C}$ that are morphisms of corings over A such that the composition $\mathcal{C} \longrightarrow \mathcal{D} \longrightarrow \mathcal{C}$ is the identity; let $\mathcal{D}_+$ be the cokernel of the map $\mathcal{C} \longrightarrow \mathcal{D}$. Assume that the multiple cotensor products $\mathcal{D} \,\square_{\mathcal{C}} \cdots \square_{\mathcal{C}} \mathcal{D}$, where $\mathcal{D}$ is endowed with a $\mathcal{C}$-$\mathcal{C}$-bimodule structure via the morphism $\mathcal{D} \longrightarrow \mathcal{C}$, are associative. We will denote by $\mathrm{Cob}^{\bullet}(\mathcal{D}, \mathcal{C})$ the reduced cobar complex

$$ \mathcal{C} \longrightarrow \mathcal{D}_+ \longrightarrow \mathcal{D}_+ \square_{\mathcal{C}} \mathcal{D}_+ \longrightarrow \mathcal{D}_+ \square_{\mathcal{C}} \mathcal{D}_+ \square_{\mathcal{C}} \mathcal{D}_+ \longrightarrow \cdots $$

It can be also defined as the semialgebra $\bigoplus_{n=0}^{\infty} \mathcal{D}_+^{\square_{\mathcal{C}} n}$ over the coring $\mathcal{C}$ (the "cotensor semialgebra" of the $\mathcal{C}$-$\mathcal{C}$-bimodule $\mathcal{D}$) endowed with the unique grading such that the component $\mathcal{D}_+$ is situated in degree 1 and the unique semiderivation (with respect to $d_{\mathcal{C}} = 0$ and $d_A = 0$) of degree 1 whose component mapping $\mathcal{D}_+$ to $\mathcal{D}_+ \square_{\mathcal{C}} \mathcal{D}_+$ is equal to the comultiplication map $\mathcal{D}_+ \longrightarrow \mathcal{D}_+ \square_{\mathcal{C}} \mathcal{D}_+$. So $\mathrm{Cob}^{\bullet}(\mathcal{D}, \mathcal{C})$ is a DG-semialgebra over the coring $\mathcal{C}$ over the k-algebra A, where A and $\mathcal{C}$ are considered as a DG-algebra and a DG-coring concentrated in degree 0.

Now let $\mathcal{D}$ be a graded coring over a k-algebra A considered as a graded k-algebra concentrated in degree 0 and $\mathcal{C}$ be a coring over A; assume additionally that $\mathcal{D}$ is concentrated in nonnegative degrees, $\mathcal{C}$ is the component of degree 0 in $\mathcal{D}$, and the maps $\mathcal{C} \longrightarrow \mathcal{D}$ and $\mathcal{D} \longrightarrow \mathcal{C}$ are simply the embedding of and the projection to the component of degree 0. In this case there is a graded version $\mathrm{Cob}_{\mathrm{gr}}^{\bullet}(\mathcal{D}, \mathcal{C})$ of the above cobar complex, which is a bigraded object with the grading denoted by upper indices coming from the cotensor powers of $\mathcal{D}_+$ and the grading denoted by lower indices coming from the grading of $\mathcal{D}_+$ itself. Notice that the component $\mathrm{Cob}_n^i(\mathcal{D}, \mathcal{C})$ can be only nonzero when $0 \leqslant i \leqslant n$.

11.4.2 Let $\mathcal{C}$ be a coring over a k-algebra A. Assume that $\mathcal{C}$ is a flat right A-module. A graded semialgebra $\mathbf{S}$ over $\mathcal{C}$ is called *right coflat Koszul* if

(i) $\mathbf{S}$ is nonnegatively graded and the semiunit homomorphism is an isomorphism $\mathcal{C} \simeq \mathbf{S}_0$;

(ii) the components $\mathbf{S}_i$ are flat right A-modules;

(iii) the cohomology $H_n^i \,\mathrm{Bar}_{\mathrm{gr}}^{\bullet}(\mathbf{S}, \mathcal{C})$ are only nonzero on the diagonal $-i = n$; and

(iv) whenever the component $\mathrm{Bar}^\bullet_n(\mathcal{S}, \mathcal{C})$ is a complex of A-flat right $\mathcal{C}$-comodules, so the diagonal cohomology $H^{-n}_n \mathrm{Bar}^\bullet_{\mathrm{gr}}(\mathcal{S}, \mathcal{C})$ can be endowed with a right $\mathcal{C}$-comodule structure as the kernel of a morphism in the category of right $\mathcal{C}$-comodules, it is a coflat right $\mathcal{C}$-comodule.

When the ring A has a finite weak homological dimension, there is an analogous definition of a *right flat and left relatively coflat Koszul* semialgebra $\mathcal{S}$ over $\mathcal{C}$. One imposes the same conditions (i–iii) and replaces (iv) with the condition

(iv′) the diagonal cohomology $H^{-n}_n \mathrm{Bar}^\bullet_{\mathrm{gr}}(\mathcal{S}, \mathcal{C})$ is a $\mathcal{C}/A$-coflat left $\mathcal{C}$-comodule for all n.

A graded coring $\mathcal{D}$ over the k-algebra A endowed with a morphism $\mathcal{D} \longrightarrow \mathcal{C}$ of corings over A is called a *right coflat Koszul coring over $\mathcal{C}$* if

 (i) $\mathcal{D}$ is nonnegatively graded and the morphism $\mathcal{D} \longrightarrow \mathcal{C}$ vanishes on the components of positive degree in $\mathcal{D}$ and induces an isomorphism $\mathcal{D}_0 \simeq \mathcal{C}$;

 (ii) whenever a component $\mathcal{D}_n$ is a flat right A-module, it is a coflat right $\mathcal{C}$-comodule;

(iii) whenever all the multiple cotensor products entering into the construction of the component $\mathrm{Cob}^\bullet_n(\mathcal{D}, \mathcal{C})$ are associative, so this component is well defined, the cohomology $H^i \mathrm{Cob}^\bullet_n(\mathcal{D}, \mathcal{C})$ is only nonzero on the diagonal $i = n$; and

(iv) in the assumptions of (iii), the diagonal cohomology $H^n \mathrm{Cob}^\bullet_n(\mathcal{D}, \mathcal{C})$ is a flat right A-module.

When the ring A has a finite weak homological dimension, there is an analogous definition of a *right flat and left relatively coflat Koszul coring* $\mathcal{D}$ over $\mathcal{C}$. One imposes the same conditions (i–ii), (iv), and replaces (iii) with the condition

(iii′) the component $\mathcal{D}_n$ is a $\mathcal{C}/A$-coflat left $\mathcal{C}$-comodule for all n.

11.4.3 The objects of the category of right coflat Koszul semialgebras are right coflat Koszul semialgebras $\mathcal{S}$ over corings $\mathcal{C}$ over k-algebras A such that $\mathcal{C}$ is a flat right A-module. Morphisms are maps of graded semialgebras $\mathcal{S} \longrightarrow \mathcal{S}'$ compatible with maps of corings $\mathcal{C} \longrightarrow \mathcal{C}'$ and morphisms of k-algebras $A \longrightarrow A'$. Imposing the additional assumption that A has a finite weak homological dimension, one analogously defines the category of right flat and left relatively coflat Koszul semialgebras.

The objects of the category of right coflat Koszul corings are right coflat Koszul corings $\mathcal{D}$ over corings $\mathcal{C}$ over k-algebras A such that $\mathcal{C}$ is a flat right A-module. Morphisms are maps of graded corings $\mathcal{D} \longrightarrow \mathcal{D}'$ compatible with morphisms of k-algebras $A \longrightarrow A'$. Imposing the additional assumption that A has a finite weak homological dimension, one analogously defines the category of right flat and left relatively Koszul corings.

Theorem. *The category of right coflat Koszul semialgebras is equivalent to the category of right coflat Koszul corings. Analogously, the category of right flat and left relatively coflat Koszul semialgebras is equivalent to the category of right flat*

and left relatively coflat Koszul corings. In both cases, the mutually inverse equiv-
alences are provided by the functor assigning to a Koszul semialgebra $\mathbf{S}$ *the coring*
of cohomology of the graded DG-coring $\mathrm{Bar}^\bullet_{\mathrm{gr}}(\mathbf{S}, \mathcal{C})$ *and the functor assigning to*
a Koszul coring $\mathcal{D}$ *the semialgebra of cohomology of the graded DG-semialgebra*
$\mathrm{Cob}^\bullet_{\mathrm{gr}}(\mathcal{D}, \mathcal{C})$.

Proof. The assertions of the theorem follow from Propositions 1 and 2 below.
To check the conditions of 11.2 needed for the coring of cohomology and the
semialgebra of cohomology to be defined, use Lemma 1.2.2 and Proposition 1.2.5.
$\qquad\qquad\qquad\qquad\qquad\qquad\qquad\qquad\qquad\qquad\qquad\qquad\qquad\qquad\qquad\qquad$ $\square$

Let $\mathcal{C}$ be a coring over a k-algebra A.

Proposition 1.

(a) *Assume that* $\mathcal{C}$ *is a flat right* A-module. *Then a graded semialgebra* $\mathbf{S}$ *over*
$\mathcal{C}$ *is right coflat Koszul if and only if*

    (i) $\mathbf{S}$ *is nonnegatively graded and the semiunit map is an isomorphism*
$\mathcal{C} \simeq \mathbf{S}_0$;

    (ii) *for any* $n \geqslant 1$ *the natural map from the quotient* k-module *of the coten-*
sor power $\mathbf{S}_1^{\square_{\mathcal{C}} n}$ *by the sum of the kernels of its maps to cotensor prod-*
ucts $\mathbf{S}_1^{\square_{\mathcal{C}} i-1} \square_{\mathcal{C}} \mathbf{S}_2 \square_{\mathcal{C}} \mathbf{S}_1^{\square_{\mathcal{C}} n-i-1}$, $i = 1, \ldots, n-1$, *to the component* $\mathbf{S}_n$
is an isomorphism;

    (iii) *the lattice of submodules of the* k-module $\mathbf{S}_i^{\square_{\mathcal{C}} n}$ *generated by these* $n-1$
kernels is distributive;

    (iv) *all the quotient modules of embedded submodules belonging to the men-*
tioned lattice are flat right A-modules *in their natural right* A-module
structures; and

    (v) *all the quotient modules of embedded submodules belonging to this lattice*
are coflat right $\mathcal{C}$-comodules *in their right* $\mathcal{C}$-comodule *structures that*
are well defined in view of (iv).

(b) *Assume that* $\mathcal{C}$ *is a flat right* A-module *and* A *has a finite weak homological*
dimension. Then a graded semialgebra $\mathbf{S}$ *over* $\mathcal{C}$ *is right flat and left relatively*
coflat Koszul if and only if it satisfies the conditions (i–iv) *of* (a) *and the*
condition

    (v′) *all the quotient modules of embedded submodules belonging to the lattice*
under consideration are $\mathcal{C}/A$-coflat *left* $\mathcal{C}$-comodules *in their natural left*
$\mathcal{C}$-comodule *structures.*

Proposition 2.

(a) *Assume that* $\mathcal{C}$ *is a flat right* A-module. *Then a graded coring* $\mathcal{D}$ *endowed*
with a morphism $\mathcal{D} \longrightarrow \mathcal{C}$ *of corings over* A *is a right coflat Koszul coring*
over $\mathcal{C}$ *if and only if*

(i) $\mathcal{D}$ is nonpositively graded and the morphism $\mathcal{D} \longrightarrow \mathcal{C}$ vanishes on the components of positive degrees in $\mathcal{D}$ and induces an isomorphism $\mathcal{D}_0 \simeq \mathcal{C}$;

(ii) for any $n \geqslant 1$ the natural map from the component $\mathcal{D}_n$ to the intersection of images of the maps from cotensor products $\mathcal{D}_1^{\square_{\mathcal{C}} i-1} \square_{\mathcal{C}} \mathcal{D}_2 \square_{\mathcal{C}} \mathcal{D}_1^{\square_{\mathcal{C}} n-i-1}$, $i = 1, \ldots, n-1$, to the cotensor power $\mathcal{D}_1^{\square_{\mathcal{C}} n}$ is an isomorphism;

(iii) the lattice of submodules of the k-module $\mathcal{D}_1^{\square_{\mathcal{C}} n}$ generated by these $n-1$ images is distributive;

(iv) all the quotient modules of the embedded submodules belonging to the mentioned lattice are flat right A-modules in their natural right A-module structures; and

(v) all the quotient modules of embedded submodules belonging to this lattice are coflat right $\mathcal{C}$-comodules in their right $\mathcal{C}$-comodule structures that are well defined in view of (iv).

(b) Assume that $\mathcal{C}$ is a flat right A-module and A has a finite weak homological dimension. Then a graded coring $\mathcal{D}$ endowed with a morphism $\mathcal{D} \longrightarrow \mathcal{C}$ of corings over A is a right flat and left relatively coflat Koszul coring over $\mathcal{C}$ if and only if it satisfies the conditions (i–iv) of (a) and the condition

(v′) all the quotient modules of embedded submodules belonging to the lattice under consideration are $\mathcal{C}/A$-coflat left $\mathcal{C}$-comodules in their natural left $\mathcal{C}$-comodule structures.

Proof of Propositions 1 and 2. Both propositions follow by induction in the internal degree n from Lemma 1.2.2, Proposition 1.2.5, and the next Lemma 1 (parts (a)$\Longleftrightarrow$(c), (a)$\Longleftrightarrow$(c*)), and the final assertion) and Lemma 2. $\square$

Lemma 1. *Let W be a k-module and $X_1, \ldots, X_{n-1} \subset W$ be a collection of submodules such that any proper subset $X_1, \ldots, \widehat{X}_k, \ldots, X_{n-1}$ generates a distributive lattice of submodules in W. Then the following conditions are equivalent:*

(a) the collection of submodules $X_1, \ldots, X_{n-1}$ generates a distributive lattice of submodules in W;

(b) the following complex of k-modules $K_\bullet(W; X_1, \ldots, X_{n-1})$ is exact

$$0 \longrightarrow X_1 \cap \cdots \cap X_{n-1} \longrightarrow X_2 \cap \cdots \cap X_{n-1} \longrightarrow X_3 \cap \cdots \cap X_{n-1}/X_1 \longrightarrow$$

$$\cdots \longrightarrow \bigcap_{s=i+1}^{n-1} X_s / \sum_{t=1}^{i-1} X_t \longrightarrow \cdots \longrightarrow$$

$$X_{n-1}/(X_1 + \cdots + X_{n-3}) \longrightarrow W/(X_1 + \cdots + X_{n-2})$$
$$\longrightarrow W/(X_1 + \cdots + X_{n-1}) \longrightarrow 0,$$

where we denote $Y/Z = Y/Y \cap Z$;

(c) *the following complex of k-modules* $B_\bullet(W; X_1, \ldots, X_{n-1})$

$$W \longrightarrow \bigoplus_t W/X_t \longrightarrow \cdots \longrightarrow \bigoplus_{t_1 < \cdots < t_{n-i}} W/\sum_{s=1}^{n-i} X_{t_s} \longrightarrow$$
$$\cdots \longrightarrow W/\sum_s X_s \longrightarrow 0$$

is exact everywhere except for the leftmost term;

(c*) *the following complex of k-modules* $B^\bullet(W; X_1, \ldots, X_{n-1})$

$$0 \longrightarrow \bigcap_s X_s \longrightarrow \cdots \longrightarrow \bigoplus_{t_1 < \cdots < t_{n-i}} \bigcap_{s=1}^{n-i} X_{t_s} \longrightarrow \cdots \longrightarrow \bigoplus_t X_t \longrightarrow W$$

is exact everywhere except for the rightmost term.

Besides, the complex in (c) *is always exact at its two rightmost nontrivial terms, and the complex in* (c*) *is always exact at its two leftmost nontrivial terms.*

Proof. See the proof of [75, Proposition 7.2 of Chapter 1]. $\square$

Assume that the coring $\mathcal{C}$ is a flat right A-module.

Lemma 2. *Let W be a k-module and $X_1, \ldots, X_{n-1} \subset W$ be a collection of submodules generating a distributive lattice of submodules in W.*

(a) *Suppose that W is a right A-module and X_s are its A-submodules. Then all the subquotient modules in the lattice of submodules generated by X_s are flat right A-modules if and only if for any $1 \leqslant t_1 < \cdots < t_{m-1} \leqslant n - 1$ the quotient module $W/(X_{t_1} + \cdots + X_{t_{m-1}})$ is a flat right A-module.*

(b) *Suppose that W is a left $\mathcal{C}$-comodule and X_s are its $\mathcal{C}$-subcomodules. Then all the subquotient modules in the lattice of submodules generated by X_s are $\mathcal{C}/A$-coflat left $\mathcal{C}$-comodules if and only if for any $1 \leqslant t_1 < \cdots < t_{m-1} \leqslant n-1$ the submodule $X_{t_1} \cap \cdots \cap X_{t_{m-1}}$ is a $\mathcal{C}/A$-coflat left $\mathcal{C}$-comodule.*

(c) *Suppose that W is a right $\mathcal{C}$-comodule and X_s are its $\mathcal{C}$-subcomodules such that all the subquotient modules in the lattice of submodules generated by X_s are flat right A-modules. Then all these subquotient modules are coflat right $\mathcal{C}$-comodules if and only if for any $1 \leqslant t_1 < \cdots < t_{m-1} \leqslant n-1$ the submodule $X_{t_1} \cap \cdots \cap X_{t_{m-1}}$ is a coflat right $\mathcal{C}$-comodule.*

Proof. Part (a): proceed by induction in n. Since the lattice is distributive, any subquotient module can be presented as an iterated extension of subquotient modules of the form $\bigcap_{j \in J} X_j / \bigcap_{j \in J} X_j \cap \sum_{i \notin J} X_i$, where $J \subset \{1, \ldots, n-1\}$. Whenever the inclusion $J \subset \{1, \ldots, n-1\}$ is proper, this subquotient module can be presented as an element of a smaller lattice generated by the submodules $X_j / X_j \cap \sum_{i \notin J} X_i$ in the quotient module $W/\sum_{i \notin J} X_i$. It follows from the induction hypothesis that all the submodules belonging to this smaller lattice are flat right A-modules. It remains to show that the submodule $X_1 \cap \cdots \cap X_{n-1}$ is a flat right A-module. But the latter submodule is the only nonzero cohomology module at the leftmost term of the complex of flat right A-modules $B_\bullet(W; X_1, \ldots, X_{n-1})$ from Lemma 1(c). The proofs of parts (b) and (c) are completely analogous, except for the use of Lemma 1.2.2. (See also [75, Proposition 7.1 of Chapter 6].) $\qquad\square$

A right coflat (right flat and left relatively coflat) Koszul semialgebra and a right coflat (right flat and left relatively coflat) Koszul coring corresponding to each other under the equivalence of categories from the above theorem are called *quadratic dual* to each other.

11.4.4 Let $\mathbf{S}$ be a right coflat (right flat and left relatively coflat) Koszul semialgebra over a coring $\mathcal{C}$ and $\mathcal{D}$ be the right coflat (right flat and left relatively coflat) Koszul coring over $\mathcal{C}$ quadratic dual to $\mathbf{S}$. Then on the cotensor products $\mathbf{S} \,\square_\mathcal{C}\, \mathcal{D}$ and $\mathcal{D} \,\square_\mathcal{C}\, \mathbf{S}$ there are structures of graded complexes whose differentials are the compositions

$$\mathbf{S}_i \,\square_\mathcal{C}\, \mathcal{D}_j \longrightarrow \mathbf{S}_i \,\square_\mathcal{C}\, \mathcal{D}_1 \,\square_\mathcal{C}\, \mathcal{D}_{j-1} \simeq \mathbf{S}_i \,\square_\mathcal{C}\, \mathbf{S}_1 \,\square_\mathcal{C}\, \mathcal{D}_{j-1} \longrightarrow \mathbf{S}_{i+1} \,\square_\mathcal{C}\, \mathcal{D}_{j-1}$$

of the maps induced by the comultiplication in $\mathcal{D}$ and the maps induced by the semimultiplication in $\mathbf{S}$ (and analogously for $\mathcal{D} \,\square_\mathcal{C}\, \mathbf{S}$). These complexes are called the *Koszul complexes* of the semialgebra $\mathbf{S}$ and the coring $\mathcal{D}$. All the grading components of the Koszul complexes with respect to the grading $i + j$, except the component of degree $i + j = 0$, are acyclic. This follows from Lemma 11.4.3.1 ((a)$\Longleftrightarrow$(b)).

Note added in proof. After this manuscript had been prepared the author learned that the Koszul property essentially does not depend on the basic (co)ring. In particular, a graded coring $\mathcal{D} = \mathcal{C} \oplus \mathcal{D}_1 \oplus \mathcal{D}_2 \oplus \cdots$ is right coflat (right flat and left relatively coflat) Koszul over a coring $\mathcal{C}$ over A if and only if the graded coring $A \oplus \mathcal{D}_1 \oplus \mathcal{D}_2 \oplus \cdots$ has the same property over the ring A considered as a coring over itself, provided that $\mathcal{D}_i$ are coflat right $\mathcal{C}$-comodules whenever they are flat right A-modules ($\mathcal{D}_i$ are $\mathcal{C}/A$-coflat left $\mathcal{C}$-comodules), and $\mathcal{C}$ and A satisfy the appropriate flatness and homological dimension assumptions above. Analogously, assuming that $\mathcal{C} = A$, a graded (semi)algebra $\mathbf{S} = A \oplus \mathbf{S}_1 \oplus \mathbf{S}_2 \oplus \cdots$ is right (co)flat Koszul over A if and only if the graded (semi)algebra $k \oplus \mathbf{S}_1 \oplus \mathbf{S}_2 \oplus \cdots$ is right (co)flat Koszul over k, provided that $\mathbf{S}_i$ are flat right A-modules and flat right k-modules for $i \geqslant 1$. One can prove this using the characterizations of Koszulity in terms of distributive lattices given in Propositions 1–2 and the results of [75, Section 6 of Chapter 1].

11.5 Central element theorem

Let $\mathcal{C}$ be a coring over a k-algebra A. Assume that $\mathcal{C}$ is a flat right A-module.

A *right coflat increasing filtration* F on a semialgebra $\mathbf{S}^\sim$ over a coring $\mathcal{C}$ is a family of $\mathcal{C}$-$\mathcal{C}$-bicomodules $F_n\mathbf{S}^\sim$ endowed with injective morphisms of $\mathcal{C}$-$\mathcal{C}$-bicomodules $F_{n-1}\mathbf{S}^\sim \longrightarrow F_n\mathbf{S}^\sim$ and an isomorphism of $\mathcal{C}$-$\mathcal{C}$-bicomodules $\varinjlim F_n\mathbf{S}^\sim \simeq \mathbf{S}^\sim$ such that

(i) $F_i \mathbf{S}^\sim = 0$ for $i < 0$, $F_0 \mathbf{S}^\sim = \mathcal{C}$, and the map $F_0 \mathbf{S}^\sim \longrightarrow \mathbf{S}^\sim$ is the semiunit map;

(ii) the compositions $F_i \mathbf{S}^\sim \square_{\mathcal{C}} F_j \mathbf{S}^\sim \longrightarrow \mathbf{S} \square_{\mathcal{C}} \mathbf{S} \longrightarrow \mathbf{S}$ of the maps induced by the injections $F_n \mathbf{S}^\sim \longrightarrow \mathbf{S}$ and the semimultiplication map factorize through $F_{i+j} \mathbf{S}^\sim$;

(iii) the successive quotients $F_n \mathbf{S}^\sim / F_{n-1} \mathbf{S}^\sim$ are flat right A-modules; and

(iv) the filtration components $F_n \mathbf{S}^\sim$ are coflat right $\mathcal{C}$-comodules (then the successive quotients are also coflat right $\mathcal{C}$-comodules).

Assuming that A has a finite weak homological dimension, one analogously defines *right flat and left relatively coflat increasing filtrations* by replacing the condition (iv) with the condition

(iv′) the filtration components $F_n \mathbf{S}^\sim$ are $\mathcal{C}/A$-coflat left $\mathcal{C}$-comodules (then the successive quotients are also $\mathcal{C}/A$-coflat left $\mathcal{C}$-comodules).

Theorem. *Let $\mathbf{S}^\sim$ be a semialgebra over a coring $\mathcal{C}$ endowed with a right coflat (right flat and left relatively coflat) increasing filtration F. Then the graded semialgebra $\mathfrak{J} = \bigoplus_n F_n \mathbf{S}^\sim$ over the coring $\mathcal{C}$ is right coflat (right flat and left relatively coflat) Koszul if and only if the graded semialgebra $\mathbf{S} = \bigoplus_n F_n \mathbf{S}^\sim / F_{n-1} \mathbf{S}^\sim$ over the coring $\mathcal{C}$ is right coflat (right flat and left relatively coflat) Koszul.*

Proof. Consider the reduced bar resolution $\cdots \longrightarrow \mathfrak{J}_+ \square_{\mathcal{C}} \mathfrak{J}_+ \square_{\mathcal{C}} \mathfrak{J} \longrightarrow \mathfrak{J}_+ \square_{\mathcal{C}} \mathfrak{J} \longrightarrow \mathfrak{J}$ of the right $\mathfrak{J}$-semimodule $\mathcal{C}$ and denote by $\mathcal{X}^\bullet$ its semitensor product

$$\cdots \longrightarrow \mathfrak{J}_+ \square_{\mathcal{C}} \mathfrak{J}_+ \square_{\mathcal{C}} \mathbf{S} \longrightarrow \mathfrak{J}_+ \square_{\mathcal{C}} \mathbf{S} \longrightarrow \mathbf{S}$$

with the left $\mathfrak{J}$-semimodule $\mathbf{S}$. Denote by $\mathcal{Y}^\bullet$ the two-term complex of graded right $\mathbf{S}$-semimodules $\mathfrak{J}_1 \longrightarrow \mathbf{S}_0 \oplus \mathbf{S}_1$, where $\mathfrak{J}_1$ is endowed with a right $\mathbf{S}$-semimodule structure via the augmentation of $\mathbf{S}$ and $\mathbf{S}_0 \oplus \mathbf{S}_1$ is the quotient semimodule $\mathbf{S} / \bigoplus_{n \geqslant 2} \mathbf{S}_n$; the components of the differential in this complex are the zero map $\mathfrak{J}_1 \longrightarrow \mathbf{S}_0$ and the projection $\mathfrak{J}_1 \longrightarrow \mathbf{S}_1$. There is a natural morphism of complexes of graded right $\mathbf{S}$-semimodules $\mathcal{X}^\bullet \longrightarrow \mathcal{Y}^\bullet$ whose components are the projections $\mathfrak{J}_+ \square_{\mathcal{C}} \mathbf{S} \longrightarrow \mathfrak{J}_1 \square_{\mathcal{C}} \mathbf{S}_0 \simeq \mathfrak{J}_1$ and $\mathbf{S} \longrightarrow \mathbf{S}_0 \oplus \mathbf{S}_1$.

  All the three complexes $\mathcal{X}^\bullet$, $\mathcal{Y}^\bullet$, and $\ker(\mathcal{X}^\bullet \to \mathcal{Y}^\bullet)$ are complexes of $\mathcal{C}$-coflat right $\mathbf{S}$-semimodules (A-flat right $\mathbf{S}$-semimodules). Let us show that the complex $\ker(\mathcal{X}^\bullet \to \mathcal{Y}^\bullet)$ is is coacyclic with respect to the exact category of coflat graded right $\mathcal{C}$-comodules (A-flat right $\mathcal{C}$-comodules). Indeed, denote by $\mathcal{Z}^\bullet$ the kernel of the map from the reduced bar resolution of the right $\mathfrak{J}$-semimodule $\mathcal{C}$ (written down above) to $\mathcal{C}$ itself. The complex of graded $\mathfrak{J}$-semimodules $\mathcal{Z}^\bullet$ has a natural endomorphism z of internal degree 1 and cohomological degree 0 induced by the endomorphism of the reduced bar resolution acting by the identity on the cotensor factors $\mathfrak{J}_+$ and by the natural injections $\mathfrak{J}_{n-1} \to \mathfrak{J}_n$ on the cotensor factors $\mathfrak{J}$. Since $\mathcal{Z}^\bullet$ is a contractible complex of coflat graded right $\mathcal{C}$-comodules (A-flat right $\mathcal{C}$-comodules), the endomorphism z is injective, and its cokernel is a complex of coflat right $\mathcal{C}$-comodules (A-flat right $\mathcal{C}$-comodules), this cokernel is coacyclic with respect to the exact category of coflat graded right $\mathcal{C}$-comodules (A-flat right $\mathcal{C}$-comodules). Now the kernel $\ker(\mathcal{X}^\bullet \to \mathcal{Y}^\bullet)$ is isomorphic as a complex of right $\mathcal{C}$-comodules to the kernel of a surjective morphism from $\operatorname{coker}(z)$ to the contractible two-term complex of coflat right $\mathcal{C}$-comodules (A-flat right $\mathcal{C}$-comodules) $\mathcal{J}_1 \longrightarrow \mathcal{J}_1$.

Since the semitensor product $\mathcal{X}^\bullet \lozenge_\mathcal{S} \mathcal{C}$ is isomorphic to $\operatorname{Bar}^\bullet_{\mathrm{gr}}(\mathcal{J}, \mathcal{C})$, it represents the object $\operatorname{SemiTor}^\mathcal{S}_{\mathrm{gr}}(\mathcal{C}, \mathcal{C})$ in the derived category of graded k-modules (see 11.3). On the other hand, since $\mathcal{X}^\bullet$ is a bounded from above complex whose terms considered as one-term complexes are semiflat complexes of graded right $\mathcal{S}$-semimodules ($\mathcal{S}/\mathcal{C}/A$-semiflat complexes of graded right $\mathcal{S}$-semimodules), $\mathcal{X}^\bullet$ is a semiflat complex of graded right $\mathcal{S}$-semimodules ($\mathcal{S}/\mathcal{C}/A$-semiflat complex of graded right $\mathcal{S}$-semimodules). The cone of the morphism $\mathcal{X}^\bullet \longrightarrow \mathcal{Y}^\bullet$ is coacyclic with respect to the exact category of coflat graded right $\mathcal{C}$-comodules (A-flat graded right $\mathcal{C}$-comodules), so the semitensor product $\mathcal{X}^\bullet \lozenge_\mathcal{S} \mathcal{C}$ represents also the object $\operatorname{SemiTor}^\mathcal{J}_{\mathrm{gr}}(\mathcal{Y}^\bullet, \mathcal{C})$ in the derived category of graded k-modules.

In the semiderived category of graded $\mathcal{C}$-coflat (A-flat) right $\mathcal{S}$-semimodules there is a distinguished triangle $\mathcal{C}(-1)[1] \longrightarrow \mathcal{Y}^\bullet \longrightarrow \mathcal{C} \longrightarrow \mathcal{C}(-1)[2]$ (where the number in round brackets denotes the shift of internal grading $M(1)_n = M_{n+1}$). It follows from the induced long exact sequence of cohomology of the objects $\operatorname{SemiTor}^\mathcal{S}_{\mathrm{gr}}(-, \mathcal{C})$ by induction in the internal degree that $\operatorname{Bar}^\bullet_{\mathrm{gr}}(\mathcal{S}, \mathcal{C})$ has nonzero cohomology on the diagonal $-i = n$ only if and only if $\operatorname{Bar}^\bullet_{\mathrm{gr}}(\mathcal{J}, \mathcal{C})$ has nonzero cohomology on the diagonal $-i = n$ only. Assume that this is so; then there are short exact sequences

$$0 \longrightarrow H^{-n+1}_{n-1} \operatorname{Bar}^\bullet_{\mathrm{gr}}(\mathcal{S}, \mathcal{C}) \longrightarrow H^{-n}_n \operatorname{Bar}^\bullet_{\mathrm{gr}}(\mathcal{J}, \mathcal{C}) \longrightarrow H^{-n}_n \operatorname{Bar}^\bullet_{\mathrm{gr}}(\mathcal{S}, \mathcal{C}) \longrightarrow 0.$$

Furthermore, the diagonal cohomology $H^{-n}_n \operatorname{Bar}^\bullet_{\mathrm{gr}}(\mathcal{J}, \mathcal{C})$ and $H^{-n}_n \operatorname{Bar}^\bullet_{\mathrm{gr}}(\mathcal{S}, \mathcal{C})$ are flat right A-modules by Lemma 11.4.3.2(a), and so are endowed with $\mathcal{C}$-$\mathcal{C}$-bicomodule structures. The maps $H^{-n}_n \operatorname{Bar}^\bullet_{\mathrm{gr}}(\mathcal{J}, \mathcal{C}) \longrightarrow H^{-n}_n \operatorname{Bar}^\bullet_{\mathrm{gr}}(\mathcal{S}, \mathcal{C})$ in the short exact sequences above are induced by the morphism of semialgebras $\mathcal{J} \longrightarrow \mathcal{S}$, hence they are morphisms of $\mathcal{C}$-$\mathcal{C}$-bicomodules.

Let us describe the compositions

$$H^{-n}_n \operatorname{Bar}^\bullet_{\mathrm{gr}}(\mathcal{J}, \mathcal{C}) \longrightarrow H^{-n}_n \operatorname{Bar}^\bullet_{\mathrm{gr}}(\mathcal{S}, \mathcal{C}) \longrightarrow H^{-n-1}_{n+1} \operatorname{Bar}^\bullet_{\mathrm{gr}}(\mathcal{J}, \mathcal{C}),$$

which will be denoted by ∂_n. Let $t \colon \mathcal{C} \longrightarrow \mathcal{J}_1$ be the natural injection. Consider the endomorphism $\partial_\mathcal{X}$ of internal degree 1 and cohomological degree -1 of the complex of graded right $\mathcal{S}$-semimodules $\mathcal{X}^\bullet$ that is defined by the following formulas: the component $\mathcal{S}$ maps to $\mathcal{J}_+ \square_\mathcal{C} \mathcal{S}$ by $t \square \operatorname{id}$, the component $\mathcal{J}_+ \square_\mathcal{C} \mathcal{S}$ maps to $\mathcal{J}_+ \square_\mathcal{C} \mathcal{J}_+ \square_\mathcal{C} \mathcal{S}$ by $t \square \operatorname{id} \square \operatorname{id} - \operatorname{id} \square t \square \operatorname{id}$, etc. Consider also the endomorphism $\partial_\mathcal{Y}$ of internal degree 1 and cohomological degree -1 of the complex of graded right $\mathcal{S}$-semimodules $\mathcal{Y}$ mapping $\mathcal{S}_0 \oplus \mathcal{S}_1$ to $\mathcal{J}_1$ by the composition of the projection $\mathcal{S}_0 \oplus \mathcal{S}_1 \to \mathcal{C}$ and the embedding t. Then the endomorphisms $\partial_\mathcal{X}$ and $\partial_\mathcal{Y}$ form a commutative diagram with the morphism $\mathcal{X}^\bullet \longrightarrow \mathcal{Y}^\bullet$.

Since the endomorphism ∂_y represents in the semiderived category of $\mathcal{C}$-coflat (A-flat) graded $\mathbf{S}$-semimodules the composition of morphisms $\mathbf{\mathcal{Y}}^\bullet \to \mathcal{C} \to \mathbf{\mathcal{Y}}^\bullet(1)[-1]$ from the distinguished triangle above, the desired maps ∂_n are induced by the endomorphism ∂_{Bar} of the bar complex $\mathrm{Bar}^\bullet_{\mathrm{gr}}(\mathbf{\mathcal{J}}, \mathcal{C}) = \mathbf{\mathcal{X}}^\bullet \lozenge_{\mathbf{S}} \mathcal{C}$ that is induced by the endomorphism $\partial_{\mathbf{\mathcal{X}}}$ of the complex $\mathbf{\mathcal{X}}^\bullet$. The endomorphism ∂_{Bar} maps the component $\mathcal{C}$ to $\mathbf{\mathcal{J}}_+$ by t, the component $\mathbf{\mathcal{J}}_+$ to $\mathbf{\mathcal{J}}_+ \square_\mathcal{C} \mathbf{\mathcal{J}}_+$ by $t \square_\mathcal{C} \mathrm{id} - \mathrm{id} \square_\mathcal{C} t$, etc. Since ∂_{Bar} is an endomorphism of complexes of $\mathcal{C}$-$\mathcal{C}$-bi-comodules, ∂_n are also endomorphisms of $\mathcal{C}$-$\mathcal{C}$-bicomodules. Hence the maps $H^{-n+1}_{n-1} \mathrm{Bar}^\bullet_{\mathrm{gr}}(\mathbf{S}, \mathcal{C}) \longrightarrow H^{-n}_n \mathrm{Bar}^\bullet_{\mathrm{gr}}(\mathbf{\mathcal{J}}, \mathcal{C})$ in the short exact sequences above are morphisms of $\mathcal{C}$-$\mathcal{C}$-bicomodules. Now it follows easily by induction using Lemma 1.2.2 that all $H^{-n}_n \mathrm{Bar}^\bullet_{\mathrm{gr}}(\mathbf{\mathcal{J}}, \mathcal{C})$ are coflat right $\mathcal{C}$-comodules ($\mathcal{C}/A$-coflat left $\mathcal{C}$-comodules) if and only if all $H^{-n}_n \mathrm{Bar}^\bullet_{\mathrm{gr}}(\mathbf{S}, \mathcal{C})$ are coflat right $\mathcal{C}$-comodules ($\mathcal{C}/A$-coflat left $\mathcal{C}$-comodules). $\qquad\square$

A semialgebra $\mathbf{S}^\sim$ over a coring $\mathcal{C}$ endowed with a right coflat (right flat and left relatively coflat) increasing filtration F is called a *right coflat (right flat and left relatively coflat) nonhomogeneous Koszul semialgebra* over $\mathcal{C}$ if the equivalent conditions of Theorem 11.5 are satisfied for it, i.e., the graded semialgebras $\bigoplus_n F_n \mathbf{S}^\sim$ and $\bigoplus_n F_n \mathbf{S}^\sim / F_{n-1} \mathbf{S}^\sim$ are right coflat (right flat and left relatively coflat) Koszul semialgebras over $\mathcal{C}$.

11.6   Poincaré–Birkhoff–Witt theorem

Let $\mathcal{C}$ be a coring over a k-algebra A; assume that C is a flat right A-module. A quasi-differential coring $\mathcal{D}^\sim$ over A concentrated in the nonnegative degrees and endowed with an isomorphism $\mathcal{C} \simeq \mathcal{D}^\sim_0$ is called *right coflat (right flat and left relatively coflat) Koszul* over $\mathcal{C}$ if the graded coring $\mathcal{D}^\sim / \mathrm{im}\, \partial$ is right coflat (right flat and left relatively coflat) Koszul over $\mathcal{C}$.

Lemma. *Let $\mathbf{\mathcal{J}}$ be a right coflat (right flat and left relatively coflat) Koszul semialgebra over $\mathcal{C}$ and $\mathcal{E}$ be the quadratic dual right coflat (right flat and left relatively coflat) Koszul coring over $\mathcal{C}$. Then a $\mathcal{C}$-$\mathcal{C}$-bicomodule morphism $\mathcal{C} \longrightarrow \mathbf{\mathcal{J}}_1 \simeq \mathcal{E}_1$ can be extended to a graded $\mathbf{\mathcal{J}}$-$\mathbf{\mathcal{J}}$-bisemimodule morphism $\mathbf{\mathcal{J}} \longrightarrow \mathbf{\mathcal{J}}$ of degree 1 (i.e., represents a "central element" of $\mathbf{\mathcal{J}}$) if and only if it can be extended to a coderivation $\mathcal{E} \longrightarrow \mathcal{E}$ of degree 1 of the coring $\mathcal{E}$ (with respect to the zero coderivation of A). Both the $\mathbf{\mathcal{J}}$-$\mathbf{\mathcal{J}}$-bisemimodule morphism and the coderivation of $\mathcal{E}$ with the given component $\mathcal{C} \longrightarrow \mathbf{\mathcal{J}}_1 \simeq \mathcal{E}_1$ are unique if they exist; the coderivation always has a zero square.*

Proof. Both conditions hold if and only if the difference of the two maps

$$\mathbf{\mathcal{J}}_1 \simeq \mathcal{C} \square_\mathcal{C} \mathbf{\mathcal{J}}_1 \longrightarrow \mathbf{\mathcal{J}}_1 \square_\mathcal{C} \mathbf{\mathcal{J}}_1 \quad \text{and} \quad \mathbf{\mathcal{J}}_1 \simeq \mathbf{\mathcal{J}}_1 \square_\mathcal{C} \mathcal{C} \longrightarrow \mathbf{\mathcal{J}}_1 \square_\mathcal{C} \mathbf{\mathcal{J}}_1$$

induced by our map $\mathcal{C} \to \mathbf{\mathcal{J}}_1$ factorizes through the injection

$$\mathcal{E}_2 \longrightarrow \mathcal{E}_1 \square_\mathcal{C} \mathcal{E}_1 \simeq \mathbf{\mathcal{J}}_1 \square_\mathcal{C} \mathbf{\mathcal{J}}_1. \qquad\square$$

The objects of the category of right coflat nonhomogeneous Koszul semial-gebras are right coflat nonhomogeneous Koszul semialgebras $(\mathbf{S}^\sim, F)$ over corings $\mathcal{C}$ over k-algebras A such that $\mathcal{C}$ is a flat right A-module. Morphisms are maps of semialgebras $\mathbf{S}^\sim \longrightarrow \mathbf{S}^{\sim\prime}$ compatible with maps of corings $\mathcal{C} \longrightarrow \mathcal{C}'$ and mor-phisms of k-algebras $A \longrightarrow A'$ which map the filtration components $F_n\mathbf{S}^\sim$ into the filtration components $F'_n\mathbf{S}^{\sim\prime}$. Imposing the additional assumption that A has a finite weak homological dimension, one analogously defines the category of right flat and left relatively coflat nonhomogeneous Koszul semialgebras.

The objects of the category of right coflat Koszul quasi-differential corings are right coflat Koszul quasi-differential corings $\mathcal{D}^\sim$ over corings $\mathcal{C}$ over k-algebras A such that $\mathcal{C}$ is a flat right A-module. Morphisms are maps of graded corings $\mathcal{D}^\sim \longrightarrow \mathcal{D}^{\sim\prime}$ compatible with morphisms of k-algebras $A \longrightarrow A'$ and making a commutative diagram with the coderivations ∂ and ∂'. Imposing the additional assumption that A has finite weak homological dimension, one analogously defines the category of right flat and left relatively coflat Koszul quasi-differential corings.

Theorem. *The category of right coflat (right flat and left relatively coflat) nonho-mogeneous Koszul semialgebras is equivalent to the category of right coflat (right flat and left relatively coflat) Koszul quasi-differential corings. If a filtered semial-gebra $\mathbf{S}^\sim$ over a coring $\mathcal{C}$ and a quasi-differential coring $\mathcal{D}^\sim$ correspond to each other under this duality, then the graded semialgebra $\mathbf{J} = \bigoplus_n F_n\mathbf{S}^\sim$ and the graded coring $\mathcal{D}^\sim$ are quadratic dual right coflat (right flat and left relatively coflat) Koszul semialgebra and coring over $\mathcal{C}$; the graded semialgebra $\mathbf{S} = \bigoplus_n F_n\mathbf{S}^\sim/F_{n-1}\mathbf{S}^\sim$ and the graded coring $\mathcal{D} = \mathcal{D}^\sim/\operatorname{im}\partial$ are quadratic dual right coflat (right flat and left relatively coflat) Koszul semialgebra and coring over $\mathcal{C}$; the related iso-morphisms $F_1\mathbf{S}^\sim \simeq \mathcal{D}_1^\sim$ and $F_1\mathbf{S}^\sim/F_0\mathbf{S}^\sim \simeq \mathcal{D}_1^\sim/\partial_0\mathcal{D}_0^\sim$ are compatible with each other; and the injection $F_0\mathbf{S}^\sim \longrightarrow F_1\mathbf{S}^\sim$ corresponds to the coderivation compo-nent $\partial_0\colon \mathcal{D}_0^\sim \longrightarrow \mathcal{D}_1^\sim$ under the isomorphisms $F_0\mathbf{S}^\sim \simeq \mathcal{C} \simeq \mathcal{D}_0^\sim$ and $F_1\mathbf{S}^\sim \simeq \mathcal{D}_1^\sim$.*

Proof. It follows from the lemma that the category of right coflat (right flat and left relatively coflat) Koszul semialgebras $\mathbf{J}$ endowed with a $\mathbf{J}$-$\mathbf{J}$-bisemimodule morphism $\mathbf{J} \longrightarrow \mathbf{J}$ of degree 1 is equivalent to the category of right coflat (right flat and left relatively coflat) Koszul corings $\mathcal{E}$ endowed with a coderivation of degree 1. It remains to prove that semialgebras $\mathbf{J}$ with maps $\mathbf{J} \longrightarrow \mathbf{J}$ of degree 1 coming from right coflat (right flat and left relatively coflat) nonhomogeneous Koszul semialgebras $\mathbf{S}^\sim$ correspond under this equivalence to right coflat (right flat and left relatively coflat) Koszul quasi-differential corings $\mathcal{D}^\sim = \mathcal{E}$ and vice versa. Besides, we will have to show that whenever for a quasi-differential coring $\mathcal{D}^\sim$ the graded coring $\mathcal{D}^\sim/\operatorname{im}\partial$ is a right coflat (left relatively coflat) Koszul coring over a coring $\mathcal{C}$, the graded coring $\mathcal{D}^\sim$ is also a right coflat (left flat and right relatively coflat) Koszul coring over $\mathcal{C}$.

According to the proof of Theorem 11.5, for any right coflat (right flat and left relatively coflat) nonhomogeneous Koszul semialgebra $\mathbf{S}^\sim$ there is a right coflat (right flat and left relatively coflat) Koszul quasi-differential coring $\mathcal{D}^\sim$. Indeed, set $\mathcal{D}^\sim = \bigoplus_n H_n^{-n} \operatorname{Bar}_{\mathrm{gr}}^\bullet(\mathcal{T}, \mathcal{C})$, where $\mathcal{T} = \bigoplus_n F_n \mathcal{S}^\sim$; then the endomorphism ∂ of the $\mathcal{C}$-$\mathcal{C}$-bicomodule $\mathcal{D}^\sim$ induced by the endomorphism ∂_{Bar} of the reduced bar construction $\operatorname{Bar}_{\mathrm{gr}}^\bullet(\mathcal{T}, \mathcal{C})$ is a coderivation of degree 1 (with respect to the zero coderivation of A) and its restriction to $\mathcal{D}_0^\sim$ coincides with the injection $\mathcal{D}_0^\sim \simeq \mathcal{T}_0 \longrightarrow \mathcal{T}_1 \simeq \mathcal{D}_1^\sim$. It also follows from this proof that the right coflat (right flat and left relatively coflat) Koszul semialgebra $\mathcal{S} = \bigoplus F_n \mathcal{S}^\sim / F_{n-1} \mathcal{S}^\sim$ is quadratic dual to the coring $\mathcal{D} = \mathcal{D}^\sim / \operatorname{im} \partial$, which is therefore right coflat (right flat and left relatively coflat) Koszul over $\mathcal{C}$.

Let us now construct the nonhomogeneous Koszul semialgebra corresponding to a right coflat (right coflat and left relatively coflat) Koszul quasi-differential coring $\mathcal{D}^\sim$ over a coring $\mathcal{C}$. Set $\mathcal{D} = \mathcal{D}^\sim / \operatorname{im} \partial$. Consider the bigraded coring $\mathcal{K}$ over the k-algebra A (which is considered as a bigraded k-algebra concentrated in the bidegree $(0,0)$) with the components $\mathcal{K}^{p,q} = \mathcal{D}_{q-p}^\sim$ for $p \leqslant 0$, $q \leqslant 0$ and $\mathcal{K}^{p,q} = 0$ otherwise. The coring $\mathcal{K}$ considered as a graded coring in the total grading $p + q$ has a coderivation $\partial_\mathcal{K}$ (with respect to the zero coderivation of A) mapping the component $\mathcal{K}^{p,q}$ to $\mathcal{K}^{p,q+1}$ by ∂_{q-p}; one has $\partial_\mathcal{K}^2 = 0$. There is a morphism of bigraded corings $\mathcal{K} \longrightarrow \mathcal{D}$ inducing an isomorphism of the corings of cohomology, where the coring $\mathcal{D}$ is placed in the bigrading $\mathcal{D}^{p,0} = \mathcal{D}_{-p}$ and endowed with the zero differential.

Denote by $\mathcal{K}_+$ the cokernel of the injection $\mathcal{C} \simeq \mathcal{K}^{0,0} \longrightarrow \mathcal{K}$. Let $\mathcal{R} = \bigoplus_{r=0}^\infty \mathcal{K}_+^{\square_\mathcal{C} r}$ be the "cotensor semialgebra" of the bigraded $\mathcal{C}$-$\mathcal{C}$-bicomodule $\mathcal{K}_+$. By the definition, $\mathcal{R}$ is a trigraded semialgebra over the coring $\mathcal{C}$ (which is considered as a trigraded coring concentrated in the tridegree $(0,0,0)$) with the gradings p and q inherited from the bigrading of $\mathcal{K}_+$ and the additional grading r by the number of cotensor factors. We will consider $\mathcal{R}$ as a graded semialgebra in the total grading $p + q + r$. The semialgebra $\mathcal{R}$ is endowed with three semiderivations (with respect to the zero derivation of the coring $\mathcal{C}$) of total degree 1, which we will now introduce.

Let $\partial_\mathcal{R}$ be the only semiderivation of $\mathcal{R}$ which preserves $\mathcal{K}_+ \subset \mathcal{R}$ (embedded as the part of degree $r = 1$) and whose restriction to $\mathcal{K}_+$ is equal to $-\partial_\mathcal{K}$. Let $d_\mathcal{R}$ be the only semiderivation of $\mathcal{R}$ which maps $\mathcal{K}_+$ to $\mathcal{K}_+ \square_\mathcal{C} \mathcal{K}_+$ by the composition of the comultiplication map $\mathcal{K}_+ \longrightarrow \mathcal{K}_+ \square_\mathcal{C} \mathcal{K}_+$ and the sign automorphism of $\mathcal{K}_+ \square_\mathcal{C} \mathcal{K}_+$ acting on the component $\mathcal{K}^{p',q'} \square_\mathcal{C} \mathcal{K}^{p'',q''}$ as $(-1)^{p'+q'}$. Finally, let $\delta_\mathcal{R}$ be the only semiderivation of $\mathcal{R}$ whose restriction to $\mathcal{K}_+$ is the identity map of the component $\mathcal{K}^{-1,-1} \simeq \mathcal{C}$ to the semiunit component $\mathcal{R}^{0,0,0} = \mathcal{C}$ and zero on all the remaining components of $\mathcal{K}_+$. All the three differentials are constructed so that they satisfy the super-Leibniz rule in the parity $p + q + r$. The semiderivations $\partial_\mathcal{R}$, $d_\mathcal{R}$, and $\delta_\mathcal{R}$ have tridegrees $(0,1,0)$, $(0,0,1)$, and $(1,1,-1)$, respectively, in the trigrading (p,q,r). All the three semiderivations have zero squares, and they pairwise anti-commute.

There is a right coflat (right flat and left relatively coflat) increasing filtration F on the graded semialgebra $\mathcal{R}$ whose component $F_n \mathcal{R}$ is the direct sum of all trigrading components $\mathcal{R}^{p,q,r}$ with $-p \leqslant n$. This filtration is compatible with the differentials $\partial_{\mathcal{R}}$, $d_{\mathcal{R}}$, and $\delta_{\mathcal{R}}$; the semialgebra $\bigoplus_n F_n\mathcal{R}/F_{n-1}\mathcal{R}$ with the differential induced by $\partial_{\mathcal{R}} + d_{\mathcal{R}} + \delta_{\mathcal{R}}$ is naturally isomorphic to the semialgebra $\mathcal{R}$ with the differential $\partial_{\mathcal{R}} + d_{\mathcal{R}}$.

Consider the following sign-modified version of the cobar construction $\mathrm{Cob}(\mathcal{D}, \mathcal{C})$. Define $'\mathrm{Cob}(\mathcal{D}, \mathcal{C})$ as the "tensor semialgebra" $\bigoplus_r \mathcal{D}_+^{\square_{\mathcal{C}} r}$ of the $\mathcal{C}$-$\mathcal{C}$-bicomodule $\mathcal{D}_+$ and endow it with the grading p coming from the grading $\mathcal{D}^p = \mathcal{D}_{-p}$ of $\mathcal{D}_+$ and the grading r by the number of cotensor factors. We will consider $'\mathrm{Cob}(\mathcal{D}, \mathcal{C})$ as a graded semialgebra over $\mathcal{C}$ in the total grading $p + r$. Let d'_{Cob} be the only coderivation of $'\mathrm{Cob}(\mathcal{D}, \mathcal{C})$ which maps $\mathcal{D}_+ \subset {}'\mathrm{Cob}(\mathcal{D}, \mathcal{C})$ to $\mathcal{D}_+ \square_{\mathcal{C}} \mathcal{D}_+$ by the composition of the comultiplication map $\mathcal{D}_+ \longrightarrow \mathcal{D}_+ \square_{\mathcal{C}} \mathcal{D}_+$ and the sign automorphism of $\mathcal{D}_+ \square_{\mathcal{C}} \mathcal{D}_+$ acting on the component $\mathcal{D}^{p'} \square_{\mathcal{C}} \mathcal{D}^{p''}$ as $(-1)^{p'}$. Then one has $d'^2_{\mathrm{Cob}} = 0$. Notice that the differential d'_{Cob} satisfies the super-Leibniz rule in the parity $p + r$, while the differential d_{Cob} of the cobar construction $\mathrm{Cob}^\bullet_{\mathrm{gr}}(\mathcal{D}, \mathcal{C})$ satisfies the super-Leibniz rule in the parity r. The automorphism of $\bigoplus_r \mathcal{D}_+^{\square_{\mathcal{C}} r}$ acting on the component $\mathcal{D}^{p_1} \square_{\mathcal{C}} \cdots \square_{\mathcal{C}} \mathcal{D}^{p_r}$ by minus one to the power $\sum_{s=1}^r p_s(p_s + 1)/2 + \sum_{1 \leqslant s < t \leqslant r} p_s(p_t + 1)$ transforms d_{Cob} to d'_{Cob}, so the semialgebras of cohomology of the DG-semialgebras $'\mathrm{Cob}(\mathcal{D}, \mathcal{C})$ and $\mathrm{Cob}^\bullet_{\mathrm{gr}}(\mathcal{D}, \mathcal{C})$ are naturally isomorphic in the Koszul case.

Consider the morphism of DG-semialgebras $(\mathcal{R}, \partial_{\mathcal{R}} + d_{\mathcal{R}}) \to ({}'\mathrm{Cob}(\mathcal{D}, \mathcal{C}), d'_{\mathrm{Cob}})$ induced by the morphism of corings $\mathcal{K} \longrightarrow \mathcal{D}$. This morphism of DG-semialgebras induces an isomorphism of the semialgebras of cohomology. Indeed, the components of fixed grading p of the DG-semialgebra $(\mathcal{R}, \partial_{\mathcal{R}} + d_{\mathcal{R}})$ are the total components of finite bicomplexes whose components of fixed grading r are multiple cotensor products of components of fixed grading p of the DG-coring $\mathcal{K}$, and the natural maps from these multiple cotensor products to the corresponding multiple cotensor products of components of the coring $\mathcal{D}$ are quasi-isomorphisms. Hence $H^0_{\partial_{\mathcal{R}} + d_{\mathcal{R}}}(\mathcal{R}) \simeq \mathbf{S}$ and $H^i_{\partial_{\mathcal{R}} + d_{\mathcal{R}}}(\mathcal{R}) = 0$ for $i \neq 0$, where $\mathbf{S}$ denotes the right coflat (right flat and left relatively coflat) Koszul semialgebra quadratic dual to $\mathcal{D}$. Analogously, the morphism of DG-semialgebras $(\mathcal{R}, \partial_{\mathcal{R}} + d_{\mathcal{R}}) \longrightarrow ({}'\mathrm{Cob}(\mathcal{D}, \mathcal{C}), d'_{\mathrm{Cob}})$ induces quasi-isomorphisms of the tensor and cotensor products related to these DG-semialgebras that were listed in (i) and (iii) of 11.2.3.

The associated graded quotient complexes to the tensor and cotensor product of the DG-semialgebra $(\mathcal{R}, \partial_{\mathcal{R}} + d_{\mathcal{R}} + \delta_{\mathcal{R}})$ listed in (i) and (iii) of 11.2.3 with respect to the filtrations induced by the filtration F are naturally isomorphic to the corresponding tensor and cotensor products of the DG-semialgebra $(\mathcal{R}, \partial_{\mathcal{R}} + d_{\mathcal{R}})$. Therefore, the associated graded modules of the cohomology of these tensor and cotensor products of the DG-semialgebra $(\mathcal{R}, \partial_{\mathcal{R}} + d_{\mathcal{R}} + \delta_{\mathcal{R}})$ are isomorphic to the cohomology of the corresponding tensor and cotensor products of the DG-semialgebra $(\mathcal{R}, \partial_{\mathcal{R}} + d_{\mathcal{R}})$. In particular, set $\mathbf{S}^\sim = H^0_{\partial_{\mathcal{R}} + d_{\mathcal{R}} + \delta_{\mathcal{R}}}(\mathcal{R})$; then $\mathbf{S}^\sim$ is endowed with an increasing filtration F such that $\bigoplus_n F_n \mathbf{S}^\sim / F_{n-1} \mathbf{S}^\sim \simeq \mathbf{S}$, while $H^i_{\partial_{\mathcal{R}} + d_{\mathcal{R}} + \delta_{\mathcal{R}}}(\mathcal{R}) = 0$ for $i \neq 0$. Since $\mathbf{S}$ is a coflat right $\mathcal{C}$-comodule (a flat right A-module and a $\mathcal{C}/A$-coflat left $\mathcal{C}$-comodule), the associated graded quotient modules to the tensor and cotensor products under consideration of the cohomology module $\mathbf{S}^\sim$ are isomorphic to the corresponding tensor and cotensor products of $\mathbf{S}$. Thus $\mathbf{S}^\sim$ is a semialgebra over $\mathcal{C}$ and F is its right coflat (right flat and left relatively coflat) increasing filtration.

Since the semialgebra $\mathbf{S}$ is right coflat (right flat and left relatively coflat) Koszul, so is the semialgebra $\mathbf{J} = \bigoplus_n F_n\mathbf{S}^\sim$. Let $\mathcal{D}^{\sim\prime}$ be the right coflat (right flat and left relatively coflat) coring quadratic dual to $\mathbf{J}$; then $\mathcal{D}^{\sim\prime}$ is endowed with a coderivation ∂', making it a right coflat (right flat and left relatively coflat) Koszul quasi-differential coring, as we have already proven. Moreover, the cokernel $\mathcal{D}'$ of the coderivation ∂' is quadratic dual to $\mathbf{S}$, hence there is a natural isomorphism of graded corings $\mathcal{D} \simeq \mathcal{D}'$. Furthermore, the embedding of the component $\mathcal{D}_1^\sim = \mathcal{R}_{-1,0,1} \longrightarrow \mathcal{R}$ induces an isomorphism $\mathcal{D}_1^\sim \simeq F_1\mathbf{S}^\sim$. The composition $\mathcal{D}_2^\sim \longrightarrow \mathcal{D}_1^\sim \square_\mathcal{C} \mathcal{D}_1^\sim \simeq F_1\mathbf{S}^\sim \square_\mathcal{C} F_1\mathbf{S}^\sim \longrightarrow F_2\mathbf{S}^\sim$ of the comultiplication and semimultiplication maps vanishes, so there is a natural morphism of graded corings $\mathcal{D}^\sim \longrightarrow \mathcal{D}^{\sim\prime}$. Since the embedding $F_0\mathbf{S}^\sim \longrightarrow F_1\mathbf{S}^\sim$ corresponds to the map $\partial_0 \colon \mathcal{D}_0^\sim \longrightarrow \mathcal{D}_1^\sim$ under the isomorphisms $F_0\mathbf{S}^\sim \simeq \mathcal{C} \simeq \mathcal{D}_0^\sim$ and $F_1\mathbf{S}^\sim \simeq \mathcal{D}_1^\sim$, the morphism $\mathcal{D}^\sim \longrightarrow \mathcal{D}^{\sim\prime}$ forms a commutative diagram with the differentials ∂ in $\mathcal{D}^\sim$ and ∂' in $\mathcal{D}^{\sim\prime}$. The induced morphism $\mathcal{D}^\sim/\operatorname{im}\partial \longrightarrow \mathcal{D}^{\sim\prime}/\operatorname{im}\partial'$ coincides with the natural isomorphism $\mathcal{D} \longrightarrow \mathcal{D}'$ on the components of degree 1, and consequently on the other components as well. Hence the morphism of corings $\mathcal{D}^\sim \longrightarrow \mathcal{D}^{\sim\prime}$ is also an isomorphism. Thus the coring $\mathcal{D}^\sim$ is right coflat (right flat and left relatively coflat) Koszul over $\mathcal{C}$, and the semialgebra $\mathbf{J}$ quadratic dual to it together with its $\mathbf{J}$-$\mathbf{J}$-bicomodule endomorphism of degree 1 comes from the right coflat (right flat and left relatively coflat) nonhomogeneous Koszul semialgebra $\mathbf{S}^\sim$.   □

A right coflat (right flat and left relatively coflat) nonhomogeneous Koszul semialgebra and a right coflat (right flat and left relatively coflat) Koszul quasi-differential coring corresponding to each other under the equivalence of categories from Theorem 11.6 are called *nonhomogeneous quadratic dual* to each other.

All the definitions and results of 11.4–11.6 have their obvious analogues for the coflatness conditions replaced with coprojectivity ones. So, when $\mathcal{C}$ is a projective left A-module one can speak of left coprojective Koszul semialgebras and corings. When $\mathcal{C}$ is a flat right A-module and A has a finite left homological dimension, one can define right flat and left relatively projective Koszul semialgebras and corings. When $\mathcal{C}$ is a projective left A-module and A has a finite left homological dimension, one can consider left projective and right relatively coflat Koszul semialgebras and corings. Of course, when $\mathcal{C}$ is a flat left A-module, one can define left coflat Koszul semialgebras and corings, etc.

Remark. All the results of 11.4–11.6 have their analogues for semimodules and semicontramodules over semialgebras, comodules and contramodules over corings. In particular, for any right coflat Koszul semialgebra $\mathbf{S}$ and the right coflat Koszul coring $\mathcal{D}$ quadratic dual to $\mathbf{S}$ there is a natural equivalence between the categories of Koszul left $\mathbf{S}$-semimodules and Koszul left $\mathcal{D}$-comodules given by the functors of cohomology of the reduced bar and cobar constructions with coefficients in the semimodules and comodules. No (co)flatness conditions need to be imposed on the semimodules and comodules in this setting. For any left coprojective Koszul semi-algebra $\mathcal{S}$ and the left coprojective Koszul coring $\mathcal{D}$ quadratic dual to $\mathcal{S}$ there is an equivalence between the categories of Koszul left $\mathcal{S}$-semicontramodules and Koszul left $\mathcal{D}$-contramodules (which are nonpositively graded). For any right flat and left relatively coflat Koszul semialgebra $\mathcal{S}$ and the right flat and left relatively coflat Koszul coring $\mathcal{D}$ quadratic dual to $\mathcal{S}$ there is an equivalence between the categories of A-flat Koszul right $\mathcal{S}$-semimodules and A-flat Koszul right $\mathcal{D}$-comodules, etc. Furthermore, for a right coflat nonhomogeneous Koszul semialgebra $\mathcal{S}^\sim$ and a left semimodule $\mathcal{M}^\sim$ over $\mathcal{S}^\sim$ endowed with an increasing filtration F compatible with the filtration of $\mathcal{S}^\sim$, the semimodule $\bigoplus_n F_n \mathcal{M}^\sim$ is Koszul over the semialgebra $\bigoplus_n F_n \mathcal{S}^\sim$ if and only if the semimodule $\bigoplus_n F_n \mathcal{M}^\sim / F_{n-1} \mathcal{M}^\sim$ is Koszul over the semialgebra $\bigoplus_n F_n \mathcal{S}^\sim / F_{n-1} \mathcal{S}^\sim$. A filtered semimodule $\mathcal{M}^\sim$ satisfying these conditions can be called a nonhomogeneous Koszul semimodule over $\mathcal{S}^\sim$. The Koszul $\mathcal{D}$-comodule quadratic dual to the second of these graded semimodules is naturally isomorphic to the Koszul $\mathcal{D}^\sim$-comodule quadratic dual to the first semimodule with the induced $\mathcal{D}$-comodule structure. A Koszul quasi-differential left $\mathcal{D}^\sim$-comodule is a graded left $\mathcal{D}^\sim$-comodule that is Koszul as a $\mathcal{D}$-comodule; then it is also Koszul as a $\mathcal{D}^\sim$-comodule. There is a natural equivalence between the categories of nonhomogeneous Koszul left semimodules over $\mathcal{S}^\sim$ and Koszul quasi-differential comodules over $\mathcal{D}^\sim$. When $\mathcal{S}^\sim$ is a left coprojective nonhomogeneous Koszul semi-algebra, there is an analogous equivalence of categories of nonhomogeneous Koszul left semicontramodules over $\mathcal{S}^\sim$ and Koszul quasi-differential contramodules over $\mathcal{D}^\sim$ (where nonhomogeneous Koszul semicontramodules are semicontramodules endowed with complete decreasing filtrations).

11.7 Quasi-differential comodules and contramodules

11.7.1 Let $(\mathcal{D}^\sim, \partial)$ be a quasi-differential coring over a k-algebra A; set $\mathcal{D} = \mathcal{D}^\sim / \operatorname{im} \partial$. Assume that $\mathcal{D}$ is a flat graded right A-module.

A *quasi-differential left comodule* over $\mathcal{D}^\sim$ is just a graded left $\mathcal{D}^\sim$-comodule (without any differential). The DG-category of quasi-differential left $\mathcal{D}^\sim$-comodules $\mathsf{DG}(\mathcal{D}^\sim\text{-qcmd})$ is defined as follows. The objects of $\mathsf{DG}(\mathcal{D}^\sim\text{-qcmd})$ are quasi-differential left $\mathcal{D}^\sim$-comodules. Let us construct the complex of morphisms in the category $\mathsf{DG}(\mathcal{D}^\sim\text{-qcmd})$ between quasi-differential left $\mathcal{D}^\sim$-comodules $\mathcal{L}$ and $\mathcal{M}$, denoted by $\operatorname{Hom}_{\mathcal{D}}^{\bullet}(\mathcal{L}, \mathcal{M})$. The component $\operatorname{Hom}_{\mathcal{D}}^n(\mathcal{L}, \mathcal{M})$ of this complex is the k-module of all homogeneous maps $\mathcal{L} \longrightarrow \mathcal{M}$ of degree $-n$ supercommuting with the coaction of $\mathcal{D}$ in $\mathcal{L}$ and $\mathcal{M}$. This means that an element $f \in \operatorname{Hom}_{\mathcal{D}}^n(\mathcal{L}, \mathcal{M})$ should satisfy the equation $\overline{f(x)}_{(-1)} \otimes f(x)_{(0)} = (-1)^{n|x_{(-1)}|}\overline{x_{(-1)}} \otimes f(x_{(0)})$ in Sweedler's notation [82], where $z \longmapsto z_{(-1)} \otimes z_{(0)}$ denotes the left coaction maps, $\overline{p}$ is the image of an element $p \in \mathcal{D}^\sim$ in $\mathcal{D}$, and $|p|$ is the degree of a homogeneous element p. To define the differential $d(f)$ of an element f, consider the supercommutator of the coaction maps $\mathcal{L} \longrightarrow \mathcal{D}^{\sim} \square_{\mathcal{D}} \mathcal{L}$ and $\mathcal{M} \longrightarrow \mathcal{D}^{\sim} \square_{\mathcal{D}} \mathcal{M}$ with f, that is the map $\delta_f \colon \mathcal{L} \longrightarrow \mathcal{D}^{\sim} \square_{\mathcal{D}} \mathcal{M}$ given by the formula $x \longmapsto f(x)_{(-1)} \square f(x)_{(0)} - (-1)^{n|x_{(-1)}|} x_{(-1)} \square f(x_{(0)})$. For any $f \in \operatorname{Hom}_{\mathcal{D}}^n(\mathcal{L}, \mathcal{M})$, the map δ_f factorizes through the injection $\mathcal{M} \simeq \mathcal{D} \square_{\mathcal{D}} \mathcal{M} \longrightarrow \mathcal{D}^{\sim} \square_{\mathcal{D}} \mathcal{M}$ induced by the homogeneous morphism $\bar{\partial} \colon \mathcal{D} \longrightarrow \mathcal{D}^{\sim}$ of degree 1 given by $\bar{\partial}(\bar{p}) = \partial(p)$, hence the desired map $d(f) \colon \mathcal{L} \longrightarrow \mathcal{M}$ of degree $-n-1$.

Since the map δ_f and the morphism $\bar{\partial}$ supercommute with the left coactions of $\mathcal{D}$, so does the map $d(f)$. Let us check that $d^2(f) = 0$, in other words, that $d(f)$ supercommutes with the left coactions of $\mathcal{D}^{\sim}$ in $\mathcal{L}$ and $\mathcal{M}$. Consider the two homogeneous maps $\mathcal{L} \longrightarrow \mathcal{D}^{\sim} \square_{\mathcal{D}} \mathcal{M}$ given by the formulas $x \longmapsto (df)(x)_{(-1)} \square (df)(x)_{(0)}$ and $x \longmapsto (-1)^{(n+1)|x_{(-1)}|} x_{(-1)} \square (df)(x_{(0)})$; we have to check that these two maps coincide. Consider the image of the former map under the map $\operatorname{Hom}_{\mathcal{D}}^{n+1}(\mathcal{L}, \mathcal{D}^{\sim} \square_{\mathcal{D}} \mathcal{M}) \longrightarrow \operatorname{Hom}_{\mathcal{D}}^{n+1}(\mathcal{L}, \mathcal{D}^{\sim} \square_{\mathcal{D}} \mathcal{D}^{\sim} \square_{\mathcal{D}} \mathcal{M})$ given by the formula $g \longmapsto (x \mapsto (-1)^{(n+1)|x_{(-1)}|} \partial(x_{(-1)}) \square g(x_{(0)}))$ and the image of the second map under the map $\operatorname{Hom}_{\mathcal{D}}^{n+1}(\mathcal{L}, \mathcal{D}^{\sim} \square_{\mathcal{D}} \mathcal{M}) \longrightarrow \operatorname{Hom}_{\mathcal{D}}^{n+1}(\mathcal{L}, \mathcal{D}^{\sim} \square_{\mathcal{D}} \mathcal{D}^{\sim} \square_{\mathcal{D}} \mathcal{M})$ given by the formula $g \longmapsto (x \mapsto (-1)^{|g(x)_1|} g(x)_1 \square \partial(g(x)_{2(-1)}) \square g(x)_{2(0)})$, where $y = y_1 \square y_2$ is a notation for an element $y \in \mathcal{D}^{\sim} \square_{\mathcal{D}} \mathcal{M}$. The sum of these two elements of $\operatorname{Hom}_{\mathcal{D}}^{n+1}(\mathcal{L}, \mathcal{D}^{\sim} \square_{\mathcal{D}} \mathcal{D}^{\sim} \square_{\mathcal{D}} \mathcal{M})$ is equal to the image of the element δ_f under the map $\operatorname{Hom}_{\mathcal{D}}^{n+1}(\mathcal{L}, \mathcal{D}^{\sim} \square_{\mathcal{D}} \mathcal{M}) \longrightarrow \operatorname{Hom}_{\mathcal{D}}^{n+1}(\mathcal{L}, \mathcal{D}^{\sim} \square_{\mathcal{D}} \mathcal{D}^{\sim} \square_{\mathcal{D}} \mathcal{M})$ induced by the comultiplication map $\mathcal{D}^{\sim} \longrightarrow \mathcal{D}^{\sim} \square_{\mathcal{D}} \mathcal{D}^{\sim}$. There is a commutative square formed by the diagonal embedding $\mathcal{D}^{\sim} \longrightarrow \mathcal{D}^{\sim} \oplus \mathcal{D}^{\sim}$, the morphism $\mathcal{D}^{\sim} \oplus \mathcal{D}^{\sim} \longrightarrow \mathcal{D}^{\sim} \square_{\mathcal{D}} \mathcal{D}^{\sim}$ given by the formula $(x, y) \longmapsto \partial(x_{(1)}) \square x_{(2)} + (-1)^{|y_{(1)}|} y_{(1)} \square \partial(y_{(2)})$, the morphism $\partial \colon \mathcal{D}^{\sim} \longrightarrow \mathcal{D}^{\sim}$, and the comultiplication morphism $\mathcal{D}^{\sim} \longrightarrow \mathcal{D}^{\sim} \square_{\mathcal{D}} \mathcal{D}^{\sim}$. Considering the filtrations originating from the two-term filtration $\partial(\mathcal{D}^{\sim}) \subset \mathcal{D}^{\sim}$, one can check that this square is Cartesian. It remains Cartesian after applying the functors $- \square_{\mathcal{D}} \mathcal{M}$ and $\operatorname{Hom}_{\mathcal{D}}^{n+1}(\mathcal{L}, -)$, so we are done.

Let $\mathcal{M}$ be a quasi-differential left $\mathcal{D}^{\sim}$-comodule and $q \colon \mathcal{M} \longrightarrow \mathcal{M}$ be an element of $\operatorname{Hom}_{\mathcal{D}}^1(\mathcal{M}, \mathcal{M})$ satisfying the Maurer–Cartan equation $d(q) + q^2 = 0$. The quasi-differential left $\mathcal{D}^{\sim}$-comodule structure on $\mathcal{M}$ twisted with q is constructed as follows. First of all, the structure of a graded $\mathcal{D}$-comodule on $\mathcal{M}$ does not change under twisting. Next, the twisted coaction map $\mathcal{M} \longrightarrow \mathcal{D}^{\sim} \square_{\mathcal{D}} \mathcal{M}$ is the sum of the original coaction map and the composition $\mathcal{M} \longrightarrow \mathcal{M} \longrightarrow \mathcal{D}^{\sim} \square_{\mathcal{D}} \mathcal{M} \longrightarrow \mathcal{D}^{\sim} \square_{\mathcal{D}} \mathcal{M}$ of the map $q \colon \mathcal{M} \longrightarrow \mathcal{M}$, the coaction map $\mathcal{M} \longrightarrow \mathcal{D}^{\sim} \square_{\mathcal{D}} \mathcal{M}$, and the map $\mathcal{D}^{\sim} \square_{\mathcal{D}} \mathcal{M} \longrightarrow \mathcal{D}^{\sim} \square_{\mathcal{D}} \mathcal{M}$ induced by the morphism $\partial \colon \mathcal{D}^{\sim} \longrightarrow \mathcal{D}^{\sim}$. Denote the quasi-differential $\mathcal{D}^{\sim}$-comodule we have constructed by $\mathcal{M}(q)$. For any quasi-differential left $\mathcal{D}^{\sim}$-comodule $\mathcal{L}$ the differential in the complex $\operatorname{Hom}_{\mathcal{D}}^{\bullet}(\mathcal{L}, \mathcal{M}(q))$ differs from the differential in the complex $\operatorname{Hom}_{\mathcal{D}}^{\bullet}(\mathcal{L}, \mathcal{M})$ according to the formula $d_q(f) = d(f) + q \circ f$.

Since infinite direct sums and shifts of objects clearly exist in the DG-category $\mathsf{DG}(\mathcal{D}^{\sim}\text{-qcmd})$, it follows from the above construction, in particular, that cones exist in it. Therefore, the homotopy category $\mathsf{Hot}(\mathcal{D}^{\sim}\text{-qcmd})$, whose objects are the objects of $\mathsf{DG}(\mathcal{D}^{\sim}\text{-qcmd})$ and morphisms are the zero cohomology of the complexes of morphisms in $\mathsf{DG}(\mathcal{D}^{\sim}\text{-qcmd})$, is naturally triangulated. Furthermore, for any complex of quasi-differential left $\mathcal{D}^\sim$-comodules one can define the total graded quasi-differential left $\mathcal{D}^\sim$-comodule such that the corresponding graded left $\mathcal{D}$-comodule will coincide with the infinite direct sum of the shifts of the terms of the complex considered as graded left $\mathcal{D}$-comodules. In particular, one can speak about the total quasi-differential left $\mathcal{D}^\sim$-comodules of exact triples of quasi-differential left $\mathcal{D}^\sim$-comodules. This allows us to define the *coderived category* of quasi-differential left $\mathcal{D}^\sim$-comodules $\mathsf{D}^{\mathsf{co}}(\mathcal{D}^\sim\text{-qcmd})$ as the quotient category of the homotopy category $\mathsf{Hot}(\mathcal{D}^\sim\text{-qcmd})$ by its minimal triangulated subcategory containing such objects associated to exact triples and closed under infinite direct sums. The objects of the latter subcategory are called *coacyclic* quasi-differential comodules.

11.7.2 Let $(\mathcal{D}^\sim, \partial)$ be a quasi-differential coring over A; assume that $\mathcal{D} = \mathcal{D}^\sim/\mathrm{im}\,\partial$ is a flat graded left A-module.

A *quasi-differential right comodule* over $\mathcal{D}^\sim$ is just a graded right $\mathcal{D}^\sim$-comodule. Let us define the DG-category of quasi-differential right comodules $\mathsf{DG}(\text{qcmd-}\mathcal{D}^\sim)$ over $\mathcal{D}^\sim$. The objects of $\mathsf{DG}(\text{qcmd-}\mathcal{D}^\sim)$ are quasi-differential right $\mathcal{D}^\sim$-comodules. The complex of morphisms $\mathrm{Hom}^\bullet_{\mathcal{D}^\sim}(\mathcal{R}, \mathcal{N})$ in the category $\mathsf{DG}(\text{qcmd-}\mathcal{D}^\sim)$ between quasi-differential right $\mathcal{D}^\sim$-comodules $\mathcal{R}$ and $\mathcal{N}$ is constructed as follows. The component $\mathrm{Hom}^n_{\mathcal{D}^\sim}(\mathcal{R}, \mathcal{N})$ of this complex is the k-module of all homogeneous maps $\mathcal{R} \longrightarrow \mathcal{N}$ of degree $-n$ commuting with the $\mathcal{D}$-comodule structures (without any signs). To define the differential of an element $f \in \mathrm{Hom}^\bullet_{\mathcal{D}}(\mathcal{R}, \mathcal{N})$, consider the map $\delta_f \colon \mathcal{R} \longrightarrow \mathcal{N} \square_{\mathcal{D}} \mathcal{D}^\sim$ given by the formula $x \longmapsto f(x)_{(0)} \square f(x)_{(1)} - f(x_{(0)}) \square x_{(1)}$, where $z \longmapsto z_{(0)} \otimes z_{(1)}$ denotes the right coaction maps. The map δ_f factorizes through the injection $\mathcal{N} \longrightarrow \mathcal{N} \square_{\mathcal{D}} \mathcal{D}^\sim$ given by the formula $y \longmapsto (-1)^{|y_{(0)}|} y_{(0)} \square \partial(y_{(1)})$, hence the desired map $d(f) \colon \mathcal{R} \longrightarrow \mathcal{N}$.

Let $\mathcal{N}$ be a quasi-differential right $\mathcal{D}^\sim$-comodule and $q \in \mathrm{Hom}^1_{\mathcal{D}}(\mathcal{N}, \mathcal{N})$ be an element satisfying the equation $d(q) + q^2 = 0$. To define quasi-differential right $\mathcal{D}^\sim$-comodule structure on $\mathcal{N}$ twisted with q, set the new coaction map $\mathcal{N} \longrightarrow \mathcal{N} \square_{\mathcal{D}} \mathcal{D}^\sim$ to be the sum of the original coaction map and the composition $\mathcal{N} \longrightarrow \mathcal{N} \longrightarrow \mathcal{D}^\sim \square_{\mathcal{D}} \mathcal{N}$ of the map $q \colon \mathcal{N} \longrightarrow \mathcal{N}$ and the map $\mathcal{N} \longrightarrow \mathcal{N} \square_{\mathcal{D}} \mathcal{D}^\sim$ given by the formula $y \longmapsto (-1)^{|y_{(0)}|} y_{(0)} \square \partial(y_{(1)})$. Denote the quasi-differential $\mathcal{D}^\sim$-comodule so constructed by $\mathcal{N}(q)$; for any quasi-differential right $\mathcal{D}^\sim$-comodule $\mathcal{R}$ the differential in the complex $\mathrm{Hom}^\bullet_{\mathcal{D}}(\mathcal{R}, \mathcal{N}(q))$ differs from the differential in the complex $\mathrm{Hom}^\bullet_{\mathcal{D}}(\mathcal{R}, \mathcal{N})$ by the rule $d_q(f) = d(f) + q \circ f$.

The definitions of the homotopy category of quasi-differential right $\mathcal{D}^\sim$-comodules $\mathsf{Hot}(\text{qcmd-}\mathcal{D}^\sim)$ and the coderived category of quasi-differential right $\mathcal{D}^\sim$-comodules $\mathsf{D}^{\mathsf{co}}(\text{qcmd-}\mathcal{D}^\sim)$ are the same as in the left quasi-differential comodule case.

11.7.3 Let $(\mathcal{D}^\sim, \partial)$ be a quasi-differential coring over A; assume that $\mathcal{D} = \mathcal{D}^\sim/\mathrm{im}\,\partial$ is a projective graded left A-module.

A *quasi-differential left contramodule* over $\mathcal{D}^\sim$ is just a graded left $\mathcal{D}^\sim$-contramodule. Let us define the DG-category of quasi-differential left $\mathcal{D}^\sim$-contramodules $\mathsf{DG}(\mathcal{D}^\sim\text{-qcntr})$. The objects of $\mathsf{DG}(\mathcal{D}^\sim\text{-qcntr})$ are quasi-differential left $\mathcal{D}^\sim$-contramodules. The complex of morphisms $\mathrm{Hom}^{\mathcal{D},\bullet}(\mathfrak{P}, \mathfrak{Q})$ in the category $\mathsf{DG}(\mathcal{D}^\sim\text{-qcntr})$ between quasi-differential left $\mathcal{D}^\sim$-contramodules $\mathfrak{P}$ and $\mathfrak{Q}$ is constructed as follows. The component $\mathrm{Hom}^{\mathcal{D},n}(\mathfrak{P}, \mathfrak{Q})$ of this complex is the k-module of all homogeneous maps $\mathfrak{P} \longrightarrow \mathfrak{Q}$ of degree $-n$ supercommuting with the $\mathcal{D}$-contramodule structures. This means that an element $f \in \mathrm{Hom}^{\mathcal{D},n}(\mathfrak{P}, \mathfrak{Q})$ should satisfy the equation $\pi_{\mathfrak{P}}(f \circ x) = (-1)^{mn} f(\pi_{\mathfrak{P}}(x))$ for any $x \in \mathrm{Hom}_A(\mathcal{D}_m, \mathfrak{P})$, where $\pi_{\mathfrak{P}}$ denotes the contraaction map. To define the differential of an element $f \in \mathrm{Hom}^{\mathcal{D},n}(\mathfrak{P}, \mathfrak{Q})$, consider the map $\delta_f\colon \mathrm{Cohom}_{\mathcal{D}}(\mathcal{D}^\sim, \mathfrak{P}) \longrightarrow \mathfrak{Q}$ given by the formula $\overline{x} \longmapsto \pi_{\mathfrak{P}}(f \circ x) - (-1)^{mn} f(\pi_{\mathfrak{P}}(x))$ for $x \in \mathrm{Hom}_A(\mathcal{D}_m^\sim, \mathfrak{P})$, where $\overline{x}$ denotes the class of x in $\mathrm{Cohom}_{\mathcal{D}}(\mathcal{D}^\sim, \mathfrak{P})$. The map δ_f factorizes through the surjection $\mathrm{Cohom}_{\mathcal{D}}(\mathcal{D}^\sim, \mathfrak{P}) \longrightarrow \mathrm{Cohom}_{\mathcal{D}}(\mathcal{D}, \mathfrak{P}) \simeq \mathfrak{P}$ induced by the morphism $\overline{\partial}\colon \mathcal{D} \longrightarrow \mathcal{D}^\sim$, hence the desired map $d(f)\colon \mathfrak{P} \longrightarrow \mathfrak{Q}$.

Let $\mathfrak{P}$ be a quasi-differential left $\mathcal{D}^\sim$-contramodule and $q \in \mathrm{Hom}^{\mathcal{D},1}(\mathfrak{P}, \mathfrak{P})$ be an element satisfying the equation $d(q) + q^2 = 0$. To define the quasi-differential left $\mathcal{D}^\sim$-contramodule structure on $\mathfrak{P}$ twisted with q, set the new contraaction map $\mathrm{Cohom}_{\mathcal{D}}(\mathcal{D}^\sim, \mathfrak{P}) \longrightarrow \mathfrak{P}$ to be the sum of the original contraaction map and the composition of the map $\mathrm{Cohom}_{\mathcal{D}}(\mathcal{D}^\sim, \mathfrak{P}) \longrightarrow \mathfrak{P}$ induced by $\overline{\partial}\colon \mathcal{D} \longrightarrow \mathcal{D}^\sim$ and the map $q\colon \mathfrak{P} \longrightarrow \mathfrak{P}$. Denote the quasi-differential left $\mathcal{D}^\sim$-contramodule so constructed by $\mathfrak{P}(q)$; for any quasi-differential left $\mathcal{D}^\sim$-contramodule $\mathfrak{Q}$ the differential in the complex $\mathrm{Hom}^{\mathcal{D},\bullet}(\mathfrak{Q}, \mathfrak{P}(q))$ differs from the differential in the complex $\mathrm{Hom}^{\mathcal{D},\bullet}(\mathfrak{Q}, \mathfrak{P})$ by the rule $d_q(f) = d(f) + q \circ f$.

The definitions of the homotopy category of quasi-differential left $\mathcal{D}^\sim$-contramodules $\mathsf{Hot}(\mathcal{D}^\sim\text{-qcntr})$, its triangulated subcategory of *contraacyclic* quasi-differential $\mathcal{D}^\sim$-contramodules, and the *contraderived category* of quasi-differential left $\mathcal{D}^\sim$-contramodules $\mathsf{D}^{\mathrm{ctr}}(\mathcal{D}^\sim\text{-qcntr})$ are completely analogous to the corresponding definitions in the comodule case; the only difference is that one considers infinite products instead of infinite direct sums.

Remark. One can define the DG-categories of CDG-comodules and CDG-contramodules over a CDG-coring (see 11.2.2 and 0.4.4) and identify them with the DG-categories of quasi-differential comodules and contramodules in the case when a quasi-differential coring corresponds to a CDG-coring. More generally, let $(\mathcal{D}^\sim, \partial)$ be a quasi-differential coring over a k-algebra A such that the components $\mathcal{D}_i$ of the coring $\mathcal{D} = \mathcal{D}^\sim/\mathrm{im}\,\partial$ are projective left A-modules. Then there is a natural structure of quasi-differential k-algebra on the graded A-A-bimodule with the components $R^{n\sim} = \mathrm{Hom}_A(\mathcal{D}_n^\sim, A)$. Let (R, d, h) be a CDG-algebra over k corresponding to $R^\sim$. In this situation the DG-category of quasi-differential right $\mathcal{D}^\sim$-comodules is isomorphic to a full subcategory of the DG-category of right CDG-modules over (R, d, h), and there is a forgetful functor from the DG-category of quasi-differential left $\mathcal{D}^\sim$-contramodules to the DG-category of left CDG-modules over (R, d, h).

11.8 Koszul duality

Let $\mathcal{C}$ be a coring over a k-algebra A.

Theorem.

(a) *Assume that $\mathcal{C}$ is a flat right A-module. Let $\mathbf{S}^\sim$ be a right coflat nonhomogeneous Koszul semialgebra over the coring $\mathcal{C}$ and $\mathcal{D}^\sim$ be the right coflat Koszul quasi-differential coring over $\mathcal{C}$ nonhomogeneous quadratic dual to $\mathbf{S}^\sim$. Then the semiderived category of left $\mathbf{S}^\sim$-semimodules is naturally equivalent to the coderived category of quasi-differential left $\mathcal{D}^\sim$-comodules.*

(b) *Assume that $\mathcal{C}$ is a flat left A-module. Let $\mathbf{S}^\sim$ be a left coflat nonhomogeneous Koszul semialgebra over the coring $\mathcal{C}$ and $\mathcal{D}^\sim$ be the left coflat Koszul quasi-differential coring over $\mathcal{C}$ nonhomogeneous quadratic dual to $\mathbf{S}^\sim$. Then the semiderived category of right $\mathbf{S}^\sim$-semimodules is naturally equivalent to the coderived category of quasi-differential right $\mathcal{D}^\sim$-comodules.*

(c) *Assume that $\mathcal{C}$ is a projective left A-module. Let $\mathbf{S}^\sim$ be a left coprojective nonhomogeneous Koszul semialgebra over the coring $\mathcal{C}$ and $\mathcal{D}^\sim$ be the left coprojective Koszul quasi-differential coring over $\mathcal{C}$ nonhomogeneous quadratic dual to $\mathbf{S}^\sim$. Then the semiderived category of left $\mathbf{S}^\sim$-semicontramodules is naturally equivalent to the contraderived category of quasi-differential left $\mathcal{D}^\sim$-contramodules.*

Proof. Part (a): let us construct a pair of adjoint functors between the DG-category of complexes of left $\mathbf{S}^\sim$-semimodules and the DG-category of quasi-differential left $\mathcal{D}^\sim$-comodules. The functor Ξ assigns to a left $\mathbf{S}^\sim$-semimodule $\mathcal{M}$ the graded left $\mathcal{D}$-comodule $\mathcal{D} \,\square_\mathcal{C}\, \mathcal{M}$ endowed with the following left $\mathcal{D}^\sim$-comodule structure. Consider the map $\mathcal{D}_n^\sim \,\square_\mathcal{C}\, \mathcal{M} \longrightarrow \mathcal{D}_n^\sim \,\square_\mathcal{C}\, \mathcal{M}$ equal to the sum of the identity map and $(-1)^n$ times the composition $\mathcal{D}_n^\sim \,\square_\mathcal{C}\, \mathcal{M} \longrightarrow \mathcal{D}_{n-1}^\sim \,\square_\mathcal{C}\, \mathcal{D}_1^\sim \,\square_\mathcal{C}\, \mathcal{M} \longrightarrow \mathcal{D}_{n-1}^\sim \,\square_\mathcal{C}\, \mathcal{M} \longrightarrow \mathcal{D}_n^\sim \,\square_\mathcal{C}\, \mathcal{M}$ of the map induced by the comultiplication morphism $\mathcal{D}_n^\sim \longrightarrow \mathcal{D}_{n-1}^\sim \,\square_\mathcal{C}\, \mathcal{D}_1^\sim$, the map induced by the semiaction morphism $F_1\mathbf{S}^\sim \,\square_\mathcal{C}\, \mathcal{M} \longrightarrow \mathcal{M}$, and the map induced by the morphism $\partial_{n-1}\colon \mathcal{D}_{n-1}^\sim \longrightarrow \mathcal{D}_n^\sim$. This map factorizes through the surjection $\mathcal{D}_n^\sim \,\square_\mathcal{C}\, \mathcal{M} \longrightarrow \mathcal{D}_n \,\square_\mathcal{C}\, \mathcal{M}$, since its composition with the map $\mathcal{D}_{n-1}^\sim \,\square_\mathcal{C}\, \mathcal{M} \longrightarrow \mathcal{D}_n^\sim \,\square_\mathcal{C}\, \mathcal{M}$ induced by the morphism ∂_{n-1} vanishes. So we obtain a natural map

$$\mathcal{D}_n \,\square_\mathcal{C}\, \mathcal{M} \longrightarrow \mathcal{D}_n^\sim \,\square_\mathcal{C}\, \mathcal{M}.$$

Now the compositions $\mathcal{D}_{i+j} \,\square_\mathcal{C}\, \mathcal{M} \longrightarrow \mathcal{D}_{i+j}^\sim \,\square_\mathcal{C}\, \mathcal{M} \longrightarrow \mathcal{D}_i^\sim \,\square_\mathcal{C}\, \mathcal{D}_j^\sim \,\square_\mathcal{C}\, \mathcal{M} \longrightarrow \mathcal{D}_i^\sim \,\square_\mathcal{C}\, \mathcal{D}_j \,\square_\mathcal{C}\, \mathcal{M}$ of the maps $\mathcal{D}_{i+j} \,\square_\mathcal{C}\, \mathcal{M} \longrightarrow \mathcal{D}_{i+j}^\sim \,\square_\mathcal{C}\, \mathcal{M}$ we have constructed with the maps induced by the comultiplication maps $\mathcal{D}_{i+j}^\sim \longrightarrow \mathcal{D}_i^\sim \,\square_\mathcal{C}\, \mathcal{D}_j^\sim$ and the maps induced by the natural surjections $\mathcal{D}_j^\sim \longrightarrow \mathcal{D}_j$ define the desired graded $\mathcal{D}^\sim$-comodule structure on $\mathcal{D} \,\square_\mathcal{C}\, \mathcal{M}$. To a complex of left $\mathbf{S}^\sim$-semimodules $\mathcal{M}^\bullet$ the functor Ξ assigns the total quasi-differential $\mathcal{D}^\sim$-comodule of the complex of quasi-differential $\mathcal{D}^\sim$-comodules $\mathcal{D} \,\square_\mathcal{C}\, \mathcal{M}^\bullet$.

The functor Υ assigns to a quasi-differential left $\mathcal{D}^\sim$-comodule $\mathcal{L}$ the complex of left $\mathbf{S}^\sim$-semimodules $\Upsilon^\bullet(\mathcal{L}) = \mathbf{S}^\sim \,\square_\mathcal{C}\, \mathcal{L}$ with the terms $\Upsilon^i(\mathcal{L}) = \mathbf{S}^\sim \,\square_\mathcal{C}\, \mathcal{L}_{-i}$ and the differential defined as the composition

$$\mathbf{S}^{\sim} \,\square_{\mathfrak{C}}\, \mathcal{L} \longrightarrow \mathbf{S}^{\sim} \,\square_{\mathfrak{C}}\, \mathcal{D}_1^{\sim} \,\square_{\mathfrak{C}}\, \mathcal{L} \longrightarrow \mathbf{S}^{\sim} \,\square_{\mathfrak{C}}\, \mathcal{L}$$

of the map induced by the coaction morphism $\mathcal{L} \longrightarrow \mathcal{D}_1^{\sim} \,\square_{\mathfrak{C}}\, \mathcal{L}$ and the map induced by the semimultiplication morphism $\mathbf{S}^{\sim} \,\square_{\mathfrak{C}}\, F_1\mathbf{S}^{\sim} \longrightarrow \mathbf{S}^{\sim}$. The functor Ξ is right adjoint to the functor Υ, since both complexes $\operatorname{Hom}_{\mathcal{D}}^{\bullet}(\mathcal{L}, \Xi(\mathbf{M}^{\bullet}))$ and $\operatorname{Hom}_{\mathbf{S}^{\sim}}(\Upsilon^{\bullet}(\mathcal{L}), \mathbf{M}^{\bullet})$ are naturally isomorphic to the total complex of the bicomplex $\operatorname{Hom}_{\mathfrak{C}}(\mathcal{L}_i, \mathbf{M}^j)$, one of whose differentials is induced by the differential in $\mathbf{M}^{\bullet}$ and the other one assigns to a $\mathfrak{C}$-comodule morphism $f\colon \mathcal{L}_i \longrightarrow \mathbf{M}^j$ the composition

$$\mathcal{L}_{i+1} \longrightarrow \mathcal{D}_1^{\sim} \,\square_{\mathfrak{C}}\, \mathcal{L}_i \longrightarrow \mathcal{D}_1^{\sim} \,\square_{\mathfrak{C}}\, \mathbf{M}^j \simeq F_1\mathbf{S}^{\sim} \,\square_{\mathfrak{C}}\, \mathbf{M}^j \longrightarrow \mathbf{M}^j$$

of the coaction map, the map induced by the morphism f, and the semiaction map.

Let us show that the functors Ξ and Υ induce mutually inverse equivalences between the semiderived category $\mathsf{D}^{\mathsf{si}}(\mathbf{S}^{\sim}\text{-simod})$ and the coderived category $\mathsf{D}^{\mathsf{co}}(\mathcal{D}^{\sim}\text{-qcmd})$. Firstly, the functor Ξ sends $\mathfrak{C}$-coacyclic complexes of $\mathbf{S}$-semimodules to coacyclic quasi-differential $\mathcal{D}^{\sim}$-comodules. Indeed, for any complex of left $\mathbf{S}$-semimodules $\mathbf{M}^{\bullet}$ the quasi-differential $\mathcal{D}^{\sim}$-comodule $\Xi(\mathbf{M}^{\bullet}) = \mathcal{D} \,\square_{\mathfrak{C}}\, \mathbf{M}^{\bullet}$ has an increasing filtration by quasi-differential $\mathcal{D}^{\sim}$-subcomodules defined by the formula

$$F_n(\mathcal{D} \,\square_{\mathfrak{C}}\, \mathbf{M}^{\bullet}) = \bigoplus_{i \leqslant n} \mathcal{D}_i \,\square_{\mathfrak{C}}\, \mathbf{M}^{\bullet}.$$

The associated graded quotient quasi-differential comodule to this filtration is described as follows. There is a functor from the DG-category of complexes of $\mathfrak{C}$-comodules to the DG-category of quasi-differential $\mathcal{D}^{\sim}$-comodules assigning to a complex of $\mathfrak{C}$-comodules the total quasi-differential $\mathcal{D}^{\sim}$-comodule of the complex of quasi-differential $\mathcal{D}^{\sim}$-comodules whose terms are the terms of the original complex of $\mathfrak{C}$-comodules endowed with the graded $\mathcal{D}^{\sim}$-comodule structure via the embedding $\mathfrak{C} \simeq \mathcal{D}_0^{\sim} \longrightarrow \mathcal{D}^{\sim}$. (This functor can be also described in terms of the quasi-differential subcoring in $\mathcal{D}^{\sim}$ whose components are $\mathcal{D}_0^{\sim}$ and $\partial_0(\mathcal{D}_0^{\sim}) \subset \mathcal{D}_1^{\sim}$.) Clearly, this functor sends coacyclic complexes of $\mathfrak{C}$-comodules to coacyclic quasi-differential $\mathcal{D}^{\sim}$-comodules. Now the quasi-differential $\mathcal{D}^{\sim}$-comodules $F_n\Xi(\mathbf{M}^{\bullet})/F_{n-1}\Xi(\mathbf{M}^{\bullet})$ are isomorphic to the images of the $\mathfrak{C}$-comodules $\mathcal{D}_n \,\square_{\mathfrak{C}}\, \mathbf{M}^{\bullet}$ under this functor, and are, therefore, coacyclic whenever $\mathbf{M}^{\bullet}$ is $\mathfrak{C}$-coacyclic.

Secondly, the functor Υ sends coacyclic quasi-differential $\mathcal{D}^{\sim}$-comodules to complexes of $\mathbf{S}^{\sim}$-semimodules that are coacyclic not only over $\mathfrak{C}$, but even over $\mathbf{S}^{\sim}$.

Thirdly, let us check that for any complex of left $\mathbf{S}$-semimodules $\mathbf{M}^{\bullet}$ the cone of the natural morphism of complexes of $\mathbf{S}$-semimodules $\Upsilon^{\bullet}\Xi(\mathbf{M}^{\bullet}) \longrightarrow \mathbf{M}^{\bullet}$ is coacyclic as a complex of left $\mathfrak{C}$-comodules. The complex of $\mathfrak{C}$-comodules $\Upsilon^{\bullet}\Xi(\mathbf{M}^{\bullet}) = \mathbf{S}^{\sim} \,\square_{\mathfrak{C}}\, \mathcal{D} \,\square_{\mathfrak{C}}\, \mathbf{M}^{\bullet}$ has an increasing filtration given by the formula

$$F_n\Upsilon^{\bullet}\Xi(\mathbf{M}^{\bullet}) = \sum_{i+j \leqslant n} F_i\mathbf{S}^{\sim} \,\square_{\mathfrak{C}}\, \mathcal{D}_j \,\square_{\mathfrak{C}}\, \mathbf{M}^{\bullet} \subset \mathbf{S}^{\sim} \,\square_{\mathfrak{C}}\, \mathcal{D} \,\square_{\mathfrak{C}}\, \mathbf{M}^{\bullet}.$$

The cone of the morphism $\Upsilon^{\bullet}\Xi(\mathbf{M}^{\bullet}) \longrightarrow \mathbf{M}^{\bullet}$ has an induced filtration F whose components are the cones of the morphisms $F_n\Upsilon^{\bullet}\Xi(\mathbf{M}^{\bullet}) \longrightarrow \mathbf{M}^{\bullet}$. The quotient complex $\operatorname{cone}(F_n \Upsilon^\bullet \Xi(\mathcal{M}^\bullet) \to \mathcal{M}^\bullet)/\operatorname{cone}(F_{n-1}\Upsilon^\bullet\Xi(\mathcal{M}^\bullet) \to \mathcal{M}^\bullet)$ is isomorphic to the cone of the identity endomorphism of $\mathcal{M}^\bullet$ for $n = 0$ and to the cotensor product of a positive-degree component of the Koszul complex $\mathcal{S} \,\square_\mathfrak{C}\, \mathcal{D}$ and the complex $\mathcal{M}^\bullet$ for $n > 0$ (where, as always, $\mathcal{S} = \bigoplus_n F_n\mathcal{S}^\sim/F_{n-1}\mathcal{S}^\sim$). Thus in both cases the quotient complex is coacyclic.

Fourthly, it remains to check that for any quasi-differential left $\mathcal{D}^\sim$-comodule $\mathcal{L}$ the cone of the natural morphism of quasi-differential $\mathcal{D}^\sim$-comodules $\mathcal{L} \longrightarrow \Xi\Upsilon^\bullet(\mathcal{L})$ is coacyclic. First let us show that it suffices to consider the case when $\mathcal{L}$ is a graded $\mathfrak{C}$-comodule endowed with a graded $\mathcal{D}^\sim$-comodule structure via the embedding of corings $\mathcal{D}_0^\sim \longrightarrow \mathcal{D}^\sim$. To this end, consider the increasing filtration of $\mathcal{L}$ by quasi-differential $\mathcal{D}^\sim$-subcomodules

$$G_n(\mathcal{L}) = \nu_L^{-1}(\bigoplus_{i \leqslant n} \mathcal{D}^\sim_i \,\square_\mathfrak{C}\, \mathcal{L}),$$

where $\nu_L \colon \mathcal{L} \longrightarrow \mathcal{D}^\sim \,\square_\mathfrak{C}\, \mathcal{L}$ denotes the coaction map. The quotient quasi-differential comodules $G_n(\mathcal{L})/G_{n-1}(\mathcal{L})$ are graded $\mathcal{D}^\sim$-comodules originating from graded $\mathfrak{C}$-comodules; the filtration G induces a filtration on the cone of the morphism $\mathcal{L} \longrightarrow \Xi\Upsilon^\bullet(\mathcal{L})$ whose components are the cones of the morphisms $G_n(\mathcal{L}) \longrightarrow \Xi\Upsilon^\bullet(G_n(\mathcal{L}))$; and the associated quotient quasi-differential comodules of the latter filtration are the cones of the morphisms $G_n(\mathcal{L})/G_{n-1}(\mathcal{L}) \longrightarrow \Xi\Upsilon^\bullet(G_n(\mathcal{L})/G_{n-1}(\mathcal{L}))$.

Now assume that $\mathcal{L}$ is a graded $\mathfrak{C}$-comodule with the induced graded $\mathcal{D}^\sim$-comodule structure, or even a complex of $\mathfrak{C}$-comodules with the induced quasi-differential $\mathcal{D}^\sim$-comodule structure. In this case, the quasi-differential $\mathcal{D}^\sim$-comodule $\Xi\Upsilon^\bullet(\mathcal{L}) = \mathcal{D} \,\square_\mathfrak{C}\, \mathcal{S}^\sim \,\square_\mathfrak{C}\, \mathcal{L}$ has an increasing filtration by quasi-differential subcomodules given by the formula

$$F_n\Xi\Upsilon^\bullet(\mathcal{L}) = \sum_{i+j \leqslant n} \mathcal{D}_j \,\square_\mathfrak{C}\, F_i\mathcal{S}^\sim \,\square_\mathfrak{C}\, \mathcal{L} \subset \mathcal{D} \,\square_\mathfrak{C}\, \mathcal{S}^\sim \,\square_\mathfrak{C}\, \mathcal{L}.$$

The cone of the morphism $\mathcal{L} \longrightarrow \Xi\Upsilon^\bullet(\mathcal{L})$ has the induced filtration F whose components are the cones of the morphisms $\mathcal{L} \longrightarrow F_n\Xi\Upsilon^\bullet(\mathcal{L})$. The associated quotient comodules of the latter filtration are coacyclic quasi-differential $\mathcal{D}^\sim$-comodules. Indeed, the component $F_0 \operatorname{cone}(\mathcal{L} \longrightarrow \Xi\Upsilon^\bullet(\mathcal{L}))$ is isomorphic to the cone of the identity endomorphism of $\mathcal{L}$, while the quotient quasi-differential comodules with $n > 0$ are isomorphic to cotensor products of positive-degree components of the Koszul complex $\mathcal{D} \,\square_\mathfrak{C}\, \mathcal{S}$ and the $\mathfrak{C}$-comodule $\mathcal{L}$, endowed with the quasi-differential $\mathcal{D}^\sim$-comodule structures originating from their structures of complexes of $\mathfrak{C}$-comodules. Thus all these quotient quasi-differential comodules are coacyclic.

Part (b): we will only construct a pair of adjoint functors between the DG-category of complexes of right $\mathcal{S}^\sim$-semimodules and the DG-category of quasi-differential right $\mathcal{D}^\sim$-comodules; the rest of the proof is identical to that of part (a). The functor Ξ assigns to a right $\mathcal{S}$-semimodule $\mathcal{N}$ the graded right $\mathcal{D}$-comodule $\mathcal{N} \,\square_\mathfrak{C}\, \mathcal{D}$ endowed with a right $\mathcal{D}^\sim$-comodule structure in terms of the following maps $\mathcal{N} \,\square_\mathfrak{C}\, \mathcal{D}_n \longrightarrow \mathcal{N} \,\square_\mathfrak{C}\, \mathcal{D}_n^\sim$. Consider the map $\mathcal{N} \,\square_\mathfrak{C}\, \mathcal{D}_n \longrightarrow \mathcal{N} \,\square_\mathfrak{C}\, \mathcal{D}_n$ equal to the difference of the identity map and the composition $\mathcal{N} \,\square_\mathfrak{C}\, \mathcal{D}_n^\sim \longrightarrow$

$\mathcal{N} \,\square_{\mathcal{C}}\, \mathcal{D}_1^{\sim} \,\square_{\mathcal{C}}\, \mathcal{D}_{n-1}^{\sim} \longrightarrow \mathcal{N} \,\square_{\mathcal{C}}\, \mathcal{D}_{n-1}^{\sim} \longrightarrow \mathcal{N} \,\square_{\mathcal{C}}\, \mathcal{D}_n^{\sim}$ of the map induced by the comultiplication morphism, the map induced by the semiaction morphism, and the map induced by the morphism ∂_{n-1}. This difference factorizes through the surjection $\mathcal{N} \,\square_{\mathcal{C}}\, \mathcal{D}_n^{\sim} \longrightarrow \mathcal{N} \,\square_{\mathcal{C}}\, \mathcal{D}_n$, hence the desired map. The functor Υ assigns to a quasi-differential right $\mathcal{D}^{\sim}$-comodule $\mathcal{R}$ the complex of right $\mathbf{S}^{\sim}$-semimodules $\Upsilon^{\bullet}(\mathcal{R}) = \mathcal{R} \,\square_{\mathcal{C}}\, \mathbf{S}^{\sim}$ with the terms $\Upsilon^i(\mathcal{R}) = \mathcal{R}_{-i} \,\square_{\mathcal{C}}\, \mathbf{S}^{\sim}$ and the differentials d^i defines as $(-1)^i$ times the composition $\mathcal{R}_{-i} \,\square_{\mathcal{C}}\, \mathbf{S}^{\sim} \longrightarrow \mathcal{R}_{-i-1} \,\square_{\mathcal{C}}\, \mathcal{D}_1^{\sim} \,\square_{\mathcal{C}}\, \mathbf{S}^{\sim} \longrightarrow \mathcal{R}_{-i-1} \,\square_{\mathcal{C}}\, \mathbf{S}^{\sim}$ of the map induced by the coaction morphism and the map induced by the semimultiplication morphism.

Part (c): let us construct a pair of adjoint functors between the DG-category of complexes of left $\mathbf{S}^{\sim}$-semicontramodules and the DG-category of quasi-differential left $\mathcal{D}^{\sim}$-contramodules. The functor Ξ assigns to a left $\mathbf{S}$-semicontramodule $\mathfrak{P}$ the graded left $\mathcal{D}$-contramodule $\mathrm{Cohom}_{\mathcal{C}}(\mathcal{D}, \mathfrak{P})$ endowed with a left $\mathcal{D}^{\sim}$-contramodule structure in terms of the following maps $\mathrm{Cohom}_{\mathcal{C}}(\mathcal{D}_n^{\sim}, \mathfrak{P}) \longrightarrow \mathrm{Cohom}_{\mathcal{C}}(\mathcal{D}_n, \mathfrak{P})$. Consider the map $\mathrm{Cohom}_{\mathcal{C}}(\mathcal{D}_n^{\sim}, \mathfrak{P}) \longrightarrow \mathrm{Cohom}_{\mathcal{C}}(\mathcal{D}_n^{\sim}, \mathfrak{P})$ equal to the difference of the identity map and the composition $\mathrm{Cohom}_{\mathcal{C}}(\mathcal{D}_n^{\sim}, \mathfrak{P}) \longrightarrow \mathrm{Cohom}_{\mathcal{C}}(\mathcal{D}_{n-1}^{\sim}, \mathfrak{P}) \longrightarrow \mathrm{Cohom}_{\mathcal{C}}(\mathcal{D}_{n-1}^{\sim}, \mathrm{Cohom}_{\mathcal{C}}(\mathcal{D}_1^{\sim}, \mathfrak{P})) \longrightarrow \mathrm{Cohom}_{\mathcal{C}}(\mathcal{D}_n^{\sim}, \mathfrak{P})$ of the map induced by the morphism ∂_{n-1}, the map induced by the semicontraaction morphism $\mathfrak{P} \longrightarrow \mathrm{Cohom}_{\mathcal{C}}(F_1 \mathbf{S}^{\sim}, \mathfrak{P})$, and the map induced by the comultiplication morphism $\mathcal{D}_n^{\sim} \longrightarrow \mathcal{D}_1^{\sim} \,\square_{\mathcal{C}}\, \mathcal{D}_{n-1}^{\sim}$. This difference factorizes through the injection $\mathrm{Cohom}_{\mathcal{C}}(\mathcal{D}_n, \mathfrak{P}) \longrightarrow \mathrm{Cohom}_{\mathcal{C}}(\mathcal{D}_n^{\sim}, \mathfrak{P})$, hence the desired map. The functor Υ right adjoint to Ξ assigns to a quasi-differential left $\mathcal{D}^{\sim}$-contramodule $\mathfrak{Q}$ the complex of left $\mathbf{S}^{\sim}$-semicontramodules $\Upsilon^{\bullet}(\mathfrak{Q}) = \mathrm{Cohom}_{\mathcal{C}}(\mathbf{S}^{\sim}, \mathfrak{Q})$ with the terms $\Upsilon^i(\mathfrak{Q}) = \mathrm{Cohom}_{\mathcal{C}}(\mathbf{S}^{\sim}, \mathfrak{Q}_{-i})$ and the differential defined as the composition $\mathrm{Cohom}_{\mathcal{C}}(\mathbf{S}^{\sim}, \mathfrak{Q}) \longrightarrow \mathrm{Cohom}_{\mathcal{C}}(F_1 \mathbf{S}^{\sim} \,\square_{\mathcal{C}}\, \mathbf{S}^{\sim}, \mathfrak{Q}) \longrightarrow \mathrm{Cohom}_{\mathcal{C}}(\mathbf{S}^{\sim}, Q)$ of the map induced by the semimultiplication morphism $F_1 \mathbf{S}^{\sim} \,\square_{\mathcal{C}}\, \mathbf{S}^{\sim} \longrightarrow \mathbf{S}^{\sim}$ and the map induced by the contraaction morphism $\mathrm{Cohom}_{\mathcal{C}}(\mathcal{D}_1^{\sim}, \mathfrak{Q}) \longrightarrow \mathfrak{Q}$.

The rest of the proof is analogous to that of part (a), with the exception of the argument related to the filtration G (the first step of the fourth part of the proof). The problem here is that the decreasing filtration G of a graded $\mathcal{D}^{\sim}$-contramodule $\mathfrak{Q}$ whose components are the images $G^n \mathfrak{Q}$ of the contraaction maps $\mathrm{Cohom}_{\mathcal{C}}(\mathcal{D}^{\sim}/ \bigoplus_{i \leqslant n} \mathcal{D}_i^{\sim}, \mathfrak{Q}) \longrightarrow \mathfrak{Q}$ is not in general separated, i.e., the intersection of $G^n \mathfrak{Q}$ may be nonzero (see Appendix A). What one should do is replace an arbitrary quasi-differential left $\mathcal{D}^{\sim}$-contramodule $\mathfrak{Q}$ with the total quasi-differential contramodule $\mathfrak{R}$ of its bar resolution $\cdots \longrightarrow \mathrm{Cohom}_{\mathcal{C}}(\mathcal{D}^{\sim} \,\square_{\mathcal{C}}\, \mathcal{D}^{\sim} \,\square_{\mathcal{C}}\, \mathcal{D}^{\sim}, \mathfrak{Q}) \longrightarrow \mathrm{Cohom}_{\mathcal{C}}(\mathcal{D}^{\sim} \,\square_{\mathcal{C}}\, \mathcal{D}^{\sim}, \mathfrak{Q}) \longrightarrow \mathrm{Cohom}_{\mathcal{C}}(\mathcal{D}^{\sim}, \mathfrak{Q})$. Since the cone of the natural morphism of quasi-differential $\mathcal{D}^{\sim}$-contramodules $\mathfrak{R} \longrightarrow \mathfrak{Q}$ is contraacyclic, one can consider the quasi-differential $\mathcal{D}^{\sim}$-contramodule $\mathfrak{R}$ instead of $\mathfrak{Q}$.

In addition to the filtration G introduced above, consider also the decreasing filtration $'G$ of a graded $\mathcal{D}^{\sim}$-contramodule $\mathfrak{Q}$ whose components are the images $'G^n \mathfrak{Q}$ of the contraaction maps $\mathrm{Cohom}_{\mathcal{C}}(\mathcal{D}/ \bigoplus_{i \leqslant n} \mathcal{D}_i, \mathfrak{Q}) \longrightarrow \mathfrak{Q}$. It is clear that $\mathfrak{R} \simeq \varprojlim_n \mathfrak{R}/'G^n \mathfrak{R}$, since the graded $\mathcal{D}$-contramodule $\mathfrak{R}$ is simply the infinite product of the terms of the above bar resolution. Next one can either show that

'G is a filtration by graded $\mathcal{D}^{\sim}$-subcontramodules and use the filtration 'G of $\mathfrak{R}$, or show that the filtrations G and 'G are commensurable, ${}'G^n\mathfrak{R} \subset G^n\mathfrak{R} \subset {}'G^{n-1}\mathfrak{R}$, and use the filtration G of $\mathfrak{R}$. (The quotient quasi-differential $\mathcal{D}^{\sim}$-contramodules $G^n\mathfrak{Q}/G^{n+1}\mathfrak{Q}$ originate from graded $\mathcal{C}$-contramodules, while the quotient quasi-differential $\mathcal{D}^{\sim}$-contramodules ${}'G^n\mathfrak{Q}/{}'G^{n+1}\mathfrak{Q}$ originate from complexes of $\mathcal{C}$-contramodules, which is also sufficient.) Both assertions for an arbitrary graded $\mathcal{D}^{\sim}$-contramodule $\mathfrak{Q}$ follow from the fact that the composition $\mathcal{D}_i^{\sim} \longrightarrow \mathcal{D}_1^{\sim} \square_{\mathcal{C}} \mathcal{D}_{i-1}^{\sim} \longrightarrow \mathcal{D}_1^{\sim} \square_{\mathcal{C}} \mathcal{D}_{i-1}$ of the comultiplication map and the map induced by the natural surjection $\mathcal{D}_{i-1}^{\sim} \longrightarrow \mathcal{D}_{i-1}$ is injective and its cokernel, being isomorphic to the cokernel of the comultiplication map $\mathcal{D}_i \longrightarrow \mathcal{D}_1 \square_{\mathcal{C}} \mathcal{D}_{i-1}$, is a coprojective left $\mathcal{C}$-comodule. To check the latter, consider the composition of the map $\mathcal{D}_i^{\sim} \longrightarrow \mathcal{D}_1^{\sim} \square_{\mathcal{C}} \mathcal{D}_{i-1}$ in question with the map ∂_{i-1}.

Alternatively, one can replace an arbitrary quasi-differential $\mathcal{D}^{\sim}$-contramodule $\mathfrak{Q}$ with the cone of the morphism $\ker(\mathrm{Cohom}_{\mathcal{C}}(\mathcal{D}^{\sim}, \mathfrak{Q}) \to \mathfrak{Q}) \longrightarrow \mathrm{Cohom}_{\mathcal{C}}(\mathcal{D}^{\sim}, \mathfrak{Q})$ and use the appropriate generalization of Lemma A.2.3. $\square$

Remark. Notice that no homological dimension condition on the k-algebra A is assumed in Theorem 11.8. In particular, when $\mathcal{C} = A$, so $\mathbf{S}^{\sim}$ is just a filtered k-algebra, Theorem 11.8 provides a description of certain semiderived categories of $\mathbf{S}^{\sim}$-modules relative to $F_0\mathbf{S}^{\sim} = A$. A description of the conventional derived category can also be obtained. Namely, in the assumptions of part (a) of the theorem the conventional derived category of left $\mathbf{S}^{\sim}$-semimodules is equivalent to the quotient category of the coderived category of quasi-differential left $\mathcal{D}^{\sim}$-comodules by its minimal triangulated subcategory containing all the quasi-differential $\mathcal{D}^{\sim}$-comodules originating from acyclic complexes of left $\mathcal{C}$-comodules and closed under infinite direct sums. This is so because for any acyclic complex of $\mathbf{S}^{\sim}$-semimodules $\mathbf{M}^{\bullet}$ the quasi-differential $\mathcal{D}^{\sim}$-comodules $F_n\Xi(\mathbf{M}^{\bullet})/F_{n-1}\Xi(\mathbf{M}^{\bullet})$ originate from acyclic complexes of $\mathcal{C}$-comodules, and conversely, for any quasi-differential $\mathcal{D}^{\sim}$-comodule $\mathcal{L}$ originating from an acyclic complex of $\mathcal{C}$-comodules the complex of $\mathbf{S}^{\sim}$-semimodules $\Upsilon^{\bullet}(\mathcal{L})$ is acyclic. The analogous result holds for right $\mathbf{S}^{\sim}$-semimodules in the assumptions of part (b); and in the assumptions of part (c) the conventional derived category of left $\mathbf{S}^{\sim}$-semicontramodules is equivalent to the quotient category of the contraderived category of quasi-differential left $\mathcal{D}^{\sim}$-contramodules by its minimal triangulated subcategory containing all the quasi-differential contramodules originating from acyclic complexes of left $\mathcal{C}$-contramodules and closed under infinite products.

11.9 SemiTor and Cotor, SemiExt and Coext

11.9.1 Let $(\mathcal{D}^{\sim}, \partial)$ be a quasi-differential coring over a k-algebra A; assume that $\mathcal{D} = \mathcal{D}^{\sim}/\mathrm{im}\,\partial$ is a flat left and right A-module.

Let $\mathcal{N}$ be a quasi-differential right $\mathcal{D}^{\sim}$-comodule and $\mathcal{M}$ be a quasi-differential left $\mathcal{D}^{\sim}$-comodule. Assume that one of the graded A-modules $\mathcal{N}$ and $\mathcal{M}$ is flat. Then on the cotensor product $\mathcal{N} \square_{\mathcal{D}} \mathcal{M}$ of the graded comodules $\mathcal{N}$ and $\mathcal{M}$ over the graded coring $\mathcal{D}$ there is a natural differential with zero square, which is defined as follows.

Consider the map $\delta\colon \mathcal{N}\,\square_{\mathcal{D}}\,\mathcal{M} \longrightarrow \mathcal{N}\,\square_{\mathcal{D}}\,\mathcal{D}^{\sim}\,\square_{\mathcal{D}}\,\mathcal{M}$ given by the formula $x\square y \longmapsto -x_{(0)}\square x_{(1)}\square y + x\square y_{(-1)}\square y_{(0)}$. This map factorizes through the injection $\mathcal{N}\square_{\mathcal{D}}\mathcal{M} \longrightarrow \mathcal{N}\square_{\mathcal{D}}\mathcal{D}^{\sim}\square_{\mathcal{D}}\mathcal{M}$ given by the formula $x\square y \longmapsto (-1)^{|x_{(0)}|}x_{(0)}\square\partial(x_{(1)})\square y = (-1)^{|x|}x\square\partial(y_{(-1)})\square y_{(0)}$, hence the desired map $d\colon \mathcal{N}\,\square_{\mathcal{D}}\,\mathcal{M} \longrightarrow \mathcal{N}\,\square_{\mathcal{D}}\,\mathcal{M}$. Let us check that $d^2 = 0$, that is the image of d is contained in $\mathcal{N}\,\square_{\mathcal{D}^{\sim}}\,\mathcal{M}$. Set $d(x\square y) = x'\square y'$. Consider the two elements $x'\square y'_{(-1)}\square y'_{(0)}$ and $x'_{(0)}\square x'_{(1)}\square y$ of the cotensor product $\mathcal{N}\square_{\mathcal{D}}\mathcal{D}^{\sim}\square_{\mathcal{D}}\mathcal{M}$; we have to check that these two elements coincide. Consider the image of the former element under the map $\mathcal{N}\,\square_{\mathcal{D}}\,\mathcal{D}^{\sim}\,\square_{\mathcal{D}}\,\mathcal{M} \longrightarrow \mathcal{N}\square_{\mathcal{D}}\mathcal{D}^{\sim}\square_{\mathcal{D}}\mathcal{D}^{\sim}\square_{\mathcal{D}}\mathcal{M}$ given by the formula $u\square b\square v \longmapsto (-1)^{|u_{(0)}|}u_{(0)}\square\partial(u_{(1)})\square b\square v$ and the image of the latter element under the map $\mathcal{N}\square_{\mathcal{D}}\mathcal{D}^{\sim}\square_{\mathcal{D}}\mathcal{M} \longrightarrow \mathcal{N}\square_{\mathcal{D}}\mathcal{D}^{\sim}\square_{\mathcal{D}}\mathcal{D}^{\sim}\,\square_{\mathcal{D}}\,\mathcal{M}$ given by the formula $u\square b\square v \longmapsto (-1)^{|u|+|b|}u\square b\square\partial(v_{(-1)})\square v_{(0)}$. The sum of these two elements of $\mathcal{N}\,\square_{\mathcal{D}}\,\mathcal{D}^{\sim}\,\square_{\mathcal{D}}\,\mathcal{D}^{\sim}\,\square_{\mathcal{D}}\,\mathcal{M}$ is equal to the image of the element $\delta(x\square y)$ under the map $\mathcal{N}\square_{\mathcal{D}}\mathcal{D}^{\sim}\square_{\mathcal{D}}\mathcal{M} \longrightarrow \mathcal{N}\square_{\mathcal{D}}\mathcal{D}^{\sim}\square_{\mathcal{D}}\mathcal{D}^{\sim}\square_{\mathcal{D}}\mathcal{M}$ induced by the comultiplication map $\mathcal{D}^{\sim} \longrightarrow \mathcal{D}^{\sim}\,\square_{\mathcal{D}}\,\mathcal{D}^{\sim}$. It remains to notice that the Cartesian square formed by the maps $\mathcal{D}^{\sim} \longrightarrow \mathcal{D}^{\sim}\oplus\mathcal{D}^{\sim}$, $\mathcal{D}^{\sim}\oplus\mathcal{D}^{\sim} \longrightarrow \mathcal{D}^{\sim}\square_{\mathcal{D}}\mathcal{D}^{\sim}$, $\mathcal{D}^{\sim} \longrightarrow \mathcal{D}^{\sim}$, and $\mathcal{D}^{\sim} \longrightarrow \mathcal{D}^{\sim}\square_{\mathcal{D}}\mathcal{D}^{\sim}$ constructed in 11.7.1 remains Cartesian after taking the cotensor product with $\mathcal{N}$ and $\mathcal{M}$. We will denote the complex we have constructed by $\mathcal{N}\square_{\mathcal{D}}^{\bullet}\,\mathcal{M}$; its terms are $\mathcal{N}\square_{\mathcal{D}}^{n}\,\mathcal{M} = (\mathcal{N}\,\square_{\mathcal{D}}\,\mathcal{M})_{-n}$.

Now assume that the ring A has a finite weak homological dimension. In order to define the double-sided derived functor of the functor $\square_{\mathcal{D}}^{\bullet}$, we will show that the coderived category of quasi-differential $\mathcal{D}^{\sim}$-comodules is equivalent to the quotient category of the homotopy category of $\mathcal{D}$-coflat quasi-differential $\mathcal{D}^{\sim}$-comodules by its intersection with the thick subcategory of coacyclic quasi-differential $\mathcal{D}^{\sim}$-comodules.

The argument is analogous to that of either Theorem 2.5 or Theorem 2.6. First let us construct for any quasi-differential left $\mathcal{D}^{\sim}$-comodule $\mathcal{K}$ a morphism into it from an A-flat quasi-differential left $\mathcal{D}^{\sim}$-comodule $\mathbb{L}_1(\mathcal{K})$ with a coacyclic cone. Use the graded version of Lemma 1.1.3 to obtain a finite resolution $0 \longrightarrow \mathcal{Z} \longrightarrow \mathcal{P}_{d-1}(\mathcal{K}) \longrightarrow \cdots \longrightarrow \mathcal{P}_0(\mathcal{K}) \longrightarrow \mathcal{K}$ of a graded $\mathcal{D}^{\sim}$-comodule $\mathcal{K}$ consisting of A-flat graded $\mathcal{D}^{\sim}$-comodules. The total quasi-differential $\mathcal{D}^{\sim}$-comodule of the complex of quasi-differential $\mathcal{D}^{\sim}$-comodules

$$\mathcal{Z} \longrightarrow \mathcal{P}_{d-1}(\mathcal{K}) \longrightarrow \cdots \longrightarrow \mathcal{P}_0(\mathcal{K})$$

is an A-flat quasi-differential $\mathcal{D}^{\sim}$-comodule whose morphism into $\mathcal{K}$ has a coacyclic cone. Indeed, the total quasi-differential $\mathcal{D}^{\sim}$-comodule of any acyclic complex of quasi-differential $\mathcal{D}^{\sim}$-comodules bounded from below is coacyclic, since it has an increasing filtration by quasi-differential $\mathcal{D}^{\sim}$-subcomodules such that the associated quotient $\mathcal{D}^{\sim}$-comodules are isomorphic to cones of identity endomorphisms of certain quasi-differential $\mathcal{D}^{\sim}$-comodules.

Now let us construct for any A-flat quasi-differential left $\mathcal{D}^{\sim}$-comodule $\mathcal{L}$ a morphism from it into a $\mathcal{D}$-coflat quasi-differential left $\mathcal{D}^{\sim}$-comodule $\mathbb{R}_2(\mathcal{L})$ with a coacyclic cone. Consider the cobar construction

$$\mathcal{D}^\sim \otimes_A \mathcal{L} \longrightarrow \mathcal{D}^\sim \otimes_A \mathcal{D}^\sim \otimes_A \mathcal{L} \longrightarrow \cdots .$$

Notice that $\mathcal{D}^\sim$ is a coflat graded left $\mathcal{D}$-comodule, since there is an exact triple of left $\mathcal{D}$-comodules $\mathcal{D}(-1) \longrightarrow \mathcal{D}^\sim \longrightarrow \mathcal{D}$ (where $\mathcal{D}(-1)_i = \mathcal{D}_{i-1}$). Hence the total quasi-differential $\mathcal{D}^\sim$-comodule of this cobar complex of quasi-differential $\mathcal{D}^\sim$-comodules is a $\mathcal{D}$-coflat quasi-differential $\mathcal{D}^\sim$-comodule such that the map into it from the quasi-differential $\mathcal{D}^\sim$-comodule $\mathcal{L}$ has a coacyclic cone.

It is easy to see that the cotensor product of a quasi-differential right $\mathcal{D}^\sim$-comodule and a quasi-differential left $\mathcal{D}^\sim$-comodule is an acyclic complex whenever one of the two quasi-differential $\mathcal{D}^\sim$-comodules is coacyclic and the other one is $\mathcal{D}$-coflat. The double-sided derived functor

$$\mathrm{Cotor}_\mathsf{q}^{\mathcal{D}^\sim} : \mathsf{D}^{\mathrm{co}}(\mathrm{qcmd}\text{-}\mathcal{D}^\sim) \times \mathsf{D}^{\mathrm{co}}(\mathcal{D}^\sim\text{-}\mathrm{qcmd}) \longrightarrow \mathsf{D}(k\text{-mod})$$

is defined by restricting the functor $\square_\mathcal{D}^\bullet$ to the Cartesian product of the homotopy category of quasi-differential right $\mathcal{D}^\sim$-comodules and the homotopy category of $\mathcal{D}$-coflat quasi-differential left $\mathcal{D}^\sim$-comodules or to the Cartesian product of the homotopy category of $\mathcal{D}$-coflat quasi-differential right $\mathcal{D}^\sim$-comodules and the homotopy category of quasi-differential left $\mathcal{D}^\sim$-comodules, and composing it with the localization functor $\mathsf{Hot}(k\text{-mod}) \longrightarrow \mathsf{D}(k\text{-mod})$.

11.9.2 Let $(\mathcal{D}^\sim, \partial)$ be a quasi-differential coring over a k-algebra A; assume that $\mathcal{D} = \mathcal{D}^\sim/\mathrm{im}\,\partial$ is a projective left and a flat right A-module.

Let $\mathcal{M}$ be a quasi-differential left $\mathcal{D}^\sim$-comodule and $\mathfrak{P}$ be a quasi-differential left $\mathcal{D}^\sim$-contramodule. Assume that either the graded A-module $\mathcal{M}$ is projective, or the graded A-module $\mathfrak{P}$ is injective. Then on the graded k-module of cohomomorphisms $\mathrm{Cohom}_\mathcal{D}(\mathcal{M}, \mathfrak{P})$ from the graded comodule $\mathcal{M}$ to the graded contramodule $\mathfrak{P}$ over the graded coring $\mathcal{D}$ there is a natural differential with zero square, which is defined as follows. Consider the map
$$\delta \colon \mathrm{Cohom}_\mathcal{D}(\mathcal{M}, \mathrm{Cohom}_\mathcal{D}(\mathcal{D}^\sim, \mathfrak{P})) \simeq \mathrm{Cohom}_\mathcal{D}(\mathcal{D}^\sim \square_\mathcal{D} \mathcal{M}, \mathfrak{P}) \longrightarrow \mathrm{Cohom}_\mathcal{D}(\mathcal{M}, \mathfrak{P})$$
defined by the formula $f \longmapsto \pi_\mathfrak{P} \circ f - f \circ \nu_\mathcal{M}$ (where $\pi_\mathfrak{P}$ and $\nu_\mathcal{M}$ denote the contraaction and coaction morphisms). This map factorizes through the surjection $\mathrm{Cohom}_\mathcal{D}(\mathcal{D}^\sim \square_\mathcal{D} \mathcal{M}, \mathfrak{P}) \longrightarrow \mathrm{Cohom}_\mathcal{D}(\mathcal{M}, \mathfrak{P})$ induced by the morphism $\bar\partial \colon \mathcal{D} \longrightarrow \mathcal{D}^\sim$, hence the map $d \colon \mathrm{Cohom}_\mathcal{D}(\mathcal{M}, \mathfrak{P}) \longrightarrow \mathrm{Cohom}_\mathcal{D}(\mathcal{M}, \mathfrak{P})$. We will denote the complex we have constructed by $\mathrm{Cohom}_\mathcal{D}^\bullet(\mathcal{M}, \mathfrak{P})$; its terms are $\mathrm{Cohom}_\mathcal{D}^n(\mathcal{M}, \mathfrak{P}) = \mathrm{Cohom}_\mathcal{D}(\mathcal{M}, \mathfrak{P})_{-n}$.

Assume that the ring A has a finite left homological dimension. Then the coderived category of quasi-differential left $\mathcal{D}^\sim$-comodules is equivalent to the quotient category of the homotopy category of $\mathcal{D}$-coprojective quasi-differential left $\mathcal{D}^\sim$-comodules by its intersection with the thick subcategory of coacyclic quasi-differential $\mathcal{D}^\sim$-comodules. Analogously, the contraderived category of quasi-differential left $\mathcal{D}^\sim$-contramodules is equivalent to the quotient category of the homotopy category of $\mathcal{D}$-coinjective quasi-differential left $\mathcal{D}^\sim$-contramodules by its

intersection with the thick subcategory of contraacyclic quasi-differential $\mathcal{D}^{\sim}$-con-
tramodules. The double-sided derived functor

$$\mathrm{Coext}^{\mathsf{q}}_{\mathcal{D}^{\sim}} : \mathsf{D}^{\mathrm{co}}(\mathcal{D}^{\sim}\text{-qcmd})^{\mathrm{op}} \times \mathsf{D}^{\mathrm{ctr}}(\mathcal{D}^{\sim}\text{-qcntr}) \longrightarrow \mathsf{D}(k\text{-mod})$$

is defined by restricting the functor $\mathrm{Cohom}^{\bullet}_{\mathcal{D}}$ to the Cartesian product of the
homotopy category of $\mathcal{D}$-coprojective quasi-differential left $\mathcal{D}^{\sim}$-comodules and the
homotopy category of quasi-differential left $\mathcal{D}^{\sim}$-contramodules or to the Cartesian
product of the homotopy category of quasi-differential left $\mathcal{D}^{\sim}$-comodules and the
homotopy category of $\mathcal{D}$-coinjective quasi-differential left $\mathcal{D}^{\sim}$-contramodules, and
composing it with the localization functor $\mathsf{Hot}(k\text{-mod}) \longrightarrow \mathsf{D}(k\text{-mod})$.

11.9.3 Let $\mathcal{C}$ be a coring over a k-algebra A. Assume that $\mathcal{C}$ is a flat left and
right A-module and A has a finite weak homological dimension. Let $\mathbf{S}^{\sim}$ be a left
and right coflat nonhomogeneous Koszul semialgebra over $\mathcal{C}$, and $\mathcal{D}^{\sim}$ be the left
and right coflat Koszul quasi-differential coring nonhomogeneous quadratic dual
to $\mathbf{S}^{\sim}$.

Corollary.
(a) *The equivalences of categories*

$$\mathsf{D}^{\mathsf{si}}(\text{simod-}\mathbf{S}^{\sim}) \simeq \mathsf{D}^{\mathrm{co}}(\text{qcmd-}\mathcal{D}^{\sim}) \quad and \quad \mathsf{D}^{\mathsf{si}}(\mathbf{S}^{\sim}\text{-simod}) \simeq \mathsf{D}^{\mathrm{co}}(\mathcal{D}^{\sim}\text{-qcmd})$$

transform the derived functor $\mathrm{SemiTor}^{\mathbf{S}^{\sim}}$ into the derived functor $\mathrm{Cotor}^{\mathcal{D}^{\sim}}_{\mathsf{q}}$.
(b) *Assume additionally that $\mathcal{C}$ is a projective left A-module, A has a finite left
homological dimension, and $\mathbf{S}^{\sim}$ is a left coprojective nonhomogeneous Koszul
semialgebra. Then the equivalences of categories*

$$\mathsf{D}^{\mathsf{si}}(\mathbf{S}^{\sim}\text{-simod}) \simeq \mathsf{D}^{\mathrm{co}}(\mathcal{D}^{\sim}\text{-qcmd}) \quad and \quad \mathsf{D}^{\mathsf{si}}(\mathbf{S}^{\sim}\text{-sicntr}) \simeq \mathsf{D}^{\mathrm{ctr}}(\mathcal{D}^{\sim}\text{-qcntr})$$

transform the derived functor $\mathrm{SemiExt}_{\mathbf{S}^{\sim}}$ into the derived functor $\mathrm{Coext}^{\mathsf{q}}_{\mathcal{D}^{\sim}}$.

Proof. Part (a): for any complex of right $\mathbf{S}^{\sim}$-semimodules $\mathbf{N}^{\bullet}$ and any quasi-
differential left $\mathcal{D}^{\sim}$-comodule $\mathcal{L}$ there is a natural isomorphism of complexes of
k-modules

$$\Xi(\mathbf{N}^{\bullet}) \,\square^{\bullet}_{\mathcal{D}}\, \mathcal{L} \simeq \mathbf{N}^{\bullet} \,\Diamond_{\mathbf{S}^{\sim}}\, \Upsilon^{\bullet}(\mathcal{L}).$$

Indeed, both complexes are isomorphic to the total complex of the bicomplex
$\mathbf{N}^i \,\square_{\mathcal{C}}\, \mathcal{L}_j$, one of whose differentials is induced by the differential in $\mathbf{N}^{\bullet}$ and the
other is equal to the composition

$$\mathbf{N}^i \,\square_{\mathcal{C}}\, \mathcal{L}_j \longrightarrow \mathbf{N}^i \,\square_{\mathcal{C}}\, \mathcal{D}^{\sim}_1 \,\square_{\mathcal{C}}\, \mathcal{L}_{j-1} \longrightarrow \mathbf{N}^i \,\square_{\mathcal{C}}\, \mathcal{L}_{j-1}$$

of the map induced by the $\mathcal{D}^{\sim}$-coaction in $\mathcal{L}$ and the map induced by the
$\mathbf{S}^{\sim}$-semiaction in $\mathbf{N}^i$. Now let $\mathbf{N}^{\bullet}$ be a semiflat complex of $\mathcal{C}$-coflat right
$\mathbf{S}^{\sim}$-semimodules and $\mathbf{M}^{\bullet}$ be a complex of left $\mathbf{S}^{\sim}$-semimodules. Then there is
an isomorphism $\Xi(\mathbf{N}^{\bullet}) \,\square^{\bullet}_{\mathcal{D}}\, \Xi(\mathbf{M}^{\bullet}) \simeq \mathbf{N}^{\bullet} \,\Diamond_{\mathbf{S}^{\sim}}\, \Upsilon^{\bullet}\Xi(\mathbf{M}^{\bullet})$ and a quasi-isomorphism

$\mathbf{N}^\bullet \lozenge_{\mathbf{S}\sim} \Upsilon^\bullet \Xi(\mathbf{M}^\bullet) \longrightarrow \mathbf{N}^\bullet \lozenge_{\mathbf{S}\sim} \mathbf{M}^\bullet$. Analogously, for a complex of right $\mathbf{S}^\sim$-semi-modules $\mathbf{N}^\bullet$ and a semiflat complex of $\mathcal{C}$-coflat left $\mathbf{S}^\sim$-semimodules $\mathbf{M}^\bullet$ there is an isomorphism $\Xi(\mathbf{N}^\bullet) \square^\bullet_{\mathcal{D}} \Xi(\mathbf{M}^\bullet) \simeq \Upsilon^\bullet \Xi(\mathbf{N}^\bullet) \lozenge_{\mathbf{S}\sim} \mathbf{M}^\bullet$ and a quasi-isomorphism $\Upsilon^\bullet \Xi(\mathbf{N}^\bullet) \lozenge_{\mathbf{S}\sim} \mathbf{M}^\bullet \longrightarrow \mathbf{N}^\bullet \lozenge_{\mathbf{S}\sim} \mathbf{M}^\bullet$. It is easy to check that the square diagram formed by these maps is commutative. The proof of part (b) is completely analogous.   □

Question. Can one construct a comodule-contramodule correspondence (equivalence between the coderived and contraderived categories) for quasi-differential comodules and contramodules? Also, is there a natural closed model category structure on the category of quasi-differential comodules (contramodules)?

Appendices

A Contramodules over Coalgebras over Fields

Let $\mathcal{C}$ be a coassociative coalgebra with counit over a field k. It is well known (see [82, Theorems 2.1.3 and 2.2.1] or [68, 5.1.1]) that $\mathcal{C}$ is the union of its finite-dimensional subcoalgebras and any $\mathcal{C}$-comodule is a union of finite-dimensional comodules over finite-dimensional subcoalgebras of $\mathcal{C}$. The dual assertion for $\mathcal{C}$-contramodules is *not* true: for the most common of noncosemisimple infinite-dimensional coalgebras $\mathcal{C}$ there exist $\mathcal{C}$-contramodules $\mathfrak{P}$ such that the intersection of the images of $\mathrm{Hom}_k(\mathcal{C}/\mathcal{U}, \mathfrak{P})$ in $\mathfrak{P}$ over all finite-dimensional subcoalgebras $\mathcal{U} \subset \mathcal{C}$ is nonzero. A weaker statement holds, however: if the contraaction map

$$\mathrm{Hom}_k(\mathcal{C}/\mathcal{U}, \mathfrak{P}) \longrightarrow \mathfrak{P}$$

is surjective for every finite-dimensional subcoalgebra $\mathcal{U}$ of $\mathcal{C}$, then $\mathfrak{P} = 0$. Besides, even though adic filtrations of contramodules are not in general separated, they are always *complete*. Using the related techniques we show that any contraflat $\mathcal{C}$-contramodule is projective, generalizing the well-known result that any flat module over a finite-dimensional algebra is projective [7].

A.1 Counterexamples

A.1.1 Let $\mathcal{C}$ be the coalgebra for which the dual algebra $\mathcal{C}^*$ is isomorphic to the algebra of formal power series $k[[x]]$. Then a $\mathcal{C}$-contramodule $\mathfrak{P}$ can be equivalently defined as a k-vector space endowed with the following operation of summation of sequences of vectors with formal coefficients x^n: for any elements $p_0, p_1, \ldots$ in $\mathfrak{P}$, an element of $\mathfrak{P}$ denoted by $\sum_{n=0}^{\infty} x^n p_n$ is defined. This operation should satisfy the following equations:

$$\sum_{n=0}^{\infty} x^n (a p_n + b q_n) = a \sum_{n=0}^{\infty} x^n p_n + b \sum_{n=0}^{\infty} x^n q_n$$

for $a, b \in k$, $p_n, q_n \in \mathfrak{P}$ (linearity);

$$\sum_{n=0}^{\infty} x^n p_n = p_0 \quad \text{when } p_1 = p_2 = \cdots = 0$$

(counity); and

$$\sum_{i=0}^{\infty} x^i \left(\sum_{j=0}^{\infty} x^j p_{ij} \right) = \sum_{n=0}^{\infty} x^n \left(\sum_{i+j=n} p_{ij} \right)$$

for any $p_{ij} \in \mathfrak{P}$, $i, j = 0, 1, \ldots$ (contraassociativity). In the latter equation, the interior summation sign in the right-hand side denotes the conventional finite sum of elements of a vector space, while the three other summation signs refer to the contramodule infinite summation operation.

The following examples of $\mathcal{C}$-contramodules are revealing. Let $\mathfrak{E}$ denote the free $\mathcal{C}$-contramodule generated by the sequence of symbols e_0, e_1, $\ldots$; its elements can be represented as formal sums $\sum_{i=0}^{\infty} a_i(x)e_i$, where $a_i(x)$ are formal power series in x such that $a_i(x) \to 0$ in the topology of $k[[x]]$ as $i \to \infty$. Let $\mathfrak{F}$ denote the free $\mathcal{C}$-contramodule generated by the sequence of symbols f_1, f_2, $\ldots$; then $\mathcal{C}$-contramodule homomorphisms from $\mathfrak{F}$ to $\mathfrak{E}$ correspond bijectively to sequences of elements of $\mathfrak{E}$ that are images of the elements f_i. We are interested in the map $g\colon \mathfrak{F} \longrightarrow \mathfrak{E}$ sending f_i to $x^i e_i - e_0$; in other words, an element $\sum_{i=1}^{\infty} b_i(x)f_i$ of $\mathfrak{F}$ is mapped to the element $\sum_{i=1}^{\infty} x^i b_i(x)e_i - \left(\sum_{i=1}^{\infty} b_i(x)\right)e_0$. It is clear from this formula that the element $e_0 \in \mathfrak{E}$ does not belong to the image of g. Let $\mathfrak{P}$ denote the cokernel of the morphism g and p_i denote the images of the elements e_i in $\mathfrak{P}$. Then one has $p_0 = x^n p_n$ in $\mathfrak{P}$; in other words, the element p_0 belongs to the image of $\operatorname{Hom}_k(\mathcal{C}/\mathcal{U}, \mathfrak{P})$ under the contraaction map $\operatorname{Hom}_k(\mathcal{C}, \mathfrak{P}) \longrightarrow \mathfrak{P}$ for any finite-dimensional subcoalgebra $\mathcal{U} = (k[[x]]/x^n)^*$ of $\mathcal{C}$.

Now let $\mathfrak{E}'$ be the free $\mathcal{C}$-contramodule generated by the symbols e_1, e_2, $\ldots$, $\mathfrak{P}'$ denote the cokernel of the map $g'\colon \mathfrak{F} \longrightarrow \mathfrak{E}'$ sending f_i to $x_i e_i$, and p_i' denote the images of e_i' in $\mathfrak{P}'$. Then the result of the contramodule infinite summation $\sum_{n=1}^{\infty} x^n p_n'$ is nonzero in $\mathfrak{P}'$, even though every element $x^n p_n'$ is equal to zero. Therefore, *the contramodule summation operation cannot be understood as any kind of limit of finite partial sums*. Actually, the $\mathcal{C}$-contramodule $\mathfrak{P}'$ is just the direct sum of the contramodules $k[[x]]/x^n k[[x]]$ over $n = 1$, 2, $\ldots$ in the category of $\mathcal{C}$-contramodules. Notice that the element $\sum_{n=1}^{\infty} x^n p_n'$ also belongs to the image of $\operatorname{Hom}_k(\mathcal{C}/\mathcal{U}, \mathfrak{P}')$ in $\mathfrak{P}'$ for any finite-dimensional subcoalgebra $\mathcal{U} \subset \mathcal{C}$.

Remark. In the above notation, a $\mathcal{C}$-contramodule structure on a k-vector space $\mathfrak{P}$ is uniquely determined by the underlying structure of a module over the algebra of polynomials $k[x]$; the natural functor $\mathcal{C}$-contra $\longrightarrow k[x]$-mod is fully faithful. Indeed, for any p_0, p_1, $\ldots$ in $\mathfrak{P}$ the sequence $q_m = \sum_{n=0}^{\infty} x^n p_{m+n} \in \mathfrak{P}$, $m = 0$, 1, $\ldots$ is the unique solution of the system of equations $q_m = p_m + x q_{m+1}$. The image of this functor is a full abelian subcategory closed under kernels, cokernels, extensions, and infinite products; it consists of all $k[x]$-modules P such that $\operatorname{Ext}_{k[x]}^i(k[x, x^{-1}], P) = 0$ for $i = 0$, 1. It follows that if $\mathcal{D}$ is a coalgebra for which the dual algebra $\mathcal{D}^*$ is isomorphic to a quotient algebra of the algebra of formal power series $k[[x_1, \ldots, x_m]]$ in a finite number of (commuting) variables by a closed ideal, then the natural functor $\mathcal{D}$-contra $\longrightarrow k[x_1, \ldots, x_m]$-mod is fully faithful.

A.1.2 Now let us give an example of *finite-dimensional* (namely, two-dimensional) contramodule $\mathfrak{P}$ over a coalgebra $\mathcal{C}$ such that the intersection of the images of $\operatorname{Hom}_k(\mathcal{C}/\mathcal{U}, \mathfrak{P})$ in $\mathfrak{P}$ is nonzero. Notice that for any coalgebra $\mathcal{C}$ there are natural left $\mathcal{C}^*$-module structures on any left $\mathcal{C}$-comodule and any left $\mathcal{C}$-contramodule; that is there are natural faithful functors

$$\mathcal{C}\text{-comod} \longrightarrow \mathcal{C}^*\text{-mod} \quad \text{and} \quad \mathcal{C}\text{-contra} \longrightarrow \mathcal{C}^*\text{-mod}$$

(where $\mathcal{C}^*$ is considered as an abstract algebra without any topology). The functor $\mathcal{C}$-comod $\longrightarrow$ $\mathcal{C}^*$-mod is fully faithful, while the functor $\mathcal{C}$-contra $\longrightarrow$ $\mathcal{C}^*$-mod is fully faithful on finite-dimensional contramodules.

Let V be a vector space and $\mathcal{C}$ be the coalgebra such that the dual algebra $\mathcal{C}^*$ has the form $ki_2 \oplus i_2 V^* i_1 \oplus ki_1$, where i_1 and i_2 are idempotent elements with $i_1 i_2 = i_2 i_1 = 0$ and $i_1 + i_2 = 1$. Then left $\mathcal{C}^*$-modules are essentially pairs of k-vector spaces M_1, M_2 endowed with a map

$$V^* \otimes_k M_1 \longrightarrow M_2,$$

left $\mathcal{C}$-comodules are pairs of vector spaces $\mathcal{M}_1$, $\mathcal{M}_2$ endowed with a map

$$\mathcal{M}_1 \longrightarrow V \otimes_k \mathcal{M}_2,$$

and left $\mathcal{C}$-contramodules are pairs of vector spaces $\mathfrak{P}_1$, $\mathfrak{P}_2$ endowed with a map

$$\mathrm{Hom}_k(V, \mathfrak{P}_1) \longrightarrow \mathfrak{P}_2.$$

In particular, the functor $\mathcal{C}$-contra $\longrightarrow$ $\mathcal{C}^*$-mod is not surjective on morphisms of infinite-dimensional objects, while the functor $\mathcal{C}$-comod $\longrightarrow$ $\mathcal{C}^*$-mod is not surjective on the isomorphism classes of finite-dimensional objects. (Neither is in general the functor $\mathcal{C}$-contra $\longrightarrow$ $\mathcal{C}^*$-mod, as one can see in the example of an analogous coalgebra with three idempotent linear functions instead of two and three vector spaces instead of one; when k is a finite field and $\mathcal{C}$ is the countable direct sum of copies of the coalgebra k, there even exists a one-dimensional $\mathcal{C}^*$-module which comes from no $\mathcal{C}$-comodule or $\mathcal{C}$-contramodule.)

Let $\mathfrak{P}$ be the $\mathcal{C}$-contramodule with $\mathfrak{P}_1 = k = \mathfrak{P}_2$ corresponding to a linear function $V^* \longrightarrow k$ coming from no element of V. Then the intersection of the images of $\mathrm{Hom}_k(\mathcal{C}/\mathcal{U}, \mathfrak{P})$ in $\mathfrak{P}$ over all finite-dimensional subcoalgebras $\mathcal{U} \subset \mathcal{C}$ is equal to $\mathfrak{P}_2$.

More generally, for any coalgebra $\mathcal{C}$ any finite-dimensional left $\mathcal{C}$-comodule $\mathcal{M}$ has a natural left $\mathcal{C}$-contramodule structure given by the composition

$$\mathrm{Hom}_k(\mathcal{C}, \mathcal{M}) \simeq \mathcal{C}^* \otimes_k \mathcal{M} \longrightarrow \mathcal{C}^* \otimes_k \mathcal{C} \otimes_k \mathcal{M} \longrightarrow \mathcal{M}$$

of the map induced by the $\mathcal{C}$-coaction in $\mathcal{M}$ and the map induced by the pairing $\mathcal{C}^* \otimes_k \mathcal{C} \longrightarrow k$. The category of finite-dimensional left $\mathcal{C}$-comodules is isomorphic to a full subcategory of the category of finite-dimensional left $\mathcal{C}$-contramodules; a finite-dimensional $\mathcal{C}$-contramodule comes from a $\mathcal{C}$-comodule if and only if it comes from a contramodule over a finite-dimensional subcoalgebra of $\mathcal{C}$. We will see below that every *irreducible* $\mathcal{C}$-contramodule is a finite-dimensional contramodule over a finite-dimensional subcoalgebra of $\mathcal{C}$; it follows that the above functor provides a bijective correspondence between irreducible left $\mathcal{C}$-comodules and irreducible left $\mathcal{C}$-contramodules.

Comparing the cobar complex for comodules with the bar complex for contramodules, one discovers that for any finite-dimensional left $\mathcal{C}$-comodules $\mathcal{L}$ and $\mathcal{M}$ there is a natural isomorphism

$$\mathrm{Ext}^{\mathcal{C},i}(\mathcal{L}, \mathcal{M}) \simeq \mathrm{Ext}^i_{\mathcal{C}}(\mathcal{L}, \mathcal{M})^{**}.$$

In other words, the Ext spaces between finite-dimensional $\mathcal{C}$-comodules in the category of arbitrary $\mathcal{C}$-contramodules are the completions of the Ext spaces in the category of finite-dimensional $\mathcal{C}$-comodules with respect to the profinite-dimensional topology.

A.2 Nakayama's Lemma

The exposition below is based on the structure theory of coalgebras, see [82, Chapters VIII–IX] or [68, Chapter 5].

A coalgebra is called *cosimple* if it has no nontrivial proper subcoalgebras. A coalgebra $\mathcal{C}$ is called *cosemisimple* if it is a union of finite-dimensional coalgebras dual to semisimple k-algebras, or equivalently, if the abelian category of (left or right) $\mathcal{C}$-comodules is semisimple. Any cosemisimple coalgebra can be decomposed into an (infinite) direct sum of cosimple coalgebras in a unique way. For any coalgebra $\mathcal{C}$, let $\mathcal{C}^{\mathrm{ss}}$ denote its maximal cosemisimple subcoalgebra; it contains all other cosemisimple subcoalgebras of $\mathcal{C}$.

Lemma 1. *Let $\mathcal{C}$ be a coalgebra over a field k and $\mathfrak{P}$ be a nonzero left $\mathcal{C}$ contramodule. Then the image of the space $\mathrm{Hom}_k(\mathcal{C}/\mathcal{C}^{\mathrm{ss}}, \mathfrak{P})$ under the contraaction map $\mathrm{Hom}_k(\mathcal{C}, \mathfrak{P}) \longrightarrow \mathfrak{P}$ is not equal to $\mathfrak{P}$.*

Proof. Notice that the coalgebra without counit $\mathcal{D} = \mathcal{C}/\mathcal{C}^{\mathrm{ss}}$ is *conilpotent*, that is any element $d \in \mathcal{D}$ is annihilated by the iterated comultiplication map $\mathcal{D} \longrightarrow \mathcal{D}^{\otimes n}$ with a large enough n (dependent on d). We will show that for any contramodule $\mathfrak{P}$ over a conilpotent coalgebra $\mathcal{D}$ surjectivity of the contraaction map $\pi_{\mathfrak{P}} \colon \mathrm{Hom}_k(\mathcal{D}, \mathfrak{P}) \longrightarrow \mathfrak{P}$ implies vanishing of $\mathfrak{P}$. Indeed, assume that $\pi_{\mathfrak{P}}$ is surjective. Let p be an element of $\mathfrak{P}$; it is equal to $\pi_{\mathfrak{P}}(f_1)$ for a certain map $f_1 \colon \mathcal{D} \longrightarrow \mathfrak{P}$. Since the map $\pi_{\mathfrak{P}}$ is surjective, the map f_1 can be lifted to a certain map

$$\mathcal{D} \longrightarrow \mathrm{Hom}_k(\mathcal{D}, \mathfrak{P}),$$

which supplies a map $f_2 \colon \mathcal{D} \otimes_k \mathcal{D} \longrightarrow \mathfrak{P}$. So one constructs a sequence of maps

$$f_i \colon \mathcal{D}^{\otimes i} \longrightarrow \mathfrak{P}$$

such that $f_{i-1} = \pi_{\mathfrak{P},1}(f_i)$, where $\pi_{\mathfrak{P},1}$ signifies the application of $\pi_{\mathfrak{P}}$ at the first tensor factor of $\mathcal{D}^{\otimes i}$. Set

$$g_i = \mu_{\mathcal{D},2..i}(f_i) = f_i \circ \mu_{\mathcal{D},2..i}, \quad g_i \colon \mathcal{D} \otimes \mathcal{D} \longrightarrow \mathfrak{P}, \quad i \geqslant 2,$$

where $\mu_{\mathcal{D},2..i}\colon \mathcal{D} \otimes \mathcal{D} \longrightarrow \mathcal{D}^{\otimes i}$ denotes the tensor product of the identity map $\mathcal{D} \longrightarrow \mathcal{D}$ with the iterated comultiplication map $\mathcal{D} \longrightarrow \mathcal{D}^{\otimes i-1}$. We have

$$\pi_{\mathfrak{P},1}(g_i) = \mu_{\mathcal{D},1..i-1}(f_{i-1}) \quad \text{and} \quad \mu_{\mathcal{D}}(g_i) = \mu_{\mathcal{D},1..i}(f_i).$$

Notice that by conilpotency of the coalgebra $\mathcal{D}$ the series $\sum_{i=2}^{\infty} g_i$ converges in the sense of point-wise limit of functions $\mathcal{D} \otimes_k \mathcal{D} \longrightarrow \mathfrak{P}$, and even of functions $\mathcal{D} \longrightarrow \mathrm{Hom}_k(\mathcal{D}, \mathfrak{P})$. (As always, we presume the identification $\mathrm{Hom}_k(U \otimes_k V, W) = \mathrm{Hom}_k(V, \mathrm{Hom}_k(U, W))$ when we consider left contramodules.) Therefore,

$$\pi_{\mathfrak{P},1}\left(\sum_{i=2}^{\infty} g_i\right) = \sum_{i=2}^{\infty} \mu_{\mathcal{D},1..i-1}(f_{i-1})$$

and

$$\mu_{\mathcal{D}}\left(\sum_{i=2}^{\infty} g_i\right) = \sum_{i=2}^{\infty} \mu_{\mathcal{D},1..i}(f_i),$$

hence

$$\pi_{\mathfrak{P},1}\left(\sum_{i=2}^{\infty} g_i\right) - \mu_{\mathcal{D}}\left(\sum_{i=2}^{\infty} g_i\right) = f_1.$$

By the contraassociativity equation, it follows that $p = \pi_{\mathfrak{P}}(f_1) = 0$. $\square$

Lemma 2. *Let coalgebra $\mathcal{C}$ be the direct sum of a family of coalgebras $\mathcal{C}_\alpha$. Then any left contramodule $\mathfrak{P}$ over $\mathcal{C}$ is the product of a uniquely defined family of left contramodules $\mathfrak{P}_\alpha$ over $\mathcal{C}_\alpha$.*

Proof. Uniqueness and functoriality is clear, since the component $\mathfrak{P}_\alpha$ can be recovered as the image of the projector corresponding to the linear function on $\mathcal{C}$ that is equal to the counit on $\mathcal{C}_\alpha$ and vanishes on $\mathcal{C}_\beta$ for all $\beta \neq \alpha$. Existence is obvious for a free $\mathcal{C}$-contramodule. Now suppose that a $\mathcal{C}$-contramodule $\mathfrak{Q}$ is the product of $\mathcal{C}_\alpha$-contramodules $\mathfrak{Q}_\alpha$; let us show that any subcontramodule $\mathfrak{R} \subset \mathfrak{Q}$ is the product of its images $\mathfrak{R}_\alpha$ under the projections $\mathfrak{Q} \longrightarrow \mathfrak{Q}_\alpha$. Let r_α be a family of elements of $\mathfrak{R}$. Consider the linear map $f\colon \mathcal{C} \longrightarrow \mathfrak{R}$ whose restriction to $\mathcal{C}_\alpha$ is equal to the composition

$$\mathcal{C}_\alpha \longrightarrow k \longrightarrow \mathfrak{R}$$

of the counit of $\mathcal{C}_\alpha$ and the map sending $1 \in k$ to r_α. Set $r = \pi_{\mathfrak{R}}(f)$. Then the image of the element r under the projection $\mathfrak{R} \longrightarrow \mathfrak{R}_\alpha$ is equal to the image of r_α under this projection. Thus $\mathfrak{R}$ is identified with the product of $\mathfrak{R}_\alpha$. It remains to notice that any $\mathcal{C}$-contramodule is isomorphic to the quotient contramodule of a free contramodule by one of its subcontramodules. $\square$

Corollary. *For any coalgebra $\mathcal{C}$ and any nonzero contramodule $\mathfrak{P}$ over $\mathcal{C}$ there exists a finite-dimensional (and even cosimple) subcoalgebra $\mathfrak{U} \subset \mathcal{C}$ such that the image of $\mathrm{Hom}_k(\mathcal{C}/\mathfrak{U}, \mathfrak{P})$ under the contraaction map $\mathrm{Hom}_k(\mathcal{C}, \mathfrak{P}) \longrightarrow \mathfrak{P}$ is not equal to $\mathfrak{P}$.*

Proof. By Lemma 1, the image of the map $\operatorname{Hom}_k(\mathcal{C}/\mathcal{C}^{\mathrm{ss}}, \mathfrak{P}) \longrightarrow \mathfrak{P}$ is not equal to $\mathfrak{P}$. Denote this image by $\mathfrak{Q}$; it is a subcontramodule of $\mathfrak{P}$ and the quotient contramodule $\mathfrak{P}/\mathfrak{Q}$ is a contramodule over $\mathcal{C}^{\mathrm{ss}}$. By Lemma 2, there exists a cosimple subcoalgebra $\mathcal{C}_\alpha$ of $\mathcal{C}^{\mathrm{ss}}$ such that $\mathfrak{P}/\mathfrak{Q}$ has a nonzero quotient which is a contramodule over $\mathcal{C}_\alpha$. $\qquad\square$

Lemma 3. *Let $\mathcal{C}_0 \subset \mathcal{C}_1 \subset \mathcal{C}_2 \subset \cdots \subset \mathcal{C}$ be a coalgebra with an increasing sequence of subcoalgebras. For a $\mathcal{C}$-contramodule $\mathfrak{P}$, denote by $G^i\mathfrak{P}$ the image of the contraaction map $\operatorname{Hom}_k(\mathcal{C}/\mathcal{C}_i, \mathfrak{P}) \longrightarrow \mathfrak{P}$. Then for any $\mathcal{C}$-contramodule $\mathfrak{P}$ the natural map $\mathfrak{P} \longrightarrow \varprojlim_i \mathfrak{P}/G^i\mathfrak{P}$ is surjective.*

Proof. The assertion is obvious for a free $\mathcal{C}$-contramodule. Represent an arbitrary $\mathcal{C}$-contramodule $\mathfrak{P}$ as the quotient contramodule $\mathfrak{Q}/\mathfrak{K}$ of a free $\mathcal{C}$-contramodule $\mathfrak{Q}$. Since the maps $G^i\mathfrak{Q} \longrightarrow G^i\mathfrak{P}$ are surjective, there are short exact sequences

$$0 \longrightarrow \mathfrak{K}/\mathfrak{K}\cap G^i\mathfrak{Q} \longrightarrow \mathfrak{Q}/G^i\mathfrak{Q} \longrightarrow \mathfrak{P}/G^i\mathfrak{P} \longrightarrow 0.$$

Passing to the projective limits, we see that the map $\varprojlim_i \mathfrak{Q}/G^i\mathfrak{Q} \longrightarrow \varprojlim_i \mathfrak{P}/G^i\mathfrak{P}$ is surjective. $\qquad\square$

When $\mathcal{C} = \bigcup_i \mathcal{C}_i$, it follows, in particular, that $\mathfrak{K} \simeq \varprojlim_i \mathfrak{K}/G^i\mathfrak{K}$ for any $\mathcal{C}$-contramodule $\mathfrak{K}$ which is a subcontramodule of a projective $\mathcal{C}$-contramodule.

A.3 Contraflat contramodules

Lemma. *Let $\mathcal{C}$ be a coalgebra over a field k. Then a left $\mathcal{C}$-contramodule is contraflat if and only if it is projective.*

Proof. For any $\mathcal{C}$-contramodule $\mathfrak{Q}$ and any subcoalgebra $\mathcal{V} \subset \mathcal{C}$ denote by

$$^\mathcal{V}\mathfrak{Q} = \operatorname{coker}(\operatorname{Hom}_k(\mathcal{C}/\mathcal{V}, \mathfrak{Q}) \to \mathfrak{Q}) \simeq \operatorname{Cohom}_{\mathcal{C}}(\mathcal{V}, \mathfrak{Q})$$

the maximal quotient contramodule of $\mathfrak{Q}$ that is a contramodule over $\mathcal{V}$. The key step is to construct for any $\mathcal{C}^{\mathrm{ss}}$-contramodule $\mathfrak{R}$ a projective $\mathcal{C}$-contramodule $\mathfrak{Q}$ such that $^{\mathcal{C}^{\mathrm{ss}}}\mathfrak{Q} \simeq \mathfrak{R}$. By Lemma A.2.2, $\mathfrak{R}$ is a product of contramodules over cosimple components $\mathcal{C}_\alpha$ of $\mathcal{C}^{\mathrm{ss}}$. Any contramodule over $\mathcal{C}_\alpha$ is, in turn, a direct sum of copies of the unique irreducible $\mathcal{C}_\alpha$-contramodule. Hence it suffices to consider the case of an irreducible $\mathcal{C}_\alpha$-contramodule $\mathfrak{R}$.

Let e_α be an idempotent element of the algebra $\mathcal{C}_\alpha^*$ such that $\mathfrak{R}$ is isomorphic to $\mathcal{C}_\alpha^* e_\alpha$. Consider the idempotent linear function e_{ss} on $\mathcal{C}^{\mathrm{ss}}$ equal to e_α on $\mathcal{C}_\alpha$ and zero on $\mathcal{C}_\beta$ for all $\beta \neq \alpha$. It is well known that for any surjective map of rings $A \longrightarrow B$ whose kernel is a nil ideal in A any idempotent element of B can be lifted to an idempotent element of A. Using this fact for finite-dimensional algebras and Zorn's Lemma, one can show that any idempotent linear function on $\mathcal{C}^{\mathrm{ss}}$ can be extended to an idempotent linear function on $\mathcal{C}$. Let e be an idempotent linear function on $\mathcal{C}$ extending e_{ss}; set $\mathfrak{Q} = \mathcal{C}^* e$. Then one has

$$\mathcal{C}^{ss}\mathfrak{Q} \simeq (\mathcal{C}^{ss}\mathcal{C}^*)e \simeq \mathcal{C}^{ss*}e_{ss} \simeq \mathfrak{R}$$

as desired.

Now let $\mathfrak{P}$ be a contraflat left $\mathcal{C}$-contramodule. Consider a projective left $\mathcal{C}$-contramodule $\mathfrak{Q}$ such that $\mathcal{C}^{ss}\mathfrak{Q} \simeq \mathcal{C}^{ss}\mathfrak{P}$. Since $\mathfrak{Q}$ is projective, the map $\mathfrak{Q} \longrightarrow \mathcal{C}^{ss}\mathfrak{P}$ can be lifted to a $\mathcal{C}$-contramodule morphism

$$f \colon \mathfrak{Q} \longrightarrow \mathfrak{P}.$$

Since $\mathcal{C}^{ss}(\operatorname{coker} f) = \operatorname{coker}(\mathcal{C}^{ss}f) = 0$, it follows from Lemma A.2.1 that the morphism f is surjective; it remains to show that f is injective.

For any right comodule $\mathcal{N}$ over a subcoalgebra $\mathcal{U} \subset \mathcal{C}$ there is a natural isomorphism

$$\mathcal{N} \odot_{\mathcal{C}} \mathfrak{P} \simeq \mathcal{N} \odot_{\mathcal{U}} {}^{\mathcal{U}}\mathfrak{P},$$

hence the $\mathcal{U}$-contramodule ${}^{\mathcal{U}}\mathfrak{P}$ is contraflat. Now let $\mathcal{U}$ be a finite-dimensional subcoalgebra; then ${}^{\mathcal{U}}\mathfrak{P}$ is a flat left $\mathcal{U}^*$-module. Denote by K the kernel of the map ${}^{\mathcal{U}}f \colon {}^{\mathcal{U}}\mathfrak{Q} \longrightarrow {}^{\mathcal{U}}\mathfrak{P}$. For any right $\mathcal{U}^*$-module N we have a short exact sequence

$$0 \longrightarrow N \otimes_{\mathcal{U}^*} K \longrightarrow N \otimes_{\mathcal{U}^*} {}^{\mathcal{U}}\mathfrak{Q} \longrightarrow N \otimes_{\mathcal{U}^*} {}^{\mathcal{U}}\mathfrak{P} \longrightarrow 0.$$

Since for any cosimple subcoalgebra $\mathcal{U}_\alpha \subset \mathcal{U}$ the map $\mathcal{U}_\alpha^* \otimes_{\mathcal{U}^*} {}^{\mathcal{U}}f = {}^{\mathcal{U}_\alpha}f$ is an isomorphism, we can conclude that the module $\mathcal{U}_\alpha^* \otimes_{\mathcal{U}^*} K = {}^{\mathcal{U}_\alpha}K$ is zero. It follows that $K = 0$ and the map ${}^{\mathcal{U}}f$ is an isomorphism.

Finally, let $\mathfrak{K}$ be the kernel of the map $\mathfrak{Q} \longrightarrow \mathfrak{P}$. Since ${}^{\mathcal{U}}f$ is an isomorphism, the subcontramodule $\mathfrak{K} \subset \mathfrak{Q}$ is contained in the image of $\operatorname{Hom}_k(\mathcal{C}/\mathcal{U}, \mathfrak{Q})$ in $\mathfrak{Q}$ for any finite-dimensional subcoalgebra $\mathcal{U} \subset \mathcal{C}$. But the intersection of such images is zero, because the $\mathfrak{Q}$ is a projective $\mathcal{C}$-contramodule. $\qquad\square$

Remark. Much more generally, one can define left contramodules over an arbitrary complete and separated topological ring R where open right ideals form a base of neighborhoods of zero (cf. [8]). Namely, for any set X let $R[[X]]$ denote the set of all formal linear combinations of elements of X with coefficients converging to zero in R, i.e., the set of all formal sums $\sum_{x \in X} r_x x$, with $r_x \in R$ such that for any neighborhood of zero $U \subset R$ one has $r_x \in U$ for all but a finite number of elements $x \in X$. Then for any set X there is a natural map of "opening the parentheses" $R[[R[[X]]]] \longrightarrow R[[X]]$ assigning a formal linear combination to a formal linear combination of formal linear combinations; it is well defined in view of our condition on R. There is also a map $X \longrightarrow R[[X]]$ defined in terms of the unit element of R; taken together, these two natural maps make the functor $X \longmapsto R[[X]]$ a monad on the category of sets. Left contramodules over R are, by the definition, modules over this monad. One can see that the category of left R-contramodules is abelian and there is a forgetful functor from it to the category of R-modules; this functor is exact and preserves infinite products. For example, when $R = \mathbb{Z}_l$ is the topological ring of l-adic integers, the category of R-contramodules is isomorphic to the category of weakly l-complete abelian groups in the sense of Jannsen [54], i.e., abelian groups P such that $\mathrm{Ext}^i_{\mathbb{Z}}(\mathbb{Z}[l^{-1}], P) = 0$ for $i = 0$, 1. When R is a topological algebra over a field, the above definition of an R-contramodule is equivalent to the definition given in D.5.2. There is an obvious way to define the contratensor product of a discrete right R-module and a left R-contramodule. Now if T is a topological ring without unit satisfying the above condition, and T is pronilpotent, that is for any neighborhood of zero $U \subset T$ there exists n such that $T^n \subset U$, then any left T-contramodule P such that the contraaction map $T[[P]] \longrightarrow P$ is surjective vanishes. Besides, any left contramodule over the topological product R of a family of rings (with units) R_α satisfying the above condition is naturally the product of a family of left R_α-contramodules. Finally, let R be a topological ring satisfying the above condition and endowed with a decreasing filtration $R \supset G^1 R \supset G^2 R \supset \cdots$ by closed left ideals such that any neighborhood of zero in R contains $G^i R$ for large i. For any left R-contramodule P, denote by $G^i P$ the image of the contraaction map $G^i R[[P]] \longrightarrow P$; then the natural map $P \longrightarrow \varprojlim_i P/G^i P$ is surjective. The proofs of these assertions are analogous to those of Lemmas A.2.1–A.2.3. When open two-sided ideals form a base of neighborhoods of zero in R and the discrete quotient rings of R are right Artinian, a left R-contramodule P is projective if and only if for any open two-sided ideal $J \subset R$ the cokernel of the contraaction map $J[[P]] \longrightarrow P$ is a projective left R/J-module. The proof of this result is analogous to that of Lemma A.3. It follows that the class of projective left R-contramodules is closed under infinite products under these assumptions. For a profinite ring R (defined equivalently as a projective limit of finite rings endowed with the topology of projective limit or as a profinite abelian group endowed with a continuous associative multiplication with unit) one can even obtain the comodule-contramodule correspondence. Namely, the coderived category of discrete left R-modules is equivalent to the contraderived category of left R-contramodules; this equivalence is constructed in a way analogous to 0.2.6 with the role of a coalgebra $\mathcal{C}$ played by the discrete R-R-bimodule of continuous abelian group homomorphisms $R \longrightarrow \mathbb{Q}/\mathbb{Z}$. More generally, let R be a topological ring where open two-sided ideals form a base of neighborhoods of zero and the discrete quotient rings are right Artinian. A pseudo-compact right R-module [43, IV.3] is a topological right R-module whose open submodules form a base of neighborhoods of zero and discrete quotient modules have finite length. The category of pseudo-compact right R-modules is an abelian category with exact functors of infinite products and enough projectives; the projective pseudo-compact right R-modules are the direct summands of infinite products of copies of the pseudo-compact right R-module R. There is a natural anti-equivalence between the contraderived categories of pseudo-compact right R-modules and left R-contramodules provided by the derived functors of pseudo-compact module and contramodule homomorphisms into R.

B Comparison with Arkhipov's Ext$^{\infty/2+*}$ and Sevostyanov's Tor$_{\infty/2+*}$

Semi-infinite cohomology of associative algebras was introduced by S. Arkhipov [2, 3]; later A. Sevostyanov studied it in [79]. The constructions of derived functors SemiTor and SemiExt in this monograph are based on three key ideas which were not known in the 1990s, namely:

(i) the functors of semitensor product and semihomomorphisms;

(ii) the constructions of adjusted objects from Lemmas 1.3.3 and 3.3.3; and

(iii) the definitions of semiderived categories.

We have discussed already Sevostyanov's substitute for (i) in 10.4 and mentioned Arkhipov's substitute for (ii) in 10.3.5. Here we consider Arkhipov's substitute for (i) and suggest an Arkhipov and Sevostyanov-style substitute for (iii). Combining these together, we obtain comparison results relating our SemiExt to Arkhipov's Ext$^{\infty/2+*}$ and our SemiTor to Sevostyanov's Tor$_{\infty/2+*}$.

Throughout this appendix we will freely use the notation and remarks of 11.1.

B.1 Algebras R and $R^{\#}$

B.1.1 Let R be a graded associative algebra over a field k endowed with a pair of subalgebras K and $B \subset R$. Assume that all the components K_i are finite dimensional, $K_i = 0$ for i large negative, and $B_i = 0$ for i large positive. Set $\mathcal{C}_i = K^*_{-i}$ and $\mathcal{C} = \bigoplus_i \mathcal{C}_i$; this is the coalgebra graded dual to the algebra K. The coalgebra structure on $\mathcal{C}$ exists due to the conditions imposed on the grading of K. There is a natural pairing $\phi \colon \mathcal{C} \otimes_k K \longrightarrow k$ satisfying the conditions of 10.1.2.

Notice that a structure of graded (left or right) $\mathcal{C}$-comodule on a graded k-vector space M with $M_i = 0$ for $i \gg 0$ is the same as a structure of graded (left or right) K-module on M. Analogously, a structure of graded (left or right) $\mathcal{C}$-contramodule on a graded k-vector space P with $P_i = 0$ for $i \ll 0$ is the same as a structure of graded (left or right) K-module on P. Indeed, one has $\mathrm{Hom}_k^{\mathrm{gr}}(\mathcal{C}, P) \simeq K \otimes_k P$.

Furthermore, assume that the multiplication map $K \otimes_k B \longrightarrow R$ is an isomorphism of graded vector spaces. The algebra R is uniquely determined by the algebras K and B and the "permutation" map

$$B \otimes_k K \longrightarrow K \otimes_k B$$

obtained by restricting the multiplication map $R \otimes_k R \longrightarrow R \simeq K \otimes_k B$ to the subspace $B \otimes_k K \subset R \otimes_k R$. Transferring the tensor factors K to the other sides of this arrow, one obtains a map

$$\mathcal{C} \otimes_k B \longrightarrow \mathrm{Hom}_k^{\mathrm{gr}}(K, B), \qquad c \otimes b \longmapsto (k' \mapsto (\phi \otimes \mathrm{id}_B)(c \otimes bk')),$$

where the graded Hom space in the right-hand side is defined, as always, as direct sum of the spaces of homogeneous maps of various degrees. By the conditions imposed on the gradings of K and B, we have $B \otimes_k \mathcal{C} \simeq \mathrm{Hom}_k^{\mathrm{gr}}(K, B)$, so we get a homogeneous map

$$\psi \colon \mathcal{C} \otimes_k B \longrightarrow B \otimes_k \mathcal{C}.$$

One can check that the map ψ is a right entwining structure for the graded coalgebra $\mathcal{C}$ and the graded algebra B over k.

Conversely, if the map ψ corresponding to a "permutation" map $B \otimes_k K \longrightarrow K \otimes_k B$ satisfies the entwining structure equations, then the latter map can be extended to an associative algebra structure on $R = K \otimes_k B$ with subalgebras K and $B \subset R$. However, *not every homogeneous map* $\mathcal{C} \otimes_k B \longrightarrow B \otimes_k \mathcal{C}$ *comes from a homogeneous map* $B \otimes_k K \longrightarrow K \otimes_k B$.

In the described situation the constructions of 10.2 and 10.3 produce the same graded semialgebra

$$\mathcal{C} \otimes_K R = \mathbf{S} \simeq \mathcal{C} \otimes_k B.$$

The pairing $\phi \colon \mathcal{C} \otimes_k K \longrightarrow k$ is nondegenerate in $\mathcal{C}$, so the functor Δ_ϕ is fully faithful and in order to show that the construction of 10.2 works one only has to check that there exists a right $\mathcal{C}$-comodule structure on $\mathcal{C} \otimes_K R$ inducing the given right K-module structure. This is so because $\mathbf{S}_i = 0$ for $i \gg 0$ according to the conditions imposed on the gradings of K and B.

B.1.2 Now suppose that we are given two graded algebras R and $R^{\#}$ with the same two graded subalgebras K, $B \subset R$ and K, $B \subset R^{\#}$ such that the multiplication maps

$$K \otimes_k B \longrightarrow R \quad \text{and} \quad B \otimes_k K \longrightarrow R^{\#}$$

are isomorphisms of vector spaces. Assume that $\dim_k K_i < \infty$ for all i, $K_i = 0$ for $i \ll 0$, and $B_i = 0$ for $i \gg 0$. Furthermore, assume that the right entwining structure $\psi \colon \mathcal{C} \otimes_k B \longrightarrow B \otimes_k \mathcal{C}$ coming from the "permutation" map in R and the left entwining structure $\psi^{\#} \colon B \otimes_k \mathcal{C} \longrightarrow \mathcal{C} \otimes_k B$ coming from the "permutation" map in $R^{\#}$ are inverse to each other.

Then there are isomorphisms of graded semialgebras

$$\mathbf{S} = \mathcal{C} \otimes_K R \simeq \mathcal{C} \otimes_k B \simeq B \otimes_k \mathcal{C} \simeq R^{\#} \otimes_K \mathcal{C} = \mathbf{S}^{\#},$$

which allow one to describe left and right $\mathbf{S}$-semimodules and $\mathbf{S}$-semicontramodules in terms of left and right R-modules and $R^{\#}$-modules. In particular, $\mathbf{S}$ has a natural structure of graded $R^{\#}$-R-bimodule.

By the graded version of the result of 10.2.2, a structure of graded right
$\mathbf{S}$-semimodule on a graded k-vector space N with $N_i = 0$ for $i \gg 0$ is the same
as a structure of graded right R-module on N. A structure of graded left $\mathbf{S}$-semi-
module on a graded k-vector space M with $M_i = 0$ for $i \gg 0$ is the same as a
structure of graded left $R^\#$-module on M. A structure of graded left $\mathbf{S}$-semicon-
tramodule on a graded k-vector space P with $P_i = 0$ for $i \ll 0$ is the same as a
structure of graded left R-module on P. In other words, there are isomorphisms
of the corresponding categories of graded modules and homogeneous morphisms
between them.

Besides, for any graded right R-module N with $N_i = 0$ for $i \gg 0$ and any
graded left R-module P with $P_i = 0$ for $i \ll 0$ there is a natural isomorphism
$N \odot_{\mathbf{S}}^{\mathrm{gr}} P \simeq N \otimes_R P$. Indeed, one has $N \odot_{\mathcal{C}}^{\mathrm{gr}} P \simeq N \otimes_K P$ and $(N \square_{\mathcal{C}}^{\mathrm{gr}} \mathbf{S}) \odot_{\mathcal{C}}^{\mathrm{gr}} P \simeq$
$N \otimes_K R \otimes_K P$.

B.1.3 A graded K-module M with $M_i = 0$ for $i \gg 0$ is injective as a graded
$\mathcal{C}$-comodule if and only if it is injective as a graded K-module and if and only if it
is injective in the category of graded K-modules with the same restriction on the
grading. Analogously, a graded K-module P with $P_i = 0$ for $i \ll 0$ is projective as
a graded $\mathcal{C}$-contramodule if and only if it is projective as a graded K-module and
if and only if it is projective in the category of graded K-modules with the same
restriction on the grading.

By the graded version of Proposition 6.2.1(a), for any graded right R-mod-
ule N with $N_i = 0$ for $i \gg 0$ and any K-injective graded left $R^\#$-module M with
$M_i = 0$ for $i \gg 0$ there are natural isomorphisms

$$N \lozenge_{\mathbf{S}}^{\mathrm{gr}} M \simeq N \odot_{\mathbf{S}}^{\mathrm{gr}} \Psi_{\mathbf{S}}^{\mathrm{gr}}(M) \simeq N \odot_{\mathbf{S}}^{\mathrm{gr}} \mathrm{Hom}_{R^\#}^{\mathrm{gr}}(\mathbf{S}, M).$$

Analogously, for any K-injective graded right R-module N with $N_i = 0$ for $i \gg 0$
and any graded left $R^\#$-module M with $M_i = 0$ for $i \gg 0$ there is a natural
isomorphism

$$N \lozenge_{\mathbf{S}}^{\mathrm{gr}} M \simeq M \odot_{\mathbf{S}^{\mathrm{op}}}^{\mathrm{gr}} \mathrm{Hom}_{R^{\mathrm{op}}}^{\mathrm{gr}}(\mathbf{S}, N).$$

The contratensor products in the right-hand sides of these formulas *cannot* be in
general replaced by the tensor product over R and $R^\#$, as the graded $\mathbf{S}$-semicon-
tramodule $\Psi_{\mathbf{S}}^{\mathrm{gr}}(M)$ does not have zero components in large negative degrees. In
this situation the contratensor product is a certain quotient space of the tensor
product.

By the graded version of Proposition 6.2.3(a), for any K-injective graded left
$R^\#$-module M with $M_i = 0$ for $i \gg 0$ and any graded left R-module P with $P_i = 0$
for $i \ll 0$ there are natural isomorphisms

$$\mathrm{SemiHom}_{\mathbf{S}}^{\mathrm{gr}}(M, P) \simeq \mathrm{Hom}_{\mathrm{gr}}^{\mathbf{S}}(\Psi_{\mathbf{S}}^{\mathrm{gr}}(M), P) \simeq \mathrm{Hom}_{\mathrm{gr}}^{\mathbf{S}}(\mathrm{Hom}_{R^\#}^{\mathrm{gr}}(\mathbf{S}, M), P).$$

Here the homomorphisms of graded $\mathbf{S}$-semicontramodules again *cannot* be replaced by homomorphisms of graded left R-modules. The former homomorphism
spaces are certain subspaces of the latter ones.

By the graded version of Proposition 6.2.2(a), for any graded left $R^{\#}$-module M with $M_i = 0$ for $i \gg 0$ and any K-projective graded left R-module P with $P_i = 0$ for $i \ll 0$ there are natural isomorphisms

$$\mathrm{SemiHom}_{\mathbf{S}}^{\mathrm{gr}}(M, P) \simeq \mathrm{Hom}_{\mathbf{S}}^{\mathrm{gr}}(M, \Phi_{\mathbf{S}}^{\mathrm{gr}}(P)) \simeq \mathrm{Hom}_{R^{\#}}^{\mathrm{gr}}(M, \mathbf{S} \otimes_R^{\mathrm{gr}} P).$$

Here the homomorphisms of graded left $\mathbf{S}$-semimodules *can* be replaced by the homomorphisms of graded left $R^{\#}$-modules, since the functor Δ_ϕ is fully faithful, and consequently so is the functor $\Delta_{\phi,f}$.

All of these formulas except the last one have ungraded versions:

$$N \lozenge_{\mathbf{S}} M \simeq N \odot_{\mathbf{S}} \mathrm{Hom}_{R^{\#}}(\mathbf{S}, M), \qquad N \lozenge_{\mathbf{S}} M \simeq M \odot_{\mathbf{S}^{\mathrm{op}}} \mathrm{Hom}_{R^{\mathrm{op}}}(\mathbf{S}, N),$$

$$\mathrm{SemiHom}_{\mathbf{S}}(M, P) \simeq \mathrm{Hom}^{\mathbf{S}}(\mathrm{Hom}_{R^{\#}}(\mathbf{S}, M), P)$$

under the appropriate K-injectivity conditions.

B.2   Finite-dimensional case

When the subalgebra $K \subset R$ is finite-dimensional, the algebra $R^{\#}$ can be constructed without any reference to the grading or the complementary subalgebra B.

B.2.1 Let K be a finite-dimensional k-algebra and $\mathcal{C} = K^*$ be the coalgebra dual to K. Then the categories of left $\mathcal{C}$-comodules and left $\mathcal{C}$-contramodules are isomorphic to the category of left K-modules and the category of right $\mathcal{C}$-comodules is isomorphic to the category of right K-modules.

The adjoint functors $\Phi_{\mathcal{C}}$ and $\Psi_{\mathcal{C}}$ can be therefore considered as adjoint endofunctors on the category of left K-modules defined by the formulas

$$P \longmapsto \mathcal{C} \otimes_K P \quad \text{and} \quad M \longmapsto K \square_{\mathcal{C}} M \simeq \mathrm{Hom}_K(\mathcal{C}, M).$$

The restrictions of these functors define an equivalence between the categories of projective and injective left K-modules.

By Proposition 1.2.3(a-b), the mutually inverse equivalences $P \longmapsto \mathcal{C} \otimes_K P$ and $M \longmapsto K \square_{\mathcal{C}} M$ between the category of K-K-bicomodules that are projective as left K-modules and the category of K-K-bicomodules that are injective as left K-modules transform the functor of tensor product over K in the former category into the functor of cotensor product over $\mathcal{C}$ in the latter one. In other words, these two tensor categories are equivalent, and therefore there is a correspondence between ring objects in the former and the latter tensor category.

B.2.2 Let K be a finite-dimensional k-algebra and $K \longrightarrow R$ be a morphism of k-algebras. By the above argument, if R is a projective left K-module, then the tensor product $\mathbf{S} = \mathcal{C} \otimes_K R$ has a natural structure of semialgebra over $\mathcal{C}$. Furthermore, if $\mathbf{S}$ is an injective right K-module, then the cotensor product $R^{\#} = \mathbf{S} \square_{\mathcal{C}} K$ has a natural structure of k-algebra endowed with a k-algebra morphism

$K \longrightarrow R^{\#}$. In this case the semialgebra $\mathbf{S}$ can be also obtained as the tensor product $R^{\#} \otimes_K \mathcal{C}$.

By the result of 10.2.2, a structure of right $\mathbf{S}$-semimodule on a k-vector space N is the same as a structure of right R-module on N. A structure of left $\mathbf{S}$-semimodule on a k-vector space M is the same as a structure of left $R^{\#}$-module on M. A structure of left $\mathbf{S}$-semicontramodule on a k-vector space P is the same as a structure of left R-module on P. In other words, the corresponding categories are isomorphic. Besides, for any right R-module N and any left R-module P there is a natural isomorphism $N \circledcirc_{\mathbf{S}} P \simeq N \otimes_R P$ (see 10.2.3).

Remark. The case of a Frobenius algebra K is of special interest. In this case the k-algebra $R^{\#}$ is isomorphic to the k-algebra R, but the k-algebra morphisms $K \longrightarrow R$ and $K \longrightarrow R^{\#}$ differ by the Frobenius automorphism of K.

B.2.3 By Proposition 6.2.1(a), for any right R-module N and any K-injective left R-module M there are natural isomorphisms

$$N \lozenge_{\mathbf{S}} M \simeq N \circledcirc_{\mathbf{S}} \Psi_{\mathbf{S}}(M) \simeq N \otimes_R \mathrm{Hom}_{R^{\#}}(\mathbf{S}, M).$$

Analogously, for any K-injective right R-module N and any left $R^{\#}$-module M there is a natural isomorphism

$$N \lozenge_{\mathbf{S}} M \simeq \mathrm{Hom}_{R^{\mathrm{op}}}(\mathbf{S}, N) \otimes_{R^{\#}} M.$$

By Proposition 6.2.3(a), for any K-injective left $R^{\#}$-module M and any left R-module P there are natural isomorphisms

$$\mathrm{SemiHom}_{\mathbf{S}}(M, P) \simeq \mathrm{Hom}^{\mathbf{S}}(\Psi_{\mathbf{S}}(M), P) \simeq \mathrm{Hom}_R(\mathrm{Hom}_{R^{\#}}(\mathbf{S}, M), P).$$

By Proposition 6.2.2(a), for any left $R^{\#}$-module M and any K-projective left R-module P there are natural isomorphisms

$$\mathrm{SemiHom}_{\mathbf{S}}(M, P) \simeq \mathrm{Hom}_{\mathbf{S}}(M, \Phi_{\mathbf{S}}(P)) \simeq \mathrm{Hom}_{R^{\#}}(M, \mathbf{S} \otimes_R P).$$

All of these formulas have obvious graded versions.

B.3 Semijective complexes

Let $\mathbf{S}$ be a graded semialgebra over a graded coalgebra $\mathcal{C}$ over a field k. Suppose that $\mathbf{S}_i = 0 = \mathcal{C}_i$ for $i > 0$ and $\mathcal{C}_0 = k$. Assume also that $\mathbf{S}$ is an injective left and right graded $\mathcal{C}$-comodule. Let $\mathcal{C}\text{-comod}^{\downarrow}$ and $\text{comod}^{\downarrow}\text{-}\mathcal{C}$ denote the categories of $\mathcal{C}$-comodules graded by nonpositive integers, $\mathcal{C}\text{-contra}^{\uparrow}$ denote the category of left $\mathcal{C}$-comodules graded by nonnegative integers, $\mathbf{S}\text{-simod}^{\downarrow}$, $\text{simod}^{\downarrow}\text{-}\mathbf{S}$, and $\mathbf{S}\text{-sicntr}^{\uparrow}$ denote the categories of graded $\mathbf{S}$-semimodules and $\mathbf{S}$-semicontramodules with analogously bounded grading.

Any acyclic complex over $\mathcal{C}\text{-comod}^{\downarrow}$ is coacyclic with respect to $\mathcal{C}\text{-comod}^{\downarrow}$. Analogously, any acyclic complex over $\mathcal{C}\text{-contra}^{\uparrow}$ is contraacyclic with respect to

$\mathcal{C}$-contra$^\uparrow$. Indeed, let $\mathcal{K}^\bullet$ be an acyclic complex of nonpositively graded $\mathcal{C}$-comodules. As before, we denote by upper indices the homological grading and by lower indices the internal grading. Introduce an increasing filtration on $\mathcal{K}^\bullet$ by the complexes of graded subcomodules $F_n\mathcal{K}^j = \bigoplus_{i \geqslant -n} \mathcal{K}^j_i$. Then the acyclic complexes of trivial $\mathcal{C}$-comodules $F_n\mathcal{K}^\bullet/F_{n-1}\mathcal{K}^\bullet$ are clearly coacyclic.

So we have $\mathsf{D}^{\mathsf{si}}(\mathsf{S}\text{-simod}^\downarrow) = \mathsf{D}(\mathsf{S}\text{-simod}^\downarrow)$ and $\mathsf{D}^{\mathsf{si}}(\mathsf{S}\text{-sicntr}^\uparrow) = \mathsf{D}(\mathsf{S}\text{-sicntr}^\uparrow)$.

A complex $\mathcal{M}^\bullet$ over $\mathcal{C}$-comod$^\downarrow$ is called *injective* if the complex of homogeneous homomorphisms into $\mathcal{M}^\bullet$ from any acyclic complex over $\mathcal{C}$-comod$^\downarrow$ is acyclic. In this case the complex of homogeneous homomorphisms into $\mathcal{M}^\bullet$ from any acyclic complex over $\mathcal{C}$-comod^{gr} is also acyclic. Analogously, a complex $\mathfrak{P}^\bullet$ over $\mathcal{C}$-contra$^\uparrow$ is called *projective* if the complex of homogeneous homomorphisms from $\mathfrak{P}^\bullet$ into any acyclic complex over $\mathcal{C}$-contra$^\uparrow$ is acyclic. In this case the complex of homogeneous homomorphisms from $\mathfrak{P}^\bullet$ into any acyclic complex over $\mathcal{C}$-contra^{gr} is also acyclic.

By Lemma 5.4, any complex of injective objects in $\mathcal{C}$-comod$^\downarrow$ is injective and any complex of projective objects in $\mathcal{C}$-contra$^\uparrow$ is projective.

A complex $\mathcal{M}^\bullet$ over S-simod$^\downarrow$ is called *quite $\mathsf{S}/\mathcal{C}$-projective* if the complex of homogeneous homomorphisms from $\mathcal{M}^\bullet$ into any $\mathcal{C}$-contractible complex over S-simod$^\downarrow$ is acyclic. Equivalently, $\mathcal{M}^\bullet$ should belong to the minimal triangulated subcategory of $\mathsf{Hot}(\mathsf{S}\text{-simod}^\downarrow)$ containing the complexes of graded semimodules induced from complexes over $\mathcal{C}$-comod$^\downarrow$ and closed under infinite direct sums. Indeed, any quite $\mathsf{S}/\mathcal{C}$-projective complex of graded S-semimodules is homotopy equivalent to the total complex of its bar resolution.

Analogously, a complex $\mathfrak{P}^\bullet$ over S-sicntr$^\uparrow$ is called *quite $\mathsf{S}/\mathcal{C}$-injective* if the complex of homogeneous homomorphisms into $\mathfrak{P}^\bullet$ from any $\mathcal{C}$-contractible complex over S-sicntr$^\uparrow$ is acyclic. Equivalently, $\mathfrak{P}^\bullet$ should belong to the minimal triangulated subcategory of $\mathsf{Hot}(\mathsf{S}\text{-sicntr}^\uparrow)$ containing the complexes of graded semicontramodules coinduced from complexes over $\mathcal{C}$-contra$^\uparrow$ and closed under infinite products.

A complex $\mathcal{M}^\bullet$ over S-simod$^\downarrow$ is called *semijective* if it is $\mathcal{C}$-injective and quite $\mathsf{S}/\mathcal{C}$-projective. Analogously, a complex $\mathfrak{P}^\bullet$ over S-sicntr$^\uparrow$ is called *semijective* if it is $\mathcal{C}$-projective and quite $\mathsf{S}/\mathcal{C}$-injective. Clearly, any acyclic semijective complex of semimodules or semicontramodules is contractible.

By the graded version of Remark 6.4, any semiprojective complex of nonpositively graded $\mathcal{C}$-injective S-semimodules is semijective and any semiinjective complex of nonnegatively graded $\mathcal{C}$-projective S-semicontramodules is semijective. Hence the homotopy category of semijective complexes over S-simod$^\downarrow$ or S-sicntr$^\uparrow$ is equivalent to the derived category $\mathsf{D}(\mathsf{S}\text{-simod}^\downarrow)$ or $\mathsf{D}(\mathsf{S}\text{-sicntr}^\uparrow)$, any semijective complex is semiprojective or semiinjective, and one can use semijective complexes to compute the derived functors SemiTor$^\mathsf{S}$ and SemiExt$_\mathsf{S}$.

B.4 Explicit resolutions

Let us return to the situation of B.1.2, but make the stronger assumptions that $\dim_k K_i < \infty$ for all i, $K_i = 0$ for $i < 0$, $K_0 = k$, and $B_i = 0$ for $i > 0$. Set $\mathcal{C}_i = K^*_{-i}$ and $\mathcal{S} = \mathcal{C} \otimes_K R \simeq R^\# \otimes_K \mathcal{C} = \mathcal{S}^\#$.

B.4.1 For any complex of nonnegatively graded left R-modules $P^\bullet$ denote by $\mathbb{L}_2(P^\bullet)$ the total complex of the reduced relative bar complex

$$\cdots \longrightarrow R \otimes_B R/B \otimes_B R/B \otimes_B P^\bullet \longrightarrow R \otimes_B R/B \otimes_B P^\bullet \longrightarrow R \otimes_B P^\bullet.$$

It does not matter whether we construct this total complex by taking infinite direct sums or infinite products in the category of graded R-modules, as the two total complexes coincide. The complex $\mathbb{L}_2(P^\bullet)$ is a complex of K-projective nonnegatively graded left R-modules quasi-isomorphic to the complex $P^\bullet$.

For any complex of nonpositively graded left $R^\#$-modules $M^\bullet$ denote by $\mathbb{L}_3(M^\bullet)$ the total complex of the reduced relative bar complex

$$\cdots \longrightarrow R^\# \otimes_K R^\#/K \otimes_K R^\#/K \otimes_K M^\bullet$$
$$\longrightarrow R^\# \otimes_K R^\#/K \otimes_K M^\bullet \longrightarrow R^\# \otimes_K M^\bullet,$$

constructed by taking infinite direct sums along the diagonals. The complex $\mathbb{L}_3(M^\bullet)$ is a quite $\mathcal{S}/\mathcal{C}$-projective complex of nonpositively graded left $\mathcal{S}$-semimodules quasi-isomorphic to the complex $M^\bullet$.

By 4.8 and B.1.3, the complex

$$\operatorname{Hom}^{\mathrm{gr}}_{R^\#}(\mathbb{L}_3(M^\bullet),\ \mathcal{S} \otimes^{\mathrm{gr}}_R \mathbb{L}_2(P^\bullet))$$

represents the object $\operatorname{SemiExt}^{\mathrm{gr}}_{\mathcal{S}}(M^\bullet, P^\bullet)$ in $\mathsf{D}(k\text{-vect}^{\mathrm{gr}})$. We have reproduced Arkhipov's explicit complex [2, 3] computing $\operatorname{Ext}^{\infty/2+*}_R(M^\bullet, P^\bullet)$.

B.4.2 For any complex of nonpositively graded left $R^\#$-modules $M^\bullet$ denote by $\mathbb{R}_2(M^\bullet)$ the total complex of the reduced relative cobar complex

$$\operatorname{Hom}^{\mathrm{gr}}_B(R^\#, M^\bullet) \longrightarrow \operatorname{Hom}^{\mathrm{gr}}_B(R^\#/B \otimes_B R,\ M^\bullet)$$
$$\longrightarrow \operatorname{Hom}^{\mathrm{gr}}_B(R^\#/B \otimes_B R^\#/B \otimes_B R,\ M^\bullet) \longrightarrow \cdots$$

It does not matter whether we construct this total complex by taking infinite direct sums or infinite products in the category of graded $R^\#$-modules, as the two total complexes coincide. The complex $\mathbb{R}_2(M^\bullet)$ is a complex of K-injective nonpositively graded left $R^\#$-modules quasi-isomorphic to the complex $M^\bullet$.

For any complex of K-injective nonpositively graded left $R^\#$-modules $M^\bullet$ the complex $\mathbb{L}_3(M^\bullet)$ defined in B.4.1 is a semiprojective complex of $\mathcal{C}$-injective left $\mathcal{S}$-semimodules, since it is isomorphic to the total complex of the reduced bar complex

$$\cdots \longrightarrow \mathcal{S} \,\square_{\mathcal{C}}\, \mathcal{S}/\mathcal{C} \,\square_{\mathcal{C}}\, \mathcal{S}/\mathcal{C} \,\square_{\mathcal{C}}\, M^\bullet \longrightarrow \mathcal{S} \,\square_{\mathcal{C}}\, \mathcal{S}/\mathcal{C} \,\square_{\mathcal{C}}\, M^\bullet \longrightarrow \mathcal{S} \,\square_{\mathcal{C}}\, M^\bullet$$

and the left $\mathcal{C}$-comodule $\mathcal{S}/\mathcal{C}$ is injective in our assumptions.

Let $N^\bullet$ be a complex of nonpositively graded right R-modules and $M^\bullet$ be a complex of nonpositively graded left $R^\#$-modules. By 10.4, the complex

$$N^\bullet \otimes_B^{\mathcal{C}} \mathbb{L}_3\mathbb{R}_2(M^\bullet)$$

represents the object $\mathrm{SemiTor}^{\mathcal{S}}_{\mathrm{gr}}(N^\bullet, M^\bullet)$ in $\mathsf{D}(k\text{-vect}^{\mathrm{gr}})$. We have reproduced Sevostyanov's explicit complex [79] computing $\mathrm{Tor}^R_{\infty/2+*}(N^\bullet, M^\bullet)$.

B.5   Explicit resolutions for a finite-dimensional subalgebra

Let us consider the situation of an associative algebra R endowed with a pair of subalgebras K and $B \subset R$ such that the multiplication map $K \otimes_k B \longrightarrow R$ is an isomorphism of vector spaces and K is a finite-dimensional algebra. Let $\mathcal{C} = K^*$ be the coalgebra dual to K. Then the construction of B.1.1–B.1.2 is applicable, e.g., with R endowed by the trivial grading, and whenever the entwining map

$$\psi \colon B \otimes_k \mathcal{C} \longrightarrow \mathcal{C} \otimes_k B$$

turns out to be invertible, this construction produces an algebra $R^\#$ with subalgebras K and B and isomorphisms of semialgebras $\mathcal{S} = \mathcal{C} \otimes_K R \simeq \mathcal{C} \otimes_k B \simeq B \otimes_k \mathcal{C} \simeq R^\# \otimes_K \mathcal{C} = \mathcal{S}^\#$.

B.5.1   For any complex of right R-modules $N^\bullet$ denote by $\mathbb{R}_2(N^\bullet)$ the total complex of the reduced relative cobar complex

$$\mathrm{Hom}_{B^{\mathrm{op}}}(R, N^\bullet) \longrightarrow \mathrm{Hom}_{B^{\mathrm{op}}}(R \otimes_B R/B, \, N^\bullet)$$
$$\longrightarrow \mathrm{Hom}_{B^{\mathrm{op}}}(R \otimes_B R/B \otimes_B R/B, \, N^\bullet) \longrightarrow \cdots,$$

constructed by taking infinite direct sums along the diagonals. The complex $\mathbb{R}_2(N^\bullet)$ is a complex of K-injective right R-modules and the cone of the morphism $N^\bullet \longrightarrow \mathbb{R}_2(N^\bullet)$ is K-coacyclic (and even R-coacyclic). For any complex of left $R^\#$-modules $M^\bullet$ the complex $\mathbb{R}_2(M^\bullet)$ is constructed in an analogous way.

For any complex of left R-modules $P^\bullet$ denote by $\mathbb{L}_2(P^\bullet)$ the total complex of the reduced relative bar complex

$$\cdots \longrightarrow R \otimes_B R/B \otimes_B R/B \otimes_B P^\bullet \longrightarrow R \otimes_B R/B \otimes_B P^\bullet \longrightarrow R \otimes_B P^\bullet,$$

constructed by taking infinite products along the diagonals. The complex $\mathbb{L}_2(P^\bullet)$ is a complex of K-projective left R-modules and the cone of the morphism $\mathbb{L}_2(P^\bullet) \longrightarrow P^\bullet$ is K-contraacyclic (and even R-contraacyclic).

For any complex of right R-modules $N^\bullet$ denote by $\mathbb{L}_3(N^\bullet)$ the total complex of the reduced relative bar complex

$$\cdots \longrightarrow N^\bullet \otimes_K R/K \otimes_K R/K \otimes_K R \longrightarrow N^\bullet \otimes_K R/K \otimes_K R \longrightarrow N^\bullet \otimes_K R,$$

constructed by taking infinite direct sums along the diagonals. The complex $\mathbb{L}_3(N^\bullet)$ is a quite $\mathcal{S}/\mathcal{C}$-projective complex of right $\mathcal{S}$-semimodules and the cone of the morphism $\mathbb{L}_3(N^\bullet) \longrightarrow N^\bullet$ is $\mathcal{C}$-contractible. Whenever $N^\bullet$ is a complex of $\mathcal{C}$-injective right $\mathcal{S}$-semimodules, $\mathbb{L}_3(N^\bullet)$ is a semiprojective complex of $\mathcal{C}$-injective right $\mathcal{S}$-semimodules, as it was explained in B.4.2. For a complex of left $R^\#$-modules $M^\bullet$ the complex $\mathbb{L}_3(M^\bullet)$ is constructed in an analogous way.

For any complex of left R-modules $P^\bullet$ denote by $\mathbb{R}_3(P^\bullet)$ the total complex of the reduced relative cobar complex

$$\mathrm{Hom}_K(R, P^\bullet) \longrightarrow \mathrm{Hom}_K(R/K \otimes_K R, \, P^\bullet)$$
$$\longrightarrow \mathrm{Hom}_K(R/K \otimes_K R/K \otimes_K R, \, M^\bullet) \longrightarrow \cdots,$$

constructed by taking infinite products along the diagonals. The complex $\mathbb{R}_3(P^\bullet)$ is a quite $\mathcal{S}/\mathcal{C}$-injective complex of left $\mathcal{S}$-semicontramodules and the cone of the morphism $P^\bullet \longrightarrow \mathbb{R}_3(P^\bullet)$ is $\mathcal{C}$-contractible. Whenever $P^\bullet$ is a complex of $\mathcal{C}$-projective left $\mathcal{S}$-semicontramodules, $\mathbb{R}_3(P^\bullet)$ is a semiinjective complex of $\mathcal{C}$-projective right $\mathcal{S}$-semimodules, since it is isomorphic to the total complex of the reduced cobar complex

$$\mathrm{Cohom}_\mathcal{C}(\mathcal{S}, P^\bullet) \longrightarrow \mathrm{Cohom}_\mathcal{C}(\mathcal{S}/\mathcal{C} \,\square_\mathcal{C}\, \mathcal{S}, \, P^\bullet)$$
$$\longrightarrow \mathrm{Cohom}_\mathcal{C}(\mathcal{S}/\mathcal{C} \,\square_\mathcal{C}\, \mathcal{S}/\mathcal{C} \,\square_\mathcal{C}\, \mathcal{S}, \, P^\bullet) \longrightarrow \cdots$$

and the right $\mathcal{C}$-comodule $\mathcal{S}/\mathcal{C}$ is injective in our assumptions.

B.5.2 One can use these resolutions in various ways to compute the derived functors $\mathrm{SemiTor}^\mathcal{S}$, $\mathrm{SemiExt}_\mathcal{S}$, $\Psi_\mathcal{S}$, $\Phi_\mathcal{S}$, $\mathrm{Ext}_\mathcal{S}$, $\mathrm{Ext}^\mathcal{S}$, and $\mathrm{CtrTor}^\mathcal{S}$.

Specifically, for any complex of right R-modules $N^\bullet$ and any complex of left $R^\#$-modules $M^\bullet$ the object $\mathrm{SemiTor}^\mathcal{S}(N^\bullet, M^\bullet)$ in $\mathsf{D}(k\text{-vect})$ is represented by either of the four complexes

$$N^\bullet \otimes_R \mathrm{Hom}_{R^\#}(\mathcal{S}, \mathbb{L}_3\mathbb{R}_2(M^\bullet)), \quad \mathbb{L}_3(N^\bullet) \otimes_R \mathrm{Hom}_{R^\#}(\mathcal{S}, \mathbb{R}_2(M^\bullet)),$$
$$\mathrm{Hom}_{R^{\mathrm{op}}}(\mathcal{S}, \mathbb{L}_3\mathbb{R}_2(N^\bullet)) \otimes_{R^\#} M^\bullet, \quad \mathrm{Hom}_{R^{\mathrm{op}}}(\mathcal{S}, \mathbb{R}_2(N^\bullet)) \otimes_{R^\#} \mathbb{L}_3(M^\bullet)$$

according to the formulas of B.2.3 and the results of 2.8. For any complex of left $R^\#$-modules $M^\bullet$ and any complex of left R-modules $P^\bullet$ the object $\mathrm{SemiExt}_\mathcal{S}(M^\bullet, P^\bullet)$ in $\mathsf{D}(k\text{-vect})$ is represented by any of the four complexes

$$\mathrm{Hom}_R(\mathrm{Hom}_{R^\#}(\mathcal{S}, \mathbb{L}_3\mathbb{R}_2(M^\bullet)), P^\bullet), \quad \mathrm{Hom}_R(\mathrm{Hom}_{R^\#}(\mathcal{S}, \mathbb{R}_2(M^\bullet)), \mathbb{R}_3(P^\bullet)),$$
$$\mathrm{Hom}_{R^\#}(M^\bullet, \, \mathcal{S} \otimes_R \mathbb{R}_3\mathbb{L}_2(P^\bullet)), \quad \mathrm{Hom}_{R^\#}(\mathbb{L}_3(M^\bullet), \, \mathcal{S} \otimes_R \mathbb{L}_2(P^\bullet))$$

according to the formulas of B.2.3 and the results of 4.8.

One can also use the constructions of 10.4 instead of the formulas B.2.3.

For any complex of left $R^\#$-modules $M^\bullet$ the object $\Psi_\mathcal{S}(M^\bullet)$ in $\mathsf{D}^{\mathrm{si}}(\mathcal{S}\text{-sicntr})$ is represented by the complex of left R-modules

$$\mathrm{Hom}_{R^\#}(\mathcal{S}, \mathbb{R}_2(M^\bullet)).$$

For any complex of left R-modules $P^\bullet$ the object $\Phi_{\mathbf{S}}(P^\bullet)$ in $\mathsf{D}^{\mathsf{si}}(\mathbf{S}\text{-simod})$ is represented by the complex of left $R^{\#}$-modules

$$\mathbf{S} \otimes_R \mathbb{L}_2(P^\bullet).$$

For any complexes of left $R^{\#}$-modules $L^\bullet$ and $M^\bullet$ the object $\mathrm{Ext}_{\mathbf{S}}(L^\bullet, M^\bullet)$ in $\mathsf{D}(k\text{-vect})$ is represented by the complex

$$\mathrm{Hom}_{R^{\#}}(\mathbb{L}_3(L^\bullet), \mathbb{R}_2(M^\bullet)).$$

For any complexes of left R-modules $P^\bullet$ and $Q^\bullet$ the object $\mathrm{Ext}^{\mathbf{S}}(P^\bullet, Q^\bullet)$ in $\mathsf{D}(k\text{-vect})$ is represented by the complex

$$\mathrm{Hom}_R(\mathbb{L}_2(P^\bullet), \mathbb{R}_3(Q^\bullet)).$$

For any complex of right R-modules $N^\bullet$ and any complex of left R-modules $P^\bullet$ the object $\mathrm{CtrTor}^{\mathbf{S}}(N^\bullet, P^\bullet)$ in $\mathsf{D}(k\text{-vect})$ is represented by the complex

$$\mathbb{L}_3(N^\bullet) \otimes_R \mathbb{L}_2(P^\bullet).$$

These assertions follow from the results of 6.5.

In the situation of B.2.2 (with no complementary subalgebra B) one may have to use the constructions of resolutions $\mathbb{R}_2(N)$, $\mathbb{R}_2(M)$, and $\mathbb{L}_2(P)$ from the proofs of Theorems 2.6 and 4.6 instead of the constructions of B.5.1. The alternative is simply to replace "B" with "k" in the formulas defining the resolutions $\mathbb{R}_2$ and $\mathbb{L}_2$ in B.5.1.

C Semialgebras Associated to Harish-Chandra Pairs

by Leonid Positselski and Dmitriy Rumynin

Recall that in 10.2 we described the categories of right semimodules and left semi-contramodules over a semialgebra of the form $\mathcal{S} = \mathcal{C} \otimes_K R$, but no satisfactory description of the category of left semimodules over $\mathcal{S}$ was found. Here we consider the situation when $\mathcal{C}$ and K are Hopf algebras over a field k, and, under certain assumptions, construct a Morita equivalence between the semialgebras $\mathcal{C} \otimes_K R$ and $R \otimes_K \mathcal{C}$. So left semimodules over $\mathcal{C} \otimes_K R$ can be described. This includes the case of an algebraic Harish-Chandra pair $(\mathfrak{g}, H)$ with a smooth affine algebraic group H.

C.1 Two semialgebras

C.1.1 Let K and $\mathcal{C}$ be two Hopf algebras over a field k with invertible antipodes s. Using Sweedler's notation [82], we will denote the multiplications in K and $\mathcal{C}$ by $x \otimes y \longmapsto xy$ and the comultiplications by $x \longmapsto x_{(1)} \otimes x_{(2)}$; the units will be denoted by e and the counits by ε, so that one has $s(x_{(1)})x_{(2)} = \varepsilon(x)e = x_{(1)}s(x_{(2)})$ for $x \in K$ or $\mathcal{C}$. Let $\langle \, , \, \rangle \colon K \otimes_k \mathcal{C} \longrightarrow k$ be a pairing between K and $\mathcal{C}$ such that

$$\langle xy, c \rangle = \langle x, c_{(1)} \rangle \langle y, c_{(2)} \rangle \quad \text{and} \quad \langle x, cd \rangle = \langle x_{(1)}, c \rangle \langle x_{(2)}, d \rangle$$

for x, $y \in K$, c, $d \in \mathcal{C}$; besides, one should have $\langle x, e \rangle = \varepsilon(x)$ and $\langle e, c \rangle = \varepsilon(c)$. Then it follows that the pairing $\langle \, , \, \rangle$ is also compatible with the antipodes, $\langle s(x), c \rangle = \langle x, s(c) \rangle$.

Finally, suppose that we are given an "adjoint" right coaction of $\mathcal{C}$ in K denoted by $x \mapsto x_{[0]} \otimes x_{[1]}$, satisfying the equations

$$x_{(1)} y s(x_{(2)}) = \langle x, y_{[1]} \rangle y_{[0]}$$

and $(xy)_{[0]} \otimes (xy)_{[1]} = x_{[0]} y_{[0]} \otimes x_{[1]} y_{[1]}$; besides, assume that $e_{[0]} \otimes e_{[1]} = e \otimes e$. This coaction should also satisfy the equations of compatibility with the squares of antipodes $(s^2 x)_{[0]} \otimes (s^2 x)_{[1]} = s^2(x_{[0]}) \otimes s^{-2}(x_{[1]})$ and compatibility with the pairing $\langle s^{-1}(x_{[0]}), c_{(2)} \rangle s(c_{(1)}) c_{(3)} x_{[1]} = \langle s^{-1}(x), c \rangle e$. When the pairing $\langle \, , \, \rangle$ is non-degenerate in $\mathcal{C}$, the latter four equations follow from the first one and the previous assumptions.

Indeed, one has

$$\langle y, s^2(x)_{[1]}\rangle\langle s^2 x\rangle_{[0]} = y_{(1)}s^2(x)s(y_{(2)}) = s^2((s^{-2}y)_{(1)}xs((s^{-2}y)_{(2)}))$$
$$= \langle s^{-2}(y), x_{[1]}\rangle s^2(x_{[0]}) = \langle y, s^{-2}(x_{[1]})\rangle s^2(x_{[0]})$$

and

$$\langle s^{-1}(x_{[0]}), c_{(2)}\rangle\langle y, s(c_{(1)})c_{(3)}x_{[1]}\rangle$$
$$= \langle s^{-1}(x_{[0]}), c_{(2)}\rangle\langle y_{(3)}, x_{[1]}\rangle\langle y_{(1)}, s(c_{(1)})\rangle\langle y_{(2)}, c_{(3)}\rangle$$
$$= \langle s^{-1}(y_{(3)}xs(y_{(4)})), c_{(2)}\rangle\langle s(y_{(1)}), c_{(1)}\rangle\langle y_{(2)}, c_{(3)}\rangle$$
$$= \langle s(y_{(1)})y_{(4)}s^{-1}(x)s^{-1}(y_{(3)})y_{(2)}, c\rangle = \langle s^{-1}(x), c\rangle\varepsilon(y);$$

analogously for the second and the third equations.

C.1.2 Let R be an associative algebra over k endowed with a morphism of algebras $f\colon K \longrightarrow R$ and a right coaction of the coalgebra $\mathcal{C}$, which we will denote by $u \longmapsto u_{[0]} \otimes u_{[1]}$, $u \in R$. Assume that f is a morphism of right $\mathcal{C}$-comodules and the right coaction of $\mathcal{C}$ in R satisfies the equations

$$f(x)uf(s(x)) = \langle x, u_{[1]}\rangle u_{[0]}$$

and $(uv)_{[0]} \otimes (uv)_{[1]} = u_{[0]}v_{[0]} \otimes u_{[1]}v_{[1]}$ for u, $v \in R$, $x \in K$.
Define a pairing $\phi_r\colon \mathcal{C} \otimes_k K \longrightarrow k$ by the formula

$$\phi_r(c, x) = \langle s^{-1}(x), c\rangle.$$

The pairing ϕ_r satisfies the conditions of 10.1.2; in particular, it induces a right action of K in $\mathcal{C}$ given by the formula

$$c \leftarrow x = \langle s^{-1}(x), c_{(2)}\rangle c_{(1)}.$$

Assume that the morphism of k-algebras f makes R a flat left K-module. We will now apply the construction of 10.2.1 in order to obtain a semialgebra structure on the tensor product $\mathbf{S}^r = \mathcal{C} \otimes_K R$.
Define a right $\mathcal{C}$-comodule structure on $\mathcal{C} \otimes_K R$ by the formula

$$c \otimes_K u \longmapsto c_{(1)} \otimes_K u_{[0]} \otimes c_{(2)}u_{[1]}.$$

First let us check that this coaction is well defined.
We have $(c \leftarrow x) \otimes u = \langle s^{-1}x, c_{(2)}\rangle c_{(1)} \otimes u \longmapsto c_{(1)} \otimes_K u_{[0]} \otimes \langle s^{(-1)}x, c_{(3)}\rangle c_{(2)}u_{[1]}$ and $c \otimes f(x)u \longmapsto c_{(1)} \otimes_K f(x_{[0]})u_{[0]} \otimes c_{(2)}x_{[1]}u_{[1]} = (c_{(1)} \leftarrow x_{[0]}) \otimes_K u_{[0]} \otimes c_{(2)}x_{[1]}u_{[1]} = c_{(1)} \otimes_K u_{[0]} \otimes \langle c_{(2)}, s^{-1}(x_{[0]})\rangle c_{(3)}x_{[1]}u_{[1]}$; now $\langle s^{-1}(x), d_{(2)}\rangle d_{(1)} = \langle s^{-1}(x_{[0]}), d_{(3)}\rangle d_{(1)}s(d_{(2)})d_{(4)}x_{[1]} = \langle s^{-1}(x_{[0]}), d_{(1)}\rangle d_{(2)}x_{[1]}$ for $d \in \mathcal{C}$, $x \in K$. Furthermore, this right $\mathcal{C}$-comodule structure agrees with the right K-module structure on $\mathcal{C} \otimes_K R$, since $\langle s^{-1}(x), c_{(2)}u_{[1]}\rangle c_{(1)} \otimes_K u_{[0]} = \langle s^{-1}(x_{(2)}), c_{(2)}\rangle c_{(1)} \otimes_K \langle s^{-1}(x_{(1)}), u_{[1]}\rangle u_{[0]} = (c \leftarrow x_{(3)}) \otimes_K f(s^{-1}(x_{(2)}))uf(x_{(1)}) = c \otimes_K uf(x)$. It is easy to see that this right coaction of $\mathcal{C}$ commutes with the left coaction of $\mathcal{C}$ and that

the semiunit map $\mathcal{C} \longrightarrow \mathcal{C} \otimes_K R$ is a morphism of right $\mathcal{C}$-comodules. Finally, to show that the semimultiplication map $(\mathcal{C} \otimes_K R) \,\square_{\mathcal{C}}\, (\mathcal{C} \otimes_K R) \longrightarrow \mathcal{C} \otimes_K R$ is a morphism of right $\mathcal{C}$-comodules, one defines a right $\mathcal{C}$-comodule structure on $\mathcal{C} \otimes_K R \otimes_K R$ by the formula $c \otimes_K u \otimes_K v \longmapsto c_{(1)} \otimes_K u_{[0]} \otimes_K v_{[0]} \otimes c_{(2)} u_{[1]} v_{[1]}$ and checks that both the isomorphism $\mathcal{C} \otimes_K R \otimes_K R \simeq (\mathcal{C} \otimes_K R) \,\square_{\mathcal{C}}\, (\mathcal{C} \otimes_K R)$ and the map $\mathcal{C} \otimes_K R \otimes_K R \longrightarrow \mathcal{C} \otimes_K R$ are morphisms of right $\mathcal{C}$-comodules.

C.1.3 Define a pairing $\phi_l \colon K \otimes_k \mathcal{C} \longrightarrow k$ by the formula

$$\phi_l(x, c) = \langle s(x), c \rangle.$$

The pairing ϕ_l induces a left action of K in $\mathcal{C}$ given by the formula

$$x \rightarrow c = \langle s(x), c_{(1)} \rangle c_{(2)}.$$

Assume that the morphism of k-algebras f makes R a flat right K-module. We will apply the opposite version of the construction of 10.2.1 in order to obtain a semialgebra structure on $\mathbf{S}^l = R \otimes_K \mathcal{C}$.

Define a left $\mathcal{C}$-comodule structure on $R \otimes_K \mathcal{C}$ by the formula

$$u \otimes_K c \longmapsto c_{(1)} s^{-1}(u_{[1]}) \otimes u_{[0]} \otimes_K c_{(2)}.$$

This coaction is well defined, since one has

$$u \otimes (x \rightarrow c) = u \otimes \langle s(x), c_{(1)} \rangle c_{(2)} \longmapsto \langle s(x), c_{(1)} \rangle c_{(2)} s^{-1}(u_{[1]}) \otimes u_{[0]} \otimes_K c_{(3)}$$

and

$$uf(x) \otimes c \longmapsto c_{(1)} s^{-1}(u_{[1]} x_{[1]}) \otimes u_{[0]} f(x_{[0]}) \otimes_K c_{(2)}$$
$$= c_{(1)} s^{-1}(x_{[1]}) s^{-1}(u_{[1]}) \otimes u_{[0]} \otimes_K (x_{[0]} \rightarrow c_{(2)})$$
$$= \langle s(x_{[0]}), c_{(2)} \rangle c_{(1)} s^{-1}(x_{[1]}) s^{-1}(u_{[1]}) \otimes u_{[0]} \otimes_K c_{(3)};$$

now the identity $\langle s(x), c_{(1)} \rangle c_{(2)} = \langle s(x_{[0]}), c_{(2)} \rangle c_{(1)} s^{-1}(x_{[1]})$ follows from the identities $\langle s^{-1}(x), d_{(2)} \rangle d_{(1)} = \langle s^{-1}(x_{[0]}), d_{(1)} \rangle d_{(2)} x_{[1]}$ and $(s^2 x)_{[0]} \otimes (s^2 x)_{[1]} = s^2(x_{[0]}) \otimes s^{-2}(x_{[1]})$. This left $\mathcal{C}$-comodule structure agrees with the left K-module structure on $R \otimes_K \mathcal{C}$, since $\langle s(x), c_{(1)} s^{-1}(u_{[1]}) \rangle \rangle u_{[0]} \otimes_K c_{(2)} = \langle s(x_{(1)}), s^{-1}(u_{[1]}) \rangle \otimes_K \langle s(x_{(2)}), c_{(1)} \rangle c_{(2)} = \langle x_{(1)}, u_{[1]} \rangle u_{[0]} \otimes_K (x_{(2)} \rightarrow c) = x_{(1)} u s(x_{(2)}) x_{(3)} \otimes_K c = xu \otimes_K c$. The rest is analogous to C.1.2; the left $\mathcal{C}$-comodule structure on $R \otimes_K R \otimes_K \mathcal{C}$ is defined by the formula $u \otimes_K v \otimes_K c \longmapsto c_{(1)} s^{-1}(v_{[1]}) s^{-1}(u_{[1]}) \otimes u_{[0]} \otimes_K v_{[0]} \otimes_K c_{(2)}$.

C.1.4 According to 10.2.2, the category of right $\mathbf{S}^r$-semimodules is isomorphic to the category of k-vector spaces $\mathbf{N}$ endowed with right $\mathcal{C}$-comodule and right R-module structures such that

$$\langle s^{-1}(x), n_{(1)} \rangle n_{(0)} = nf(x) \quad \text{and} \quad (nr)_{(0)} \otimes (nr)_{(1)} = n_{(0)} r_{[0]} \otimes n_{(1)} r_{[1]}$$

for $n \in \mathbf{N}$, $x \in K$, $r \in R$, where $n \longmapsto n_{(0)} \otimes n_{(1)}$ denotes the right $\mathcal{C}$-coaction map and $n \otimes r \longmapsto nr$ denotes the right R-action map.

Assuming that R is a projective left K-module, the category of left $\boldsymbol{S}^r$-semi-contramodules is isomorphic to the category of k-vector spaces $\mathfrak{P}$ endowed with left $\mathcal{C}$-contramodule and left R-module structures such that

$$\pi_{\mathfrak{P}}(c \mapsto \langle s^{-1}(x), c \rangle p) = f(x)p$$

and

$$\pi_{\mathfrak{P}}(c \mapsto r_{[0]}g(cr_{[1]})) = r\pi_{\mathfrak{P}}(g)$$

for $p \in \mathfrak{P}$, $x \in K$, $c \in \mathcal{C}$, $r \in R$, $g \in \mathrm{Hom}_k(\mathcal{C}, \mathfrak{P})$, where $\pi_{\mathfrak{P}}$ denotes the contraaction map and $r \otimes p \longrightarrow rp$ denotes the left action map.

The category of left $\boldsymbol{S}^l$-semimodules is isomorphic to the category of k-vector spaces $\mathcal{M}$ endowed with left $\mathcal{C}$-comodule and left R-module structures such that

$$\langle s(x), m_{(-1)} \rangle m_{(0)} = f(x)m$$

and

$$(rm)_{(-1)} \otimes (rm)_{(0)} = m_{(-1)}s^{-1}(r_{[1]}) \otimes r_{[0]}m_{[0]}$$

for $m \in \mathcal{M}$, $x \in K$, $r \in R$, where $m \longmapsto m_{(-1)} \otimes m_{(0)}$ denotes the left $\mathcal{C}$-coaction map and $r \otimes m \longmapsto rm$ denotes the left R-action map.

C.2   Morita equivalence

C.2.1   Let $\mathcal{C}$ be a k-vector space endowed with a $\mathcal{C}$-$\mathcal{C}$-bicomodule structure and a right $\mathcal{C}$-module structure satisfying the equation

$$(jc)_{(-1)} \otimes (jc)_{(0)} \otimes (jc)_{(1)} = j_{(-1)}c_{(1)} \otimes j_{(0)}c_{(2)} \otimes j_{(1)}c_{(3)}$$

for $j \in \mathcal{E}$, $c \in \mathcal{C}$, where $j \longmapsto j_{(-1)} \otimes j_{(0)} \otimes j_{(1)}$ denotes the bicoaction map and $j \otimes c \longmapsto jc$ denotes the right action map.

In particular, $\mathcal{E}$ is a right Hopf module (see [82, 4.1] or [68, 1.9]) over $\mathcal{C}$, hence $\mathcal{E}$ is isomorphic to the tensor product $E \otimes_k \mathcal{C}$ as a right $\mathcal{C}$-comodule and a right $\mathcal{C}$-module, where E is a k-vector space which can be defined as the subspace in $\mathcal{E}$ consisting of all $i \in \mathcal{E}$ such that $i_{(0)} \otimes i_{(1)} = i \otimes e$. One can see that E is a left $\mathcal{C}$-sub-comodule in $\mathcal{E}$, so $\mathcal{E}$ can be identified with the tensor product $E \otimes_k \mathcal{C}$ endowed with the bicoaction $(i \otimes c)_{(-1)} \otimes (i \otimes c)_{(0)} \otimes (i \otimes c)_{(1)} = i_{(-1)}c_{(1)} \otimes (i_{(0)} \otimes c_{(2)}) \otimes c_{(3)}$ and the right action $(i \otimes c)d = i \otimes cd$. Besides, the isomorphism $E \otimes_k \mathcal{C} \simeq \mathcal{C} \otimes_k E$ given by the formulas $i \otimes c \longmapsto i_{(-1)}c \otimes i_{(0)}$ and $c \otimes i \longmapsto i_{(0)} \otimes s^{-1}(i_{(-1)})c$ identifies $\mathcal{E}$ with the tensor product $\mathcal{C} \otimes_k E$ endowed with the bicoaction $(c \otimes i)_{(-1)} \otimes (c \otimes i)_{(0)} \otimes (c \otimes i)_{(1)} = c_{(1)} \otimes (c_{(2)} \otimes i_{(0)}) \otimes s^{-1}(i_{(-1)})c_{(3)}$ and the right action $(c \otimes i)d = cd \otimes i$.

C.2.2   The pairings ϕ_l and ϕ_r induce left and right actions of K in $\mathcal{E}$ given by the formulas $x \rightarrow j = \langle s(x), j_{(-1)} \rangle j_{(0)}$ and $j \leftarrow x = \langle s^{-1}(x), j_{(1)} \rangle j_{(0)}$ for $j \in \mathcal{E}$, $x \in K$. Assume that these two actions satisfy the equation

$$x_{[0]} \rightarrow (jx_{[1]}) = j \leftarrow x, \quad \text{or equivalently,} \quad x \rightarrow j = (js^{-1}(x_{[1]})) \leftarrow x_{[0]}. \qquad (*)$$

Let us construct an isomorphism

$$\mathcal{E} \otimes_K R \simeq R \otimes_K \mathcal{E}.$$

Set the map $\mathcal{E} \otimes_K R \longrightarrow R \otimes_K \mathcal{E}$ to be given by the formula

$$j \otimes_K u \longmapsto u_{[0]} \otimes_K ju_{[1]}$$

and the map $R \otimes_K \mathcal{E} \longrightarrow \mathcal{E} \otimes_K R$ to be given by the formula

$$u \otimes_K j \longmapsto js^{-1}(u_{[1]}) \otimes_K u_{[0]}$$

for $j \in \mathcal{E}$, $u \in R$. We have to check that these maps are well defined.

One has

$$x \rightarrow (jc) = \langle s(x), j_{(-1)}c_{(1)} \rangle j_{(0)}c_{(1)}$$
$$= \langle s(x_{(2)}), j_{(-1)} \rangle \langle s(x_{(1)})c_{(1)} \rangle j_{(0)}c_{(1)} = (x_{(2)} \rightarrow j)(x_{(1)} \rightarrow c).$$

Therefore,

$$\langle x_{(1)}, d_{(1)} \rangle x_{(2)} \rightarrow (jd_{(2)}) = (x_{(3)} \rightarrow j)(x_{(2)} \rightarrow (s^{-1}(x_{(1)}) \rightarrow d)) = (x \rightarrow j)d.$$

Now we have $(j \leftarrow x) \otimes u \longmapsto u_{[0]} \otimes_K (j \leftarrow x)u_{[1]}$ and

$$j \otimes f(x)u \longmapsto f(x_{[0]})u_{[0]} \otimes_K jx_{[1]}u_{[1]}$$
$$= f(x_{[0](1)})u_{[0]}f(s(x_{[0](2)}))f(x_{[0](3)}) \otimes_K jx_{[1]}u_{[1]}$$
$$= \langle x_{[0](1)}, u_{[1]} \rangle u_{[0]}f(x_{[0](2)}) \otimes_K jx_{[1]}u_{[2]}$$
$$= u_{[0]} \otimes_K \langle x_{[0](1)}, u_{[1]} \rangle x_{[0](2)} \rightarrow (jx_{[1]}u_{(2)}) = u_{[0]} \otimes_K (x_{[0]} \rightarrow (jx_{[1]}))u_{[1]}.$$

Analogously, one has

$$(jc) \leftarrow x = \langle s^{-1}(x), j_{(1)}c_{(2)} \rangle j_{(0)}c_{(1)}$$
$$= \langle s^{-1}(x_{(2)}), j_{(1)} \rangle \langle s^{-1}(x_{(1)}), c_{(2)} \rangle j_{(0)}c_{(1)} = (j \leftarrow x_{(2)})(c \leftarrow x_{(1)}).$$

Therefore,

$$\langle s^{-1}(x_{(1)}), d_{(1)} \rangle (js^{-1}(d_{(2)})) \leftarrow x_{(2)} = \langle x_{(1)}, s^{-1}(d_{(1)}) \rangle (js^{-1}(d_{(2)})) \leftarrow x_{(2)}$$
$$= (j \leftarrow x_{(3)})((s^{-1}(d) \leftarrow s(x_{(1)})) \leftarrow x_{(2)}) = (j \leftarrow x)s^{-1}(d).$$

Now we have

$$u \otimes (x \rightarrow j) \longmapsto (x \rightarrow j)s^{-1}(u_{[1]}) \otimes_K u_{[0]}$$

and

$$uf(x) \otimes j \longmapsto js^{-1}(x_{[1]})s^{-1}(u_{[1]}) \otimes_K u_{[0]}f(x_{[0]})$$
$$= js^{-1}(x_{[1]})s^{-1}(u_{[1]}) \otimes_K f(x_{[0](3)})f(s^{-1}(x_{[0](2)}))u_{[0]}f(x_{[0](1)})$$
$$= js^{-1}(x_{[1]})s^{-1}(u_{[2]}) \otimes_K \langle s^{-1}(x_{[0](1)}), u_{[1]} \rangle f(x_{[0](2)})u_{[0]}$$
$$= \langle s^{-1}(x_{[0](1)}), u_{[1]} \rangle (js^{-1}(x_{[1]})s^{-1}(u_{[2]})) \leftarrow x_{[0](2)} \otimes_K u_{[0]}$$
$$= ((js^{-1}(x_{[1]})) \leftarrow x_{[0]})s^{-1}(u_{[1]}) \otimes_K u_{[0]}.$$

Checking that these two maps are mutually inverse is easy.

C.2.3   Assume that R is a flat left and right K-module. Then the vector space

$$\mathcal{E} \otimes_K R \simeq \mathcal{E} \,\square_{\mathcal{C}}\, (\mathcal{C} \otimes_K R)$$

is endowed with the structures of left $\mathcal{C}$-comodule, right $\mathbf{S}^r$-semimodule, and right R-module. The vector space

$$R \otimes_K \mathcal{E} \simeq (R \otimes_K \mathcal{C}) \,\square_{\mathcal{C}}\, \mathcal{E}$$

is endowed with the structures of right $\mathcal{C}$-comodule, left $\mathbf{S}^l$-semimodule, and left R-module.

Let us check that the isomorphism $\mathcal{E} \otimes_K R \simeq R \otimes_K \mathcal{E}$ preserves the $\mathcal{C}$-$\mathcal{C}$-bicomodule structures. Indeed, one has $(j \otimes_K u)_{(-1)} \otimes (j \otimes_K u)_{(0)} \otimes (j \otimes_K u)_{(1)} = j_{(-1)} \otimes (j_{(0)} \otimes_K u_{[0]}) \otimes j_{(1)} u_{[1]}$ and $(u \otimes_K j)_{(-1)} \otimes (u \otimes_K j)_{(0)} \otimes (u \otimes_K j)_{(1)} = j_{(-1)} s^{-1}(u_{[1]}) \otimes (u_{[0]} \otimes_K j_{(0)}) \otimes j_{(1)}$, hence $(u_{[0]} \otimes_K j u_{[1]})_{(-1)} \otimes (u_{[0]} \otimes_K j u_{[1]})_{(-0)} \otimes (u_{[0]} \otimes_K j u_{[1]})_{(1)} = j_{(-1)} u_{[2]} s^{-1}(u_{[1]}) \otimes (u_{[0]} \otimes_K j_{(0)} u_{[3]}) \otimes j_{(1)} u_{[4]} = j_{(-1)} \otimes (u_{[0]} \otimes_K j_{(0)} u_{[1]}) \otimes j_{[1]} u_{[2]}$.

Set $\mathcal{E} \otimes_K R = \mathcal{E} \simeq R \otimes_K \mathcal{E}$. The left and right actions of R in $\mathcal{E}$ commute; indeed, the left and the induced right actions of R in $R \otimes_K \mathcal{E}$ are given by the formulas $w(u \otimes_K j) = wu \otimes_K j$ and $(u \otimes_K j)v = uv_{[0]} \otimes_K jv_{[1]}$.

It follows easily that

$$\mathcal{E} \,\square_{\mathcal{C}}\, \mathbf{S}^r \simeq \mathcal{E} \simeq \mathbf{S}^l \,\square_{\mathcal{C}}\, \mathcal{E}$$

is an $\mathbf{S}^l$-$\mathbf{S}^r$-bisemimodule.

C.2.4   Now assume that the $\mathcal{C}$-$\mathcal{C}$-bicomodule $\mathcal{E}$ can be included into a Morita autoequivalence $(\mathcal{E}, \mathcal{E}^\vee)$ of $\mathcal{C}$. This means that a $\mathcal{C}$-$\mathcal{C}$-bicomodule $\mathcal{E}^\vee$ is given together with isomorphisms of $\mathcal{C}$-$\mathcal{C}$-bicomodules $\mathcal{E} \,\square_{\mathcal{C}}\, \mathcal{E}^\vee \simeq \mathcal{C} \simeq \mathcal{E}^\vee \,\square_{\mathcal{C}}\, \mathcal{E}$ such that the two induced isomorphisms $\mathcal{E} \,\square_{\mathcal{C}}\, \mathcal{E}^\vee \,\square_{\mathcal{C}}\, \mathcal{E} \Rightarrow \mathcal{E}$ coincide and the two induced isomorphisms $\mathcal{E}^\vee \,\square_{\mathcal{C}}\, \mathcal{E} \,\square_{\mathcal{C}}\, \mathcal{E}^\vee \Rightarrow \mathcal{E}^\vee$ coincide (see 7.5). The Morita equivalence $(\mathcal{E}, \mathcal{E}^\vee)$ is unique if it exists, and it exists if and only if the left $\mathcal{C}$-comodule E is one-dimensional. In the latter case, the bicomodule $\mathcal{E}^\vee$ is constructed as follows.

The left $\mathcal{C}$-coaction in E has the form $i_{(-1)} \otimes i_{(0)} = c_E \otimes i$ for a certain element $c_E \in \mathcal{C}$ such that $c_{E(1)} \otimes c_{E(2)} = c_E \otimes c_E$ and $\varepsilon(c_E) = 1$. Set $E^\vee = \mathrm{Hom}_k(E, k)$ and define a left coaction of $\mathcal{C}$ in $E^\vee$ by the formula $\check{i}_{(-1)} \otimes \check{i}_{(0)} = s(c_E) \otimes \check{i}$. Take $\mathcal{E}^\vee = E^\vee \otimes_k \mathcal{C}$ and define the $\mathcal{C}$-$\mathcal{C}$-bicomodule structure on $\mathcal{E}^\vee$ by the formula $(\check{i} \otimes c)_{(-1)} \otimes (\check{i} \otimes c)_{(0)} \otimes (\check{i} \otimes c)_{(1)} = \check{i}_{(-1)} c_{(1)} \otimes (\check{i}_{(0)} \otimes c_{(2)}) \otimes c_{(3)}$. Then one has $\mathcal{E} \,\square_{\mathcal{C}}\, \mathcal{E}^\vee \simeq E \otimes_k E^\vee \otimes_k \mathcal{C} \simeq \mathcal{C}$ and $\mathcal{E}^\vee \,\square_{\mathcal{C}}\, \mathcal{E} \simeq E^\vee \otimes_k E \otimes_k \mathcal{C} \simeq \mathcal{C}$. There is also an isomorphism $E^\vee \otimes_k \mathcal{C} \simeq \mathcal{C} \otimes_k E^\vee$ given by the formulas analogous to C.2.1.

C.2.5   Taking the cotensor product of the isomorphism $\mathcal{E} \,\square_{\mathcal{C}}\, \mathbf{S}^r \simeq \mathbf{S}^l \,\square_{\mathcal{C}}\, \mathcal{E}$ with the $\mathcal{C}$-$\mathcal{C}$-bicomodule $\mathcal{E}^\vee$ on the left, we obtain an isomorphism

$$\mathbf{S}^r \simeq \mathcal{E}^\vee \,\square_{\mathcal{C}}\, \mathbf{S}^l \,\square_{\mathcal{C}}\, \mathcal{E}.$$

Define a semialgebra structure on $\mathcal{E}^\vee \,\square_{\mathcal{C}}\, \mathbf{S}^l \,\square_{\mathcal{C}}\, \mathcal{E}$ in terms of the semialgebra structure on $\mathbf{S}^l$ and the isomorphisms $\mathcal{E} \,\square_{\mathcal{C}}\, \mathcal{E}^\vee \simeq \mathcal{C} \simeq \mathcal{E}^\vee \,\square_{\mathcal{C}}\, \mathcal{E}$ (see 8.4.1). Using the facts that $\mathcal{E} \,\square_{\mathcal{C}}\, \mathbf{S}^r \simeq \mathbf{S}^l \,\square_{\mathcal{C}}\, \mathcal{E}$ is an $\mathbf{S}^l\text{-}\mathbf{S}^r$-bisemimodule and the isomorphism $\mathcal{E} \,\square_{\mathcal{C}}\, \mathbf{S}^r \simeq \mathbf{S}^l \,\square_{\mathcal{C}}\, \mathcal{E}$ forms a commutative diagram with the maps $\mathcal{E} \longrightarrow \mathcal{E} \,\square_{\mathcal{C}}\, \mathbf{S}^r$ and $\mathcal{E} \longrightarrow \mathbf{S}^l \,\square_{\mathcal{C}}\, \mathcal{E}$ induced by the semiunit morphisms of $\mathbf{S}^r$ and $\mathbf{S}^l$, one can check that $\mathbf{S}^r \simeq \mathcal{E}^\vee \,\square_{\mathcal{C}}\, \mathbf{S}^l \,\square_{\mathcal{C}}\, \mathcal{E}$ is an isomorphism of semialgebras over $\mathcal{C}$.

It follows that $\mathbf{S}^l$ and $\mathbf{S}^r$ are left and right coflat $\mathcal{C}$-comodules. Set $\mathbf{S}^r \square_{\mathcal{C}} \mathcal{E}^\vee = \mathcal{E}^\vee \simeq \mathcal{E}^\vee \,\square_{\mathcal{C}}\, \mathbf{S}^l$; then $\mathcal{E}^\vee$ is an $\mathbf{S}^r\text{-}\mathbf{S}^l$-bisemimodule and the pair $(\mathcal{E}, \mathcal{E}^\vee)$ is a left and right coflat Morita equivalence between $\mathbf{S}^l$ and $\mathbf{S}^r$ (see 8.4.5).

The category of left $\mathbf{S}^r$-semimodules can be now described. Namely, it is equivalent to the category of left $\mathbf{S}^l$-semimodules; this equivalence assigns to a left $\mathbf{S}^r$-semimodule $\mathcal{L}$ the left $\mathbf{S}^l$-semimodule $\mathcal{E} \,\square_{\mathcal{C}}\, \mathcal{L}$ and to a left $\mathbf{S}^l$-semimodule $\mathcal{M}$ the left $\mathbf{S}^r$-semimodule $\mathcal{E}^\vee \,\square_{\mathcal{C}}\, \mathcal{M}$. On the level of $\mathcal{C}$-comodules, one has $\mathcal{E} \,\square_{\mathcal{C}}\, \mathcal{L} \simeq E \otimes_k \mathcal{L}$ and $\mathcal{E}^\vee \,\square_{\mathcal{C}}\, \mathcal{M} \simeq E^\vee \otimes_k \mathcal{M}$; the left $\mathcal{C}$-coaction in $E \otimes_k \mathcal{L}$ and $E^\vee \otimes_k \mathcal{M}$ is given by the formulas $(i \otimes l)_{(-1)} \otimes (i \otimes l)_{(0)} = c_E l_{(-1)} \otimes (i \otimes l_{(0)})$ and $(\check{i} \otimes m)_{(-1)} \otimes (\check{i} \otimes m)_{(0)} = s(c_E) m_{(-1)} \otimes (\check{i} \otimes m_{(0)})$.

C.2.6 In particular, when the left and right actions of K in $\mathcal{C}$ satisfy the equation

$$x_{[0]} \to (c x_{[1]}) = c \leftarrow x, \quad \text{or equivalently,} \quad x \to c = (c s^{-1}(x_{[1]})) \leftarrow x_{[0]}, \qquad (**)$$

one can take $\mathcal{E} = \mathcal{C} = \mathcal{E}^\vee$. Thus the semialgebras $\mathbf{S}^r$ and $\mathbf{S}^l$ are isomorphic in this case, the isomorphism being given by the formulas $c \otimes_K u \longmapsto u_{[0]} \otimes_K c u_{[1]}$ and $u \otimes_K c \longmapsto c s^{-1}(u_{[1]}) \otimes_K u_{[0]}$ for $c \in \mathcal{C}$ and $u \in R$.

Remark. One can construct an isomorphism between versions of the semialgebras $\mathbf{S}^r$ and $\mathbf{S}^l$ in slightly larger generality. Namely, let $\chi_r, \chi_l \colon K \longrightarrow k$ be k-algebra homomorphisms satisfying the equations

$$\chi(x_{[0]}) x_{[1]} = \chi(x) e \quad \text{and} \quad \chi(x_{[0](2)}) x_{[0](1)} \otimes x_{[1]} = \chi(x_{(2)}) x_{(1)[0]} \otimes x_{(1)[1]}.$$

These equations hold automatically when the pairing $\langle \, , \, \rangle$ is nondegenerate in $\mathcal{C}$ (apply $\mathrm{id} \otimes \chi$ to the identity

$$\langle y, x_{[1]} \rangle x_{[0](1)} \otimes x_{[0](2)} = (\langle y, x_{[1]} \rangle x_{[0]})_{(1)} \otimes \langle y, x_{[1]} \rangle x_{[0]})_{(2)}$$
$$= (y_{(1)} x s(y_{(2)}))_{(1)} \otimes (y_{(1)} x s(y_{(2)}))_{(2)} = y_{(1)} x_{(1)} s(y_{(4)}) \otimes y_{(2)} x_{(2)} s(y_{(3)})).$$

Define pairings $\phi_r \colon \mathcal{C} \otimes_k K \longrightarrow k$ and $\phi_l \colon K \otimes_k \mathcal{C} \longrightarrow k$ by the formulas $\phi_r(c, x) = \chi_r(x_{(2)}) \langle s^{-1}(x_{(1)}), c \rangle$ and $\phi_l(x, c) = \chi_l(x_{(2)}) \langle s(x_{(1)}), c \rangle$, and modify the definitions of the right and left actions of K in $\mathcal{C}$ accordingly, $c \leftarrow x = \phi_r(c_{(2)}, x) c_{(1)}$ and $x \to c = \phi_l(x, c_{(1)}) c_{(2)}$. Then the tensor products $\mathcal{C} \otimes_K R$ and $R \otimes_K \mathcal{C}$ are semialgebras over $\mathcal{C}$ with the $\mathcal{C}$-$\mathcal{C}$-bicomodule structures given by the same formulas as in C.1.2–C.1.3. Assuming that the modified left and right actions of K in $\mathcal{C}$ satisfy the above equation, the maps $c \otimes_K u \longmapsto u_{[0]} \otimes_K c u_{[1]}$ and $u \otimes_K c \longmapsto c s^{-1}(u_{[1]}) \otimes_K u_{[0]}$ are mutually inverse isomorphisms between these two semialgebras.

C.2.7   The *opposite coring* $\mathcal{D}^{\mathrm{op}}$ to a coring $\mathcal{D}$ over a k-algebra B and the *opposite semialgebra* $\mathcal{T}^{\mathrm{op}}$ to a semialgebra $\mathcal{T}$ over $\mathcal{D}$ are defined in the obvious way; $\mathcal{D}^{\mathrm{op}}$ is a coring over B^{op} and $\mathcal{T}^{\mathrm{op}}$ is a semialgebra over $\mathcal{D}^{\mathrm{op}}$.

In the above assumptions, notice the identity $s(x{\to}c) = s(c){\leftarrow}s(x)$ for $x \in K$, $c \in \mathcal{C}$. Suppose that the k-algebra R is endowed with an anti-endomorphism s satisfying the equations $f(s(x)) = s(f(x))$ and $c_{(1)} \otimes_K (s(u))_{[0]} \otimes c_{(2)}(s(u))_{[1]} = c_{(1)} \otimes_K s(u_{[0]}) \otimes u_{[1]}c_{(2)}$; the second equation follows from the first one if the pairing $\langle\,,\rangle$ is nondegenerate in $\mathcal{C}$. Then there is a map of semialgebras $s\colon (R \otimes_K \mathcal{C})^{\mathrm{op}} \longrightarrow \mathcal{C} \otimes_K R$ compatible with the isomorphism of coalgebras $s\colon \mathcal{C}^{\mathrm{op}} \simeq \mathcal{C}$; it is defined by the formula $s(u \otimes_K c) = s(c) \otimes_K s(u)$.

Suppose that we are given a map $s\colon \mathcal{E} \longrightarrow \mathcal{E}$ satisfying the equation $(s(j))_{(0)} \otimes (s(j))_{(1)} = s(j_{(0)}) \otimes s(j_{(-1)})$. Then one has $s(x \to j) = s(j) \leftarrow s(x)$. The induced map $s\colon R \otimes_K \mathcal{E} \longrightarrow \mathcal{E} \otimes_K R$ given by the formula $s(u \otimes_K j) = s(j) \otimes_K s(u)$ is a map of right semimodules compatible with the isomorphism of coalgebras $s\colon \mathcal{C}^{\mathrm{op}} \simeq \mathcal{C}$ and the map of semialgebras $s\colon \mathbf{S}^{l\,\mathrm{op}} \longrightarrow \mathbf{S}^r$, where the right $\mathbf{S}^{l\,\mathrm{op}}$-semimodule structure on $R \otimes_K \mathcal{E}$ corresponds to its left $\mathbf{S}^l$-semimodule structure.

Now assume that $\mathcal{C}$ is commutative, K is cocommutative, and our data satisfy the equations $s^2(u) = u$, $s^2(j) = j$, $s(jc) = s(j)s(c)$, and $(s(u))_{[0]} \otimes (s(u))_{[1]} = s(u_{[0]}) \otimes u_{[1]}$; the latter equation holds automatically when the pairing $\langle\,,\rangle$ is nondegenerate in $\mathcal{C}$. Then the composition of the isomorphism of $\mathbf{S}^l$-$\mathbf{S}^r$-bisemimodules $\mathcal{E} \otimes_K R \simeq R \otimes_K \mathcal{E}$ and the map $s\colon R \otimes_K \mathcal{E} \longrightarrow \mathcal{E} \otimes_K R$ in an involution of the bisemimodule $\mathcal{E}$ transforming its left $\mathbf{S}^l$-semimodule and right $\mathbf{S}^r$-semimodule structures into each other in a way compatible with the isomorphism of coalgebras $s\colon \mathcal{C}^{\mathrm{op}} \longrightarrow \mathcal{C}$ and the isomorphism of semialgebras $s\colon \mathbf{S}^{l\,\mathrm{op}} \simeq \mathbf{S}^r$. In particular, in the situation of C.2.6 the map $s\colon R \otimes_K \mathcal{C} \longrightarrow \mathcal{C} \otimes_K R$ becomes an involutive anti-automorphism of the semialgebra $\mathbf{S}^l \simeq \mathbf{S}^r$ compatible with the anti-automorphism s of the coalgebra $\mathcal{C}$.

C.3   Semitensor product and semihomomorphisms, SemiTor and SemiExt

Let us return to the assumptions of C.1.1–C.2.4.

C.3.1   Let $\mathcal{N}$ be a right $\mathcal{C}$-comodule and $\mathcal{M}$ be a left $\mathcal{C}$-comodule. Then one can easily check that the two injections $\mathcal{N} \,\square_\mathcal{C}\, \mathcal{E}^{\vee} \,\square_\mathcal{C}\, \mathcal{M} \simeq \mathcal{N} \,\square_\mathcal{C}\, (E^{\vee} \otimes_k \mathcal{C}) \,\square_\mathcal{C}\, \mathcal{M} \simeq \mathcal{N} \,\square_\mathcal{C}\, (E^{\vee} \otimes_k \mathcal{M}) \longrightarrow \mathcal{N} \otimes_k E^{\vee} \otimes_k \mathcal{M}$ and $\mathcal{N} \,\square_\mathcal{C}\, \mathcal{E}^{\vee} \,\square_\mathcal{C}\, \mathcal{M} \simeq \mathcal{N} \,\square_\mathcal{C}\, (\mathcal{C} \otimes_k E^{\vee}) \,\square_\mathcal{C}\, \mathcal{M} \simeq (\mathcal{N} \otimes_k E^{\vee}) \,\square_\mathcal{C}\, \mathcal{M} \longrightarrow \mathcal{N} \otimes_k E^{\vee} \otimes_k \mathcal{M}$ coincide.

Let $\mathcal{N}$ be a right $\mathbf{S}^r$-semimodule and $\mathcal{M}$ be a left $\mathbf{S}^l$-semimodule (see C.1.4). Then the isomorphism $(\mathcal{N} \otimes_K R) \,\square_\mathcal{C}\, (E^{\vee} \otimes_k \mathcal{M}) \simeq \mathcal{N} \,\square_\mathcal{C}\, (\mathcal{C} \otimes_K R) \,\square_\mathcal{C}\, \mathcal{E}^{\vee} \,\square_\mathcal{C}\, \mathcal{M} \simeq \mathcal{N} \,\square_\mathcal{C}\, \mathcal{E}^{\vee} \,\square_\mathcal{C}\, (R \otimes_K \mathcal{C}) \,\square_\mathcal{C}\, \mathcal{M} \simeq (\mathcal{N} \otimes_k E^{\vee}) \,\square_\mathcal{C}\, (R \otimes_K \mathcal{M})$ induced by the isomorphism $(\mathcal{C} \otimes_K R) \,\square_\mathcal{C}\, \mathcal{E}^{\vee} \simeq \mathcal{E}^{\vee} \,\square_\mathcal{C}\, (R \otimes_K \mathcal{C})$ can be computed as follows.

There is an isomorphism $(\mathcal{C} \otimes_k R) \,\square_\mathcal{C}\, \mathcal{E}^{\vee} \simeq \mathcal{E}^{\vee} \,\square_\mathcal{C}\, (R \otimes_K \mathcal{C})$ defined by the same formulas as the isomorphism $\mathbf{S}^r \,\square_\mathcal{C}\, \mathcal{E}^{\vee} \simeq \mathcal{E}^{\vee} \,\square_\mathcal{C}\, \mathbf{S}^l$ ($\otimes_K$ being replaced with

$\otimes$). Hence the induced isomorphism $(\mathcal{N} \otimes_k R) \,\square_{\mathcal{C}}\, (E^{\vee} \otimes_k \mathcal{M}) \simeq (\mathcal{N} \otimes_k E^{\vee}) \,\square_{\mathcal{C}}\, (R \otimes_k \mathcal{M})$, which is given by the simple formula $n \otimes r \otimes \breve{\imath} \otimes m \longmapsto n \otimes \breve{\imath} \otimes r \otimes m$. The isomorphisms $(\mathcal{N} \otimes_K R) \,\square_{\mathcal{C}}\, (E^{\vee} \otimes_k \mathcal{M}) \simeq (\mathcal{N} \otimes_k E^{\vee}) \,\square_{\mathcal{C}}\, (R \otimes_K \mathcal{M})$ and $(\mathcal{N} \otimes_k R) \,\square_{\mathcal{C}}\, (E^{\vee} \otimes_k \mathcal{M}) \simeq (\mathcal{N} \otimes_k E^{\vee}) \,\square_{\mathcal{C}}\, (R \otimes_k \mathcal{M})$ form a commutative diagram with the natural maps $(\mathcal{N} \otimes_k R) \,\square_{\mathcal{C}}\, (E^{\vee} \otimes_k \mathcal{M}) \longrightarrow (\mathcal{N} \otimes_K R) \,\square_{\mathcal{C}}\, (E^{\vee} \otimes_k \mathcal{M})$ and $(\mathcal{N} \otimes_k E^{\vee}) \,\square_{\mathcal{C}}\, (R \otimes_k \mathcal{M}) \longrightarrow (\mathcal{N} \otimes_k E^{\vee}) \,\square_{\mathcal{C}}\, (R \otimes_K \mathcal{M})$. This provides a description of the first isomorphism whenever the latter two maps are surjective – in particular, when either $\mathcal{N}$ or $\mathcal{M}$ is a coflat $\mathcal{C}$-comodule. To compute the desired isomorphism in the general case, it suffices to represent either $\mathcal{N}$ or $\mathcal{M}$ as the kernel of a morphism of $\mathcal{C}$-coflat semimodules (both sides of this isomorphism preserve kernels, since R is a flat left and right K-module).

C.3.2 Let $\mathcal{M}$ be a left $\mathcal{C}$-comodule and $\mathfrak{P}$ be a left $\mathcal{C}$-contramodule. Then one can check that the two surjections $\mathrm{Hom}_k(E^{\vee} \otimes_k \mathcal{M}, \mathfrak{P}) \to \mathrm{Cohom}_{\mathcal{C}}(E^{\vee} \otimes_k \mathcal{M}, \mathfrak{P}) \simeq \mathrm{Cohom}_{\mathcal{C}}(\mathcal{E}^{\vee}\square_{\mathcal{C}}\mathcal{M}, \mathfrak{P})$ and $\mathrm{Hom}_k(\mathcal{M}, \mathrm{Hom}_k(E^{\vee}, \mathfrak{P})) \to \mathrm{Cohom}_{\mathcal{C}}(\mathcal{M}, \mathrm{Hom}_k(E^{\vee}, \mathfrak{P})) \simeq \mathrm{Cohom}_{\mathcal{C}}(\mathcal{M}, \mathrm{Cohom}_{\mathcal{C}}(\mathcal{E}^{\vee}, \mathfrak{P}))$ coincide.

Let $\mathcal{M}$ be a left $\mathcal{S}^l$-semimodule and $\mathfrak{P}$ be a left $\mathcal{S}^r$-semicontramodule. Assuming that R is a projective left K-module, the isomorphism $\mathrm{Cohom}_{\mathcal{C}}(E^{\vee} \otimes_k \mathcal{M}, \mathrm{Hom}_K(R, \mathfrak{P})) \simeq \mathrm{Cohom}_{\mathcal{C}}(\mathcal{E}^{\vee} \,\square_{\mathcal{C}}\, \mathcal{M}, \mathrm{Cohom}_{\mathcal{C}}(\mathcal{C} \otimes_K R, \mathfrak{P})) \simeq \mathrm{Cohom}_{\mathcal{C}}((R \otimes_K \mathcal{C}) \,\square_{\mathcal{C}}\, \mathcal{M}, \mathrm{Cohom}_{\mathcal{C}}(\mathcal{E}^{\vee} \mathfrak{P})) \simeq \mathrm{Cohom}_{\mathcal{C}}(R \otimes_K \mathcal{M}, \mathrm{Hom}_k(E^{\vee}, \mathfrak{P}))$ induced by the isomorphism $(\mathcal{C} \otimes_K R) \,\square_{\mathcal{C}}\, \mathcal{E}^{\vee} \simeq \mathcal{E}^{\vee} \,\square_{\mathcal{C}}\, (R \otimes_K \mathcal{C})$ can be computed as follows.

The isomorphism $\mathrm{Cohom}_{\mathcal{C}}(E^{\vee} \otimes_k \mathcal{M}, \mathrm{Hom}_k(R, \mathfrak{P})) \simeq \mathrm{Cohom}_{\mathcal{C}}(R \otimes_k \mathcal{M}, \mathrm{Hom}_k(E^{\vee}, \mathfrak{P}))$ induced by the isomorphism $(\mathcal{C} \otimes_k R) \,\square_{\mathcal{C}}\, \mathcal{E}^{\vee} \simeq \mathcal{E}^{\vee} \,\square_{\mathcal{C}}\, (R \otimes_k \mathcal{C})$ is given by the simple formula $g \longmapsto h$, $h(r \otimes m)(\breve{\imath}) = g(\breve{\imath} \otimes m)(r)$. The isomorphisms $\mathrm{Cohom}_{\mathcal{C}}(E^{\vee}\otimes_k\mathcal{M}, \mathrm{Hom}_K(R, \mathfrak{P})) \simeq \mathrm{Cohom}_{\mathcal{C}}(R\otimes_K\mathcal{M}, \mathrm{Hom}_k(E^{\vee}, \mathfrak{P}))$ and $\mathrm{Cohom}_{\mathcal{C}}(E^{\vee}\otimes_k\mathcal{M}, \mathrm{Hom}_k(R, \mathfrak{P})) \simeq \mathrm{Cohom}_{\mathcal{C}}(R\otimes_K\mathcal{M}, \mathrm{Hom}_k(E^{\vee}, \mathfrak{P}))$ form a commutative diagram with the natural maps $\mathrm{Cohom}_{\mathcal{C}}(E^{\vee} \otimes_k \mathcal{M}, \mathrm{Hom}_K(R, \mathfrak{P})) \longrightarrow \mathrm{Cohom}_{\mathcal{C}}(E^{\vee} \otimes_k \mathcal{M}, \mathrm{Hom}_k(R, \mathfrak{P}))$ and $\mathrm{Cohom}_{\mathcal{C}}(R \otimes_K \mathcal{M}, \mathrm{Hom}_k(E^{\vee}, \mathfrak{P})) \longrightarrow \mathrm{Cohom}_{\mathcal{C}}(R \otimes_k \mathcal{M}, \mathrm{Hom}_k(E^{\vee}, \mathfrak{P}))$. This provides a description of the first isomorphism in the case when the latter two maps are injective – in particular, when either $\mathcal{M}$ is a coprojective $\mathcal{C}$-comodule, or $\mathfrak{P}$ is a coinjective $\mathcal{C}$-contramodule. To compute the desired isomorphism in the general case, it suffices to either represent $\mathcal{M}$ as the kernel of a morphism of $\mathcal{C}$-coprojective semimodules, or represent $\mathfrak{P}$ as the cokernel of a morphism of $\mathcal{C}$-coinjective semicontramodules.

C.3.3 Assume that the k-algebra R is endowed with a Hopf algebra structure $u \longmapsto u_{(1)} \otimes u_{(2)}$, $u \longmapsto \varepsilon(u)$ with invertible antipode s such that $f\colon K \longrightarrow R$ is a Hopf algebra morphism. Let $\mathcal{N}$ be a right $\mathcal{S}^r$-semimodule and $\mathcal{M}$ be a left $\mathcal{S}^l$-semimodule; assume that either $\mathcal{N}$ or $\mathcal{M}$ is a coflat $\mathcal{C}$-comodule. Define right R-module and right $\mathcal{C}$-comodule structures on the tensor product $\mathcal{N} \otimes_k E^{\vee} \otimes_k \mathcal{M}$ by the formulas $(n \otimes \breve{\imath} \otimes m)r = nr_{(2)} \otimes \breve{\imath} \otimes s^{-1}(r_{(1)})m$ and $(n \otimes \breve{\imath} \otimes m)_{(0)} \otimes (n \otimes \breve{\imath} \otimes m)_{(1)} = (n_{(0)} \otimes \breve{\imath} \otimes m_{(0)}) \otimes s^{-1}(m_{(-1)})c_E n_{(1)}$. Then the semitensor product $\mathcal{N} \lozenge_{\mathcal{S}^r} \mathcal{E}^{\vee} \lozenge_{\mathcal{S}^l} \mathcal{M}$ (which is easily seen to be associative) is uniquely determined by these right R-module and right $\mathcal{C}$-comodule structures on $\mathcal{N} \otimes_k E^{\vee} \otimes_k \mathcal{M}$.

Indeed, the subspace $\mathbf{N} \square_{\mathcal{C}} (E^\vee \otimes_k \mathbf{M}) \simeq \mathbf{N} \square_{\mathcal{C}} \mathcal{E}^\vee \square_{\mathcal{C}} \mathbf{M} \simeq (\mathbf{M} \otimes_k E^\vee) \square_{\mathcal{C}} \mathbf{M}$ of the space $\mathbf{N} \otimes_k E^\vee \otimes_k \mathbf{M}$ can be defined by the equation $n_{(0)} \otimes \check{\imath} \otimes m_{(0)} \otimes s^{-1}(m_{(-1)})c_E n_{(0)} = n \otimes \check{\imath} \otimes m \otimes e$. The isomorphism $\mathbf{N} \otimes_k R \otimes_k E^\vee \otimes_k \mathbf{M} \simeq \mathbf{N} \otimes_k E^\vee \otimes_k \mathbf{M} \otimes_k R$ given by the formulas $n \otimes r \otimes \check{\imath} \otimes m \longmapsto n \otimes \check{\imath} \otimes r_{(1)}m \otimes r_{(2)}$ and $n \otimes \check{\imath} \otimes m \otimes r \longmapsto n \otimes r_{(2)} \otimes \check{\imath} \otimes s^{-1}(r_{(1)})m$ transforms the pair of maps $\mathbf{N} \otimes_k R \otimes_k E^\vee \otimes_k \mathbf{M} \rightrightarrows \mathbf{N} \otimes_k E^\vee \otimes_k \mathbf{M}$ given by the formulas $n \otimes r \otimes \check{\imath} \otimes m \longmapsto nr \otimes \check{\imath} \otimes m$, $n \otimes \check{\imath} \otimes rm$ into the pair of maps $\mathbf{N} \otimes_k E^\vee \otimes_k \mathbf{M} \otimes_k R \rightrightarrows \mathbf{N} \otimes_k E^\vee \otimes_k \mathbf{M}$ given by the formulas $n \otimes \check{\imath} \otimes m \otimes r \longmapsto nr_{(2)} \otimes \check{\imath} \otimes s^{-1}(r_{(1)})m$, $\varepsilon(r)n \otimes \check{\imath} \otimes m$. This isomorphism also transforms the subspace $(\mathbf{N} \otimes_k R) \square_{\mathcal{C}} (E^\vee \otimes_k \mathbf{M})$ of $\mathbf{N} \otimes_k R \otimes_k E^\vee \otimes_k \mathbf{M}$, which can be defined by the equation $n_{(0)} \otimes r_{[0]} \otimes \check{\imath} \otimes m_{(0)} \otimes s^{-1}(m_{(-1)})c_E n_{(1)}r_{[1]} = n \otimes r \otimes \check{\imath} \otimes m \otimes e$, into the subspace of $\mathbf{N} \otimes_k E^\vee \otimes_k \mathbf{M} \otimes_k R$ defined by the equation $n_{(0)} \otimes \check{\imath} \otimes r_{(2)[0](1)}(s^{-1}(r_{(1)}))_{[0]}m_{(0)} \otimes r_{(2)[0](2)} \otimes s^{-2}((s^{-1}(r_{(1)}))_{[1]})s^{-1}(m_{(-1)})c_E n_{(1)}r_{(2)[1]} = n \otimes \check{\imath} \otimes m \otimes r \otimes e$. Finally, the same isomorphism transforms the quotient space $\mathbf{N} \otimes_K R \otimes_k E^\vee \otimes_k \mathbf{M}$ of the space $\mathbf{N} \otimes_k R \otimes_k E^\vee \otimes_k \mathbf{M}$ into the quotient space $(\mathbf{N} \otimes_k E^\vee \otimes_k \mathbf{M}) \otimes_K R$ of the space $\mathbf{N} \otimes_k E^\vee \otimes_k \mathbf{M} \otimes_k R$, as one can check using the isomorphism $\mathbf{N} \otimes_k K \otimes_k R \otimes_k E^\vee \otimes_k \mathbf{M} \simeq \mathbf{N} \otimes_k E^\vee \otimes_k \mathbf{M} \otimes_k K \otimes_k R$ given by the formulas $n \otimes x \otimes r \otimes \check{\imath} \otimes m \longmapsto n \otimes \check{\imath} \otimes x_{(1)}r_{(1)}m \otimes x_{(2)} \otimes r_{(2)}$ and $n \otimes \check{\imath} \otimes m \otimes x \otimes r \longmapsto n \otimes x_{(2)} \otimes r_{(2)} \otimes \check{\imath} \otimes s^{-1}(r_{(1)})s^{-1}(x_{(1)})m$.

C.3.4 Let $\mathbf{M}$ be a left $\mathbf{S}^l$-semimodule and $\mathbf{\mathfrak{P}}$ be a left $\mathbf{S}^r$-semicontramodule; assume that either $\mathbf{M}$ is a coprojective $\mathcal{C}$-comodule, or $\mathbf{\mathfrak{P}}$ is a coinjective $\mathcal{C}$-contramodule. Define left R-module and left $\mathcal{C}$-contramodule structures on the space $\mathrm{Hom}_k(E^\vee \otimes_k \mathbf{M}, \mathbf{\mathfrak{P}})$ by the formulas $rg(\check{\imath} \otimes m) = r_{(2)}g(\check{\imath} \otimes s^{-1}(r_{(1)})m)$ and $\pi(h)(\check{\imath} \otimes m) = \pi_{\mathbf{\mathfrak{P}}}(c \mapsto h(s^{-1}(m_{(-1)})c_E c)(\check{\imath} \otimes m_{(0)}))$ for $g \in \mathrm{Hom}_k(E^\vee \otimes_k \mathbf{M}, \mathbf{\mathfrak{P}})$ and $h \in \mathrm{Hom}_k(\mathcal{C}, \mathrm{Hom}_k(E^\vee \otimes_k \mathbf{M}, \mathbf{\mathfrak{P}}))$, where $\pi_{\mathbf{\mathfrak{P}}}$ denotes the $\mathcal{C}$-contraaction in $\mathbf{\mathfrak{P}}$. Then the semihomomorphism space $\mathrm{SemiHom}_{\mathbf{S}^r}(\mathcal{E} \lozenge_{\mathbf{S}^l} \mathbf{M}, \mathbf{\mathfrak{P}})$ is uniquely determined by these R-module and $\mathcal{C}$-contramodule structures on $\mathrm{Hom}_k(E^\vee \otimes_k \mathbf{M}, \mathbf{\mathfrak{P}})$.

This is established in a way analogous to C.3.3 using the isomorphism $\mathrm{Hom}_k(R \otimes_k E^\vee \otimes_k \mathbf{M}, \mathbf{\mathfrak{P}}) \simeq \mathrm{Hom}_k(R, \mathrm{Hom}_k(E^\vee \otimes_k \mathbf{M}, \mathbf{\mathfrak{P}}))$ given by the formulas $g \longmapsto h$, $g(r \otimes \check{\imath} \otimes m) = h(r_{(2)})(\check{\imath} \otimes r_{(1)})$, $h(r)(\check{\imath} \otimes m) = g(r_{(2)} \otimes \check{\imath} \otimes s^{-1}(r_{(1)})m)$.

C.3.5 Now assume that $\mathcal{C}$ is commutative, K is cocommutative, and the equations $(s(u))_{[0]} \otimes (s(u))_{[1]} = s(u_{[0]}) \otimes u_{[1]}$, $\varepsilon(u_{[0]})u_{[1]} = \varepsilon(u)e$, and $u_{(1)[0]} \otimes u_{(2)[0]} \otimes u_{(1)[1]}u_{(2)[1]} = u_{[0](1)} \otimes u_{[0](2)} \otimes u_{[1]}$ are satisfied for $u \in R$; when the pairing $\langle\,,\,\rangle$ is nondegenerate in $\mathcal{C}$, these equations hold automatically.

Let $\mathbf{N}$ be a right $\mathbf{S}^r$-semimodule and $\mathbf{M}$ be a left $\mathbf{S}^l$-semimodule. Define right R-module and right $\mathcal{C}$-comodule structures on the tensor product $\mathbf{N} \otimes_k \mathbf{M}$ by the formulas $(n \otimes m)r = nr_{(2)} \otimes s^{-1}(r_{(1)})m$ and $(n \otimes m)_{(0)} \otimes (n \otimes m)_{(1)} = (n_{(0)} \otimes m_{(0)}) \otimes s^{-1}(m_{(-1)})n_{(1)}$. These right action and right coaction satisfy the equations of C.1.4, so they define a right $\mathbf{S}^r$-semimodule structure on $\mathbf{N} \otimes_k \mathbf{M}$. The ground field k, endowed with the trivial left R-module and left $\mathcal{C}$-comodule structures $ra = \varepsilon(r)a$ and $a_{(-1)} \otimes a_{(0)} = e \otimes a$ for $a \in k$, becomes a left $\mathbf{S}^l$-semimodule. Then there is a natural isomorphism

$$\mathbf{N} \lozenge_{\mathbf{S}^r} \mathcal{E}^\vee \lozenge_{\mathbf{S}^l} \mathbf{M} \simeq (\mathbf{N} \otimes_k \mathbf{M}) \lozenge_{\mathbf{S}^r} \mathcal{E}^\vee \lozenge_{\mathbf{S}^l} k.$$

Indeed, let us first assume that either $\mathcal{N}$ or $\mathcal{M}$ is a coflat $\mathcal{C}$-comodule; notice that $\mathcal{N} \otimes_k \mathcal{M}$ is then a coflat $\mathcal{C}$-comodule, too. The isomorphism $\mathcal{N} \square_{\mathcal{C}} (E^\vee \otimes_k \mathcal{M}) \simeq (\mathcal{N} \otimes_k \mathcal{M}) \square_{\mathcal{C}} E^\vee$ given by the formula $n \otimes \check{\imath} \otimes m \longmapsto n \otimes m \otimes \check{\imath}$ and the isomorphism $(\mathcal{N} \otimes_K R) \square_{\mathcal{C}} (E^\vee \otimes_k \mathcal{M}) \simeq ((\mathcal{N} \otimes_k \mathcal{M}) \otimes_K R) \square_{\mathcal{C}} E^\vee$ constructed in C.3.3 transform the pair of maps whose cokernel is $\mathcal{N} \lozenge_{\mathbf{S}^r} \mathcal{E}^\vee \lozenge_{\mathbf{S}^l} \mathcal{M}$ into the pair of maps whose cokernel is $(\mathcal{N} \otimes_k \mathcal{M}) \lozenge_{\mathbf{S}^r} \mathcal{E}^\vee \lozenge_{\mathbf{S}^l} k$. In the general case, represent $\mathcal{N}$ or $\mathcal{M}$ as the kernel of a morphism of $\mathcal{C}$-coflat semimodules; then the pair of maps whose cokernel is $\mathcal{N} \lozenge_{\mathbf{S}^r} \mathcal{E}^\vee \lozenge_{\mathbf{S}^l} \mathcal{M}$ and the pair of maps whose cokernel is $(\mathcal{N} \otimes_k \mathcal{M}) \lozenge_{\mathbf{S}^r} \mathcal{E}^\vee \lozenge_{\mathbf{S}^l} k$ become the kernels of isomorphic morphisms of pairs of maps.

C.3.6 Let $\mathcal{M}$ be a left $\mathbf{S}^l$-semimodule and $\mathfrak{P}$ be a left $\mathbf{S}^r$-semicontramodule. Define left R-module and left $\mathcal{C}$-contramodule structures on the space $\mathrm{Hom}_k(\mathcal{M}, \mathfrak{P})$ by the formulas $rg(m) = r_{(2)}g(s^{-1}(r_{(1)})m)$ and $\pi(h)(m) = \pi_{\mathfrak{P}}(c \mapsto h(s^{-1}(m_{(-1)})c)(m_{(0)}))$ for $g \in \mathrm{Hom}_k(\mathcal{M}, \mathfrak{P})$ and $h \in \mathrm{Hom}_k(\mathcal{C}, \mathrm{Hom}_k(\mathcal{M}, \mathfrak{P}))$. These left action and left contraaction satisfy the equations of C.1.4, so they define a left $\mathbf{S}^r$-semicontramodule structure on $\mathrm{Hom}_k(\mathcal{M}, \mathfrak{P})$. Then there is a natural isomorphism

$$\mathrm{SemiHom}_{\mathbf{S}^r}(\mathcal{E}^\vee \lozenge_{\mathbf{S}^l} \mathcal{M}, \mathfrak{P}) \simeq \mathrm{SemiHom}_{\mathbf{S}^r}(\mathcal{E}^\vee \lozenge_{\mathbf{S}^l} k, \mathrm{Hom}_k(\mathcal{M}, \mathfrak{P})).$$

C.3.7 Let $\mathcal{N}^\bullet$ be a complex of right $\mathbf{S}^r$-semimodules and $\mathcal{M}^\bullet$ be a complex of left $\mathbf{S}^l$-semimodules. Then there are natural isomorphisms

$$\mathrm{SemiTor}^{\mathbf{S}^r}(\mathcal{N}^\bullet, \mathcal{E}^\vee \lozenge_{\mathbf{S}^l} \mathcal{M}^\bullet) \simeq \mathrm{SemiTor}^{\mathbf{S}^l}(\mathcal{N}^\bullet \lozenge_{\mathbf{S}^r} \mathcal{E}^\vee, \mathcal{M}^\bullet)$$
$$\simeq \mathrm{SemiTor}^{\mathbf{S}^r}(\mathcal{N}^\bullet \otimes_k \mathcal{M}^\bullet, \mathcal{E}^\vee \lozenge_{\mathbf{S}^l} k)$$

in the derived category of k-vector spaces. The isomorphism between the first two objects is provided by the results of 8.4.3, and the isomorphism between either of the first two objects and the third one follows from C.3.5.

Indeed, assume that the complex $\mathcal{N}^\bullet$ is semiflat. Then the complex $\mathcal{N}^\bullet \otimes_k \mathcal{M}^\bullet$ is also semiflat, since $(\mathcal{N}^\bullet \otimes_k \mathcal{M}^\bullet) \lozenge_{\mathbf{S}^r} \mathcal{L}^\bullet \simeq (\mathcal{N}^\bullet \otimes_k \mathcal{M}^\bullet) \lozenge_{\mathbf{S}^r} \mathcal{E}^\vee \lozenge_{\mathbf{S}^l} \mathcal{E} \lozenge_{\mathbf{S}^r} \mathcal{L}^\bullet \simeq (\mathcal{N}^\bullet \otimes_k \mathcal{M}^\bullet \otimes_k (\mathcal{E} \lozenge_{\mathbf{S}^r} \mathcal{L}^\bullet)) \lozenge_{\mathbf{S}^r} \mathcal{E}^\vee \lozenge_{\mathbf{S}^l} k \simeq \mathcal{N}^\bullet \lozenge_{\mathbf{S}^r} \mathcal{E}^\vee \lozenge_{\mathbf{S}^l} (\mathcal{M}^\bullet \otimes_k (\mathcal{E} \lozenge_{\mathbf{S}^r} \mathcal{L}^\bullet))$ for any complex of left $\mathbf{S}^r$-semimodules $\mathcal{L}^\bullet$.

C.3.8 Let $\mathcal{M}^\bullet$ be a complex of left $\mathbf{S}^l$-semimodules and $\mathfrak{P}^\bullet$ be a complex of left $\mathbf{S}^r$-semicontramodules. Then there are natural isomorphisms

$$\mathrm{SemiExt}_{\mathbf{S}^r}(\mathcal{E}^\vee \lozenge_{\mathbf{S}^l} \mathcal{M}^\bullet, \mathfrak{P}^\bullet) \simeq \mathrm{SemiExt}_{\mathbf{S}^l}(\mathcal{M}^\bullet, \mathrm{SemiHom}_{\mathbf{S}^r}(\mathcal{E}^\vee, \mathfrak{P}^\bullet))$$
$$\simeq \mathrm{SemiExt}_{\mathbf{S}^r}(\mathcal{E}^\vee \lozenge_{\mathbf{S}^l} k, \mathrm{Hom}_k(\mathcal{M}^\bullet, \mathfrak{P}^\bullet)).$$

C.4 Harish-Chandra pairs

For a discussion of terminology related to Harish-Chandra pairs and Harish-Chandra modules, see D.2.5.

C.4.1 Let $(\mathfrak{g}, H)$ be an algebraic Harish-Chandra pair over a field k, that is H is an algebraic group, which we will assume to be affine, $\mathfrak{g}$ is a Lie algebra into which the Lie algebra $\mathfrak{h}$ of the algebraic group H is embedded, and an action of H by Lie algebra automorphisms of $\mathfrak{g}$ is given. Two conditions should be satisfied: $\mathfrak{h}$ is an H-submodule in $\mathfrak{g}$ where H acts by the adjoint action of H in $\mathfrak{h}$, and the action of $\mathfrak{h}$ in $\mathfrak{g}$ obtained by differentiating the action of H in $\mathfrak{g}$ coincides with the adjoint action of $\mathfrak{h}$ in $\mathfrak{g}$. Notice that the dimension of $\mathfrak{h}$ is presumed to be finite, though the dimension of $\mathfrak{g}$ may be infinite.

Set $K = U(\mathfrak{h})$ and $R = U(\mathfrak{g})$ to be the universal enveloping algebras of $\mathfrak{h}$ and $\mathfrak{g}$, and $f\colon K \longrightarrow R$ to be the morphism induced by the embedding $\mathfrak{h} \longrightarrow \mathfrak{g}$. Let $\mathcal{C} = \mathcal{C}(H)$ be the coalgebra of functions on H. Then $\mathcal{C}$, K, and R are Hopf algebras; the adjoint action of H in $\mathfrak{h}$ and the given action of H in $\mathfrak{g}$ provide us with right coactions $x \longmapsto x_{[0]} \otimes x_{[1]}$ and $u \longmapsto u_{[0]} \otimes u_{[1]}$ of $\mathcal{C}$ in K and R; and there is a natural pairing $\langle\,,\,\rangle\colon K \otimes_k \mathcal{C} \longrightarrow k$ such that the equations of C.1.1–C.1.2 and C.3.5 are satisfied.

C.4.2 So we obtain two opposite semialgebras $\boldsymbol{\mathcal{S}}^l = \boldsymbol{\mathcal{S}}^l(\mathfrak{g}, H)$ and $\boldsymbol{\mathcal{S}}^r = \boldsymbol{\mathcal{S}}^r(\mathfrak{g}, H)$ such that the categories of left $\boldsymbol{\mathcal{S}}^l$-semimodules and right $\boldsymbol{\mathcal{S}}^r$-semimodules are isomorphic to the category of Harish-Chandra modules over $(\mathfrak{g}, H)$. Recall that a Harish-Chandra module $\mathbf{N}$ over $(\mathfrak{g}, H)$ is a k-vector space endowed with $\mathfrak{g}$-module and H-module structures such that the two induced $\mathfrak{h}$-module structures coincide and the action map $\mathfrak{g} \otimes_k \mathbf{N} \longrightarrow \mathbf{N}$ is a morphism of H-modules. The assertion follows from C.1.4; indeed, it suffices to notice that the equations of C.1.4 hold whenever they hold for x and r belonging to some sets of generators of the algebras K and R.

Analogously, the category of left $\boldsymbol{\mathcal{S}}^r(\mathfrak{g}, H)$-semicontramodules is isomorphic to the category of k-vector spaces $\mathfrak{P}$ endowed with $\mathfrak{g}$-module and $\mathcal{C}(H)$-contramodule structures such that the two induced $\mathfrak{h}$-module structures coincide and the action map $\mathfrak{P} \longrightarrow \operatorname{Hom}_k(\mathfrak{g}, \mathfrak{P})$ is a morphism of $\mathcal{C}(H)$-contramodules. Here a left $\mathcal{C}$-contramodule structure induces an $\mathfrak{h}$-module structure by the formula $xp = -\pi_{\mathfrak{P}}(c \mapsto \langle x, c\rangle p)$ for $p \in \mathfrak{P}$, $x \in \mathfrak{h}$, $c \in \mathcal{C}$; for a left $\mathcal{C}$-comodule $\mathcal{M}$ and a left $\mathcal{C}$-contramodule $\mathfrak{P}$, the left $\mathcal{C}$-contramodule structure on $\operatorname{Hom}_k(\mathcal{M}, \mathfrak{P})$ is defined by the formula $\pi(g)(m) = \pi_{\mathfrak{P}}(c \mapsto g(s^{-1}(m_{(-1)})c)(m_{(0)}))$ for $m \in \mathcal{M}$, $g \in \operatorname{Hom}_k(\mathcal{C}, \operatorname{Hom}_k(\mathcal{M}, \mathfrak{P}))$. Vector spaces $\mathfrak{P}$ with such structures may be called *Harish-Chandra contramodules* over $(\mathfrak{g}, H)$ (see D.2.8).

C.4.3 Now assume that the algebraic group H is smooth (i.e., reduced). Let $\mathcal{E}$ be the $\mathcal{C}$-$\mathcal{C}$-bicomodule and right $\mathcal{C}$-module of differential top forms on H, with the bicomodule structure coming from the action of H on itself by left and right shifts and the module structure given by the multiplication of top forms with functions. Let $\mathcal{E}^\vee$ be the $\mathcal{C}$-$\mathcal{C}$-bicomodule of top polyvector fields on H. Then the equation of C.2.1 is clearly satisfied, and one can see that $(\mathcal{E}, \mathcal{E}^\vee)$ is a Morita autoequivalence of $\mathcal{C}$. The left $\mathcal{C}$-comodules E and $E^\vee$ can be identified with the top exterior powers of the vector spaces $\operatorname{Hom}_k(\mathfrak{h}, k)$ and $\mathfrak{h}$, respectively; c_E is the modular character of H. The inverse element anti-automorphism of H induces a map $s\colon \mathcal{E} \longrightarrow \mathcal{E}$ satisfying the equations of C.2.7.

Let us show that the equation $(*)$ of C.2.2 holds for $\mathcal{E}$. First let us check that it suffices to prove the desired equation for all x belonging to a set of generators of the algebra K. Indeed, one has $(xy)_{[0]} \to (j(xy)_{[1]}) = x_{[0]}y_{[0]} \to (jx_{[1]}y_{[1]}) = x_{[0](1)}y_{[0]}s(x_{[0](2)})x_{[0](3)} \to (jx_{[1]}y_{[1]}) = \langle x_{[0](1)}, y_{[1]} \rangle y_{[0]}x_{[0](2)} \to (jx_{[1]}y_{[2]}) = y_{[0]} \to (x_{[0](2)} \to (\langle x_{[0](1)}, y_{[1]} \rangle jx_{[1]}y_{[2]})) = y_{[0]} \to ((x_{[0]} \to (jx_{[1]}))y_{[1]})$ for $j \in \mathcal{E}$, $x, y \in K$, since $x_{(2)} \to (\langle x_{(1)}, c_{(1)} \rangle jc_{(2)}) = x_{(2)} \to (j(s^{-1}(x_{(1)}) \to c)) = (x_{(3)} \to j)(x_{(2)} \to (s^{-1}(x_{(1)}) \to c)) = (x \to j)c$ for $j \in \mathcal{E}$, $x \in K$, $c \in \mathcal{C}$. So it remains to check that the equation holds for $x \in \mathfrak{h} \subset K$.

For $h \in \mathfrak{h}$, let r_h and l_h denote the left- and right-invariant vector fields on H corresponding to h. Then one has $r_h = h_{[1]}l_{h_{[0]}}$, hence $\omega \leftarrow h = \mathrm{Lie}_{r_h} \omega = \mathrm{Lie}_{l_{h_{[0]}}}(\omega h_{[1]}) = h_{[0]} \to (\omega h_{[1]})$ for $\omega \in \mathcal{E}$, where $\mathrm{Lie}_v \omega$ denotes the Lie derivative of a top form ω along a vector field v.

C.4.4 Thus there is a left and right coflat Morita equivalence $(\mathcal{E}(\mathfrak{g}, H), \mathcal{E}^\vee(\mathfrak{g}, H))$ between the semialgebras $\mathbf{S}^r(\mathfrak{g}, H)$ and $\mathbf{S}^l(\mathfrak{g}, H)$. The functors of semitensor product with $\mathcal{E}(\mathfrak{g}, H)$ and $\mathcal{E}^\vee(\mathfrak{g}, H)$ provide mutually inverse equivalences between the category of left $\mathbf{S}^r$-semimodules and the category of left $\mathbf{S}^l$-semimodules. Notice that the underlying $\mathcal{C}$-comodule structures of semimodules are not preserved by this equivalence, but get twisted with the one-dimensional modular representation of H. The $\mathbf{S}^l$-$\mathbf{S}^r$-bisemimodule $\mathcal{E}(\mathfrak{g}, H)$ is endowed with an involutive automorphism transforming its two semimodule structures into each other in a way compatible with the antipode isomorphisms $\mathcal{C}^\mathrm{op} \simeq \mathcal{C}$ and $\mathbf{S}^{l\,\mathrm{op}} \simeq \mathbf{S}^r$. When the algebraic group H is unimodular (i.e., admits a biinvariant differential top form), the equation $(**)$ of C.2.6 is satisfied, so the semialgebras $\mathbf{S}^l(\mathfrak{g}, H)$ and $\mathbf{S}^r(\mathfrak{g}, H)$ are naturally isomorphic and endowed with an involutive anti-automorphism.

Remark. Let us assume for simplicity that the field k has characteristic 0 and the Harish-Chandra pair $(\mathfrak{g}, H)$ originates from an embedding of affine algebraic groups $H \subset G$. Then the bisemimodule $\mathcal{E}(\mathfrak{g}, H)$ can be interpreted geometrically as the Harish-Chandra bimodule of distributions on G, supported on H and regular along H. Technically, the desired vector space of distributions can be defined as the direct image of the right Diff_H-module of top forms on H under the closed embedding $H \longrightarrow G$, where Diff denotes the rings of differential operators [14]. This vector space has two commuting structures of a Harish-Chandra module over $(\mathfrak{g}, H)$, one given by the action of H by left shifts and the action of $\mathfrak{g}$ by right invariant vector fields, the other in the opposite way; so it can be considered as an $\mathbf{S}^l$-$\mathbf{S}^r$-bisemimodule. The desired map from the vector space $\mathcal{E} \simeq \mathcal{E} \otimes_K R$ to the space of distributions can be defined as the unique map forming a commutative diagram with the embeddings of the space of top forms $\mathcal{E}$ into both vector spaces and preserving the right R-module structures. To prove that this map is an isomorphism, it suffices to consider the filtration of $\mathcal{E}$ induced by the natural filtration of the universal enveloping algebra R and the filtration of the space of

distributions induced by the filtration of Diff_G by the order of differential operators. When the algebraic group H is unimodular, one can identify the semialgebra
$\mathbf{S}^l \simeq \mathbf{S} = \mathbf{S}^r$ itself with the above vector space of distributions by choosing a
nonzero biinvariant top form ω on H. The semiunit and semimultiplication in $\mathbf{S}$
are then described as follows. Given a function on $\mathcal{C}$, one has to multiply it with
ω and take the push-forward with respect to the closed embedding $H \longrightarrow G$ to
obtain the corresponding distribution under the semiunit map. To describe the
semimultiplication, denote by $G \times^H G$ the quotient variety of the Cartesian product $G \times G$ by the equivalence relation $(g'h, g'') \sim (g', hg'')$. Then the pull-back of
distributions with respect to the smooth map $G \times G \longrightarrow G \times^H G$ using the relative
top form ω identifies $\mathbf{S} \square_{\mathcal{C}} \mathbf{S}$ with the space of distributions on $G \times^H G$ supported
in $H \subset G \times^H G$ and regular along H. The push-forward of distributions with
respect to the multiplication map $G \times^H G \longrightarrow G$ provides the semimultiplication
in $\mathbf{S}$. (Cf. Appendix F.)

C.5 Semiinvariants and semicontrainvariants

C.5.1 Let $\mathfrak{h} \subset \mathfrak{g}$ be a Lie algebra with a finite-dimensional subalgebra; let N be
a $\mathfrak{g}$-module. Then there is a natural map

$$(\det(\mathfrak{h}) \otimes_k \mathfrak{g}/\mathfrak{h} \otimes_k N)^{\mathfrak{h}} \longrightarrow (\det(\mathfrak{h}) \otimes_k N)^{\mathfrak{h}},$$

where $\det(V)$ is the top exterior power of a finite-dimensional vector space V and
the superindex $\mathfrak{h}$ denotes the $\mathfrak{h}$-invariants. This natural map is constructed as
follows.

   Tensoring the action map $\mathfrak{g} \otimes_k N \longrightarrow N$ with $\det(\mathfrak{h})$ and passing to the $\mathfrak{h}$-in-
variants, we obtain a map $(\det(\mathfrak{h}) \otimes_k \mathfrak{g} \otimes_k N)^{\mathfrak{h}} \longrightarrow (\det(\mathfrak{h}) \otimes_k N)^{\mathfrak{h}}$. Let us check
that the composition

$$(\det(\mathfrak{h}) \otimes_k \mathfrak{h} \otimes_k N)^{\mathfrak{h}} \longrightarrow (\det(\mathfrak{h}) \otimes_k \mathfrak{g} \otimes_k N)^{\mathfrak{h}} \longrightarrow (\det(\mathfrak{h}) \otimes_k N)^{\mathfrak{h}}$$

vanishes. Notice that this composition only depends on the $\mathfrak{h}$-module structure
on N. Let n be an $\mathfrak{h}$-invariant element of $\det(\mathfrak{h}) \otimes_k \mathfrak{h} \otimes_k N$; it can be also considered
as an $\mathfrak{h}$-module map $n^* \colon (\det(\mathfrak{h}) \otimes_k \mathfrak{h})^* \longrightarrow N$, where V^* denotes the dual vector
space $\mathrm{Hom}_k(V, k)$. Let t denote the trace element of the tensor product $(\det(\mathfrak{h}) \otimes_k$
$\mathfrak{h}) \otimes_k (\det(\mathfrak{h}) \otimes_k \mathfrak{h})^*$; then t is an $\mathfrak{h}$-invariant element and one has $(\mathrm{id} \otimes n^*)(t) = n$.
So it remains to check that the image of t under the action map

$$(\det(\mathfrak{h}) \otimes_k \mathfrak{h}) \otimes_k (\det(\mathfrak{h}) \otimes_k \mathfrak{h})^* \longrightarrow \det(\mathfrak{h}) \otimes_k (\det(\mathfrak{h}) \otimes_k \mathfrak{h})^* \simeq \mathfrak{h}^*$$

vanishes; this is straightforward.

   We have constructed a map

$$(\det(\mathfrak{h}) \otimes_k \mathfrak{g} \otimes_k N)^{\mathfrak{h}}/(\det(\mathfrak{h}) \otimes_k \mathfrak{h} \otimes_k N)^{\mathfrak{h}} \longrightarrow (\det(\mathfrak{h}) \otimes_k N)^{\mathfrak{h}}.$$

When N is an injective $U(\mathfrak{h})$-module, this provides the desired map $(\det(\mathfrak{h}) \otimes_k \mathfrak{g}/\mathfrak{h} \otimes_k N)^{\mathfrak{h}} \longrightarrow (\det(\mathfrak{h}) \otimes_k N)^{\mathfrak{h}}$. To construct the latter map in the general case, it suffices to represent N as the kernel of a morphism of $U(\mathfrak{h})$-injective $U(\mathfrak{g})$-modules (notice that any injective $U(\mathfrak{g})$-module is an injective $U(\mathfrak{h})$-module). Indeed, both the left- and the right-hand sides of the desired map preserve kernels.

The vector space of $(\mathfrak{g}, \mathfrak{h})$-*semiinvariants* $N_{\mathfrak{g}, \mathfrak{h}}$ of a $\mathfrak{g}$-module N is defined as the cokernel of the map $(\det(\mathfrak{h}) \otimes_k \mathfrak{g}/\mathfrak{h} \otimes_k N)^{\mathfrak{h}} \longrightarrow (\det(\mathfrak{h}) \otimes_k N)^{\mathfrak{h}}$ that we have obtained. The $(\mathfrak{g}, \mathfrak{h})$-semiinvariants are a mixture of invariants along $\mathfrak{h}$ and coinvariants in the direction of $\mathfrak{g}$ relative to $\mathfrak{h}$.

C.5.2 Let P be another $\mathfrak{g}$-module. Then there is a natural map

$$\operatorname{Hom}_k(\det(\mathfrak{h}), P)_{\mathfrak{h}} \longrightarrow \operatorname{Hom}_k(\det(\mathfrak{h}) \otimes_k \mathfrak{g}/\mathfrak{h}, \, P)_{\mathfrak{h}},$$

where the subindex $\mathfrak{h}$ denotes the $\mathfrak{h}$-coinvariants. This map is constructed as follows.

Tensoring the action map $P \longrightarrow \operatorname{Hom}_k(\mathfrak{g}, P)$ with $\det(\mathfrak{h})^*$ and passing to the $\mathfrak{h}$-coinvariants, we obtain a map $\operatorname{Hom}_k(\det(\mathfrak{h}), P)_{\mathfrak{h}} \longrightarrow \operatorname{Hom}_k(\det(\mathfrak{h}) \otimes_k \mathfrak{g}, P)_{\mathfrak{h}}$. Let us check that the composition

$$\operatorname{Hom}_k(\det(\mathfrak{h}), P)_{\mathfrak{h}} \longrightarrow \operatorname{Hom}_k(\det(\mathfrak{h}) \otimes_k \mathfrak{g}, \, P)_{\mathfrak{h}} \longrightarrow \operatorname{Hom}_k(\det(\mathfrak{h}) \otimes_k \mathfrak{h}, \, P)_{\mathfrak{h}}$$

vanishes. Notice that this composition only depends on the $\mathfrak{h}$-module structure on P. Let p be an $\mathfrak{h}$-invariant map $\operatorname{Hom}_k(\det(\mathfrak{h}) \otimes_k \mathfrak{h}, P) \longrightarrow k$; it can be also considered as an $\mathfrak{h}$-module map $p^*: P \longrightarrow \det(\mathfrak{h}) \otimes \mathfrak{h}$. Then the map p factorizes through the map

$$\operatorname{Hom}_k(\det(\mathfrak{h}) \otimes_k \mathfrak{h}, \, P) \longrightarrow \operatorname{Hom}_k(\det(\mathfrak{h}) \otimes_k \mathfrak{h}, \, \det(\mathfrak{h}) \otimes \mathfrak{h})$$

induced by p^*. So it suffices to consider the case of a finite-dimensional $\mathfrak{h}$-module $P = \det(\mathfrak{h}) \otimes \mathfrak{h}$, when the assertion follows by duality from the result of C.5.1.

We have constructed a map from $\operatorname{Hom}_k(\det(\mathfrak{h}), P)_{\mathfrak{h}}$ to the kernel of the map $\operatorname{Hom}_k(\det(\mathfrak{h}) \otimes_k \mathfrak{g}, P)_{\mathfrak{h}} \longrightarrow \operatorname{Hom}_k(\det(\mathfrak{h}) \otimes_k \mathfrak{h}, P)_{\mathfrak{h}}$. When P is a projective $U(h)$-module, this provides the desired map $\operatorname{Hom}_k(\det(\mathfrak{h}), P)_{\mathfrak{h}} \longrightarrow \operatorname{Hom}_k(\det(\mathfrak{h}) \otimes_k \mathfrak{g}/\mathfrak{h}, \, P)_{\mathfrak{h}}$. In the general case, represent P as the cokernel of a morphism of $U(\mathfrak{h})$-projective $U(\mathfrak{g})$-modules and notice that both the left- and the right-hand side of the desired map preserve cokernels.

The vector space of $(\mathfrak{g}, \mathfrak{h})$-*semicontrainvariants* $P^{\mathfrak{g}, \mathfrak{h}}$ of a $\mathfrak{g}$-module P is defined as the kernel of the map $\operatorname{Hom}_k(\det(\mathfrak{h}), P)_{\mathfrak{h}} \longrightarrow \operatorname{Hom}_k(\det(\mathfrak{h}) \otimes_k \mathfrak{g}/\mathfrak{h}, P)_{\mathfrak{h}}$ that we have obtained. The $(\mathfrak{g}, \mathfrak{h})$-semicontrainvariants are a mixture of coinvariants along $\mathfrak{h}$ and invariants in the direction of $\mathfrak{g}$ relative to $\mathfrak{h}$.

C.5.3 Now let $(\mathfrak{g}, H)$ be an algebraic Harish-Chandra pair. Let $\mathcal{N}$ be a right $\mathcal{S}^r(\mathfrak{g}, H)$-semimodule, that is a Harish-Chandra module over $(\mathfrak{g}, H)$. Then the action map $\mathfrak{g} \otimes_k \mathcal{N} \longrightarrow \mathcal{N}$ is a morphism of Harish-Chandra modules. Tensoring it with $\det(\mathfrak{h})$ and passing to the H-invariants, we obtain a map

$$(\det(\mathfrak{h}) \otimes_k \mathfrak{g} \otimes_k \mathcal{N})^H \longrightarrow (\det(\mathfrak{h}) \otimes_k \mathcal{N})^H.$$

By the result of C.5.1, the composition

$$(\det(\mathfrak{h}) \otimes_k \mathfrak{h} \otimes_k \mathbf{N})^H \longrightarrow (\det(\mathfrak{h}) \otimes_k \mathfrak{g} \otimes_k \mathbf{N})^H \longrightarrow (\det(\mathfrak{h}) \otimes_k \mathbf{N})^H$$

vanishes. When $\mathbf{N}$ is a coflat $\mathcal{C}(H)$-comodule, this provides a natural map

$$(\det(\mathfrak{h}) \otimes_k \mathfrak{g}/\mathfrak{h} \otimes_k \mathbf{N})^H \longrightarrow (\det(\mathfrak{h}) \otimes_k \mathbf{N})^H;$$

to define this map in the general case, it suffices to represent $\mathbf{N}$ as the kernel of a morphism of $\mathcal{C}$-coflat $\mathbf{S}^r$-semimodules (see Lemma 1.3.3).

The vector space of $(\mathfrak{g}, H)$-*semiinvariants* $\mathbf{N}_{\mathfrak{g},H}$ is defined as the cokernel of the map $(\det(\mathfrak{h}) \otimes_k \mathfrak{g}/\mathfrak{h} \otimes_k \mathbf{N})^H \longrightarrow (\det(\mathfrak{h}) \otimes_k \mathbf{N})^H$ that we have constructed.

C.5.4 Let $\mathfrak{P}$ be a left $\mathbf{S}^r(\mathfrak{g}, H)$-semicontramodule (see C.4.2). Then the action map $\mathfrak{P} \longrightarrow \operatorname{Hom}_k(\mathfrak{g}, \mathfrak{P})$ is a morphism of $\mathbf{S}^r$-semicontramodules. Applying to it the functor $\operatorname{Hom}_k(\det(\mathfrak{h}), -)$, we get a morphism of $\mathcal{C}(H)$-contramodules. Passing to the H-coinvariants, i.e., the maximal quotient $\mathcal{C}$-contramodules with the trivial contraaction, we obtain a map

$$\operatorname{Hom}_k(\det(\mathfrak{h}), \mathfrak{P})_H \longrightarrow \operatorname{Hom}_k(\det(\mathfrak{h}) \otimes_k \mathfrak{g}, \mathfrak{P})_H.$$

By the result of C.5.2, the composition

$$\operatorname{Hom}_k(\det(\mathfrak{h}), \mathfrak{P})_H \longrightarrow \operatorname{Hom}_k(\det(\mathfrak{h}) \otimes_k \mathfrak{g}, \mathfrak{P})_H \longrightarrow \operatorname{Hom}_k(\det(\mathfrak{h}) \otimes_k \mathfrak{h}, \mathfrak{P})_H$$

vanishes. When $\mathfrak{P}$ is a coinjective $\mathcal{C}$-contramodule, this provides a natural map

$$\operatorname{Hom}_k(\det(\mathfrak{h}), \mathfrak{P})_H \longrightarrow \operatorname{Hom}_k(\det(\mathfrak{h}) \otimes_k \mathfrak{g}/\mathfrak{h}, \mathfrak{P})_H;$$

to define this map in the general case, it suffices to represent $\mathfrak{P}$ as the cokernel of a morphism of $\mathcal{C}$-coinjective $\mathbf{S}^r$-semicontramodules (see Lemma 3.3.3).

The vector space of $(\mathfrak{g}, H)$-*semicontrainvariants* $\mathfrak{P}^{\mathfrak{g},H}$ is defined as the kernel of the map $\operatorname{Hom}_k(\det(\mathfrak{h}), \mathfrak{P})_H \longrightarrow \operatorname{Hom}_k(\det(\mathfrak{h}) \otimes_k \mathfrak{g}/\mathfrak{h}, \mathfrak{P})_H$ that we have constructed.

C.5.5 Let $\mathbf{N}$ be a right $\mathbf{S}^r(\mathfrak{g}, H)$-semimodule and $\mathbf{M}$ be a left $\mathbf{S}^l(\mathfrak{g}, H)$-semimodule; assume that either $\mathbf{N}$ or $\mathbf{M}$ is a coflat $\mathcal{C}(H)$-comodule. Then there is a natural isomorphism

$$\mathbf{N} \lozenge_{\mathbf{S}^r} \mathcal{E}^\vee \lozenge_{\mathbf{S}^r} \mathbf{M} \simeq (\mathbf{N} \otimes_k \mathbf{M})_{\mathfrak{g},H},$$

where $\mathbf{N} \otimes_k \mathbf{M}$ is considered as the tensor product of Harish-Chandra modules $\mathbf{N}$ and $\mathbf{M}$.

Indeed, introduce an increasing filtration F of the k-algebra $R = U(\mathfrak{g})$ whose component $F_t R$, $t = 0, 1, \ldots$ is the linear span of all products of elements of $\mathfrak{g}$ where at most t factors do not belong to $\mathfrak{h}$. In particular, we have $F_0 R \simeq K = U(\mathfrak{h})$. Set

$$F_t \mathbf{S}^r = \mathcal{C} \otimes_K F_t R;$$

then we have $F_0\boldsymbol{S}^r \simeq \mathcal{C}$, the natural maps $F_{t-1}\boldsymbol{S}^r \longrightarrow F_t\boldsymbol{S}^r$ are injective, their cokernels are coflat left and right $\mathcal{C}$-comodules, $\boldsymbol{S}^r \simeq \varinjlim F_t\boldsymbol{S}^r$, and the semimultiplication map $F_p\boldsymbol{S}^r \square_\mathcal{C} F_q\boldsymbol{S}^r \longrightarrow \boldsymbol{S}\square_\mathcal{C}\boldsymbol{S} \longrightarrow \boldsymbol{S}$ factorizes through $F_{p+q}\boldsymbol{S}^r$. Moreover, the maps $F_p\boldsymbol{S}^r \square_\mathcal{C} F_q\boldsymbol{S}^r \longrightarrow F_{p+q}\boldsymbol{S}^r$ are surjective and their kernels are coflat left and right $\mathcal{C}$-comodules. (Cf. 11.5.)

Let $\mathcal{N}$ be a right $\boldsymbol{S}^r$-semimodule and $\mathcal{L}$ be a left $\boldsymbol{S}^r$-semimodule such that either $\mathcal{N}$ or $\mathcal{L}$ is a coflat $\mathcal{C}$-comodule. Denote by

$$\eta_t \colon \mathcal{N}\square_\mathcal{C} F_t\boldsymbol{S}^r \square_\mathcal{C} \mathcal{L} \longrightarrow \mathcal{N}\square_\mathcal{C} \mathcal{L}$$

the map equal to the difference of the map induced by the semiaction map $F_t\boldsymbol{S}^r \square_\mathcal{C} \mathcal{L} \longrightarrow \mathcal{L}$ and the map induced by the semiaction map $\mathcal{N}\square_\mathcal{C} F_t\boldsymbol{S}^r \longrightarrow \mathcal{N}$. Let us show that the images of η_t coincide for $t \geqslant 1$. Let $p, q \geqslant 1$; then the map

$$\mathcal{N}\square_\mathcal{C} F_p\boldsymbol{S}^r \square_\mathcal{C} F_q\boldsymbol{S}^r \square_\mathcal{C} \mathcal{L} \longrightarrow \mathcal{N}\square_\mathcal{C} F_{p+q}\boldsymbol{S}^r \square_\mathcal{C} \mathcal{L}$$

is surjective in view of our assumption on $\mathcal{N}$ and $\mathcal{L}$. The composition of the map $\mathcal{N}\square_\mathcal{C} F_p\boldsymbol{S}^r \square_\mathcal{C} F_q\boldsymbol{S}^r \square_\mathcal{C} \mathcal{L} \longrightarrow \mathcal{N}\square_\mathcal{C} F_{p+q}\boldsymbol{S}^r \square_\mathcal{C} \mathcal{L}$ with the map η_{p+q} is equal to the sum of the composition of the map $\mathcal{N}\square_\mathcal{C} F_p\boldsymbol{S}^r \square_\mathcal{C} F_q\boldsymbol{S}^r \square_\mathcal{C} \mathcal{L} \longrightarrow \mathcal{N}\square_\mathcal{C} F_p\boldsymbol{S}^\sim \square_\mathcal{C} \mathcal{L}$ induced by the semiaction map $F_q\boldsymbol{S}^r \square_\mathcal{C} \mathcal{L} \longrightarrow \mathcal{L}$ and the map η_p, and the composition of the map $\mathcal{N}\square_\mathcal{C} F_p\boldsymbol{S}^r \square_\mathcal{C} F_q\boldsymbol{S}^r \square_\mathcal{C} \mathcal{L} \longrightarrow \mathcal{N}\square_\mathcal{C} F_q\boldsymbol{S}^r \square_\mathcal{C} \mathcal{L}$ induced by the semiaction map $\mathcal{N}\square_\mathcal{C} F_p\boldsymbol{S}^r \longrightarrow \mathcal{N}$ and the map η_q. So the assertion follows by induction. Therefore, the semitensor product $\mathcal{N}\lozenge_{\boldsymbol{S}} \mathcal{L}$ is isomorphic to the cokernel of the map η_1.

On the other hand, the map η_0 vanishes. In view of our assumption on $\mathcal{N}$ and $\mathcal{L}$, the quotient space $(\mathcal{N}\square_\mathcal{C} F_1\boldsymbol{S}^r \square_\mathcal{C} \mathcal{L})/(\mathcal{N}\square_\mathcal{C} F_0\boldsymbol{S}^r \square_\mathcal{C} \mathcal{L})$ is isomorphic to $\mathcal{N}\square_\mathcal{C} F_1\boldsymbol{S}^r/F_0\boldsymbol{S}^r \square_\mathcal{C} \mathcal{L}$. Hence the semitensor product $\mathcal{N}\lozenge_{\boldsymbol{S}} \mathcal{L}$ is isomorphic to the cokernel of the induced map

$$\bar\eta_1 \colon \mathcal{N}\square_\mathcal{C} (F_1\boldsymbol{S}^r/F_0\boldsymbol{S}^r)\square_\mathcal{C} \mathcal{L} \longrightarrow \mathcal{N}\square_\mathcal{C} \mathcal{L}.$$

Now when $\mathcal{L} = \mathcal{E}^\vee \lozenge_{\boldsymbol{S}} \mathcal{M}$ for a left $\boldsymbol{S}^l$-semimodule $\mathcal{M}$, the natural isomorphisms $\mathcal{N}\square_\mathcal{C} \mathcal{E}^\vee \square_\mathcal{C} \mathcal{M} \simeq (\mathcal{N}\otimes_k E^\vee \otimes_k \mathcal{M})^H \simeq (E^\vee \otimes_k \mathcal{N}\otimes_k \mathcal{M})^H$ and $\mathcal{N}\square_\mathcal{C} (F_1\boldsymbol{S}^r/F_0\boldsymbol{S}^r)\square_\mathcal{C} \mathcal{E}^\vee \square_\mathcal{C} \mathcal{M} \simeq \mathcal{N}\square_\mathcal{C} (\mathcal{C}\otimes_k \mathfrak{g}/\mathfrak{h})\square_\mathcal{C} \mathcal{E}^\vee \square_\mathcal{C} \mathcal{M} \simeq (\mathcal{N}\otimes_k \mathfrak{g}/\mathfrak{h}\otimes_k E^\vee \otimes_k \mathcal{M})^H \simeq (E^\vee \otimes_k \mathfrak{g}/\mathfrak{h}\otimes_k \mathcal{N}\otimes_k \mathcal{M})^H$ given by the formulas $i\otimes n\otimes m' \longmapsto n\otimes i\otimes m$ and $n\otimes \bar z\otimes i\otimes m \longmapsto i\otimes \bar z\otimes n\otimes m$ identify the map $\bar\eta_1$ with the map whose cokernel is, by the definition, the space of semiinvariants $(\mathcal{N}\otimes_k \mathcal{M})_{\mathfrak{g},H}$.

C.5.6 Let $\mathcal{M}$ be a left $\boldsymbol{S}^l(\mathfrak{g}, H)$-semimodule and $\mathfrak{P}$ be a left $\boldsymbol{S}^r(\mathfrak{g}, H)$-semicontramodule; assume that either $\mathcal{M}$ is an coprojective $\mathcal{C}(H)$-comodule, or $\mathcal{N}$ is a coinjective $\mathcal{C}(H)$-contramodule. Then there is a natural isomorphism

$$\mathrm{SemiHom}_{\boldsymbol{S}^r}(\mathcal{E}^\vee \lozenge_{\boldsymbol{S}^l} \mathcal{M}, \mathfrak{P}) \simeq \mathrm{Hom}_k(\mathcal{M}, \mathfrak{P})^{\mathfrak{g},H},$$

where the structure of left $\boldsymbol{S}^r$-semicontramodule on $\mathrm{Hom}_k(\mathcal{M}, \mathfrak{P})$ was introduced in C.3.6. The proof is analogous to that of C.5.5.

D Tate Harish-Chandra Pairs and Tate Lie Algebras

by Sergey Arkhipov and Leonid Positselski

In order to formulate the comparison theorem relating the functors SemiTor and SemiExt to the semi-infinite (co)homology of Tate Lie algebras, one has to consider Harish-Chandra pairs $(\mathfrak{g}, H)$ with a Tate Lie algebra $\mathfrak{g}$ and a proalgebraic group H corresponding to a compact open subalgebra $\mathfrak{h} \subset \mathfrak{g}$. In such a situation, the construction of a Morita equivalence from Appendix C no longer works; instead, there is an isomorphism of "left" and "right" semialgebras corresponding to different central charges. The proof of this isomorphism is based on the non-homogeneous quadratic duality theory developed in Chapter 11 (see also 0.4). Once the isomorphism of semialgebras is constructed and the standard semi-infinite (co)homological complexes are introduced, the proof of the comparison theorem becomes pretty straightforward. The equivalence between the semiderived categories of Harish-Chandra modules and Harish-Chandra contramodules with complementary (or rather, shifted) central charges follows immediately from the isomorphism of semialgebras.

D.1 Continuous coactions

D.1.1 Let k be a fixed ground field. A *linear topology* on a vector space over k is a topology compatible with the vector space structure for which open vector subspaces form a base of neighborhoods of zero. In the sequel, by a topological vector space we will mean a k-vector space endowed with a complete and separated linear topology. Equivalently, a topological vector space is a filtered projective limit of discrete vector spaces with its projective limit topology. Accordingly, the (separated) completion of a vector space endowed with a linear topology is just the projective limit of its quotient spaces by open vector subspaces.

The category of topological vector spaces and continuous linear maps between them has an exact category structure in which a triple of topological vector spaces $V' \longrightarrow V \longrightarrow V''$ is exact if it is an exact triple of vector spaces strongly compatible with the topologies, i.e., the map $V' \longrightarrow V$ is closed and the map $V \longrightarrow V''$ is open. Any open surjective map of topological vector spaces is an admissible epimorphism. Any closed injective map from a topological vector space admitting a countable base of neighborhoods of zero is a split admissible monomorphism.

A topological vector space is called (*linearly*) *compact* if it has a base of neighborhoods of zero consisting of vector subspaces of finite codimension. Equivalently, a topological vector space is compact if it is a projective limit of finite-dimensional discrete vector spaces. A *Tate vector space* is a topological vector space admitting a compact open subspace. Equivalently, a topological vector space is a Tate vector space if it is topologically isomorphic to the direct sum of a compact vector space and a discrete vector space. The *dual Tate vector space* $V^\vee$ to a Tate vector space V is defined as the space of continuous linear functions $V \longrightarrow k$ endowed with the topology where annihilators of compact open subspaces of V form a base of neighborhoods of zero. In particular, the dual Tate vector spaces to compact vector spaces are discrete and vice versa; for any Tate vector space V, the natural map $V \longrightarrow (V^\vee)^\vee$ is a topological isomorphism.

D.1.2 The projective limit of a projective system of topological vector spaces endowed with the topology of projective limit is a topological vector space. This is called the *topological projective limit*.

The inductive limit of an inductive system of topological vector spaces can be endowed with the topology of inductive limit of vector spaces with linear topologies; we will call the inductive limit endowed with this topology the *uncompleted inductive limit*. The *completed inductive limit* is the (separated) completion of the uncompleted inductive limit. For any countable filtered inductive system formed by closed embeddings of topological vector spaces the uncompleted and completed inductive limits coincide. Moreover, let V_α be a filtered inductive system of topological vector spaces satisfying the following condition. For any increasing sequence of indices $\alpha_1 \leqslant \alpha_2 \leqslant \cdots$ the uncompleted inductive limit of V_{α_i} is a direct summand of the uncompleted inductive limit of V_α considered as an object of the category of vector spaces endowed with noncomplete linear topologies. Then the uncompleted and completed inductive limits of V_α coincide.

D.1.3 We will consider three operations of tensor product of topological vector spaces [8]. For any two topological vector spaces V and W, denote by $V \otimes^! W$ the completion of the tensor product $V \otimes_k W$ with respect to the topology with a base of neighborhoods of zero consisting of the vector subspaces $V' \otimes W + V \otimes W'$, where $V' \subset V$ and $W' \subset W$ are open vector subspaces in V and W.

Furthermore, denote by $V \otimes^* W$ the completion of $V \otimes_k W$ with respect to the topology formed by the subspaces of $V \otimes W$ satisfying the following conditions: a vector subspace $T \subset V \otimes W$ is open if

(i) there exist open subspaces $V' \subset V$, $W' \subset W$ such that $V' \otimes W' \subset T$,

(ii) for any vector $v \in V$ there exists a subspace $W'' \subset W$ such that $v \otimes W'' \subset T$, and

(iii) for any vector $w \in W$ there exists a subspace $V'' \subset V$ such that $V'' \otimes w \subset T$.

Finally, denote by $V \overrightarrow{\otimes} W$ the completion of $V \otimes_k W$ with respect to the topology formed by the subspaces satisfying the following conditions: a vector subspace $T \subset V \otimes_k W$ is open if

(i) there exists an open subspace $W' \subset W$ such that $V \otimes_k W' \subset T$, and

(ii) for any vector $w \in W$ there exists an open subspace $V'' \subset V$ such that $V'' \otimes w \subset T$.

Set $W \overleftarrow{\otimes} V = V \overrightarrow{\otimes} W$.

The topological tensor products $\otimes^!$ and $\otimes^*$ define two structures of associative and commutative tensor category on the category of topological vector spaces. The topological tensor product $\overrightarrow{\otimes}$ defines a structure of associative, but not commutative tensor category on the category of topological vector spaces. For any topological vector spaces $V_1, \ldots, V_n$ and W the vector space of continuous polylinear maps $V_1 \times \cdots \times V_n \longrightarrow W$ is naturally isomorphic to the vector space of continuous linear maps $V_1 \otimes^* \cdots \otimes^* V_n \longrightarrow W$. When both topological vector spaces V and W are compact (discrete), the topological tensor product $V \otimes^* W \simeq V \overrightarrow{\otimes} W \simeq V \overleftarrow{\otimes} W \simeq V \otimes^! W$ is also compact (discrete). The functor $\otimes^!$ preserves topological projective limits. The functor $\otimes^*$ preserves (uncompleted or completed) inductive limits of filtered inductive systems of open injections. The topological tensor product $V \overrightarrow{\otimes} W$ is the topological projective limit of $\overrightarrow{\otimes}$-products of V with discrete quotient spaces of W. The functor $(V, W) \longmapsto V \overrightarrow{\otimes} W$ preserves completed inductive limits in its second argument W. The underlying vector space of the topological tensor product $V \overrightarrow{\otimes} W$ is determined by (the topological vector space W and) the underlying vector space of the topological vector space V.

For Tate vector spaces $V_1, \ldots, V_n$ and a topological vector space U, consider the vector space of continuous polylinear maps $\times_i V_i \longrightarrow U$ endowed with the topology with a base of neighborhoods of zero formed by the subspaces of all polylinear maps mapping the Cartesian product of a collection of compact subspaces $V_i' \subset V_i$ into an open subspace $U' \subset U$ (the "compact-open" topology). This vector space is naturally topologically isomorphic to the topological tensor product $V_1^\vee \otimes^! \cdots \otimes^! V_n^\vee \otimes^! W$ [10, 3.8.17]. For any topological vector spaces U, W and Tate vector space V, the vector space of continuous linear maps $V \otimes^* W \longrightarrow U$ is naturally isomorphic to the vector space of continuous linear maps $W \longrightarrow V^\vee \otimes^! U$.

D.1.4 Let $\mathcal{C}$ be a coalgebra over the field k and V be a topological vector space. A *continuous right coaction* of $\mathcal{C}$ in V is a continuous linear map

$$V \longrightarrow V \otimes^! \mathcal{C},$$

where $\mathcal{C}$ is considered as a discrete vector space, satisfying the coassociativity and counity equations. Namely, the map $V \longrightarrow V \otimes^! \mathcal{C}$ should have equal compositions with the two maps $V \otimes^! \mathcal{C} \rightrightarrows V \otimes^! \mathcal{C} \otimes^! \mathcal{C}$ induced by the map $V \longrightarrow V \otimes^! \mathcal{C}$ and the comultiplication in $\mathcal{C}$, and the composition of the map $V \longrightarrow V \otimes^! \mathcal{C}$ with the map $V \otimes^! \mathcal{C} \longrightarrow V$ induced by the counit of $\mathcal{C}$ should be equal to the identity map. Equivalently, a continuous right coaction of $\mathcal{C}$ in V can be defined as a continuous linear map

$$V \otimes^* \mathcal{C}^\vee \longrightarrow V,$$

where $\mathcal{C}^\vee$ is considered as a compact vector space, satisfying the associativity and unity equations. *Continuous left coactions* are defined in an analogous way.

A closed subspace $W \subset V$ of a topological vector space V endowed with a continuous right coaction of a coalgebra $\mathcal{C}$ is said to be invariant with respect to the continuous coaction (or $\mathcal{C}$-invariant) if the image of W under the continuous coaction map $V \longrightarrow V \otimes^! \mathcal{C}$ is contained in the closed subspace $W \otimes^! \mathcal{C} \subset V \otimes^! \mathcal{C}$. It follows from the next lemma that any topological vector space with a continuous coaction of a coalgebra $\mathcal{C}$ is a filtered projective limit of discrete vector spaces endowed with $\mathcal{C}$-comodule structures.

Lemma. *For any topological vector space V endowed with a continuous coaction $V \longrightarrow V \otimes^! \mathcal{C}$ of a coalgebra $\mathcal{C}$, open subspaces of V invariant under the continuous coaction form a base of neighborhoods of zero in V.*

Proof. Let $U \subset V$ be an open subspace; then the full preimage U' of the open subspace $U \otimes^! \mathcal{C} \subset V \otimes^! \mathcal{C}$ under the continuous coaction map $V \longrightarrow V \otimes^! \mathcal{C}$ is an invariant open subspace in V contained in U. To check that U' is $\mathcal{C}$-invariant, use the fact the functor of $\otimes^!$-product preserves kernels in the category of topological vector spaces, and in particular, the $\otimes^!$-product with $\mathcal{C}$ preserves the kernel of the composition $V \longrightarrow V \otimes^! \mathcal{C} \longrightarrow V/U \otimes^! \mathcal{C}$. To check that U' is contained in U, use the counity equation for the continuous coaction. $\qquad\square$

The category of topological vector spaces endowed with a continuous coaction of a coalgebra $\mathcal{C}$ has an exact category structure such that a triple of topological vector spaces with continuous coactions of $\mathcal{C}$ is exact if and only if it is exact as a triple of topological vector spaces.

If V is a Tate vector space with a continuous right coaction of $\mathcal{C}$, then the dual Tate vector space $V^\vee$ is endowed with a continuous left coaction of $\mathcal{C}$.

Let V be a topological vector space with a continuous right coaction of a coalgebra $\mathcal{C}$ and W be a topological vector space with a continuous coaction of a coalgebra $\mathcal{D}$. Then all the three topological tensor products $V \otimes^! W$, $V \otimes^* W$, and $V \overrightarrow{\otimes} W$ are endowed with continuous right coactions of the coalgebra $\mathcal{C} \otimes_k \mathcal{D}$. To construct the continuous coaction on $V \otimes^! W$, one uses the natural isomorphism

$$(V \otimes^! \mathcal{C}) \otimes^! (W \otimes^! \mathcal{D}) \simeq (V \otimes^! W) \otimes^! (\mathcal{C} \otimes_k \mathcal{D}).$$

The continuous coaction on $V \otimes^* W$ is defined in terms of the natural continuous map

$$(V \otimes^! \mathcal{C}) \otimes^* (W \otimes^! \mathcal{D}) \longrightarrow (V \otimes^* W) \otimes^! (\mathcal{C} \otimes_k \mathcal{D}),$$

which exists for any topological vector spaces V, W and any discrete vector spaces $\mathcal{C}$, $\mathcal{D}$. The continuous coaction on $V \overrightarrow{\otimes} W$ is defined in terms of the natural continuous map

$$(V \otimes^! \mathcal{C}) \overrightarrow{\otimes} (W \otimes^! \mathcal{D}) \longrightarrow (V \overrightarrow{\otimes} W) \otimes^! (\mathcal{C} \otimes_k \mathcal{D}).$$

It follows that for a commutative Hopf algebra $\mathcal{C}$ the topological tensor products $V \otimes^! W$, $V \otimes^* W$, and $V \vec{\otimes} W$ of topological vector spaces with continuous right coactions of $\mathcal{C}$ are also endowed with continuous right coactions of $\mathcal{C}$. Besides, one can transform a continuous left coaction of $\mathcal{C}$ in V into a continuous right coaction using the antipode.

Now let W, U be topological vector spaces and V be a Tate vector space; suppose that W, U, and V are endowed with continuous coactions of a commutative Hopf algebra $\mathcal{C}$. Let $f \colon V \otimes^* W \longrightarrow U$ and $g \colon W \longrightarrow V^\vee \otimes^! U$ be continuous linear maps corresponding to each other under the isomorphism from D.1.3; then f preserves the continuous coactions of $\mathcal{C}$ if and only if g does.

D.1.5 A topological Lie algebra $\mathfrak{g}$ is a topological vector space endowed with a Lie algebra structure such that the bracket is a continuous bilinear map $\mathfrak{g} \times \mathfrak{g} \longrightarrow \mathfrak{g}$. Topological associative algebras are defined in an analogous way. For example, let V be a Tate vector space. Denote by $\mathrm{End}(V)$ the associative algebra of continuous endomorphisms of V endowed with the compact-open topology and by $\mathfrak{gl}(V)$ the Lie algebra corresponding to $\mathrm{End}(V)$. Then $\mathrm{End}(V)$ is a topological associative algebra and $\mathfrak{gl}(V)$ is a topological Lie algebra.

Let U, V, W be topological vector spaces endowed with continuous coactions of a commutative Hopf algebra $\mathcal{C}$. Then a continuous bilinear map $V \times W \longrightarrow U$ is called compatible with the continuous coactions of $\mathcal{C}$ if the corresponding linear map $V \otimes^* W \longrightarrow U$ preserves the continuous coactions of $\mathcal{C}$. So one can speak about compatibility of continuous pairings, Lie or associative algebra structures, Lie or associative actions, etc., with continuous coactions of a commutative Hopf algebra.

Explicitly, a bilinear map $V \times W \longrightarrow U$ is continuous and compatible with the continuous coactions of $\mathcal{C}$ if and only if the following condition holds. For any $\mathcal{C}$-invariant open subspace $U' \subset U$ and any finite-dimensional subspaces $E \subset V$, $F \subset W$ there should exist invariant open subspaces $V' \subset V'' \subset V$, $W' \subset W'' \subset W$ such that $E \subset V''$, $F \subset W''$, the map $V'' \otimes_k W'' \longrightarrow U/U'$ factorizes through $V''/V' \otimes_k W''/W'$, and the induced map $V''/V' \otimes_k W''/W' \longrightarrow U/U'$ is a morphism of $\mathcal{C}$-comodules.

D.1.6 For any Tate vector spaces V and W, there is a split exact triple of topological vector spaces

$$V \otimes^* W \longrightarrow V \vec{\otimes} W \oplus W \vec{\otimes} V \longrightarrow V \otimes^! W,$$

where the first map is the sum of the natural maps $V \otimes^* W \longrightarrow V \vec{\otimes} W$, $V \otimes^* W \longrightarrow W \vec{\otimes} V$, while the second map is the difference of the natural maps $V \vec{\otimes} W \longrightarrow V \otimes^! W$, $W \vec{\otimes} V \longrightarrow V \otimes^! W$. Let us take $W = V^\vee$. Then $V \otimes^! V^\vee$ is naturally isomorphic to $\mathfrak{gl}(V)$; the spaces $V \vec{\otimes} V^\vee$ and $V^\vee \vec{\otimes} V$ can be identified with the subspaces in $\mathfrak{gl}(V)$ formed by the linear operators with open kernel and compact closure of image, respectively; and $V \otimes^* V^\vee$ is the intersection of $V \vec{\otimes} V^\vee$ and $V^\vee \vec{\otimes} V$ in $\mathfrak{gl}(V)$.

Taking the push-forward of the exact triple $V \otimes^* V^\vee \longrightarrow V \vec{\otimes} V^\vee \oplus V^\vee \vec{\otimes} V \longrightarrow V \otimes^! V^\vee$ with respect to the natural trace map tr : $V \otimes^* V^\vee \longrightarrow k$ corresponding to the pairing $V \times V^\vee \longrightarrow k$, one obtains an exact triple of topological vector spaces

$$k \longrightarrow \mathfrak{gl}(V)^\sim \longrightarrow \mathfrak{gl}(V).$$

This is also an exact triple of $\mathfrak{gl}(V)$-modules, which allows one to define [10, 2.7.8] a Lie algebra structure on $\mathfrak{gl}(V)^\sim$ making it a central extension of the Lie algebra $\mathfrak{gl}(V)$. The anti-commutativity and the Jacobi identity follow from the fact that the commutator of an operator with open kernel and an operator with compact closure of image has zero trace.

Now assume that a Tate vector space V is endowed with a continuous coaction of a commutative Hopf algebra $\mathcal{C}$. Then $V \otimes^* V^\vee \longrightarrow V \vec{\otimes} V^\vee \oplus V^\vee \vec{\otimes} V \longrightarrow V \otimes^! V^\vee$ is an exact triple of topological vector spaces endowed with continuous coactions of $\mathcal{C}$; the trace map also preserves the continuous coactions. Thus the topological vector space $\mathfrak{gl}(V)^\sim$ acquires a continuous coaction of $\mathcal{C}$.

D.1.7 Here is another construction of the Lie algebra $\mathfrak{gl}(V)^\sim$ (see [10, 3.8.17–18]). Consider the quotient space of the vector space $V \otimes_k V^\vee \oplus V^\vee \otimes_k V \oplus k$ by the relation $v \otimes g + g \otimes v = \langle g, v \rangle$, where $\langle \, , \, \rangle$ denotes the pairing of $V^\vee$ with V. This vector space is a Lie subalgebra of the Clifford algebra $\mathrm{Cl}(V \oplus V^\vee)$ of the vector space $V \oplus V^\vee$ with the symmetric bilinear form given by the pairing $\langle \, , \, \rangle$; the Lie bracket on this subalgebra is given by the formulas

$$[v_1 \otimes g_1, \, v_2 \otimes g_2] = \langle g_1, v_2 \rangle v_1 \otimes g_2 - \langle g_2, v_1 \rangle v_2 \otimes g_1, \qquad [v \otimes g, \, 1] = 0.$$

This Lie algebra acts in the vector space V by the formulas $(v \otimes g)(v') = \langle g, v' \rangle v$, $1(v) = 0$. There is a separated topology on this Lie algebra with a base of neighborhoods of zero formed by the Lie subalgebras $V \otimes W' + V' \otimes V^\vee$, where $V' \subset V$ and $W' \subset V^\vee$ are open subspaces such that $\langle W', V' \rangle = 0$. The completion of this Lie algebra with respect to this topology can be easily identified with the Lie algebra $\mathfrak{gl}(V)^\sim$ defined above.

Hence the Lie bracket on $\mathfrak{gl}(V)^\sim$ is continuous. In addition, we need to check that when V is endowed with a continuous coaction of a commutative Hopf algebra $\mathcal{C}$, the Lie bracket is compatible with the continuous coaction of $\mathcal{C}$ in $\mathfrak{gl}(V)^\sim$. The latter follows from the existence of a well-defined commutator map

$$\mathrm{Hom}(X_4, X_3, X_1; \, X/X_1, X_4/X_1, X_2/X_1)^{\otimes 2} \longrightarrow \mathfrak{gl}(X)^\sim / (X \otimes X_3^\perp + X_2 \otimes X^\vee)$$

for any flag of finite-dimensional vector spaces $X_1 \subset X_2 \subset X_3 \subset X_4 \subset X$, where the Hom space in the left-hand side consists of all maps $X_4 \longrightarrow X/X_1$ sending X_3 to X_4/X_1 and X_1 to X_2/X_1, and $Y^\perp \subset X^\vee$ denotes the orthogonal complement to a vector subspace $Y \subset X$. .

D.1.8 A *Tate Lie algebra* is a Tate vector space endowed with a topological Lie algebra structure. Let $\mathfrak{g}$ be a Tate Lie algebra endowed with a continuous coaction of a commutative Hopf algebra $\mathcal{C}$ such that the Lie algebra structure is compatible with the continuous coaction. Then $\mathcal{C}$-invariant compact open subalgebras form a base of neighborhoods of zero in $\mathfrak{g}$.

Indeed, choose a $\mathcal{C}$-invariant compact open subspace $U \subset \mathfrak{g}$; let $\mathfrak{h}$ be the normalizer of U in $\mathfrak{g}$, i.e., the subspace of all $x \in \mathfrak{g}$ such that $[x, U] \subset U$. Then $\mathfrak{h}$ is a $\mathcal{C}$-invariant open subalgebra in $\mathfrak{g}$, since it is the kernel of the adjoint action map $\mathfrak{g} \longrightarrow \mathrm{Hom}_k(U, \mathfrak{g}/U)$. Therefore, the intersection $\mathfrak{h} \cap U$ is a $\mathcal{C}$-invariant compact open subalgebra in $\mathfrak{g}$ contained in U.

The canonical central extension $\mathfrak{g}^\sim$ of a Tate Lie algebra $\mathfrak{g}$ is defined as the fibered product of $\mathfrak{g}$ and $\mathfrak{gl}(\mathfrak{g})^\sim$ over $\mathfrak{gl}(\mathfrak{g})$, where $\mathfrak{g}$ maps to $\mathfrak{gl}(\mathfrak{g})$ by the adjoint representation. The vector space $\mathfrak{g}^\sim$ is endowed with the topology of fibered product; this makes $\mathfrak{g}^\sim$ a Tate Lie algebra. The central extension $\mathfrak{g}^\sim \longrightarrow \mathfrak{g}$ splits canonically and continuously over any compact open Lie subalgebra $\mathfrak{h} \subset \mathfrak{g}$. Indeed, the image of $\mathfrak{h}$ in $\mathfrak{gl}(\mathfrak{g})$ is contained in the open Lie subalgebra $\mathfrak{gl}(\mathfrak{g}, \mathfrak{h}) \subset \mathfrak{gl}(\mathfrak{g})$ of endomorphisms preserving $\mathfrak{h}$, and $\mathfrak{gl}(\mathfrak{g}, \mathfrak{h})$ can be embedded into $\mathfrak{gl}(\mathfrak{g})^\sim$ as the completion of the subspace

$$\mathfrak{g} \otimes_k \mathfrak{h}^\perp + \mathfrak{g}^\vee \otimes_k \mathfrak{h}.$$

The natural continuous coaction of $\mathcal{C}$ in $\mathfrak{g}^\sim$ is constructed as the fibered product of the coactions in $\mathfrak{g}$ and $\mathfrak{gl}(\mathfrak{g})^\sim$; it is clear that the Lie algebra structure on $\mathfrak{g}^\sim$ is compatible with the continuous coaction. If $\mathfrak{h} \subset \mathfrak{g}$ is a $\mathcal{C}$-invariant compact open subalgebra, then the canonical splitting $\mathfrak{h} \longrightarrow \mathfrak{g}^\sim$ preserves the continuous coactions.

When a Tate vector space V is decomposed into a direct sum $V \simeq E \oplus F$ of a compact vector space E and a discrete vector space F, there is a natural section $\mathfrak{gl}(V) \longrightarrow \mathfrak{gl}(V)^\sim$ of the central extension $\mathfrak{gl}(V)^\sim \longrightarrow \mathfrak{gl}(V)$; the image of this section is the completion of the subspace

$$V \otimes F^\vee + F \otimes V^\vee + V^\vee \otimes E + E^\vee \otimes V.$$

Consequently, when a Tate Lie algebra $\mathfrak{g}$ is decomposed into a direct sum $\mathfrak{g} \simeq \mathfrak{h} \oplus \mathfrak{b}$ of a compact open Lie subalgebra $\mathfrak{h}$ and a discrete vector subspace $\mathfrak{b}$, there is a natural section $\mathfrak{g} \longrightarrow \mathfrak{g}^\sim$ of the central extension $\mathfrak{g}^\sim \longrightarrow \mathfrak{g}$; this section agrees with the natural splitting $\mathfrak{h} \longrightarrow \mathfrak{g}^\sim$.

D.2 Construction of semialgebra

D.2.1 We will sometimes use Sweedler's notation [82] $c \longmapsto c_{(1)} \otimes c_{(2)}$ for the comultiplication map in a coassociative coalgebra $\mathcal{C}$. The analogous notation for coactions of $\mathcal{C}$ in a right $\mathcal{C}$-comodule $\mathcal{N}$ and a left $\mathcal{C}$-comodule $\mathcal{M}$ is $n \longmapsto n_{(0)} \otimes n_{(1)}$ and $m \longmapsto m_{(-1)} \otimes m_{(0)}$, where $n, n_{(0)} \in \mathcal{N}$, $m, m_{(0)} \in \mathcal{M}$, and $n_{(1)}, m_{(-1)} \in \mathcal{C}$.

A *Lie coalgebra* $\mathcal{L}$ is a k-vector space endowed with a k-linear map $\mathcal{L} \longrightarrow \bigwedge_k^2 \mathcal{L}$ from $\mathcal{L}$ to the second exterior power of $\mathcal{L}$ denoted by $l \longmapsto l_{\{1\}} \wedge l_{\{2\}}$, which should satisfy the dual version of Jacobi identity

$$l_{\{1\}\{1\}} \wedge l_{\{1\}\{2\}} \wedge l_{\{2\}} = l_{\{1\}} \wedge l_{\{2\}\{1\}} \wedge l_{\{2\}\{2\}},$$

where $l' \wedge l'' \wedge l'''$ denotes an element of $\bigwedge_k^3 \mathcal{L}$. A *comodule* $\mathcal{M}$ over a Lie coalgebra $\mathcal{L}$ is a k-vector space endowed with a k-linear map $\mathcal{M} \longrightarrow \mathcal{L} \otimes \mathcal{M}$ denoted by $m \longmapsto m_{\{-1\}} \otimes m_{\{0\}}$ satisfying the equation

$$m_{\{-1\}} \wedge m_{\{0\}\{-1\}} \otimes m_{\{0\}\{0\}} = m_{\{-1\}\{1\}} \wedge m_{\{-1\}\{2\}} \otimes m_{\{0\}},$$

where $l' \wedge l'' \otimes m$ denotes an element of $\bigwedge_k^2 \mathcal{L} \otimes_k \mathcal{M}$.

A *Tate Harish-Chandra pair* $(\mathfrak{g}, \mathcal{C})$ is a set of data consisting of a Tate Lie algebra $\mathfrak{g}$, a commutative Hopf algebra $\mathcal{C}$, a continuous coaction of $\mathcal{C}$ in $\mathfrak{g}$ such that the Lie algebra structure on $\mathfrak{g}$ is compatible with the continuous coaction, a $\mathcal{C}$-invariant compact open subalgebra $\mathfrak{h} \subset \mathfrak{g}$, and a continuous pairing $\psi \colon \mathcal{C} \times \mathfrak{h} \longrightarrow k$, where $\mathcal{C}$ is considered with the discrete topology. This data should satisfy the following conditions (cf. [12, Section 3.1]):

(i) The pairing ψ is compatible with the multiplication and comultiplication in $\mathcal{C}$, i.e., the map $\check{\psi} \colon \mathcal{C} \longrightarrow \mathfrak{h}^{\vee}$ corresponding to ψ is a morphism of Lie coalgebras such that

$$\check{\psi}(c'c'') = \varepsilon(c')\check{\psi}(c'') + \varepsilon(c'')\check{\psi}(c')$$

for $c', c'' \in \mathcal{C}$. Here the Lie coalgebra structure on $\mathcal{C}$ is defined by the formula $c \longmapsto c_{(1)} \wedge c_{(2)}$ and the Lie coalgebra structure on $\mathfrak{h}^{\vee}$ is given by the formula

$$\langle x^*, [x', x''] \rangle = \langle x^*_{\{1\}}, x'' \rangle \langle x^*_{\{2\}}, x' \rangle - \langle x^*_{\{1\}}, x' \rangle \langle x^*_{\{2\}}, x'' \rangle$$

for $x^* \in \mathfrak{h}^{\vee}$, $x', x'' \in \mathfrak{h}$. By ε we denote the counit of $\mathcal{C}$.

(ii) The pairing ψ is compatible with the continuous coaction of $\mathcal{C}$ in $\mathfrak{h}$ obtained by restricting the coaction in $\mathfrak{g}$ and the adjoint coaction of $\mathcal{C}$ in itself. The latter is defined by the formula

$$c \longmapsto c_{[0]} \otimes c_{[1]} = c_{(2)} \otimes s(c_{(1)})c_{(3)},$$

where s denotes the antipode map of the Hopf algebra $\mathcal{C}$ (the square brackets are used to avoid ambiguity of notation). The compatibility means that the continuous linear map $\mathcal{C} \otimes^* \mathfrak{h} \longrightarrow k$ corresponding to ψ preserves the continuous coactions, or equivalently, the map $\check{\psi}$ is a morphism of $\mathcal{C}$-comodules.

(iii) The action of $\mathfrak{h}$ in $\mathfrak{g}$ induced by the continuous coaction of $\mathcal{C}$ in $\mathfrak{g}$ and the pairing ψ coincides with the adjoint action of $\mathfrak{h}$ in $\mathfrak{g}$. Here the former action is constructed as the projective limit of the actions of $\mathfrak{h}$ in quotient spaces of $\mathfrak{g}$ by $\mathcal{C}$-invariant open subspaces; for a right $\mathcal{C}$-comodule $\mathcal{N}$, the $\mathfrak{h}$-module structure on $\mathcal{N}$ induced by the pairing ψ is defined by the formula

$$xn = -\psi(n_{(1)}, x)n_{(0)} \quad \text{for } x \in \mathfrak{h}, \ n \in \mathcal{N}.$$

Given a Tate Harish-Chandra pair $(\mathfrak{g}, \mathcal{C})$, one can construct a Tate Harish-Chandra pair $(\mathfrak{g}^\sim, \mathcal{C})$ with the same Lie subalgebra $\mathfrak{h}$, where $\mathfrak{g}^\sim$ is the canonical central extension of a Tate Lie algebra $\mathfrak{g}$. A continuous coaction of $\mathcal{C}$ in $\mathfrak{g}^\sim$ and a canonical embedding of $\mathfrak{h}$ into $\mathfrak{g}$ preserving the continuous coactions of $\mathcal{C}$ were constructed above; it remains to check the condition (iii). Here it suffices to notice that the adjoint action of $\mathfrak{gl}(\mathfrak{g})$ in $\mathfrak{gl}(\mathfrak{g})^\sim$ coincides with the action of $\mathfrak{gl}(\mathfrak{g})$ in $\mathfrak{gl}(\mathfrak{g})^\sim$ induced by the action of $\mathfrak{gl}(\mathfrak{g})$ in $\mathfrak{g}$, hence the adjoint action of $\mathfrak{h}$ in $\mathfrak{gl}(\mathfrak{g})^\sim$ coincides with the action of $\mathfrak{h}$ in $\mathfrak{gl}(\mathfrak{g})^\sim$ induced by the coaction of $\mathcal{C}$ in $\mathfrak{gl}(\mathfrak{g})^\sim$ and the pairing ψ.

D.2.2 Let $(\mathfrak{g}', \mathcal{C})$ be a Tate Harish-Chandra pair such that the Tate Lie algebra $\mathfrak{g}'$ is a central extension of a Tate Lie algebra $\mathfrak{g}$ with the kernel identified with k; assume that $\mathcal{C}$ coacts trivially on $k \subset \mathfrak{g}'$ and the Lie subalgebra $\mathfrak{h} \subset \mathfrak{g}'$ that is a part of the Tate Harish-Chandra pair structure does not contain k. Then $(\mathfrak{g}, \mathcal{C})$ is naturally also a Tate Harish-Chandra pair with the induced continuous coaction of $\mathcal{C}$ in $\mathfrak{g}$ and the Lie subalgebra $\mathfrak{h} \subset \mathfrak{g}$ defined as the image of $\mathfrak{h}$ in $\mathfrak{g}$. In this case, we will say that $(\mathfrak{g}', \mathcal{C}) \longrightarrow (\mathfrak{g}, \mathcal{C})$ is a central extension of Tate Harish-Chandra pairs with the kernel k. One example of a central extension of Tate Harish-Chandra pairs is the canonical central extension $(\mathfrak{g}^\sim, \mathcal{C}) \longrightarrow (\mathfrak{g}, \mathcal{C})$.

Let $\varkappa \colon (\mathfrak{g}', \mathcal{C}) \longrightarrow (\mathfrak{g}, \mathcal{C})$ be a central extension of Tate Harish-Chandra pairs with the kernel k. Consider the tensor product

$$\mathbf{S}^r_{\varkappa}(\mathfrak{g}, \mathcal{C}) = \mathcal{C} \otimes_{U(\mathfrak{h})} U_{\varkappa}(\mathfrak{g}),$$

where $U(\mathfrak{h})$ and $U(\mathfrak{g}')$ denote the universal enveloping algebras of the Lie algebras $\mathfrak{h}$ and $\mathfrak{g}'$ considered as Lie algebras without any topologies, $U_{\varkappa}(\mathfrak{g}) = U(\mathfrak{g}')/(1_{U(\mathfrak{g}')} - 1_{\mathfrak{g}'})$ is the modification of the universal enveloping algebra of $\mathfrak{g}$ corresponding to the central extension $k \longrightarrow \mathfrak{g}' \longrightarrow \mathfrak{g}$, and $1_{U(\mathfrak{g}')}$ and $1_{\mathfrak{g}'}$ denote the unit elements of the algebra $U(\mathfrak{g}')$ and the vector subspace $k \subset \mathfrak{g}'$, respectively. The structure of right $U(\mathfrak{h})$-module on $\mathcal{C}$ comes from the pairing $\phi \colon \mathcal{C} \otimes_k U(\mathfrak{h}) \longrightarrow k$ corresponding to the algebra morphism $U(\mathfrak{h}) \longrightarrow \mathcal{C}^\vee$ induced by the Lie algebra morphism $\check{\psi} \colon \mathfrak{h} \longrightarrow \mathcal{C}^\vee$, where the multiplication on $\mathcal{C}^*$ is defined by the formula $\langle c'^* c''^*, c \rangle = \langle c'^*, c_{(2)} \rangle \langle c''^*, c_{(1)} \rangle$ for $c'^*, c''^* \in \mathcal{C}^*$, $c \in \mathcal{C}$ and the Lie bracket is given by the formula $[c'^*, c''^*] = c'^* c''^* - c''^* c'^*$.

We claim that the vector space $\mathbf{S}^r_{\varkappa}(\mathfrak{g}, \mathcal{C})$ has a natural structure of semialgebra over the coalgebra $\mathcal{C}$ provided by the general construction of 10.2.1. The construction of this semialgebra structure becomes a little simpler if one assumes that

(iv) the pairing $\phi \colon \mathcal{C} \otimes_k U(\mathfrak{h}) \longrightarrow k$ is nondegenerate in $\mathcal{C}$,

but this is not necessary. When $\mathfrak{h}$ is the Lie algebra of the proalgebraic group H corresponding [30] to $\mathcal{C}$, and the characteristic of the field k is zero, the condition (iv) simply means that H is connected [78].

D.2.3 To construct a right $\mathcal{C}$-comodule structure on $\mathbf{S}_{\varkappa}^r(\mathfrak{g}, \mathcal{C})$, we will have to approximate this vector space by finite-dimensional spaces. Let $V_1, \ldots, V_t$ be a sequence of $\mathcal{C}$-invariant compact open subspaces of $\mathfrak{g}'$ containing $\mathfrak{h}$ and k such that $V_i + [V_i, V_i] \subset V_{i-1}$. Let $\mathcal{N}$ be a finite-dimensional right $\mathcal{C}$-comodule. Choose a $\mathcal{C}$-invariant compact open subspace $W_1 \subset \mathfrak{h}$ such that the $\mathcal{C}$-comodule $\mathcal{N}$ is annihilated by the action of W_1 obtained by restricting the action of $\mathfrak{h}$ induced by the pairing ψ. For each $i = 2, \ldots, t$ choose a $\mathcal{C}$-invariant compact open subspace $W_i \subset \mathfrak{h}$ such that $W_i + [V_i, W_i] \subset W_{i-1}$. Denote by $\mathbf{S}_{\varkappa}^r(\mathcal{N}; V_1, \ldots, V_t)$ the quotient space of the vector space

$$\mathcal{N} \otimes_k (k \oplus V_1/W_1 \oplus \cdots \oplus (V_t/W_t)^{\otimes t})$$

by the obvious relations imitating the relations in the enveloping algebra $U_{\varkappa}(\mathfrak{g})$ and its tensor product with $\mathcal{N}$ over $U(\mathfrak{h})$. It is easy to see that this quotient space does not depend on the choice of the subspaces W_i. In other words, denote by $R(V_1, \ldots, V_t)$ the subspace

$$U(\mathfrak{h})(k + V_1 + \cdots + V_t^t) \subset U_{\varkappa}(\mathfrak{g});$$

it is an $U(\mathfrak{h})$-$U(\mathfrak{h})$-submodule of $U_{\varkappa}(\mathfrak{g})$ and a free left $U(\mathfrak{h})$-module. There is a natural isomorphism

$$\mathbf{S}_{\varkappa}^r(\mathcal{N}; V_1, \ldots, V_t) \simeq \mathcal{N} \otimes_{U(\mathfrak{h})} R(V_1, \ldots, V_t).$$

This is an isomorphism of right $U(\mathfrak{h})$-modules; when $\mathcal{N} = \mathcal{D}$ is a finite-dimensional subcoalgebra of $\mathcal{C}$, this is also an isomorphism of left $\mathcal{C}$-comodules. Clearly, the inductive limit of $\mathbf{S}_{\varkappa}^r(\mathcal{D}; V_1, \ldots, V_t)$ over increasing t, V_i, and finite-dimensional subcoalgebras $\mathcal{D} \subset \mathcal{C}$ is naturally isomorphic to $\mathbf{S}_{\varkappa}^r(\mathfrak{g}, \mathcal{C})$.

Now the vector space $\mathbf{S}_{\varkappa}^r(\mathcal{N}; V_1, \ldots, V_t)$ has a right $\mathcal{C}$-comodule structure induced by the right $\mathcal{C}$-comodule structure on $\mathcal{N} \otimes_k (k \oplus V_1/W_1 \oplus \cdots \oplus (V_t/W_t)^{\otimes t})$ obtained by taking the tensor product of the $\mathcal{C}$-comodule structures on V_i/W_i and the right $\mathcal{C}$-comodule structure on $\mathcal{N}$. The inductive limit of these $\mathcal{C}$-comodule structures for $\mathcal{N} = \mathcal{D}$ provides the desired right $\mathcal{C}$-comodule structure on $\mathbf{S}_{\varkappa}^r(\mathfrak{g}, \mathcal{C})$. It commutes with the left $\mathcal{C}$-comodule structure on $\mathbf{S}_{\varkappa}^r(\mathfrak{g}, \mathcal{C})$ and agrees with the right $U(\mathfrak{h})$-module structure, since such commutativity and agreement hold on the level of the spaces $\mathbf{S}_{\varkappa}^r(\mathcal{D}; V_1, \ldots, V_t)$. Furthermore, by the (classical) Poincaré–Birkhoff–Witt theorem $U_{\varkappa}(\mathfrak{g})$ is a free left $U(\mathfrak{h})$-module. If the condition (iv) holds, the construction of the semialgebra $\mathbf{S}_{\varkappa}^r(\mathfrak{g}, \mathcal{C})$ is finished; otherwise, we still have to check that the semiunit map $\mathcal{C} \longrightarrow \mathbf{S}_{\varkappa}^r(\mathfrak{g}, \mathcal{C})$ and the semimultiplication map $\mathbf{S}_{\varkappa}^r(\mathfrak{g}, \mathcal{C}) \,\square_{\mathcal{C}}\, \mathbf{S}_{\varkappa}^r(\mathfrak{g}, \mathcal{C}) \longrightarrow \mathbf{S}_{\varkappa}^r(\mathfrak{g}, \mathcal{C})$ are morphisms of right $\mathcal{C}$-comodules.

The former is clear, and the latter can be proven in the following way. Any finite-dimensional $\mathcal{C}$-comodule $\mathcal{N}$ is a comodule over a finite-dimensional subcoalgebra $\mathcal{E} \subset \mathcal{C}$. There is a natural isomorphism $\mathcal{N} \otimes_{U(\mathfrak{h})} R(V_1, \ldots, V_t) \simeq \mathcal{N} \square_{\mathcal{C}} (\mathcal{E} \otimes_{U(\mathfrak{h})} R(V_1, \ldots, V_t))$. The corresponding isomorphism $\mathbf{S}_{\varkappa}^r(\mathcal{N}; V_1, \ldots, V_t) \simeq \mathcal{N} \square_{\mathcal{C}} \mathbf{S}_{\varkappa}^r(\mathcal{E}; V_1, \ldots, V_t)$, which is induced by the isomorphism $\mathcal{N} \simeq \mathcal{N} \square_{\mathcal{C}} \mathcal{E}$, preserves the right $\mathcal{C}$-comodule structures. All of this is applicable to the case of

$\mathcal{N} = \mathbf{S}^r_\varkappa(V'_1, \ldots, V'_t; \mathcal{D})$, where $V'_1, \ldots, V'_t$ is another sequence of subspaces of $\mathfrak{g}'$ satisfying the above conditions. Now let $V''_1, \ldots, V''_{2t} \subset \mathfrak{g}'$ be a sequence of subspaces satisfying the above conditions and such that V'_i, $V_i \subset V''_{t+i}$. The map $\mathcal{C} \otimes_{U(\mathfrak{h})} U_\varkappa(\mathfrak{g}) \otimes_{U(\mathfrak{h})} U_\varkappa(\mathfrak{g}) \longrightarrow \mathcal{C} \otimes_{U(\mathfrak{h})} U_\varkappa(\mathfrak{g})$ induced by the multiplication map $U_\varkappa(\mathfrak{g}) \otimes_{U(\mathfrak{h})} U_\varkappa(\mathfrak{g}) \longrightarrow U_\varkappa(\mathfrak{g})$ is the inductive limit of the maps

$$\mathcal{D} \otimes_{U(\mathfrak{h})} R(V'_1, \ldots, V'_t) \otimes_{U(\mathfrak{h})} R(V_1, \ldots, V_t) \longrightarrow \mathcal{D} \otimes_{U(\mathfrak{h})} R(V''_1, \ldots, V''_{2t})$$

over increasing t, V_i, V'_i, V''_i, and $\mathcal{D}$. The corresponding map

$$\mathbf{S}^r_\varkappa(\mathbf{S}^r_\varkappa(\mathcal{D}; V'_1, \ldots, V'_t); V_1, \ldots, V_t) \longrightarrow \mathbf{S}^r_\varkappa(\mathcal{D}; V''_1, \ldots, V''_{2t})$$

is induced by the map

$$\mathcal{D} \otimes_k (k \oplus V'_1/W'_1 \oplus \cdots \oplus (V'_t/W'_t)^{\otimes t}) \otimes_k (k \oplus V_1/W_1 \oplus \cdots \oplus (V_t/W_t)^{\otimes t})$$
$$\longrightarrow \mathcal{D} \otimes_k (k \oplus V''_1/W''_1 \oplus \cdots \oplus (V''_{2t}/W''_{2t})^{\otimes 2t}),$$

where the sequences of subspaces W'_i, W_i, W''_i satisfy the above conditions with respect to the sequences of subspaces V'_i, V_i, V''_i, and the right $\mathcal{C}$-comodules $\mathcal{D}$, $\mathcal{D} \otimes_k (k \oplus V'_1/W'_1 \oplus \cdots \oplus (V'_t/W'_t)^{\otimes t})$, $\mathcal{D}$, respectively, and the additional condition that W'_i, $W_i \subset W''_{t+i}$. One can easily see that the latter map is a morphism of right $\mathcal{C}$-comodules. The semialgebra $\mathbf{S}^r_\varkappa(\mathfrak{g}, \mathcal{C})$ over the coalgebra $\mathcal{C}$ is constructed.

D.2.4 Analogously one defines a semialgebra structure on the tensor product $\mathbf{S}^l_\varkappa(\mathfrak{g}, \mathcal{C}) = U_\varkappa(\mathfrak{g}) \otimes_{U(\mathfrak{h})} \mathcal{C}$. The semialgebras $\mathbf{S}^r_\varkappa = \mathbf{S}^r_\varkappa(\mathfrak{g}, \mathcal{C})$ and $\mathbf{S}^l_\varkappa = \mathbf{S}^l_\varkappa(\mathfrak{g}, \mathcal{C})$ are essentially opposite to each other (see C.2.7) up to replacing $\varkappa$ with $-\varkappa$. More precisely, the antipode anti-automorphisms of $U(\mathfrak{g}')$ and $\mathcal{C}$ induce a natural isomorphism of semialgebras $\mathbf{S}^r_\varkappa \simeq \mathbf{S}^{l\,\mathrm{op}}_{-\varkappa}$ compatible with the isomorphism of coalgebras $\mathcal{C}^{\mathrm{op}} \simeq \mathcal{C}$, where $-\varkappa$ is defined as the central extension of Tate Harish-Chandra pairs with the kernel k that is obtained from the central extension $\varkappa$ by multiplying the embedding $k \longrightarrow \mathfrak{g}'$ with -1.

D.2.5 A *discrete module* M over a topological Lie algebra $\mathfrak{g}$ is a $\mathfrak{g}$-module such that the action map $\mathfrak{g} \times M \longrightarrow M$ is continuous with respect to the discrete topology of M. Equivalently, a $\mathfrak{g}$-module M is discrete if the annihilator of any element of M is an open Lie subalgebra in $\mathfrak{g}$. In particular, if $\psi\colon \mathcal{C} \times \mathfrak{h} \longrightarrow k$ is a continuous pairing between a compact Lie algebra $\mathfrak{h}$ and a coalgebra $\mathcal{C}$ such that the map $\check{\psi}\colon \mathcal{C} \longrightarrow \mathfrak{h}^\vee$ is a morphism of Lie coalgebras, then the $\mathfrak{h}$-module structure induced by a $\mathcal{C}$-comodule structure by the formula of D.2.1 (iii) is always discrete.

Let $\varkappa\colon (\mathfrak{g}', \mathcal{C}) \longrightarrow (\mathfrak{g}, \mathcal{C})$ be a central extension of Tate Harish-Chandra pairs with the kernel k. Then the category of left semimodules over $\mathbf{S}^l_\varkappa(\mathfrak{g}, \mathcal{C})$ is isomorphic to the category of k-vector spaces $\mathfrak{M}$ endowed with $\mathcal{C}$-comodule and discrete $\mathfrak{g}'$-module structures such that the induced discrete $\mathfrak{h}$-module structures coincide, the action map $\mathfrak{g}/U \otimes_k \mathcal{L} \longrightarrow \mathfrak{M}$ is a morphism of $\mathcal{C}$-comodules for any finite-dimensional $\mathcal{C}$-subcomodule $\mathcal{L} \subset \mathfrak{M}$ and any $\mathcal{C}$-invariant compact open subspace

$U \subset \mathfrak{g}$ annihilating $\mathcal{L}$, and the unit element of $k \subset \mathfrak{g}'$ acts by the identity in $\mathcal{M}$. The second of these three conditions can be reformulated as follows: for any $\mathcal{C}$-invariant compact subspace $V \subset \mathfrak{g}'$, the natural Lie coaction map $\mathcal{M} \longrightarrow V^\vee \otimes_k \mathcal{M}$ is a morphism of $\mathcal{C}$-comodules. When the assumption (iv) of D.2.2 is satisfied, the second condition is redundant.

Abusing terminology, we will call vector spaces $\mathcal{M}$ endowed with such a structure *Harish-Chandra modules* over $(\mathfrak{g}, \mathcal{C})$ *with the central charge* $\varkappa$. Analogously, the category of right semimodules over $\mathbf{S}^r_\varkappa(\mathfrak{g}, \mathcal{C})$ is isomorphic to the category of Harish-Chandra modules over $(\mathfrak{g}, \mathcal{C})$ with central charge $-\varkappa$. The category of Harish-Chandra modules over $(\mathfrak{g}, \mathcal{C})$ with the central charge $\varkappa$ will be denoted by $\mathsf{O}_\varkappa(\mathfrak{g}, \mathcal{C})$.

Remark. The terminology related to Harish-Chandra modules is a hopeless mess. In classical representation theory, by "Harish-Chandra modules" people meant modules over a Lie algebra over the field of real numbers integrable to a maximal compact subgroup of the corresponding Lie group [32]. Later came abstract Harish-Chandra pairs (Lie algebra; subgroup), algebraic Harish-Chandra pairs (with an algebraic group in the role of subgroup), and modules over these [9]. What we might now call "Harish-Chandra modules over a semisimple Lie algebra relative to its Borel subgroup" came to be known as "the Bernstein–Gelfand–Gelfand category O" [53]. We reflect this terminology in the notation above. What we call "Tate Harish-Chandra pairs" (with a proalgebraic subgroup) were first introduced, under the name of "Harish-Chandra pairs", by Beilinson–Feigin–Mazur in [12] (the exposition in *loc. cit.* is restricted to the case of the ground field of complex numbers). In our approach, some results are applicable to arbitrary Tate Harish-Chandra pairs, or just those satisfying a weak assumption (iv). This includes Theorem D.3.1 and Corollary D.3.1. And Tate Harish-Chandra pairs can be thought of as analogues of the classical Harish-Chandra pairs in that the subgroup corresponds to a compact Lie subalgebra $\mathfrak{h}$, although this is an entirely different notion of compactness. Still the case that interests us most is that of a *prounipotent* subgroup (we are not concerned about this prounipotent subgroup being maximal in any sense, but only about it corresponding to a compact open Lie subalgebra). So what we work with in D.6 is a variation of the classical "category O" rather than the classical "Harish-Chandra modules". An additional detail is that Verma modules, which were among the most important objects of the "category O" from the classical point of view, are hardly visible in our approach at all. Dual, or *contragredient*, Verma modules are important for us instead (see Remark D.3.1). Neither do we have any duality functor acting from our category $\mathsf{O}_\varkappa(\mathfrak{g}, \mathcal{C})$ to itself. Instead, we consider duality functors $\mathsf{O}_\varkappa(\mathfrak{g}, \mathcal{C})^{\mathrm{op}} \longrightarrow \mathsf{O}^{\mathrm{ctr}}_{-\varkappa}(\mathfrak{g}, \mathcal{C})$ transforming Harish-Chandra modules into Harish-Chandra contramodules (see D.2.8).

D.2.6 For a topological vector space V and a vector space P, denote by $V \otimes^\widehat{} P$ the tensor product $V \otimes^! P = V \overleftarrow{\otimes} P$ *considered as a vector space without any topology*, where P is endowed with the discrete topology for the purpose of making the topological tensor product. In other words, one has

$$V \otimes^\wedge P = \varprojlim_U V/U \otimes_k P,$$

where the projective limit is taken over all open subspaces $U \subset V$.

For a topological vector space V, denote by $\bigwedge^{*,2}(V)$ the completion of $\bigwedge_k^2(V)$ with respect to the topology with the base of neighborhoods of zero formed by all the subspaces $T \subset \bigwedge_k^2(V)$ such that there exists an open subspace $V' \subset V$ for which $\bigwedge_k^2(V') \subset T$ and for any vector $v \in V$ there exists an open subspace $V'' \subset V$ for which $v \wedge V'' \subset T$. For any topological vector spaces V and W, the vector space of continuous skew-symmetric bilinear maps $V \times V \longrightarrow W$ is naturally isomorphic to the vector space of continuous linear maps $\bigwedge^{2,*}(V) \longrightarrow W$. The space $\bigwedge^{*,2}(V)$ is a closed subspace of the space $V \otimes^* V$; the skew-symmetrization map $V \otimes^* V \longrightarrow V \otimes^* V$ factorizes through $\bigwedge^{*,2}(V)$.

Let $\mathfrak{g}$ be a topological Lie algebra. A *contramodule* over $\mathfrak{g}$ is a vector space P endowed with a linear map $\mathfrak{g} \otimes^\wedge P \longrightarrow P$ satisfying the following version of Jacobi equation. Consider the vector space $\bigwedge^{*,2}(\mathfrak{g}) \otimes^\wedge P$. There is a natural map $\bigwedge^{*,2}(\mathfrak{g}) \otimes^\wedge P \longrightarrow \mathfrak{g} \otimes^\wedge P$ induced by the bracket map $\bigwedge^{*,2}(\mathfrak{g}) \longrightarrow \mathfrak{g}$. Furthermore, there is a natural map

$$(\mathfrak{g} \otimes^* \mathfrak{g}) \otimes^\wedge P \longrightarrow \mathfrak{g} \otimes^\wedge (\mathfrak{g} \otimes^\wedge P),$$

which is constructed as follows. For any open subspace $U \subset \mathfrak{g}$ there is a natural surjection $(\mathfrak{g} \otimes^* \mathfrak{g}) \otimes^\wedge P \longrightarrow (\mathfrak{g}/U \otimes^* \mathfrak{g}) \otimes^\wedge P$ and for any discrete vector space F there is a natural isomorphism

$$(F \otimes^* \mathfrak{g}) \otimes^\wedge P \simeq F \otimes_k (\mathfrak{g} \otimes^\wedge P),$$

so the desired map is obtained as the projective limit over U. Composing the map $\bigwedge^{*,2}(\mathfrak{g}) \otimes^\wedge P \longrightarrow (\mathfrak{g} \otimes^* \mathfrak{g}) \otimes^\wedge P$ induced by the embedding $\bigwedge^{*,2}(\mathfrak{g}) \longrightarrow \mathfrak{g} \otimes^* \mathfrak{g}$ with the map $(\mathfrak{g} \otimes^* \mathfrak{g}) \otimes^\wedge P \longrightarrow \mathfrak{g} \otimes^\wedge (\mathfrak{g} \otimes^\wedge P)$ that we have constructed and with the map $\mathfrak{g} \otimes^\wedge (\mathfrak{g} \otimes^\wedge P) \longrightarrow \mathfrak{g} \otimes^\wedge P$ induced by the contraaction map $\mathfrak{g} \otimes^\wedge P \longrightarrow P$, we obtain a second map $\bigwedge^{*,2}(\mathfrak{g}) \otimes^\wedge P \longrightarrow P$. Now the contramodule Jacobi equation claims that the two maps $\bigwedge^{*,2}(\mathfrak{g}) \otimes^\wedge P \rightrightarrows \mathfrak{g} \otimes^\wedge P$ should have equal compositions with the contraaction map $\mathfrak{g} \otimes^\wedge P \longrightarrow P$.

Alternatively, the map $(\mathfrak{g} \otimes^* \mathfrak{g}) \otimes^\wedge P \longrightarrow \mathfrak{g} \otimes^\wedge (\mathfrak{g} \otimes^\wedge P)$ can be constructed as the composition

$$(\mathfrak{g} \otimes^* \mathfrak{g}) \otimes^\wedge P \longrightarrow (\mathfrak{g} \overset{\leftarrow}{\otimes} \mathfrak{g}) \otimes^\wedge P \simeq \mathfrak{g} \overset{\leftarrow}{\otimes} \mathfrak{g} \overset{\leftarrow}{\otimes} P \simeq \mathfrak{g} \otimes^\wedge (\mathfrak{g} \otimes^\wedge P)$$

of the map induced by the natural continuous map $\mathfrak{g} \otimes^* \mathfrak{g} \longrightarrow \mathfrak{g} \overset{\leftarrow}{\otimes} \mathfrak{g}$ and the natural isomorphisms whose existence follows from the fact that the topological tensor product $W \overset{\leftarrow}{\otimes} V$ considered as a vector space without any topology does not depend on the topology of V. The following comparison between the definitions of a discrete $\mathfrak{g}$-module and a $\mathfrak{g}$-contramodule can be made: a discrete $\mathfrak{g}$-module structure on a vector space M is given by a continuous linear map

$$\mathfrak{g} \otimes M \simeq \mathfrak{g} \otimes^* M = \mathfrak{g} \vec{\otimes} M \longrightarrow M,$$

while a $\mathfrak{g}$-contramodule structure on a vector space P is given by a discontinuous linear map

$$\mathfrak{g} \otimes^{\wedge} P \simeq \mathfrak{g} \otimes^! P = \mathfrak{g} \overleftarrow{\otimes} P \longrightarrow P,$$

where M and P are endowed with discrete topologies. For any topological Lie algebra $\mathfrak{g}$, the category of $\mathfrak{g}$-contramodules is abelian (cf. D.5.2). There is a natural exact forgetful functor from the category of $\mathfrak{g}$-contramodules to the category of modules over the Lie algebra $\mathfrak{g}$ considered without any topology.

For any discrete $\mathfrak{g}$-module M and any vector space E there is a natural structure of $\mathfrak{g}$-contramodule on the space of linear maps $\mathrm{Hom}_k(M, E)$. The contraaction map $\mathfrak{g} \otimes^{\wedge} \mathrm{Hom}_k(M, E) \longrightarrow \mathrm{Hom}_k(M, E)$ is constructed as the projective limit over all open subspaces $U \subset \mathfrak{g}$ of the maps

$$\mathfrak{g}/U \otimes_k \mathrm{Hom}_k(M, E) \longrightarrow \mathrm{Hom}_k(M^U, E)$$

given by the formula $\bar{z} \otimes g \longmapsto (m \longmapsto -g(\bar{z}m))$ for $\bar{z} \in \mathfrak{g}/U$, $g \in \mathrm{Hom}_k(M, E)$, and $m \in M^U$, where $M^U \subset M$ denotes the subspace of all elements of M annihilated by U.

More generally, for any discrete module M over a topological Lie algebra $\mathfrak{g}_1$ and any contramodule P over a topological Lie algebra $\mathfrak{g}_2$ there is a natural $(\mathfrak{g}_1 \oplus \mathfrak{g}_2)$-contramodule structure on $\mathrm{Hom}_k(M, P)$ with the contraaction map $(\mathfrak{g}_1 \oplus \mathfrak{g}_2) \otimes^{\wedge} \mathrm{Hom}_k(M, P) \longrightarrow \mathrm{Hom}_k(M, P)$ defined as the sum of two commuting contraactions of $\mathfrak{g}_1$ and $\mathfrak{g}_2$ in $\mathrm{Hom}_k(M, P)$, one of which is introduced above and the other one is given by the composition

$$\mathfrak{g}_2 \otimes^{\wedge} \mathrm{Hom}_k(M, P) \longrightarrow \mathrm{Hom}_k(M, \mathfrak{g}_2 \otimes^{\wedge} P) \longrightarrow \mathrm{Hom}_k(M, P)$$

of the natural map $\mathfrak{g}_2 \otimes^{\wedge} \mathrm{Hom}_k(M, P) \longrightarrow \mathrm{Hom}_k(M, \mathfrak{g}_2 \otimes^{\wedge} P)$ and the map $\mathrm{Hom}_k(M, \mathfrak{g}_2 \otimes^{\wedge} P) \longrightarrow \mathrm{Hom}_k(M, P)$ induced by the $\mathfrak{g}_2$-contraaction in P. Hence for any discrete $\mathfrak{g}$-module M and any $\mathfrak{g}$-contramodule P there is a natural $\mathfrak{g}$-contramodule structure on $\mathrm{Hom}_k(M, P)$ induced by the diagonal embedding of Lie algebras $\mathfrak{g} \longrightarrow \mathfrak{g} \oplus \mathfrak{g}$.

D.2.7   When $\mathfrak{g}$ is a Tate Lie algebra, a $\mathfrak{g}$-contramodule P can be also defined as a k-vector space endowed with a linear map

$$\mathrm{Hom}_k(V^{\vee}, P) \longrightarrow P$$

for every compact open subspace $V \subset \mathfrak{g}$. These linear maps should satisfy the following two conditions: when $U \subset V \subset \mathfrak{g}$ are compact open subspaces, the maps $\mathrm{Hom}_k(U^{\vee}, P) \longrightarrow P$ and $\mathrm{Hom}_k(V^{\vee}, P) \longrightarrow P$ should form a commutative diagram with the map $\mathrm{Hom}_k(U^\vee, P) \longrightarrow \mathrm{Hom}_k(V^\vee, P)$ induced by the natural surjection $V^\vee \longrightarrow U^\vee$, and for any compact open subspaces V', V'', $W \subset \mathfrak{g}$ such that $[V', V''] \subset W$ the composition

$$\mathrm{Hom}_k(V''^\vee \otimes_k V'^\vee,\, P) \longrightarrow \mathrm{Hom}_k(W^\vee, P) \longrightarrow P$$

of the map induced by the cobracket map $W^\vee \longrightarrow V''^\vee \otimes_k V'^\vee$ with the contraaction map should be equal to the difference of the iterated contraaction map

$$\mathrm{Hom}_k(V''^\vee \otimes_k V'^\vee,\, P) \simeq \mathrm{Hom}_k(V'^\vee, \mathrm{Hom}_k(V''^\vee, P)) \longrightarrow \mathrm{Hom}_k(V'^\vee, P) \longrightarrow P$$

and the composition of the isomorphism $\mathrm{Hom}_k(V''^\vee \otimes_k V'^\vee,\, P) \simeq \mathrm{Hom}_k(V'^\vee \otimes_k V''^\vee,\, P)$ induced by the isomorphism $V'^\vee \otimes_k V''^\vee \simeq V''^\vee \otimes_k V'^\vee$ with the iterated contraaction map

$$\mathrm{Hom}_k(V'^\vee \otimes_k V''^\vee,\, P) \simeq \mathrm{Hom}_k(V''^\vee, \mathrm{Hom}_k(V'^\vee, P)) \longrightarrow \mathrm{Hom}_k(V''^\vee, P) \longrightarrow P.$$

The following example illustrates the notion of a contramodule over a Tate Lie algebra as a kind of module with infinite summation operations (cf. A.1). Let $\mathfrak{g} = k((t))d/dt$ be the Lie algebra of vector fields on the formal circle with its natural topology. Denote by $L_i = t^{i+1}d/dt$ its topological basis; then the Lie bracket in $\mathfrak{g}$ is given by the formula $[L_i, L_j] = (j - i)L_{i+j}$. A $\mathfrak{g}$-contramodule P is a k-vector space endowed with the infinite summation operations

$$(p_{-N}, p_{-N+1}, \dots) \longmapsto \sum_{i=-N}^{+\infty} L_i p_i$$

defined for any $N \in \mathbb{Z}$ and assigning to a sequence of vectors $p_i \in P$ the corresponding infinite sum, which must be also an element of P. These summation operations should satisfy the following equations:

$$\sum_{i=-N}^{+\infty} L_i p_i = \sum_{i=-M}^{+\infty} L_i p_i$$

whenever $-N < -M$ and $p_{-N} = \dots = p_{-M+1} = 0$ (agreement);

$$\sum_{i=-N}^{+\infty} L_i(ap_i + bq_i) = a \sum_{i=-N}^{+\infty} L_i p_i + b \sum_{i=-N}^{+\infty} L_i q_i$$

for any a, $b \in k$ and p_i, $q_i \in P$ (linearity); and

$$\sum_{i=-N}^{+\infty} L_i\left(\sum_{j=-M}^{+\infty} L_j p_{ij}\right) - \sum_{j=-M}^{+\infty} L_j\left(\sum_{i=-N}^{+\infty} L_i p_{ij}\right)$$
$$= \sum_{n=-N-M}^{+\infty} L_n\left(\sum_{i+j=n} (j - i)p_{i+j}\right)$$

for any $p_{ij} \in P$, $i \geqslant -N$, $j \geqslant -M$ (contra-Jacobi equation).

For a Tate Lie algebra $\mathfrak{g}$, a discrete $\mathfrak{g}$-module M, and a $\mathfrak{g}$-contramodule P, the structure of $\mathfrak{g}$-contramodule on $\mathrm{Hom}_k(M, P)$ defined above is given by the formula $\pi(g)(m) = \pi_P(x^* \mapsto g(x^*)(m)) - g(m_{\{-1\}})(m_{\{0\}})$ for a compact open subspace $V \subset \mathfrak{g}$, a linear map $g \in \operatorname{Hom}_k(V^\vee, \operatorname{Hom}_k(M, P))$, and elements $x^* \in V^\vee$, $m \in M$, where $m \mapsto m_{\{-1\}} \otimes m_{\{0\}}$ denotes the map $M \longrightarrow V^\vee \otimes_k M$ corresponding to the $\mathfrak{g}$-action map $V \times M \longrightarrow M$ and π_P denotes the $\mathfrak{g}$-contraaction map $\operatorname{Hom}_k(V^\vee, P) \longrightarrow P$.

If $\psi \colon \mathcal{C} \times \mathfrak{h} \longrightarrow k$ is a continuous pairing between a coalgebra $\mathcal{C}$ and a compact Lie algebra $\mathfrak{h}$ such that the map $\check{\psi} \colon \mathcal{C} \longrightarrow \mathfrak{h}^\vee$ is a morphism of Lie coalgebras, then for any left $\mathcal{C}$-contramodule $\mathfrak{P}$ the induced contraaction of $\mathfrak{h}$ in $\mathfrak{P}$ is defined as the composition $\operatorname{Hom}_k(\mathfrak{h}^\vee, \mathfrak{P}) \longrightarrow \operatorname{Hom}_k(\mathcal{C}, \mathfrak{P}) \longrightarrow \mathfrak{P}$ of the map induced by the map $\check{\psi}$ and the $\mathcal{C}$-contraaction map.

D.2.8　Let $\varkappa \colon (\mathfrak{g}', \mathcal{C}) \longrightarrow (\mathfrak{g}, \mathcal{C})$ be a central extension of Tate Harish-Chandra pairs with the kernel k. Then the category of left semicontramodules over the semialgebra $\mathcal{S}^r_\varkappa(\mathfrak{g}, \mathcal{C})$ is isomorphic to the category of k-vector spaces $\mathfrak{P}$ endowed with a left $\mathcal{C}$-contramodule and a $\mathfrak{g}'$-contramodule structures such that the induced $\mathfrak{h}$-contramodule structures coincide, for any $\mathcal{C}$-invariant compact open subspace $V \subset \mathfrak{g}$ the $\mathfrak{g}$-contraaction map

$$\operatorname{Hom}_k(V^\vee, \mathfrak{P}) \longrightarrow \mathfrak{P}$$

is a morphism of $\mathcal{C}$-contramodules, and the unit element of $k \subset \mathfrak{g}'$ acts by the identity in $\mathfrak{P}$. Here the left $\mathcal{C}$-contramodule structure on the vector space $\operatorname{Hom}_k(M, \mathfrak{P})$ for a left $\mathcal{C}$-comodule M and a left $\mathcal{C}$-contramodule $\mathfrak{P}$ is defined by the formula $\pi(g)(m) = \pi_{\mathfrak{P}}(c \mapsto g(s(m_{(-1)})c)(m_{(0)}))$ for $m \in M$, $g \in \operatorname{Hom}_k(\mathcal{C}, \operatorname{Hom}_k(M, \mathfrak{P}))$.

Indeed, according to 10.2.2, a left $\mathcal{S}^r_\varkappa$-semicontramodule structure on $\mathfrak{P}$ is the same that a left $\mathcal{C}$-contramodule and a left $U_\varkappa(\mathfrak{g})$-module structures such that the induced $U(\mathfrak{h})$-module structures on $\mathfrak{P}$ coincide and the (semicontra)action map

$$\mathfrak{P} \longrightarrow \operatorname{Hom}_{U(\mathfrak{h})}(U_\varkappa(\mathfrak{g}), \mathfrak{P}) \simeq \operatorname{Cohom}_{\mathcal{C}}(\mathcal{S}^r_\varkappa, \mathfrak{P})$$

is a morphism of $\mathcal{C}$-contramodules. The latter condition is equivalent to the map $\mathfrak{P} \longrightarrow \operatorname{Hom}_{U(\mathfrak{h})}(U(\mathfrak{h}) \cdot V, \mathfrak{P}) \simeq \operatorname{Cohom}_{\mathcal{C}}(\mathcal{C} \otimes_{U(\mathfrak{h})} U(\mathfrak{h}) \cdot V, \mathfrak{P})$ being a morphism of $\mathcal{C}$-contramodules for any compact $\mathcal{C}$-invariant subspace $\mathfrak{h} \oplus k \subset V \subset \mathfrak{g}'$, where $U(\mathfrak{h}) \cdot V \subset U_\varkappa(\mathfrak{g})$. Given this data, one can use the short exact sequences

$$\mathfrak{h} \otimes_k \mathfrak{P} \longrightarrow \mathfrak{h} \otimes^\wedge \mathfrak{P} \oplus V \otimes_k \mathfrak{P} \longrightarrow V \otimes^\wedge \mathfrak{P}$$

to construct the Lie contraaction maps $V \otimes^\wedge \mathfrak{P} \longrightarrow \mathfrak{P}$. Then the map $\mathfrak{P} \longrightarrow \operatorname{Hom}_{U(\mathfrak{h})}(U(\mathfrak{h}) \cdot V, \mathfrak{P})$ is a morphism of $\mathcal{C}$-contramodules if and only if the map $\operatorname{Hom}_k(V^\vee, \mathfrak{P}) \longrightarrow \mathfrak{P}$ is a morphism of $\mathcal{C}$-contramodules. To check this, one can express the first condition in terms of the equality of two appropriate maps $\operatorname{Hom}_k(\mathcal{C}, \mathfrak{P}) \rightrightarrows \operatorname{Hom}_k(V, \mathfrak{P})$ and the second condition in terms of the equality of two maps $\operatorname{Hom}_k(V^\vee \otimes_k \mathcal{C}, \mathfrak{P}) \rightrightarrows \mathfrak{P}$. These two pairs of maps correspond to each other under a natural isomorphism $V^\vee \otimes_k \mathcal{C} \simeq \mathcal{C} \otimes_k V^\vee$ depending on the $\mathcal{C}$-comodule structure on $V^\vee$. In particular, our maps $\operatorname{Hom}_k(V^\vee, \mathfrak{P}) \longrightarrow \mathfrak{P}$ are morphisms of $\mathfrak{h}$-contramodules, and it follows that they define a $\mathfrak{g}'$-contramodule structure.

We will call vector spaces $\mathfrak{P}$ endowed with such a structure *Harish-Chandra contramodules over* $(\mathfrak{g}, \mathcal{C})$ *with the central charge* $\varkappa$. The abelian category of Harish-Chandra contramodules over $(\mathfrak{g}, \mathcal{C})$ with the central charge $\varkappa$ will be denoted by $\mathsf{O}^{\mathrm{ctr}}_{\varkappa}(\mathfrak{g}, \mathcal{C})$. If $\varkappa_1 \colon (\mathfrak{g}', \mathcal{C}) \longrightarrow (\mathfrak{g}, \mathcal{C})$ and $\varkappa_2 \colon (\mathfrak{g}'', \mathcal{C}) \longrightarrow (\mathfrak{g}, \mathcal{C})$ are two central extensions of Tate Harish-Chandra pairs with the kernels k, and $\mathcal{M}$ and $\mathfrak{P}$ are a Harish-Chandra module and a Harish-Chandra contramodule over $(\mathfrak{g}, \mathcal{C})$ with the central charges $\varkappa_1$ and $\varkappa_2$, respectively, then the vector space $\mathrm{Hom}_k(\mathcal{M}, \mathfrak{P})$ has a natural structure of Harish-Chandra contramodule with the central charge $\varkappa_2 - \varkappa_1$. Here $\varkappa_2 - \varkappa_1 \colon (\mathfrak{g}''', \mathcal{C}) \longrightarrow (\mathfrak{g}, \mathcal{C})$ denotes the Baer difference of the central extensions $\varkappa_2$ and $\varkappa_1$. This Harish-Chandra contramodule structure consists of the $\mathfrak{g}'''$-contramodule and $\mathcal{C}$-contramodule structures on $\mathrm{Hom}_k(\mathcal{M}, \mathfrak{P})$ defined by the above rules.

D.3 Isomorphism of semialgebras

D.3.1 For any two central extensions of Tate Harish-Chandra pairs $\varkappa' \colon (\mathfrak{g}', \mathcal{C}) \longrightarrow (\mathfrak{g}, \mathcal{C})$ and $\varkappa'' \colon (\mathfrak{g}'', \mathcal{C}) \longrightarrow (\mathfrak{g}, \mathcal{C})$ with the kernels identified with k we denote by $\varkappa' + \varkappa''$ their Baer sum, i.e., the central extension of Tate Harish-Chandra pairs $(\mathfrak{g}''', \mathcal{C}) \longrightarrow (\mathfrak{g}, \mathcal{C})$ with $\mathfrak{g}''' = \ker(\mathfrak{g}' \oplus \mathfrak{g}'' \to \mathfrak{g})/\mathrm{im}\, k$, where the map $\mathfrak{g}' \oplus \mathfrak{g}'' \longrightarrow \mathfrak{g}$ is the difference of the maps $\mathfrak{g}' \longrightarrow \mathfrak{g}$ and $\mathfrak{g}'' \longrightarrow \mathfrak{g}$, and the map $k \longrightarrow \mathfrak{g}' \oplus \mathfrak{g}''$ is the difference of the maps $k \longrightarrow \mathfrak{g}'$ and $k \longrightarrow \mathfrak{g}''$. The canonical central extension $(\mathfrak{g}^{\sim}, \mathcal{C}) \longrightarrow (\mathfrak{g}, \mathcal{C})$ will be denoted by $\varkappa_0$.

We claim that for any central extension of Tate Harish-Chandra pairs $\varkappa \colon (\mathfrak{g}', \mathcal{C}) \longrightarrow (\mathfrak{g}, \mathcal{C})$ with the kernel k such that the condition (iv) of D.2.2 is satisfied there is a natural isomorphism

$$\mathsf{S}^r_{\varkappa+\varkappa_0}(\mathfrak{g}, \mathcal{C}) \simeq \mathsf{S}^l_{\varkappa}(\mathfrak{g}, \mathcal{C})$$

of semialgebras over the coalgebra $\mathcal{C}$. This isomorphism is characterized by the following three properties.

(a) Consider the increasing filtration F of the k-algebra $U_{\varkappa}(\mathfrak{g})$ with the components

$$F_i U_{\varkappa}(\mathfrak{g}) = (k + \mathfrak{g}' + \cdots + \mathfrak{g}'^i)U(\mathfrak{h}) = U(\mathfrak{h})(k + \mathfrak{g}' + \cdots + \mathfrak{g}'^i)$$

and the induced filtration

$$F_i \mathsf{S}^l_{\varkappa} = F_i U_{\varkappa}(\mathfrak{g}) \otimes_{U(\mathfrak{h})} \mathcal{C}$$

of the semialgebra $\mathsf{S}^l_{\varkappa} = \mathsf{S}^l_{\varkappa}(\mathfrak{g}, \mathcal{C})$. Then we have $F_0 \mathsf{S}^l_{\varkappa} \simeq \mathcal{C}$, $\mathsf{S}^l_{\varkappa} \simeq \varinjlim F_i \mathsf{S}^l_{\varkappa}$, and the semimultiplication maps $F_i \mathsf{S}^l_{\varkappa} \square_{\mathcal{C}} F_j \mathsf{S}^l_{\varkappa} \longrightarrow \mathsf{S}^l_{\varkappa} \square_{\mathcal{C}} \mathsf{S}^l_{\varkappa} \longrightarrow \mathsf{S}^l_{\varkappa}$ factorize through $F_{i+j} \mathsf{S}^l_{\varkappa}$. There is an analogous filtration

$$F_i \mathsf{S}^r_{\varkappa+\varkappa_0} = \mathcal{C} \otimes_{U(\mathfrak{h})} F_i U_{\varkappa+\varkappa_0}(\mathfrak{g})$$

of the semialgebra $\mathbf{S}^r_{\varkappa+\varkappa_0} = \mathbf{S}^r_{\varkappa+\varkappa_0}(\mathfrak{g}, \mathcal{C})$. The desired isomorphism $\mathbf{S}^r_{\varkappa+\varkappa_0} \simeq \mathbf{S}^l_{\varkappa}$ preserves the filtrations F.

(b) The natural maps $F_{i-1}\mathbf{S}^l_{\varkappa} \longrightarrow F_i\mathbf{S}^l_{\varkappa}$ are injective and their cokernels are coflat left and right $\mathcal{C}$-comodules, so the associated graded quotient semialgebra

$$\operatorname{gr}_F \mathbf{S}^l_{\varkappa} = \bigoplus_i F_i\mathbf{S}^l_{\varkappa}/F_{i-1}\mathbf{S}^l_{\varkappa}$$

is defined (cf. 11.5). The semialgebra $\operatorname{gr}_F \mathbf{S}^l_{\varkappa}$ is naturally isomorphic to the tensor product $\operatorname{Sym}_k(\mathfrak{g}/\mathfrak{h}) \otimes_k \mathcal{C}$ of the symmetric algebra $\operatorname{Sym}_k(\mathfrak{g}/\mathfrak{h})$ of the k-vector space $\mathfrak{g}/\mathfrak{h}$ and the coalgebra $\mathcal{C}$, endowed with the semialgebra structure corresponding to the left entwining structure

$$\operatorname{Sym}_k(\mathfrak{g}/\mathfrak{h}) \otimes_k \mathcal{C} \longrightarrow \mathcal{C} \otimes_k \operatorname{Sym}_k(\mathfrak{g}/\mathfrak{h})$$

for the coalgebra $\mathcal{C}$ and the algebra $\operatorname{Sym}_k(\mathfrak{g}/\mathfrak{h})$ (see 10.3). Here the entwining map is given by the formula $u \otimes c \longmapsto cu_{(-1)} \otimes u_{(0)}$, where $u \longmapsto u_{(-1)} \otimes u_{(0)}$ denotes the $\mathcal{C}$-coaction in $\operatorname{Sym}_k(\mathfrak{g}/\mathfrak{h})$ induced by the $\mathcal{C}$-coaction in $\mathfrak{g}/\mathfrak{h}$. Analogously, the semialgebra $\operatorname{gr}_F \mathbf{S}^r_{\varkappa+\varkappa_0}$ is naturally isomorphic to the tensor product $\mathcal{C} \otimes_k \operatorname{Sym}_k(\mathfrak{g}/\mathfrak{h})$ endowed with the semialgebra structure corresponding to the right entwining structure

$$\mathcal{C} \otimes_k \operatorname{Sym}_k(\mathfrak{g}/\mathfrak{h}) \longrightarrow \operatorname{Sym}_k(\mathfrak{g}/\mathfrak{h}) \otimes_k \mathcal{C}.$$

Here the entwining map is given by the formula $c \otimes u \longmapsto u_{(0)} \otimes cu_{(1)}$, where the right coaction $u \longmapsto u_{(0)} \otimes u_{(1)}$ is obtained from the above left coaction $u \longmapsto u_{(-1)} \otimes u_{(0)}$ by applying the antipode. These left and right entwining maps are inverse to each other, hence there is a natural isomorphism of semialgebras

$$\operatorname{gr}_F \mathbf{S}^l_{\varkappa} \simeq \operatorname{gr}_F \mathbf{S}^r_{\varkappa+\varkappa_0}.$$

This isomorphism can be obtained by passing to the associated graded quotient semialgebras in the desired isomorphism $\mathbf{S}^l_{\varkappa} \simeq \mathbf{S}^r_{\varkappa+\varkappa_0}$.

(c) Choose a section $b': \mathfrak{g}/\mathfrak{h} \longrightarrow \mathfrak{g}'$ of the natural surjection $\mathfrak{g}' \longrightarrow \mathfrak{g}'/(\mathfrak{h} \oplus k) \simeq \mathfrak{g}/\mathfrak{h}$. Composing b' with the surjection $\mathfrak{g}' \longrightarrow \mathfrak{g}$, we obtain a section b of the natural surjection $\mathfrak{g} \longrightarrow \mathfrak{g}/\mathfrak{h}$, hence a direct sum decomposition $\mathfrak{g} \simeq \mathfrak{h} \oplus b(\mathfrak{g}/\mathfrak{h})$. So there is the corresponding section $\mathfrak{g} \longrightarrow \mathfrak{g}^\sim$ of the canonical central extension $\mathfrak{g}^\sim \longrightarrow \mathfrak{g}$; denote by $\tilde{b}$ the composition $\mathfrak{g}/\mathfrak{h} \longrightarrow \mathfrak{g} \longrightarrow \mathfrak{g}^\sim$ of the section b and the section $\mathfrak{g} \longrightarrow \mathfrak{g}^\sim$. The Baer sum of the sections b' and $\tilde{b}$ provides a section $b'': \mathfrak{g}/\mathfrak{h} \longrightarrow \mathfrak{g}''$, where $(\mathfrak{g}'', \mathcal{C}) \longrightarrow (\mathfrak{g}, \mathcal{C})$ denotes the central extension $\varkappa + \varkappa_0$. Now the composition

$$\mathfrak{g}/\mathfrak{h} \otimes_k \mathcal{C} \longrightarrow \mathfrak{g}' \otimes_k \mathcal{C} \simeq F_1 U_\varkappa(\mathfrak{g}) \otimes_k \mathcal{C} \longrightarrow F_1\mathbf{S}^l_{\varkappa}$$

of the map induced by the map b', the isomorphism induced by the natural isomorphism $\mathfrak{g}' \simeq F_1 U_\varkappa(\mathfrak{g})$, and the surjection $U_\varkappa(\mathfrak{g}) \otimes_k \mathcal{C} \longrightarrow U_\varkappa(\mathfrak{g}) \otimes_{U(\mathfrak{h})} \mathcal{C}$

provides a section of the natural surjection $F_1 \mathcal{S}^l_{\varkappa} \longrightarrow F_1 \mathcal{S}^l_{\varkappa} / F_0 \mathcal{S}^l_{\varkappa} \simeq \mathfrak{g}/\mathfrak{h} \otimes_k \mathcal{C}$. This section is a morphism of right $\mathcal{C}$-comodules. Hence the corresponding retraction $F_1 \mathcal{S}^l_{\varkappa} \longrightarrow F_0 \mathcal{S}^l_{\varkappa} \simeq \mathcal{C}$ is also a morphism of right $\mathcal{C}$-comodules. Analogously, the composition

$$\mathcal{C} \otimes_k \mathfrak{g}/\mathfrak{h} \longrightarrow \mathcal{C} \otimes_k \mathfrak{g}'' \simeq \mathcal{C} \otimes_k F_1 U_{\varkappa + \varkappa_0}(\mathfrak{g}) \longrightarrow F_1 \mathcal{S}^r_{\varkappa + \varkappa_0},$$

where the first morphism is induced by the map b'', is a section of the natural surjection $F_1 \mathcal{S}^r_{\varkappa + \varkappa_0} \longrightarrow F_1 \mathcal{S}^r_{\varkappa + \varkappa_0} / F_0 \mathcal{S}^r_{\varkappa + \varkappa_0} \simeq \mathcal{C} \otimes_k \mathfrak{g}/\mathfrak{h}$; this section is a morphism of left $\mathcal{C}$-comodules. Hence so is the corresponding retraction $F_1 \mathcal{S}^r_{\varkappa + \varkappa_0} \longrightarrow F_0 \mathcal{S}^r_{\varkappa + \varkappa_0} \simeq \mathcal{C}$. The desired isomorphism $F_1 \mathcal{S}^l_{\varkappa} \simeq F_1 \mathcal{S}^r_{\varkappa + \varkappa_0}$ identifies the compositions

$$F_1 \mathcal{S}^l_{\varkappa} \longrightarrow \mathcal{C} \longrightarrow k \quad \text{and} \quad F_1 \mathcal{S}^r_{\varkappa + \varkappa_0} \longrightarrow \mathcal{C} \longrightarrow k$$

of the retractions $F_1 \mathcal{S}^l_{\varkappa} \longrightarrow \mathcal{C}$ and $F_1 \mathcal{S}^r_{\varkappa + \varkappa_0} \longrightarrow \mathcal{C}$ with the counit map $\mathcal{C} \longrightarrow k$. This condition holds for all sections b'.

Theorem. *Assuming the condition* (iv), *there exists a unique isomorphism of semialgebras* $\mathcal{S}^r_{\varkappa + \varkappa_0}(\mathfrak{g}, \mathcal{C}) \simeq \mathcal{S}^l_{\varkappa}(\mathfrak{g}, \mathcal{C})$ *over* $\mathcal{C}$ *satisfying the above properties* (a–c).

Proof. Uniqueness is clear, since a morphism from a $\mathcal{C}$-$\mathcal{C}$-bicomodule to the bicomodule $\mathcal{C}$ is determined by its composition with the counit map $\mathcal{C} \longrightarrow k$. The proof of existence occupies subsections D.3.2–D.3.7.

The next result is obtained by specializing the semimodule-semicontramodule correspondence theorem to the case of Harish-Chandra modules and contramodules.

Corollary. *There is a natural equivalence* $\mathbb{R}\Psi_{\mathcal{S}^l_{\varkappa}} = \mathbb{L}\Phi^{-1}_{\mathcal{S}^r_{\varkappa + \varkappa_0}}$ *between the semiderived category of Harish-Chandra modules with the central charge* $\varkappa$ *over* $(\mathfrak{g}, \mathcal{C})$ *and the semiderived category of Harish-Chandra contramodules with the central charge* $\varkappa + \varkappa_0$ *over* $(\mathfrak{g}, \mathcal{C})$. *Here the semiderived category of Harish-Chandra modules is defined as the quotient category of the homotopy category of complexes of Harish-Chandra modules by the thick subcategory of* $\mathcal{C}$-coacyclic *complexes; the semiderived category of Harish-Chandra contramodules is analogously defined as the quotient category by the thick subcategory of* $\mathcal{C}$-contraacyclic *complexes.*

Proof. This follows from the results of D.2.5 and D.2.8, Theorem D.3.1, and Corollary 6.3. □

Remark. The main property of the equivalence of semiderived categories provided by Corollary D.3.1 is that it transforms the Harish-Chandra modules $\mathfrak{M}$ that, considered as $\mathcal{C}$-comodules, are the cofree comodules $\mathcal{C} \otimes_k E$ cogenerated by a vector space E, into the Harish-Chandra contramodules $\mathfrak{P}$ that, considered as $\mathcal{C}$-contramodules, are the free contramodules $\mathrm{Hom}_k(\mathcal{C}, E)$ generated by the same vector space E, and vice versa. A similar assertion holds for any complexes of $\mathcal{C}$-cofree Harish-Chandra modules and $\mathcal{C}$-free Harish-Chandra contramodules. The above corollary is a way to formulate the classical duality between Harish-Chandra modules with the complementary central charges $\varkappa$ and $-\varkappa - \varkappa_0$ [39, 77]. Of course, there is no hope of establishing an *anti-equivalence* between any kinds of exotic derived categories of *arbitrary* Harish-Chandra modules over $(\mathfrak{g}, \mathcal{C})$ with the complementary central charges, as the derived category of vector spaces is not anti-equivalent to itself. At the very least, one would have to impose some finiteness conditions on the Harish-Chandra modules. The introduction of contramodules allows us to resolve this problem. Still one can use the functor $\Phi_{\mathbf{S}}$ to construct a *contravariant functor* between the semiderived categories of Harish-Chandra modules with the complementary central charges. Choose a vector space U; for example, $U = k$. Consider the functor $\mathcal{N} \longmapsto \mathrm{Hom}_k(\mathcal{N}, U)$ acting from the semiderived category of Harish-Chandra modules over $(\mathfrak{g}, \mathcal{C})$ with the central charge $-\varkappa - \varkappa_0$ to the semiderived category of Harish-Chandra contramodules over $(\mathfrak{g}, \mathcal{C})$ with the central charge $\varkappa + \varkappa_0$. Composing this functor $\mathrm{Hom}_k(-, U)$ with the functor $\mathbb{L}\Phi_{\mathbf{S}^r_{\varkappa+\varkappa_0}}$, one obtains a contravariant functor $\mathsf{D}^{\mathsf{si}}(\mathsf{O}_{-\varkappa-\varkappa_0}(\mathfrak{g}, \mathcal{C})) = \mathsf{D}^{\mathsf{si}}(\text{simod-}\mathbf{S}^r_{\varkappa+\varkappa_0}) \longrightarrow \mathsf{D}^{\mathsf{si}}(\mathbf{S}^l_\varkappa\text{-simod}) = \mathsf{D}^{\mathsf{si}}(\mathsf{O}_\varkappa(\mathfrak{g}, \mathcal{C}))$. The latter functor transforms the Harish-Chandra modules that as $\mathcal{C}$-comodules are cofreely cogenerated by a vector space E into the Harish-Chandra modules that as $\mathcal{C}$-comodules are cofreely cogenerated by the vector space $\mathrm{Hom}_k(E, U)$, and similarly for complexes of $\mathcal{C}$-cofree Harish-Chandra modules. One cannot avoid using the exotic derived categories in this construction, because the functor $\mathbb{L}\Phi_{\mathbf{S}}$ does not preserve acyclicity, in general (see 0.2.7).

D.3.2   The semialgebras $\mathbf{S}^l_\varkappa$ and $\mathbf{S}^r_{\varkappa+\varkappa_0}$ endowed with the increasing filtrations F are left and right coflat nonhomogeneous Koszul semialgebras over the coalgebra $\mathcal{C}$ (see 11.5). In other words, the graded semialgebras $\mathrm{gr}_F\mathbf{S}^l_\varkappa$ and $\mathrm{gr}_F\mathbf{S}^r_{\varkappa+\varkappa_0}$ over $\mathcal{C}$ are left and right coflat Koszul in the sense of 11.4. Indeed, there are natural isomorphisms of complexes of $\mathcal{C}$-$\mathcal{C}$-bicomodules

$$\mathrm{Bar}^\bullet_{\mathrm{gr}}(\mathrm{gr}_F\mathbf{S}^l_\varkappa, \mathcal{C}) \simeq \mathrm{Bar}^\bullet_{\mathrm{gr}}(\mathrm{Sym}_k(\mathfrak{g}/\mathfrak{h}), k) \otimes_k \mathcal{C}$$

and $\mathrm{Bar}^\bullet_{\mathrm{gr}}(\mathrm{gr}_F\mathbf{S}^r_{\varkappa+\varkappa_0}, \mathcal{C}) \simeq \mathcal{C} \otimes_k \mathrm{Bar}^\bullet_{\mathrm{gr}}(\mathrm{Sym}_k(\mathfrak{g}/\mathfrak{h}), k)$, and the k-algebra $\mathrm{Sym}_k(\mathfrak{g}/\mathfrak{h})$ is Koszul. Here the left $\mathcal{C}$-coaction in $\mathrm{Bar}^\bullet_{\mathrm{gr}}(\mathrm{Sym}_k(\mathfrak{g}/\mathfrak{h}), k)\otimes_k\mathcal{C}$ is the tensor product of the $\mathcal{C}$-coaction in $\mathrm{Bar}^\bullet_{\mathrm{gr}}(\mathrm{Sym}_k(\mathfrak{g}/\mathfrak{h}), k)$ induced by the $\mathcal{C}$-coaction in $\mathfrak{g}/\mathfrak{h}$ and the left $\mathcal{C}$-coaction in $\mathcal{C}$, while the right $\mathcal{C}$-coaction in $\mathrm{Bar}^\bullet_{\mathrm{gr}}(\mathrm{Sym}_k(\mathfrak{g}/\mathfrak{h}), k) \otimes_k \mathcal{C}$ is induced by the right $\mathcal{C}$-coaction in $\mathcal{C}$. The $\mathcal{C}$-$\mathcal{C}$-bicomodule structure on $\mathcal{C} \otimes_k \mathrm{Bar}^\bullet_{\mathrm{gr}}(\mathrm{Sym}_k(\mathfrak{g}/\mathfrak{h}), k)$ is defined in an analogous way (with the left and right sides switched).

   The left and right coflat Koszul coalgebras $\mathcal{D}^l$ and $\mathcal{D}^r$ over $\mathcal{C}$ quadratic dual to the left and right coflat Koszul semialgebras $\mathrm{gr}_F\mathbf{S}^l_\varkappa$ and $\mathrm{gr}_F\mathbf{S}^r_{\varkappa+\varkappa_0}$ are described as follows. One has

$$\mathcal{D}^l \simeq \textstyle\bigwedge_k(\mathfrak{g}/\mathfrak{h}) \otimes_k \mathcal{C},$$

where $\bigwedge_k(\mathfrak{g}/\mathfrak{h})$ denotes the exterior coalgebra of the k-vector space $\mathfrak{g}/\mathfrak{h}$, i.e., the coalgebra quadratic dual to the symmetric algebra $\mathrm{Sym}_k(\mathfrak{g}/\mathfrak{h})$. The counit of $\bigwedge_k(\mathfrak{g}/\mathfrak{h}) \otimes_k \mathcal{C}$ is the tensor product of the counits of $\bigwedge_k(\mathfrak{g}/\mathfrak{h})$ and $\mathcal{C}$, while the co-multiplication in $\bigwedge_k(\mathfrak{g}/\mathfrak{h}) \otimes_k \mathcal{C}$ is constructed as the composition $\bigwedge_k(\mathfrak{g}/\mathfrak{h}) \otimes_k \mathcal{C} \longrightarrow \bigwedge_k(\mathfrak{g}/\mathfrak{h}) \otimes_k \bigwedge_k(\mathfrak{g}/\mathfrak{h}) \otimes_k \mathcal{C} \otimes_k \mathcal{C} \longrightarrow \bigwedge_k(\mathfrak{g}/\mathfrak{h}) \otimes_k \mathcal{C} \otimes_k \bigwedge_k(\mathfrak{g}/\mathfrak{h}) \otimes_k \mathcal{C}$ of the map induced by the comultiplications in $\bigwedge_k(\mathfrak{g}/\mathfrak{h})$ and $\mathcal{C}$ and the map induced by the "permutation" map $\bigwedge_k(\mathfrak{g}/\mathfrak{h}) \otimes_k \mathcal{C} \longrightarrow \mathcal{C} \otimes_k \bigwedge_k(\mathfrak{g}/\mathfrak{h})$. The latter map is given by the formula $u \otimes c \longmapsto cu_{[-1]} \otimes u_{[0]}$ for $u \in \bigwedge_k(\mathfrak{g}/\mathfrak{h})$ and $c \in \mathcal{C}$, where $u \longmapsto u_{[-1]} \otimes u_{[0]}$ denotes the $\mathcal{C}$-coaction in $\bigwedge_k(\mathfrak{g}/\mathfrak{h})$ induced by the $\mathcal{C}$-coaction in $\mathfrak{g}/\mathfrak{h}$.

Analogously, one has $\mathcal{D}^r \simeq \mathcal{C} \otimes_k \bigwedge_k(\mathfrak{g}/\mathfrak{h})$, where the counit of $\mathcal{C} \otimes_k \bigwedge_k(\mathfrak{g}/\mathfrak{h})$ is the tensor product of the counits of $\bigwedge_k(\mathfrak{g}/\mathfrak{h})$ and $\mathcal{C}$, while the comultiplication in $\mathcal{C} \otimes_k \bigwedge_k(\mathfrak{g}/\mathfrak{h})$ is defined in terms of the "permutation" map $\mathcal{C} \otimes_k \bigwedge_k(\mathfrak{g}/\mathfrak{h}) \longrightarrow \bigwedge_k(\mathfrak{g}/\mathfrak{h}) \otimes_k \mathcal{C}$. The latter map is given by the formula $c \otimes u \longmapsto u_{[0]} \otimes cu_{[1]}$, where the right coaction $u \longmapsto u_{[0]} \otimes u_{[1]}$ is obtained from the left coaction $u \longmapsto u_{[-1]} \otimes u_{[0]}$ by applying the antipode. Both coalgebras $\bigwedge_k(\mathfrak{g}/\mathfrak{h}) \otimes_k \mathcal{C}$ and $\mathcal{C} \otimes_k \bigwedge_k(\mathfrak{g}/\mathfrak{h})$ have grad-ings induced by the grading of $\bigwedge_k(\mathfrak{g}/\mathfrak{h})$. The two "permutation" maps are inverse to each other, and they provide an isomorphism of graded coalgebras $\mathcal{D}^l \simeq \mathcal{D}^r$.

Now recall that we have assumed the condition (iv) of D.2.2. Denote by $\cdots \subset V^2\mathcal{C} \subset V^1\mathcal{C} \subset V^0\mathcal{C} = \mathcal{C}$ the decreasing filtration of $\mathcal{C}$ orthogonal to the natural increasing filtration of the universal enveloping algebra $U(\mathfrak{h})$, that is

$$V^i\mathcal{C} = \{c \in \mathcal{C} \mid \phi(c, x) = 0 \text{ for all } x \in k + \mathfrak{h} + \cdots + \mathfrak{h}^{i-1} \subset U(\mathfrak{h})\}.$$

Notice that the decreasing filtration V is compatible with both the coalgebra and algebra structures on $\mathcal{C}$; in particular, it is a filtration by ideals with respect to the multiplication. The subspace $V^1\mathcal{C}$ is the kernel of the counit map $\mathcal{C} \longrightarrow k$; the subspace $V^2\mathcal{C}$ is the kernel of the map $\mathcal{C} \longrightarrow \mathfrak{h}^\vee \oplus k$ which is the sum of the map $\check{\psi}$ and the counit map.

Define decreasing filtrations V on the coalgebras $\mathcal{D}^l$ and $\mathcal{D}^r$ by the formulas

$$V^i\mathcal{D}^l \simeq \bigwedge_k(\mathfrak{g}/\mathfrak{h}) \otimes_k V^i\mathcal{C} \quad \text{and} \quad V^i\mathcal{D}^r \simeq V^i\mathcal{C} \otimes_k \bigwedge_k(\mathfrak{g}/\mathfrak{h});$$

these filtrations are compatible with the coalgebra structures on $\mathcal{D}^l$ and $\mathcal{D}^r$, and correspond to each other under the isomorphism $\mathcal{D}^l \simeq \mathcal{D}^r$. Set $\mathcal{D}^l \simeq \mathcal{D} \simeq \mathcal{D}^r$. The coalgebra $\mathcal{D}$ is cogenerated by the maps $\mathcal{D} \longrightarrow \mathcal{D}_0/V^2\mathcal{D}_0$ and $\mathcal{D} \longrightarrow \mathcal{D}_1/V^1\mathcal{D}_1$, i.e., the iterated comultiplication map from $\mathcal{D}$ to the direct product of all tensor powers of $\mathcal{D}_0/V^2\mathcal{D}_0 \oplus \mathcal{D}_1/V^1\mathcal{D}_1$ is injective. Moreover, the decreasing filtration V on $\mathcal{D}$ is cogenerated by the filtrations on $\mathcal{D}_0/V^2\mathcal{D}_0$ and $\mathcal{D}_1/V^1\mathcal{D}_1$, i.e., the subspaces $V^i\mathcal{D}$ are the full preimages of the subspaces of the induced filtration on the product of all tensor powers of $\mathcal{D}_0/V^2\mathcal{D}_0 \oplus \mathcal{D}_1/V^1\mathcal{D}_1$ under the iterated comultiplication map.

D.3.3 Composing the equivalences of categories from 11.2.2 and Theorem 11.6, we obtain an equivalence between the category of left (right) coflat nonhomoge-neous Koszul semialgebras over $\mathcal{C}$ and the category of left (right) coflat Koszul

CDG-coalgebras over $\mathcal{C}$. Here a CDG-coalgebra $(\mathcal{D}, d, h)$ is called Koszul over $\mathcal{C}$ if the underlying graded coalgebra $\mathcal{D}$ is Koszul over $\mathcal{C}$. Recall that for a left (right) coflat nonhomogeneous Koszul semialgebra $\mathbf{S}^\sim$ and the corresponding quasi-differential coalgebra $\mathcal{D}^\sim$ one has $F_1\mathbf{S}^\sim \simeq \mathcal{D}_1^\sim$, so to construct a specific CDG-coalgebra $(\mathcal{D}, d, h)$ corresponding to a given filtered semialgebra $\mathbf{S}^\sim$ one has to choose a linear map

$$\delta \colon F_1\mathbf{S}^\sim \longrightarrow k$$

such that the composition of the injection $\mathcal{C} \simeq F_0\mathbf{S}^\sim \longrightarrow F_1\mathbf{S}^\sim$ with δ coincides with the counit map of $\mathcal{C}$.

Choose a section $b' \colon \mathfrak{g}/\mathfrak{h} \longrightarrow \mathfrak{g}'$ and construct the related section $b'' \colon \mathfrak{g}/\mathfrak{h} \longrightarrow \mathfrak{g}''$; denote by

$$\delta_{b'}^l \colon F_1\mathbf{S}_\varkappa^l \longrightarrow k \quad \text{and} \quad \delta_{b''}^r \colon F_1\mathbf{S}_{\varkappa+\varkappa_0}^r \longrightarrow k$$

the corresponding linear functions constructed in (c) of D.3.1. In order to obtain an isomorphism of left and right coflat nonhomogeneous Koszul semialgebras $\mathbf{S}_\varkappa^l \simeq \mathbf{S}_{\varkappa+\varkappa_0}^r$, we will construct an isomorphism between the CDG-coalgebras

$$(\mathcal{D}^l, d_{b'}^l, h_{b'}^l) \quad \text{and} \quad (\mathcal{D}^r, d_{b''}^r, h_{b''}^r)$$

corresponding to the filtered semialgebras $\mathbf{S}_\varkappa^l$ and $\mathbf{S}_{\varkappa+\varkappa_0}^r$ endowed with the linear functions $\delta_{b'}^l$ and $\delta_{b''}^r$. For the corresponding isomorphism of semialgebras to identity the linear functions $\delta_{b'}^l$ and $\delta_{b''}^r$, we need this isomorphism of CDG-coalgebras to have the form $(f, 0)$, i.e., the change-of-connection linear function has to vanish. The isomorphism of coalgebras $\mathcal{D}^l \simeq \mathcal{D}^r$ is already defined; all we have to do is to check that it identifies $d_{b'}^l$ with $d_{b''}^r$ and $h_{b'}^l$ with $h_{b''}^r$.

Besides, we need to show that the isomorphism $\mathbf{S}_\varkappa^l \simeq \mathbf{S}_{\varkappa+\varkappa_0}^r$ so obtained does not depend on the choice of b'. Here it suffices to check that changing the section b' to b_1' leads to isomorphisms of CDG-coalgebras

$$(\mathrm{id}, a^l) \colon (\mathcal{D}^l, d_{b'}^r, h_{b'}^r) \longrightarrow (\mathcal{D}^l, d_{b_1'}^r, h_{b_1'}^r)$$

and

$$(\mathrm{id}, a^r) \colon (\mathcal{D}^r, d_{b''}^r, h_{b''}^r) \longrightarrow (\mathcal{D}^r, d_{b_1''}^r, h_{b_1''}^r)$$

with the linear functions a^l and a^r being identified by the isomorphism $\mathcal{D}^l \simeq \mathcal{D}^r$.

Since the coalgebra $\mathcal{D}^l = \mathcal{D} \simeq \mathcal{D}^r$ is cogenerated by the maps

$$\mathcal{D} \longrightarrow \mathcal{D}_0/V^2\mathcal{D}_0 \subset \mathfrak{h}^\vee \oplus k \quad \text{and} \quad \mathcal{D} \longrightarrow \mathcal{D}_1/V_1\mathcal{D}_1 \simeq \mathfrak{g}/\mathfrak{h},$$

it suffices to check that the compositions of $d_{b'}^l$ and $d_{b''}^r$ with these two maps coincide in order to show that $d_{b'}^l = d_{b''}^r$. We will also see that these compositions factorize through $\mathcal{D}_1/V^2\mathcal{D}_1$ and $\mathcal{D}_2/V^1\mathcal{D}_2$, respectively, and the induced map $\mathcal{D}_1/V^2\mathcal{D}_1 \longrightarrow \mathcal{D}_0/V^2\mathcal{D}_0$ preserves the images of V^1 (actually, even maps the whole of $\mathcal{D}_1/V^2\mathcal{D}_1$ into $V^1\mathcal{D}_0/V^2\mathcal{D}_1$), hence it will follow that the differential $d_{b'}^l = d_{b''}^r$ preserves the decreasing filtration V. Besides, we will see that the linear function $h_{b'}^l = h_{b''}^r$ annihilates the subspace $V^2\mathcal{D}_2$ and the linear function $a^l = a^r$ corresponding to a change of section b' annihilates the subspace $V^2\mathcal{D}_1$.

D.3.4 Let us introduce notation for the components of the commutator map with respect to the direct sum decomposition $\mathfrak{g}' \simeq \mathfrak{h} \oplus b'(\mathfrak{g}/\mathfrak{h}) \oplus k$. As above, the Lie coalgebra structure on $\mathfrak{h}^\vee$ is denoted by $x^* \longmapsto x^*_{\{1\}} \wedge x^*_{\{2\}}$. Denote the Lie coaction of $\mathfrak{h}^\vee$ in $\mathfrak{g}/\mathfrak{h}$, i.e., the map $\mathfrak{g}/\mathfrak{h} \longrightarrow \mathfrak{h}^\vee \otimes_k \mathfrak{g}/\mathfrak{h}$ corresponding to the commutator map $\mathfrak{h} \times \mathfrak{g}/\mathfrak{h} \longrightarrow \mathfrak{g}/\mathfrak{h}$, by $u \longmapsto u_{\{-1\}} \otimes u_{\{0\}}$. These two maps do not depend on the choice of the section b'; the rest of them do.

Denote by $u \otimes x^* \longmapsto u(x^*)$ the map $\mathfrak{g}/\mathfrak{h} \otimes_k \mathfrak{h}^\vee \longrightarrow \mathfrak{h}^\vee$ corresponding to the projection of the commutator map $b'(\mathfrak{g}/\mathfrak{h}) \times \mathfrak{h} \longrightarrow \mathfrak{g}' \longrightarrow \mathfrak{h}$. Denote by $u \wedge v \longmapsto \{u, v\}$ the map $\bigwedge^2_k(\mathfrak{g}/\mathfrak{h}) \longrightarrow \mathfrak{g}/\mathfrak{h}$ corresponding to the commutator map $\bigwedge^2_k b'(\mathfrak{g}/\mathfrak{h}) \longrightarrow \mathfrak{g}/\mathfrak{h}$. Denote by $u \wedge v \otimes x^* \longmapsto (u, v)_{x^*}$ the map $\bigwedge^2_k(\mathfrak{g}/\mathfrak{h}) \otimes_k \mathfrak{h}^\vee \longrightarrow k$ corresponding to the projection of the commutator map $\bigwedge^2_k b'(\mathfrak{g}/\mathfrak{h}) \longrightarrow \mathfrak{g}' \longrightarrow \mathfrak{h}$.

The above five maps only depend on the Lie algebra $\mathfrak{g}$ with the subalgebra $\mathfrak{h}$ and the section $b \colon \mathfrak{g}/\mathfrak{h} \longrightarrow \mathfrak{g}$, but the following two will depend essentially on $\mathfrak{g}'$ and b'. Denote by $\rho' \colon \mathfrak{g}/\mathfrak{h} \longrightarrow \mathfrak{h}^\vee$ the map corresponding to the projection of the commutator map $b'(\mathfrak{g}/\mathfrak{h}) \times \mathfrak{h} \longrightarrow \mathfrak{g}' \longrightarrow k$. Denote by $\sigma' \colon \bigwedge^2_k(\mathfrak{g}/\mathfrak{h}) \longrightarrow k$ the map corresponding to the projection of the commutator map $\bigwedge^2_k b'(\mathfrak{g}/\mathfrak{h}) \longrightarrow \mathfrak{g}' \longrightarrow k$. Denote by $\tilde{\rho}, \tilde{\sigma}$ and ρ'', σ'' the analogous maps corresponding to the central extensions $\mathfrak{g}^\sim \longrightarrow \mathfrak{g}$ and $\mathfrak{g}'' \longrightarrow \mathfrak{g}$ with the sections $\tilde{b}$ and b''. Clearly, we have $\rho'' = \rho' + \tilde{\rho}$ and $\sigma'' = \sigma' + \tilde{\sigma}$.

Set $\mathfrak{b} = b(\mathfrak{g}/\mathfrak{h}) \subset \mathfrak{g}$. The composition θ of the commutator map in $\mathfrak{gl}(\mathfrak{g})^\sim$ with the projection $\mathfrak{gl}(\mathfrak{g})^\sim \longrightarrow k$ corresponding to the section $\mathfrak{gl}(\mathfrak{g}) \longrightarrow \mathfrak{gl}(\mathfrak{g})^\sim$ coming from the direct sum decomposition $\mathfrak{g} \simeq \mathfrak{h} \oplus \mathfrak{b}$ is written explicitly as follows. For any continuous linear operator $A \colon \mathfrak{g} \longrightarrow \mathfrak{g}$ denote by $A_{\mathfrak{h} \to \mathfrak{b}} \colon \mathfrak{h} \longrightarrow \mathfrak{b}$, $A_{\mathfrak{b} \to \mathfrak{h}} \colon \mathfrak{b} \longrightarrow \mathfrak{h}$, etc., its components with respect to our direct sum decomposition. Then the cocycle θ is given by the formula $\theta(A \wedge B) = \operatorname{tr}(A_{\mathfrak{b} \to \mathfrak{h}} B_{\mathfrak{h} \to \mathfrak{b}}) - \operatorname{tr}(B_{\mathfrak{b} \to \mathfrak{h}} A_{\mathfrak{h} \to \mathfrak{b}})$, where tr denotes the trace of a linear operator $\mathfrak{h} \longrightarrow \mathfrak{h}$ with an open kernel.

Using this formula, one can find that, in the above notation, $\tilde{\rho}(u) = -u_{\{0\}}(u_{\{-1\}})$ and $\tilde{\sigma}(u \wedge v) = -(u, v_{\{0\}})v_{\{-1\}} + (v, u_{\{0\}})u_{\{-1\}}$.

D.3.5 We have $\mathcal{D}^l_0 \simeq \mathcal{C}$, $\mathcal{D}^l_1 \simeq \mathfrak{g}/\mathfrak{h} \otimes_k \mathcal{C}$, and $\mathcal{D}^l_2 \simeq \bigwedge^2_k(\mathfrak{g}/\mathfrak{h}) \otimes_k \mathcal{C}$. The composition of the map $d^l_{b'} \colon \mathfrak{g}/\mathfrak{h} \otimes_k \mathcal{C} \longrightarrow \mathcal{C}$ with the counit map $\varepsilon \colon \mathcal{C} \longrightarrow k$ vanishes, since $d^l_{b'}$ is a coderivation. Let us start with computing the composition of the map $d^l_{b'}$ with the map $\psi \colon \mathcal{C} \longrightarrow \mathfrak{h}^\vee$.

The class of an element $u \otimes c \in \mathfrak{g}/\mathfrak{h} \otimes_k \mathcal{C}$ can be represented by the element $b'(u) \otimes_{U(\mathfrak{h})} c \in F_1 U_\varkappa(\mathfrak{g}) \otimes_{U(\mathfrak{h})} \mathcal{C} \simeq \mathcal{D}^l_1{}^\sim$ in the quasi-differential coalgebra $\mathcal{D}^l{}^\sim$ corresponding to the filtered semialgebra $\mathcal{S}^l_\varkappa$. Denote the image of $b'(u) \otimes_{U(\mathfrak{h})} c$ under the comultiplication map $\mathcal{D}^l_1{}^\sim \longrightarrow \mathcal{C} \otimes_k \mathcal{D}^l_1{}^\sim$ by $c_1 \otimes (z \otimes_{U(\mathfrak{h})} c_2)$, where $z \in F_1 U_\varkappa(\mathfrak{g})$. The total comultiplication of $b'(u) \otimes_{U(\mathfrak{h})} c$ is then equal to $c_1 \otimes (z \otimes_{U(\mathfrak{h})} c_2) + (b'(u) \otimes_{U(\mathfrak{h})} c_{(1)}) \otimes c_{(2)}$. We have

$$d^l_{b'}(u \otimes c) = \delta^l_{b'}(b'(u) \otimes_{U(\mathfrak{h})} c_{(1)})c_{(2)} - \delta^l_{b'}(z \otimes_{U(\mathfrak{h})} c_2)c_1 = -\delta^l_{b'}(z \otimes_{U(\mathfrak{h})} c_2)c_1.$$

Furthermore,

$$
\begin{aligned}
\psi(d_{b'}^l(u \otimes c), x) &= -\psi(c_1, x)\delta_{b'}^l(z \otimes_{U(\mathfrak{h})} c_2) = -\delta_{b'}^l(xb'(u) \otimes_{U(\mathfrak{h})} c) \\
&= -\delta_{b'}^l([x, b'(u)] \otimes_{U(\mathfrak{h})} c) - \delta_{b'}^l(b'(u) \otimes_{U(\mathfrak{h})} xc) \\
&= \delta_{b'}^l([b'(u), x] \otimes_{U(\mathfrak{h})} c) = \langle x, u(\breve{\psi}(c))\rangle + \langle x, \rho'(u)\rangle\varepsilon(c)
\end{aligned}
$$

for $x \in \mathfrak{h}$, since $\psi(x, c_1)z \otimes_{U(\mathfrak{h})} c_2 = xb'(u) \otimes_{U(\mathfrak{h})} c$.

So the composition of the map $d_{b'}^l \colon \mathfrak{g}/\mathfrak{h} \otimes_k \mathcal{C} \longrightarrow \mathcal{C}$ with the map $\breve{\psi} \colon \mathcal{C} \longrightarrow \mathfrak{h}^\vee$ is equal to the composition of the map $\mathrm{id} \otimes (\breve{\psi}, \varepsilon) \colon \mathfrak{g}/\mathfrak{h} \otimes_k \mathcal{C} \longrightarrow \mathfrak{g}/\mathfrak{h} \otimes_k (\mathfrak{h}^\vee \oplus k)$ with the map $\mathfrak{g}/\mathfrak{h} \otimes_k (\mathfrak{h}^\vee \oplus k) \longrightarrow \mathfrak{h}^\vee$ given by the formula $u \otimes x^* + v \longmapsto u(x^*) + \rho'(v)$.

Now let us compute the composition of the map $d_{b'}^l \colon \bigwedge_k^2(\mathfrak{g}/\mathfrak{h}) \otimes_k \mathcal{C} \longrightarrow \mathfrak{g}/\mathfrak{h} \otimes_k \mathcal{C}$ with the map $\mathrm{id} \otimes \varepsilon \colon \mathfrak{g}/\mathfrak{h} \otimes_k \mathcal{C} \longrightarrow \mathfrak{g}/\mathfrak{h}$. The vector space $\mathcal{D}_2^l{}^\sim$ is the kernel of the semimultiplication map $F_1\mathcal{S}_\varkappa^l \square_{\mathcal{C}} F_1\mathcal{S}_\varkappa^l \longrightarrow F_2\mathcal{S}_\varkappa^l$, which can be identified with the kernel of the map $F_1U_\varkappa(\mathfrak{g}) \otimes_{U(\mathfrak{h})} F_1U_\varkappa(\mathfrak{g}) \otimes_{U(\mathfrak{h})} \mathcal{C} \longrightarrow F_2U_\varkappa(\mathfrak{g}) \otimes_{U(\mathfrak{h})} \mathcal{C}$ induced by the multiplication map $F_1U_\varkappa(\mathfrak{g}) \otimes_{U(\mathfrak{h})} F_1U_\varkappa(\mathfrak{g}) \longrightarrow F_2U_\varkappa(\mathfrak{g})$. The class of an element $u \wedge v \otimes c \in \bigwedge_k^2(\mathfrak{g}/\mathfrak{h}) \otimes_k \mathcal{C}$ can be represented by the element $b'(u) \otimes_{U(\mathfrak{h})} b'(v) \otimes_{U(\mathfrak{h})} c - b'(v) \otimes_{U(\mathfrak{h})} b'(u) \otimes_{U(\mathfrak{h})} c - [b'(u), b'(v)] \otimes_{U(\mathfrak{h})} 1 \otimes_{U(\mathfrak{h})} c$ in the latter kernel.

Denote the image of $b'(v) \otimes_{U(\mathfrak{h})} c$ under the comultiplication map $\mathcal{D}_1^l{}^\sim \longrightarrow \mathcal{C} \otimes_k \mathcal{D}_1^l{}^\sim$ by $c_1 \otimes (z \otimes_{U(\mathfrak{h})} c_2)$, where $z \in F_1U_\varkappa(\mathfrak{g})$; then the image of $b'(u) \otimes_{U(\mathfrak{h})} b'(v) \otimes_{U(\mathfrak{h})} c$ under the comultiplication map $\mathcal{D}_2^l{}^\sim \longrightarrow \mathcal{D}_1^l{}^\sim \otimes_k \mathcal{D}_1^l{}^\sim$ is equal to $(b'(u) \otimes_{U(\mathfrak{h})} c_1) \otimes (z \otimes_{U(\mathfrak{h})} c_2)$. The image of $[b'(u), b'(v)] \otimes_{U(\mathfrak{h})} 1 \otimes_{U(\mathfrak{h})} c$ under the same map $\mathcal{D}_2^l{}^\sim \longrightarrow \mathcal{D}_1^l{}^\sim \otimes_k \mathcal{D}_1^l{}^\sim$ is equal to $([b'(u), b'(v)] \otimes_{U(\mathfrak{h})} c_{(1)}) \otimes (1 \otimes_{U(\mathfrak{h})} c_{(2)})$. We have $\delta_{b'}^l(b'(u) \otimes_{U(\mathfrak{h})} c_1)\overline{z \otimes_{U(\mathfrak{h})} c_2} = 0$ and $\delta_{b'}^l(z \otimes_{U(\mathfrak{h})} c_2)(\mathrm{id} \otimes \varepsilon)\overline{b'(u) \otimes_{U(\mathfrak{h})} c_1} = \varepsilon(c_1)\delta_{b'}^l(z \otimes_{U(\mathfrak{h})} c_2)u = \delta_{b'}^l(b'(v) \otimes_{U(\mathfrak{h})} c)u = 0$, where $\overline{p} \in \mathcal{D}_1^l$ denotes the image of an element $p \in \mathcal{D}_1^l{}^\sim$. Furthermore, $\delta_{b'}^l([b'(u), b'(v)] \otimes_{U(\mathfrak{h})} c_{(1)})\overline{1 \otimes_{U(\mathfrak{h})} c_{(2)}} = 0$. Hence $(\mathrm{id} \otimes \varepsilon)d_{b'}^l(u \wedge v \otimes c) = -\delta_{b'}^l(1 \otimes_{U(\mathfrak{h})} c_{(2)})(\mathrm{id} \otimes \varepsilon)\overline{[b'(u), b'(v)] \otimes_{U(\mathfrak{h})} c_{(1)}} = -(\mathrm{id} \otimes \varepsilon)\overline{[b'(u), b'(v)] \otimes_{U(\mathfrak{h})} c} = -\{u, v\}\varepsilon(c)$.

So the composition of the map $d_{b'}^l \colon \bigwedge_k^2(\mathfrak{g}/\mathfrak{h}) \otimes_k \mathcal{C} \longrightarrow \mathfrak{g}/\mathfrak{h} \otimes_k \mathcal{C}$ with the map $\mathrm{id} \otimes \varepsilon \colon \mathfrak{g}/\mathfrak{h} \otimes_k \mathcal{C} \longrightarrow \mathfrak{g}/\mathfrak{h}$ is equal to the composition of the map $\mathrm{id} \otimes \varepsilon \colon \bigwedge_k^2(\mathfrak{g}/\mathfrak{h}) \otimes_k \mathcal{C} \longrightarrow \bigwedge_k^2(\mathfrak{g}/\mathfrak{h})$ with the map $\bigwedge_k^2(\mathfrak{g}/\mathfrak{h}) \longrightarrow \mathfrak{g}/\mathfrak{h}$ given by the formula $u \wedge v \longmapsto -\{u, v\}$.

Finally, let us compute the linear function $h_{b'}^l \colon \bigwedge_k^2(\mathfrak{g}/\mathfrak{h}) \otimes_k \mathcal{C} \longrightarrow k$. We have $\delta_{b'}^l(b'(u) \otimes_{U(\mathfrak{h})} c_1)\delta_{b'}^l(z \otimes_{U(\mathfrak{h})} c_2) = 0$, hence $h(u \wedge v \otimes c) = -\delta_{b'}^l([b'(u), b'(v)] \otimes_{U(\mathfrak{h})} c_{(1)})\delta_{b'}^l(1 \otimes_{U(\mathfrak{h})} c_{(2)}) = -\delta_{b'}^l([b'(u), b'(v)] \otimes_{U(\mathfrak{h})} c) = -(u, v)_{\breve{\psi}(c)} - \sigma'(u \wedge v)\varepsilon(c)$. So the linear function $h_{b'}^l$ is equal to the composition of the map $\mathrm{id} \otimes (\breve{\psi}, \varepsilon) \colon \bigwedge_k^2(\mathfrak{g}/\mathfrak{h}) \otimes_k \mathcal{C} \longrightarrow \bigwedge_k^2(\mathfrak{g}/\mathfrak{h}) \otimes_k (\mathfrak{h}^\vee \oplus k)$ and the linear function $\bigwedge_k^2(\mathfrak{g}/\mathfrak{h}) \otimes_k (\mathfrak{h}^\vee \oplus k) \longrightarrow k$ given by the formula $u_1 \wedge v_1 \otimes x^* + u \wedge v \longmapsto -(u_1, v_1)_{x^*} - \sigma'(u \wedge v)$.

D.3.6 Analogously, we have $\mathcal{D}_0^r \simeq \mathcal{C}$, $\mathcal{D}_1^r \simeq \mathcal{C} \otimes_k \mathfrak{g}/\mathfrak{h}$, and $\mathcal{D}_2^r \simeq \mathcal{C} \otimes_k \bigwedge_k^2(\mathfrak{g}/\mathfrak{h})$. The composition of the map $d_{b''}^r \colon \mathcal{C} \otimes_k \mathfrak{g}/\mathfrak{h} \longrightarrow \mathcal{C}$ with the map $\varepsilon \colon \mathcal{C} \longrightarrow k$ vanishes. The composition of the map $d_{b''}^r \colon \mathcal{C} \otimes_k \mathfrak{g}/\mathfrak{h} \longrightarrow \mathcal{C}$ with the map $\breve{\psi} \colon \mathcal{C} \longrightarrow \mathfrak{h}^\vee$ is equal to the composition of the map $(\check{\psi}, \varepsilon) \otimes \mathrm{id} \colon \mathcal{C} \otimes_k \mathfrak{g}/\mathfrak{h} \longrightarrow (\mathfrak{h}^\vee \oplus k) \otimes_k \mathfrak{g}/\mathfrak{h}$ and the map $(\mathfrak{h}^\vee \oplus k) \otimes_k \mathfrak{g}/\mathfrak{h} \longrightarrow \mathfrak{h}^\vee$ given by the formula $x^* \otimes u + v \longmapsto u(x^*) + \rho''(v)$. The composition of the map $d_{b''}^r \colon \mathcal{C} \otimes_k \bigwedge_k^2(\mathfrak{g}/\mathfrak{h}) \longrightarrow \mathcal{C} \otimes_k \mathfrak{g}/\mathfrak{h}$ with the map $\varepsilon \otimes \mathrm{id} \colon \mathcal{C} \otimes_k \mathfrak{g}/\mathfrak{h} \longrightarrow \mathfrak{g}/\mathfrak{h}$ is equal to the composition of the map $\varepsilon \otimes \mathrm{id} \colon \mathcal{C} \otimes_k \bigwedge_k^2(\mathfrak{g}/\mathfrak{h}) \longrightarrow \bigwedge_k^2(\mathfrak{g}/\mathfrak{h})$ and the map $\bigwedge_k^2(\mathfrak{g}/\mathfrak{h}) \longrightarrow \mathfrak{g}/\mathfrak{h}$ given by the formula $u \wedge v \longmapsto -\{u, v\}$. The linear function $h_{b''}^r \colon \mathcal{C} \otimes_k \bigwedge_k^2(\mathfrak{g}/\mathfrak{h}) \longrightarrow k$ is equal to the composition of the map $(\check{\psi}, \varepsilon) \otimes \mathrm{id} \colon \mathcal{C} \otimes_k \bigwedge_k^2(\mathfrak{g}/\mathfrak{h}) \longrightarrow (\mathfrak{h}^\vee \oplus k) \otimes_k \bigwedge_k^2(\mathfrak{g}/\mathfrak{h})$ and the linear function $(\mathfrak{h}^\vee \oplus k) \otimes_k \bigwedge_k^2(\mathfrak{g}/\mathfrak{h}) \longrightarrow k$ given by the formula $x^* \otimes u_1 \wedge v_1 + u \wedge v \longmapsto -(u_1, v_1)_{x^*} - \sigma''(u \wedge v)$.

The isomorphism $\mathfrak{g}/\mathfrak{h} \otimes_k \mathcal{C} \simeq \mathcal{C} \otimes_k \mathfrak{g}/\mathfrak{h}$ forms a commutative diagram with the map $\mathfrak{g}/\mathfrak{h} \otimes_k \mathcal{C} \longrightarrow \mathfrak{g}/\mathfrak{h} \otimes_k (\mathfrak{h}^\vee \oplus k)$, the map $\mathcal{C} \otimes_k \mathfrak{g}/\mathfrak{h} \longrightarrow (\mathfrak{h}^\vee \oplus k) \otimes_k \mathfrak{g}/\mathfrak{h}$, and the isomorphism $\mathfrak{g}/\mathfrak{h} \otimes_k (\mathfrak{h}^\vee \oplus k) \simeq (\mathfrak{h}^\vee \oplus k) \otimes_k \mathfrak{g}/\mathfrak{h}$ given by the formula $u \otimes x^* + v \longmapsto x^* \otimes u + v_{\{-1\}} \otimes v_{\{0\}} + v$. Analogously, the isomorphism $\bigwedge_k^2(\mathfrak{g}/\mathfrak{h}) \otimes_k \mathcal{C} \simeq \mathcal{C} \otimes_k \bigwedge_k^2(\mathfrak{g}/\mathfrak{h})$ forms a commutative diagram with the map $\bigwedge_k^2(\mathfrak{g}/\mathfrak{h}) \otimes_k \mathcal{C} \longrightarrow \bigwedge_k^2(\mathfrak{g}/\mathfrak{h}) \otimes_k (\mathfrak{h}^\vee \oplus k)$, the map $\mathcal{C} \otimes_k \bigwedge_k^2(\mathfrak{g}/\mathfrak{h}) \longrightarrow (\mathfrak{h}^\vee \oplus k) \otimes_k \bigwedge_k^2(\mathfrak{g}/\mathfrak{h})$, and the isomorphism $\bigwedge_k^2(\mathfrak{g}/\mathfrak{h}) \otimes_k (\mathfrak{h}^\vee \oplus k) \simeq (\mathfrak{h}^\vee \oplus k) \otimes_k \bigwedge_k^2(\mathfrak{g}/\mathfrak{h})$ given by the formula $u_1 \wedge v_1 \otimes x^* + u \wedge v \longmapsto x^* \otimes u_1 \wedge v_1 + u_{\{-1\}} \otimes u_{\{0\}} \wedge v + v_{\{-1\}} \otimes u \wedge v_{\{0\}} + u \wedge v$.

Now it is straightforward to check that the isomorphism $\mathcal{D}^l \simeq \mathcal{D}^r$ identifies $d_{b'}^l$ with $d_{b''}^r$ modulo $V^2 \mathcal{D}_0 \oplus V^1 \mathcal{D}_1 \oplus \mathcal{D}_2 \oplus \mathcal{D}_3 \oplus \cdots$ and $h_{b'}^l$ with $h_{b''}^r$. Indeed, one has $u(x^*) + v_{\{0\}}(v_{\{-1\}}) + \rho''(v) = u(x^*) + \rho'(v)$ and $-(u_1, v_1)_{x^*} - (u_{\{0\}}, v)_{u_{\{-1\}}} - (u, v_{\{0\}})_{v_{\{-1\}}} - \sigma''(u \wedge v) = -(u_1, v_1)_{x^*} - \sigma'(u \wedge v)$.

D.3.7 Finally, let $b_1' \colon \mathfrak{g}/\mathfrak{h} \longrightarrow \mathfrak{g}'$ be another section of the surjection $\mathfrak{g}' \longrightarrow \mathfrak{g}'/(\mathfrak{h} \oplus k) \simeq \mathfrak{g}/\mathfrak{h}$. Then we can write $b_1' = b + t + t'$ with $t \colon \mathfrak{g}/\mathfrak{h} \longrightarrow \mathfrak{h}$ and $t' \colon \mathfrak{g}/\mathfrak{h} \longrightarrow k$. Analogously, the sections $\tilde{b}_1 \colon \mathfrak{g}/\mathfrak{h} \longrightarrow \mathfrak{g}^\sim$ and $b_1'' \colon \mathfrak{g}/\mathfrak{h} \longrightarrow \mathfrak{g}''$ corresponding to b_1' have the form $\tilde{b}_1 = \tilde{b} + t + \tilde{t}$ and $b_1'' = b'' + t + t''$ with $t'' = t' + \tilde{t}$.

Denote by τ, $\tau_1 \colon \mathfrak{gl}(\mathfrak{g}) \longrightarrow \mathfrak{gl}(\mathfrak{g})^\sim$ the sections corresponding to direct sum decompositions $\mathfrak{g} = \mathfrak{h} \oplus b(\mathfrak{g}/\mathfrak{h})$ and $\mathfrak{g} = \mathfrak{h} \oplus b_1(\mathfrak{g}/\mathfrak{h})$ with $b_1 = b + t$, $t \colon \mathfrak{g}/\mathfrak{h} \longrightarrow \mathfrak{h}$. Then one has $\tau_1(A) - \tau(A) = \mathrm{tr}(tA_{\mathfrak{h} \to \mathfrak{g}/\mathfrak{h}})$ for any $A \in \mathfrak{gl}(\mathfrak{g})$, where $A_{\mathfrak{h} \to \mathfrak{g}/\mathfrak{h}}$ denotes the composition $\mathfrak{h} \longrightarrow \mathfrak{g} \longrightarrow \mathfrak{g} \longrightarrow \mathfrak{g}/\mathfrak{h}$ of the endomorphism A with the injection $\mathfrak{h} \longrightarrow \mathfrak{g}$ and the surjection $\mathfrak{g} \longrightarrow \mathfrak{g}/\mathfrak{h}$.

Using this formula, one can find that $\tilde{t}(u) = -\langle u_{\{-1\}}, t(u_{\{0\}})\rangle$, where $\langle \,,\, \rangle$ denotes the natural pairing $\mathfrak{h}^\vee \times \mathfrak{h} \longrightarrow k$.

The natural isomorphism $(\mathrm{id}, a^l) \colon (\mathcal{D}^l, d_{b'}^l, h_{b'}^l) \longrightarrow (\mathcal{D}^l, d_{b_1'}^l, h_{b_1'}^l)$ between the CDG-coalgebras corresponding to the sections b' and b_1' can be computed easily; the linear function $a^l \colon \mathcal{D}_1^l \longrightarrow k$ is the composition of the map $\mathrm{id} \otimes (\check{\psi}, \varepsilon) \colon \mathfrak{g}/\mathfrak{h} \otimes_k \mathcal{C} \longrightarrow \mathfrak{g}/\mathfrak{h} \otimes_k (\mathfrak{h}^\vee \oplus k)$ and the linear function $\mathfrak{g}/\mathfrak{h} \otimes_k (\mathfrak{h}^\vee \oplus k) \longrightarrow k$ given by the formula $u \otimes x^* + v \longmapsto -\langle x^*, t(u)\rangle - t'(v)$. Analogously, the linear function a^r in the natural isomorphism $(\mathrm{id}, a^r) \colon (\mathcal{D}^r, d_{b''}^r, h_{b''}^r) \longrightarrow (\mathcal{D}^r, d_{b_1''}^r, h_{b_1''}^r)$ between the CDG-coalgebras corresponding to the sections b'' and b_1'' is the composition of the map $(\check{\psi}, \varepsilon) \otimes \mathrm{id} \colon \mathcal{C} \otimes_k \mathfrak{g}/\mathfrak{h} \longrightarrow (\mathfrak{h}^\vee \oplus k) \otimes_k \mathfrak{g}/\mathfrak{h}$ and the linear function $(\mathfrak{h}^\vee \oplus k) \otimes_k \mathfrak{g}/\mathfrak{h} \longrightarrow k$ given by the formula $x^* \otimes u + v \longmapsto -\langle x^*, t(u)\rangle - t''(v)$.

Now it is straightforward to check that the isomorphism $\mathcal{D}^l \simeq \mathcal{D}^r$ identifies a^l with a^r. Indeed, $-\langle x^*, t(u)\rangle - \langle v_{\{-1\}}, t(v_{\{0\}})\rangle - t''(v) = -\langle x^*, t(u)\rangle - t'(v)$.

Theorem D.3.1 is proven. $\qquad\qquad\qquad\qquad\qquad\qquad\qquad\qquad\qquad\square$

D.4   Semiinvariants and semicontrainvariants

D.4.1   Let $\mathfrak{g}$ be a Tate Lie algebra, $\mathfrak{g}^\sim \longrightarrow \mathfrak{g}$ be the canonical central extension, and $\mathfrak{h} \subset \mathfrak{g}$ be a compact open subalgebra; recall that the central extension $\mathfrak{g}^\sim \longrightarrow \mathfrak{g}$ splits canonically over $\mathfrak{h}$. Let N be a discrete $\mathfrak{g}^\sim$-module where the unit element of $k \subset \mathfrak{g}^\sim$ acts by minus the identity. We would like to construct a natural map

$$(\mathfrak{g}/\mathfrak{h} \otimes_k N)^\mathfrak{h} \longrightarrow N^\mathfrak{h},$$

where the superindex $\mathfrak{h}$ denotes the $\mathfrak{h}$-invariants and the action of $\mathfrak{h}$ in N is defined in terms of the canonical splitting $\mathfrak{h} \longrightarrow \mathfrak{g}^\sim$.

Choose a section $b \colon \mathfrak{g}/\mathfrak{h} \longrightarrow \mathfrak{g}$ of the surjection $\mathfrak{g} \longrightarrow \mathfrak{g}/\mathfrak{h}$. The direct sum decomposition $\mathfrak{g} \simeq \mathfrak{h} \oplus b(\mathfrak{g}/\mathfrak{h})$ leads to a section of the central extension $\mathfrak{gl}(\mathfrak{g})^\sim \longrightarrow \mathfrak{gl}(\mathfrak{g})$, and consequently to a section of the central extension $\mathfrak{g}^\sim \longrightarrow \mathfrak{g}$. Composing the section b with the latter section, we get a section $\tilde{b} \colon \mathfrak{g}/\mathfrak{h} \longrightarrow \mathfrak{g}^\sim$ of the surjection $\mathfrak{g}^\sim \longrightarrow \mathfrak{g}^\sim/(\mathfrak{h} \oplus k) \simeq \mathfrak{g}/\mathfrak{h}$.

Consider the composition

$$(\mathfrak{g}/\mathfrak{h} \otimes_k N)^\mathfrak{h} \longrightarrow \mathfrak{g}/\mathfrak{h} \otimes_k N \longrightarrow \mathfrak{g}^\sim \otimes_k N \longrightarrow N$$

of the natural injection $(\mathfrak{g}/\mathfrak{h} \otimes_k N)^\mathfrak{h} \longrightarrow \mathfrak{h}$, the map $\mathfrak{g}/\mathfrak{h} \otimes_k N \longrightarrow \mathfrak{g}^\sim \otimes_k N$ induced by the section $\tilde{b} \colon \mathfrak{g}/\mathfrak{h} \longrightarrow \mathfrak{g}^\sim$, and the $\mathfrak{g}^\sim$-action map $\mathfrak{g}^\sim \otimes_k N \longrightarrow N$. Let us check that this composition does not depend on the choice of b and its image lies in the subspace of invariants $N^\mathfrak{h} \subset N$, so it provides the desired natural map $(\mathfrak{g}/\mathfrak{h} \otimes_k N)^\mathfrak{h} \longrightarrow N^\mathfrak{h}$.

Let $u \otimes n$ be a formal notation for an element of $\mathfrak{g}/\mathfrak{h} \otimes_k N$. Denote by $n \longmapsto n_{\{-1\}} \otimes n_{\{0\}}$ the map $N \longrightarrow \mathfrak{h}^\vee \otimes_k N$ corresponding to the $\mathfrak{h}$-action map $\mathfrak{h} \times N \longrightarrow N$. Rewriting the identity $x\tilde{b}(u)n = \tilde{b}(u)xn + [x, \tilde{b}(u)]n$ for $x \in \mathfrak{h}$ in the notation of D.3.4, we obtain the identity $(\tilde{b}(u)n)_{\{-1\}} \otimes (\tilde{b}(u)n)_{\{0\}} = n_{\{-1\}} \otimes \tilde{b}(u)n_{\{0\}} + u_{\{-1\}} \otimes \tilde{b}(u_{\{0\}})n - u(n_{\{-1\}}) \otimes n_{\{0\}} - u_{\{0\}}(u_{\{-1\}}) \otimes n$. Now whenever $u \otimes n$ is an $\mathfrak{h}$-invariant element of $\mathfrak{g}/\mathfrak{h} \otimes_k N$ one has $n_{\{-1\}} \otimes u \otimes n_{\{0\}} + u_{\{-1\}} \otimes u_{\{0\}} \otimes n = 0$, hence $(\tilde{b}(u)n)_{\{-1\}} \otimes (\tilde{b}(u)n)_{\{0\}} = 0$ and $\tilde{b}(u)n$ is an $\mathfrak{h}$-invariant element of N.

Let $b_1 \colon \mathfrak{g}/\mathfrak{h} \longrightarrow \mathfrak{h}$ be another section of the surjection $\mathfrak{g} \longrightarrow \mathfrak{g}/\mathfrak{h}$ and $\tilde{b}_1 \colon \mathfrak{g}/\mathfrak{h} \longrightarrow \mathfrak{g}^\sim$ be the corresponding section of the surjection $\mathfrak{g}^\sim \longrightarrow \mathfrak{g}/\mathfrak{h}$. According to D.3.7, we have $\tilde{b}_1 = \tilde{b} + t + \tilde{t}$ with a map $t = b_1 - b \colon \mathfrak{g}/\mathfrak{h} \longrightarrow \mathfrak{h}$ and the linear function $\tilde{t} \colon \mathfrak{g}/\mathfrak{h} \longrightarrow k$ given by the formula $\tilde{t}(u) = -\langle u_{\{-1\}}, t(u_{\{0\}})\rangle$. Let $u \otimes n$ be an $\mathfrak{h}$-invariant element of $\mathfrak{g}/\mathfrak{h} \otimes N$; then the equation $n_{\{-1\}} \otimes u \otimes n_{\{0\}} + u_{\{-1\}} \otimes u_{\{0\}} \otimes n = 0$ implies $\langle n_{\{-1\}}, t(u)\rangle n_{\{0\}} + \langle u_{\{-1\}}, t(u_{\{0\}})\rangle n = 0$ and $t(u)n - \tilde{t}(u)n = 0$.

The cokernel $N_{\mathfrak{g},\mathfrak{h}}$ of the natural map $(\mathfrak{g}/\mathfrak{h} \otimes_k N)^{\mathfrak{h}} \longrightarrow N^{\mathfrak{h}}$ that we have constructed is called the space of $(\mathfrak{g},\mathfrak{h})$-*semiinvariants* of a discrete $\mathfrak{g}$-module N. The $(\mathfrak{g},\mathfrak{h})$-semiinvariants are a mixture of $\mathfrak{h}$-invariants and "coinvariants along $\mathfrak{g}/\mathfrak{h}$".

D.4.2 For a topological Lie algebra $\mathfrak{h}$ and an $\mathfrak{h}$-contramodule P the space of $\mathfrak{h}$-coinvariants $P_{\mathfrak{h}}$ is defined as the maximal quotient contramodule of P where $\mathfrak{h}$ contraacts by zero, i.e., the cokernel of the contraaction map $\mathfrak{h} \otimes^{\wedge} P \longrightarrow P$.

Let $\mathfrak{g}$ be a Tate Lie algebra with a compact open subalgebra $\mathfrak{h}$. Let P be a $\mathfrak{g}^{\sim}$-contramodule where the unit element of $k \subset \mathfrak{g}^{\sim}$ acts by the identity. We would like to construct a natural map

$$P_{\mathfrak{h}} \longrightarrow \operatorname{Hom}_k(\mathfrak{g}/\mathfrak{h}, P)_{\mathfrak{h}},$$

where the $\mathfrak{h}$-contraaction in $\operatorname{Hom}_k(\mathfrak{g}/\mathfrak{h}, P)$ is induced by the discrete action of $\mathfrak{h}$ in $\mathfrak{g}/\mathfrak{h}$ and the $\mathfrak{h}$-contraaction in P as explained in D.2.6–D.2.7.

As above, choose a section $b \colon \mathfrak{g}/\mathfrak{h} \longrightarrow \mathfrak{g}$ and construct the corresponding section $\tilde{b} \colon \mathfrak{g}/\mathfrak{h} \longrightarrow \mathfrak{g}^{\sim}$. Consider the composition

$$P \longrightarrow \operatorname{Hom}_k(\mathfrak{g}^{\sim}, P) \longrightarrow \operatorname{Hom}_k(\mathfrak{g}/\mathfrak{h}, P) \longrightarrow \operatorname{Hom}_k(\mathfrak{g}/\mathfrak{h}, P)_{\mathfrak{h}}$$

of the map $P \longrightarrow \operatorname{Hom}_k(\mathfrak{g}^{\sim}, P)$ corresponding to the action of $\mathfrak{g}^{\sim}$ in P induced by the contraaction of $\mathfrak{g}^{\sim}$ in P, the map $\operatorname{Hom}_k(\mathfrak{g}^{\sim}, P) \longrightarrow \operatorname{Hom}_k(\mathfrak{g}/\mathfrak{h}, P)$ induced by the section $\tilde{b}$, and the natural surjection $\operatorname{Hom}_k(\mathfrak{g}/\mathfrak{h}, P) \longrightarrow \operatorname{Hom}_k(\mathfrak{g}/\mathfrak{h}, P)_{\mathfrak{h}}$. Let us check that this composition factorizes through the natural surjection $P \longrightarrow P_{\mathfrak{h}}$ and does not depend on the choice of b, so it defines the desired map $P_{\mathfrak{h}} \longrightarrow \operatorname{Hom}_k(\mathfrak{g}/\mathfrak{h}, P)_{\mathfrak{h}}$.

Let f a linear function $\mathfrak{h}^{\vee} \longrightarrow P$ and $\pi_P(f) \in P$ be its image under the contraaaction map. The image of $\pi_P(f)$ under the composition

$$P \longrightarrow \operatorname{Hom}_k(\mathfrak{g}^{\sim}, P) \longrightarrow \operatorname{Hom}_k(\mathfrak{g}/\mathfrak{h}, P)$$

is given by the formula

$$u \longmapsto \tilde{b}(u)\pi_P(f) = \pi_P(x^* \mapsto \tilde{b}(u)f(x^*)) - \tilde{b}(u_{\{0\}})f(u_{\{-1\}})$$
$$+ \pi_P(x^* \mapsto f(u(x^*))) - f(u_{\{0\}}(u_{\{-1\}}))$$

in the notation of D.3.4. This element of $\operatorname{Hom}_k(\mathfrak{g}/\mathfrak{h}, P)$ is the image of the element $g \in \operatorname{Hom}_k(\mathfrak{h}^{\vee}, \operatorname{Hom}_k(\mathfrak{g}/\mathfrak{h}, P))$ given by the formula $g(x^*)(u) = \tilde{b}(u)f(x^*) + f(u(x^*))$ under the contraaction map.

If $b_1 \colon \mathfrak{g}/\mathfrak{h} \longrightarrow \mathfrak{g}$ is a different section, then

$$\tilde{b}_1(u) = b_1(u) + t(u) - \langle u_{\{-1\}}, t(u_{\{0\}}) \rangle$$

and for any $p \in P$ the element of $\operatorname{Hom}_k(\mathfrak{g}/\mathfrak{h}, P)$ given by the formula $u \longmapsto t(u)p - \langle u_{\{-1\}}, t(u_{\{0\}}) \rangle p$ is the image of the element $g \in \operatorname{Hom}_k(\mathfrak{h}^{\vee}, \operatorname{Hom}_k(\mathfrak{g}/\mathfrak{h}, P))$ given by the formula $g(x^*)(u) = \langle x^*, t(u) \rangle p$ under the contraaction map.

The kernel $P^{\mathfrak{g},\mathfrak{h}}$ of the natural map $P_{\mathfrak{h}} \longrightarrow \mathrm{Hom}_k(\mathfrak{g}/\mathfrak{h}, P)_{\mathfrak{h}}$ is called the space of $(\mathfrak{g}, \mathfrak{h})$-*semicontrainvariants* of a $\mathfrak{g}$-contramodule P. The $(\mathfrak{g}, \mathfrak{h})$-semicontrainvariants are a mixture of $\mathfrak{h}$-coinvariants and "invariants along $\mathfrak{g}/\mathfrak{h}$".

Remark. The above definitions of $(\mathfrak{g}, \mathfrak{h})$-semiinvariants and $(\mathfrak{g}, \mathfrak{h})$-semicontrainvariants agree with the definitions from C.5.1–C.5.2 up to twists with a one-dimensional vector space $\det(\mathfrak{h})$, essentially for the following reason. When $\mathfrak{g}$ is a discrete Lie algebra, the central extension $\mathfrak{g}^{\sim} \longrightarrow \mathfrak{g}$ has a canonical splitting induced by the canonical splitting of the central extension $\mathfrak{gl}(\mathfrak{g})^{\sim} \longrightarrow \mathfrak{gl}(\mathfrak{g})$. When $\mathfrak{g}$ is a Tate Lie algebra and $\mathfrak{h} \subset \mathfrak{g}$ is a compact open Lie subalgebra, the central extension $\mathfrak{g}^{\sim} \longrightarrow \mathfrak{g}$ has a canonical splitting over $\mathfrak{h}$. When $\mathfrak{g}$ is a discrete Lie algebra and $\mathfrak{h} \subset \mathfrak{g}$ is a finite-dimensional Lie subalgebra, these two splittings do *not* agree over $\mathfrak{h}$; instead, they differ by the modular character of the Lie algebra $\mathfrak{h}$.

D.4.3 Let $\mathcal{C}$ be a coalgebra endowed with a coaugmentation (morphism of coalgebras) $e\colon k \longrightarrow \mathcal{C}$, $\mathfrak{h}$ be a compact Lie algebra, and $\psi\colon \mathcal{C} \times \mathfrak{h} \longrightarrow k$ be a pairing such that the map $\check{\psi}\colon \mathfrak{h}^{\vee} \longrightarrow \mathcal{C}$ is a morphism of coalgebras and ψ annihilates $e(k)$. For any right $\mathcal{C}$-comodule $\mathcal{N}$, the maximal subcomodule of $\mathcal{N}$ where the coaction of $\mathcal{C}$ is trivial can be described as the cotensor product $\mathcal{N} \square_{\mathcal{C}} k$. Here a coaction of $\mathcal{C}$ is called trivial if it is induced by e; the vector space k is endowed with the trivial coaction. There is a natural injective map $\mathcal{N} \square_{\mathcal{C}} k \longrightarrow \mathcal{N}^{\mathfrak{h}}$, which is an isomorphism provided that the assumption (iv) of D.2.2 holds.

Analogously, for any left $\mathcal{C}$-contramodule $\mathfrak{P}$ the maximal quotient $\mathcal{C}$-contramoduleof $\mathfrak{P}$ with the trivial contraaction can be described as the space of cohomomorphisms $\mathrm{Cohom}_{\mathcal{C}}(k, \mathfrak{P})$. There is a natural surjective map $\mathfrak{P}_{\mathfrak{h}} \longrightarrow \mathrm{Cohom}_{\mathcal{C}}(k, \mathfrak{P})$, which is also an isomorphism provided that the condition (iv) holds. Indeed, it suffices to consider the case when $\mathfrak{P} = \mathrm{Hom}_k(\mathcal{C}, E)$ is an induced $\mathcal{C}$-contramodule; in this case one only has to check that the kernel of the composition $\mathcal{C} \longrightarrow \mathcal{C} \otimes_k \mathcal{C} \longrightarrow \mathcal{C} \otimes_k \mathfrak{h}^{\vee}$ of the comultiplication map and the map induced by the map $\check{\psi}$ coincides with $e(k)$.

Let $\mathcal{C}$ be a commutative Hopf algebra. Then for any right $\mathcal{C}$-comodule $\mathcal{N}$ and left $\mathcal{C}$-comodule $\mathcal{M}$ there is a natural isomorphism $\mathcal{N} \square_{\mathcal{C}} \mathcal{M} \simeq (\mathcal{N} \otimes_k \mathcal{M}) \square_{\mathcal{C}} k$, where the right $\mathcal{C}$-comodule structure on $\mathcal{N} \square_{\mathcal{C}} \mathcal{M}$ is defined using the antipode and multiplication in $\mathcal{C}$. Analogously, for any left $\mathcal{C}$-comodule $\mathcal{M}$ and left $\mathcal{C}$-contramodule $\mathfrak{P}$ there is a natural isomorphism $\mathrm{Cohom}_{\mathcal{C}}(\mathcal{M}, \mathfrak{P}) \simeq \mathrm{Cohom}_{\mathcal{C}}(k, \mathrm{Hom}_k(\mathcal{M}, \mathfrak{P}))$.

D.4.4 Now let $\varkappa\colon (\mathfrak{g}', \mathcal{C}) \longrightarrow (\mathfrak{g}, \mathcal{C})$ be a central extension of Tate Harish-Chandra pairs with the kernel k satisfying the assumption (iv) of D.2.2 and $\mathbf{S}^l_{\varkappa} = \mathbf{S} \simeq \mathbf{S}^r_{\varkappa + \varkappa_0}$ be the corresponding semialgebra over $\mathcal{C}$.

Lemma.

(a) *Let* $\mathbf{N}$ *be a Harish-Chandra module with the central charge* $-\varkappa - \varkappa_0$ *and* $\mathcal{M}$ *be a Harish-Chandra module with the central charge* $\varkappa$ *over* $(\mathfrak{g}, \mathcal{C})$; *in other words,* $\mathbf{N}$ *is a right* $\mathbf{S}^r_{\varkappa + \varkappa_0}$-*semimodule and* $\mathcal{M}$ *is a left* $\mathbf{S}^l_{\varkappa}$-*semimodule. Assume that either* $\mathbf{N}$ *or* $\mathcal{M}$ *is a coflat* $\mathcal{C}$-*comodule. Then there is a natural isomorphism*

$$\mathbf{N} \lozenge_{\mathbf{S}} \mathbf{M} \simeq (\mathbf{N} \otimes_k \mathbf{M})_{\mathfrak{g},\mathfrak{h}},$$

where the tensor product $\mathbf{N} \otimes_k \mathbf{M}$ *is a Harish-Chandra module with the central charge* $-\varkappa_0$.

(b) *Let* $\mathbf{M}$ *be a Harish-Chandra module with the central charge* $\varkappa$ *and* $\mathfrak{P}$ *be a Harish-Chandra contramodule with the central charge* $\varkappa + \varkappa_0$ *over* $(\mathfrak{g}, \mathbb{C})$; *in other words,* $\mathbf{M}$ *is a left* $\mathbf{S}^l_\varkappa$-*semimodule and* $\mathfrak{P}$ *is a left* $\mathbf{S}^r_{\varkappa+\varkappa_0}$-*semicontramodule. Assume that either* $\mathbf{M}$ *is a coprojective* $\mathbb{C}$-*comodule, or* $\mathfrak{P}$ *is a coinjective* $\mathbb{C}$-*contramodule. Then there is a natural isomorphism*

$$\mathrm{SemiHom}_{\mathbf{S}}(\mathbf{M}, \mathfrak{P}) \simeq \mathrm{Hom}_k(\mathbf{M}, \mathfrak{P})^{\mathfrak{g},\mathfrak{h}},$$

where the space $\mathrm{Hom}_k(\mathbf{M}, \mathfrak{P})$ *is a Harish-Chandra contramodule with the central charge* $\varkappa_0$.

Proof. Part (a): denote by $\eta_i \colon \mathbf{N} \, \square_{\mathbb{C}} \, F_i \mathbf{S} \, \square_{\mathbb{C}} \, \mathbf{M} \longrightarrow \mathbf{N} \, \square_{\mathbb{C}} \, \mathbf{M}$ the map equal to the difference of the map induced by the semiaction map $F_i \mathbf{S} \, \square_{\mathbb{C}} \, \mathbf{M} \longrightarrow \mathbf{M}$ and the map induced by the semiaction map $\mathbf{N} \, \square_{\mathbb{C}} \, F_1 \mathbf{S} \longrightarrow \mathbf{N}$. The map η_0 vanishes and the quotient space $(\mathbf{N} \, \square_{\mathbb{C}} \, F_1 \mathbf{S} \, \square_{\mathbb{C}} \, \mathbf{M})/(\mathbf{N} \, \square_{\mathbb{C}} \, F_0 \mathbf{S} \, \square_{\mathbb{C}} \, \mathbf{M})$ is isomorphic to $\mathbf{N} \, \square_{\mathbb{C}} \, (F_1 \mathbf{S}/F_0 \mathbf{S}) \, \square_{\mathbb{C}} \, \mathbf{M}$, hence the induced map

$$\bar{\eta}_1 \colon \mathbf{N} \, \square_{\mathbb{C}} \, (F_1 \mathbf{S}/F_0 \mathbf{S}) \, \square_{\mathbb{C}} \, \mathbf{M} \longrightarrow \mathbf{N} \, \square_{\mathbb{C}} \, \mathbf{M}.$$

The cokernel of the map $\bar{\eta}_1$ coincides with the semitensor product $\mathbf{N} \lozenge_{\mathbf{S}} \mathbf{M}$ for the reasons explained in C.5.5. The cotensor product

$$\mathbf{N} \, \square_{\mathbb{C}} \, (F_1 \mathbf{S}/F_0 \mathbf{S}) \, \square_{\mathbb{C}} \, \mathbf{M} \simeq \mathbf{N} \, \square_{\mathbb{C}} \, (\mathfrak{g}/\mathfrak{h} \otimes_k \mathbb{C}) \, \square_{\mathbb{C}} \, \mathbf{M} \simeq \mathbf{N} \, \square_{\mathbb{C}} \, (\mathfrak{g}/\mathfrak{h} \otimes_k \mathbf{M})$$

is isomorphic to the space of invariants $(\mathfrak{g}/\mathfrak{h} \otimes_k \mathbf{M} \otimes_k \mathbf{N})^{\mathfrak{h}}$ in view of the assumption (iv); this isomorphism coincides with the isomorphism

$$\mathbf{N} \, \square_{\mathbb{C}} \, (F_1 \mathbf{S}/F_0 \mathbf{S}) \, \square_{\mathbb{C}} \, \mathbf{M} \simeq (\mathfrak{g}/\mathfrak{h} \otimes_k \mathbf{M} \otimes_k \mathbf{N})^{\mathfrak{h}}$$

induced by the isomorphism $F_1 \mathbf{S}/F_0 \mathbf{S} \simeq \mathbb{C} \otimes_k \mathfrak{g}/\mathfrak{h}$. Let us check that this isomorphism identifies the map $\bar{\eta}_1$ with the map whose cokernel is, by the definition, the space of semiinvariants $(\mathbf{N} \otimes_k \mathbf{M})_{\mathfrak{g},\mathfrak{h}}$.

   Choose a section $b' \colon \mathfrak{g}/\mathfrak{h} \longrightarrow \mathfrak{g}'$ and consider the corresponding section $b'' \colon \mathfrak{g}/\mathfrak{h} \longrightarrow \mathfrak{g}''$. There is an isomorphism of right $\mathbb{C}$-comodules $\mathbb{C} \oplus \mathfrak{g}/\mathfrak{h} \otimes_k \mathbb{C} \simeq F_1 \mathbf{S}^l_\varkappa$ given by the formula $c'_1 + u' \otimes c' \longmapsto 1 \otimes_{U(\mathfrak{h})} c'_1 + b'(u') \otimes_{U(\mathfrak{h})} c'$ and an analogous isomorphism of left $\mathbb{C}$-comodules $\mathbb{C} \oplus \mathbb{C} \otimes_k \mathfrak{g}/\mathfrak{h} \simeq F_1 \mathbf{S}^r_{\varkappa+\varkappa_0}$ given by the formula $c''_1 + c'' \otimes u'' \longmapsto c''_1 \otimes_{U(\mathfrak{h})} 1 + c'' \otimes_{U(\mathfrak{h})} b''(u'')$. The induced isomorphism $\mathbf{M} \oplus \mathfrak{g}/\mathfrak{h} \otimes_k \mathbf{M} \simeq F_1 \mathbf{S}^l_\varkappa \, \square_{\mathbb{C}} \, \mathbf{M} \simeq F_1 U_\varkappa(\mathfrak{g}) \otimes_{U(\mathfrak{h})} \mathbf{M}$ is given by the formula $m_1 + u \otimes m \longmapsto 1 \otimes_{U(\mathfrak{h})} m_1 + b'(u) \otimes_{U(\mathfrak{h})} m$. Now let $z = n \otimes (c'_1 + u' \otimes_{U(\mathfrak{h})} c') \otimes m = n \otimes (c''_1 + c'' \otimes_{U(\mathfrak{h})} u'') \otimes m$ be an element of $\mathbf{N} \, \square_{\mathbb{C}} \, F_1 \mathbf{S} \, \square_{\mathbb{C}} \, \mathbf{M}$. Then the corresponding element of $\mathbf{N} \, \square_{\mathbb{C}} \, F_1 U_\varkappa(\mathfrak{g}) \otimes_{U(\mathfrak{h})} \mathbf{M}$ can be written as

$\varepsilon(c_1')n \otimes 1 \otimes_{U(\mathfrak{h})} m + \varepsilon(c')n \otimes b'(u') \otimes_{U(\mathfrak{h})} m$, hence the image of z under the map $\mathcal{N} \square_{\mathfrak{e}} F_1 \mathcal{S} \square_{\mathfrak{e}} \mathcal{M} \longrightarrow \mathcal{N} \square_{\mathfrak{e}} \mathcal{M}$ induced by the semiaction map $F_1 \mathcal{S} \square_{\mathfrak{e}} \mathcal{M} \longrightarrow \mathcal{M}$ is equal to $\varepsilon(c_1')n \otimes m + \varepsilon(c')n \otimes b'(u')m$. Analogously, the image of z under the map $\mathcal{N} \square_{\mathfrak{e}} F_1 \mathcal{S} \square_{\mathfrak{e}} \mathcal{M} \longrightarrow \mathcal{N} \square_{\mathfrak{e}} \mathcal{M}$ induced by the semiaction map $\mathcal{N} \square_{\mathfrak{e}} F_1 \mathcal{S} \longrightarrow \mathcal{N}$ is equal to $\varepsilon(c_1'')n \otimes m - \varepsilon(c'')b''(u'')n \otimes m$. One has $\varepsilon(c_1') = \varepsilon(c_1'')$ by the condition (c) of D.3.1. Thus $\eta_1(z) = n \otimes b'(u)m + b''(u)n \otimes m = \tilde{b}(u)(n \otimes m)$, where $u = \varepsilon(c')u' = \varepsilon(c'')u''$. Part (a) is proven; the proof of part (b) is completely analogous. $\square$

D.5 Semi-infinite homology and cohomology

D.5.1 A *discrete right module* N over a topological associative algebra R is a right R-module such that the action map $N \times R \longrightarrow N$ is continuous with respect to the discrete topology of N. Equivalently, a right R-module N is discrete if the annihilator of any element of N is an open right ideal in R.

Let A and B be topological associative algebras in which open right ideals form bases of neighborhoods of zero. Then the topological tensor product $A \otimes^! B$ has a natural structure of topological associative algebra with the same property. The tensor product of a discrete right A-module and a discrete right B-module is naturally a discrete right $A \otimes^! B$-module.

Let $\varkappa \colon \mathfrak{g}' \longrightarrow \mathfrak{g}$ be a central extension of topological Lie algebras with the kernel k. Then the modified enveloping algebra $U_\varkappa(\mathfrak{g}) = U(\mathfrak{g}')/(1_{U(\mathfrak{g}')} - 1_{\mathfrak{g}'})$ can be endowed with the topology where right ideals generated by open subspaces of $\mathfrak{g}'$ form a base of neighborhoods of zero (see [10, 3.8.17] and [8, 2.4], where left ideals are considered in place of our right ideals). Denote the completion of $U_\varkappa(\mathfrak{g})$ with respect to this topology by $U_\varkappa^{\frown}(\mathfrak{g})$; this is a topological associative algebra. The category of discrete $\mathfrak{g}'$-modules where the unit element of $k \subset \mathfrak{g}'$ acts by minus the identity is isomorphic to the category of discrete right $U_\varkappa^{\frown}(\mathfrak{g})$-modules.

D.5.2 Let R be a topological associative algebra where open right ideals form a base of neighborhoods of zero. Then for any k-vector space P there is a natural map

$$R \otimes^{\frown}(R \otimes^{\frown} P) \longrightarrow R \otimes^{\frown} P$$

induced by the multiplication in R; it is constructed as the projective limit over all open right ideals $U \subset R$ of the maps $R/U \otimes_k (R \otimes^{\frown} P) \longrightarrow R/U \otimes_k P$ induced by the discrete right action of R in R/U. A *left contramodule* over R is a vector space P endowed with a linear map $R \otimes^{\frown} P \longrightarrow P$ satisfying the following contraassociativity and unity equations. First, the two maps $R \otimes^{\frown}(R \otimes^{\frown} P) \rightrightarrows R \otimes^{\frown} P$, one induced by the multiplication in R and the other induced by the contraaction map $R \otimes^{\frown} P \longrightarrow P$, should have equal compositions with the contraaction map. Second, the composition $P \longrightarrow R \otimes^{\frown} P \longrightarrow P$ of the map induced by the unit of R and the contraaction map should be equal to the identity endomorphism of P.

The category of left R-contramodules is abelian and there is a natural exact forgetful functor from it to the category of left modules over the algebra R

considered without any topology (cf. Remark A.3). Notice also the isomorphisms

$$R \otimes^\wedge (R \otimes^\wedge P) \simeq R \overleftarrow{\otimes} R \overleftarrow{\otimes} P \simeq (R \overleftarrow{\otimes} R) \otimes^\wedge P,$$

demonstrating the similarity of the above definition with the definition of a contra-module over a Lie algebra given in D.2.6. The above natural map $R \otimes^\wedge (R \otimes^\wedge P) \longrightarrow R \otimes^\wedge P$ is induced by the continuous multiplication map $R \overleftarrow{\otimes} R \longrightarrow R$, which exists for any topological associative algebra R where open right ideals form a base of neighborhoods of zero [8]. Just as for Lie algebras, a structure of a discrete right R-module on a vector space N is given by a continuous linear map

$$N \otimes R \simeq N \otimes^* R = N \overleftarrow{\otimes} R \longrightarrow N,$$

while a structure of a left R-contramodule on a vector space P is given by a discontinuous linear map

$$R \otimes^\wedge P \simeq R \otimes^! P = R \overleftarrow{\otimes} P \longrightarrow P,$$

where N and P are endowed with discrete topologies.

For any discrete right R-module N and any k-vector space E, the vector space $\mathrm{Hom}_k(N, E)$ has a natural structure of left contramodule over R. The contraaction map $R \otimes^\wedge \mathrm{Hom}_k(N, E) \longrightarrow \mathrm{Hom}_k(N, E)$ is constructed as the projective limit over all open right ideals $U \subset R$ of the maps

$$R/U \otimes_k \mathrm{Hom}_k(N, E) \longrightarrow \mathrm{Hom}_k(N^U, E)$$

given by the formulas $\bar{r} \otimes_k g \longmapsto (n \longmapsto g(n\bar{r}))$ for $\bar{r} \in R/U$, $g \in \mathrm{Hom}_k(N, E)$, and $n \in N^U$, where $N^U \subset N$ denotes the subspace of all elements of N annihilated by U.

More generally, let A and B be topological associative algebras where open right ideals form bases of neighborhoods of zero, N be a discrete right B-module, and P be a left A-contramodule. Then the vector space $\mathrm{Hom}_k(N, P)$ has a natural structure of left contramodule over $A \otimes^! B$. The contraaction map $(A \otimes^! B) \otimes^\wedge \mathrm{Hom}_k(N, E) \longrightarrow \mathrm{Hom}_k(N, E)$ is constructed as the projective limit over all open right ideals $U \subset B$ of the compositions

$$A \otimes^\wedge (B/U \otimes_k \mathrm{Hom}_k(N, P)) \longrightarrow A \otimes^\wedge \mathrm{Hom}_k(N^U, P)$$
$$\longrightarrow \mathrm{Hom}_k(N^U, A \otimes^\wedge P) \longrightarrow \mathrm{Hom}_k(N^U, P),$$

where the first map is induced by the right B-action in N and the third map is induced by the A-contraaction in P.

D.5.3 Let $\varkappa\colon \mathfrak{g}' \longrightarrow \mathfrak{g}$ be a central extension of topological Lie algebras with the kernel k. Assume that the topological Lie algebra $\mathfrak{g}'$ has a countable base of neighborhoods of zero consisting of open Lie subalgebras (concerning the second part of this condition, cf. [8]).

Theorem. *The category of $\mathfrak{g}'$-contramodules where the unit element of $k \subset \mathfrak{g}'$ acts by the identity is isomorphic to the category of left contramodules over the topological associative algebra $U_{\varkappa}^{\widehat{\ }}(\mathfrak{g})$.*

Proof. It is easy to see that the composition $\mathfrak{g}' \otimes^{\widehat{\ }} P \longrightarrow U_{\varkappa}^{\widehat{\ }}(\mathfrak{g}) \otimes^{\widehat{\ }} P \longrightarrow P$ defines a $\mathfrak{g}'$-contramodule structure on any left $U_{\varkappa}^{\widehat{\ }}(\mathfrak{g})$-contramodule P (so, in particular, $U_{\varkappa}^{\widehat{\ }}(\mathfrak{g})$ itself is a $\mathfrak{g}'$-contramodule). Let us construct the functor in the opposite direction.

The standard homological Chevalley complex

$$\cdots \longrightarrow \textstyle\bigwedge_k^2(\mathfrak{g}') \otimes_k U_{\varkappa}(\mathfrak{g}) \longrightarrow \mathfrak{g}' \otimes_k U_{\varkappa}(\mathfrak{g}) \longrightarrow U_{\varkappa}(\mathfrak{g}) \longrightarrow 0$$

is acyclic. For any open Lie subalgebra $\mathfrak{h} \subset \mathfrak{g}'$ not containing $k \subset \mathfrak{g}'$, the complex

$$\cdots \longrightarrow \textstyle\bigwedge_k^2(\mathfrak{h}) \otimes_k U_{\varkappa}(\mathfrak{g}) \longrightarrow \mathfrak{h} \otimes_k U_{\varkappa}(\mathfrak{g}) \longrightarrow \mathfrak{h} U_{\varkappa}(\mathfrak{g}) \longrightarrow 0$$

is an acyclic subcomplex of the previous complex. Taking the quotient complex and passing to the projective limit over $\mathfrak{h}$, we obtain a split exact complex of topological vector spaces

$$\cdots \longrightarrow \textstyle\bigwedge^{s,2}(\mathfrak{g}') \otimes^! U_{\varkappa}(\mathfrak{g}) \longrightarrow \mathfrak{g}' \otimes^! U_{\varkappa}(\mathfrak{g}) \longrightarrow U_{\varkappa}^{\widehat{\ }}(\mathfrak{g}) \longrightarrow 0,$$

where we denote by $\bigwedge^{s,i}(\mathfrak{g}')$ the completion of $\bigwedge_k^i(\mathfrak{g}')$ with respect to the topology with a base of neighborhoods of zero formed by the subspaces $\bigwedge_k^i(\mathfrak{h})$ and the enveloping algebra $U_{\varkappa}(\mathfrak{g})$ is considered as a discrete topological vector space. Applying the functor $\otimes^{\widehat{\ }} P$, we obtain an exact sequence of vector spaces

$$\textstyle\bigwedge^{s,2}(\mathfrak{g}') \otimes^{\widehat{\ }}(U_{\varkappa}(\mathfrak{g}) \otimes_k P) \longrightarrow \mathfrak{g}' \otimes^{\widehat{\ }}(U_{\varkappa}(\mathfrak{g}) \otimes_k P) \longrightarrow U_{\varkappa}^{\widehat{\ }}(\mathfrak{g}) \otimes^{\widehat{\ }} P \longrightarrow 0$$

for any k-vector space P.

Now let P be a $\mathfrak{g}'$-contramodule where the unit element of $k \subset \mathfrak{g}'$ acts by the identity; then, in particular, P is a $\mathfrak{g}'$-module and a $U_{\varkappa}(\mathfrak{g})$-module. It is clear from the above exact sequence that the composition

$$\mathfrak{g}' \otimes^{\widehat{\ }}(U_{\varkappa}(\mathfrak{g}) \otimes_k P) \longrightarrow \mathfrak{g}' \otimes^{\widehat{\ }} P \longrightarrow P$$

of the map induced by the $U_{\varkappa}(\mathfrak{g})$-action map and the $\mathfrak{g}'$-contraaction map factorizes through $U_{\varkappa}^{\widehat{\ }}(\mathfrak{g}) \otimes^{\widehat{\ }} P$, providing the desired contraaction map $U_{\varkappa}^{\widehat{\ }}(\mathfrak{g}) \otimes^{\widehat{\ }} P \longrightarrow P$.

Let us check that this contraaction map satisfies the contraassociativity equation. Any element z of $U_{\varkappa}^{\widehat{\ }}(\mathfrak{g}) \otimes^{\widehat{\ }} P$ can be presented in the form $z = \sum_{i=0}^{\infty} u_i \otimes p_i$ with $u_i \in U_{\varkappa}(\mathfrak{g})$ and $p_i \in P$, where $u_i \to 0$ in $U_{\varkappa}(\mathfrak{g})$ as $i \to \infty$ and the infinite sum is understood as the limit in the topology of $U_{\varkappa}^{\widehat{\ }}(\mathfrak{g}) \otimes^! P$. Let us denote the image of the element $\sum_i u_i \otimes p_i$ under the contraaction map $U_{\varkappa}^{\widehat{\ }}(\mathfrak{g}) \otimes^{\widehat{\ }} P \longrightarrow P$ by $\sum_i u_i p_i \in P$. In this notation, the $U_{\varkappa}^{\widehat{\ }}(\mathfrak{g})$-contraaction map is defined by the formula $\sum_i (x_{i_1} x_{i_2} \cdots x_{i_{k_i}}) p_i = \sum_i x_{i_1}(x_{i_2} \cdots x_{i_{k_i}} p_i)$ for any $x_{i_t} \in \mathfrak{g}'$ and $p_i \in P$ such that $x_{i_1} \to 0$ in $\mathfrak{g}'$ as $i \to \infty$. We have to show that $\sum_i u_i \sum_j v_{ij} p_{ij} =$

$\sum_{i,j}(u_iv_{ij})p_{ij}$ for any u_i, $v_{ij} \in U_\varkappa(\mathfrak{g})$ and $p_{ij} \in P$ such that $u_i \to 0$ as $i \to \infty$ and $v_{ij} \to 0$ as $j \to \infty$ for any i.

Let us first check that $\sum_i x_i \sum_j y_{ij}p_{ij} = \sum_{i,j}(x_iy_{ij})p_{ij}$ for any x_i, $y_{ij} \in \mathfrak{g}'$ and $p_{ij} \in P$ such that $x_i \to 0$ in $\mathfrak{g}'$ as $i \to \infty$ and $y_{ij} \to 0$ in $\mathfrak{g}'$ as $j \to \infty$ for any i. Choose an integer j_i for each i such that $\{y_{ij} \mid j > j_i\}$ converges to zero in $\mathfrak{g}'$ as $i + j \to \infty$. Then we have

$$\sum_{i,j}(x_iy_{ij})p_{ij} = \sum_{j \leqslant j_i} x_i(y_{ij}p_{ij}) + \sum_{j > j_i} y_{ij}(x_ip_{ij}) + \sum_{j > j_i}[x_i, y_{ij}]p_{ij}.$$

To check that

$$\sum_i x_i \sum_{j > j_i} y_{ij}p_{ij} = \sum_{j > j_i} y_{ij}(x_ip_{ij}) + \sum_{j > j_i}[x_i, y_{ij}]p_{ij},$$

apply the equation on the contraaction map of a contramodule over a topological Lie algebra to the element $\sum_{j > j_i} x_i \wedge y_{ij} \otimes p_{ij}$ of the vector space $\bigwedge^{*,2}(\mathfrak{g}') \otimes^\frown P$.

It follows that $\sum_i x_i \sum_j v_{ij}p_{ij} = \sum_{i,j}(x_iv_{ij})p_{ij}$ for any $x_i \in \mathfrak{g}'$, $v_{ij} \in U_\varkappa(\mathfrak{g})$, and $p_{ij} \in P$ such that $x_i \to 0$ in $\mathfrak{g}'$ as $i \to \infty$ and $v_{ij} \to 0$ in $U_\varkappa^\frown(\mathfrak{g})$ as $j \to \infty$ for any i. Indeed, assuming that $v_{ij} = y_{ij1}y_{ij2}\cdots y_{ijk_{ij}}$, where $y_{ijt} \in \mathfrak{g}'$ and $y_{ij1} \to 0$ in $\mathfrak{g}'$ as $j \to \infty$, we have

$$\sum_i x_i \sum_j (y_{ij1}y_{ij2}\cdots y_{ijk_{ij}})p_{ij} = \sum_i x_i \sum_j y_{ij1}(y_{ij2}\cdots y_{ijk_{ij}}p_{ij})$$
$$= \sum_{i,j}(x_iy_{ij1})(y_{ij2}\cdots y_{ijk_{ij}}p_{ij}) = \sum_{i,j}(x_iy_{ij1}y_{ij2}\cdots y_{ijk_{ij}})p_{ij}.$$

Furthermore, it follows that $x_1\cdots x_s \sum_j v_jp_j = \sum_j(x_1\cdots x_sv_j)p_j$ for any $x_t \in \mathfrak{g}'$, $v_j \in U_\varkappa(\mathfrak{g})$, and $p_j \in P$ such that $v_j \to 0$ in $U_\varkappa^\frown(\mathfrak{g})$ as $j \to \infty$. Now to check that $\sum_i u_i \sum_j v_{ij}p_{ij} = \sum_{i,j}(u_iv_{ij})p_{ij}$, we can assume that $u_i = x_{i_1}x_{i_2}\cdots x_{i_{k_i}}$, where $x_{i_t} \in \mathfrak{g}'$ and $x_{i_1} \to 0$ in $\mathfrak{g}'$ as $i \to \infty$. Then we have

$$\sum_i(x_{i_1}x_{i_2}\cdots x_{i_{k_i}})\sum_j v_{ij}p_{ij} = \sum_i x_{i_1}(x_{i_2}\cdots x_{i_{k_i}}\sum_j v_{ij}p_{ij})$$
$$= \sum_i x_{i_1} \sum_j(x_{i_2}\cdots x_{i_{k_i}}v_{ij}p_{ij}) = \sum_{i,j}(x_{i_1}x_{i_2}\cdots x_{i_{k_i}}v_{ij})p_{ij}. \qquad \square$$

Question. Can one construct an isomorphism between the categories of "$\mathfrak{g}$-contra-modules with central charge $\varkappa$" and left $U_\varkappa^\frown(\mathfrak{g})$-contramodules without the count-ability assumption on the topology of $\mathfrak{g}'$?

D.5.4 The following weaker version of Theorem D.5.3 holds without the count-ability assumption. Let $\varkappa\colon \mathfrak{g}' \longrightarrow \mathfrak{g}$ be a central extension of topological Lie algebras with the kernel k; assume that open subalgebras form a base of neighborhoods of zero in $\mathfrak{g}'$. Let B be a topological associative algebra where open right ideals form a base of neighborhoods of zero, N be a discrete right B-module, and P be a $\mathfrak{g}'$-contramodule where the unit element of $k \subset \mathfrak{g}'$ acts by the identity. Then one can define the contraaction map $(\mathfrak{g}' \otimes^! B) \otimes^\frown \mathrm{Hom}_k(N, P) \longrightarrow \mathrm{Hom}_k(N, P)$ as in D.5.2.

Consider the iterated contraaction map

$$((\mathfrak{g}' \otimes^! B) \overleftarrow{\otimes} (\mathfrak{g}' \otimes^! B)) \otimes^\wedge \operatorname{Hom}_k(N, P)$$
$$\simeq (\mathfrak{g}' \otimes^! B) \otimes^\wedge ((\mathfrak{g}' \otimes^! B) \otimes^\wedge \operatorname{Hom}_k(N, P)) \longrightarrow \operatorname{Hom}_k(N, P).$$

It was noticed in [8] that a topological associative algebra A has the property that open right ideals form a base of neighborhoods of zero if and only if the multiplication map $A \otimes^* A \longrightarrow A$ factorizes through $A \overleftarrow{\otimes} A$. Let K denote the kernel of the multiplication map

$$(\mathfrak{g}' \otimes^! B) \overleftarrow{\otimes} (\mathfrak{g}' \otimes^! B) \longrightarrow U_\varkappa^\wedge(\mathfrak{g}) \otimes^! B.$$

We claim that the composition of the injection

$$K \otimes^\wedge \operatorname{Hom}_k(N, P) \longrightarrow ((\mathfrak{g}' \otimes^! B) \overleftarrow{\otimes} (\mathfrak{g}' \otimes^! B)) \otimes^\wedge \operatorname{Hom}_k(N, P)$$

and the iterated contraaction map

$$((\mathfrak{g}' \otimes^! B) \overleftarrow{\otimes} (\mathfrak{g}' \otimes^! B)) \otimes^\wedge \operatorname{Hom}_k(N, P) \longrightarrow \operatorname{Hom}_k(N, P)$$

vanishes.

For any topological vector spaces U, V, X, Y there is a natural map

$$(U \otimes^! X) \overleftarrow{\otimes} (V \otimes^! Y) \longrightarrow (U \overleftarrow{\otimes} V) \otimes^! (X \overleftarrow{\otimes} Y).$$

The composition $(\mathfrak{g}' \otimes^! B) \overleftarrow{\otimes} (\mathfrak{g}' \otimes^! B) \longrightarrow (\mathfrak{g}' \overleftarrow{\otimes} \mathfrak{g}') \otimes^! (B \overleftarrow{\otimes} B) \longrightarrow (\mathfrak{g}' \overleftarrow{\otimes} \mathfrak{g}') \otimes^! B$ induces the map $((\mathfrak{g}' \otimes^! B) \overleftarrow{\otimes} (\mathfrak{g}' \otimes^! B)) \otimes^\wedge \operatorname{Hom}_k(N, P) \longrightarrow ((\mathfrak{g}' \overleftarrow{\otimes} \mathfrak{g}') \otimes^! B) \otimes^\wedge \operatorname{Hom}_k(N, P)$. A contraaction map $((\mathfrak{g}' \overleftarrow{\otimes} \mathfrak{g}') \otimes^! B) \otimes^\wedge \operatorname{Hom}_k(N, P) \longrightarrow \operatorname{Hom}_k(N, P)$ can be defined in terms of the discrete right action of B in N and the iterated contraaction map $(\mathfrak{g}' \overleftarrow{\otimes} \mathfrak{g}') \otimes^\wedge P \longrightarrow P$. The iterated contraaction map $((\mathfrak{g}' \otimes^! B) \overleftarrow{\otimes} (\mathfrak{g}' \otimes^! B)) \otimes^\wedge \operatorname{Hom}_k(N, P) \longrightarrow \operatorname{Hom}_k(N, P)$ is equal to the composition $((\mathfrak{g}' \otimes^! B) \overleftarrow{\otimes} (\mathfrak{g}' \otimes^! B)) \otimes^\wedge \operatorname{Hom}_k(N, P) \longrightarrow ((\mathfrak{g}' \overleftarrow{\otimes} \mathfrak{g}') \otimes^! B) \otimes^\wedge \operatorname{Hom}_k(N, P) \longrightarrow \operatorname{Hom}_k(N, P)$ of the above induced map and contraaction map. Let Q denote the kernel of the multiplication map

$$\mathfrak{g}' \overleftarrow{\otimes} \mathfrak{g}' \longrightarrow U_\varkappa^\wedge(\mathfrak{g}).$$

The image of K under the map $(\mathfrak{g}' \otimes^! B) \overleftarrow{\otimes} (\mathfrak{g}' \otimes^! B) \longrightarrow (\mathfrak{g}' \overleftarrow{\otimes} \mathfrak{g}') \otimes^! B$ is contained in $Q \otimes^! B$. So it suffices to check that the composition of the injection $Q \otimes^\wedge P \longrightarrow (\mathfrak{g}' \overleftarrow{\otimes} \mathfrak{g}') \otimes^\wedge P$ and the iterated contraaction map $(\mathfrak{g}' \overleftarrow{\otimes} \mathfrak{g}') \otimes^\wedge P \longrightarrow P$ vanishes.

The topological vector space Q is the topological projective limit of the kernels of multiplication maps $\mathfrak{g}'/\mathfrak{h} \otimes^* \mathfrak{g}' \longrightarrow U_\varkappa(\mathfrak{g})/\mathfrak{h} U_\varkappa(\mathfrak{g})$ over all open subalgebras $\mathfrak{h} \subset \mathfrak{g}'$ not containing $k \subset \mathfrak{g}'$. Since the intersection of $\mathfrak{h} U_\varkappa(\mathfrak{g})$ and $\mathfrak{g}'^2$ inside $U_\varkappa(\mathfrak{g}')$ is equal to $\mathfrak{h}\mathfrak{g}'$, the kernel of the (nontopological) multiplication map $\mathfrak{g}' \otimes_k \mathfrak{g}' \longrightarrow U_\varkappa(\mathfrak{g}')$ maps surjectively onto the kernels we are interested in. This nontopological kernel is the image of the map $\bigwedge_k^2(\mathfrak{g}') \longrightarrow \mathfrak{g}' \otimes_k \mathfrak{g}'$ given by the formula $x \wedge y \longmapsto x \otimes y - y \otimes x - 1 \otimes [x, y]$. The kernel of the composition $\Lambda_k^2(\mathfrak{g}') \longrightarrow \mathfrak{g}' \otimes_k \mathfrak{g}' \longrightarrow \mathfrak{g}'/\mathfrak{h} \otimes_k \mathfrak{g}'$ is the subspace $\Lambda_k^2(\mathfrak{h}) \subset \Lambda_k^2(\mathfrak{g}')$. Hence the kernel of the map $\mathfrak{g}'/\mathfrak{h} \otimes^* \mathfrak{g}' \longrightarrow U_\varkappa(\mathfrak{g})/\mathfrak{h} U_\varkappa(\mathfrak{g})$ is the subspace $\Lambda_k^2(\mathfrak{g}')/\Lambda_k^2(\mathfrak{h}) \subset \mathfrak{g}'/\mathfrak{h} \otimes^* \mathfrak{g}'$, embedded by the above formula, endowed with the induced topology of a closed subspace. One can easily check that this topology on $\Lambda_k^2(\mathfrak{g}')/\Lambda_k^2(\mathfrak{h})$ is the topology of the quotient space $\Lambda^{*,2}(\mathfrak{g}')/\Lambda^{*,2}(\mathfrak{h})$. Thus the topological vector space Q is isomorphic to $\Lambda^{*,2}(\mathfrak{g}')$.

D.5.5 The following constructions are due to Beilinson and Drinfeld [10, 3.8.19–22].

Let V be a Tate vector space and $E \subset V$ be a compact open subspace. The graded vector space of *semi-infinite forms*

$$\Lambda_E^{\infty/2}(V) = \bigoplus_i \Lambda_E^{\infty/2+i}(V)$$

is defined as the inductive limit of the spaces $\Lambda_k(V/U) \otimes_k \det(E/U)^\vee$ over all compact open subspaces $U \subset E$. Here $\det(X)$ denotes the top exterior power of a finite-dimensional vector space X and $\Lambda_k(W)$ denotes the direct sum of all exterior powers of a vector space W; the grading on $\Lambda_k(V/U) \otimes_k \det(E/U)^\vee$ is defined so that $\Lambda_k^j(V/U) \otimes_k \det(E/U)^\vee$ is the component of degree $j - \dim(E/U)$. The limit is taken over the maps induced by the natural maps

$$\Lambda_k^j(V/U') \otimes_k \det(U'/U'') \longrightarrow \Lambda_k^{j+m}(V/U''),$$

where $U'' \subset U'$ and $m = \dim(U'/U'')$. The spaces of semi-infinite forms corresponding to different compact open subspaces $E \subset V$, only differ by a dimensional shift and a determinantal twist: if $F \subset V$ is another compact open subspace, then there are natural isomorphisms

$$\Lambda_F^{\infty/2+i}(V) \simeq \Lambda_E^{\infty/2+i+\dim(E,F)}(V) \otimes_k \det(E, F),$$

where $\dim(E, F) = \dim(E/E \cap F) - \dim(F/E \cap F)$ and $\det(E, F) = \det(E/E \cap F) \otimes_k \det(F/E \cap F)^\vee$.

Denote by $\overline{\mathrm{Cl}}(V)$ the algebra of endomorphisms of the vector space $\Lambda_E^{\infty/2}(V)$ endowed with the topology where annihilators of finite-dimensional subspaces of $\Lambda_E^{\infty/2}(V)$ form a base of neighborhoods of zero. Clearly, the topological associative algebra $\overline{\mathrm{Cl}}(V)$ does not depend on the choice of a compact open subspace $E \subset V$; open left ideals form a base of neighborhoods of zero in $\overline{\mathrm{Cl}}(V)$. Denote by $\overline{\mathrm{Cl}}^i(V)$ the closed subspace of homogeneous endomorphisms of degree i in $\overline{\mathrm{Cl}}(V)$.

The Clifford algebra $\mathrm{Cl}(V \oplus V^\vee)$ acts naturally in $\Lambda_E^{\infty/2}(V)$, so there is a morphism of associative algebras $e \colon \mathrm{Cl}(V \oplus V^\vee) \longrightarrow \overline{\mathrm{Cl}}(V)$; in particular, the map e sends V to $\overline{\mathrm{Cl}}^1(V)$ and $V^\vee$ to $\overline{\mathrm{Cl}}^{-1}(V)$. Let $\Lambda^{!,i}(V^\vee)$ denote the completion of $\Lambda_k^i(V^\vee)$ with respect to the topology with the base of neighborhoods of zero formed by the subspaces $U \wedge \Lambda_k^{i-1}(V^\vee) \subset \Lambda_k^i(V^\vee)$, where $U \subset V^\vee$ is an open subspace. The composition $\bigwedge_k^i(V^\vee) \longrightarrow \mathrm{Cl}(V \oplus V^\vee) \longrightarrow \overline{\mathrm{Cl}}(V)$ can be extended by continuity to a map $\bigwedge^{!,i}(V^\vee) \longrightarrow \overline{\mathrm{Cl}}(V)$, which we will denote also by e. The construction of D.1.7 provides a morphism of topological Lie algebras $\mathfrak{gl}(V)^\sim \longrightarrow \overline{\mathrm{Cl}}^0(V)$.

Let $\mathfrak{g}$ be a Tate Lie algebra and $\varkappa_0 \colon \mathfrak{g}^\sim \longrightarrow \mathfrak{g}$ be its canonical central extension. Consider the topological tensor product $\mathfrak{g}^\sim \otimes^! \overline{\mathrm{Cl}}(\mathfrak{g})^{\mathrm{op}}$, where $\overline{\mathrm{Cl}}(\mathfrak{g})^{\mathrm{op}}$ denotes the topological algebra opposite to $\overline{\mathrm{Cl}}(\mathfrak{g})$; this topological tensor product is a bimodule over $\overline{\mathrm{Cl}}(\mathfrak{g})^{\mathrm{op}}$. The unit elements of $\overline{\mathrm{Cl}}(\mathfrak{g})^{\mathrm{op}}$ and $k \subset \mathfrak{g}^\sim$ induce embeddings of $\mathfrak{g}^\sim$ and $\overline{\mathrm{Cl}}(\mathfrak{g})^{\mathrm{op}}$ into $\mathfrak{g}^\sim \otimes^! \overline{\mathrm{Cl}}(\mathfrak{g})^{\mathrm{op}}$. Consider the difference of the composition $\mathfrak{g}^\sim \longrightarrow \mathfrak{gl}(\mathfrak{g})^\sim \longrightarrow \overline{\mathrm{Cl}}(\mathfrak{g}) \simeq \overline{\mathrm{Cl}}(\mathfrak{g})^{\mathrm{op}} \longrightarrow \mathfrak{g}^\sim \otimes^! \overline{\mathrm{Cl}}(\mathfrak{g})^{\mathrm{op}}$ and the embedding $\mathfrak{g}^\sim \longrightarrow \mathfrak{g}^\sim \otimes^! \overline{\mathrm{Cl}}(\mathfrak{g})^{\mathrm{op}}$; this difference maps $k \subset \mathfrak{g}^\sim$ to zero and so induces a natural map

$$l \colon \mathfrak{g} \longrightarrow \mathfrak{g}^\sim \otimes^! \overline{\mathrm{Cl}}^0(\mathfrak{g})^{\mathrm{op}}.$$

The composition of the map l with the embedding $\mathfrak{g}^\sim \otimes^! \overline{\mathrm{Cl}}^{-1}(\mathfrak{g})^{\mathrm{op}} \longrightarrow U_{\varkappa_0}^\frown(\mathfrak{g}) \otimes^! \overline{\mathrm{Cl}}^{-1}(\mathfrak{g})^{\mathrm{op}}$ is an anti-homomorphism of Lie algebras, i.e., it transforms the commutators to minus the commutators.

Denote by $\delta \colon \mathfrak{g}^\vee \longrightarrow \bigwedge^{!,2}(\mathfrak{g}^\vee)$ the continuous linear map given by the formula $\delta(x^*) = x_{\{1\}}^* \wedge x_{\{2\}}^*$, where $\langle x^*, [x', x''] \rangle = \langle x_{\{1\}}^*, x'' \rangle \langle x_{\{2\}}^*, x' \rangle - \langle x_{\{1\}}^*, x' \rangle \langle x_{\{2\}}^*, x'' \rangle$ for $x^* \in \mathfrak{g}^\vee$, $x', x'' \in \mathfrak{g}$. Define the map $\chi \colon \mathfrak{g} \otimes \mathfrak{g}^\vee \longrightarrow \mathfrak{g}^\sim \otimes^! \overline{\mathrm{Cl}}^{-1}(\mathfrak{g})^{\mathrm{op}}$ by the formula

$$\chi(x \otimes x^*) = l(x)e(x^*)^{\mathrm{op}} - e(x)^{\mathrm{op}}e(\delta(x^*))^{\mathrm{op}} = e(x^*)^{\mathrm{op}}l(x) - e(\delta(x^*))^{\mathrm{op}}e(x)^{\mathrm{op}},$$

where a^{op} denotes the element of $\mathrm{Cl}(\mathfrak{g})^{\mathrm{op}}$ corresponding to an element $a \in \mathrm{Cl}(\mathfrak{g})$; and extend this map by continuity to a map

$$\chi \colon \mathfrak{g} \otimes^! \mathfrak{g}^\vee \longrightarrow \mathfrak{g}^\sim \otimes^! \overline{\mathrm{Cl}}^{-1}(\mathfrak{g})^{\mathrm{op}}.$$

Identify $\mathfrak{g} \otimes^! \mathfrak{g}^\vee$ with $\mathrm{End}(\mathfrak{g})$ and set $\mathfrak{d} = \chi(\mathrm{id}_\mathfrak{g}) \in \mathfrak{g}^\sim \otimes^! \overline{\mathrm{Cl}}^{-1}(\mathfrak{g})^{\mathrm{op}}$.

Denote the image of $\mathfrak{d}$ under the embedding $\mathfrak{g}^\sim \otimes^! \overline{\mathrm{Cl}}^{-1}(\mathfrak{g})^{\mathrm{op}} \longrightarrow U_{\varkappa_0}^\frown(\mathfrak{g}) \otimes^! \overline{\mathrm{Cl}}^{-1}(\mathfrak{g})^{\mathrm{op}}$ also by $\mathfrak{d}$. Using the identity $[l(x), e(y)^{\mathrm{op}}] = -e([x, y])^{\mathrm{op}}$, one can check that $[\mathfrak{d}, e(x)^{\mathrm{op}}] = l(x)$ and $[[\mathfrak{d}^2, e(x)^{\mathrm{op}}], e(y)^{\mathrm{op}}] = 0$ for all $x, y \in \mathfrak{g}$, where $[\,,\,]$ denotes the supercommutator with respect to the grading in which $U_{\varkappa_0}^\frown(\mathfrak{g}) \otimes^! \overline{\mathrm{Cl}}^i(\mathfrak{g})^{\mathrm{op}}$ lies in the degree i. It is easy to see that any element of $\overline{\mathrm{Cl}}^i(\mathfrak{g})$ supercommuting with $e(x)$ for all $x \in \mathfrak{g}$ is zero when $i < 0$; hence the same applies to elements of $U_{\varkappa_0}^\frown(\mathfrak{g}) \otimes^! \overline{\mathrm{Cl}}^i(\mathfrak{g})^{\mathrm{op}}$ with $i < 0$. It follows that $\mathfrak{d}$ is the unique element of $U_{\varkappa_0}^\frown(\mathfrak{g}) \otimes^! \overline{\mathrm{Cl}}^{-1}(\mathfrak{g})^{\mathrm{op}}$ satisfying the equation $[\mathfrak{d}, e(x)^{\mathrm{op}}] = l(x)$, and that $\mathfrak{d}^2 = 0$.

D.5.6 Let $\mathfrak{g}$ be a Tate Lie algebra and $E \subset \mathfrak{g}$ be a compact open vector subspace. Let N be a discrete $\mathfrak{g}^\sim$-module where the unit element of $k \subset \mathfrak{g}^\sim$ acts by minus the identity. Then N can be considered as a discrete right $U_{\varkappa_0}^\frown(\mathfrak{g})$-module and $\bigwedge_E^{\infty/2}(\mathfrak{g})$ is a discrete right $\overline{\mathrm{Cl}}(\mathfrak{g})^{\mathrm{op}}$-module, so the tensor product $\bigwedge_E^{\infty/2}(\mathfrak{g}) \otimes_k N$ is a discrete right module over $U_{\varkappa_0}^\frown(\mathfrak{g}) \otimes^! \overline{\mathrm{Cl}}(\mathfrak{g})^{\mathrm{op}}$. The action of the element $\mathfrak{d} \in$

$U_{\varkappa_0}^{\wedge}(\mathfrak{g}) \otimes^! \overline{\mathrm{Cl}}^{-1}(\mathfrak{g})^{\mathrm{op}}$ defines a differential $d_{\infty/2}$ of degree -1 on the graded vector space $C_{\infty/2+\bullet}^{E}(\mathfrak{g}, N)$ with the components

$$C_{\infty/2+i}^{E}(\mathfrak{g}, N) = \bigwedge_{E}^{\infty/2+i}(\mathfrak{g}) \otimes_k N.$$

One has $d_{\infty/2}^2 = 0$, since $\partial^2 = 0$; so $C_{\infty/2+\bullet}^{E}(\mathfrak{g}, N)$ becomes a complex. This complex is called the *semi-infinite homological complex* and its homology is called the *semi-infinite homology* of a Tate Lie algebra $\mathfrak{g}$ with coefficients in a discrete $\mathfrak{g}^\sim$-module N. One defines the semi-infinite homology of $\mathfrak{g}$ with coefficients in a complex of discrete $\mathfrak{g}^\sim$-modules $N^\bullet$ as the homology of the total complex of the bicomplex $C_{\infty/2+\bullet}^{E}(\mathfrak{g}, N^\bullet)$ constructed by taking infinite direct sums along the diagonals.

Let P be a $\mathfrak{g}^\sim$-contramodule where the unit element of $k \subset \mathfrak{g}^\sim$ acts by the identity. First assume that $\mathfrak{g}$ has a countable base of neighborhoods of zero. Then P can be considered as a left $U_{\varkappa_0}^{\wedge}(\mathfrak{g})$-contramodule and $\bigwedge_{E}^{\infty/2}(\mathfrak{g})$ is a discrete right $\overline{\mathrm{Cl}}(\mathfrak{g})^{\mathrm{op}}$-module, so the space of homomorphisms $\mathrm{Hom}_k(\bigwedge_{E}^{\infty/2}(\mathfrak{g}), P)$ is a left contramodule over $U_{\varkappa_0}^{\wedge}(\mathfrak{g}) \otimes^! \overline{\mathrm{Cl}}(\mathfrak{g})^{\mathrm{op}}$. The action of the element $\partial \in U_{\varkappa_0}^{\wedge}(\mathfrak{g}) \otimes^! \overline{\mathrm{Cl}}^{-1}(\mathfrak{g})^{\mathrm{op}}$ defines a differential $d^{\infty/2}$ of degree 1 on the graded vector space $C_{E}^{\infty/2+\bullet}(\mathfrak{g}, P)$ with the components

$$C_{E}^{\infty/2+i}(\mathfrak{g}, P) = \mathrm{Hom}_k(\bigwedge_{E}^{\infty/2+i}(\mathfrak{g}), P).$$

One has $(d^{\infty/2})^2 = 0$, since $\partial^2 = 0$. Without the countability assumption, the element $\partial \in \mathfrak{g}^\sim \otimes^! \overline{\mathrm{Cl}}^{-1}(\mathfrak{g})^{\mathrm{op}}$ still acts on the graded vector space $C_{E}^{\infty/2+\bullet}(\mathfrak{g}, P)$ by an operator $d^{\infty/2}$ of degree 1. By the result of D.5.4, the identity $\partial^2 = 0$ in $U_{\varkappa_0}^{\wedge}(\mathfrak{g}) \otimes^! \overline{\mathrm{Cl}}(\mathfrak{g})^{\mathrm{op}}$ implies the equation $(d^{\infty/2})^2 = 0$; so $C_{E}^{\infty/2+\bullet}(\mathfrak{g}, P)$ becomes a complex. This complex is called the *semi-infinite cohomological complex* and its cohomology is called the *semi-infinite cohomology* of a Tate Lie algebra $\mathfrak{g}$ with coefficients in a $\mathfrak{g}^\sim$-contramodule P. One defines the semi-infinite cohomology of $\mathfrak{g}$ with coefficients in a complex of $\mathfrak{g}^\sim$-contramodules $P^\bullet$ as the cohomology of the total complex of the bicomplex $C_{E}^{\infty/2+\bullet}(\mathfrak{g}, P^\bullet)$ constructed by taking infinite products along the diagonals.

D.6 Comparison theorem

D.6.1 The purpose of this subsection is to partly extend to arbitrary characteristic the classical equivalence between the categories of nilpotent Lie algebras and unipotent algebraic groups in characteristic zero (see [31, Corollary IV.2.4.5], or [72, 3.3.6] and [78, Remark LG.V.4.1], or [48]). We will not obtain an equivalence of categories, but only construct a functor assigning a commutative Hopf algebra to a pronilpotent Lie algebra over a field k.

The correspondence $\mathfrak{h} \longmapsto \mathcal{L} = \mathfrak{h}^\vee$ provides an anti-equivalence between the categories of compact Lie algebras and Lie coalgebras. The correspondence between the action maps $\mathfrak{h} \times M \longrightarrow M$ and the coaction maps $M \longrightarrow \mathfrak{h}^\vee \otimes_k M$ defines an equivalence between the categories of discrete $\mathfrak{h}$-modules and $\mathcal{L}$-comodules.

A Lie coalgebra $\mathcal{L}$ is called *conilpotent* if it is a filtered inductive limit of finite-dimensional Lie coalgebras dual to finite-dimensional nilpotent Lie algebras; in other words, $\mathcal{L}$ is conilpotent if the dual compact Lie algebra $\mathfrak{h}$ is *pronilpotent*. A comodule M over a Lie coalgebra $\mathcal{L}$ is called *conilpotent* if it is an inductive limit of finite-dimensional comodules which can be represented as iterated extensions of trivial comodules (that is comodules with a zero coaction map); analogously one defines *nilpotent* discrete modules over topological Lie algebras. A coassociative coalgebra $\mathcal{C}$ endowed with a coaugmentation map $k \longrightarrow \mathcal{C}$ is called *conilpotent* if for every element c of the coalgebra without counit $\mathcal{C}/\operatorname{im} k$ there exists a positive integer i such that the iterated comultiplication map $\mathcal{C}/\operatorname{im} k \longrightarrow (\mathcal{C}/\operatorname{im} k)^{\otimes i}$ annihilates c.

Remark. The terminology related to conilpotent (coassociative) coalgebras is not well settled. Sweedler [82] calls such coalgebras "pointed irreducible", Montgomery [68] calls them "connected", and many recent publications (see, e.g., [61]) use the word "cocomplete". The latter term comes from the condition that the natural filtration (defined above) should be cocomplete (i.e., exhaustive). In our view, this terminology is misleading and the term "conilpotent" is preferable.

For a conilpotent Lie coalgebra $\mathcal{L}$, the *conilpotent coenveloping coalgebra* $\mathcal{C}(\mathcal{L})$ is constructed as follows. Consider the category of finite-dimensional conilpotent $\mathcal{L}$-comodules together with the forgetful functor from it to the category of finite-dimensional vector spaces; by [30, Proposition 2.14], this category is equivalent (actually, isomorphic) to the category of finite-dimensional left comodules over a certain uniquely defined coalgebra $\mathcal{C}(\mathcal{L})$ together with the forgetful functor from this category to the category of finite-dimensional vector spaces. Clearly, the category of (arbitrary) left $\mathcal{C}(\mathcal{L})$-comodules is isomorphic to the category of conilpotent $\mathcal{L}$-comodules. The trivial $\mathcal{L}$-comodule k defines a coaugmentation $k \longrightarrow \mathcal{C}(\mathcal{L})$; since this is the only irreducible left $\mathcal{C}(\mathcal{L})$-comodule, the coalgebra $\mathcal{C}(\mathcal{L})$ is conilpotent.

The coalgebra $\mathcal{C}(\mathcal{L})$ is the universal final object in the category of conilpotent coalgebras $\mathcal{C}$ endowed with a Lie coalgebra morphism $\mathcal{C} \longrightarrow \mathcal{L}$ such that the composition $k \longrightarrow \mathcal{C} \longrightarrow \mathcal{L}$ vanishes. Indeed, there is a morphism of Lie coalgebras $\mathcal{C}(\mathcal{L}) \longrightarrow \mathcal{L}$, since there is a natural $\mathcal{L}$-comodule structure on every left $\mathcal{C}(\mathcal{L})$-comodule, and in particular, on the left comodule $\mathcal{C}(\mathcal{L})$. Conversely, a morphism $\mathcal{C} \longrightarrow \mathcal{L}$ with the above properties defines a functor assigning to a left $\mathcal{C}$-comodule M a conilpotent $\mathcal{L}$-comodule structure on the same vector space M, hence a left $\mathcal{C}(\mathcal{L})$-comodule structure on M; this induces a coalgebra morphism $\mathcal{C} \longrightarrow \mathcal{C}(\mathcal{L})$. Since the category of finite-dimensional conilpotent $\mathcal{L}$-comodules is a tensor category with duality, the coalgebra $\mathcal{C}(\mathcal{L})$ acquires a Hopf algebra structure.

Let $\mathfrak{h}$ be the compact Lie algebra dual to $\mathcal{L}$; then the pairing $\phi \colon \mathcal{C}(\mathcal{L}) \times U(\mathfrak{h}) \longrightarrow k$ is nondegenerate in $\mathcal{C}$, since the morphism $\mathcal{C}(\mathcal{L}) \longrightarrow \mathcal{L}$ factorizes through the quotient coalgebra of $\mathcal{C}(\mathcal{L})$ by the kernel of ϕ, so a nonzero kernel would be contradict the universality property.

Let $\mathfrak{M}$ be a conilpotent $\mathcal{C}$-comodule; set $\mathcal{C} = \mathcal{C}(\mathcal{L})$ and $\mathcal{C}_+ = \mathcal{C}/\operatorname{im} k$. Then the natural surjective morphism from the reduced cobar complex

$$\mathfrak{M} \longrightarrow \mathcal{C}_+ \otimes_k \mathfrak{M} \longrightarrow \mathcal{C}_+ \otimes_k \mathcal{C}_+ \otimes_k \mathfrak{M} \longrightarrow \cdots$$

computing $\operatorname{Cotor}^{\mathcal{C}}(k, \mathfrak{M}) \simeq \operatorname{Ext}_{\mathcal{C}}(k, \mathfrak{M})$ onto the cohomological Chevalley complex

$$\mathfrak{M} \longrightarrow \mathcal{L} \otimes_k \mathfrak{M} \longrightarrow \textstyle\bigwedge^2_k \mathcal{L} \otimes_k \mathfrak{M} \longrightarrow \cdots$$

is a quasi-isomorphism. It suffices to check this for a finite-dimensional Lie coalgebra $\mathcal{L}$ dual to a finite-dimensional nilpotent Lie algebra $\mathfrak{h}$; essentially, one has to show that the fully faithful functor from the category of nilpotent $\mathfrak{h}$-modules to the category of arbitrary $\mathfrak{h}$-modules induces isomorphisms on the Ext spaces. This well-known fact can be proven by induction on the dimension of $\mathfrak{h}$ using the Serre–Hochschild spectral sequences for both types of cohomology under consideration. The key step is to check that for a Lie subcoalgebra $\mathcal{E} \subset \mathcal{L}$ the $\mathcal{C}(\mathcal{L}/\mathcal{E})$-comodule $\mathcal{C}(\mathcal{L})$ is injective and the $\mathcal{E}$-comodule $\mathcal{C}(\mathcal{E})$ is the comodule of $\mathcal{L}/\mathcal{E}$-invariants in the $\mathcal{L}$-comodule $\mathcal{C}(\mathcal{L})$; it suffices to consider the case when $\mathcal{L}/\mathcal{E}$ is one-dimensional.

D.6.2 Let $\mathfrak{g}$ be a Tate Lie algebra and $\mathfrak{h} \subset \mathfrak{g}$ be a compact open Lie subalgebra. Assume that $\mathfrak{h}$ is pronilpotent and the discrete $\mathfrak{h}$-module $\mathfrak{g}/\mathfrak{h}$ is nilpotent. Then the conilpotent coalgebra $\mathcal{C} = \mathcal{C}(\mathfrak{h}^\vee)$ coacts continuously in $\mathfrak{g}$, making $(\mathfrak{g}, \mathcal{C})$ a Tate Harish-Chandra pair. Let $k \longrightarrow \mathfrak{g}' \longrightarrow \mathfrak{g}$ be a central extension of Tate Lie algebras endowed with a splitting over $\mathfrak{h}$; then there are a Tate Harish-Chandra pair $(\mathfrak{g}', \mathcal{C})$ and a central extension of Tate Harish-Chandra pairs $\varkappa\colon (\mathfrak{g}', \mathcal{C}) \longrightarrow (\mathfrak{g}, \mathcal{C})$ with the kernel k. Denote by $\varkappa_0\colon (\mathfrak{g}^\sim, \mathcal{C}) \longrightarrow (\mathfrak{g}, \mathcal{C})$ the canonical central extension. Set $\mathbf{S}^l_\varkappa = \mathbf{S}^l_\varkappa(\mathfrak{g}, \mathcal{C}) \simeq \mathbf{S}^r_{\varkappa+\varkappa_0}(\mathfrak{g}, \mathcal{C}) = \mathbf{S}^r_{\varkappa+\varkappa_0}$.

Theorem.

(a) *Let $\mathbf{N}^\bullet$ be a complex of right $\mathbf{S}^r_{\varkappa+\varkappa_0}$-semimodules and $\mathfrak{M}^\bullet$ be a complex of left $\mathbf{S}^l_\varkappa$-semimodules; in other words, $\mathbf{N}^\bullet$ is a complex of Harish-Chandra modules with the central charge $-\varkappa - \varkappa_0$ and $\mathfrak{M}^\bullet$ is a complex of Harish-Chandra modules with the central charge $\varkappa$ over $(\mathfrak{g}, \mathcal{C})$.*

   Then the total complex of the semi-infinite homological bicomplex $C^\mathfrak{h}_{\infty/2+\bullet}(\mathfrak{g}, \mathbf{N}^\bullet \otimes_k \mathfrak{M}^\bullet)$ constructed by taking infinite direct sums along the diagonals represents the object $\operatorname{SemiTor}^{\mathbf{S}^l_\varkappa}(\mathbf{N}^\bullet, \mathfrak{M}^\bullet)$ in the derived category of k-vector spaces. Here the tensor product $\mathbf{N}^\bullet \otimes_k \mathfrak{M}^\bullet$ is a complex of Harish-Chandra modules with the central charge $-\varkappa_0$.

(b) *Let $\mathfrak{M}^\bullet$ be a complex of left $\mathbf{S}^l_\varkappa$-semimodules and $\mathfrak{P}^\bullet$ be a complex of left $\mathbf{S}^r_{\varkappa+\varkappa_0}$-semicontramodules; in other words, $\mathfrak{M}^\bullet$ is a complex of Harish-Chandra modules with the central charge $\varkappa$ and $\mathfrak{P}^\bullet$ is a complex of Harish-Chandra contramodules with the central charge $\varkappa + \varkappa_0$ over $(\mathfrak{g}, \mathcal{C})$.*

> *Then the total complex of the semi-infinite cohomological bicomplex $C_{\mathfrak{h}}^{\infty/2+\bullet}(\mathfrak{g}, \mathrm{Hom}_k(\mathbf{M}^\bullet, \mathfrak{P}^\bullet))$ constructed by taking infinite products along the diagonals represents the object $\mathrm{SemiExt}_{\mathbf{S}^l_{\varkappa}}(\mathbf{M}^\bullet, \mathfrak{P}^\bullet)$ in the derived category of k-vector spaces. Here $\mathrm{Hom}_k(\mathbf{M}^\bullet, \mathfrak{P}^\bullet)$ is a complex of Harish-Chandra contramodules with the central charge $\varkappa_0$.*

Proof. Part (a): set $\mathbf{S}^l_{-\varkappa_0} \simeq \mathbf{S} = \mathbf{S}^r_0$. Consider the semi-infinite homological complex $C_{\infty/2+\bullet}^{\mathfrak{h}}(\mathfrak{g}, \mathbf{S})$ of the $\mathfrak{g}^\sim$-module $\mathbf{S}$ with the discrete $\mathfrak{g}^\sim$-module structure originating from the left $\mathbf{S}^l_{-\varkappa_0}$-semimodule structure. The complex $C_{\infty/2+\bullet}^{\mathfrak{h}}(\mathfrak{g}, \mathbf{S})$ is a complex of right $\mathbf{S}^r_0$-semimodules. Let us check that it is a semiflat complex naturally isomorphic to the semimodule k in the semiderived category of right $\mathbf{S}^r_0$-semimodules.

Let $F_i \mathbf{S}$ denote the increasing filtration of the semialgebra $\mathbf{S}$ introduced in D.3.1. Set

$$F_i\left(\textstyle\bigwedge_{\mathfrak{h}}^{\infty/2}(\mathfrak{g})\right) = \textstyle\bigwedge_k^i(\mathfrak{g}) \wedge \bigwedge_{\mathfrak{h}}^{\infty/2}(\mathfrak{h}).$$

Denote by F the induced filtration of the tensor product $C_{\infty/2+\bullet}^{\mathfrak{h}}(\mathfrak{g}, \mathbf{S}) = \bigwedge_{\mathfrak{h}}^{\infty/2}(\mathfrak{g}) \otimes_k \mathbf{S}$; this is an increasing filtration of the complex of right $\mathbf{S}^r_0$-semimodules $C_{\infty/2+\bullet}^{\mathfrak{h}}(\mathfrak{g}, \mathbf{S})$ by subcomplexes of right $\mathcal{C}$-comodules. The complex $\mathrm{gr}_F C_{\infty+\bullet}^{\mathfrak{h}}(\mathfrak{g}, \mathbf{S})$ can be identified with the total complex of the cohomological Chevalley bicomplex

$$\textstyle\bigwedge_k(\mathfrak{h}^\vee) \otimes_k \bigwedge_k(\mathfrak{g}/\mathfrak{h}) \otimes_k \mathrm{Sym}_k(\mathfrak{g}/\mathfrak{h}) \otimes_k \mathcal{C}$$

of the complex of $\mathfrak{h}^\vee$-comodules $\bigwedge_k(\mathfrak{g}/\mathfrak{h}) \otimes_k \mathrm{Sym}_k(\mathfrak{g}/\mathfrak{h}) \otimes_k \mathcal{C}$ obtained as the tensor product of the Koszul complex $\bigwedge_k(\mathfrak{g}/\mathfrak{h}) \otimes_k \mathrm{Sym}_k(\mathfrak{g}/\mathfrak{h})$ with the coaction of $\mathfrak{h}^\vee$ induced by the coaction in $\mathfrak{g}/\mathfrak{h}$ and the left $\mathcal{C}$-comodule $\mathcal{C}$ with the induced $\mathfrak{h}^\vee$-comodule structure. It follows that the cone of the injection $F_0 C_{\infty/2+\bullet}^{\mathfrak{h}}(\mathfrak{g}, \mathbf{S}) \longrightarrow C_{\infty/2+\bullet}^{\mathfrak{h}}(\mathfrak{g}, \mathbf{S})$ is a coacyclic complex of right $\mathcal{C}$-comodules.

The complex $F_0 C_{\infty/2+\bullet}^{\mathfrak{h}}(\mathfrak{g}, \mathbf{S})$ is naturally isomorphic to the cohomological Chevalley complex $\bigwedge_k(\mathfrak{h}^\vee) \otimes_k \mathcal{C}$; it is a complex of right $\mathcal{C}$-comodules bounded from below and endowed with a quasi-isomorphism of complexes of right $\mathcal{C}$-comodules $k \longrightarrow F_0 C_{\infty/2+\bullet}^{\mathfrak{h}}(\mathfrak{g}, \mathbf{S})$. Let us check that the right $\mathbf{S}^r_0$-semimodule structure on $H_0 C_{\infty/2+\bullet}^{\mathfrak{h}}(\mathfrak{g}, \mathbf{S}) \simeq k$ corresponds to the trivial $\mathfrak{g}$-module structure. The unit element of this homology group can be represented by the cycle $\lambda \otimes 1 \in C_{\infty/2+0}^{\mathfrak{h}}(\mathfrak{g}, \mathbf{S})$, where λ denotes the unit element of $k \simeq \bigwedge_{\mathfrak{h}}^{\infty/2+0}(\mathfrak{h}) \subset \bigwedge_{\mathfrak{h}}^{\infty/2+0}(\mathfrak{g})$ and $1 \in \mathcal{C} \subset \mathbf{S}$ is the unit (coaugmentation) element of $\mathcal{C}$. Then for any $z \in \mathfrak{g}$ one has $(\lambda \otimes 1)z = \lambda \otimes (1 \otimes_{U(\mathfrak{h})} z) = \lambda \otimes (\tilde{b}(\overline{z}_{(0)}) \otimes_{U(\mathfrak{h})} s(\overline{z}_{(-1)})) = d_{\infty/2}((\overline{z}_{(0)} \wedge \lambda) \otimes s(\overline{z}_{(-1)}))$, where $\overline{z}$ denotes the image of z in $\mathfrak{g}/\mathfrak{h}$ and $\tilde{b}: \mathfrak{g}/\mathfrak{h} \longrightarrow \mathfrak{g}^\sim$ is the section corresponding to any section $b: \mathfrak{g}/\mathfrak{h} \longrightarrow \mathfrak{g}$. The second equation holds, since the elements $1 \otimes_{U(\mathfrak{h})} z$ and $\tilde{b}(\overline{z}_{(0)}) \otimes_{U(\mathfrak{h})} s(\overline{z}_{(-1)})$ of $F_1 \mathbf{S}$ have the same images in $F_1 \mathbf{S}/F_0 \mathbf{S}$ and are both annihilated by the left action of $\mathfrak{h}$ and the map $\delta_b^l = \delta_b^r$. To check the third equation, one can use the supercommutation relation $[\partial, e(y)^{\mathrm{op}}] = l(y)$ for $y \in \mathfrak{g}$.

Now let $\tau_{\geqslant 0} C_{\infty/2+\bullet}^{\mathfrak{h}}(\mathfrak{g}, \mathcal{S})$ denote the quotient complex of canonical truncation of the complex of right $\mathcal{S}_0^r$-semimodules $C_{\infty/2+\bullet}^{\mathfrak{h}}(\mathfrak{g}, \mathcal{S})$ concentrated in the nonnegative cohomological (nonpositive homological) degrees; then there are natural morphisms of complexes of right $\mathcal{S}_0^r$-semimodules

$$k \longrightarrow \tau_{\geqslant 0} C_{\infty/2+\bullet}^{\mathfrak{h}}(\mathfrak{g}, \mathcal{S}) \longleftarrow C_{\infty/2+\bullet}^{\mathfrak{h}}(\mathfrak{g}, \mathcal{S})$$

with $\mathcal{C}$-coacyclic cones. Indeed, recall that any acyclic complex bounded from below is coacyclic. The embedding $F_0 C_{\infty/2+\bullet}^{\mathfrak{h}}(\mathfrak{g}, \mathcal{S}) \longrightarrow C_{\infty/2+\bullet}^{\mathfrak{h}}(\mathfrak{g}, \mathcal{S})$ has a $\mathcal{C}$-coacyclic cone, as has the composition $F_0 C_{\infty/2+\bullet}^{\mathfrak{h}}(\mathfrak{g}, \mathcal{S}) \longrightarrow C_{\infty/2+\bullet}^{\mathfrak{h}}(\mathfrak{g}, \mathcal{S}) \longrightarrow \tau_{\geqslant 0} C_{\infty/2+\bullet}^{\mathfrak{h}}(\mathfrak{g}, \mathcal{S})$, so the cone of the map $C_{\infty/2+\bullet}^{\mathfrak{h}}(\mathfrak{g}, \mathcal{S}) \longrightarrow \tau_{\geqslant 0} C_{\infty/2+\bullet}^{\mathfrak{h}}(\mathfrak{g}, \mathcal{S})$ is also $\mathcal{C}$-coacyclic.

For any complex of left $\mathcal{S}_{-\varkappa_0}^l$-semimodules $\mathcal{K}^\bullet$, the semitensor product $C_{\infty/2+\bullet}^{\mathfrak{h}}(\mathfrak{g}, \mathcal{S}) \lozenge_{\mathcal{S}} \mathcal{K}^\bullet$ is naturally isomorphic to the total complex of the bicomplex $C_{\infty/2+\bullet}^{\mathfrak{h}}(\mathfrak{g}, \mathcal{K}^\bullet)$, constructed by taking infinite direct sums along the diagonals.

Consider the increasing filtration of the total complex of $C_{\infty/2+\bullet}^{\mathfrak{h}}(\mathfrak{g}, \mathcal{K}^\bullet) = \bigwedge_{\mathfrak{h}}^{\infty/2}(\mathfrak{g}) \otimes_k \mathcal{K}^\bullet$ induced by the above filtration F of $\bigwedge_{\mathfrak{h}}^{\infty/2}(\mathfrak{g})$. The associated graded quotient complex of this filtration can be identified with the total complex of the cohomological Chevalley bicomplex

$$\bigwedge\nolimits_k(\mathfrak{h}^\vee) \otimes_k \bigwedge\nolimits_k(\mathfrak{g}/\mathfrak{h}) \otimes_k \mathcal{K}^\bullet$$

of the tensor product of the graded $\mathfrak{h}^\vee$-comodule $\bigwedge_k(\mathfrak{g}/\mathfrak{h})$ and the complex $\mathcal{K}^\bullet$ with the $\mathfrak{h}^\vee$-comodule structure induced by the left $\mathcal{C}$-comodule structure. It follows that the complex $C_{\infty/2+\bullet}^{\mathfrak{h}}(\mathfrak{g}, \mathcal{S}) \lozenge_{\mathcal{S}} \mathcal{K}^\bullet$ is acyclic whenever a complex of left $\mathcal{S}_{-\varkappa_0}^l$-semimodules $\mathcal{K}^\bullet$ is $\mathcal{C}$-coacyclic, so the complex of right $\mathcal{S}_0^r$-semimodules $C_{\infty/2+\bullet}^{\mathfrak{h}}(\mathfrak{g}, \mathcal{S})$ is semiflat.

The tensor product $C_{\infty/2+\bullet}^{\mathfrak{h}}(\mathfrak{g}, \mathcal{S}) \otimes_k N^\bullet$ of the complex of Harish-Chandra modules $C_{\infty/2+\bullet}^{\mathfrak{h}}(\mathfrak{g}, \mathcal{S})$ with central charge 0 and the complex of Harish-Chandra modules $N^\bullet$ with central charge $-\varkappa - \varkappa_0$ is a complex of Harish-Chandra modules with the central charge $-\varkappa - \varkappa_0$. This complex of right $\mathcal{S}_{\varkappa+\varkappa_0}^r$-semimodules is semiflat and naturally isomorphic to $N^\bullet$ in the semiderived category of right $\mathcal{S}_{\varkappa+\varkappa_0}^r$-semimodules. The latter is clear, and to check the former, notice the isomorphisms of Lemma D.4.4(a)

$$(C_{\infty/2+\bullet}^{\mathfrak{h}}(\mathfrak{g}, \mathcal{S}) \otimes_k N^\bullet) \lozenge_{\mathcal{S}_\varkappa^l} \mathcal{L}^\bullet \simeq (C_{\infty/2+\bullet}^{\mathfrak{h}}(\mathfrak{g}, \mathcal{S}) \otimes_k N^\bullet \otimes_k \mathcal{L}^\bullet)_{\mathfrak{g}, \mathfrak{h}}$$

$$\simeq C_{\infty/2+\bullet}^{\mathfrak{h}}(\mathfrak{g}, \mathcal{S}) \lozenge_{\mathcal{S}_0^r} (N^\bullet \otimes_k \mathcal{L}^\bullet)$$

for any complex of left $\mathbf{S}^l_{\varkappa}$-semimodules $\mathcal{L}^{\bullet}$. Now the object $\mathrm{SemiTor}^{\mathbf{S}^l_{\varkappa}}(\mathbf{N}^{\bullet}, \mathcal{M}^{\bullet})$ is represented by the complex

$$(C^{\mathfrak{h}}_{\infty/2+\bullet}(\mathfrak{g}, \mathbf{S}) \otimes_k \mathbf{N}^{\bullet}) \lozenge_{\mathbf{S}^l_{\varkappa}} \mathcal{M}^{\bullet} \simeq C^{\mathfrak{h}}_{\infty/2+\bullet}(\mathfrak{g}, \mathbf{S}) \lozenge_{\mathbf{S}^r_0} (\mathbf{N}^{\bullet} \otimes_k \mathcal{M}^{\bullet})$$

$$\simeq C^{\mathfrak{h}}_{\infty/2+\bullet}(\mathfrak{g}, \mathbf{N}^{\bullet} \otimes_k \mathcal{M}^{\bullet})$$

in the derived category of k-vector spaces.

Another way to identify $\mathrm{SemiTor}^{\mathbf{S}^l_{\varkappa}}(\mathbf{N}^{\bullet}, \mathcal{M}^{\bullet})$ with $C^{\mathfrak{h}}_{\infty/2+\bullet}(\mathfrak{g}, \mathbf{N}^{\bullet} \otimes_k \mathcal{M}^{\bullet})$ is to consider the semiflat complex of left $\mathbf{S}^l_{\varkappa}$-semimodules $C^{\mathfrak{h}}_{\infty/2+\bullet}(\mathfrak{g}, \mathbf{S}) \otimes_k \mathcal{M}^{\bullet}$ naturally isomorphic to $\mathcal{M}^{\bullet}$ in the semiderived category of left $\mathbf{S}^l_{\varkappa}$-semimodules. To check that these two identifications coincide, represent the images of $\mathbf{N}^{\bullet}$ and $\mathcal{M}^{\bullet}$ in the semiderived categories of semimodules by arbitrary semiflat complexes. The proof of part (b) is completely analogous.   □

Question. Can one obtain the semi-infinite homology of arbitrary discrete modules over a Tate Lie algebra with a fixed compact open subalgebra (rather than only Harish-Chandra modules under the nilpotency conditions) as a kind of double-sided derived functor of the functor of semiinvariants on an appropriate exotic derived category of discrete modules? Notice that the cohomology of the Chevalley complex $\mathcal{M} \longrightarrow \mathcal{L} \otimes_k \mathcal{M} \longrightarrow \bigwedge_k^2 \mathcal{L} \otimes_k \mathcal{M} \longrightarrow \cdots$ for a comodule $\mathcal{M}$ over a Lie coalgebra $\mathcal{L}$ is indeed the right derived functor of the functor of $\mathcal{L}$-invariants on the abelian category of $\mathcal{L}$-comodules, since the category of $\mathcal{L}$-comodules has enough injectives and the cohomology of the Chevalley complex is an effaceable cohomological functor. The former holds since the category of discrete modules over a compact Lie algebra $\mathfrak{h} = \mathcal{L}^{\vee}$ has exact functors of filtered inductive limits preserved by the forgetful functor to the category of k-vector spaces, and the discrete $\mathfrak{h}$-modules $U(\mathfrak{h}) \otimes_{U(\mathfrak{a})} k$ induced from trivial modules over open subalgebras $\mathfrak{a} \subset \mathfrak{h}$ form a set of generators, so the forgetful functor even has a right adjoint. To check the latter, one can represent cocycles in the cohomological Chevalley complex by discrete $\mathfrak{h}$-module morphisms into $\mathcal{M}$ from the relative homological complexes $\cdots \longrightarrow \bigwedge_k^2(\mathfrak{h}/\mathfrak{a}) \otimes_{U(\mathfrak{a})} U\mathfrak{h} \longrightarrow \mathfrak{h}/\mathfrak{a} \otimes_{U(\mathfrak{a})} U(\mathfrak{h}) \longrightarrow U(\mathfrak{h})$, which are quotient complexes of the Chevalley homological complex of the $\mathfrak{h}$-module $U(\mathfrak{h})$ and finite discrete $\mathfrak{h}$-module resolutions of the trivial $\mathfrak{h}$-module k. Furthermore, the semiderived category of discrete modules over a Tate Lie algebra $\mathfrak{g}$, defined as the quotient category of the homotopy category of discrete $\mathfrak{g}$-modules by the thick subcategory of complexes coacyclic as complexes of discrete $\mathfrak{h}$-modules, does not depend on the choice of an open compact subalgebra $\mathfrak{h} \subset \mathfrak{g}$. This can be demonstrated by considering the tensor product over k of the above relative homological complex with a complex of discrete $\mathfrak{h}$-modules coacyclic over $\mathfrak{a}$. Notice that in the above proof we have essentially shown that the semi-infinite homology is a functor on the semiderived category of discrete $\mathfrak{g}$-modules.

D.6.3 We keep the assumptions and notation of D.6.2, and also use the notation of Corollary D.3.1. The following result makes use of the semimodule-semicontramodule correspondence in order to express the semi-infinite homology and cohomology in terms of compositions of one-sided derived functors.

Corollary.
(a) *Let $\mathcal{M}^{\bullet}$ be a complex of Harish-Chandra modules with the central charge $\varkappa$ and $\mathfrak{P}^{\bullet}$ be a complex of Harish-Chandra contramodules with the central charge $\varkappa + \varkappa_0$ over $(\mathfrak{g}, \mathcal{C})$. Then the semi-infinite cohomological complex $C_{\mathfrak{h}}^{\infty/2+\bullet}(\mathfrak{g}, \mathrm{Hom}_k(\mathcal{M}^{\bullet}, \mathfrak{P}^{\bullet}))$ represents the object*

$$\mathrm{Ext}_{\mathcal{S}_{\varkappa}^l}(\mathcal{M}^{\bullet}, \mathbb{L}\Phi_{\mathcal{S}_{\varkappa+\varkappa_0}^r}(\mathfrak{P}^{\bullet})) \simeq \mathrm{Ext}^{\mathcal{S}_{\varkappa+\varkappa_0}^r}(\mathbb{R}\Psi_{\mathcal{S}_{\varkappa}^l}(\mathcal{M}^{\bullet}), \mathfrak{P}^{\bullet})$$

in the derived category of k-vector spaces.

(b) *Let $\mathcal{M}^{\bullet}$ be a complex of Harish-Chandra modules with the central charge $\varkappa$ and $\mathcal{N}^{\bullet}$ be a complex of Harish-Chandra modules with the central charge $-\varkappa - \varkappa_0$ over $(\mathfrak{g}, \mathcal{C})$. Then the semi-infinite homological complex $C^{\mathfrak{h}}_{\infty/2+\bullet}(\mathfrak{g}, \mathcal{N}^{\bullet} \otimes_k \mathcal{M}^{\bullet})$ represents the object*

$$\mathrm{Ctrtor}^{\mathcal{S}_{\varkappa+\varkappa_0}^r}(\mathcal{N}^{\bullet}, \mathbb{R}\Psi_{\mathcal{S}_{\varkappa}^l}(\mathcal{M}^{\bullet})) \simeq \mathrm{Ctrtor}^{\mathcal{S}_{-\varkappa}^r}(\mathcal{M}^{\bullet}, \mathbb{R}\Psi_{\mathcal{S}_{-\varkappa-\varkappa_0}^l}(\mathcal{N}^{\bullet}))$$

in the derived category of k-vector spaces.

Proof. This follows from Theorem D.6.2 and Corollary 6.6. $\qquad\square$

Remark. Set $\mathcal{S}_0^l = \mathcal{S} \simeq \mathcal{S}_{\varkappa_0}^r$ and consider the complex of left $\mathcal{S}$-semimodules $\mathcal{R}^{\bullet} = \mathcal{S} \otimes_k \bigwedge_{\infty/2+\bullet}^{\mathfrak{h}}(\mathfrak{g})$. This is a semiprojective complex of semiprojective left $\mathcal{S}$-semimodules isomorphic to the trivial $\mathcal{S}$-semimodule k in $\mathsf{D}^{\mathsf{si}}(\mathcal{S}\text{-simod})$. Assume that the pronilpotent Lie algebra $\mathfrak{h}$ is infinite-dimensional (cf. 0.2.7). Then the complex of left $\mathcal{S}$-semicontramodules $\Psi_{\mathcal{S}}(\mathcal{R}^{\bullet})$ is acyclic. Indeed, it suffices to check that the complex of left $\mathcal{C}$-contramodules obtained by applying the functor $\Psi_{\mathcal{C}}$ to the cohomological Chevalley complex $\mathcal{C} \otimes_k \bigwedge_k(\mathfrak{h}^{\vee})$ is acyclic; one can reduce this problem to the case of an abelian Lie algebra by considering the decreasing filtration $\mathfrak{h} \supset [\mathfrak{h}, \mathfrak{h}] \supset [\mathfrak{h}, [\mathfrak{h}, \mathfrak{h}]] \supset \cdots$ on $\mathfrak{h}$ and the induced increasing filtrations on $\mathfrak{h}^{\vee}$ and $\mathcal{C}$. The complex $\Psi_{\mathcal{S}}(\mathcal{R}^{\bullet})$ is also a projective complex of projective left $\mathcal{S}$-semicontramodules (see Remark 6.5 and Section 9.2); it can be thought of as the "projective $\mathcal{S}$-semicontramodule resolution of a (nonexistent) one-dimensional left $\mathcal{S}$-semicontramodule placed in the degree $+\infty$". It is a complex over $\mathcal{O}_{\varkappa_0}^{\mathrm{ctr}}(\mathfrak{g}, \mathcal{C})$. For any complex of right $\mathcal{S}$-semimodules $\mathcal{N}^{\bullet}$, the contratensor product complex $\mathcal{N}^{\bullet} \odot_{\mathcal{S}} \Psi_{\mathcal{S}}(\mathcal{R}^{\bullet})$ computes the semi-infinite homology of $\mathfrak{g}$ with coefficients in $\mathcal{N}^{\bullet}$. For any complex of left $\mathcal{S}$-semicontramodules $\mathfrak{P}^{\bullet}$, the complex of semicontramodule homomorphisms $\mathrm{Hom}^{\mathcal{S}}(\Psi_{\mathcal{S}}(\mathcal{R}^{\bullet}), \mathfrak{P}^{\bullet})$ computes the semi-infinite cohomology of $\mathfrak{g}$ with coefficients in $\mathfrak{P}^{\bullet}$. (Cf. [84, Subsection 3.11.4].)

E Groups with Open Profinite Subgroups

To a locally compact totally disconnected topological group G and a commutative ring k one associates a family of left and right semiprojective Morita equivalent semialgebras $\mathcal{S} = \mathcal{S}_k(G, H)$ numbered by open profinite subgroups $H \subset G$. As explained in 8.4.5, Morita equivalences of semialgebras do not have to preserve the semiderived categories or the derived functors SemiTor and SemiExt, and this is indeed the case here: $\mathrm{SemiTor}^{\mathcal{S}}$ and $\mathrm{SemiExt}_{\mathcal{S}}$ depend very essentially on H. For a complex of smooth G-modules $N^\bullet$ over k and a complex of k-flat smooth G-modules $M^\bullet$ over k, we show that $\mathrm{SemiTor}^{\mathcal{S}_k(G,H)}(N^\bullet, M^\bullet)$ only depends on the complex of smooth G-modules $N^\bullet \otimes_k M^\bullet$, and analogously for $\mathrm{SemiExt}_{\mathcal{S}_k(G,H)}$.

When k is a field of zero characteristic, one can climb one step higher and assign to a "good enough" group object $\mathbb{G}$ in the category of ind-pro-topological spaces with a subgroup object $\mathbb{H}$ belonging to the category of pro-topological groups a coring object in the tensor category of representations of $\mathbb{H} \times \mathbb{H}$ in pro-vector spaces over k. So there is a functor of cotensor product [45] on certain categories of representations of central extensions of $\mathbb{G}$; it has a double-sided derived functor ProCotor.

E.1 Morita equivalent semialgebras

E.1.1 In the sequel, all topological spaces and topological groups are presumed to be locally compact and totally disconnected (see [20, II.4.4 and III.4.6]).

For a topological space X and an abelian group A, denote by $A(X)$ the abelian group of locally constant compactly supported A-valued functions on X. For any proper map of topological spaces $X \longrightarrow Y$, the pull-back map $A(Y) \longrightarrow A(X)$ is defined. For any étale map (local homeomorphism) of topological spaces $X \longrightarrow Y$, the push-forward map $A(X) \longrightarrow A(Y)$ is defined.

For any topological spaces X and Y and an abelian group A, there is a natural isomorphism $A(X \times Y) \simeq A(X)(Y)$. For any topological space X, a commutative ring k, and a k-module A, there is a natural isomorphism $A(X) \simeq A \otimes_k k(X)$.

For a topological space X and an abelian group A, denote by $A[[X]]$ the abelian group of finitely-additive compactly supported A-valued measures defined on the open-closed subsets of X. For any map of topological spaces $X \longrightarrow Y$, the push-forward map $A[[X]] \longrightarrow A[[Y]]$ is defined.

For any compact topological space X, a commutative ring k, and a k-module A, there is a natural isomorphism $A[[X]] \simeq \mathrm{Hom}_k(k(X), A)$. When the space X is discrete, we use the notation $A[X]$ instead of $A[[X]]$.

E.1.2   Let H be a profinite group and k be a commutative ring. Then the module of locally constant functions $k(H)$ has a natural structure of coring over k where the left and right actions of k coincide (or a "coalgebra over k" in the more traditional terminology of [23]). This coring structure, which we denote by $\mathcal{C} = \mathcal{C}_k(H)$, is defined as follows. The counit map $k(H) \longrightarrow k$ is the evaluation at the unit element $e \in H$. The comultiplication map $k(H) \longrightarrow k(H) \otimes_k k(H)$ is provided by the pull-back map $k(H) \longrightarrow k(H \times H)$ induced by the multiplication map $H \times H \longrightarrow H$ together with the identification $k(H \times H) \simeq k(H) \otimes_k k(H)$.

Let G be a topological group and $H \subset G$ be an open profinite subgroup. Then the module of locally constant compactly supported functions $k(G)$ has a natural structure of semialgebra over the coring $\mathcal{C}_k(H)$. This semialgebra structure, which we denote by $\mathbf{S} = \mathbf{S}_k(G, H)$, is defined as follows. The bicoaction map

$$k(G) \longrightarrow k(H) \otimes_k k(G) \otimes_k k(H) \simeq k(H \times G \times H)$$

is the pull-back map induced by the multiplication map $H \times G \times H \longrightarrow G$. The semiunit map $k(H) \longrightarrow k(G)$ is the push-forward map induced by the injection $H \longrightarrow G$. Denote by $G \times^H G$ the quotient space of the Cartesian square $G \times G$ by the equivalence relation $(g'h, g'') \sim (g', hg'')$ for g', $g'' \in G$ and $h \in H$. The pull-back map $k(G \times^H G) \longrightarrow k(G \times G)$ induced by the natural surjection $G \times G \longrightarrow G \times^H G$ identifies $k(G \times^H G)$ with the cotensor product

$$k(G) \,\square_{\mathcal{C}(H)}\, k(G) \subset k(G) \otimes_k k(G) \simeq k(G \times G).$$

The semimultiplication map $k(G) \,\square_{\mathcal{C}}\, k(G) \longrightarrow k(G)$ is the push-forward map induced by the multiplication map $G \times^H G \longrightarrow G$.

The involutions $k(H) \longrightarrow k(H)$ and $k(G) \longrightarrow k(G)$ induced by the inverse element maps $H \longrightarrow H$ and $G \longrightarrow G$ provide the isomorphism of semialgebras $\mathbf{S}_k(G, H)^{\mathrm{op}} \simeq \mathbf{S}_k(G, H)$ compatible with the isomorphism of corings $\mathcal{C}_k(H)^{\mathrm{op}} \simeq \mathcal{C}_k(H)$ over k (in the notation of C.2.7).

Now let H_1, $H_2 \subset G$ be two open profinite subgroups of G. Then the k-module $k(G) = \mathbf{S}_k(G, H_1)$ has a natural left $\mathbf{S}_k(G, H_1)$-semimodule structure and at the same time $k(G) = \mathbf{S}_k(G, H_2)$ has a natural right $\mathbf{S}_k(G, H_2)$-semimodule structure. Obviously, these two semimodule structures commute; we denote this bisemimodule structure on $k(G)$ by $\mathbf{S}_k(G, H_1, H_2)$. For any three open profinite subgroups H_1, H_2, $H_3 \subset G$, there is a natural isomorphism $\mathbf{S}_k(G, H_1, H_2) \lozenge_{S_k(G, H_2)} \mathbf{S}_k(G, H_2, H_3) \simeq \mathbf{S}_k(G, H_2) \lozenge_{S_k(G, H_2)} \mathbf{S}_k(G, H_2) \simeq \mathbf{S}_k(G, H_2) \simeq \mathbf{S}_k(G, H_1, H_3)$; this is an isomorphism of $\mathbf{S}_k(G, H_1)$-$\mathbf{S}_k(G, H_3)$-bisemimodules.

One can check that $k(H)$ is a projective k-module. Clearly, $\mathbf{S}_k(G, H)$ is a coprojective left and right $\mathcal{C}_k(H)$-comodule. So the pair

$$(\mathbf{S}_k(G, H_1, H_2), \mathbf{S}_k(G, H_2, H_1))$$

is a left and right semiprojective Morita equivalence between the semialgebras $\mathbf{S}_k(G, H_1)$ and $\mathbf{S}_k(G, H_2)$.

E.1.3 The semialgebra $\mathbf{S}_k(G, H)$ can be also obtained by the construction of 10.2.1.

Denote by $k[H]$ and $k[G]$ the group k-algebras of the groups H and G considered as groups without any topology. There is a pairing $\phi \colon \mathcal{C}_k(H) \otimes_k k[H] \longrightarrow k$ satisfying the conditions of 10.1.2, given by the formula $(c, h) \longmapsto c(h^{-1})$ for any $c \in k(H)$ and $h \in H$. The induced functor $\Delta_\phi \colon \mathsf{comod\text{-}}\mathcal{C}_k(H) \longrightarrow \mathsf{mod\text{-}}k[H]$ is fully faithful; its image is described as follows.

A module M over a topological group G is called *smooth* (discrete), if the action map $G \times M \longrightarrow M$ is continuous with respect to the discrete topology of M; equivalently, M is smooth if the stabilizer of every its element is an open subgroup in G. The functor Δ_ϕ identifies the category of right $\mathcal{C}_k(H)$-comodules with the category of smooth H-modules over k.

The tensor product $\mathcal{C}_k(H) \otimes_{k[H]} k[G]$ is a smooth G-module with respect to the action of G by right multiplications, so it becomes a semialgebra over $\mathcal{C}_k(H)$. This semialgebra can be identified with $\mathbf{S}_k(G, H)$ by the formula $(c \otimes g) \longmapsto (g' \longmapsto c(g'g^{-1}))$, where a locally constant function $c \colon H \longrightarrow k$ is presumed to be extended to G by zero. By the result of 10.2.2, the category of right $\mathbf{S}_k(G, H)$-semimodules is isomorphic to the category of smooth G-modules over k.

One also obtains the following description of the category of left $\mathbf{S}_k(G, H)$-semicontramodules. For a topological group G and a commutative ring k, a *G-contramodule over k* is a k-module P endowed with a k-linear map $P[[G]] \longrightarrow P$ satisfying the following conditions. First, the point measure supported in the unit element $e \in G$ and corresponding to an element $p \in P$ should map to the element p. Second, the composition $P[[G \times G]] \longrightarrow P[[G]][[G]] \longrightarrow P$ of the natural map

$$P[[G \times G]] \longrightarrow P[[G]][[G]]$$

and the iterated contraaction map $P[[G]][[G]] \longrightarrow P[[G]] \longrightarrow P$ should be equal to the composition $P[[G \times G]] \longrightarrow P[[G]] \longrightarrow P$ of the push-forward map $P[[G \times G]] \longrightarrow P[[G]]$ induced by the multiplication map $G \times G \longrightarrow G$ and the contraaction map $P[[G]] \longrightarrow P$. The images of the point measures under the contraaction map define the forgetful functor from the category of G-contramodules over k to the category of (nontopological) G-modules over k.

The category of left $\mathbf{S}_k(G, H)$-semicontramodules is isomorphic to the category of G-contramodules over k. Indeed, the result of 10.2.2 describes left $\mathbf{S}_k(G, H)$-semicontramodules as vector spaces endowed with compatible structures of a left $\mathcal{C}_k(H)$-contramodule and a left $k[G]$-module, or equivalently, a left H-contramodule and a left G-module over k. One checks that such a pair of compatible structures can be extended to a G-contramodule structure in a unique way.

E.1.4 For any smooth G-module M over k and any k-module E there is a natural G-contramodule structure on the space of k-linear maps $\mathrm{Hom}_k(M, E)$. The contraaction map $\mathrm{Hom}_k(M, E)[[G]] \longrightarrow \mathrm{Hom}_k(M, E)$ is constructed as the projective limit over all open subgroups $U \subset G$ of the compositions

$$\mathrm{Hom}_k(M, E)[[G]] \longrightarrow \mathrm{Hom}_k(M, E)[G/U] \longrightarrow \mathrm{Hom}_k(M^U, E)$$

of the maps $\mathrm{Hom}_k(M, E)[[G]] \longrightarrow \mathrm{Hom}_k(M, E)[G/U]$ induced by the surjections $G \longrightarrow G/U$ and the maps $\mathrm{Hom}_k(M, E)[G/U] \longrightarrow \mathrm{Hom}_k(M^U, E)$ induced by the action maps $G/U \times M^U \longrightarrow M$, where M^U denotes the k-submodule of U-invariants in M and G/U is the set of all left cosets of G modulo U.

More generally, let G_1 and G_2 be topological groups. Then for any smooth G_1-module M over k and G_2-contramodule P over k there is a natural $G_1 \times G_2$-contramodule structure on $\mathrm{Hom}_k(M, P)$ with the contraaction map

$$\mathrm{Hom}_k(M, P)[[G_1 \times G_2]] \longrightarrow \mathrm{Hom}_k(M, P)$$

defined as either of the compositions

$$\mathrm{Hom}_k(M, P)[[G_1 \times G_2]] \longrightarrow \mathrm{Hom}_k(M, P)[[G_1]][[G_2]]$$
$$\longrightarrow \mathrm{Hom}_k(M, P)[[G_2]] \longrightarrow \mathrm{Hom}_k(M, P)$$

or

$$\mathrm{Hom}_k(M, P)[[G_1 \times G_2]] \longrightarrow \mathrm{Hom}_k(M, P)[[G_2]][[G_1]]$$
$$\longrightarrow \mathrm{Hom}_k(M, P)[[G_1]] \longrightarrow \mathrm{Hom}_k(M, P),$$

where the G_1-contraaction map $\mathrm{Hom}_k(M, P)[[G_1]] \longrightarrow \mathrm{Hom}_k(M, P)$ is defined above and the G_2-contraaction map $\mathrm{Hom}_k(M, P)[[G_2]] \longrightarrow \mathrm{Hom}_k(M, P)$ is constructed as the composition

$$\mathrm{Hom}_k(M, P)[[G_2]] \longrightarrow \mathrm{Hom}_k(M, P[[G_2]]) \longrightarrow \mathrm{Hom}_k(M, P).$$

Hence for any smooth G-module M over k and any G-contramodule P over k there is a natural G-contramodule structure on $\mathrm{Hom}_k(M, P)$ induced by the diagonal map of topological groups $G \longrightarrow G \times G$.

E.2 Semiinvariants and semicontrainvariants

E.2.1 Let G be a topological group and $H \subset G$ be an open profinite subgroup. For a smooth H-module M over k, let $\mathrm{Ind}_H^G M$ denote the induced G-module $k[G] \otimes_{k[H]} M$. For any smooth G-module N over k we will construct a pair of maps

$$(\mathrm{Ind}_H^G N)^H \rightrightarrows N^H,$$

where the superindex H denotes the k-submodule of H-invariants. Namely, the first map $(\mathrm{Ind}_H^G N)^H \longrightarrow N^H$ is obtained by applying the functor of H-invariants to the map $(\mathrm{Ind}_H^G N) \longrightarrow N$ given by the formula $g \otimes n \longmapsto g(n)$. The second map $(\mathrm{Ind}_H^G N)^H \longrightarrow N^H$ only depends on the H-module structure on N.

To define this second map, identify the induced representation $\operatorname{Ind}_H^G N$ with the k-module of all compactly supported functions $G \longrightarrow N$ transforming the right action of H in G into the action of H in N; this identification assigns to a function $g \longmapsto n_g$ the formal linear combination $\sum_{g \in G/H} g \otimes n_g$, where G/H denotes the set of all left cosets of G modulo H. An H-invariant element of $\operatorname{Ind}_H^G N$ is then represented by a compactly supported function $G \longrightarrow N$ denoted by $g \longmapsto n_g$ and satisfying the equations $n_{gh} = h^{-1}(n_g)$ and $n_{hg} = n_g$ for $g \in G$, $h \in H$. The second map $(\operatorname{Ind}_H^G N)^H \longrightarrow N^H$ sends a function $g \longmapsto n_g$ to the element $\sum_{g \in H \backslash G} n_g \in N$, where $H \backslash G$ denotes the set of all right cosets of G modulo H.

The cokernel $N_{G,H}$ of this pair of maps $(\operatorname{Ind}_H^G N)^H \rightrightarrows N^H$ is called the module of (G, H)-*semiinvariants* of a smooth G-module N. The (G, H)-semiinvariants are a mixture of H-invariants and coinvariants along G relative to H.

E.2.2 For an H-contramodule Q over k, let $\operatorname{Coind}_H^G(Q)$ denote the coinduced G-contramodule $\operatorname{Hom}_{k[H]}(k[G], Q) \simeq \operatorname{Cohom}_{\mathfrak{C}_k(H)}(\mathfrak{S}_k(G, H), Q)$. For any G-contramodule P over k we will construct a pair of maps

$$P_H \rightrightarrows \operatorname{Coind}_H^G(P)_H,$$

where the subindex H denotes the k-module of H-coinvariants, i.e., the maximal quotient H-contramodule with the trivial contraaction. Namely, the first map $P_H \longrightarrow \operatorname{Coind}_H^G(P)_H$ is obtained by applying the functor of H-coinvariants to the (semicontraaction) map $P \longrightarrow \operatorname{Coind}_H^G(P)$ given by the formula $p \longmapsto (g \longmapsto g(p))$.

The second map $P_H \longrightarrow \operatorname{Coind}_H^G(P)_H$ only depends on the H-contramodule structure on P. It is given by the formula $p \longmapsto \sum_{g \in G/H} g * p$ for $p \in P$, where $g*p \colon k[G] \longrightarrow P$ is the $k[H]$-linear map defined by the rules $hg \longmapsto h(p)$ for $h \in H$ and $g' \longmapsto 0$ for all $g' \in G$ not belonging to the right coset Hg. Clearly, the infinite sum over G/H converges element-wise on $k[G]$ for any choice of representatives of the left cosets; its image in the module of coinvariants does not depend on this choice and is determined by the image of p in P_H.

The kernel $P^{G,H}$ of this pair of maps $P_H \rightrightarrows \operatorname{Coind}_H^G(P)_H$ is called the module of (G, H)-*semicontrainvariants* of a G-contramodule P. The (G, H)-semicontra-invariants are a mixture of H-coinvariants and invariants along G relative to H.

E.2.3 Denote by $\chi \colon G \longrightarrow \mathbb{Q}^*$ the modular character of G, i.e., the character with which G acts by left shifts on the one-dimensional $\mathbb{Q}$-vector space of $\mathbb{Q}_\chi$ of right invariant $\mathbb{Q}$-valued measures defined on the open compact subsets of G. Equivalently, $\chi(g)$ is equal to the ratio of the number of left cosets contained in the double coset HgH to the number of right cosets contained in it. Whenever the commutative ring k contains $\mathbb{Q}$, there is a natural isomorphism

$$N_{G,H} \simeq (N \otimes_{\mathbb{Q}} \mathbb{Q}_\chi)_G$$

for any smooth G-module N over k, where the subindex G denotes the k-module of G-coinvariants.

Indeed, the composition $M^H \longrightarrow M \longrightarrow M_H$ is an isomorphism for any smooth H-module M over k, so in particular there are isomorphisms $N^H \simeq N_H$ and $(\mathrm{Ind}_H^G N)^H \longrightarrow (\mathrm{Ind}_H^G)_H$. These isomorphisms transform the above pair of maps $(\mathrm{Ind}_H^G N)^H \rightrightarrows N^H$ into the pair of maps $(\mathrm{Ind}_H^G N)_H \rightrightarrows N_H$ given by the formulas $g \otimes n \longmapsto g(n)$ and $g \otimes n \longmapsto \chi^{-1}(g)n$.

E.2.4 For any G-contramodule P over k denote by P^G the k-module of G-invariants in P defined as the submodule of all $p \in P$ such that for any measure $m \in k[[G]]$ the image of the measure $pm \in P[[G]]$ under the contraaction map $P[[G]] \longrightarrow P$ is equal to the value $m(G)$ of m at G.

Assuming that the commutative ring k contains $\mathbb{Q}$, the composition $Q^H \longrightarrow Q \longrightarrow Q_H$ is an isomorphism for any H-contramodule Q over k, as one can show using the action of the Haar measure of the profinite group H in the contramodule Q. One can also use the Haar measure to check that Q^H is the maximal subcontramodule of Q with the trivial contraaction of H, and it follows that P^G is the maximal subcontramodule of P with the trivial contraaction of G, under our assumption. Finally, when $k \supset \mathbb{Q}$ there is a natural isomorphism

$$P^{G,H} \simeq \mathrm{Hom}_{\mathbb{Q}}(\mathbb{Q}_\chi, P)^G.$$

E.2.5 Let N be a left $\mathbf{S}_k(G, H)$-semimodule and M be a right $\mathbf{S}_k(G, H)$-semimodule; in other words, N and M are smooth G-modules over k. Then there is a natural isomorphism

$$N \lozenge_{\mathbf{S}_k(G,H)} M \simeq (N \otimes_k M)_{G,H},$$

where $N \otimes_k M$ is considered as a smooth G-module over k. Here the semitensor product is well defined by Proposition 1.2.5(f).

Indeed, there is an obvious isomorphism $N \square_{\mathcal{C}_k(H)} M \simeq (N \otimes_k M)^H$. The k-module $N\square_{\mathcal{C}_k(H)} \mathbf{S}_k(G, H)\square_{\mathcal{C}_k(H)} M$ can be identified with the module of locally constant compactly supported functions $f \colon G \longrightarrow N\otimes_k M$ satisfying the equations $f(hg) = hf(g)$ and $f(gh) = f(g)h$ for $g \in G$, $h \in H$, where $(g, a) \longmapsto ga$ and $(a, g) \longmapsto ag$ denote the actions of G in $N \otimes_k M$ induced by the actions in N and M, respectively. At the same time, the k-module $(\mathrm{Ind}_H^G(N \otimes_k M))^H$ can be identified with the module of locally constant compactly supported functions $f' \colon G \longrightarrow N \otimes_k M$ satisfying the equations $f(hg) = f(g)$ and $f(gh) = h^{-1}f(g)h$. The formula $f'(g) = g^{-1}f(g)$ defines an isomorphism between these two k-modules transforming the pair of maps whose cokernel is $N \lozenge_{\mathbf{S}_k(G,H)} M$ into the pair of maps whose cokernel is $(N \otimes_k M)_{G,H}$.

E.2.6 Let M be a left $\mathbf{S}_k(G, H)$-semimodule and P be a left $\mathbf{S}_k(G, H)$-semicontramodule. Then there is a natural isomorphism

$$\mathrm{SemiHom}_{\mathbf{S}_k(G,H)}(M, P) \simeq \mathrm{Hom}_k(M, P)^{G,H},$$

where $\mathrm{Hom}_k(M, P)$ is considered as a G-contramodule over k and the semihomomorphism module is well defined by Proposition 3.2.5(j).

Indeed, the quotient modules $\mathrm{Cohom}_{\mathcal{C}_k(H)}(M, P)$ and $\mathrm{Hom}_k(M, P)_H$ of the k-module $\mathrm{Hom}_k(M, P)$ coincide. There are two commuting contraactions of G in $\mathrm{Hom}_k(M, P)$ induced by the smooth action in M and the contraaction in P; denote these contraaction maps by π_M and π_P, and the corresponding actions of G in $\mathrm{Hom}_k(M, P)$ by $(g, x) \longmapsto g_M(x)$ and $(g, x) \longmapsto g_P(x)$.

The k-module $\mathrm{Cohom}_{\mathcal{C}_k(H)}(\mathbf{S}_k(G, H) \square_{\mathcal{C}_k(H)} M, P)$ can be identified with the quotient module of the module of all finitely-additive measures defined on compact open subsets of G and taking values in $\mathrm{Hom}_k(M, P)$ by the submodule generated by measures of the form

$$ U \longmapsto \pi_M(W \mapsto \mu(U \times W^{-1})) - \mu(\{(g, h) \mid gh \in U\}) $$

and

$$ U \longmapsto \pi_P(W \mapsto \nu(U \times W)) - \nu(\{(h, g) \mid hg \in U\}), $$

where μ and ν are finitely-additive measures defined on compact open subsets of $G \times H$ and $H \times G$, respectively, and taking values in $\mathrm{Hom}_k(M, P)$. Here U denotes a compact open subset of G and W is an open-closed subset of H; by W^{-1} we denote the (pre)image of W under the inverse element map.

At the same time, the k-module $\mathrm{Coind}_H^G(\mathrm{Hom}_k(M, P))_H$ can be identified with the quotient module of the same module of measures on G by the submodule generated by measures of the form

$$ U \longmapsto \mu'(U \times H) - \mu'(\{(g, h) \mid gh \in U\}) $$

and

$$ U \longmapsto \pi_P(W_2 \mapsto \pi_M(W_1 \mapsto \nu'((W_1 \cap W_2) \times U))) - \nu'(\{(h, g) \mid hg \in U\}), $$

where μ' and ν' are finitely-additive measures defined on compact open subsets of $G \times H$ and $H \times G$, respectively, and taking values in $\mathrm{Hom}_k(M, P)$. Here W_1 and W_2 denote open-closed subsets of H. The formulas

$$ m'(U) = \pi_M(V \longmapsto m(U \cap V^{-1})) \quad \text{and} \quad m(U) = \pi_M(V \longmapsto m'(U \cap V)), $$

where V denotes an open-closed subset of G, define an isomorphism between these two quotient spaces of measures.

The pair of maps

$$ \mathrm{Cohom}_{\mathcal{C}_k(H)}(M, P) \rightrightarrows \mathrm{Cohom}_{\mathcal{C}_k(H)}(\mathbf{S}_k(G, H) \square_{\mathcal{C}_k(H)} M, P) $$

whose kernel is $\mathrm{SemiHom}_{\mathbf{S}_k(G, H)}(M, P)$ is given by the rules

$$ x \longmapsto \sum\nolimits_{g \in G/H} g_M^{-1}(x)\delta_g \quad \text{and} \quad x \longmapsto \sum\nolimits_{g \in H \backslash G} g_P(x)\delta_g, $$

where $y\delta_g$ denotes the $\mathrm{Hom}_k(M, P)$-valued measure defined on compact open subsets of G that is supported in the point $g \in G$ and corresponds to an element

$y \in \operatorname{Hom}_k(M, P)$. The pair of maps $\operatorname{Hom}_k(M, P)_H \rightrightarrows \operatorname{Coind}_H^G(\operatorname{Hom}_k(M, P))_H$ whose kernel is $\operatorname{Hom}_k(M, P)^{G,H}$ is given by the rules $x \longmapsto \sum_{g \in G/H} x \delta_g$ and $x \longmapsto \sum_{g \in H \backslash G} g_P g_M(x) \delta_g$. The above isomorphism between two quotient spaces of measures transforms one of these two pairs of maps into the other.

E.2.7   Let $H_1 \subset H_2$ be two open profinite subgroups of a topological group G and N be a smooth G-module over k. Then, at least, in the following two situations there is a natural isomorphism $N_{G,H_1} \simeq N_{G,H_2}$: when N as a G-module is induced from a smooth H_1-module over k, and when N as an H_2-module over k is coinduced from a module over the trivial subgroup $\{e\}$ (i.e., N is a coinduced $\mathfrak{C}_k(H_2)$-co-module).

These isomorphisms are constructed as follows. In the first case, one shows that the triple semitensor product $N \lozenge_{\mathbf{S}_k(G,H_1)} \mathbf{S}_k(G, H_1, H_2) \lozenge_{\mathbf{S}_k(G,H_2)} k$ is associative in the sense that the conclusion of Proposition 1.4.4 applies to it.

In the second case, one shows that the triple semitensor product $N \lozenge_{\mathbf{S}_k(G,H_2)} \mathbf{S}_k(G, H_2, H_1) \lozenge_{\mathbf{S}_k(G,H_1)} k$ is associative in the similar sense. In both cases, the argument is analogous to that of Proposition 1.4.4.

E.2.8   Let $H_1 \subset H_2$ be two open profinite subgroups of G and P be a G-contra-module over k. Then, at least, in the following two situations there is a natural isomorphism $P^{G,H_1} \simeq P^{G,H_2}$: when P is coinduced from an H_1-contramodule over k, and when P as an H_2-contramodule over k is induced from a contramodule over the trivial subgroup $\{e\}$ (i.e., P is an induced $\mathfrak{C}_k(H_2)$-contramodule).

E.3   SemiTor and SemiExt

E.3.1   Assume that the ring k has a finite weak homological dimension. Then for any complexes of smooth G-modules $N^\bullet$ and $M^\bullet$ over k the object $\operatorname{SemiTor}^{\mathbf{S}_k(G,H)}(N^\bullet, M^\bullet)$ in the derived category of k-modules is defined. Furthermore, whenever either $N^\bullet$ or $M^\bullet$ is a complex of k-flat smooth G-modules over k there is a natural isomorphism

$$\operatorname{SemiTor}^{\mathbf{S}_k(G,H)}(N^\bullet, M^\bullet) \simeq \operatorname{SemiTor}^{\mathbf{S}_k(G,H)}(N^\bullet \otimes_k M^\bullet, k)$$

in $\mathsf{D}(k\text{-mod})$.

Indeed, assume that $M^\bullet$ is a complex of flat k-modules. If $N^\bullet$ is a semiflat complex of right $\mathbf{S}_k(G, H)$-semimodules, then so is the tensor product $N^\bullet \otimes_k M^\bullet$, since $(N^\bullet \otimes_k M^\bullet) \lozenge_{\mathbf{S}_k(G,H)} L^\bullet \simeq (N^\bullet \otimes_k M^\bullet \otimes_k L^\bullet)_{G,H} \simeq N^\bullet \lozenge_{\mathbf{S}_k(G,H)} (M^\bullet \otimes_k L^\bullet)$ for any complex of smooth G-modules $L^\bullet$ over k, and the complex of left $\mathfrak{C}_k(H)$-co-modules $M^\bullet \otimes_k L^\bullet$ is coacyclic whenever the complex of left $\mathfrak{C}_k(H)$-comodules $L^\bullet$ is. It remains to use the natural isomorphism $N^\bullet \lozenge_{\mathbf{S}_k(G,H)} M^\bullet \simeq (N^\bullet \otimes_k M^\bullet)_{G,H}$.

E.3.2   Let $k' \to k$ be a morphism of commutative rings of finite weak homological dimension. Then for any complex of smooth G-modules $N^\bullet$ over k the image of the object $\operatorname{SemiTor}^{\mathbf{S}_k(G,H)}(N^\bullet, k)$ under the restriction of scalars functor $\mathsf{D}(k\text{-mod}) \to$

$D(k'\text{-mod})$ is naturally isomorphic to the object $\mathrm{SemiTor}^{\mathcal{S}_{k'}(G,H)}(N^\bullet, k')$. This follows from Corollary 8.3.3(a). In this sense, the object $\mathrm{SemiTor}^{\mathcal{S}_k(G,H)}(N^\bullet, k)$ does not depend essentially on k, but only on G, H, and $N^\bullet$. The cohomology k-modules of the complex

$$\mathrm{SemiTor}^{\mathcal{S}_k(G,H)}(N^\bullet, k)$$

are called the *semi-infinite homology* of G relative to H with coefficients in $N^\bullet$.

Remark. When G is a discrete group, $H = \{e\}$ is the trivial subgroup, and N is a G-module over k, the cohomology of the object $\mathrm{SemiTor}^{\mathcal{S}_k(G,\{e\})}(k, N)$ coincides with the discrete group homology $H_*(G, N)$ and is concentrated in the nonpositive cohomological degrees. When G is a profinite group, $H = G$ is the whole group, and N is a smooth (discrete) G-module over k, the cohomology of $\mathrm{SemiTor}^{\mathcal{S}_k(G,G)}(k, N)$ coincides with the profinite group cohomology $H^*(G, N)$ and is concentrated in the nonnegative cohomological degrees. More generally, when $G = G_1 \times G_2$ is the product of a discrete group G_1 and a profinite group G_2, $H = G_2 \subset G$, k is a field, and $N = N_1 \otimes_k N_2$ is the tensor product of a G_1-module and a smooth G_2-module, the cohomology of $\mathrm{SemiTor}^{\mathcal{S}_k(G,H)}(k, N)$ is isomorphic to the tensor product $H_*(G_1, N_1) \otimes_k H^*(G_2, N_2)$. Applying these observations to the case of a finite group G, one can see that the semi-infinite homology of a topological group G does depend essentially on the choice of an open profinite subgroup $H \subset G$.

E.3.3 Assume that the ring k has a finite homological dimension. Then for any complex of smooth G-modules $M^\bullet$ over k and any complex of G-contramodules $P^\bullet$ over k the object $\mathrm{SemiExt}_{\mathcal{S}_k(G,H)}(M^\bullet, P^\bullet)$ in the derived category of k-modules is defined. Furthermore, whenever either $M^\bullet$ is a complex of projective k-modules or $P^\bullet$ is a complex of injective k-modules there is a natural isomorphism

$$\mathrm{SemiExt}_{\mathcal{S}_k(G,H)}(M^\bullet, P^\bullet) \simeq \mathrm{SemiExt}_{\mathcal{S}_k(G,H)}(k, \mathrm{Hom}_k(M^\bullet, P^\bullet))$$

in $D(k\text{-mod})$.

E.3.4 Let $k' \longrightarrow k$ be a morphism of commutative rings of finite homological dimension. Then for any complex of G-contramodules $P^\bullet$ over k the image of the object $\mathrm{SemiExt}_{\mathcal{S}_k(G,H)}(k, P^\bullet)$ under the restriction of scalars functor $D(k\text{-mod}) \longrightarrow D(k'\text{-mod})$ is naturally isomorphic to the object $\mathrm{SemiExt}_{\mathcal{S}_{k'}(G,H)}(k', P^\bullet)$. The cohomology k-modules of the complex

$$\mathrm{SemiExt}_{\mathcal{S}_k(G,H)}(k, P^\bullet)$$

are called the *semi-infinite cohomology* of G relative to H with coefficients in $P^\bullet$.

E.4 Remarks on the Gaitsgory–Kazhdan construction

This section contains some comments on the papers [44, 45].

E.4.1 Let G be a topological group and k be a field of characteristic 0. Then the category of discrete $G \times G$-modules over k has a tensor category structure with the tensor product of two modules $K' \otimes_G K''$ defined as the module of coinvariants of the action of G in $K' \otimes_k K''$ induced by the action of the second copy of G in K' and the action of the first copy of G in K''. The category of discrete G-modules over k has structures of a left and a right module category over this tensor category; and the functor

$$(N, M) \longmapsto N \otimes_G M = (N \otimes_k M)_G$$

defines a pairing between these two module categories taking values in the category of k-vector spaces. A finitely-additive k-valued measure defined on compact open subsets of G is called *smooth* if it is equal to the product of a locally constant k-valued function on G and a (left or right invariant) Haar measure on G. The $G \times G$-module $\mathcal{A}(G)$ of compactly supported smooth k-valued measures on G is the unit object of the above tensor category.

Notice that the $G \times G$-module $\mathcal{A}(G)$ has a natural k-algebra structure, given by the convolution of measures. However, this algebra has no unit and the category of (left or right) modules over it contains the category of smooth G-modules over k as a proper full subcategory.

E.4.2 For any category C, denote by $\mathsf{Pro}\, C$ and $\mathsf{Ind}\, C$ the categories of pro-objects and ind-objects in C. Let $\mathsf{Set}_{\mathsf{fin}}$ denote the category of finite sets. We will identify the category of compact topological spaces with the category $\mathsf{Pro}\,\mathsf{Set}_{\mathsf{fin}}$, the category of (discrete) sets with the category $\mathsf{Ind}\,\mathsf{Set}_{\mathsf{fin}}$, and the category of topological spaces with a full subcategory of the category $\mathsf{Ind}\,\mathsf{Pro}\,\mathsf{Set}_{\mathsf{fin}}$ formed by the inductive systems of profinite sets and their open embeddings. In addition, we will consider the category $\mathsf{Ind}\,\mathsf{Pro}\,\mathsf{Set}_{\mathsf{fin}}$ as a full subcategory in $\mathsf{Pro}\,\mathsf{Ind}\,\mathsf{Pro}\,\mathsf{Set}_{\mathsf{fin}}$ and the latter category as a full subcategory in $\mathsf{Ind}\,\mathsf{Pro}\,\mathsf{Ind}\,\mathsf{Pro}\,\mathsf{Set}_{\mathsf{fin}}$. All of these categories have symmetric monoidal structures coming from the Cartesian product of sets in $\mathsf{Set}_{\mathsf{fin}}$.

Let $\mathbb{H}$ be a group object in $\mathsf{Pro}\,\mathsf{Ind}\,\mathsf{Pro}\,\mathsf{Set}_{\mathsf{fin}}$ such that $\mathbb{H}$ can be represented by a projective system of topological groups in $\mathsf{Ind}\,\mathsf{Pro}\,\mathsf{Set}_{\mathsf{fin}}$ and open surjective morphisms between them. A representation of $\mathbb{H}$ in k-vect is just a smooth G-module over k, where G is a quotient group object of $\mathbb{H}$ that is a topological group in $\mathsf{Ind}\,\mathsf{Pro}\,\mathsf{Set}_{\mathsf{fin}}$. The category $\mathsf{Rep}_k(\mathbb{H})$ of representations of $\mathbb{H}$ in $\mathsf{Pro}(k\text{-vect})$, defined in [44], is equivalent to the category of pro-objects in the category of representations of $\mathbb{H}$ in k-vect. So the category $\mathsf{Rep}_k(\mathbb{H} \times \mathbb{H})$ has a natural tensor category structure with the unit object $\mathcal{A}(\mathbb{H})$ given by the projective system formed by the modules $\mathcal{A}(G)$ of smooth compactly supported measures on topological groups G that are quotient groups of $\mathbb{H}$ and the push-forward maps between the spaces of measures. The category $\mathsf{Rep}_k(\mathbb{H})$ is a left and a right module category over this tensor category, and there is a pairing between these left and right module categories taking values in $\mathsf{Pro}(k\text{-vect})$.

E.4.3 Let $\mathbb{G}$ be a group object in $\mathsf{Ind}\,\mathsf{Pro}\,\mathsf{Ind}\,\mathsf{Pro}\,\mathsf{Set}_{\mathsf{fin}}$ and let $\mathbb{H} \subset \mathbb{G}$ be a subgroup object which belongs to $\mathsf{Pro}\,\mathsf{Ind}\,\mathsf{Pro}\,\mathsf{Set}_{\mathsf{fin}}$. Assume that the object $\mathbb{G}$ is given by an inductive system of objects $\mathbb{G}_\alpha \in \mathsf{Pro}\,\mathsf{Ind}\,\mathsf{Pro}\,\mathsf{Set}_{\mathsf{fin}}$ and the group object $\mathbb{H}$ is given by a projective system of its quotient group objects $\mathbb{H}/\mathbb{H}^i \in \mathsf{Ind}\,\mathsf{Pro}\,\mathsf{Set}_{\mathsf{fin}}$ satisfying the following conditions. It is convenient to assume that $\mathbb{G}_0 = \mathbb{H} = \mathbb{H}^0$ is the initial object in the inductive system of $\mathbb{G}_\alpha$ and the final object in the projective system of $\mathbb{H}^i$. As the notation suggests, $\mathbb{H}^i$ are normal subgroup objects of $\mathbb{H}$. The group object $\mathbb{H}$ acts in $\mathbb{G}_\alpha$ in a way compatible with the action of $\mathbb{H}$ in $\mathbb{G}$ by right multiplications. The quotient objects $\mathbb{G}_\alpha/\mathbb{H}^i$ are topological spaces in $\mathsf{Ind}\,\mathsf{Pro}\,\mathsf{Set}_{\mathsf{fin}}$. The morphisms $\mathbb{G}_\alpha/\mathbb{H}^i \longrightarrow \mathbb{G}_\beta/\mathbb{H}^i$ are closed embeddings of topological spaces; the morphisms $\mathbb{G}_\alpha/\mathbb{H}^i \longrightarrow \mathbb{G}_\alpha/\mathbb{H}^j$ are principal $\mathbb{H}^i/\mathbb{H}^j$-bundles. Finally, the quotient objects $\mathbb{G}_\alpha/\mathbb{H}$ are compact topological spaces, i.e., belong to $\mathsf{Pro}\,\mathsf{Set}_{\mathsf{fin}} \subset \mathsf{Ind}\,\mathsf{Pro}\,\mathsf{Set}_{\mathsf{fin}}$.

Let $\mathbb{G}'$ be a group object in $\mathsf{Ind}\,\mathsf{Pro}\,\mathsf{Ind}\,\mathsf{Pro}\,\mathsf{Set}_{\mathsf{fin}}$ endowed with a central subgroup object identified with the multiplicative group k^*, which is considered as a discrete topological group, that is a group object in $\mathsf{Ind}\,\mathsf{Set}_{\mathsf{fin}} \subset \mathsf{Pro}\,\mathsf{Ind}\,\mathsf{Pro}\,\mathsf{Set}_{\mathsf{fin}}$. Suppose that the quotient group object $\mathbb{G}'/k^*$ is identified with $\mathbb{G}$ and the central extension $\mathbb{G}' \longrightarrow \mathbb{G}$ is endowed with a splitting over $\mathbb{H}$, i.e., $\mathbb{H}$ is a subgroup object in $\mathbb{G}'$. Moreover, denote by $\mathbb{G}'_\alpha$ of the preimages of $\mathbb{G}_\alpha$ in $\mathbb{G}'$ and assume that the morphisms $\mathbb{G}'_\alpha/\mathbb{H}^i \longrightarrow \mathbb{G}_\alpha/\mathbb{H}^i$ are principal k^*-bundles of topological spaces in $\mathsf{Ind}\,\mathsf{Pro}\,\mathsf{Set}_{\mathsf{fin}}$.

E.4.4 One example of such a group $\mathbb{G}'$ is provided by the canonical central extension $\mathbb{G}^\sim$ of the group $\mathbb{G}$ with the kernel k^*, which is constructed as follows. For each α let us choose i such that $\mathrm{Ad}_{\mathbb{G}^{-1}}(\mathbb{H}^i) \subset \mathbb{H}$ and j such that $\mathrm{Ad}_{\mathbb{G}_\alpha}(\mathbb{H}^j) \subset \mathbb{H}^i$. For any topological group G denote by $\mu(G)$ the one-dimensional vector space of left invariant finitely-additive k-valued Haar measures defined on compact open subsets of G. An element of the topological space $\mathbb{G}^\sim_\alpha/\mathbb{H}$ is a pair consisting of an element of $g \in \mathbb{G}_\alpha/\mathbb{H}$ and an isomorphism

$$\mu(\mathbb{H}^i/\mathrm{Ad}_g(\mathbb{H}^j)) \simeq \mu(\mathbb{H}^i/\mathbb{H}^j).$$

The topology on $\mathbb{G}^\sim_\alpha/\mathbb{H}$ is defined by the condition that the following set of sections of the k^*-torsor $\mathbb{G}^\sim_\alpha/\mathbb{H} \longrightarrow \mathbb{G}_\alpha/\mathbb{H}$ consists of continuous maps. Choose m such that $\mathrm{Ad}_{\mathbb{G}^{-1}_\alpha}(\mathbb{H}^m) \subset \mathbb{H}^j$, a compact open subset U in the quotient group $\mathbb{H}^i/\mathbb{H}^m$, and an element $a \in k^*$; for each $g \in \mathbb{G}_\alpha/\mathbb{H}$ define the isomorphism $\mu(\mathbb{H}^i/\mathrm{Ad}_g(\mathbb{H}^j)) \simeq \mu(\mathbb{H}^i/\mathbb{H}^j)$ so that the left-invariant measure on $\mathbb{H}^i/\mathrm{Ad}_g(\mathbb{H}^j)$ for which the measure of the image of U is equal to 1 corresponds to the left-invariant measure on $\mathbb{H}^i/\mathbb{H}^j$ for which the measure of the image of U is equal to a. The ratio of any two such sections is a locally constant function. Now the object $\mathbb{G}^\sim_\alpha$ of $\mathsf{Pro}\,\mathsf{Ind}\,\mathsf{Pro}\,\mathsf{Set}_{\mathsf{fin}}$ is the fibered product of $\mathbb{G}^\sim_\alpha/\mathbb{H}$ and $\mathbb{G}_\alpha$ over $\mathbb{G}_\alpha/\mathbb{H}$; it is easy to define the group structure on $\mathbb{G}^\sim$ (one should first check that the construction of $\mathbb{G}^\sim_\alpha/\mathbb{H}$ does not depend on the choice of $\mathbb{H}^i$ and $\mathbb{H}^j$).

E.4.5 Let $c' \colon \mathbb{G}' \longrightarrow \mathbb{G}$ be a central extension satisfying the above conditions. Denote by $\mathsf{Rep}_{c'}(\mathbb{G})$ the category of representations of $\mathbb{G}'$ in $\mathsf{Pro}(k\text{-vect})$ in which the central subgroup $k^* \subset \mathbb{G}'$ acts tautologically by automorphisms proportional to the identity, as defined in [44]. Then the forgetful functor $\mathsf{Rep}_{c'}(\mathbb{G}) \longrightarrow \mathsf{Rep}(\mathbb{H})$ admits a right adjoint functor, which can be described as the functor of tensor product over $\mathbb{H}$ with a certain representation of $\mathbb{G}' \times \mathbb{H}$ in $\mathsf{Pro}(k\text{-vect})$. The underlying pro-vector space of this representation, denoted by $\mathcal{C}_{c'}(\mathbb{G}, \mathbb{H})$, is the space of "pro-semimeasures on $\mathbb{G}$ relative to $\mathbb{H}$ on the level c'"; it is given by the projective system formed by the vector spaces

$$k_{c'}(\mathbb{G}_\alpha/\mathbb{H}^i) \otimes_k \mu(\mathbb{H}/\mathbb{H}^i),$$

where the first factor is the space of locally constant compactly supported functions on $\mathbb{G}'_\alpha/\mathbb{H}^i$ which transform by the tautological character under the action of k^*. The morphism in this projective system corresponding to a change of α is the pullback map with respect to a closed embedding, while the morphism corresponding to a change of i is the map of integration along the fibers of a principal bundle. Hence the representation $\mathcal{C}_{c'}(\mathbb{G}, \mathbb{H})$ considered as an object of the tensor category $\mathsf{Rep}(\mathbb{H} \times \mathbb{H})$ is endowed with a structure of coring (with a counit $\mathcal{C}_{c'}(\mathbb{G}, \mathbb{H}) \longrightarrow \mathcal{A}(\mathbb{H})$), and it follows from Theorem 7.4.1 that the category $\mathsf{Rep}_{c'}(\mathbb{G})$ is equivalent to the category of left comodules over $\mathcal{C}_{c'}(\mathbb{G}, \mathbb{H})$ in $\mathsf{Rep}(\mathbb{H})$.

Now let $c'' \colon \mathbb{G}'' \longrightarrow \mathbb{G}$ be the central extension satisfying the same conditions and complementary to c', i.e., the Baer sum $c' + c''$ is identified with minus the canonical central extension $c_0 \colon \mathbb{G}^\sim \longrightarrow \mathbb{G}$. Gaitsgory and Kazhdan noticed that one can extend the right action of $\mathbb{H}$ in $\mathcal{C}_{c'}(\mathbb{G}, \mathbb{H})$ to an action of $\mathbb{G}''$ commuting with the left action of $\mathbb{G}'$, with the central subgroup k^* of $\mathbb{G}''$ acting tautologically. Moreover, in [45] there is a construction of a natural anti-isomorphism of corings

$$\mathcal{C}_{c'}(\mathbb{G}, \mathbb{H}) \simeq \mathcal{C}_{c''}(\mathbb{G}, \mathbb{H})$$

permuting the left and right actions of $\mathbb{G}'$ and $\mathbb{G}''$. Thus the category $\mathsf{Rep}_{c''}(\mathbb{G})$ is equivalent to the category of right comodules over $\mathcal{C}_{c'}(\mathbb{G}, \mathbb{H})$ in $\mathsf{Rep}(\mathbb{H})$.

E.4.6 So there is the functor of cotensor product

$$\square_{\mathcal{C}_{c'}(\mathbb{G}, \mathbb{H})} \colon \mathsf{Rep}_{c''}(\mathbb{G}) \times \mathsf{Rep}_{c'}(\mathbb{G}) \longrightarrow \mathsf{Pro}(k\text{-vect}),$$

which is called "semi-invariants" in [45]. This functor is neither left, nor right exact in general. One can construct its double-sided derived functor in a way analogous to that of Remark 2.7, at least when the set of indices i is countable.

The semiderived category $\mathsf{D}^{\mathsf{si}}(\mathsf{Rep}_{c'}(\mathbb{G}))$ is defined as the quotient category of the homotopy category $\mathsf{Hot}(\mathsf{Rep}_{c'}(\mathbb{G}))$ by the thick subcategory of complexes that are contraacyclic as complexes over the abelian category $\mathsf{Rep}(\mathbb{H})$. Then Lemma 2.7 allows us to define the double-sided derived functor

$$\mathsf{ProCotor}^{\mathcal{C}_{c'}(\mathbb{G}, \mathbb{H})} \colon \mathsf{D}^{\mathsf{si}}(\mathsf{Rep}_{c''}(\mathbb{G})) \times \mathsf{D}^{\mathsf{si}}(\mathsf{Rep}_{c'}(\mathbb{G})) \longrightarrow \mathsf{D}(\mathsf{Pro}(k\text{-vect}))$$

in terms of coflat complexes in $\mathsf{Hot}(\mathsf{Rep}_{c''}(\mathbb{G}))$ and $\mathsf{Hot}(\mathsf{Rep}_{c'}(\mathbb{G}))$. The key step is to construct for any object of $\mathsf{Rep}_{c'}(\mathbb{G})$ a surjective map onto it from an object of $\mathsf{Rep}_{c'}(\mathbb{G})$ that is flat as a representation of $\mathbb{H}$. This construction is dual to that of Lemma 1.3.3; it is based on the fact that any module over a topological group G induced from the trivial module k over a compact open subgroup $H \subset G$ is flat with respect to the tensor product of discrete G-modules over k.

Question. Can the cotensor product $\mathcal{N} \,\square_{\mathfrak{e}_{c'}(\mathbb{G},\mathbb{H})}\, \mathcal{M}$ of an object $\mathcal{N} \in \mathsf{Rep}_{c''}(\mathbb{G})$ and an object $\mathcal{M} \in \mathsf{Rep}_{c'}(\mathbb{G})$ be recovered from the tensor product $\mathcal{N} \otimes_k^{\mathrm{p}} \mathcal{M}$ in the category of pro-vector spaces, considered as a representation of $\mathbb{G}^\sim$ with the diagonal action?

F Algebraic Groupoids with Closed Subgroupoids

To any smooth affine groupoid (M, H) one can associate a coring $\mathcal{C}(H)$ over a ring $A(M)$ and a natural left and right coflat Morita autoequivalence $(\mathcal{E}, \mathcal{E}^\vee)$ of $\mathcal{C}(H)$; it has the form

$$\mathcal{C}(H) \otimes_{A(M)} E \simeq \mathcal{E} \simeq E \otimes_{A(M)} \mathcal{C}(H)$$

and

$$E^\vee \otimes_{A(M)} \mathcal{C}(H) \simeq \mathcal{E}^\vee \simeq \mathcal{C}(H) \otimes_{A(M)} E^\vee,$$

where $(E, E^\vee)$ is a certain pair of mutually inverse invertible $A(M)$-modules. To any groupoid (M, G) containing (M, H) as a closed subgroupoid, one can assign two opposite semialgebras $\mathbf{S}^l(G, H)$ and $\mathbf{S}^r(G, H)$ over $\mathcal{C}(H)$ together with a natural left and right semiflat Morita equivalence $(\mathcal{E}, \mathcal{E}^\vee)$ between them formed by the bisemimodules

$$\mathbf{S}^l(G, H) \,\square_{\mathcal{C}(H)}\, \mathcal{E} \simeq \mathcal{E} \simeq \mathcal{E} \,\square_{\mathcal{C}(H)}\, \mathbf{S}^r(G, H)$$

and

$$\mathcal{E}^\vee \,\square_{\mathcal{C}(H)}\, \mathbf{S}^l(G, H) \simeq \mathcal{E}^\vee \simeq \mathbf{S}^r(G, H) \,\square_{\mathcal{C}(H)}\, \mathcal{E}^\vee.$$

To obtain these results, we will have to assume the existence of a quotient variety G/H.

In this appendix, by a variety we mean a smooth algebraic variety (smooth separated scheme) over a fixed ground field k of zero characteristic. The structure sheaf of a variety X is denoted by $O = O_X$ and the sheaf of differential top forms by $\Omega = \Omega_X$; for any invertible sheaf L over X, its tensor powers are denoted by L^n for $n \in \mathbb{Z}$.

F.1 Coring associated to affine groupoid

A (*smooth*) *groupoid* (M, G) is a set of data consisting of two varieties M and G, two smooth morphisms $s, t: G \rightrightarrows M$ of *source* and *target*, a *unit* morphism $e: M \longrightarrow G$, a *multiplication* morphism $m: G \times_M G \longrightarrow G$ (where the first factor G in the fibered product $G \times_M G$ maps to M by the morphism s and the second factor G maps to M by the morphism t), and an *inverse element* morphism $i: G \longrightarrow G$. The following equations should be satisfied: first, $se = \mathrm{id}_M = te$ and $m(e \times \mathrm{id}_G) = \mathrm{id}_G = m(\mathrm{id}_G \times e)$ (unity); second, $tm = tp_1$, $sm = sp_2$, and $m(m \times \mathrm{id}_G) = m(\mathrm{id}_G \times m)$, where p_1 and p_2 denote the canonical projections of the fibered product $G \times_M G$ to the first and the second factors, respectively (associativity); third, $si = t$, $ti = s$, $m(i \times \mathrm{id}_G)\Delta_t = es$, and $m(\mathrm{id}_G \times i)\Delta_s = et$, where Δ_s and Δ_t denote the diagonal embeddings of G into the fibered squares of G over M with respect to the morphisms s and t, respectively (inverseness). It follows from these equations that $i^2 = \mathrm{id}_G$.

A groupoid (M, H) is said to be *affine* if M and H are affine varieties. In the sequel (M, H) denotes an affine groupoid; its structure morphisms are denoted by the same letters s, t, e, m, i.

Let $A = A(M) = O(M)$ and $\mathcal{C} = \mathcal{C}(H) = O(H)$ be the rings of functions on M and H, respectively. The maps of source and target s, $t\colon H \rightrightarrows M$ induce two maps of rings $A \rightrightarrows \mathcal{C}$, which endow $\mathcal{C}$ with two structures of A-module; we will consider the A-module structure on $\mathcal{C}$ coming from the morphism t as a left module structure and the A-module structure on $\mathcal{C}$ coming from the morphism s as a right module structure. Then there is a natural isomorphism $O(H \times_M H) \simeq \mathcal{C} \otimes_A \mathcal{C}$, hence the multiplication morphism $m\colon H \times_M H \longrightarrow H$ induces a comultiplication map $\mathcal{C} \longrightarrow \mathcal{C} \otimes_A \mathcal{C}$. Besides, the unit map $e\colon M \longrightarrow H$ induces a counit map $\mathcal{C} \longrightarrow A$. It follows from the associativity and unity equations of the groupoid (M, H) that these comultiplication and counit maps are morphisms of A-A-bimodules satisfying the coassociativity and couinty equations; so $\mathcal{C}$ is a coring over A. Clearly, $\mathcal{C}$ is a coflat left and right A-module.

F.2   Canonical Morita autoequivalence

Consider the invertible sheaf

$$V = V_H = \Omega_H \otimes s^*(\Omega_M^{-1}) \otimes t^*(\Omega_M^{-1})$$

on H. Let q_1 and q_2 denote the canonical projections of the fibered product $H \times_M H$ to the first and the second factors, respectively. Then there are natural isomorphisms

$$q_1^*(V) \simeq m^*(V) \simeq q_2^*(V)$$

of invertible sheaves on $H \times_M H$. Indeed, one has $q_1^*(\Omega_H) \simeq \Omega_{H \times_M H} \otimes q_2^*(\Omega_H^{-1}) \otimes q_2^* t^*(\Omega_M)$ and $m^*(\Omega_H) \simeq \Omega_{H \times_M H} \otimes q_2^*(\Omega_H^{-1}) \otimes q_2^* s^*(\Omega_M)$. Now denote by U the invertible sheaf

$$e^*\Omega_H \otimes \Omega_M^{-2}$$

on M. Applying the functors of inverse image with respect to the morphisms $e \times \mathrm{id}_H$, $\mathrm{id}_H \times e\colon H \rightrightarrows H \times_M H$ to the above isomorphisms, one obtains natural isomorphisms of invertible sheaves

$$t^*U \simeq V \simeq s^*U.$$

Set $\mathcal{E} = V(H)$ and $\mathcal{E}^\vee = V^{-1}(H)$. Then $\mathcal{E}$ and $\mathcal{E}^\vee$ are $\mathcal{C}$-modules, and consequently A-A-bimodules. The pull-back map $V(H) \longrightarrow m^*(V)(H \times_M H)$

with respect to the multiplication morphism m together with the isomorphisms

$$m^*(V)(H \times_M H) \simeq q_1^*(V)(H \times_M H) \simeq \mathcal{E} \otimes_A \mathcal{C}$$

and $m^*(V)(H \times_M H) \simeq q_2^*(V)(H \times_M H) \simeq \mathcal{C} \otimes_A \mathcal{E}$ defines the right and left coactions of $\mathcal{C}$ in $\mathcal{E}$. It follows from the associativity equation of H that these coactions commute; so $\mathcal{E}$ is a $\mathcal{C}$-$\mathcal{C}$-bicomodule. Analogously one defines a $\mathcal{C}$-$\mathcal{C}$-bi-module structure on $\mathcal{E}^\vee$. Set $E = U(M)$ and $E^\vee = U^{-1}(M)$; then there are natural isomorphisms of $\mathcal{C}$-comodules $\mathcal{C} \otimes_A E \simeq \mathcal{E} \simeq E \otimes_A \mathcal{C}$ and $E^\vee \otimes_A \mathcal{C} \simeq \mathcal{E}^\vee \simeq \mathcal{C} \otimes_A E^\vee$. These isomorphisms have the property that two maps $\mathcal{E} \simeq \mathcal{C} \otimes_A E \longrightarrow E$ and $\mathcal{E} \simeq E \otimes_A \mathcal{C} \longrightarrow E$ induced by the counit map $\mathcal{C} \longrightarrow A$ coincide, and analogously for $\mathcal{E}^\vee$. Besides, there are obvious isomorphisms $E \otimes_A E^\vee \simeq A \simeq E^\vee \otimes_A E$.

It follows that $\mathcal{E} \,\square_{\mathcal{C}}\, \mathcal{E}^\vee \simeq (E \otimes_A \mathcal{C}) \,\square_{\mathcal{C}}\, (\mathcal{C} \otimes_A E^\vee) \simeq E \otimes_A \mathcal{C} \otimes_A E^\vee \simeq \mathcal{C}$, and analogously $\mathcal{E}^\vee \,\square_{\mathcal{C}}\, \mathcal{E} \simeq \mathcal{C}$. So the pair $(\mathcal{E}, \mathcal{E}^\vee)$ is a left and right coflat Morita equivalence (see 7.5) between $\mathcal{C}$ and itself. Since the bicomodules $\mathcal{E}$ and $\mathcal{E}^\vee$ can be expressed in the above form in terms of the A-modules E and $E^\vee$, it follows that there are natural isomorphisms of corings

$$E \otimes_A \mathcal{C} \otimes_A E^\vee \simeq \mathcal{C} \simeq E^\vee \otimes_A \mathcal{C} \otimes_A E.$$

F.3 Distributions and generalized sections

Let $X \supset Z$ be a variety with a (smooth) closed subvariety and L be a locally constant sheaf on X. The sheaf L^Z of generalized sections of L, supported in Z and regular along Z, can be defined as the image of L with respect to the d-th right derived functor of the functor assigning to any quasi-coherent sheaf on X its maximal subsheaf supported set-theoretically in Z, where $d = \dim X - \dim Z$. The sheaf L^Z is a quasi-coherent sheaf on X supported set-theoretically in Z.

There is a natural isomorphism $L^Z \simeq L \otimes_{O_X} O_X^Z$. The sheaf Ω_X^Z can be alternatively defined as the direct image of the constant right module Ω_Z over the sheaf of differential operators Diff_Z under the closed embedding $Z \longrightarrow X$ (see [14]); this makes Ω_X^Z not only an O_X-module, but even a Diff_X-module. The sheaf Ω_X^Z is called the sheaf of distributions on X, supported in Z and regular along Z.

Let $g \colon Y \longrightarrow X$ be a morphism of varieties and $Z \subset X$ be a closed subvariety. Assume that the fibered product $Z \times_X Y$ is smooth and $\dim Y - \dim Z \times_X Y = \dim X - \dim Z$ if $Z \times_X Y$ is nonempty. Then there is a natural isomorphism $g^*(L^Z) \simeq (g^*L)^{Z \times_X Y}$ of quasi-coherent sheaves on Y. In particular, there is a natural pull-back map of the modules of global generalized sections

$$g^+ \colon L^Z(X) \longrightarrow (g^*L)^{Z \times_X Y}(Y).$$

Let $h \colon W \longrightarrow X$ be a morphism of varieties and $Z \subset W$ be a closed subvari-ety such that the composition $Z \longrightarrow W \longrightarrow X$ is also a closed embedding. Then there is a natural push-forward map $h_*(\Omega_W^Z) \longrightarrow \Omega_X^Z$ of quasi-coherent sheaves on X. Consequently, for any locally constant sheaves L' on X and L'' on W endowed with an isomorphism $L'' \otimes \Omega_W^{-1} \simeq h^*(L' \otimes \Omega_X^{-1})$ there is a push-forward map $h_*(L''^Z) \longrightarrow L'^Z$. In particular, there is a natural push-forward map of the modules of generalized sections

$$h_+ : L''^Z(W) \longrightarrow L'^Z(X).$$

Let $g : Y \longrightarrow X$ and $h : W \longrightarrow X$ be morphisms of varieties satisfying the above conditions with respect to a closed subvariety $Z \subset W$. Assume also that the fibered product $W \times_X Y$ is smooth and $\dim W \times_X Y + \dim X = \dim W + \dim Y$ if $W \times_X Y$ is nonempty. Set $\tilde{g} = \mathrm{id}_W \times g : W \times_X Y \longrightarrow W$ and $\tilde{h} = h \times \mathrm{id}_Y : W \times_X Y \longrightarrow Y$. Then there is a natural isomorphism of invertible sheaves

$$\tilde{g}^* \Omega_W \otimes \Omega_{W \times_X Y}^{-1} \simeq \tilde{h}^*(g^* \Omega_X \otimes \Omega_Y^{-1})$$

on $W \times_X Y$, as one can see by decomposing g and h into closed embeddings followed by smooth morphisms. For any quasi-coherent sheaf F on W supported set-theoretically in Z there is a natural isomorphism $\tilde{h}_* \tilde{g}^* F \simeq g^* h_* F$ of quasi-coherent sheaves on $W \times_X Y$. The push-forward maps of the sheaves of distributions with respect to the morphisms h and $\tilde{h}$ are compatible with the pull-back isomorphisms with respect to g and $\tilde{g}$ in the obvious sense.

The sheaf of generalized sections L^Z of a locally constant sheaf L on a variety $X \supset Z$ is endowed with a natural increasing filtration by coherent subsheaves $F_n L^Z$ of generalized sections of order no greater than n. This filtration is preserved by all the above natural isomorphisms and maps. The associated graded sheaf $\mathrm{gr}_F \Omega_X^Z$ is the direct image under our closed embedding $\iota : Z \longrightarrow X$ of a sheaf of O_Z-modules naturally isomorphic to the tensor product $\Omega_Z \otimes_{O_Z} \mathrm{Sym}_{O_Z} N_{Z,X}$, where $N_{Z,X}$ is the normal bundle to Z in X and Sym denotes the symmetric algebra.

In particular, there is a natural isomorphism $\lambda_0 : \iota_* \Omega_Z \longrightarrow F_0 \Omega_X^Z$. Furthermore, there is a natural map of sheaves of k-vector spaces $\lambda_1 : \iota_* \Omega_Z \otimes_{O_X} T_X \longrightarrow F_1 \Omega_X^Z$ which induces the isomorphism $\iota_*(\Omega_Z \otimes_{O_Z} N_{Z,X}) \simeq F_1 \Omega_X^Z / F_0 \Omega_X^Z$, where $T = T_X$ denotes the tangent bundle of X. The map λ_1 satisfies the equation

$$\lambda_1(f\omega \otimes v) = \lambda_1(\omega \otimes fv) = f\lambda_1(\omega \otimes v) - \lambda_0(v(f)\omega)$$

for local sections $f \in O_X$, $\omega \in \Omega_Z$, and $v \in T_X$, where $(v, f) \longmapsto v(f)$ denotes the action of vector fields in functions.

F.4   Lie algebroid of a groupoid

A *Lie algebroid* $\mathfrak{g}$ over a commutative ring A is an A-module endowed with a Lie algebra structure and a Lie action of $\mathfrak{g}$ by derivations of A satisfying the equations $[x, ay] = a[x, y] + x(a)y$ and $(ax)(b) = a(x(b))$ for a, $b \in A$, x, $y \in \mathfrak{g}$. The *enveloping algebra* $U_A(\mathfrak{g})$ of a Lie algebroid $\mathfrak{g}$ over A is generated by A and $\mathfrak{g}$ with the relations $a \cdot b = ab$, $a \cdot x = ax$, $x \cdot a = ax + x(a)$, and $x \cdot y - y \cdot x = [x, y]$, where $(u, v) \longmapsto u \cdot v$ denotes the multiplication in $U_A(\mathfrak{g})$. The algebra $U_A(\mathfrak{g})$ is endowed with a natural increasing filtration $F_n U_A(\mathfrak{g})$ defined by the rules $F_0 U_A(\mathfrak{g}) = \operatorname{im} A$, $F_1 U_A(\mathfrak{g}) = \operatorname{im} A + \operatorname{im} \mathfrak{g}$, and $F_n U_A(\mathfrak{g}) = F_1 U_A(\mathfrak{g})^n$ for $n \geqslant 1$. When $\mathfrak{g}$ is a flat A-module, the associated graded algebra $\operatorname{gr}_F U_A(\mathfrak{g})$ is isomorphic to the symmetric algebra $\operatorname{Sym}_A(\mathfrak{g})$ of the A-module $\mathfrak{g}$.

Let (M, G) be a groupoid with an affine base variety M; set $A = A(M) = O(M)$. Then the A-module $\mathfrak{g} = N_{e(M), G}(M)$ has a natural Lie algebroid structure. To define the action of $\mathfrak{g}$ in A, consider the A-module $(e^* T_G)(M)$ of vector fields on $e(M)$ tangent to G. There are natural push-forward morphisms s_+, $t_+ \colon (e^* T_G)(M) \rightrightarrows T(M)$. Identify $\mathfrak{g}$ with the kernel of the morphism t_+; then the action of $\mathfrak{g}$ in A is defined in terms of the map $s_+ \colon \mathfrak{g} \longrightarrow T(M)$. To define the Lie algebra structure on $\mathfrak{g}$, we will embed $\mathfrak{g}$ into a certain module of generalized sections on G, supported in $e(M)$ and regular along $e(M)$.

Set

$$K^l(G) = (t^* \Omega_M^{-1} \otimes \Omega_G)^{e(M)}(G);$$

this module of generalized sections is endowed with a natural filtration F. The $O(G)$-module structure on $K^l(G)$ induces an A-A-bimodule structure; as in F.1, we consider the A-module structure coming from the morphism t as a left A-module structure and the A-module structure coming from the morphism s as a right A-module structure. There is a natural isomorphism of A-A-bimodules $A \simeq F_0 K^l(G)$. Define a k-linear map of sheaves

$$e_* N_{e(M), G} \longrightarrow (t^* \Omega_M^{-1} \otimes \Omega_G)^{e(M)}$$

locally by the formula $v \longmapsto t^* \omega^{-1} \otimes \lambda_1(\omega \otimes v)$, where v is a local vector field on $e(M)$ tangent to G such that $t_*(v) = 0$ and ω is a local nonvanishing top form on M; it is easy to show that this expression does not depend on the choice of ω. Passing to the global sections, we obtain an injective map $\mathfrak{g} \longrightarrow F_1 K^l(G)$ inducing an isomorphism $\mathfrak{g} \simeq F_1 K^l(G) / F_0 K^l(G)$. This injective map and the A-A-bimodule structure satisfy the compatibility equations $a \cdot x = ax$ and $x \cdot a = x(a) + ax$ for $x \in \mathfrak{g} \subset F_1 K^l(G)$ and $a \in A$, where $(a, x) \longmapsto ax$ denotes the action of A in $\mathfrak{g}$, while $(a, u) \longmapsto a \cdot u$ and $(u, a) \longmapsto u \cdot a$ denote the left and right actions of A in $F_1 K^l(G)$.

Let us define a k-algebra structure on $K^l(G)$. There is a natural isomorphism of invertible sheaves on $G \times_M G$,

$$p_1^*(t^* \Omega_M^{-1} \otimes \Omega_G) \otimes p_2^*(t^* \Omega_M^{-1} \otimes \Omega_G) \simeq p_1^* t^* \Omega_M^{-1} \otimes \Omega_{G \times_M G}.$$

The pull-back with respect to the closed embedding $G \times_M G \longrightarrow G \times_{\operatorname{Spec} k} G$ provides an isomorphism

$$(p_1^* t^* \Omega_M^{-1} \otimes \Omega_{G \times_M G})^{(e \times e)(M)}(G \times_M G) \simeq K^l(G) \otimes_A K^l(G),$$

and the push-forward with respect to the multiplication map $G \times_M G \longrightarrow G$ defines an associative multiplication $K^l(G) \otimes_A K^l(G) \longrightarrow K^l(G)$. The associated graded algebra $\mathrm{gr}_F K^l(G)$ is naturally isomorphic to $\mathrm{Sym}_A F_1 K^l(G)/F_0 K^l(G)$. The formula $(u, a) \longmapsto t_+(s^+(a)u)$ defines a left action of $K^l(G)$ in A; the subspace $\mathfrak{g}$ in $F_1 K^l(G)$ is characterized as the annihilator of the unit element of A under this action. Hence $\mathfrak{g}$ is a Lie subalgebra of $K^l(G)$; this makes it a Lie algebra and a Lie algebroid over A. It follows that there is a natural isomorphism

$$U_A(\mathfrak{g}) \simeq K^l(G).$$

Analogously one defines an algebra structure on the A-A-bimodule of generalized sections

$$K^r(G) = (s^* \Omega_M^{-1} \otimes \Omega_G)^{e(M)}(G);$$

then there is a natural isomorphism of k-algebras $U_A(\mathfrak{g})^{\mathrm{op}} \simeq K^r(G)$.

F.5 Two Morita equivalent semialgebras

Let $(M, H) \longrightarrow (M, G)$ be a morphism of smooth groupoids with the same base variety M such that the groupoid H is affine and the morphism of varieties $H \longrightarrow G$ is a closed embedding. Denote by $\mathfrak{h} = N_{e(M),H}(M)$ the Lie algebroid of the groupoid H; then there is a natural injective morphism $\mathfrak{h} \longrightarrow \mathfrak{g}$ of Lie algebroids over $A(M)$.

There is a natural pairing

$$\phi_l \colon K^l(H) \otimes_A \mathcal{C} \longrightarrow A, \qquad \phi_l(u, c) = t_+(i^+(c)u)$$

between the algebra $K^l(H)$ and the coring $\mathcal{C}$ satisfying the conditions of 10.1.2 with the left and right sides switched. The push-forward with respect to the closed embedding $H \longrightarrow G$ defines an injective morphism of k-algebras $K^l(H) \longrightarrow K^l(G)$; this is the enveloping algebra morphism induced by the morphism of Lie algebroids $\mathfrak{h} \longrightarrow \mathfrak{g}$. Since $\mathfrak{h}$ and $\mathfrak{g}/\mathfrak{h}$ are projective A-modules, $K^l(G)$ is a projective left and right $K^l(H)$-module. Set

$$\mathbf{S}^l = \mathbf{S}^l(G, H) = K^l(G) \otimes_{K^l(H)} \mathcal{C}$$

we will use the construction of 10.2.1 to endow $\mathbf{S}^l$ with a structure of semialgebra over $\mathcal{C}$.

Consider the cotensor product $\mathbf{S}^l \,\square_{\mathcal{C}}\, \mathcal{E} \simeq K^l(G) \otimes_{K^l(H)} \mathcal{E}$. Denote by p_1 and q_2 the projections of the fibered product $G \times_M H$ to the first and the second factors. There is a natural isomorphism

$$p_1^*(t^* \Omega_M^{-1} \otimes \Omega_G) \otimes q_2^*(V_H) \simeq p_1^* t^* \Omega_M^{-1} \otimes \Omega_{G \times_M H} \otimes q_2^* s^* \Omega_M^{-1}$$

of invertible sheaves on $G \times_M H$, where $V_H = \Omega_H \otimes s^*(\Omega_M^{-1}) \otimes t^*(\Omega_M^{-1})$ is the invertible sheaf on H defined in F.2. The pull-back with respect to the closed embedding $G \times_M H \longrightarrow G \times_{\mathrm{Spec}\,k} H$ identifies the tensor product $K^l(G) \otimes_A \mathcal{E}$ with the module of generalized sections

$$(p_1^* t^* \Omega_M^{-1} \otimes \Omega_{G \times_M H} \otimes q_2^* s^* \Omega_M^{-1})^{e(M) \times_M H}(G \times_M H).$$

The push-forward with respect to the multiplication morphism $G \times_M H \longrightarrow G$ defines a natural map $K^l(G) \otimes_A \mathcal{E} \longrightarrow \mathcal{E}$, where $\mathcal{E} = V_G^H(G)$ is the space of generalized sections of the invertible sheaf

$$V_G = \Omega_G \otimes s^*(\Omega_M^{-1}) \otimes t^*(\Omega_M^{-1})$$

on G. It follows from the associativity equation for the two iterated multiplication maps $G \times_M H \times_M H \rightrightarrows G$ that this map factorizes through $K^l(G) \otimes_{K^l(H)} \mathcal{E}$. The induced map

$$K^l(G) \otimes_{K^l(H)} \mathcal{E} \longrightarrow \mathcal{E}$$

is an isomorphism, since the associated graded map with respect to the filtrations F is.

Denote by q_1 and p_2 the projections of the fibered product $H \times_M G$ to the first and second factors. One constructs a natural isomorphism $m^*(V_G) \simeq p_2^*(V_G)$ of invertible sheaves on $H \times_M G$ in the same way as in F.2. The pull-back with respect to the multiplication map $m \colon H \times_M G \longrightarrow G$ together with this isomorphism provide a left coaction of $\mathcal{C}$ in $\mathcal{E}$. Analogously one defines a right coaction of $\mathcal{C}$ in $\mathcal{E}$; it follows from the associativity equation for $H \times_M G \times_M H \rightrightarrows G$ that these two coactions commute. We have $\mathcal{S}^l \square_{\mathcal{C}} \mathcal{E} \simeq \mathcal{E}$, thus the isomorphism $\mathcal{S}^l \simeq \mathcal{E} \square_{\mathcal{C}} \mathcal{E}^\vee$ provides a left coaction of $\mathcal{C}$ in $\mathcal{S}^l$ commuting with the natural right coaction of $\mathcal{C}$ in $\mathcal{S}^l$.

Since the natural map $\mathcal{E} \longrightarrow \mathcal{E}$ provided by the push-forward with respect to the closed embedding $H \longrightarrow G$ is a morphism of $\mathcal{C}$-$\mathcal{C}$-bicomodules, so is the semiunit map $\mathcal{C} \longrightarrow \mathcal{S}^l$. It remains to show that the semimultiplication map $\mathcal{S}^l \square_{\mathcal{C}} \mathcal{S}^l \longrightarrow \mathcal{S}^l$ is a morphism of left $\mathcal{C}$-comodules; here it suffices to check that the map $\mathcal{S}^l \square_{\mathcal{C}} \mathcal{S}^l \square_{\mathcal{C}} \mathcal{E} \longrightarrow \mathcal{S}^l \square_{\mathcal{C}} \mathcal{E}$ is a morphism of left $\mathcal{C}$-comodules. After we have done with this verification, the latter map will define a left $\mathcal{S}^l$-semimodule structure on $\mathcal{E}$.

Analogously, define the pairing

$$\phi_r \colon \mathcal{C} \otimes_A K^r(H) \longrightarrow A, \qquad \phi_r(c, u) = s_+(i^+(c)u)$$

and set

$$\mathcal{S}^r = \mathcal{S}^r(G, H) = \mathcal{C} \otimes_{K^r(H)} K^r(G).$$

The same construction makes $\mathcal{S}^r$ a semialgebra over $\mathcal{C}$ and $\mathcal{E}$ a right $\mathcal{S}^r$-semimodule. We will have to check that the left $\mathcal{S}^l$-semimodule and the right $\mathcal{S}^r$-semimodule structures on $\mathcal{E}$ commute.

After this is done, we get an $\mathcal{S}^l$-$\mathcal{S}^r$-bisemimodule

$$\mathcal{S}^l \square_{\mathcal{C}} \mathcal{E} \simeq \mathcal{E} \simeq \mathcal{E} \square_{\mathcal{C}} \mathcal{S}^r,$$

where both the maps $\mathcal{E} \rightrightarrows \mathcal{E}$ induced by the semiunit maps $\mathcal{C} \longrightarrow \mathcal{S}^l$ and $\mathcal{C} \longrightarrow \mathcal{S}^r$ coincide with the push-forward map $V(H) \longrightarrow V_G^H(G)$ under the closed embedding $H \longrightarrow G$. The isomorphisms $\mathcal{E}^\vee \square_\mathcal{C} \mathcal{S}^l \simeq \mathcal{E}^\vee \square_\mathcal{C} \mathcal{S}^l \square_\mathcal{C} \mathcal{E} \square_\mathcal{C} \mathcal{E}^\vee \simeq \mathcal{E}^\vee \square_\mathcal{C} \mathcal{E} \square_\mathcal{C} \mathcal{S}^r \square_\mathcal{C} \mathcal{E}^\vee \simeq \mathcal{S}^r \square_\mathcal{C} \mathcal{E}^\vee$ define an $\mathcal{S}^r$-$\mathcal{S}^l$-bisemimodule

$$\mathcal{E}^\vee \square_\mathcal{C} \mathcal{S}^l = \mathcal{E}^\vee \simeq \mathcal{S}^r \square_\mathcal{C} \mathcal{E}^\vee$$

endowed with bisemimodule isomorphisms $\mathcal{E} \Diamond_{\mathcal{S}^r} \mathcal{E}^\vee \simeq (\mathcal{E} \square_\mathcal{C} \mathcal{S}^r) \Diamond_{\mathcal{S}^r} (\mathcal{S}^r \square_\mathcal{C} \mathcal{E}^\vee) \simeq \mathcal{E} \square_\mathcal{C} \mathcal{S}^r \square_\mathcal{C} \mathcal{E}^\vee \simeq \mathcal{S}^l$ and $\mathcal{E}^\vee \Diamond_{\mathcal{S}^l} \mathcal{E} \simeq (\mathcal{E}^\vee \square_\mathcal{C} \mathcal{S}^l) \Diamond_{\mathcal{S}^l} (\mathcal{S}^l \square_\mathcal{C} \mathcal{E}) \simeq \mathcal{E}^\vee \square_\mathcal{C} \mathcal{S}^l \square_\mathcal{C} \mathcal{E} \simeq \mathcal{S}^r$. This provides a left and right semiflat Morita equivalence $(\mathcal{E}, \mathcal{E}^\vee)$ between the semialgebras $\mathcal{S}^l$ and $\mathcal{S}^r$, and isomorphisms of semialgebras

$$\mathcal{S}^r \simeq \mathcal{E}^\vee \square_\mathcal{C} \mathcal{S}^l \square_\mathcal{C} \mathcal{E} \quad \text{and} \quad \mathcal{S}^l \simeq \mathcal{E} \square_\mathcal{C} \mathcal{S}^r \square_\mathcal{C} \mathcal{E}^\vee.$$

(See 8.4.5 and 8.4.1 for the relevant definition and construction.)

F.6   Compatibility verifications

In order to check that the map

$$\mathcal{S}^l \square_\mathcal{C} \mathcal{E} \simeq \mathcal{S}^l \square_\mathcal{C} \mathcal{S}^l \square_\mathcal{C} \mathcal{E} \longrightarrow \mathcal{S}^l \square_\mathcal{C} \mathcal{E} \simeq \mathcal{E}$$

is a morphism of left $\mathcal{C}$-comodules, we will identify this map with a certain push-forward map of appropriate modules of generalized sections.

Here we will need to assume the existence of a variety of left cosets G/H such that $G \longrightarrow G/H$ is a smooth surjective morphism and the fibered square $G \times_{G/H} G$ can be identified with $G \times_M H$ so that the canonical projection maps $G \times_{G/H} G \rightrightarrows G$ correspond to the projection and multiplication maps $p_1, m \colon G \times_M H \rightrightarrows G$.

Actually, we are interested in the quotient variety $G \times^H G$ of $G \times_M G$ by the equivalence relation $(g'h, g'') \sim (g', hg'')$ for $g', g'' \in G$ and $h \in H$; it can be constructed as either of the fibered products

$$H \backslash G \times_M G \simeq G \times^H G = G \times_M G/H,$$

where $H \backslash G$ denotes the variety of right cosets, $H \backslash G \simeq G/H$. Analogously one can construct the quotient variety

$$G \times^H G \times^H G = G \times_M G \times_M G/\{(g'h_1, g''h_2, g''') \sim (g', h_1g'', h_2g''')\}$$

for $g', g'', g''' \in G$ and $h_1, h_2 \in H$.

We have $\mathcal{S}^l \square_\mathcal{C} \mathcal{E} \simeq \mathcal{E} \square_\mathcal{C} \mathcal{E}^\vee \square_\mathcal{C} \mathcal{E}$. Consider the natural map

$$r \colon G \times_M H \times_M G \longrightarrow G \times^H G, \qquad (g', h, g'') \longmapsto (g'h, g'') = (g', hg'').$$

Let p_1, q_2, p_3 denote the projections of the triple fibered product $G \times_M H \times_M G$ to the three factors and $n \colon G \times^H G \longrightarrow G$ denote the multiplication morphism. There are natural isomorphisms

$$p_1^* t^* (\Omega_M^{-1}) \otimes \Omega_{G \times_M H \times_M G} \otimes p_3^* s^* (\Omega_M^{-1}) \simeq p_1^* (V_G) \otimes q_2^* (\Omega_H) \otimes p_3^* (V_G)$$

and therefore

$$r^*(n^*t^*\Omega_M^{-1} \otimes \Omega_{G\times_H G} \otimes n^*s^*\Omega_M^{-1}) \simeq p_1^*(V_G) \otimes q_2^*(V_H^{-1}) \otimes p_3^*(V_G)$$

of invertible sheaves on $G \times_M H \times_M G$.

The pull-back with respect to the closed embedding

$$G \times_M H \times_M G \longrightarrow G \times_{\operatorname{Spec} k} H \times_{\operatorname{Spec} k} G$$

provides an isomorphism

$$(p_1^*V_G \otimes q_2^*V_H^{-1} \otimes p_3^*V_G)^{H\times_M H\times_M H}(G \times_M H \times_M G) \simeq \mathcal{E} \otimes_A \mathcal{E}^\vee \otimes_A \mathcal{E}.$$

The pull-back with respect to the smooth morphism r identifies the module of generalized sections

$$(n^*t^*\Omega_M^{-1} \otimes \Omega_{G\times_H G} \otimes n^*s^*\Omega_M^{-1})^{H\times_H H}(G \times_H G)$$

with the submodule $\mathcal{E} \,\square_{\mathcal{C}}\, \mathcal{E}^\vee \,\square_{\mathcal{C}}\, \mathcal{E} \subset \mathcal{E} \otimes_A \mathcal{E}^\vee \otimes_A \mathcal{E}$, as one can see by identifying the tensor product $\mathcal{E} \otimes_A \mathcal{C} \otimes_A \mathcal{E}^\vee \otimes_A \mathcal{C} \otimes_A \mathcal{E}$ with a module of generalized sections on the fibered square of $G \times_M H \times_M G$ over $G \times_H G$. Now our map

$$\mathcal{E} \,\square_{\mathcal{C}}\, \mathcal{E}^\vee \,\square_{\mathcal{C}}\, \mathcal{E} \longrightarrow \mathcal{E}$$

is identified with the push-forward map with respect to the multiplication morphism n; to check this, one can first identify the map

$$K^l(G) \otimes_A \mathcal{E} \longrightarrow K^l(G) \otimes_{K^l(H)} \mathcal{E} \simeq \mathcal{E} \,\square_{\mathcal{C}}\, \mathcal{E}^\vee \,\square_{\mathcal{C}}\, \mathcal{E}$$

with a push-forward map with respect to the morphism $G \times_M G \longrightarrow G \times_H G$. The desired compatibility with the left $\mathcal{C}$-comodule structures now follows from the commutation of the pull-back and push-forward maps of generalized sections.

To check that the left $\mathcal{S}^l$-semimodule and the right $\mathcal{S}^r$-semimodule structures on $\mathcal{E}$ commute, one can identify $\mathcal{S}^l \square_{\mathcal{C}} \mathcal{E} \square_{\mathcal{C}} \mathcal{S}^r$ with a module of generalized sections on $G \times_H G \times_H G$ and use the associativity equation for the iterated multiplication maps

$$G \times_H G \times_H G \rightrightarrows G \times_H G \longrightarrow G.$$

Bibliography

[1] M. Aguiar. Internal categories and quantum groups. Cornell Univ. Ph.D. Thesis, 1997. Available from http://www.math.tamu.edu/~maguiar/.

[2] S.M. Arkhipov. Semi-infinite cohomology of associative algebras and bar-duality. *Internat. Math. Research Notices* **1997**, #17, pp. 833–863. arXiv:q-alg/9602013

[3] S. Arkhipov. Semi-infinite cohomology of quantum groups II. Topics in quantum groups and finite-type invariants, pp. 3–42, *Amer. Math. Soc. Translations, Ser. 2* **185**, AMS, Providence, RI, 1998. arXiv:q-alg/9610020

[4] S. Arkhipov. Semiinfinite cohomology of contragradient Weyl modules over small quantum groups. Electronic preprint arXiv:math.QA/9906071.

[5] S. Arkhipov. Semiinfinite cohomology of Tate Lie algebras. *Moscow Math. Journ.* **2**, #1, pp. 35–40, 2002. arXiv:math.QA/0003015

[6] M. Barr. Coequalizers and free triples. *Math. Zeitschrift* **116**, #4, pp. 307–322, 1970.

[7] H. Bass. Finitistic dimension and a homological generalization of semi-primary rings. *Trans. Amer. Math. Soc.* **95**, #3, pp. 466–488, 1960.

[8] A. Beilinson. Remarks on topological algebras. *Moscow Math. Journ.* **8**, #1, pp. 1–20, 2008.

[9] A. Beilinson, J. Bernstein. A proof of Jantzen conjectures. *Advances in Soviet Math.* **16**, #1, pp. 1–50, 1993.

[10] A. Beilinson, V. Drinfeld. Chiral algebras. AMS Colloquium Publications, 51. AMS, Providence, RI, 2004.

[11] A. Beilinson, V. Drinfeld. Quantization of Hitchin's integrable system and Hecke eigensheaves. February 2000. Available from http://www.math.utexas.edu/~benzvi/Langlands.html.

[12] A. Beilinson, B. Feigin, B. Mazur. Notes on Conformal Field Theory (incomplete), 1991. Available from http://www.math.sunysb.edu/~kirillov/manuscripts.html.

[13] A. Beilinson, V. Ginzburg, W. Soergel. Koszul duality patterns in representation theory. *Journ. Amer. Math. Soc.* **9**, #2, pp. 473–527, 1996.

[14] J. Bernstein. Algebraic theory of D-modules. Available from http://www.math.uchicago.edu/~mitya/langlands.html.

[15] J. Bernstein, V. Lunts. Equivariant sheaves and functors. *Lecture Notes Math.* **1578**, Springer-Verlag, Berlin, 1994.

[16] R. Bezrukavnikov, M. Finkelberg, V. Schechtman. Factorizable sheaves and quantum groups. *Lecture Notes Math.* **1691**, Springer-Verlag, Berlin–Heidelberg, 1998.

[17] R. Bezrukavnikov, L. Positselski. On semi-infinite cohomology of finite-dimensional graded algebras. *Compositio Math.* **146**, #2, pp. 480–496, 2010. arXiv:0803.3252 [math.QA]

[18] A. Bondal, M. Kapranov. Representable functors, Serre functors, and mutations. *Math. USSR Izvestiya* **35**, #3, pp. 519–541, 1990.

[19] A. Bondal, M. Kapranov. Enhanced triangulated categories. *Math. USSR Sbornik* **70**, #1, pp. 93–107, 1991.

[20] N. Bourbaki. Topologie générale. Chapitres 1 à 4. Springer-Verlag Berlin–Heidelberg–New York, 2007. Réimpression inchangée de l'édition originale, Hermann, Paris, 1971.

[21] T. Brzeziński. The structure of corings: induction functors, Maschke-type theorem, and Frobenius and Galois-type properties. *Algebr. Represent. Theory* **5**, #4, pp. 389–410, 2002. arXiv:math.RA/0002105

[22] T. Brzeziński. Flat connections and (co)modules. New Techniques in Hopf Algebras and Graded Ring Theory, Universa Press, Wetteren, 2007, pp. 35–52. arXiv:math.QA/0608170

[23] T. Brzezinski, R. Wisbauer. Corings and comodules. London Mathematical Society Lecture Note Series, 309. Cambridge University Press, Cambridge, 2003.

[24] T. Brzeziński, L. El Kaoutit, J. Gomez-Torrecillas. The bicategories of corings. *Journ. Pure Appl. Algebra* **205**, #3, pp. 510–541, 2006. arXiv:math.RA/0408042

[25] T. Brzeziński, R.B. Turner. The Galois theory of matrix C-rings. *Appl. categor. struct.* **14**, #5–6, pp. 471–501, 2006. arXiv:math.RA/0512049

[26] G. Böhm, T. Brzeziński, R. Wisbauer. Monads and comonads in module categories. *Journ. Algebra* **233**, #5, pp. 1719–1747, 2009. arXiv:0804.1460 [math.RA]

[27] T. Brzeziński, A. Vazquez Marquez. Internal Kleisli categories. Electronic preprint arXiv:0911.4048 [math.CT].

[28] T. Bühler. Exact categories. *Expositiones Math.* **28**, #1, pp. 1–69, 2010. arXiv:0811.1480 [math.HO]

[29] P. Deligne. Cohomologie à supports propres. SGA4, Tome 3. *Lecture Notes Math.* **305**, Springer-Verlag, Berlin–Heidelberg–New York, 1973, pp. 250–480.

[30] P. Deligne, J.S. Milne. Tannakian categories. *Lecture Notes Math.* **900**, Springer-Verlag, Berlin–Heidelberg–New York, 1982, pp. 101–228.

[31] M. Demazure, P. Gabriel. Groupes algébriques, Tome I. North Holland, Amsterdam, 1970.

[32] J. Dixmier. Enveloping Algebras. Translated from French. North Holland, Amsterdam, 1977. Or: Graduate Studies in Math. 11, AMS, Providence, RI, 1996.

[33] S. Eilenberg, J.C. Moore. Foundations of relative homological algebra. *Memoirs Amer. Math. Soc.* **55**, 1965.

[34] S. Eilenberg, J.C. Moore. Homology and fibrations I. Coalgebras, cotensor product and its derived functors. *Comment. Math. Helvetici* **40**, #1, pp. 199–236, 1966.

[35] E.E. Enochs, O.M.G. Jenda. Relative homological algebra. De Gruyter Expositions in Mathematics, 30. De Gruyter, Berlin–New York, 2000.

[36] B.L. Feigin. The semi-infinite homology of Kac-Moody and Virasoro Lie algebras. *Russian Math. Surveys* **39**, #2, pp. 155–156, 1984.

[37] B.L. Feigin, E.V. Frenkel. Affine Kac-Moody algebras and semi-infinite flag manifolds. *Comm. Math. Phys.* **128**, #1, pp. 161–189, 1990.

[38] B. Feigin, E. Frenkel. Semi-infinite Weil complex and the Virasoro algebra. *Comm. Math. Phys.* **137**, #3, pp. 617–639, 1991. Erratum in *Comm. Math. Phys.* **147**, #3, pp. 647–648, 1992.

[39] B.L. Feigin, D.B. Fuchs. Verma modules over the Virasoro algebra. Topology (Leningrad, 1982), pp. 230–245, *Lecture Notes in Math.* **1060**, Springer-Verlag, Berlin, 1984.

[40] E. Frenkel, D. Gaitsgory. Local geometric Langlands correspondence and affine Kac-Moody algebras. Algebraic Geometry and Number Theory, In Honor of Vladimir Drinfeld's 50th Birthday, *Progress in Math.* **253**, Birkhäuser Boston, 2006, pp. 69–260. `arXiv:math.RT/0508382`

[41] I.B. Frenkel, H. Garland, G.J. Zuckerman. Semi-infinite cohomology and string theory. *Proc. Natl. Acad. Sci. USA* **83**, #22, pp. 8442–8446, 1986.

[42] I.B. Frenkel, A.M. Zeitlin. Quantum group as semi-infinite cohomology. Electronic preprint `arXiv:0812.1620 [math.RT]`, to appear in *Comm. Math. Physics*.

[43] P. Gabriel. Des catégories abéliennes. *Bull. Soc. Math. France* **90**, pp. 323–448, 1962.

[44] D. Gaitsgory, D. Kazhdan. Representations of algebraic groups over a 2-dimensional local field. *Geom. and Funct. Anal.* **14**, #3, pp. 535–574, 2004. `arXiv:math.RT/0302174`

[45] D. Gaitsgory, D. Kazhdan. Algebraic groups over a 2-dimensional local field: some further constructions. Studies in Lie theory, pp. 97-130, *Progress in Math.* **243**, Birkhäuser Boston, Boston, MA, 2006. `arXiv:math.RT/0406282`

[46] S.I. Gelfand, Yu.I. Manin. Methods of homological algebra. Translation from the 1988 Russian original. Springer-Verlag, Berlin, 1996. Or: Second edition. Springer-Verlag, 2003.

[47] G. Hochschild. Relative homological algebra. *Trans. Amer. Math. Soc.* **82**, #1, pp. 246–269, 1956.

[48] G. Hochschild. Note on algebraic Lie algebras. *Proc. Amer. Math. Soc.* **29**, #1, pp. 10–16, 1971.

[49] M. Hovey. Model categories. Mathematical Surveys and Monographs, 63. AMS, Providence, RI, 1999.

[50] M. Hovey. Cotorsion pairs, model category structures, and representation theory. *Math. Zeitschrift* **241**, #3, pp. 553–592, 2002.

[51] M. Hovey. Cotorsion pairs and model categories. *Contemporary Math.* **436**, AMS, Providence, RI, 2007, pp. 277–296. `arXiv:math.AT/0701161`

[52] D. Husemoller, J.C. Moore, J. Stasheff. Differential homological algebra and homogeneous spaces. *Journ. Pure Appl. Algebra* **5**, #2, pp. 113–185, 1974.

[53] J.E. Humphreys. Representations of semisimple Lie algebras in the BGG category $\mathcal{O}$. Graduate Studies in Math. 94, AMS, Providence, RI, 2008.

[54] U. Jannsen. Continuous étale cohomology. *Math. Annalen* **280**, #2, pp. 207–245, 1988.

[55] P. Jørgensen. The homotopy category of complexes of projective modules. *Advances Math.* **193**, #1, pp. 223–232, 2005. `arXiv:math.RA/0312088`

[56] M. Kapranov. On DG-modules over the de Rham complex and the vanishing cycles functor. *Lect. Notes Math.* **1479**, 1991, pp. 57–86.

[57] M. Kapranov. Semi-infinite symmetric powers. Electronic preprint `arXiv:math.QA/0107089`.

[58] M. Kapranov, E. Vasserot. Vertex algebras and the formal loop space. *Publ. Math. Inst. Hautes Études Sci.* **100**, pp. 209–269, 2004. `arXiv:math.AG/0107143`

[59] M. Kapranov, E. Vasserot. Formal loops IV: Chiral differential operators. Electronic preprint `arXiv:math.AG/0612371`.

[60] B. Keller. Deriving DG-categories. *Ann. Sci. École Norm. Sup.* (4) **27**, #1, pp. 63–102, 1994.

[61] B. Keller. Koszul duality and coderived categories (after K. Lefèvre). October 2003. Available from `http://www.math.jussieu.fr/~keller/publ/index.html`.

[62] M. Kontsevich, A. Rosenberg. Noncommutative smooth spaces. *The Gelfand Mathematical Seminars* 1996–1999, pp. 85–108, Birkhäuser Boston, Boston, MA, 2000. `arXiv:math.AG/9812158`

[63] M. Kontsevich, A. Rosenberg. Noncommutative spaces and flat descent. Max-Planck-Institut für Mathematik (Bonn) preprint MPIM 2004-36.

[64] H. Krause. The stable derived category of a Noetherian scheme. *Compositio Math.* **141**, #5, pp. 1128–1162, 2005. `arXiv:math.AG/0403526`

[65] S. Iyengar, H. Krause. Acyclicity versus total acyclicity for complexes over Noetherian rings. *Documenta Math.* **11**, pp. 207–240, 2006.

[66] K. Lefèvre-Hasegawa. Sur les A_∞-catégories. Thèse de doctorat, Université Denis Diderot – Paris 7, November 2003. `arXiv:math.CT/0310337`. Corrections, by B. Keller. Available from `http://people.math.jussieu.fr/~keller/lefevre/publ.html`.

[67] S. MacLane. Categories for the working mathematician. Graduate Texts in Mathematics, 5. Springer-Verlag, New York–Berlin, 1971. Or: Second edition. Springer-Verlag, 1998.

[68] S. Montgomery. Hopf algebras and their actions on rings. Regional Conference Series in Mathematics, 82. AMS, 1993.

[69] A. Neeman. The derived category of an exact category. *Journ. Algebra* **135**, #2, pp. 388–394, 1990.

[70] M. Bøkstedt, A. Neeman. Homotopy limits in triangulated categories. *Compositio Math.* **86**, #2, pp. 209–234, 1993.

[71] A. Neeman. The homotopy category of flat modules, and Grothendieck duality. *Inventiones Math.* **174**, pp. 225–308, 2008.

[72] A.L. Onishchik, E.B. Vinberg. Lie groups and algebraic groups. Translated from the Russian and with a preface by D.A. Leites. Springer Series in Soviet Mathematics, Springer-Verlag, Berlin, 1990.

[73] L. Positselski. Nonhomogeneous quadratic duality and curvature. *Funct. Anal. Appl.* **27**, #3, pp. 197–204, 1993.

[74] L. Positselski, A. Vishik. Koszul duality and Galois cohomology. *Math. Research Letters* **2**, #6, pp. 771–781, 1995. `arXiv:alg-geom/9507010`

[75] A. Polishchuk, L. Positselski. Quadratic algebras. University Lecture Series, 37. AMS, Providence, RI, 2005.

[76] L. Positselski. Two kinds of derived categories, Koszul duality, and comodule-contramodule correspondence. Electronic preprint `arXiv:0905.2621 [math.CT]`.

[77] A. Rocha-Caridi, N. Wallach. Characters of irreducible representations of the Virasoro algebra. *Math. Zeitschrift* **185**, #1, pp. 1–21, 1984.

[78] J.-P. Serre. Lie algebras and Lie groups. Second edition. *Lecture Notes Math.* **1500**, Springer-Verlag, Berlin–Heidelberg, 1992.

[79] A. Sevostyanov. Semi-infinite cohomology and Hecke algebras. *Advances Math.* **159**, #1, pp. 83–141, 2001. `arXiv:math.RA/0004139`

[80] E.G. Sklyarenko. Relative homological algebra in categories of modules. *Russian Math. Surveys* **33**, #3, pp. 97–137, 1978.

[81] N. Spaltenstein. Resolutions of unbounded complexes. *Compositio Math.* **65**, #2, pp. 121–154, 1988.

[82] M.E. Sweedler. Hopf algebras. Mathematics Lecture Note Series, W.A. Benjamin, Inc., New York, 1969.

[83] J. Tate. Residues of differentials on curves. *Ann. Sci. École Norm. Sup.* (4) **1**, #1, pp. 149–159, 1968.

[84] A. Voronov. Semi-infinite homological algebra. *Inventiones Math.* **113**, #1, pp. 103–146, 1993.

[85] C.E. Watts. Intrinsic characterizations of some additive functors. *Proc. Amer. Math. Soc.* **11**, #1, pp. 5–8, 1960.

[86] C.A. Weibel. An introduction to homological algebra. Cambridge Studies in Advanced Mathematics, 38. Cambridge University Press, 1994.

Notation

$\mathbb{Z}$ – ring of rational integers

$\mathbb{Q}$ – field of rational numbers

k – commutative ring or field

$k^{\vee}$ – injective cogenerator of k-modules

R, S, K – noncommutative rings

A, B, F – associative k-algebras

$M, N, L, P, E, U, \ldots$ – modules

K, E – bimodules

$\mathfrak{g}, \mathfrak{h}, \mathfrak{a}$ – Lie algebras

$\mathfrak{g}, \mathfrak{h}$ – Lie algebroids

G, H – topological groups, (pro)alge- braic groups, algebraic groupoids

$\mathbb{G}, \mathbb{H}$ – groups in (ind-)pro-ind-pro-finite sets, see E.4.2–E.4.3

$U, V, W, E, \ldots$ – vector spaces

$M, X, Y, Z, \ldots$ – algebraic varieties

$E, L, U, V, \ldots$ – vector bundles/locally free sheaves

P, Q – contramodules over topological rings/Lie algebras/groups

$\mathcal{C}, \mathcal{D}, \mathcal{E}, \ldots$ – coalgebras or corings

$\mathcal{L}$ – Lie coalgebra

$\mathcal{M}, \mathcal{N}, \mathcal{L}, \mathcal{R}, \mathcal{K}, \mathcal{E}, \mathcal{P}, \mathcal{Q}, \mathcal{J}, \mathcal{U}, \mathcal{Z}, \ldots$ – comodules

$\mathcal{K}, \mathcal{E}$ – bicomodules

$\mathfrak{P}, \mathfrak{Q}, \mathfrak{R}, \mathfrak{K}, \mathfrak{E}, \mathfrak{F}, \mathfrak{I}, \mathfrak{U}, \mathfrak{Z}, \ldots$ – contramodules over coalgebras or corings

$\mathcal{S}, \mathcal{T}, \mathcal{R}, \ldots$ – semialgebras

$\mathcal{K}, \mathcal{E}$ – bisemimodules

$\mathcal{M}, \mathcal{N}, \mathcal{L}, \mathcal{R}, \mathcal{K}, \mathcal{E}, \mathcal{P}, \mathcal{Q}, \mathcal{F}, \mathcal{J}, \mathcal{U}, \mathcal{Z}, \ldots$ – semimodules

$\mathfrak{P}, \mathfrak{Q}, \mathfrak{R}, \mathfrak{K}, \mathfrak{E}, \mathfrak{F}, \mathfrak{I}, \mathfrak{U}, \mathfrak{Z}, \ldots$ – semicontramodules

A, H, K, M, N, E, F, ... – categories

$\Theta, \Xi, \Delta, \Gamma, \Sigma, \Pi, \Lambda, \ldots$ – functors

$M^{\bullet}, \mathcal{M}^{\bullet}, \mathfrak{M}^{\bullet}, \ldots$ – complexes

$\mathbb{D}$ – double-sided derived functor

$\mathbb{L}$ – left derived functor

$\mathbb{R}$ – right derived functor

$\mathbb{L}_1, \mathbb{L}_2, \mathbb{L}_3$ – functorial partial left resolutions, see 2.5, 2.6, 4.5, 4.6

$\mathbb{R}_1, \mathbb{R}_2, \mathbb{R}_3$ – functorial partial right resolutions

$\mathcal{C}/A\text{-}, \mathcal{D}/B\text{-}, \mathcal{S}/\mathcal{C}\text{-}, \ldots$ – relatively

$\mathcal{S}/\mathcal{C}/A\text{-}, \mathcal{T}/\mathcal{D}/B\text{-}, \ldots$ – birelatively

$\mathrm{Hom}_R(L, M)$, $\mathrm{Hom}_{R^{\mathrm{op}}}(K, N)$ – left module, right module homomorphisms

$\mathcal{N} \,\square_{\mathcal{C}}\, \mathcal{M}$ – cotensor product

$\mathrm{Cohom}_{\mathcal{C}}(\mathcal{M}, \mathfrak{P})$ – cohomomorphisms

$\mathcal{N} \odot_{\mathcal{C}} \mathfrak{P}$ – contratensor product over coalgebra or coring

$\mathrm{Hom}_{\mathcal{C}}(\mathcal{L}, \mathcal{M})$, $\mathrm{Hom}^{\mathcal{C}}(\mathfrak{P}, \mathfrak{Q})$ – left comodule, left contramodule homomorphisms

$\mathrm{Cotor}^{\mathcal{C}}$ – right or double-sided derived functor of $\square_{\mathcal{C}}$, see 0.2.2, 1.2.2, 2.7

$\mathrm{Coext}_{\mathcal{C}}$ – left or double-sided derived functor of $\mathrm{Cohom}_{\mathcal{C}}$, see 0.2.5, 3.2.2, 4.7

$\mathrm{Ctrtor}^{\mathcal{C}}$ – left derived functor of $\odot_{\mathcal{C}}$, see 0.2.8, 5.3, 5.5

$\mathrm{Ext}_{\mathcal{C}}$ – right derived functor of $\mathrm{Hom}_{\mathcal{C}}$

$\mathrm{Ext}^{\mathcal{C}}$ – right derived functor of $\mathrm{Hom}^{\mathcal{C}}$

$\mathcal{N} \,\Diamond_{\mathcal{S}}\, \mathcal{M}$ – semitensor product

$\mathrm{SemiHom}_{\mathcal{S}}(\mathcal{M}, \mathfrak{P})$ – semihomomorphisms

$\mathcal{N} \circledS_{\mathcal{S}} \mathfrak{P}$ – contratensor product over semialgebra

$\mathrm{Hom}_{\mathcal{S}}(\mathcal{L}, \mathcal{M})$, $\mathrm{Hom}^{\mathcal{S}}(\mathfrak{P}, \mathfrak{Q})$ – semimodule, semicontramodule homomorphisms

$\mathrm{SemiTor}^{\mathcal{S}}$ – double-sided derived functor of $\Diamond_{\mathcal{S}}$, see 0.3.3, 2.7

$\mathrm{SemiExt}_{\mathcal{S}}$ – double-sided derived functor of $\mathrm{SemiHom}_{\mathcal{S}}$, see 0.3.6, 4.7

$\mathrm{CtrTor}^{\mathcal{S}}$ – left derived functor of $\circledS_{\mathcal{S}}$, see 0.3.8, 6.5

$\mathrm{Ext}_{\mathcal{S}}$ – right derived functor of $\mathrm{Hom}_{\mathcal{S}}$

$\mathrm{Ext}^{\mathcal{S}}$ – right derived functor of $\mathrm{Hom}^{\mathcal{S}}$

$\Phi_{\mathcal{C}}(\mathfrak{P}) = \mathcal{C} \odot_{\mathcal{C}} \mathfrak{P}$, $\Psi_{\mathcal{C}}(\mathcal{M}) = \mathrm{Hom}_{\mathcal{C}}(\mathcal{C}, \mathcal{M})$ – see 0.2.6, 5.1, 5.3

$\mathbb{L}\Phi_{\mathcal{C}}, \mathbb{R}\Psi_{\mathcal{C}}$ – derived comodule-contramodule correspondence functors, see 5.4

$\Phi_{\mathcal{S}}(\mathfrak{P}) = \mathcal{S} \circledS_{\mathcal{S}} \mathfrak{P}$, $\Psi_{\mathcal{S}}(\mathcal{M}) = \mathrm{Hom}_{\mathcal{S}}(\mathcal{S}, \mathcal{M})$ – see 0.3.7, 6.2

$\mathbb{L}\Phi_{\mathcal{S}}$, $\mathbb{R}\Psi_{\mathcal{S}}$ – derived semimodule-semi-contramodule correspondence, see 6.3

ProCotor$^{\mathcal{C}}$ – see 2.7, E.4.6

IndCoext$_{\mathcal{C}}$ – see 4.7

$M \longmapsto {}_B M$, $M \longmapsto M_B$, $P \longmapsto {}^B P$ – change-of-ring functors, see 7.1.2

$\mathcal{N} \longmapsto {}_{\mathcal{C}}\mathcal{N}$, $\mathcal{N} \longmapsto \mathcal{N}_{\mathcal{C}}$, $\mathfrak{Q} \longmapsto {}^{\mathcal{C}}\mathfrak{Q}$ – change-of-coring functors, see 7.1.2

$\mathcal{M} \longmapsto {}_{\mathcal{J}}\mathcal{M}$, $\mathcal{M} \longmapsto \mathcal{M}_{\mathcal{J}}$, $\mathfrak{P} \longmapsto {}^{\mathcal{J}}\mathfrak{P}$ – change-of-semialgebra functors, see 8.1.2

$\mathcal{C} \longmapsto {}_B \mathcal{C}_B$, see 7.4.1, 7.5.1

$\mathcal{J} \longmapsto {}_{\mathcal{C}}\mathcal{J}_{\mathcal{C}}$, see 8.4.1

$\mathcal{M}^{\bullet} \longmapsto {}_B^{\mathbb{L}}\mathcal{M}^{\bullet}$, $\mathcal{M}^{\bullet} \longmapsto \mathcal{M}_B^{\bullet \mathbb{L}}$, $\mathfrak{P}^{\bullet} \longmapsto {}_{\mathbb{R}}^B \mathfrak{P}^{\bullet}$ – derived push-forwards, see 7.3

$\mathcal{N}^{\bullet} \longmapsto {}_{\mathcal{C}}^{\mathbb{R}}\mathcal{N}^{\bullet}$, $\mathcal{N}^{\bullet} \longmapsto \mathcal{N}_{\mathcal{C}}^{\bullet \mathbb{R}}$, $\mathfrak{Q}^{\bullet} \longmapsto {}_{\mathbb{L}}^{\mathcal{C}}\mathfrak{Q}^{\bullet}$ – derived pull-backs, see 7.3

$\mathcal{N}^{\bullet} \longmapsto {}_{\mathcal{C}}^{\mathbb{R}}\mathcal{N}^{\bullet}$, $\mathcal{N}^{\bullet} \longmapsto \mathcal{N}_{\mathcal{C}}^{\bullet \mathbb{R}}$, $\mathfrak{Q}^{\bullet} \longmapsto {}_{\mathbb{L}}^{\mathcal{C}}\mathfrak{Q}^{\bullet}$ – derived pull-backs, see 8.3

$\mathcal{M}^{\bullet} \longmapsto {}_{\mathcal{J}}^{\mathbb{L}}\mathcal{M}^{\bullet}$, $\mathcal{M}^{\bullet} \longmapsto \mathcal{M}_{\mathcal{J}}^{\bullet \mathbb{L}}$, $\mathfrak{P}^{\bullet} \longmapsto {}_{\mathbb{R}}^{\mathcal{J}}\mathfrak{P}^{\bullet}$ – derived push-forwards, see 8.3

functors Δ_{ϕ}, Δ^{ϕ} – see 10.1.2

functors Γ_{ϕ}, Γ^{ϕ} – see 10.1.3

functors $\Delta_{\phi,f}$, $\Delta^{\phi,f}$ – see 10.2.2

functors $\Gamma_{\phi,f}$, $\Gamma^{\phi,f}$ – see 10.2.4

$\mathcal{N} \otimes_B^{\mathcal{C}} \mathcal{M}$, $\mathrm{Hom}_B^{\mathcal{C}}(\mathcal{M}, \mathfrak{P})$ – semiproduct, semimorphisms, see 10.4

$\mathrm{Ind}_H^G N$, $\mathrm{Coind}_H^G P$ – induced and coinduced representations, see E.2.1–E.2.2

$N^{\natural}$, N^H – invariants

$P_{\natural}$, P_H – coinvariants

$N_{\mathfrak{g},\natural}$, $\mathbf{N}_{\mathfrak{g},H}$, $N_{G,H}$ – semiinvariants

$P^{\mathfrak{g},\natural}$, $\mathfrak{P}^{\mathfrak{g},H}$, $P^{G,H}$ – semicontrainvariants

$\lozenge_{\mathcal{S}}^{\mathrm{gr}}$, $\mathrm{SemiHom}_{\mathcal{S}}^{\mathrm{gr}}$, $\mathrm{Hom}_{\mathrm{gr}}^{\mathcal{S}}$, $\Phi_{\mathcal{S}}^{\mathrm{gr}}$, ... – graded versions of functors, see 11.1.1, B.1.3

Σ, Π – forgetting-the-grading functors, see 11.1.1

$\mathcal{M}^{\bullet} \longmapsto \Xi(\mathcal{M}^{\bullet})$, $\mathcal{L} \longmapsto \Upsilon^{\bullet}(\mathcal{L})$ – derived Koszul duality functors, see 11.8

R^{op}, $\mathcal{C}^{\mathrm{op}}$, $\mathcal{S}^{\mathrm{op}}$ – opposite ring, coring, semialgebra

C^{op} – opposite category

$\mathsf{Ind}\,\mathsf{C}$, $\mathsf{Pro}\,\mathsf{C}$ – categories of ind-objects, pro-objects

Hot – (unbounded) homotopy category

D – (unbounded) derived category

Acycl, Acycl^{co}, Acycl^{ctr} – categories of acyclic, coacyclic, contraacyclic complexes

D^{co}, $\mathsf{D}^{\mathrm{ctr}}$, D^{si} – coderived, contraderived, semiderived category

R-mod, mod-R, k-vect – categories of left, right R-modules, k-vector spaces

$\mathcal{C}$-comod, comod-$\mathcal{C}$ – categories of left, right $\mathcal{C}$-comodules

$\mathcal{C}$-contra – category of left $\mathcal{C}$-contramodules

$\mathcal{S}$-simod, simod-$\mathcal{S}$ – categories of left, right $\mathcal{S}$-semimodules

$\mathcal{S}$-sicntr – category of left $\mathcal{S}$-semicontramodules

$\mathcal{D}^{\sim}$-qcmd, qcmd-$\mathcal{D}^{\sim}$ – categories of left, right quasi-differential $\mathcal{D}^{\sim}$-comodules

$\mathcal{D}^{\sim}$-qcntr – category of left quasi-differential $\mathcal{D}^{\sim}$-contramodules

$\mathsf{O}(\mathfrak{g}, H)$, $\mathsf{O}_{\varkappa}(\mathfrak{g}, \mathcal{C})$ – category of Harish-Chandra modules

$\mathsf{O}_{\varkappa}^{\mathrm{ctr}}(\mathfrak{g}, \mathcal{C})$ – category of Harish-Chandra contramodules

k-vect^{gr}, k-mod^{gr}, $\mathcal{C}$-comod^{gr}, $\mathcal{C}$-contra^{gr}, $\mathcal{S}$-simod^{gr}, $\mathcal{S}$-sicntr^{gr}, ... – categories of graded modules, comodules, contramodules, semimodules, etc., see 11.1

$\mathcal{S}$-simod$^{\uparrow}$, simod$^{\uparrow}$-$\mathcal{S}$, $\mathcal{S}$-sicntr$^{\downarrow}$, $\mathcal{C}$-comod$^{\downarrow}$, $\mathcal{C}$-contra$^{\uparrow}$, $\mathcal{S}$-simod$^{\downarrow}$, $\mathcal{S}$-sicntr$^{\uparrow}$, ... – categories of nonnegatively/nonpositively graded modules, see 11.1.2 and B.3

$\mathsf{Rep}_k(\mathbb{H})$ – see E.4.2

$\mathsf{Rep}_{c'}(\mathbb{G})$ – see E.4.5

$\mathsf{Set}_{\mathrm{fin}}$ – category of finite sets

$\varepsilon_{\mathcal{C}}$, $\mu_{\mathcal{C}}$, $\nu_{\mathcal{M}}$, $\pi_{\mathfrak{P}}$ – counit, comultiplication, coaction, contraaction maps

$\mathbf{e}$, $\mathbf{m}$, $\mathbf{n}$, $\mathbf{p}$ – semiunit, semimultiplication, semiaction, semicontraaction morphisms

$c \longmapsto c_{(1)} \otimes c_{(2)}$, $x \longmapsto x_{(-1)} \otimes x_{(0)}$, $x \longmapsto x_{[0]} \otimes x_{[1]}$, $l \longmapsto l_{\{1\}} \otimes l_{\{2\}}$, ... – Sweedler's notation for comultiplications and coactions, see [82, 1.2 and 2.0].

$c \longmapsto s(c)$, $x \longmapsto s(x)$ – antipode map

$\ker(X \to Y)$, $\mathrm{coker}(X \to Y)$, $\mathrm{cone}(X^\bullet \to Y^\bullet)$ – kernel, cokernel, cone of map

$\mathrm{im}\,\partial$, $\mathrm{im}\,k$ – image of map or set

id, $\mathrm{id}_{\mathcal{C}}$, id_G, ... – identity endomorphism

Id – identity endofunctor

$V \longmapsto V^*$ – dual vector space

$V \longmapsto V^\vee$ – dual Tate vector space

$\bigwedge_k(V)$ – exterior (co)algebra

$\mathrm{Sym}_k(V)$ – symmetric (co)algebra

$\det(W) = \bigwedge_k^{\dim W}(W)$ – top forms

$\bigwedge_E^{\infty/2}(V)$ – semi-infinite forms

$U(\mathfrak{g})$ – enveloping algebra

$U_\varkappa(\mathfrak{g})$ – modified for a central extension, see D.2.2

$U_A(\mathfrak{g})$ – enveloping algebra of Lie algebroid

$\mathcal{C}(\mathcal{L}) = \mathcal{C}(\mathfrak{h}^\vee)$ – conilpotent coenveloping coalgebra, see D.6.1

End – endomorphisms

$\mathrm{End}(E)$ – bundle of endomorphisms of a vector bundle

$\mathrm{End}(V)$, $\mathfrak{gl}(V)$ – continuous endomorphisms of Tate vector space, see D.1.5

$\mathfrak{gl}(V)^\sim$, $\mathfrak{g}^\sim$ – canonical central extensions, see D.1.6, D.1.8

$\mathrm{Cl}(V \oplus V^\vee)$ – Clifford algebra, see D.1.7

$\overline{\mathrm{Cl}}(V)$, see D.5.5

$V \otimes^* W$, $V \otimes^! W$, $V \overset{\rightarrow}{\otimes} W$, $V \overset{\leftarrow}{\otimes} W$ – topological tensor products

$\bigwedge^{*,i}(V)$, $\bigwedge^{s,i}(V)$, $\bigwedge^{!,i}(V)$ – topological exterior powers, see D.2.6, D.5.3, D.5.5

$V \otimes^{\widehat{\ }} P$ – completed tensor product, see D.2.6

$U_\varkappa^{\widehat{\ }}(\mathfrak{g})$ – topological enveloping algebra, see D.5.1

O_X – structure sheaf/trivial bundle

Ω_X – sheaf/bundle of top forms

$\Omega(M, E)$ – differential forms with coefficients in vector bundle E

Diff_M – differential operators on M

$\mathrm{Diff}_{M,E}$ – acting in the sections of E

T_X – tangent bundle

$N_{Z,X}$ – normal bundle to Z in X

$L(X)$ – global sections of a sheaf L on X

$G \times_M G$ – fibered product

$G \times^H G$ – see C.4.4, E.1.2, F.6

$(E, E^\vee)$, $(\mathcal{E}, \mathcal{E}^\vee)$, $(\boldsymbol{\mathcal{E}}, \boldsymbol{\mathcal{E}}^\vee)$ – Morita morphisms

$A(\mathrm{M})$ – see F.1

$\mathcal{C}(H)$ – see C.4.1 and F.1

$\mathcal{C}_k(H)$ – see E.1.2

$\boldsymbol{S}^r$, $\boldsymbol{S}^l$ – see C.1.2–C.1.3

$\boldsymbol{S}^l(\mathfrak{g}, H)$, $\boldsymbol{S}^r(\mathfrak{g}, H)$ – see C.4.2

$\boldsymbol{S}_\varkappa^r(\mathfrak{g}, \mathcal{C})$, $\boldsymbol{S}_\varkappa^l(\mathfrak{g}, \mathcal{C})$ – see D.2.2, D.2.4

$\boldsymbol{S}_k(G, H)$ – see E.1.2

$\boldsymbol{S}^r(G, H)$, $\boldsymbol{S}^l(G, H)$ – see F.5

$\mathcal{A}(G)$, $\mathcal{A}(\mathbb{H})$ – see E.4.1, E.4.2

$\mathcal{C}_{c'}(\mathbb{G}, \mathbb{H})$ – see E.4.5

$A(X)$, $A[X]$, $A[[X]]$ – functions and measures, see E.1.1

Ad_g – adjoint action of a group element g in the group

L^Z – sheaf of generalized sections of L supported in Z

g^*, h_* – inverse, direct images of quasi-coherent sheaves

g^+, h_+ – pull-back, push-forward of generalized sections, see F.3

$\psi^\# = \psi^{-1}$, see 10.3

R and $R^\#$, see B.1.2, B.2.2

$\mathcal{C} \longmapsto \mathcal{C}^{\mathrm{ss}}$ – maximal cosemisimple subcoalgebra, see A.2

$\mathrm{Bar}_{\mathrm{gr}}^\bullet(\boldsymbol{S}, \mathcal{C})$, $\mathrm{Cob}_{\mathrm{gr}}^\bullet(\mathcal{D}, \mathcal{C})$ – relative bar and cobar constructions, see 11.4.1

$F_i X$, $G_n X$, ... – increasing filtrations

$V^i X$, $G^n X$, ... – decreasing filtrations

$\mathrm{gr}_F X$ – associated graded object to filtered object (X, F)

$\mathsf{H}_1 \times \mathsf{H}_2 \dashrightarrow \mathsf{K}$ – partially defined functor (well-defined on a full subcategory)

$\mathcal{P}(\mathcal{M}) \longrightarrow \mathcal{M}$ – see 1.1.3, 1.3.2

$\mathfrak{P} \longrightarrow \mathfrak{I}(\mathfrak{P})$ – see 3.1.3, 3.3.2

$\mathcal{M} \longrightarrow \mathcal{J}(\mathcal{M})$ – see Lemma 9.1.2

$\mathfrak{F}(\mathfrak{P}) \longrightarrow \mathfrak{P}$ – see Lemma 9.1.2

$\mathcal{P}(\mathcal{M}) \longrightarrow \mathcal{M}$ – see 1.3.2

$\mathfrak{P} \longrightarrow \mathfrak{I}(\mathfrak{P})$ – see 3.3.2

$\mathcal{N} \longrightarrow \mathcal{J}(\mathcal{N})$ – see 1.3.3

$\mathfrak{F}(\mathfrak{P}) \longrightarrow \mathfrak{P}$ – see 3.3.3

$\mathcal{M} \longrightarrow \mathcal{J}(\mathcal{M})$ – see Lemma 9.2.2

$\mathcal{F}(\mathcal{L}) \longrightarrow \mathcal{L}$ – see Lemma 9.2.3

$\mathcal{P}^+(\mathcal{M})$, $\mathfrak{I}^+(\mathfrak{P})$, $\boldsymbol{\mathcal{P}}^+(\mathcal{M})$, ... – see 2.5, 2.6

Index

Monografie Matematyczne

[1] S. Banach, *Théorie des opérations linéaires*, 1932
[2] S. Saks, *Théorie de l'integrale*, 1933
[3] C. Kuratowski, *Topologie I*, 1933
[4] W. Sierpiński, *Hypothèse de continu*, 1934
[5] A. Zygmund, *Trigonometrical Series*, 1935
[6] S. Kaczmarz, H. Steinhaus, *Theorie der Orthogonalreihen*, 1935
[7] S. Saks, *Theory of the integral*, 1937
[8] S. Banach, *Mechanika*, T. I, 1947
[9] S. Banach, *Mechanika*, T. II, 1947
[10] S. Saks, A. Zygmund, *Funkcje analityczne*, 1948
[11] W. Sierpiński, *Zasady algebry wyższej*, 1946
[12] K. Borsuk, *Geometria analityczna w n wymiarach*, 1950
[13] W. Sierpiński, *Działania nieskończone*, 1948
[14] W. Sierpiński, *Rachunek różniczkowy poprzedzony badaniem funkcji elementarnych*, 1947
[15] K. Kuratowski, *Wykłady rachunku różniczkowego i całkowego*, T. I, 1948
[16] E. Otto, *Geometria wykreślna*, 1950
[17] S. Banach, *Wstęp do teorii funkcji rzeczywistych*, 1951
[18] A. Mostowski, *Logika matematyczna*, 1948
[19] W. Sierpiński, *Teoria liczb*, 1950
[20] C. Kuratowski, *Topologie I*, 1948
[21] C. Kuratowski, *Topologie II*, 1950
[22] W. Rubinowicz, *Wektory i tensory*, 1950
[23] W. Sierpiński, *Algèbre des ensembles*, 1951
[24] S. Banach, *Mechanics*, 1951
[25] W. Nikliborc, *Równania różniczkowe*, Cz. I, 1951
[26] M. Stark, *Geometria analityczna*, 1951
[27] K. Kuratowski, A. Mostowski, *Teoria mnogości*, 1952
[28] S. Saks, A. Zygmund, *Analytic functions*, 1952
[29] F. Leja, *Funkcje analityczne i harmoniczne*, Cz. I, 1952
[30] J. Mikusiński, *Rachunek operatorów*, 1953
* [31] W. Ślebodziński, *Formes extérieures et leurs applications*, 1954
[32] S. Mazurkiewicz, *Podstawy rachunku prawdopodobieństwa*, 1956
[33] A. Walfisz, *Gitterpunkte in mehrdimensionalen Kugeln*, 1957
[34] W. Sierpiński, *Cardinal and ordinal numbers*, 1965
[35] R. Sikorski, *Funkcje rzeczywiste*, 1958
[36] K. Maurin, *Metody przestrzeni Hilberta*, 1959
[37] R. Sikorski, *Funkcje rzeczywiste*, T. II, 1959
[38] W. Sierpiński, *Teoria liczb II*, 1959

* [39] J. Aczél, S. Gołąb, *Funktionalgleichungen der Theorie der geometrischen Objekte*, 1960

[40] W. Ślebodziński, *Formes extérieures et leurs applications*, II, 1963

[41] H. Rasiowa, R. Sikorski, *The mathematics of metamathematics*, 1963

[42] W. Sierpiński, *Elementary theory of numbers*, 1964

* [43] J. Szarski, *Differential inequalities*, 1965

[44] K. Borsuk, *Theory of retracts*, 1967

[45] K. Maurin, *Methods of Hilbert spaces*, 1967

[46] M. Kuczma, *Functional equations in a single variable*, 1967

[47] D. Przeworska-Rolewicz, S. Rolewicz, *Equations in linear spaces*, 1968

[48] K. Maurin, *General eigenfunction expansions and unitary representations of topological groups*, 1968

[49] A. Alexiewicz, *Analiza funkcjonalna*, 1969

* [50] K. Borsuk, *Multidimensional analytic geometry*, 1969

* [51] R. Sikorski, *Advanced calculus. Functions of several variables*, 1969

[52] W. Ślebodziński, *Exterior forms and their applications*, 1971

[53] M. Krzyżański, *Partial differential equations of second order*, vol. I, 1971

[54] M. Krzyżański, *Partial differential equations of second order*, vol. II, 1971

[55] Z. Semadeni, *Banach spaces of continuous functions*, 1971

[56] S. Rolewicz, *Metric linear spaces*, 1972

[57] W. Narkiewicz, *Elementary and analytic theory of algebraic numbers*, 1974

[58] Cz. Bessaga, A. Pełczyński, *Selected topics in infinite dimensional topology*, 1975

* [59] K. Borsuk, *Theory of shape*, 1975

[60] R. Engelking, *General topology*, 1977

[61] J. Dugundji, A. Granas, *Fixed point theory*, 1982

* [62] W. Narkiewicz, *Classical problems in number theory*, 1986

The volumes marked with * are available at the exchange department of the library of the Institute of Mathematics, Polish Academy of Sciences.

Monografie Matematyczne, New Series (MMNS)

Edited by
Przemysław Wojtaszczyk, IMPAN and Warsaw University, Poland

Starting in the 1930s with volumes written by such distinguished mathematicians as Banach, Saks, Kuratowski, and Sierpinski, the original series grew to comprise 62 excellent monographs up to the 1980s. In cooperation with the Institute of Mathematics of the Polish Academy of Sciences (IMPAN), Birkhäuser now resumes this tradition to publish high quality research monographs in all areas of pure and applied mathematics.

BIRKHÄUSER